AHDB LIBRARY
02476 478839
ACC.NO. 1385 (2)   PO NO
PRICE: £115.00   DATE: 05/11/2012
CLASS.NO. 636.085 SUT

D1555977

Mineral Nutrition of Livestock, 4th Edition

Mixed Sources
Product group from well-managed
forests and other controlled sources
www.fsc.org Cert no. SA-COC-1565
© 1996 Forest Stewardship Council

Mineral Nutrition of Livestock, 4th Edition

Neville F. Suttle

*Honorary Research Fellow
Moredun Foundation
Pentland Science Park
Bush Loan
Penicuik
Midlothian EH26 0PZ
UK*

CABI is a trading name of CAB International

CABI Head Office
Nosworthy Way
Wallingford
Oxfordshire OX10 8DE
UK

Tel: +44 (0)1491 832111
Fax: +44 (0)1491 833508
E-mail: cabi@cabi.org
Website: www.cabi.org

CABI North American Office
875 Massachusetts Avenue
7th Floor
Cambridge, MA 02139
USA

Tel: +1 617 395 4056
Fax: +1 617 354 6875
E-mail: cabi-nao@cabi.org

© N. Suttle 2010. All rights reserved. No part of this publication may be reproduced in any form or by any means, electronically, mechanically, by photocopying, recording or otherwise, without the prior permission of the copyright owners.

A catalogue record for this book is available from the British Library, London, UK.

Library of Congress Cataloging-in-Publication Data

Suttle, N. F.
  Mineral nutrition of livestock / Neville F. Suttle. -- 4th ed.
    p. cm.
  Earlier editions entered under Eric J. Underwood.
  Includes bibliographical references and index.
  ISBN 978-1-84593-472-9 (alk. paper)
 1. Minerals in animal nutrition. I. Title.

SF98.M5U5 2010
636.08'527--dc22

2009022346

ISBN-13: 978 1 84593 472 9

Commissioning editor: Sarah Hulbert
Production editor: Kate Hill

Typeset by SPi, Pondicherry, India.
Printed and bound in the UK by the MPG Books Group.

Contents

Preface		vii
1.	The Requirement for Minerals	1
2.	Natural Sources of Minerals	14
3.	Assessing and Controlling Mineral Status in Livestock	39
4.	Calcium	54
5.	Magnesium	92
6.	Phosphorus	122
7.	Potassium	168
8.	Sodium and Chloride	182
9.	Sulfur	206
10.	Cobalt	223
11.	Copper	255
12.	Iodine	306
13.	Iron	334
14.	Manganese	355
15.	Selenium	377
16.	Zinc	426
17.	Occasionally Beneficial Elements	459

18.	Potentially Toxic Elements	489
19.	Design of Supplementation Trials For Assessing Mineral Deprivation	528
20.	Minerals and Humans	540

Appendices   555

Index   565

Preface

Reviewing progress in the mineral nutrition of livestock over the last 10 years has been complicated by a small fraction of the many new papers that addressed practical problems in a logical way and were subjected to rigorous peer review. The focus has been either on fundamental studies of molecular pathways or on commercially attractive nutritional supplements, bypassing the 'middle-ground' of basic mineral nutrition. Fundamental studies have rarely produced information that changes the way mineral imbalances are diagnosed, treated or avoided. Applied studies have been dominated by commercial interests and some published conclusions have been highly misleading, even those appearing in prestigious journals. Commercially biased experiments in mineral nutrition can be easily recognized: experimental designs follow practices adopted by QCs during cross examinations, avoiding questions to which the answer is unknown and possibly damning; statistical significance conventions are ignored, trends ($P < 0.1$) becoming 'significant' and if that ruse fails 'numerical superiority' may be claimed; positive comments about supplements are always mentioned in the abstract, however trivial; negative comments are avoided. One journal of human clinical nutrition now appends a 'declaration of interest' to their published papers, with an indication of what each author has contributed to minimize biased reporting, but commercial influences are now so pervasive that a similar declaration may be needed from referees. One veterinary journal now discriminates against citations of papers not subjected to peer review, a commendable practice that can greatly shorten reference lists.

Rigorous review has led me to reject some widely held views, including those that unreservedly accredit selenomethionine and metal chelates with superior availability. While I think that Eric Underwood would have agreed with me, it seemed unfair to link his reputation to such contentious issues and a text which has continued to shift from the solid foundation that he so carefully laid in 1981. Two major changes in organization have been made: first, the complete separation of calcium from phosphorus, since they are mutual antagonists rather than bedfellows when it comes to non-ruminant nutrition; second, to provide a nutritional 'heart' to most chapters by tagging mineral requirements behind mineral composition and availability. To counter the growing separation of mineral nutrition in man and livestock, a final chapter has been added, which highlights areas of interdependence with implications for the health of both, and the sustainability of their shared environment.

After a decade that has seen a vast increase in the pace and complexity of research, I have been greatly assisted by the following chapter referees for covering my ignorance, particularly at the modelling and molecular levels: Professors Andrew Sykes (Lincoln College, NZ; magnesium), Susan Fairweather-Tait (University of East Anglia, UK; iron), John Arthur (Rowett Research Institute, Aberdeen; iodine and selenium) and Jerry Spears (North Carolina University; manganese and zinc). Old friends Bob Orskov, Dennis Scott and Chris Livesey interrupted their retirement, the latter's only just begun, to put me right on the rumenology of sulfur, physiological aspects of sodium and potassium and environmental aspects of potentially toxic elements, respectively.

A criticism of the last edition by Professor Ivan Caple, the then Dean of University of Melbourne's Veterinary School, that it lacked a quick reference section for busy vets, has resulted in the addition of three summary appendices, though the lists of cautionary footnotes are necessarily long. Readers are again invited to notify the author of errors, omissions or irrelevancies – it may be possible to respond to some of them in a subsequent paperback version.

Neville Suttle
Moredun Foundation
Penicuik
EH26 0PZ
suttle_hints@hotmail.com

1 The Requirement for Minerals

Early Discoveries

All animal and plant tissues contain widely varying amounts and proportions of mineral elements, which largely remain as oxides, carbonates, phosphates and sulfates in the ash after ignition of organic matter. In the 17th century, a Swedish chemist, Gahn, found calcium phosphate to be the major component of bone ash. In 1875, Sir Humphrey Davy identified the element potassium in the residues of incinerated wood and gave it the name 'pot ash'. The earliest hint of nutritional significance for such apparently inert substances came from Fordyce (1791), who showed that canaries on a seed diet required a supplement of 'calcareous earth' to remain healthy and produce eggs. Calcium supplements were eventually used for the prevention of rickets, a childhood disorder of bone development that had plagued people for centuries. The discovery that iron was a characteristic component of blood led Frodisch (1832) to link blood iron content with 'chlorosis' (anaemia) in people. In livestock, Boussingault (1847) showed that cattle had a dietary need for common salt, and Babcock (1905) induced calcium deficiency in dairy cows by feeding diets low in calcium. A craving of sick cattle and sheep for the bones of dead animals on the South African veldt led to the identification of phosphorus deficiency in cattle (Theiler, 1912). Chatin (1850–1854) linked environmental iodine deficiency to the incidence of endemic goitre in man and animals and, early in the next century, the iodine-rich molecule thyroxine was isolated from thyroid tissue (Harington, 1926). At the beginning of the last century, it was becoming apparent that there is more to minerals than meets the eye.

Essentiality

Between 1928 and 1931, novel studies at Wisconsin with rats given specially purified diets showed that copper, manganese and zinc were each essential for health (Underwood, 1977). The 1930s saw the extension of such studies to livestock and, in the field, animals were shown to suffer from deficiencies of copper and cobalt in North America, Australia and Europe. Further studies with rats maintained in plastic isolators to exclude atmospheric contamination extended the list of essential minerals to include selenium (Smith and Schwarz, 1967), an element previously renowned for its toxicity to livestock. By 1981, 22 mineral elements were believed to be essential for animal life: seven major or macronutrient minerals – calcium, phosphorus, potassium, sodium, chlorine, magnesium and sulfur – and 15 trace or micronutrient mineral elements – iron, iodine, zinc, copper, manganese, cobalt, molybdenum, selenium, chromium, tin, vanadium, fluorine, silicon, nickel and arsenic (Underwood, 1981). Subsequently, dietary supplements of aluminium, boron, cadmium, lithium, lead and rubidium were shown to improve growth or health in rats, goats, pigs or poultry reared in highly specialized conditions, without inducing specific abnormalities or being associated with breakdown along metabolic pathways where they have specific functions (see Chapters 17 and 18); these, and some of their predecessors that were labelled 'newer essential elements' in the previous edition of this book (notably fluorine, nickel, tin and vanadium) (Underwood and Suttle, 1999), must be re-examined in the light of the theory of hormesis (Calabrese and Baldwin, 1988). All animal tissues contain a further 20–30 mineral elements, mostly in small and variable concentrations. These are probably adventitious constituents, arising from contact with a chemically diverse environment.

Complexity

The last decade of the 20th century saw the increased application of molecular biology to studies of mineral metabolism and function, and the complex mechanisms by which minerals are safely transported across cell membranes and incorporated into functional intracellular molecules began to be clarified (O'Dell and Sunde, 1997). For potassium alone, 10 different membrane transport mechanisms were identified. Genes controlling the synthesis of key metalloproteins such as metallothionein, selenoenzymes such as glutathione peroxidases and superoxide dismutases (SODs) were the first to be isolated, and deficiencies of zinc were found to influence the expression of genes controlling the synthesis of molecules that did not contain zinc (Chesters, 1992). The induction of messenger RNA for transport and storage proteins promised to be a sensitive indicator of copper deprivation (Wang et al., 1996) but, with ever-increasing arrays to choose from, animals are proving to be highly selective in which genes they switch on and when they do it. New functions are being revealed, such as the role of zinc in alkylation reactions (see Chapter 16) and some, such as the activation of a methionine synthetase by copper (see Chapter 11), open up fresh possibilities for one element to compensate for the lack of another. Two families of zinc transporters have been identified, some of which are needed to activate the zinc enzyme alkaline phosphatase (see Chapter 16)

(Suzuki et al., 2005). The new millennium has seen an explosion of activity in this field and a new focus: the signalling mechanisms by which intracellular needs are communicated and orchestrated. Calcium and superoxide ions and selenocysteine play pivotal roles, with the selenocysteine 'altering our understanding of the genetic code' (Hatfield and Gladyshev, 2002). New journals dedicated to the subject of 'proteomics' have been launched, but proteomics has yet to impact the practical nutrition of livestock and it is largely beyond the scope of this book.

Functions

Minerals perform four broad types of function in animals:

1. *Structural*: minerals can form structural components of body organs and tissues, exemplified by minerals such as calcium, phosphorus and magnesium; silicon in bones and teeth; and phosphorus and sulfur in muscle proteins. Minerals such as zinc and phosphorus can also contribute structural stability to the molecules and membranes of which they are a part.
2. *Physiological*: minerals occur in body fluids and tissues as electrolytes concerned with the maintenance of osmotic pressure, acid–base balance, membrane permeability and transmission of nerve impulses. Sodium, potassium, chlorine, calcium and magnesium in the blood, cerebrospinal fluid and gastric juice provide examples of such functions.
3. *Catalytic*: minerals can act as catalysts in enzyme and endocrine systems, as integral and specific components of the structure of metalloenzymes and hormones or as activators (coenzymes) within those systems. The number and variety of metalloenzymes and coenzymes identified has continued to increase since the late 1990s. Activities may be anabolic or catabolic, life enhancing (oxidant) or life protecting (antioxidant).
4. *Regulatory*: minerals regulate cell replication and differentiation; for example, calcium ions influence signal transduction and selenocysteine influences gene transcription, leading to its nomination as 'the 21st amino acid'

Table 1.1. Some important metalloenzymes and metalloproteins in livestock.

Metal	Metalloenzyme or metalloprotein	Function
Fe	Hepcidin	Iron regulating hormone
	Succinate dehydrogenase	Oxidation of carbohydrates
	Haemoglobin	Oxygen transport in blood
	Catalase	Protection against hydrogen peroxide, H_2O_2
Cu	Cytochrome oxidase	Terminal oxidase
	Lysyl oxidase	Lysine oxidation
	Hephaestin	Iron absorption
	Caeruloplasmin	Copper transport
	Superoxide dismutase	Dismutation of superoxide radical, O_2^-
Mn	Pyruvate carboxylase	Pyruvate metabolism
	Superoxide dismutase	Antioxidant by removing O_2^-
	Glycosyl aminotransferases	Proteoglycan synthesis
Se	Glutathione peroxidases (four)	Removal of H_2O_2 and hydroperoxides
	Type 1 and 2 deiodinases	Conversion of tetraiodothyronine to triiodothyronine
	Selenocysteine	Selenium transport and synthesis of selenoenzymes
Zn	Carbonic anhydrase	Formation of carbon dioxide
	Alkaline phosphatase	Hydrolysis of phosphate esters
	Phospholipase A_2	Hydrolysis of phosphatidylcholine

(Hatfield and Gladyshev, 2002). The pivotal metabolic role of thyroxine has been attributed to the influence of triiodothyronine on gene transcription (Bassett et al., 2003).

An indication of the wide range and functional importance of metalloproteins is given in Table 1.1.

Multiplicity of Function

Many functions can be performed simultaneously by the same element in the same animal and many take place in both the plants on which livestock depend (e.g. glutathione peroxidase) and the microbes or parasites that infect them (e.g. MnSOD and CuZnSOD). Preoccupation with the structural functions of calcium and phosphorus in the skeleton initially drew attention away from their influence on manifold activities in soft tissues. These include the maintenance of calcium ion concentrations in extracellular fluid for the orderly transmission of nerve impulses and intracellular energy exchanges, which all rely on the making or breaking of high-energy phosphate bonds and cell signalling. Phosphorus is an integral part of regulatory proteins and nucleic acids and thus integral to transmission of the genetic code by translation and transcription.

Copper is an essential constituent of the growing number of cuproenzymes and cuproproteins with functions as diverse as electron transfer (as cytochrome oxidases), iron absorption (as hephaestin) and antioxidant defence (CuZnSOD) (see Chapter 11).

Functional Forms

In metalloenzymes, the metal is firmly attached to the protein moiety with a fixed number of metal atoms per mole of protein and cannot be removed without loss of enzyme activity. Where two metals are present in the same enzyme they may serve different purposes: in CuZnSOD, the ability of copper to change its valency facilitates dismutation of the superoxide free radical, while zinc stabilizes the molecule. Manganese can also change valency and thus serves a similar function to copper in MnSOD.

In regulatory proteins or peptides that contain more than one atom of a mineral, the precise number and/or position of atoms can determine function. Thyroxine contains four atoms of iodine, two attached to an outer ring and two to an inner tyrosine ring (T4). Removal of one atom from the outer ring creates a physiologically active molecule (triiodothyronine), while removal of an inner atom creates an inactive analogue. Deiodination is accomplished by a family of three selenium-dependent deiodinases, synthesized from the encoded selenocysteine (Beckett and Arthur, 1994). Low metalloenzyme activity and low concentrations of metalloproteins in particular cells or fluids sometimes accompany and explain specific clinical symptoms of mineral deprivation, but some serious pathological disorders cannot be explained in such biochemical terms (Chesters and Arthur, 1988).

Cobalt is a unique element in that its functional significance can be accounted for by its presence at the core of a single large molecule, vitamin B_{12}, with two different functions determined by the side chain that is attached.

Metabolism

Minerals follow labyrinthine pathways through the animal once ingested and Fig. 1.1 gives the barest of introductions. The digestive process can enhance or constrain the proportions of ingested minerals that are absorbed from the diet and occasionally change the forms in which they are absorbed (e.g. selenium). However, minerals are not broken down into metabolizable forms (i.e. 'digested') in the way that organic dietary components are. Absorption of many minerals is carefully regulated, but some share the same regulator, a divalent metal transporter (Garrick et al., 2003; Bai et al., 2008). Iron and manganese are delivered to a shared protective binding protein, transferrin, within the gut mucosa (see Chapter 13). Minerals are usually transported from the serosal side of the mucosa to the liver in free or bound forms via the portal blood stream, but they can get 'stuck' in the mucosa. From the liver, they are transported by the peripheral bloodstream to be taken up by different organs and tissues at rates determined by local transporter mechanisms in cell membranes and organelles to meet intracellular needs: a single insight is given into the intricacies of such movements in the context of zinc (see Chapter 16).

Mineral turnover rates vary from tissue to tissue, but are generally high in the intestinal mucosa and liver, intermediate in other soft tissues and slow in bone, although turnover rates are influenced by physiological (e.g. lactation) and nutritional (deficient or overloaded) state. Minerals also leave the transport pool by secretion (e.g. milk, sweat and digestive juices) and excretion (urine): those secreted into the gut prior to sites of absorption may be reabsorbed, and the resultant recycling delays the onset of mineral deprivation, as in the case of phosphorus secreted in saliva. Using computer programmes such as SAAM[27], rates of exchange of minerals between body pools or 'compartments' with different turnover rates and between the gut and bloodstream can be measured from the rates of change in specific radioactivity in selected pools after a single intravenous radioisotope dose. Compartmental analysis has thus clarified the changes in calcium accretion and resorption in the skeleton during pregnancy and lactation in sheep (Braithwaite, 1983) and two minerals with interrelated pathways, such as calcium and phosphorus, can be tracked simultaneously (Fernandez, 1995). Figure 1.1a shows a macro-model representing the dynamics of mineral metabolism for calcium at a single moment in time. Similar intracellular fluxes and sequential events take place on a much smaller nano-scale between organelles (Fig. 1.1b) and they often rely on the same transporters that facilitate absorption (see Chapter 13) (Garcia et al., 2007).

Net Requirements

The functions performed by minerals can only be fulfilled if sufficient amounts of the ingested mineral are absorbed and retained to keep pace with growth, development and reproduction and to replace minerals that are 'lost' either as products, such as milk or eggs, or insidiously during the process of living.

Maintenance

Finite amounts of all essential minerals are required to replace unavoidable losses even in non-productive livestock. The magnitude of this 'maintenance requirement' (M) varies for different elements and species, but the major component

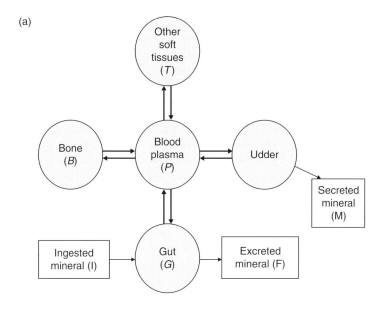

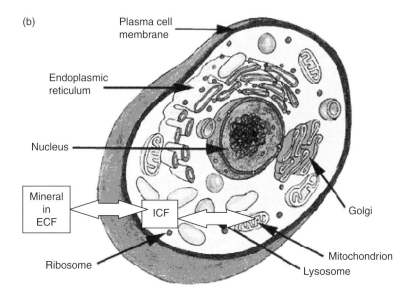

Fig. 1.1. (a) An illustration of two-way flows of mineral between pools (circles) in a lactating ewe, following ingestion (I). By labelling mineral in the diet or plasma pool with an isotope, flows can be calculated as amounts of mineral absorbed and partitioned. In one such study (Braithwaite, 1983), the major daily calcium flows (in grams) in early lactation were: I, 5.2; absorption (*GP*), 2.5; secretion into gut (*PG*), 1.4; *PB*, 1.0; *BP,* 3.1; with faecal excetion (F) at 4.1, most of the calcium secreted in milk (M, 3.0) came from bone rather than diet. (b) Whole-body fluxes are the outcome of innumerable flows of mineral into, within and from each cell, with extracellular fluid (ECF) the provider and intracellular fluid (ICF) the distributor acting in similar ways to the diet and plasma in Fig. 1.1a and the organelles (e.g. mitochondria) participating in two-way flows with mineral in ICF.

is usually the amount unavoidably lost via the faeces as sloughed mucosal cells, microbial residues and unabsorbed digestive secretions. If the faeces serve as a regulated route of excretion for a mineral absorbed in excess of need, as it does for phosphorus in ruminants and for copper and manganese in all species, it is unnecessary to replace the total faecal endogenous loss. Estimates of the minimum endogenous loss can be obtained from the intercept of the regression of faecal excretion against intake for the mineral (i.e. the faecal loss at zero mineral intake). Estimates of the maintenance requirement for phosphorus for ruminants have thus been reduced from 40 to 12 mg P kg^{-1} live weight (LW) (ARC, 1980).

An alternative approach is to feed a synthetic, mineral-free diet and assume that all the mineral excreted in faeces is of endogenous origin (e.g. Petersen and Stein, 2006). However, no animal can survive for long without essential minerals and it may be necessary to do more than replace the minimum faecal endogenous loss. Theoretically, the maintenance requirement should equate to the faecal endogenous loss measured at the minimum mineral intake needed for maximal production (or zero balance in the case of non-producing livestock) and be expressed on a kg^{-1} dry matter (DM) intake basis. In fact, the maintenance requirement is more closely related to food intake (or metabolic body weight, $LW^{0.75}$) rather than body weight per se, reflecting the increased secretion of minerals into the gut and increased 'wear and tear' on the mucosal lining at the high food intakes needed to sustain high levels of performance, something termed the 'productivity increment' in the maintenance requirement (Milligan and Summers, 1986). Lactating animals eat more food per unit of body weight than non-lactating animals, and therefore have relatively high maintenance requirements.

Work

With grazing stock, movement may raise maintenance needs for energy by 10–20% or more above those of the animal at rest. With working animals, such as horses or bullocks used for draft or transport purposes, maintenance needs for energy can be increased several fold. The extra food required for movement and work will usually supply the animal with sufficient additional minerals (even in the cow, which is increasingly used for draft purposes), except perhaps for sodium and chloride. Hard physical work, especially in hot conditions, greatly increases losses of sodium, potassium and chlorine in sweat, thereby increasing the net requirement for those elements. No other increased requirements have been reported for macro-minerals with physical work. However, there is a growing belief that increased consumption of oxygen during exercise leads to increased requirements for trace elements such as selenium that are involved in antioxidant defence (see Chapter 15) (Avellini et al., 1995).

Reproduction

The mineral requirements for reproduction in mammals are usually equated to the mineral content of the fetus and products of conception (i.e. placenta, uterus and fetal fluids), and therefore increase exponentially to reach a peak in late gestation (Fig. 1.2). For the twin-bearing ewe, the calcium requirement in late gestation actually exceeds that of lactation, leading to important contrasts in the period of vulnerability to calcium deprivation when compared with the dairy cow (see Chapter 4). There is also a small additional requirement for the growth of mammary tissue and accumulation of colostrum prior to parturition. For those minerals where the generosity of maternal nutrition determines the size of the fetal reserve (e.g. copper and selenium) (Langlands et al., 1984), it may be prudent but not essential to allow for maximal fetal retention.

Production

The net mineral requirement for production (P) is determined by the mineral content of each unit of production such as weight gain (W), milk yield (M) or fleece growth (F), and each is usually taken to remain constant. However, for elements

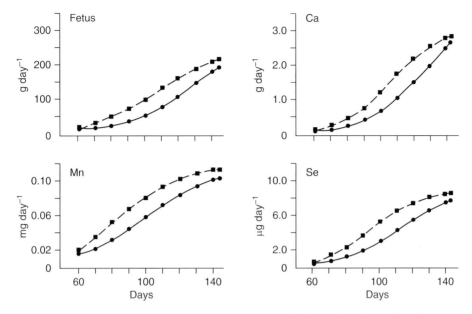

Fig. 1.2. Patterns of fetal accumulation of calcium, manganese and selenium in singleton (●) and twin-bearing ewes (■) during gestation reflect fetal weight gain (from Grace et al., 1986).

such as calcium and phosphorus, which are far richer in bone than in soft tissue, requirements for growth may diminish slightly as animals mature because bone makes a progressively smaller contribution to each unit of live-weight gain (AFRC, 1991). Table 1.2 shows that net calcium requirements for growth differ greatly between species, with the horse having a far greater need than the sheep. This difference is attributable to both a higher proportion of bone in the equine carcass (0.12 versus 0.10) and a higher proportion of dense or compact bone in the equine skeleton (see Chapter 4).

As a further example, with copper, iron and zinc concentrations in white meat less than a quarter of those in red meat (Gerber et al., 2009), W for those elements is bound to be low for broilers and turkeys compared to weaned calves and lambs.

Net requirements are affected by the productive capacity of the species or breed, the rate of production allowed by other dietary constituents – notably energy – and the environment. Animals may adjust to a suboptimal mineral intake by reducing the concentration of mineral in tissues or products. Thus, qualities such as the shell strength of eggs and the tensile strength of wool fibres may be reduced in order to conserve minerals for more essential functions. Values used for production must be sufficient to optimize the quality as well as the quantity of product. Milk is exceptional in that normal mineral concentrations are usually maintained during deficiency, with priority being given to the suckling at the expense of

Table 1.2. Net mineral requirements for production are determined by the mineral concentrations in the product. These data for calcium concentrations in the empty bodies of sheep, deer and horses at different stages of growth show that the net calcium requirement of the horse for growth is about 70% greater than that of the lamb (data mostly from Grace et al., 2008).

	Ca concentration (g kg^{-1} empty body weight)		
	Sheep	Deer	Horse
Newborn	10.1[a]	18.4	18.2
Weanling	9.6[b]	14.9	17.1
Young adult	10.5	15.0	16.7

[a]Field and Suttle (1967).
[b]Field et al. (1975).

the mother. Where the provision of excess mineral increases mineral concentrations in milk (e.g. iodine), it is unnecessary to allow for replacement of all the secreted mineral. By analogy with maintenance requirements, the requirement for milk production is the unavoidable rather than the regulated secretion of a given mineral in milk.

Gross Requirements

Net mineral requirements underestimate gross dietary needs for minerals because ingested minerals are incompletely absorbed, the degree of underestimate being *inversely* related to the efficiency with which a given mineral is absorbed from a given diet. For some minerals (e.g. sodium and potassium) absorption is virtually complete under all circumstances, but for others (e.g. copper and manganese) most of the ingested mineral can remain unabsorbed. For such elements, a range of gross requirements (GR) will usually be appropriate, inversely related to the extent to which 'conditioning' or antagonistic factors limit mineral absorption from a particular grazing or ration (Fig. 1.3). By the same reasoning, there must be a range of maximum 'tolerable' dietary levels for a given mineral, which are *directly* related to antagonist potency (Fig. 1.3). The chemical form in which a mineral is present in the diet can determine the efficiency with which it is absorbed. For example, although cereals contain high levels of phosphorus the mineral is largely present as phytate, which is poorly absorbed by pigs and poultry. There are two principal methods for estimating the needs of livestock for minerals: factorial models and dietary trials.

Factorial models

Factorial models summate the components of net requirement and divide the total by the absorbability coefficient (A) for the given mineral to allow for inefficient use of the dietary mineral supply. Thus, the calcium need of a lactating ewe is given by:

$$GR_{Ca} = (M_{Ca} + P_{Ca})/A_{Ca} = (M_{Ca} + L_{Ca} + F_{Ca})/A_{Ca}$$

where GR is the gross requirement, A_{Ca} is the absorbability coefficient for calcium and M, L and F are the net requirements of calcium for maintenance, lactation and fleece growth, respectively.

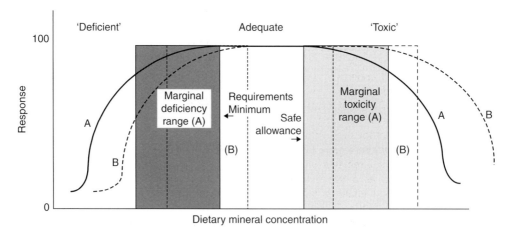

Fig. 1.3. The dose–response relationship between mineral supply and animal production showing marginal bands between adequate and inadequate or toxic dietary concentrations. For a given range of concentrations, the graph and marginal ranges move to the right as absorbability of the mineral source declines; thus A represents the more and B the lesser absorbable of two mineral sources. 'Requirements' are variously set within the central 'Adequate' band, ranging from minimum requirements to safe allowances, depending on the extent to which absorbability and other variables are taken into consideration.

The main advantage of the factorial approach is that requirements can be predicted for a wide range of production circumstances, provided that reliable data are available for each model component (ARC, 1980; Suttle, 1985; White, 1996). However, predicted requirements are heavily influenced by the value chosen for the absorbability coefficient in the model. The measurement of the absorbability coefficient can be technically difficult and varies greatly for elements such as calcium and copper in ruminants and phosphorus and zinc in non-ruminants, as explained in Chapter 2. Nevertheless, factorial models have been used to generate many of the mineral requirements tabulated in this book.

Dietary trials

Mineral requirements can be estimated by feeding groups of livestock with diets providing a range of mineral inputs above and below the minimum requirement and measuring responses in a relevant variable such as growth rate or blood composition. However, five or more different mineral input levels may be needed to precisely define the optimum requirement, and the result can still depend on the statistical model used to describe the response (Remmenga et al., 1997). A further challenge is that it is often necessary to use purified ingredients to obtain sufficiently low mineral inputs: if the diet then lacks naturally occurring antagonists, the results will underestimate requirements on natural diets (Suttle, 1983; White, 1996). Another difficulty is that it is impractical to allow for requirements that vary with time. Where the requirement for production is high relative to that for maintenance, as it is for iron in growing livestock, requirements will fall with time when expressed as a proportion of a steadily increasing food intake if the growth rate remains constant (Suttle, 1985). The dietary approach has demonstrated much higher zinc requirements for turkey poults than chicks on the same semi-purified diet (25 versus 18 mg Zn kg^{-1} DM) (Dewar and Downie, 1984). The difference may be partly attributable to the higher food conversion efficiency of the turkey poult. Dietary estimates of requirement are needed to validate factorially derived requirements.

Criteria of 'Requirement'

The criterion of adequacy is an important determinant of the estimated mineral requirement because mineral-dependent metabolic processes can vary in sensitivity to deprivation. For instance, the processes of pigmentation and keratinization of wool are the first and sometimes the only processes to be affected by a low copper status in sheep. The copper requirement of sheep is higher if wool growth rather than growth rate or blood haemoglobin content is used as criterion of adequacy. The minimum zinc requirements for spermatogenesis and testicular development in young male sheep are significantly higher than they are for body growth (Underwood and Somers, 1969). Zinc requirements are therefore lower for lambs destined for meat production than for those destined for breeding. Optimum mineral intakes are often lower when defined by production traits rather than by biochemical traits. For example, Dewar and Downie (1984) concluded that the optimum dietary zinc intakes for chicks were 18, 24 and 27 mg kg^{-1} DM, respectively, for growth, plasma zinc and tibia zinc. The adequacy of phosphorus nutrition can be assessed on the basis of growth, bone dimensions, bone composition or bone histology, with bone ash and growth plate hypertrophy giving the most sensitive estimates of requirements in turkey poults (Fig. 1.4) (Qian et al., 1996).

Requirements for Breeding Stock

Particular problems arise in defining and meeting requirements for reproduction and lactation. The mother invariably gives priority to offspring and any deficit between daily need and dietary provision is met by drawing on maternal mineral reserves. Requirements do not need to be met consistently throughout gestation or lactation. A classic experiment was performed by Kornegay et al. (1973), in which sows were fed constant daily amounts of

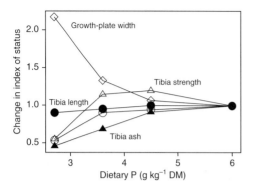

Fig. 1.4. Measures of bone quality in turkey poults vary in sensitivity to dietary phosphorus supply and assessments of 'optimal' phosphorus supply vary accordingly. Sensitivity of growth rate is shown by the open circles (values relative to those at maximal phosphorus intake) (data from Qian et al., 1996).

calcium and phosphorus (10.3 and 11.0 g day^{-1}, respectively) throughout five reproductive cycles. The allowance was grossly inadequate for lactation, but sufficient to allow some recovery of reserves between lactations. No harm was done to the offspring and even the fifth generation remained biochemically normal. The only production trait to suffer was sow longevity, which was chiefly reduced by the gradual development of lameness and other leg abnormalities. Longevity is an important trait in economic terms, and the lengthy critical experiments required to define mineral requirements for lifetime production of the breeding female have yet to be carried out for any species (including the sow) for most minerals. When a first pregnancy precedes the attainment of mature weight, the requirements for both growth and pregnancy must be met in the short term otherwise the dam or its offspring may suffer. When a cow has to conceive at peak lactation, there may be no reserves to draw on and fertility may suffer if peak daily mineral requirements are not met.

Disagreement on Mineral Requirements

Many national authorities have recommended mineral requirements to ensure that the production of native livestock is not impaired by dietary mineral imbalances (Table 1.3). However, agreement is rare, whether the recommendations are for poultry (Blair et al., 1983), sheep (Table 1.2) or other species. When one authority reviews the work of its predecessor (e.g. Agriculture & Food Research Council versus Agricultural Research Council) (ARC, 1980; AFRC, 1991) recommendations may change. Both international and national differences are attributable to the following:

- different criteria of adequacy;
- different components in factorial models; and
- different safety factors that leave no individual at risk ('safe allowances'; Fig. 1.4).

When using tables of requirements, the methods and assumptions underlying the estimates should be checked and values should not be regarded as absolute because of uncertainties surrounding precise mineral needs in specific contexts. Marginal bands or different categories will be attached to most of the mineral requirements presented in this book, rather than a single misleading average value.

Table 1.3. Variation in recommended gross mineral requirements for sheep between countries (from White, 1996). Where ranges are given, they represent the needs of specific classes of stock rather than uncertainty in the minds of their progenitors, but they should serve both purposes.

National recommendation	Ca (g kg^{-1} DM)	P (g kg^{-1} DM)	Cu (mg kg^{-1} DM)	Se (mg kg^{-1} DM)
New Zealand (1983)	2.9	2.0	5.0	0.03
USA (1985)	2.0–8.2	1.6–3.8	7–11	0.1–0.2
Australia (1990)	1.5–2.6	1.3–2.5	5.0	0.05
United Kingdom (1980)	1.4–4.5	1.2–3.5	1[a]–8.6	0.03–0.05

[a]The low requirement was a unique recognition of the distinctive mineral requirements of milk-fed animals.

Average requirements can be useful for a preliminary assessment of the adequacy of mineral supplies from forages for grazing livestock, but minimum requirements are preferable for diagnosing deficiencies because of individual variations in absorption, which also affect tolerance (Suttle, 1991). Complete avoidance of mineral deprivation may not be compatible with avoidance of toxicity, for example in providing copper and phosphorus to breeds of sheep that are vulnerable to copper poisoning (see Chapter 11) and urinary calculi (see Chapter 6), respectively. There is rarely sufficient knowledge for precise safety factors to be calculated and their deployment is seldom acknowledged. The ARC (1980) used a safety factor to allow for a low A_{Mg} in some pastures and the AFRC (1991) used one to allow for high faecal endogenous loss of phosphorus on dry roughages. Feed compounders and farmers generally apply their own safety factors regardless of the machinations of national authorities, and it is common for farm rations to provide far more than the already generous amounts recommended by some authorities (see Chapter 20).

Expression of Mineral Requirements

The requirements of animals for minerals can be expressed in several ways: in amounts per day or per unit of product, such as milk, eggs or weight gain; in proportions, such as percentage, parts per million (ppm), mass per mass (e.g. mg kg^{-1}) or moles (sometimes micro- or millimoles) per kg DM of the whole diet. Required amounts of minerals are the most precise, but they can vary with total food intake (e.g. those for calcium and phosphorus for ruminants) (AFRC, 1991). Dietary concentrations of minerals are acceptable as long as the diet is palatable and food intake is not constrained. These have the merit of simplicity, being relatively constant. However, required dietary mineral concentrations are affected by the efficiency with which organic constituents in the diet are utilized. Chicks and weaned pigs consume similar types of diet, but the faster growing broiler chick converts food to body weight the more efficiently and requires nearly twice as much calcium and manganese in its diet than the pig. Although high-yielding dairy cows require more dietary calcium and phosphorus than low-yielding cows, the dietary concentrations required increase far less because feed intakes rise as milk yield increases (AFRC, 1991). The total phosphorus requirements of hens increase with the onset of egg production but remain a constant proportion of the diet, while the required calcium concentration increases some 10-fold. Whether expressed as amounts or concentrations, requirements can be greatly influenced by factors that limit the absorption and utilization of the mineral in question.

References

AFRC (1991) Technical Committee on Responses to Nutrients Report No. 6. A reappraisal of the calcium and phosphorus requirements of sheep and cattle. *Nutrition Abstracts and Reviews* 61, 573–612.

ARC (1980) *The Nutrient Requirements of Ruminants*. Commonwealth Agricultural Bureaux, Farnham Royal, UK, pp. 184–185.

Avellini, L., Silvestrelli, M. and Gaiti, A. (1995) Training-induced modifications in some biochemical defences against free radicals in equine erythrocytes. *Veterinary Research Communications* 19, 179–184.

Babcock, S.M. (1905) The addition of salt to the ration of dairy cows. In: *University of Wisconsin Experiment Station, 22nd Annual Report*, p. 129.

Bai, S.P., Lu, L., Luo, X.G. and Liu, B. (2008) Cloning, sequencing, characterisation and expressions of divalent metal transporter one in the small intestine of broilers. *Poultry Science* 87, 768–776.

Bassett, J.H.D., Harvey, C.B. and Williams, G.R. (2003) Mechanisms of thyroid hormone receptor-specific nuclear and extra-nuclear actions. *Molecular and Cellular Endocrinology* 213, 1–11.

Beckett, G.J. and Arthur, J. (1994) The iodothyronine deiodinases and 5′ deiodination. *Baillieres Clinical Endocrinology and Metabolism* 8, 285–304.

Blair, R., Daghir, N.J., Morimoto, H., Peter, V. and Taylor, T.G. (1983) International nutrition standards for poultry. *Nutrition Abstracts and Reviews, Series B* 53, 673–703.

Boussingault, J.B. (1847) *Comptes Rendus des Séances de l'Académie des Sciences* 25, 729. Cited by McCollum, E.V. (1957) *A History of Nutrition*. Houghton Mifflin, Boston, Massachusetts.

Braithwaite, G.D. (1983) Calcium and phosphorus requirements of the ewe during pregnancy and lactation. 1. Calcium. *Journal of Agricultural Science, Cambridge* 50, 711–722.

Calabrese, E.J. and Baldwin, L.A. (1998) Hormesis as a biological hypothesis. *Environmental Health Perspectives* 106 (Suppl. 1) 357–362.

Chatin, A. (1850–1854) Recherche de l'iode dans l'air, les eaux, le sol et les produits alimentoures des Alpes de la France. *Comptes Rendus des Séances de l'Académie des Sciences* 30–39.

Chesters, J.K. (1992) Trace elements and gene expression. *Nutrition Reviews* 50, 217–223.

Chesters, J.K. and Arthur, J.R. (1988) Early biochemical defects caused by dietary trace element deficiencies. *Nutrition Research Reviews* 1, 39–56.

Dewar, W.A. and Downie, J.N. (1984) The zinc requirements of broiler chicks and turkey poults fed on purified diets. *British Journal of Nutrition* 51, 467–477.

Fernandez, J. (1995) Calcium and phosphorus metabolism in growing pigs. II Simultaneous radio-calcium and radio-phosphorus kinetics. *Livestock Production Science* 41, 243–254.

Field, A.C. and Suttle, N.F. (1967) Retention of calcium, phosphorus, magnesium, sodium and potassium by the developing sheep foetus. *Journal of Agricultural Science, Cambridge* 69, 417–423.

Field, A.C., Suttle, N.F. and Nisbet, D.I. (1975) Effects of diets low in calcium and phosphorus on the development of growing lambs. *Journal of Agricultural Science, Cambridge* 85, 435–442.

Fordyce, G. (1791) *A Treatise on the Digestion of Food*, 2nd edn. J. Johnson, London. Cited by McCollum, E. V. (1957) *A History of Nutrition*. Houghton Mifflin, Boston, Massachusetts.

Garcia, S.J., Gellein, K., Syversen, T. and Aschner, M. (2007) Iron deficient and manganese supplemented diets alter metals and transporters in developing rat brain. *Toxicological Sciences* 95, 205–214.

Garrick, M. D., Dolan, K.G., Horbinski, C., Ghoio, A.J., Higgins, D., Porcubin, M., Moore, E.G., Hainsworth, L.N., Umbreit, J.N., Conrad, M.E., Feng, L., Lis, A., Roth, J.A., Singleton, S. and Garrick, L.M. (2003) DMT 1: a mammalian transporter for multiple metals. *Biometals* 16, 41–54.

Gerber, N., Brogioli, R., Hattendof, B., Scheeder, M.R.L., Wenk, C. and Gunther, D. (2009) Variability in selected trace elements of different meat cuts determined by ICP-MS and DRC-ICPMS. *Animal* 3, 166–172.

Grace, N.D., Watkinson, J.H. and Martinson, P. (1986) Accumulation of minerals by the foetus(es) and conceptus of single- and twin-bearing ewes. *New Zealand Journal of Agricultural Research* 29, 207–222.

Grace, N.D., Castillo-Alcala, F. and Wilson, P.R. (2008) Amounts and distribution of mineral elements associated with live-weight gains of grazing red deer (*Cervus elaphus*). *New Zealand Journal of Agricultural Research* 51, 439–449.

Harington, C.R. (1926) Chemistry of thyroxine. *Biochemistry Journal* 20, 300–313.

Hatfield, D.L. and Gladyshev, V.N. (2002) How selenium has changed our understanding of the genetic code. *Molecular and Cellular Biology* 22, 3565–3576.

Kornegay, E.T., Thomas, H.R. and Meacham, T.N. (1973) Evaluation of dietary calcium and phosphorus for reproducing sows housed in total confinement on concrete or in dirt lots. *Journal of Animal Science* 37, 493–500.

Langlands, J.P., Bowles, J.E., Donald, G.E. and Smith, A.J. (1984) Deposition of copper, manganese, selenium and zinc in 'Merino' sheep. *Australian Journal of Agricultural Research* 35, 701–707.

Milligan, L.P. and Summers, M. (1986) The biological basis of maintenance and its relevance to assessing responses to nutrients. *Proceedings of the Nutrition Society* 45, 185–193.

O'Dell, B.L. and Sunde, R.A. (1997) *Introduction to Handbook of Nutritionally Essential Minerals*. Marcel Dekker Inc., New York, pp. 8–11.

Petersen, G.I. and Stein, H.H. (2006) Novel procedure for estimating endogenous losses and measurement of apparent and true digestibility of phosphorus by growing pigs. *Journal of Animal Science* 84, 2126–2132.

Qian, H., Kornegay, E.T. and Veit, H.P. (1996) Effects of supplemental phytase and phosphorus on histological, mechanical and chemical traits of tibia and performance of turkeys fed on soybean meal-based, semi-purified diets high in phytate phosphorus. *British Journal of Nutrition* 76, 263–272.

Remmenga, M.D., Milliken, G.A., Kratzer, D., Schwenke, J.R. and Rolka, H.R. (1997) Estimating the maximum effective dose in a quantitative dose–response experiment. *Journal of Animal Science* 75, 2174–2183.

Smith, J.C. and Schwarz, K. (1967) A controlled environment system for new trace element deficiencies. *Journal of Nutrition* 93, 182–188.

Suttle, N.F. (1983) Assessing the mineral and trace element status of feeds. In: Robards, G.E. and Packham, R.G. (eds) *Proceedings of the Second Symposium of the International Network of Feed Information Centres, Brisbane.* Commonwealth Agricultural Bureaux, Farnham Royal, UK, pp. 211–237.

Suttle, N.F. (1985) Estimation of requirements by factorial analysis: potential and limitations. In: Mills, C.F., Bremner, I. and Chesters, J.K. (eds) *Proceedings of the Fifth International Symposium on Trace Elements in Man and Animals.* Commonwealth Agricultural Bureaux, Farnham Royal, UK, pp. 881–883.

Suttle, N.F. (1991) Mineral supplementation of low quality roughages. In: *Proceedings of Symposium on Isotope and Related Techniques in Animal Production and Health.* International Atomic Energy Agency, Vienna, pp. 101–104.

Suzuki, T., Ishihara, K., Migaki, H., Matsuura, W., Kohda, A., Okumura, K., Nagao, M., Yamaguchi-Iwai, Y. and Kambe, T. (2005) Zinc transporters, ZnT5 and ZnT7, are required for the activation of alkaline phosphatases, zinc-requiring enzymes that are glycosylphosphatidylinositol-anchored to the cytoplasmic membrane. *Journal of Biological Chemistry* 280, 637–643.

Theiler, A. (1912) Facts and theories about styfziekte and lamziekte. In: *Second Report of the Directorate of Veterinary Research.* Onderstepoort Veterinary Institute, Pretoria, South Africa, pp. 7–78.

Underwood, E.J. (1977) *Trace Elements in Human and Animal Nutrition*, 4th edn. Academic Press, London.

Underwood, E.J. (1981) *The Mineral Nutrition of Livestock*, 2nd edn. Commonwealth Agricultural Bureaux, Farnham Royal, UK, p. 1.

Underwood, E.J. and Somers, M. (1969) Studies of zinc nutrition in sheep. I. The relation of zinc to growth, testicular development and spermatogenesis in young rams. *Australian Journal of Agricultural Research* 20, 889–897.

Underwood, E.J. and Suttle, N.F. (1999) *The Mineral Nutrition of Livestock*, 3rd edn. CAB International, Wallingford, UK.

Wang, Y.R., Wu, J.Y.J., Reaves, S.K. and Lei, K.Y. (1996) Enhanced expression of hepatic genes in copper-deficient rats detected by the messenger RNA differential display method. *Journal of Nutrition* 126, 1772–1781.

White, C.L. (1996) Understanding the mineral requirements of sheep. In: Masters, D.G. and White, C.L. (eds) *Detection and Treatment of Mineral Nutrition Problems in Grazing Sheep.* ACIAR Monograph No. 37, Canberra, pp. 15–29.

2 Natural Sources of Minerals

Livestock normally obtain most of their minerals from the feeds and forages that they consume, and their mineral intakes are influenced by the factors that determine the mineral content of plants and their seeds. The concentrations of all minerals in plants depend largely on four factors: (i) plant genotype; (ii) soil environment; (iii) climate; and (iv) stage of maturity.

The importance of a given factor varies between minerals and is influenced by interactions with the other listed factors and with aspects of crop or pasture husbandry, including the use of fertilizers, soil amendments, irrigation, crop rotation, intercropping and high-yielding cultivars.

Influence of Plant Genotype

Adaptation to extreme environments

The most striking evidence of genetic influence on mineral composition is provided by certain genera and species growing in mineral-enriched soils that carry concentrations of particular elements often several orders of magnitude higher than those of other species growing in the same extreme conditions. In saline soils, salt-bushes (*Atriplex* species) and blue-bushes (*Kochia* species) typically contain 80–140 g sodium chloride (NaCl) kg^{-1} dry matter (DM) compared to about 1 g NaCl kg^{-1} DM in common pasture plants in the same soils. Similarly, certain species of *Astragalus* growing on seleniferous soils contain >5000 mg Se kg^{-1} DM compared to <20 mg Se kg^{-1} DM in common herbage species supported by the same soils. The significance of selenium 'accumulator' plants to the problem of selenosis in grazing livestock is considered in Chapter 15. Strontium-accumulating species have also been reported (Underwood and Suttle, 1999).

Legumes

Leguminous species are generally much richer in macro-elements than grasses growing in comparable conditions, whether temperate or tropical. Minson (1990) reported mean calcium concentrations of 14.2 and 10.1 g kg^{-1} DM in temperate and tropical legumes, respectively, against 3.7 and 3.8 g kg^{-1} DM in the corresponding grasses. Many tropical legumes are exceedingly low in sodium, with half containing less than 4 g kg^{-1} DM (Minson, 1990). Mixed swards of *Lolium perenne* and *Trifolium repens* have been found to have consistently higher concentrations of calcium, magnesium and potassium than pure swards of *L. perenne*, but sodium, sulfur and phosphorus concentrations were similar with no nitrogen applied (Hopkins *et al.*, 1994). The trace elements, notably iron, copper, zinc, cobalt and nickel, are also generally higher in leguminous than in gramineous species grown in temperate climates, with copper and zinc higher in mixed than in pure grass swards (Burridge et al., 1983; Hopkins et al., 1994). However, these differences are narrowed when the soil is low in available minerals, and they can be reversed (e.g. for copper) in tropical climates (Minson, 1990). On molybdeniferous 'muck' soils, red clover can be ten times richer in molybdenum than grasses in the same pasture (Kincaid et al., 1986).

Variation among grasses and forages

Pasture species can be divided into two distinct types with regard to their sodium content: natrophilic and natrophobic. Natrophilic species have sodium-rich shoots and leaves but sodium-poor roots, while natrophobic species have the opposite distribution (Smith et al., 1978). Natrophobic species such as red clover and timothy that are grazed away from the coast could supply insufficient sodium. Grasses grown together on the same soil type and sampled at the same growth stage can vary widely in mineral composition. The following data were extracted from three studies by Underwood and Suttle (1999): total ash 40–122, calcium 0.9–5.5 and phosphorus 0.5–3.7 g kg^{-1} DM amongst 58 species in East Africa; and cobalt 0.05–0.14, copper 4.5–21.1 and manganese 96–815 mg kg^{-1} DM amongst 17 species grown in Florida. While some of the variation in cobalt and manganese may be due to soil contamination, this in turn may reflect genetic characteristics such as growth habit (upright or prostrate). Differences in mineral composition within species were highlighted by the third study in New Zealand, where short-rotation (H1) hybrid ryegrass contained one-tenth of the iodine found in perennial ryegrass (L. perenne), a difference that far outweighed the influence of soil iodine status. Botanical composition was far more important than soil origin and composition in determining the mineral content of tropical grasses in western Kenya, where the introduced species kikuyu (Pennisetum clandestinum) was richer in most macro- and trace elements than native species (Fig. 2.1). In Florida, selenium concentrations were twice as high in bermuda grass as in bahia grass, and magnesium levels were 60% higher in sorghum silage than in corn silage (Kappel et al., 1983). Leafy brassicas are naturally rich in selenium, a possible adaptation that provides protection from high sulfur concentrations in potentially toxic forms (e.g. dimethyl sulfide). Herbs generally contain higher mineral concentrations than cultivated

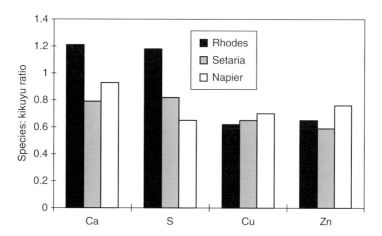

Fig. 2.1. Species differences in macro- and trace mineral composition amongst four grasses sampled during the dry season in western Kenya. Values of three species are shown as fractions of values for the fourth species, kikuyu, which were 1.4 and 1.7 g kg^{-1} DM for calcium and sulfur, respectively, and 6.0 and 30.1 mg kg^{-1} DM for copper and zinc, respectively (Jumba et al., 1996).

plants (e.g. Wilman and Derrick, 1994), but reversion to 'natural' as opposed to sown pasture does not increase the supply of phosphorus.

Variation amongst grains and seeds

Genetic differences in the mineral compositions of the vegetative parts of plants are not necessarily mirrored in their seeds. However, the seeds of leguminous plants and oilseeds are invariably richer in most minerals than the seeds of grasses and cereals, which are inherently low in calcium and sodium. Fivefold differences exist in manganese concentrations in cereal grains, with typical values as follows: wheat and oats, 35–40 mg kg^{-1} DM; barley and sorghum, 14–16 mg kg^{-1} DM; and maize 5–8 mg kg^{-1} DM (maize is also low in cobalt). Variation amongst lupin seeds is even greater: white lupins (*Lupinus albus*) contain potentially toxic levels of manganese (817–3397 mg kg^{-1} DM), 10–15 times as much as other species growing on the same sites in Western Australia (for citations, see Underwood and Suttle 1999). *L. albus* seed is also far richer in cobalt than *L. augustifolius* (240 versus 74 µg kg^{-1} DM). Zinc and selenium concentrations are similar in the two lupin species (29–30 mg kg^{-1} DM and 57–89 µg kg^{-1} DM, respectively), but much higher than in wheat grain harvested from the same sites (22 mg kg^{-1} DM and 23 µg kg^{-1} DM, respectively) (White *et al.*, 1981).

Browse species

Browse species are important sources of nutrients to smallholders and are likely to play an increasing role in more intensive intercropping systems of farming, but their mineral composition has only recently come under scrutiny. A study of nine browse species in Tanzania has revealed wide variations in both macro- and trace mineral concentrations (Table 2.1) (Mtui *et al.*, 2008). *Balanites aegyptiaca* has an exceptionally high calcium to phosphorus ratio of 35.6 and a low copper (mg) to sulfur (g) ratio of 0.66, which indicate risks of phosphorus and copper deprivation, respectively, if the species is heavily relied upon.

Effects of Soils and Fertilizers

Mineral concentrations in plants generally reflect the adequacy with which the soil can supply absorbable minerals to the roots. However, plants react to inadequate supplies of available minerals in the soil by limiting their growth, reducing the concentration of the deficient elements in their tissues or, more

Table 2.1. Data for mean mineral concentrations in the leaves of nine browse species from Tanzania show some exotic mineral profiles that could cause mineral imbalances in the browser (Mtui *et al.*, 2008).

Species	(g kg^{-1} DM)				(mg kg^{-1} DM)			
	Ca	P	Mg	S	Zn	Cu	Mn	Se
Acacia nilotica	9.5	1.2	1.0	1.3	18.9	13.8	32.8	0.19
Acacia tortilis	14.7	2.2	3.0	1.8	29.0	2.5	52.7	0.20
Balanites aegyptiaca	35.6	1.0	6.6	6.2	12.5	4.4	42.6	0.77
Cryptolepis africana	15.1	2.3	3.4	1.7	17.5	5.4	33.6	0.05
Ficus sycomorus	18.2	3.8	5.7	2.1	29.1	10.0	29.9	0.07
Lantana camara	10.2	2.5	3.1	1.4	26.4	9.6	190.3	0.02
Lawsonia inermis	11.2	4.3	3.0	2.2	47.5	9.8	72.8	0.17
Senna siamea	7.5	1.9	2.0	1.9	2.3	2.9	23.4	0.07
Trichilia emetia	13.3	2.5	3.1	1.6	41.2	5.5	21.4	0.14

commonly, by reducing growth and concentration simultaneously. The extent to which a particular response occurs varies with different minerals and different plant species or varieties and with the soil and climatic conditions.

Nevertheless, the primary reason for mineral deficiencies in grazing animals, such as those of phosphorus, sodium, cobalt, selenium and zinc, is that the soils are inherently low in plant-available minerals. These deficiencies can be mapped (e.g. zinc) (Alloway, 2004).

Large differences can exist between the mineral needs of plants and those of the animals dependent on those plants. Indeed plants do not need iodine or cobalt at all, and soil treatments would only be required to meet the requirements of livestock. Applications of magnesium compounds are sometimes necessary to raise the pasture magnesium concentration to meet the needs of cows, although this usually has no effect on pasture yield (see Chapter 5). In contrast, plant needs for potassium and manganese exceed those of animals, and applications of these elements in fertilizer can substantially boost crop or pasture yields on deficient soils that have given no signs of deficiencies in grazing stock.

Soil characteristics

Mineral uptake by plants is greatly influenced by soil pH: the effects in a grass are illustrated in Fig. 2.2 and are even more striking in legumes. Molybdenum uptake increases as soil pH rises. Pasture molybdenum can rise sufficiently on alkaline or calcareous soils (clays, shales and limestones) to induce copper-responsive disorders in livestock by impairing copper absorption (see Chapter 11). By contrast, plant uptake can be sufficiently poor on acidic soils to cause molybdenum deficiency in leguminous plants. The application of lime and sulfur can raise or lower soil pH, respectively, and so change the availability of particular minerals to plants. Liming is occasionally needed to maintain or improve soil fertility, but the associated increase in pasture molybdenum can induce copper deprivation in grazing sheep (Suttle and Jones, 1986). The absorption of nickel, cobalt and manganese by plants is

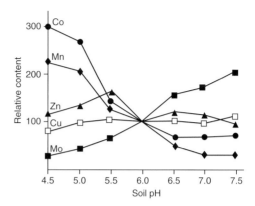

Fig. 2.2. Concentrations of cobalt and manganese in plants decrease markedly as soil pH increases, while those of molybdenum increase and those of copper and zinc show little change. Values at pH 6.0 were (mg kg^{-1} DM): Co 0.03, Cu 3.1, Mn 51.0, Mo 2.7 and Zn 23 and given a relative value of 1.0. Data are for ryegrass in Aberdeenshire (from COSAC, 1982).

favoured by acid soil conditions (Underwood and Suttle, 1999). The cobalt, molybdenum and manganese concentrations in pasture plants are all greatly increased by soil waterlogging (Burridge et al., 1983). Thus, soil conditions greatly influence the mineral composition of crops and forages and the efficacy of fertilizer applications. However, the use of soil extractants to measure 'plant-available' minerals frequently fails to predict plant mineral concentrations (Silanpaa 1982; Jumba et al., 1996), although soil extractants may be good predictors of plant yields in some cases (e.g. iron, manganese and zinc). Good prediction of herbage cobalt has been obtained using 2.5% acetic acid as the extractant and adjusting for soil pH (Suttle et al., 2003), but more sophisticated simulations of the root microenvironment are desperately needed.

Phosphorus fertilizers

Most soils in the tropics supply insufficient phosphorus for maximum crop or pasture growth and yields can be increased by applying phosphate fertilizers at 20–50 kg ha^{-1} (Jones, 1990). Herbage phosphorus can be increased from

0.2–1.2 to 0.7–2.1 g kg^{-1} DM, depending on the site, but this may not always be sufficient to meet the requirements of sheep and cattle. Australian research has established that superphosphate applications to pastures, over and above those required for maximum plant growth responses, can result in herbage of improved palatability and digestibility, giving significantly greater weight gains in sheep and cattle, wool yields and lamb and calf crops (Winks, 1990). However, expected responses do not always materialize. A further complication is that the botanical composition of leguminous tropical pasture can be changed by applying phosphatic fertilizer (Coates et al., 1990).

The position with grains and seeds is similar to that just described for whole plants. The phosphorus concentrations in wheat, barley and oat grains from soils low in available phosphorus are generally only 50–60% of those from more fertile soils. However, phosphorus levels can be raised from 2–3 to 4–5 g kg^{-1} DM by superphosphate applications, which, at the same time, increase grain yields.

Trace element and sulfur fertilizers

In some areas, small applications of molybdenum to deficient soils markedly increases legume yields and herbage molybdenum and protein concentrations. The increase in protein is usually advantageous to the grazing animal, but the increases in molybdenum are of no value, except where copper intakes are high. In the latter circumstance, copper retention is depressed and the chances of chronic copper poisoning are reduced. Where pasture copper is low, however, copper deficiency may be induced in sheep and cattle.

Zinc and selenium concentrations in grains and pastures reflect the soil status of those minerals and fertilizer usage. Wheat grain from zinc-deficient soils in Western Australia has been found to average only 16 mg Zn kg^{-1} DM compared to 35 mg kg^{-1} DM for wheat from the same soil fertilized with zinc oxide at 0.6 kg ha^{-1} (Underwood, 1977). The effect of soil selenium status on wheat grain selenium is striking: median values of 0.80 and 0.05 mg kg^{-1} DM have been reported from soils high and low in selenium, respectively, in the USA (Scott and Thompson, 1971) and even lower values (down to 0.005) occur in parts of New Zealand, Finland and Sweden where soils are extremely deficient. The application of selenium fertilizers has increased pasture selenium from 0.03 to 0.06 mg kg^{-1} DM and improved sheep production in Western Australia (see Chapter 15). The selenium concentration in sugarcane tops was found to be 6–14 times higher in seleniferous than in non-seleniferous areas in India, but was reduced from 15.2 to 5.1 mg kg^{-1} DM by applying gypsum as a fertilizer at 1 t ha^{-1} (Dhillon and Dhillon, 1991). Much smaller applications of gypsum are used to raise pasture sulfur (see Chapter 9). The nature of the soil and its treatment are thus important determinants of the value of seeds and forages as sources of minerals to animals.

Nitrogenous and potassic fertilizers

There is a widely held view that the 'improvement' of permanent pasture by reseeding and applying nitrogenous fertilizer increases the risk of mineral deficiencies in grazing livestock. However, a study involving 16 sites in England and Wales over 4 years did not support that generalization: the concentrations of three minerals (magnesium, sodium and zinc) rose, four (calcium, manganese, molybdenum and sulfur) fell and three (potassium, cobalt and copper) were not consistently changed (Hopkins et al., 1994). It was concluded that only where availability fell (for copper and magnesium) was the risk of deficiency increased by pasture improvement. However, fertilizers are applied to increase yields of forage and crop and their use increases the rate at which minerals are 'exported' in products from the site of application and thus increases the risk of mineral deficiency (see final section). Heavy applications of nitrogenous fertilizers can depress legume growth, thereby reducing the overall calcium content of a legume/grass sward. Heavy applications of potassium fertilizers can raise herbage yields and potassium contents, while at the same time depressing herbage magnesium and sodium.

The Influence of Maturity and Season

Plants mature partly in response to internal factors inherent in their genetic makeup and partly in response to external factors, either natural (e.g. season, climate) or manmade (e.g. provision of irrigation, shelter), and there are associated changes in mineral composition. Phosphorus concentrations of crop and forage plants decline markedly with advancing maturity, although the decline is less in legumes than in grasses (Coates et al., 1990). Concentrations of many other elements (e.g. cobalt, copper, iron, potassium, magnesium, manganese, molybdenum and zinc) also decline, but rarely to the same extent as those of phosphorus. Such changes often reflect increases in the proportion of stem to leaf and old to new leaves, with stems and old leaves having lower mineral concentrations than young leaves (Fig. 2.3) (Minson, 1990), although this does not apply to manganese in Fig. 2.3.

Seasonal fluctuations in calcium, phosphorus, magnesium, potassium, manganese, copper and selenium in bahia grass have been found to persist even when pasture is kept in an immature state by rotational grazing and clipping (Kappel et al., 1983). Contrasts between wet- and dry-season forages have been reported in Florida, with wet-season forage the higher in potassium, phosphorus and magnesium by 110%, 60% and 75%, respectively (Kiatoko et al., 1982). There is a negative correlation between rainfall and the selenium concentration in wheat grain (White et al., 1981). The tendency for forage concentrations of copper to increase and selenium to decrease with increasing altitude in the dry season (Jumba et al., 1996) is probably a reflection of the covariation in rainfall. Selenium concentrations in sugarcane are much higher in the tops than in the cane (5.7–9.5 versus 1.8–2.1 mg kg^{-1} DM) on seleniferous soils, but values decline with maturity (Dhillon and Dhillon, 1991). Minerals are lost with the shedding of seed and the remaining stem or straw is low in most minerals (Suttle, 1991). Furthermore, standing straw is subject to leaching of phosphorus and potassium (see Chapter 8). In a study of two Australian barley crops, the average phosphorus and potassium concentrations of straw fell from 0.7 to 0.4 and from 9 to 0.9 g kg^{-1} DM, respectively, in the 5 months following

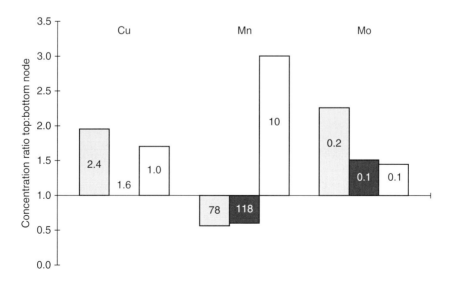

Fig. 2.3. Concentrations of minerals in plants vary from part to part and with state of maturity. Values for leaf (grey), sheath (black) and stem (white) are given in different columns. The height of the column indicates the proportionate difference between the parts from the top (youngest) and bottom (oldest) nodes of wheat plants (actual values for bottom node are given within columns in mg kg^{-1} DM) (Burridge et al., 1983).

harvest, but copper, zinc and manganese showed no significant decline (Southern, 1979). Calcium and silicon show unusual responses to season and maturity: calcium concentrations usually remain stable while those of silicon increase as the plant matures, and concentrations of both elements are higher in straw than in grain.

Analytical and Experimental Precision

The results of feed analysis must be accurate and errors have decreased with the widespread application of quality control and approved standard feed materials (e.g. from the National Bureau of Standards, USA). The main sources of error now arise from contamination during the collection and processing of samples for analysis. Milling can contaminate samples with iron, cobalt, manganese and nickel, but contamination can be reduced by the use of mills with agate-tipped blades or coffee grinders instead of laboratory mills made of steel.

When the mineral composition of diets is manipulated by adding supplements, the intended mineral concentrations may not match those attained and should always be checked before an experiment is started to spot mistakes and allow correction. In one published study involving eight different diets (Sands et al., 2001), it appeared that a phosphorus supplement was added twice to one diet and a limestone supplement omitted from another. Such errors can pass unnoticed, yet profoundly affect the validity of conclusions drawn from an experiment.

Soil Ingestion

Soil and dust contamination of herbage can at times provide a further significant source of minerals to grazing animals, especially when grazing intensity is high or when pasture availability is low. Daily soil intake can rise to 163 g in sheep in arid conditions (Vaithiyanathan and Singh, 1994) and constitute 10–25% of the total DM intake by outwintered sheep and cattle in temperate zones (Healy et al., 1974; Thornton, 1974). With elements such as cobalt, iodine and selenium, which occur in soils in concentrations usually much higher than those of the plants growing on them, soil ingestion can be beneficial to the animal (Grace et al., 1996; Grace, 2006). By contrast, the copper antagonists iron, molybdenum and zinc are biologically active in soils and their ingestion from soil contamination of herbage may be a factor in the aetiology of hypocuprosis in cattle and 'swayback' (copper deficiency) in sheep (Suttle et al., 1984). Soil ingestion can also provide an important route for toxic elements such as lead (Hill et al., 1998), cadmium (Bramley, 1990) and fluorine (Grace et al., 2003a,b) to accumulate in the tissues of grazing livestock. Contamination of forage samples with soil is a common cause of anomalously high cobalt, iron and manganese concentrations.

Atmospheric Inputs

Significant amounts of minerals can enter ecosystems from the atmosphere. A New Zealand study found that inputs of sulfur in rainfall varied from 0.5 to 15.0 kg ha^{-1}, decreasing exponentially with distance from the coast (Ledgard and Upsdell, 1991). Similar gradients exist for iodine and sodium. Inputs of sulfur from industrial sources also occur, but are declining in many countries under the influence of stricter controls of pollution. In Scandinavia, the decreasing acidity of rain is believed to be responsible for increases in soil pH, with consequent increases in plant uptake of molybdenum towards the end of the last century. Contamination with fluorine-containing dusts and fumes from industrial sources and the dispersal of wastes from mining activities can pose a further hazard to livestock in some areas (see Chapter 18). The removal of lead from petrol in Finland in 1993 was associated with a reduction in lead concentrations in the liver and kidneys of moose in 1999 compared with values for 1980 (Venlainen, 2007).

Representative Values for Mineral Composition of Feeds

Crops and forages

Given the many factors that can influence the mineral composition of a particular crop or forage, there are considerable difficulties in deriving representative values and limitations to their use. Many have used NRC values for representative purposes (McDowell, 2003), but these have long been in need of revision (Berger, 1995). Workers in Alberta derived a local database that indicated a far greater risk of copper deficiency and lesser risks of zinc deficiency in Albertan cattle than was evident from the NRC standards (Suleiman et al., 1997). Large coefficients of variation accompanied many mean values, with those for lucerne (alfalfa) hay ranging from 23.8% to 107.4% for phosphorus and manganese, respectively (over 220 samples), indicating a need for large sample sizes. The problem is summarized in Table 2.2, which includes data from the UK and USA. For barley grain, there was a general lowering of copper and increase in zinc concentration compared with the old NRC standard. For lucerne hay and corn silage there were no 'global' trends, but these may have been masked by some extreme values in the case of manganese and zinc.

Two final points should be noted:

- When population distributions are clearly skewed, median values are more representative and may show closer agreement between states or countries and with historical datasets.
- The adequacy of the mineral supply from a particular compound ration can rarely be predicted with sufficient accuracy from tables of mineral composition for the principal feed constituents.

Crop by-products

Extensive use has been made of crop residues such as oilseed cakes as sources of energy, protein and fibre in livestock diets and their inclusion generally enriches the whole diet with minerals. However, crop by-products can vary more widely in mineral composition than the whole crop (Arosemena et al., 1995). Data for three by-products are presented in Table 2.3 to illustrate the effect of the industrial process on mineral content. Differences between nutrients from the brewery and distillery industries in the UK and USA derive partly from the different grains used. The high and variable copper concentration in distillery by-products from the UK arises from the widespread use of copper stills for the distillation process. Differences between countries for soybean meal (SBM) are small because the USA is the major supplier to the UK. Note that for a given source of by-product, only phosphorus values have sufficiently low variability for useful predictions to be made of the contribution of the mineral from the by-product to the total ration. Most brans are good sources of phosphorus, but by-products such as beet pulp, citrus pulp and sorghum hulls are not. In tropical countries, by-products of the sugarcane industry such as bagasse and molasses have formed the basis of successful beef and milk production enterprises, processes that will increase as dual-purpose crops are grown for fuel as well as for foods and feeds (Leng and Preston, 1984). A watchful eye must be kept on antagonists such as goitrogens in rapeseed meals (see Chapter 12) and the feed value of by-products will only be realized if any shortcomings in mineral or antagonist content are recognized and overcome.

Drinking Water

Drinking water is not normally a major source of minerals for livestock, although there are exceptions (Shirley, 1978). Sulfur concentrations in water from deep aquifers can reach 600 mg l^{-1}, adding 3 g S kg^{-1} DM to the diet as sulfates. This is far more than any nutritional requirement and may even create problems by inducing copper deficiency (Smart et al., 1986). Iodine concentrations in drinking water can be barely detectable in east Denmark but rise as high as 139 µg l^{-1} in islands off the west coast where the groundwater filters through marine sediments, rich in iodine, on its way to the aquifers. In most

Table 2.2. The mean (standard deviation) mineral concentrations in three major foodstuffs from three different countries compared with the widely used NRC (1982) database US–Canadian Feed Composition Tables.

Feed	IFN	Source	Macro-element (g kg^{-1} DM)					Trace element (mg kg^{-1} DM)			
			Ca	P	K	Mg	Cu	Mn	Se	Zn
Barley	4–07–939	CAN	0.6 (0.2)	3.7 (0.6)	5.4 (1.8)	1.5 (0.2)	5.4 (2.4)	18.9 (6.7)	0.11 (0.09)	42.3 (10.5)
	4–00–549	USA	0.7 (0.02)	3.8 (0.07)	6.3 (0.11)	1.4 (0.02)	7.0 (2.0)	22.0 (6.0)	–	38.0 (7.0)
	4–00–549	UK	0.9 (0.6)	4.0 (0.5)	5.0 (0.7)	1.2 (0.2)	4.2 (1.5)	18.5 (3.6)	0.10 (–)	32.5 (8.5)
		NRC	0.5	3.7	4.7	1.5	9.0	18.0	–	17.1
Lucerne hay	1–00–059	CAN	17.9 (5.3)	2.1 (0.5)	17.5 (4.9)	3.1 (0.9)	6.1 (2.5)	43.7 (41.8)	0.27 (0.29)	25.4 (11.7)
	1–00–059	USA	13.1 (3.3)	3.0 (0.6)	26.3 (4.4)	2.7 (0.8)	4.4 (6.3)	22.7 (13.9)	–	13.4 (17.5)
	1–00–078	UK	15.6 (1.8)	3.1 (0.7)	27.3 (5.0)	1.7 (0.3)	–	–	–	–
		NRC	14.2	2.2	25.4	3.3	11.1	31.0	–	24.8
Maize silage	3–00–216	CAN	2.6 (1.2)	2.3 (0.5)	11.2 (2.6)	2.5 (0.9)	4.4 (2.2)	37.8 (18.8)	0.04 (0.02)	34.0 (13.1)
	3–28–250	USA	2.5 (0.14)	2.3 (0.6)	10.8 (3.3)	1.8 (0.4)	3.0 (2.7)	17.1 (20.3)	–	12.2 (18.8)
	3–02–822	UK	4.3 (2.0)	2.6 (1.2)	12.3 (4.1)	2.2 (0.7)	5.2 (0.75)	14.6 (7.2)	–	45.3 (13.9)
		NRC	2.3	2.2	9.9	1.9	10.1	30.0	–	21.0

IFN, International Food Number; CAN, data from Alberta (Suleiman et al., 1997); USA, pooled data from the states of New York, Indiana, Idaho and Arizona (Berger, 1995); UK, data from the Ministry of Agriculture, Fisheries and Food (MAFF, 1990).

Table 2.3. Mean (standard deviation) mineral composition of crop by-products can vary with the source and nature of the industrial process.

Crop by-product	IFN	Source	(g kg⁻¹ DM)		(mg kg⁻¹ DM)	
			Ca	P	Cu	Zn
Brewers' grains	5–02–141	USA (maize)	3.3 (1.2)	5.9 (0.8)	11.2 (5.0)	97.0 (16.0)
	5–00–517	USA (barley)	3.5 (1.4)	5.1 (1.0)	18.7 (9.2)	72.8 (12.5)
Distillers' dried	5–28–236	USA (maize)	2.9 (1.5)	8.3 (1.7)	9.7 (9.0)	73.5 (47.0)
grains	5–12–185	UK (barley)	1.7 (0.32)	9.6 (0.80)	40.5 (16.8)	55.4 (4.7)
Soybean meal						
Dehulled	5–04–612	USA	4.1 (2.9)	7.2 (2.8)	17.7 (7.0)	69.5 (141)
Extracted	5–04–604	UK	3.9 (1.6)	7.4 (0.44)	15.8 (2.6)	49.0 (9.9)

IFN, International Food Number; USA, data pooled from four states (Berger, 1995); UK, data collected by ADAS in England (MAFF, 1990).

natural fluorosis areas, such as those in the semi-arid interior of Argentina, the water supplies and not the feed are mainly responsible for supplying toxic quantities of fluorine (see Chapter 18). In some parts of the world the water available to animals is so saline that sodium and chlorine are ingested in quantities well beyond the requirements of those elements (see Chapter 8). Some 'hard' waters also supply significant concentrations of calcium, magnesium and sulfur and occasionally of other minerals. Individual daily water consumption is highly variable, as is the mineral composition of different drinking-water sources. A New Zealand study with grazing beef cattle reported a fourfold variation in individual water intakes (Wright et al., 1978).

Meaningful 'average' mineral intakes from drinking water are therefore impossible to calculate.

Milk

The most important animal source of minerals for species that suckle their young is milk. The mineral composition of milk varies with parity, stage of lactation, nutrition and the presence of disease. Representative values for the mineral constituents of the main milk of various species are given in Table 2.4, and they generally correlate with ash concentration. Milk is clearly a rich source of calcium, phosphorus, potassium, chlorine and zinc, but is much less satisfactory

Table 2.4. Mineral concentrations in the main milk of farm animals (representative values).

Animal	(g l⁻¹)						(mg l⁻¹)			
	Ca	P	Mg	K	Na	Cl	Zn	Fe	Cu	Mn
Cow	1.2	1.0	0.1	1.5	0.5	1.1	4.0	0.5	0.15	0.03
Ewe	1.9	1.5	0.2	1.9	–	1.4	4.0	0.5	0.25	0.04
Goat	1.4	1.2	0.2	1.7	0.4	1.5	5.5	0.4	0.15	–
Sow	2.7	1.6	–	1.0	0.03	0.9	5.0	1.5	0.70	–
Mare	0.84[b]	0.54[b]	0.06[b]	0.59[b]	0.12[b]	0.2	2.1[b]	0.34[b]	0.19[b]	0.05[b]
Buffalo	1.8	1.2	–	–	–	0.6	–	–	–	–
Llama[a]	1.7	1.2	0.15	1.2	0.27	0.7	–	–	–	–

[a]Morin et al. (1995).
[b]Grace et al. (1999).

Effects of stage of lactation, parity and species

The first milk or colostrum is usually richer in minerals than later milk, as shown for bovine colostrum in Table 2.5. In the study shown here, by the fourth calving concentrations of calcium, phosphorus, magnesium and zinc in colostrum had fallen to 72–83% of values for the first calving, but there was no effect of parity after 4 days of lactation (Kume and Tanabe, 1993). The length of time that suckled offspring remain with their mothers varies widely according to species and management practice, but would commonly be 42 days for piglets, 120 days for lambs, 150 days for kids, 180 days for foals and 300 days for beef calves. The ash content of the milk tends to decline with the lactation length of the species. The sow and the ewe secrete milk of the highest ash content (almost $10\,g\,l^{-1}$). They are followed by the goat and the cow (7–8 g l^{-1}). Mares secrete milk exceptionally low in ash (about $4\,g\,l^{-1}$), mainly due to a low concentration of the principal mineral, potassium (Table 2.5), but the volume secreted each day is high (20 l). Peak concentrations of calcium and phosphorus in mare's milk (1.25 and $0.90\,g\,l^{-1}$, respectively) were not reached until the seventh day of lactation in Grace et al.'s (1999) study. The relative growth rate of the suckled offspring declines with lactation length so that mineral concentration tends to match need across species.

Effect of diet

The effects of diet on the composition of milk vary greatly for different minerals (Kirchgessner et al., 1967). Deficiencies of calcium, phosphorus, sodium and iron are reflected in a diminished yield of milk, but not in the concentrations of these minerals in the milk that is secreted. In copper and iodine deficiencies, by contrast, there can be a marked fall in milk copper and iodine. In cobalt deprivation, concentrations of vitamin B_{12} in milk decrease and can be boosted by cobalt or vitamin B_{12} supplements (see Chapter 10). Iodine and molybdenum concentrations in milk can also be increased by raising mineral intakes, but those of copper, manganese, molybdenum and zinc cannot. In the case of selenium, the effect of dietary supplements depends on the source: inorganic selenium supplements have a limited effect on milk selenium, but organic selenium in the form of selenomethionine or brewers' yeast is preferentially partitioned to milk, giving large increases in milk selenium (Givens

Table 2.5. Mean colostrum yield and mineral concentrations in colostrum of 21 cows shortly after parturition (Kume and Tanabe, 1993). Comparative values for mare's milk (Grace et al., 1999) at 24 and 72 h are given in parentheses.

	Time after parturition (h)				
	0	12	24	72	SD
Colostrum yield (kg day^{-1})	11.7	–	14.9	21.6	2.4
Ca (g l^{-1})	2.09	1.68	1.43 (0.69)	1.25 (1.16)	0.19
P (g l^{-1})	1.75	1.43	1.25 (0.71)	1.01 (0.89)	0.17
Mg (g l^{-1})	0.31	0.21	0.15 (0.30)	0.11 (0.10)	0.04
Na (g l^{-1})	0.69	0.64	0.58 (0.56)	0.53 (0.17)	0.008
K (g l^{-1})	1.48	1.49	1.58 (0.96)	1.50 (0.98)	0.12
Fe (mg l^{-1})	2.0	1.5	1.2 (0.79)	1.1 (0.34)	0.6
Zn (mg l^{-1})	17.2	10.3	6.4 (5.5)	5.2 (2.7)	2.5
Cu (mg l^{-1})	0.12	0.09	0.08 (0.76)	0.08 (0.37)	0.03
Mn (mg l^{-1})	0.06	0.04	0.03	0.02	0.02

SD, standard deviation.

et al., 2004). Some potentially toxic elements in the mother's diet can be passed to the offspring via her milk (e.g. lead), but maternal fluorine exposure has little effect on milk fluorine (see Chapters 17 and 18).

Animal By-products

Bovine milk and the by-products of butter and cheese manufacture – skimmed milk, buttermilk and whey, in either liquid or dried forms – are valuable mineral as well as protein supplements to cereal-grain diets for pigs and poultry, particularly for calcium (Table 2.6). Skimmed milk and buttermilk differ little in mineral composition from the milk from which they are made, since little of the minerals, other than iodine, is separated with the fat in cream and butter manufacture. They also vary little from source to source, except for iron content. Whey contains less calcium and phosphorus than separated milk or buttermilk, because much of the phosphorus and a proportion of the calcium and other minerals in the milk separate with the curd in cheese manufacture, but whey is a valuable source of many minerals.

Meat and fish by-products from abattoirs and processing factories, given primarily to supply protein, also supply minerals, especially calcium, phosphorus, iron, zinc and selenium (Table 2.6). Mineral content varies greatly with the source materials and processing methods employed. Thus, blood meal, liver meal and whale meal contain only 2–3% total ash and are low in calcium and phosphorus, but they can be valuable sources of iron, copper, selenium and zinc. Commercial meat meals and fish meals, on the other hand, are usually rich in calcium, phosphorus, magnesium and zinc, depending on the proportion of bone they contain. The ash content of meat and bone meals varies from 4% to 25% and most of this ash consists of calcium and phosphorus. A typical meat meal containing 6–8% calcium (and 3–4% phosphorus), about 100 times (ten times) more than the cereal grains they commonly supplement. Fish meal often contains appreciable amounts of sodium chloride as well as calcium and phosphorus. The use of animal by-products has recently been restricted for fear of transmitting harmful pathogens, but the restriction will probably be lifted for fish meals.

Forms of Minerals in Feeds

Minerals occur in plants and foods in diverse forms, some of which may influence the efficiency which they are absorbed and utilized by livestock. Iron can be found in the storage protein ferritin in soybeans and bound to phytate in cereals. Both iron and manganese are poorly absorbed from corn/SBM diets by pigs and poultry, and manganese shares with iron an affinity for complexing agents such as ferritin (see Chapters 13 and 14). The association of iron and manganese with storage compounds

Table 2.6. Macro-mineral concentrations (g kg^{-1} DM) in animal by-products (source MAFF, 1990; NRC as cited by McDowell, 2003). Standard deviations given in parentheses are for MAFF data.

	Ca	P	Mg	Na	K	S
Battery waste	32	18	0.5	–	17	13
Broiler waste	93	25	0.6	–	23	0.2
Blood meal	0.6 (0.26)	1.5 (0.19)	0.2 (0.01)	3.6 (0.32)	1.9 (0.12)	8.4 (0.25)
Butter milk	14.4	10.1	5.2	9.0	9.0	0.9
Feather meal	5.6 (0.07)	3.1 (0.06)	0.4 (0.04)	1.4 (0.01)	1.5 (0.01)	18 (0.6)
Fish meal	56 (6.0)	38 (14.5)	2.3 (0.3)	11.2 (1.5)	10.2 (1.3)	5.0
Meat and bone meal	90 (13)	43 (6.5)	2.2 (0.2)	8.0 (1.0)	5.2 (0.6)	9.2 (0.72)
Whey	9.2	8.2	2.3	7.0	12	11.5

in grains may therefore limit their value to livestock, but the precise forms in which they occur may also be important. For example, iron is well absorbed by poultry from wheat, in which it occurs as a well-absorbed mono-ferric phytate (see Chapter 13) (Morris and Ellis, 1976). Phytate is a hexaphosphate and provides most of the phosphorus inherent in commercial pig and poultry rations (see Chapter 6), but it is poorly absorbed and forms unabsorbable complexes with calcium and several trace elements. Elements such as copper and selenium perform similar functions in plants and animals and occur in fresh plant tissue in the enzymic forms of cytochrome oxidase and glutathione peroxidase, respectively. The mineral composition of milk plays a crucial part in the development of the neonate, containing iron bound to lactoferrin, a form that can be used by the animal but not by potentially harmful gut pathogens. Milk also contains minerals in enzymic (e.g. molybdenum in xanthine oxidase) and hormonal (triiodothyronine) forms, the latter facilitating early maturation of the intestinal mucosa. Much of the zinc in milk is bound to casein, but the precise binding may influence the ability of one species to absorb zinc from the milk of another species (see Chapter 16).

Measuring the 'Mineral Value' of Feeds

The value of feeds (and supplements) as sources of minerals depends not only on their mineral content, but also on the proportion of minerals that can be utilized by the animal, a property termed 'bioavailability' (for reviews see Hazell, 1985 and Ammerman et al., 1995). Bioavailability has four components:

- accessibility – the potential access of the plant mineral to the absorptive mucosa;
- absorbability – the potential transfer of absorbable mineral across the mucosa;
- retainability – the potential retention of the transferred mineral; and
- functionality – the potential for incorporation of retained mineral into functional forms.

Accessibility is affected by the chemical forms of mineral present and their interactions with agonists or antagonists in the feed or in the gut. It is generally the major determinant of bioavailability and hence the mineral requirement of the animal. Absorbability is determined by the capacity of the mucosa to take up accessible mineral. If uptake mechanisms are non-specific, one element may decrease the uptake and therefore the absorbability of another (e.g. the reciprocal antagonism between iron and manganese; see Chapter 13). Retainability reflects the ability of circulating mineral to escape excretion via the kidneys or gut. Functionality can occasionally be affected by the forms in which the mineral is absorbed and site of retention: absorbed iron may get no further than the gut mucosa, while exposure to cadmium may cause copper to be trapped there; the selenium from selenium yeast is largely in a form (selenomethionine) that is poorly incorporated into functional selenoproteins.

The multiplicity of methods for measuring bioavailability (Suttle, 1985; Ammerman et al., 1995) gives an indication of the difficulties involved. While most methods can satisfactorily rank mineral supplements, far fewer assess the potential bioavailability of a feed source because the test animal's genotype, age, mineral status and needs can constrain absorbability, retainability and functionality.

In Vitro Determination of Mineral Accessibility

Animal influences on bioavailability can be avoided by using *in vitro* methods, but early results have generally lacked credibility and limitations are acknowledged by the introduction of the term 'bioaccessibility' to describe assessments made without animals. Reliance on simplistic attributes, such as solubility of the plant mineral in water or unphysiological extractants (e.g. citric acid), produces results that correlate poorly with *in vivo* assessments of 'mineral value', but methods have improved. Solubility after the sequential simulation of gastric and intestinal digestion can give convincing results, testing the stability of amino acid–trace element chelates in the gut (Fig. 2.4) and

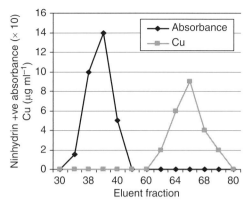

Fig. 2.4. The stability of trace element–amino acid complexes at pH conditions encountered during passage through the gut has been tested *in vitro*. The separate movements of the amino (ninhydrin-positive) and copper components of a copper–lysine complex after chromatography at a pH of 2 indicate probable breakdown of the complex in the stomach. Similar separation occurred for all 15 trace element–amino acid complexes tested, whether the pH was 2 or 5 (Brown and Zeringue, 1994), and explains the lack of efficacy of metal proteinates *in vivo*.

evaluating the accessibility of phytate phosphorus in cereals to poultry (see Chapter 6). A final dialysis stage can be added. Interlaboratory collaboration has thus yielded a standardized procedure for measuring the bioaccessibility of iron (Luten *et al.*, 1996). Approximate estimates of absorbability can be obtained by culturing cells in the dialysate (e.g. Caco 2 cells, see Chapter 12).

Under conditions of simulated digestion, the completeness of extrinsic labelling with an isotope such as zinc-65 (^{65}Zn) may predict the accuracy of a potential tracer *in vivo* for a particular feed (Schwarz *et al.*, 1982). For ruminants, the accessibility of minerals to rumen microbes is important. This has been measured *in vitro* as rates of vitamin B_{12} synthesis in continuous cultures of rumen microbes for cobalt sources as diverse as ingested soil and a novel 'chelate' (with propionate) (see Chapter 10). Increasing use will be made of such *in vitro* techniques, but account must be taken of possible interactions between feeds and results will always have to be validated *in vivo* – both of which are more difficult tasks (e.g. Menedes *et al.*, 2003).

In Vivo Measurement of 'Absorbability'

Apparent absorption

The difference between intake and faecal excretion of a mineral in a 'digestibility trial' is the amount 'apparently' absorbed and an approximate measure of accessibility plus absorbability, provided the mineral is not being absorbed by the animal according to need or excreted via the faeces when mineral intakes exceed requirement. Early techniques required animals to be housed in metabolism crates to facilitate the total collection of faeces that were free from contamination with urine and detritus. In poultry, cannulation of the ileum is required to allow disappearance of mineral from the gut to be calculated before the mineral re-enters the large bowel in urine. The need for housing and/or total faecal collection can be avoided by measuring faecal concentrations of the ingested mineral relative to those of an inert, exogenous, unabsorbable 'marker' such as chromium (see Chapter 17) or an unabsorbable constituent of the diet, either inorganic (e.g. pasture titanium) or organic (e.g. an alkane) whose total intake or dietary concentration is known. Markers that colour the faeces, such as chromic and ferric oxides, can be used to synchronize the faecal collection period with the feeding period (e.g. Pedersen *et al.*, 2007).

With minerals such as calcium, iron, manganese and zinc that are absorbed according to need, the amount apparently absorbed will be less than the available fraction since the *active* absorptive process is down-regulated and absorbability (^{max}A), the key factor in determining gross requirement (Chapter 1), can only be measured at mineral intakes at or below need (AFRC, 1991). Alfalfa is widely believed to contain poorly available calcium, but is so rich in calcium that absorbability may never have been evaluated because provision has usually exceeded need (see Chapter 4). For minerals such as copper and manganese for which the faeces can constitute the major route of endogenous excretion, the amount apparently absorbed again underestimates the 'mineral value' if animals have been given more mineral than they need.

True absorption using isotopes

Radioactive and stable isotopes can be used to measure the endogenous component of total faecal mineral excretion and thus allow the true extent of mineral absorption to be calculated in balance studies. Re-entry of absorbed mineral into the gut can be measured by labelling the pool from which gut secretions are drawn with radioisotopes. Following a single parenteral injection of radioisotope (^{45}Ca), total collection of faeces and measurements of total excretion of both radioisotope and stable mineral, Comar (1956) calculated the faecal excretion of calcium in cattle by assuming that the ratio of radioactive to stable calcium in gut secretions was the same as that in a plasma sample, once equilibrium was reached (urine or saliva samples can also be used), thus:

$$TA = I - (F - FE) \quad [2.1]$$

where TA is true absorption, I is intake, F is faecal excretion and FE is faecal endogenous loss of calcium.

Giving radioisotopes orally *and* intravenously, either to the same animals after an appropriate clearance phase or simultaneously to two matched groups (Atkinson et al., 1993), absorptive efficiency can be similarly calculated:

$$TA = O - (F_O - F_I) \quad [2.2]$$

where O is oral administration and F_O and F_I denote faecal excretion of the respective doses of oral and intravenous isotopes.

This 'comparative balance' technique is useful for elements that may contaminate the experimental environment, polluting both feed and faeces (e.g. iron, zinc and copper) and invalidating a conventional balance study. Where two isotopes of the same element are suitable (e.g. ^{55}Fe and ^{59}Fe) (Fairweather-Tait et al., 1989), their comparative balance can be assessed after simultaneous dosage by oral and intravenous routes. The absorption of as many as five elements has been assessed simultaneously in piglets (Atkinson et al., 1993). Absorption coefficients can also be derived from the kinetic models described earlier (see Chapter 1) since the sustained faecal excretion of isotope after the unabsorbed oral dose of isotope has passed through the animal is of entirely endogenous origin and the amounts lost can be calculated from their specific radioactivity in the conventional way.

Problems with isotopic markers

For an isotopic marker to give valid information on 'mineral value' it must mimic the physiological and biochemical behaviour of all forms of mineral present in the feed, from digestion through to excretion (e.g. Buckley, 1988). The lower the accessibility of a mineral in a feed, the greater the likelihood of an orally administered tracer (usually given as a soluble and highly accessible inorganic salt) overestimating true absorption. Failure to recognize the distinctive metabolism of different forms of selenium in feeds may have given aberrant values for the absorbability of selenium in hay, using an oral inorganic selenium source as the tracer (Krishnamurti et al., 1997). When the tracer is given parenterally, only the absorbable mineral is being 'traced' and errors are generally reduced, although not in the case of manganese (Davis et al., 1992) or selenomethionine (see Chapter 15). The accuracy of tracing can be improved by growing plants in isotope-enriched (radioactive or stable) nutrient solutions, a technique known as 'intrinsic labelling', but the technique is laborious and expensive to apply to large farm animals. Furthermore, results with laboratory animals are often similar to those obtained with extrinsic labelling, with selenium, chromium and cadmium being the exceptions (Johnson et al., 1991).

The natural presence of stable isotopes of certain minerals in feeds promised to add a new dimension to the assessment of their availability (Turnlund, 1989). The natural isotope ratio can be perturbed by dosing animals with the least abundant isotope and applying the same 'dilution' principle as that used for radioisotopes to calculate true absorption of the naturally abundant isotope from the feed (Fairweather-Tait et al., 1989). The stable isotope approach has the advantage of avoiding the hazards and safety constraints associated with the use of radioisotopes. However, care must be taken not to administer so much stable tracer that metabolism is disturbed. Expensive equipment (e.g. an inductively coupled 'plasma' generator and mass spectrometer) is required and stable isotope techniques have yet to be applied to the estimation of mineral absorption in livestock.

Isotope-free estimation of true absorption

There are two ways of correcting balance data for faecal endogenous loss that do not require isotopes, in addition to those already mentioned (faecal regression and the mineral-free diet, see Chapter 1). When excretion or secretion of a mineral is closely correlated to mineral intake – as is the case for magnesium in urine or molybdenum and iodine in milk – the slopes ratio method can be applied to rates of excretion or secretion. If retention and faecal endogenous loss remain constant, the slopes represent the fractional absorption of the mineral. Thus, the bioavailability of magnesium has been assessed from rates of urinary magnesium excretion in cattle (see Chapter 5). This 'comparative loss' approach has also been modified by the use of radioisotopes. Lengemann (1969) gave lactating goats both Na^{131}I and Na^{125}IO$_3$ orally, and concluded from the ^{125}I/^{131}I ratio in milk that iodate was absorbed to 0.86 of the value of iodide.

The final method requires cannulation of the portal vein (carrying absorbed mineral to the liver) and bile duct (carrying surplus absorbed mineral to the gut). After measuring portal and biliary flow rates, true absorption can be calculated by correcting the amount apparently absorbed for the fraction of the portal mineral flow that was excreted. This technique has given an absorbability of manganese in pigs on a corn/SBM diet of only 0.05 (Finley et al., 1997).

Problems with isotope-free methods

There has been a resurgence of interest in the faecal regression technique following the publication of a 'novel' method for estimating the faecal endogenous loss and true absorption of phosphorus in pigs (Fan et al., 2001), but a 10-fold variation in the resultant estimates of 'minimum' faecal endogenous loss (see Pettey et al., 2006) and extremely high true absorption of phosphorus (>0.5) for SBM and corn in pigs and poultry suggest that the technique is unreliable. The published method (Fan et al., 2001) breaks the golden rule for determining mineral requirements – that no other mineral or nutrient should be lacking – by using the feed under test as the only incremental source of calcium and protein as well as phosphorus (as little as 6% crude protein was fed and no calcium supplement given). Although the balance periods were relatively short, it is possible that gross calcium and protein deprivation at the lowest corn or SBM inclusion level impairs bone mineralization. With nowhere to go, more absorbed phosphorus than usual may be recycled and appear as 'minimal' faecal endogenous loss. Furthermore, since dietary calcium concentration largely determines how much phytate phosphorus remains unavailable (see Chapter 6), the high estimates of true absorption of phosphorus obtained for phytate-rich corn and phytate-rich SBM in calcium-deficient diets are of no practical value. While the mineral-free, semi-purified diet technique gives relatively consistent values for faecal endogenous loss of phosphorus in pigs (138–211 mg kg^{-1} DM intake) (Widmer et al., 2007), they are two- to threefold higher than those obtained by faecal regression with diets adequate in calcium (Pettey et al., 2006). Thus, the estimation of 'phosphorus value' for pig and poultry feeds by balance methods remains problematic, although not intractable (see Chapter 6).

Measurement of 'Retainability'

Whole-body retention

The 'mineral value' of feeds can be assessed via the fraction of ingested mineral that is retained in the body rather than removed from the digesta (Suttle, 1985). Assessment methods fall into three categories, each of which has disadvantages. First, retention of surrogate radioisotopes of minerals can be measured with whole-body monitors (Fig. 2.5), but the reliability of either orally or parenterally administered isotope remains questionable. For

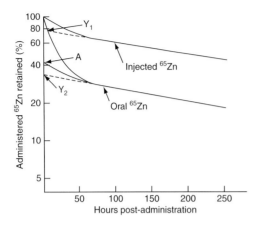

Fig. 2.5. A comparison of retention curves for zinc-65 given in the feed and by intramuscular injection. Percentage absorption (A) of oral is calculated by dividing Y_2 (the y intercept of the extrapolated retention curve for oral ^{65}Zn) by Y_1 (the y intercept of the extrapolated retention curve for injected ^{65}Zn) and multiplying by 100 (from Heth and Hoekstra, 1965).

example, if the disappearance rates of the two isotopes do not become parallel then the absorbed and parenteral isotope are being metabolized differently, invalidating the conventional correction for faecal endogenous loss (Davis et al., 1992). Second, the retention of mineral recorded in balance studies has been used, but it is often a small yet highly variable statistic bearing the cumulative errors of estimating intake, faecal excretion and urinary excretion of the mineral. Third, the mineral content of the carcass at the end of a feeding period can be compared with that of an initial slaughter group. Carcass analysis methods require the accurate homogenization and subsampling of the carcass. All retention methods are subject to the same animal influences as absorption methods and may underestimate 'retainability', but the amounts apparently absorbed and retained generally correlate well in growing animals, provided the animals are not provided with excesses of absorbed mineral that are then excreted. Retention methods can underestimate absorption by failing to detect losses of mineral from the body by sweating (of potassium and sodium) and exhalation (of selenium). However, they will overestimate the availability if the retained mineral is in poorly exchangeable forms (e.g. tissue calcification, urinary calculi, mucosal ferritin); is in a form relatively unavailable for functional purposes (e.g. selenomethionine); or if the feed provides absorbable agonists or antagonists of the retained mineral. Such contingencies are rare.

Partial retention

Comparisons of mineral accretion in the organs (e.g. liver copper), bones (e.g. tibia zinc) or bloodstream of two or more sources at two or more intakes have been widely used to derive relative availability from statistical comparisons of the linear relationships between retention and intake for the sources ('slopes ratio') (Littell et al., 1995, 1997). The technique requires accurate diet formulation, which is not always attained. If the range of mineral intake stretches to near-toxic levels, passive absorption and non-linear responses may be encountered, giving false estimates of availability, especially if comparisons are made across different ranges of intake. The values obtained reflect relative apparent absorption rather than relative true absorption, but they can be converted to true absorption by comparing the slopes of responses to oral and intravenous repletion and applying the comparative balance principle. Using this technique, the true absorption of copper was measured for the first time in ruminants (Suttle, 1974) and further applications are possible (Suttle, 2000). If a particular organ is responsible for a high proportion of total mineral accretion, its weight is predictable and it can be biopsied; as is the case with the liver in ruminants, mineral retention can be estimated from rates of increase in mineral concentration in the liver. Thus, increases in liver copper concentration can yield approximate estimates of the true absorption of copper in sheep and cattle (ARC, 1980).

Measurement of Functionality

The ability of a feed to ameliorate, prevent or potentiate signs of mineral deprivation reflects the retention of mineral in functional forms and is a complete measure of bioavailability in that it incorporates all four components (accessibility, absorbability, retainability and functionality). The relative availability of plant and animal feed sources of selenium for chicks has been assessed from their ability to prevent exudative diathesis and pancreatic atrophy (Cantor et al., 1975). In a chick growth assay using a zinc-deficient basal diet, zinc in wheat, fish meal and non-fat milk was shown to be 59%, 75% and 82%, respectively, as available as that in $ZnCO_3$ (Odell et al., 1972). Rates of recovery from or induction of anaemia have been measured as blood haemoglobin responses in iron-depleted animals, and compared to those with a standard source of high bioavailability ($FeSO_4$). The shear strength of bone is used to assess relative phosphorus availability in pigs (Ketaren et al., 1993), but may not always give the same assessment as other methods. In one study (Spencer et al., 2000), the value obtained for a conventional corn/SBM blend (9%) was far less than that obtained with a partial retention method (33%) that used plasma inorganic phosphorus concentration as the index of phosphorus status (Sands et al., 2001). Complications can arise when natural foods are added to mineral-deficient diets as the major mineral source and non-specific functional indices of relative availability, such as bone strength or growth rate, are used because the food may provide other essential nutrients or antagonists that improve or impair performance (Suttle, 1983). For example, the release of phosphorus from phytate leaves the essential nutrient *myo*-inositol behind and it can improve bone strength in poultry. However, the discrepancy mentioned above may have more to do with the application than the choice of technique.

Interactions between Food Constituents

Single foodstuffs rarely comprise the entire ration of livestock because an individual feed can rarely meet requirements for all nutrients. The assessment of 'mineral value' in individual feeds therefore presents a dilemma: if a single feed is fed, the diet will probably lack one or more nutrients, breaking the 'golden rule'. But if other ingredients are added to provide a balanced diet, the assessment becomes one of the particular blend. The need for balanced diets can be blithely ignored, but we have already seen that this may yield false values. However, in attempting to achieve dietary balance with feeds of contrasting composition, one can end up comparing the mineral value of blends that have no practical significance (e.g. Fig. 2.6) and there may be interactions between different constituents of the blend. Attempts have been made to correct data obtained with a simple practical blend for the contribution of its principal constituent, but this can introduce further difficulties. Pedersen et al. (2007) wished to determine the 'phosphorus value' in a new by-product of the ethanol industry, distillers' dried grains with solubles (DDGS), when blended with corn (the most likely practice in the USA). A diet consisting of 97% corn had an apparent absorption of phosphorus coefficient of 0.19 and that of 10 blends (48:50) with DDGS had a mean (standard deviation) apparent absorption of phosphorus of 0.51 (±0.025). On the assumption that the corn retained its 'phosphorus value' when mixed with DDGS, the latter's 'phosphorus value' was upgraded to 0.59, giving a threefold advantage over corn. The lack of calcium in both feeds was corrected by adding limestone but, for no valid reason, twice as much was added to the corn diet than to the blends, giving respective dietary calcium concentrations of 8.4 and 3.9–5.4 g kg^{-1} DM, while the corn diet contained far less protein (6.9% versus 16.1–18.5% crude protein). The most reliable statistic from the study is the uncorrected apparent absorption of phosphorus for the nutritionally imbalanced blends (0.51), but this number is twice as high as that found in balanced corn/SBM blends (see Chapter 6).

In summary, published estimates of the 'mineral value' of feeds are often a characteristic of the experiment rather than the feeds, particularly where the 'phosphorus value' is concerned.

Interactions between Minerals

Interactions between minerals are a major cause of variation in bioavailability and thus

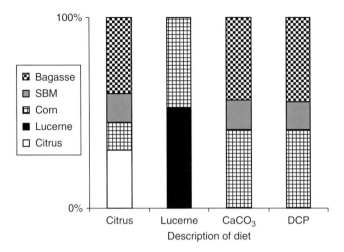

Fig. 2.6. An illustration of the difficulties of balancing diets in undertaking assessments of the 'mineral value' of feeds and supplements. The aim was to compare calcium availability in citrus pulp, lucerne hay, limestone (CaCO$_3$) and dicalcium phosphate (DCP). In succeeding in creating diets of equal energy, protein and fibre content by the variable use of bagasse, corn and soybean meal (SBM), three different rations were formed and neither 'citrus pulp' nor 'lucerne' is an accurate description of what was fed (from Roque et al., 2007).

influence both the nutritive value and potential toxicity of a particular source. An interaction is demonstrated when responses to a given element are studied at two or more levels of another element and the response to multiple supplementation is greater or less than the sum of responses to single supplements. The simplest example is provided by the 2×2 factorial experiment and an early indication of interaction is provided by arranging the results in a two-way table (Table 2.7). Striking main effects are given by the differences in row and column totals, but the interaction – indicated by differences between diagonal totals – is equally pronounced. Large groups are often needed to show significant interactions because the interaction term has few degrees of freedom in the analysis of variance (only 1 degree of freedom in a 2×2 experiment). The frequency with which physiological interactions between minerals occur has often been underestimated.

Mechanisms

The primary mechanisms whereby minerals interact to affect bioavailability are as follows:

Table 2.7. The outcome of a 2×2 factorial experiment in which the separate and combined antagonisms of iron and molybdenum towards copper were investigated using liver copper concentrations to assess status (Bremner et al., 1987). The difference in totals on the diagonals indicates the strength of the interaction equal to the main effects (i.e. row and column totals).

		Fe treatment		Mo totals
		0	+	
Mo treatment	0	1.18 (0.52)	0.35 (1.32)	1.53
	+	0.17	0.14	0.31
	Fe totals	1.35	0.49	1.89 mmol kg^{-1} DM

- Formation of unabsorbable complexes in the gut (e.g. metal phytate, see Chapters 6 and 16).
- Competition between cations for the non-specific divalent cation transporter (Fe and Mn).
- Competition between similar anions for a metabolic pathway (e.g. SO_4^{2-} and MoO_4^{2-}, see Chapter 17).
- Induction of non-specific metal-binding proteins (e.g. ferritin by iron and metallothionein by copper, zinc or cadmium, see Chapter 18).

Thus, interactions between minerals can affect each component of bioavailability and two or more components may be affected simultaneously. The net effect of an interaction is usually a change in retention or function throughout the body, but there may be only localized changes in tissue distribution in one organ (e.g. kidney) if the interactions occur post-absorptively. Interactions between minerals usually have negative effects, but they can be beneficial (e.g. small copper supplements can enhance iron utilization) and may depend on the level of supplementation (e.g. large copper supplements can increase iron requirements, see Chapter 11).

Outcome

The effects of interactions between minerals on bioavailability can sometimes be predicted, as is the case for interactions between potassium and magnesium and between copper, molybdenum and sulfur in ruminants and the influence of phytate on phosphorus and zinc absorption in non-ruminants. However, outcome is hard to predict when three mutually antagonistic elements are involved, such as copper, cadmium and zinc. Campbell and Mills (1979) anticipated that since cadmium and zinc each separately impaired copper metabolism in sheep, the addition of both elements would severely compromise copper status. However, zinc (>120 mg kg^{-1} DM) apparently protected lambs from the copper-depleting effects of cadmium (>3 mg kg^{-1} DM) and exceedingly high levels of zinc (>750 mg kg^{-1} DM) were needed to exacerbate copper depletion (Bremner and Campbell, 1980).

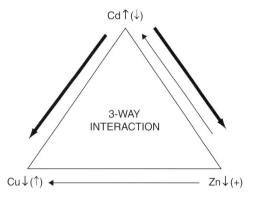

Fig. 2.7. Mutual antagonisms between copper, cadmium and zinc lead to complex three-way interactions. Raising the level of one interactant can lower the status of the other two, as shown for Cd (↑). However, inequalities in the strength of particular interactions, such as the strong antagonism of Cu by Cd, may mean that supplements of Zn (+) may raise (↑) Cu status by reducing (↓) the Cd–Cu interaction.

When each element can negatively affect the other, a supplement of any one element may have a positive net effect either by countering an antagonism against itself or neutralizing the antagonistic properties of the second element upon the third (Fig. 2.7). The clinical outcome of such three-way interactions will depend upon which element is the first to become rate-limiting. This in turn will be determined by the strength of each separate antagonism and change with the level at which each antagonist is employed. Thus, cadmium is capable of inducing either copper or zinc deprivation. Prediction of three-way interactions is also difficult if there is a shared potentiator. Thus, sulfur (as sulfide) is necessary for both the iron- and molybdenum-mediated antagonism of copper in ruminants. When iron and molybdenum are added together in a diet marginal in copper there is no additive antagonism towards copper (Table 2.7) (Bremner et al., 1987), possibly because iron and molybdenum compete for sulfide.

Mineral Cycles

There can be substantial inputs of minerals in imported feeds and/or fertilizers, withdrawals of minerals in harvested crops and/or livestock

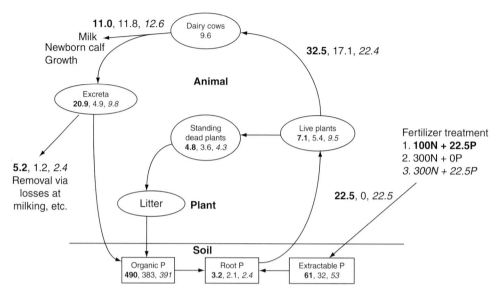

Fig. 2.8. Example of a mineral cycle showing pool sizes and flows of phosphorus (kg P ha^{-1}) for a dairy herd grazing tropical pasture under three different fertilizer regimens (after Davison *et al.*, 1997). Raising the amount of N given with P (treatment 3 versus 1) shifts the route of recycling from excreta to plant litter, a less efficient process. The application of P prevents a steady decline in extractable soil P (treatment 3 versus 2). Giving P with maximum N (treatment 3) increases the milk yield, without changing the P status of the cow, by increasing the mass of green herbage on offer.

products and recycling of mineral supplies via excreta in the farm ecosystem each year. Mineral status can therefore change with time, and changes in husbandry may also shift mineral balances substantially. Figure 2.8 illustrates the effects of nitrogen fertilizer usage on the phosphorus cycle in a dairy enterprise on tropical grass pasture. In this century, emphasis has shifted to the accumulation of phosphorus in the environment and watercourses, and different models have been conceptualized (see Turner *et al.*, 2002). The influence of fertilizer inputs of a potentially toxic mineral, cadmium, on the cycling and accumulation of that element in soil are described in Chapter 18. The net flow of utilizable mineral to the grazing animal in particular may vary widely from season to season and from year to year. In areas where mineral concentrations in herbage barely meet animal requirements, changes brought about by atmospheric, climatic or seasonal influences may determine the prevalence of deprivation. For example, the risk of 'swayback' in newborn lambs deprived of copper is high after mild winters in the UK. This is probably due to decreased use of supplementary feeds of higher available copper content and the increased ingestion of soil (containing copper antagonists) during winters of low snow cover (Suttle *et al.*, 1975). The shedding of seed can determine the incidence or severity of phosphorus deficiency states in livestock (see Chapter 6). It is important to appreciate the cyclical nature of mineral nutrition before breaking problems down into small compartments, and there is a need to frequently reassess the adequacy of mineral supplies experienced by the animal.

References

AFRC (1991) Technical Committee on Responses to Nutrients Report No. 6. A reappraisal of the calcium and phosphorus requirements of sheep and cattle. *Nutrition Abstracts and Reviews* 61, 573–612.
Alloway, B.J. (2004) *Zinc in Soils and Crop Nutrition*. IZA Publications, International Zinc Association, Brussels, pp. 1–16.

Ammerman, C.B., Baker, D.H. and Lewis, A.J. (1995) *Bioavailability of Nutrients For Animals*. Academic Press, San Diego, California.
ARC (1980) *The Nutrient Requirements of Ruminants*. Commonwealth Agricultural Bureaux, Farnham Royal, UK, pp. 184–185.
Arosemena, A., DePeters, E.J. and Fadel, J.G. (1995) Extent of variability in the nutrient composition within selected by-product feedstuffs. *Animal Feed Science and Technology* 54, 103–120.
Atkinson, S.A., Shah, J.K., Webber, C.E., Gibson, I.L. and Gibson, R.S. (1993) A multi-element tracer assessment of the fractional absorption of minerals from formula with additives of calcium, phosphorus, zinc, copper, and iron in young piglets. *Journal of Nutrition* 123, 1586–1593.
Berger, L.L. (1995) Why do we need a new NRC data base? *Animal Feed Science and Technology* 53, 99–107.
Bramley, R.G.V. (1990) Cadmium in New Zealand agriculture. *New Zealand Journal of Agricultural Research* 33, 505–519.
Bremner, I. and Campbell, T.K. (1980) The influence of dietary copper intake on the toxicity of cadmium. *Annals of the New York Academy of Sciences* 355, 319–332.
Bremner, I., Humphries, W.R., Phillippo, M., Walker, M.J. and Morrice, P.C. (1987) Iron-induced copper deficiency in calves: dose–response relationships and interactions with molybdenum and sulfur. *Animal Production* 45, 403–414.
Brown, T.F. and Zeringue, L.K. (1994) Laboratory evaluations of solubility and structural integrity of complexed and chelated trace mineral supplements. *Journal of Dairy Science* 77, 181–189.
Buckley, W.T. (1988) The use of stable isotopes in studies of mineral metabolism. *Proceedings of the Nutrition Society* 47, 407–416.
Burridge, J.C., Reith, J.W.S. and Berrow, M.L. (1983) Soil factors and treatments affecting trace elements in crops and herbage. In: Suttle, N.F., Gunn, R.G., Allen, W.M., Linklater, K.A. and Wiener, G. (eds) *Trace Elements in Animal Production and Veterinary Practice*. British Society of Animal Production Occasional Publication, No. 7, Edinburgh, pp. 77–86.
Campbell, J.K. and Mills, C.F. (1979) The toxicity of zinc to pregnant sheep. *Environmental Research* 20, 1–13.
Cantor, A.H., Scott, M.L. and Noguchi, T. (1975) Biological availability of selenium in feedstuffs and selenium compounds for prevention of exudative diathesis in chicks. *Journal of Nutrition* 105, 96–105.
Coates, D.B., Kerridge, P.C., Miller, C.P. and Winter, W.H. (1990) Phosphorus and beef production in Australia. 7. The effect of phosphorus on the composition, yield and quality of legume based pasture and their relation to animal production. *Tropical Grasslands* 24, 209–220.
Comar, C.L. (1956) Radiocalcium studies in pregnancy. *Annals of the New York Academy of Sciences* 64, 281.
COSAC (1982) Trace elements in Scottish soils and their uptake by crops, especially herbage. In: *Trace Element Deficiency in Ruminants*. Report of a Scottish Agricultural Colleges (SAC) and Research Institutes (SARI) Study Group. SAC, Edinburgh, pp. 49–50.
Davis, C.D., Wolf, T.L. and Greger, J.L. (1992) Varying levels of dietary manganese and iron effect absorption and gut endogenous losses of manganese by rats. *Journal of Nutrition* 122, 1300–1308.
Davison, T.M., Orr, W.N., Doogan, V. and Moody, P. (1997) Phosphorus fertilizer for nitrogen-fertilized dairy pastures. 2. Long-term effects on milk production and a model of phosphorus flow. *Journal of Agricultural Science, Cambridge* 129, 219–231.
Dhillon, K.S. and Dhillon, S.K. (1991) Accumulation of selenium in sugarcane (*Sacharum officinarum* Linn) in seleniferous areas of Punjab, India. *Environmental Geochemistry and Health* 13, 165–170.
Fairweather-Tait, S.J., Johnson, A., Eagles, J., Gurt, M.I., Ganatra, S. and Kennedy, H. (1989) A double label stable isotope method for measuring calcium absorption from foods. In: Southgate, D., Johnson, I. and Fenwick, G.R. (eds) *Nutrient Availability: Chemical and Biological Aspects*. Royal Society of Chemistry Special Publication No. 72, Cambridge, pp. 45–47.
Fan, M. Z., Archbold, T., Sauer, W.C., Lackeyram, D., Rideout, T., Gao, Y., de Lange, C.F.M. and Hacker, R.R. (2001) Novel methodology allows simultaneous measurement of the true phosphorus digestibility and the gastrointestinal endogenous phosphorus outputs in studies with pigs. *Journal of Nutrition* 131, 2388–2396.
Finley, J.W., Caton, J.S., Zhou, Z. and Davison, K.L. (1997) A surgical model for determination of true absorption and biliary excretion of manganese in conscious swine fed commercial diets. *Journal of Nutrition* 127, 2334–2341.

Givens, D.I., Allison, R., Cottrill, B. and Blake, J.S. (2004) Enhancing the selenium content of bovine milk through alteration of the form and concentration of selenium in the diet of the dairy cow. *Journal of the Science of Food and Agriculture* 84, 811–817.

Grace, N.D. (2006) Effect of ingestion of soil on the iodine, copper, cobalt (vitamin B_{12}) and selenium status of grazing sheep. *New Zealand Veterinary Journal* 54, 44–46.

Grace, N.D., Rounce, J.R. and Lee, J. (1996) Effect of soil ingestion on the storage of Se, vitamin B_{12}, Cu, Cd, Fe, Mn and Zn in the liver of sheep. *New Zealand Journal of Agricultural Research* 39, 325–331.

Grace, N.D., Pearce, S.G., Firth, E.C. and Fennessy, P.F. (1999) Concentrations of macro- and trace-elements in the milk of pasture-fed Thoroughbred mares. *Australian Veterinary Journal* 77, 177–180.

Grace, N.D., Loganathan, P., Hedley, M.J. and Wallace, G.C. (2003a) Ingestion of soil fluorine: its impact on the fluorine metabolism and status of grazing young sheep. *New Zealand Journal of Agricultural Research* 46, 279–286.

Grace, N.D., Loganathan, P., Hedley, M.J. and Wallace, G.C. (2003b) Ingestion of soil fluorine: its impact on the fluorine metabolism of dairy cows. *New Zealand Journal of Agricultural Research* 48, 23–27.

Hazell, T. (1985) Minerals in foods: dietary sources, chemical forms, interactions, bioavailability. *World Review of Nutrition and Dietetics* 46, 1–123.

Healy, W.B., Rankin, P.C. and Watts, H.M. (1974) Effect of soil contamination on the element composition of herbage. *New Zealand Journal of Agricultural Research* 17, 59–61.

Heth, D.A. and Hoekstra, W.G. (1965) Zinc-65 absorption and turnover in rats. 1. A procedure to determine zinc-65 absorption and the antagonistic effect of calcium in a practical diet. *Journal of Nutrition* 85, 367–374.

Hill, J., Stark, B.A., Wilkinson, J.M., Curran, M.K., Lean, I.J., Hall, J.E. and Livesey, C.T. (1998) Accumulation of potentially toxic elements by sheep given diets containing soil and sewage sludge. 2. Effect of the ingestion of soils historically treated with sewage sludge. *Animal Science* 67, 87–96.

Hopkins, A., Adamson, A.H. and Bowling, P.J. (1994) Response of permanent and reseeded grassland to fertilizer nitrogen. 2. Effects on concentrations of Ca, Mg, Na, K, S, P, Mn, Zn, Cu, Co and Mo in herbage at a range of sites. *Grass and Forage Science* 49, 9–20.

Johnson, P.E., Lykken, G.I. and Korynta, E.D. (1991) Absorption and biological half-life in humans of intrinsic and extrinsic ^{154}Mn tracers for foods of plant origin. *Journal of Nutrition* 121, 711–717.

Jones, R.J. (1990) Phosphorus and beef production in northern Australia. 1. Phosphorus and pasture productivity – a review. *Tropical Grasslands* 24, 131–139.

Jumba, I.O., Suttle, N.F., Hunter, E.A. and Wandiga, S.O. (1996) Effects of botanical composition, soil origin and composition on mineral concentrations in dry season pastures in Western Kenya. In: Appleton, J.D., Fuge, R. and McCall, G.J.H. (eds) *Environmental Geochemistry and Health.* Geological Society Special Publication No. 113, London, pp. 39–45.

Kappel, L.C., Morgan, E.B., Kilgore, L., Ingraham, R.H. and Babcock, D.K. (1983) Seasonal changes of mineral content in Southern forages. *Journal of Dairy Science* 66, 1822–1828.

Ketaren, P., Batterham, E.S., Dettman, E.B. and Farrell, D.J. (1993) Phosphorus studies in pigs 2. Assessing phosphorus availability for pigs and rats. *British Journal of Nutrition* 70, 269–288.

Kiatoko, M., McDowell, L.R., Bertrand, J.E., Chapman, H.C., Pete, F.M., Martin, F.G. and Conrad, J.H. (1982) Evaluating the nutritional status of beef cattle herds from four soil order regions of Florida. I. Macroelements, protein, carotene, vitamins A and E, haemoglobin and haematocrits. *Journal of Animal Science* 55, 28–47.

Kincaid, R.L., Gay, C.C. and Krieger, R.I. (1986) Relationship of serum and plasma copper and caeruloplasmin concentrations of cattle and the effects of whole sample storage. *American Journal of Veterinary Research* 47, 1157–1159.

Kirchgessner, M., Friesecke, H. and Koch, G. (1967) *Nutrition and the Composition of Milk.* Crosby Lockward, London, pp. 209–238.

Krishnamurti, C.R., Ramberg, C.F., Shariff, M.A. and Boston, R.C. (1997) A compartmental model depicting short-term kinetic changes in selenium metabolism in ewes fed hay containing normal or inadequate levels of selenium. *Journal of Nutrition* 127, 95–102.

Kume, S.-I. and Tanabe, S. (1993) Effect of parity on colostral mineral concentrations of Holstein cows and value of colostrum as a mineral source for newborn calves. *Journal of Dairy Science* 76, 1654–1660.

Ledgard, S.F. and Upsdell, M.P. (1991) Sulfur inputs from rainfall throughout New Zealand. *New Zealand Journal of Agricultural Research* 34, 105–111.

Leng, R.A. and Preston, T.R. (1984) Nutritional strategies for the utilisation of agroindustrial by-products by ruminants and extension of the principles and technologies to the small farmer in Asia. In: *Proceedings of the Fifth World Conference on Animal Production, Tokyo,* pp. 310–318.

Lengemann, F.W. (1969) Comparative metabolism of Na^{125}IO$_3$ and Na^{131}I in lactating cows and goats. In: *Trace Mineral Studies with Isotopes in Domestic Animals*. International Atomic Energy Agency, Vienna, pp. 113–120.

Lengemann, F.W., Comar, C.L. and Wasserman, R.H. (1957) Absorption of calcium and strontium from milk and nonmilk diets. *Journal of Nutrition* 61, 571–583.

Littell, R.C., Lewis, A.J. and Henry, P.R. (1995) Statistical evaluation of bioavailability assays. In: *Bioavailability of Nutrients For Animals*. Academic Press, San Diego, California, pp. 5–34.

Littell, R.C., Lewis, A.J., Henry, P.R. and Ammerman, C.B. (1997) Estimation of relative bioavailability of nutrients. *Journal of Animal Science* 75, 2672–2683.

Luten, J., Crews, H., Flynn, A., Dael, P.V., Kastenmayer, P. and Hurrel, R. (1996) Inter-laboratory trial on the determination of the *in vitro* iron dialysability from food. *Journal of Science in Food and Agriculture* 72, 415–424.

MAFF (1990) *UK Tables of the Nutritive Value and Chemical Composition of Foodstuffs*. In: Givens, D.I. (ed.) Rowett Research Services, Aberdeen, UK.

McDowell, L. R. (2003) *Minerals in Animal and Human Nutrition*. Elsevier, Amsterdam, p. 688.

McDowell, L. R., Conrad, J.H. and Hembry, F.G. (1993) *Minerals for Grazing Ruminants in Tropical Regions*, 2nd edn. Animal Science Department, University of Florida, Gainsville, Florida.

Minson, D.J. (1990) *Forages in Ruminant Nutrition*. Academic Press, San Diego, California, pp. 208–229.

Mitchell, R.L. (1957) The trace element content of plants. *Research, UK* 10, 357–362.

Morin, D.E., Rowan, L.L., Hurley, W.L. and Braselton, W.E. (1995) Composition of milk from llamas in the United States. *Journal of Dairy Science* 78, 1713–1720.

Morris, E.R. and Ellis, R. (1976) Isolation of monoferric phytate from wheat bran and its biological value as an iron source to the rat. *Journal of Nutrition* 106, 753–760.

Mtui, D.J., Lekule, F.P., Shem, M.N., Hayashida, M. and Fujihara, T. (2008) Mineral concentrations in leaves of nine browse species collected from Mvomero, Morogo, Tanzania. *Journal of Food, Agriculture & Environment* 6, 226–230.

Odell, B.L., Burpo, C.E. and Savage, J.E. (1972) Evaluation of zinc availability in foodstuffs of plant and animal origin. *Journal of Nutrition* 102, 653–660.

Pedersen, C., Boersma, M.G. and Stein, H.H. (2007) Digestibility of energy and phosphorus in ten samples of distillers dried grains with solubles fed to growing pigs. *Journal of Animal Science* 85, 1168–1176.

Pettey, L.A., Cromwell, G.L. and Lindemann. M.D. (2006) Estimation of endogenous phosphorus loss in growing and finishing pigs fed semi-purified diets. *Journal of Animal Science* 84, 618–626.

Pott, F.B., Henry, P.R., Ammerman, C.B., Merritt, A.M., Madison, J.B. and Miles, R.D. (1994) Relative bioavailability of copper in a copper:lysine complex for chicks and lambs. *Animal Feed Science and Technology* 45, 193–203.

Roque, A.P., Dias, R.S., Vitti, D.M.S.S., Bueno, I.C.S., Cunha, E.A., Santos, L.E. and Bueno, M.S. (2007) True digestibility of calcium from sources used in lamb finishing diets. *Small Ruminant Research* 71, 243–249.

Sands, J.S., Ragland, D., Baxter, C., Joern, B.C., Sauber, T.E. and Adeola, O. (2001) Phosphorus bioavailability, growth performance and nutrient balance in pigs fed high available phosphorus corn and phytase. *Journal of Animal Science* 79, 2134–2142.

Schwarz, R., Balko, A.Z. and Wien, E.M. (1982) An *in vitro* system for measuring intrinsic dietary mineral exchangeability: alternatives to intrinsic labelling. *Journal of Nutrition* 112, 497–504.

Scott, M.L. and Thompson, J.N. (1971) Selenium content of feedstuffs and effects of dietary selenium levels upon tissue selenium in chicks and poults. *Poultry Science* 50, 1742–1748.

Shirley, R.L. (1978) Water as a source of minerals. In: Conrad, J.H. and McDowell, L.R. (eds) *Latin American Symposium on Mineral Nutrition Research with Grazing Ruminants*. Animal Science Department, University of Florida, Gainsville, Florida, pp. 40–47.

Silanpaa, M. (1982) *Micronutrients and the Nutrient Status of Soils: A Global Study*. Food and Agriculture Organization, Rome.

Smart, M.E., Cohen, R., Christensen, D.A. and Williams, C.M. (1986) The effects of sulphate removal from the drinking water on the plasma and liver copper and zinc concentrations of beef cows and their calves. *Canadian Journal of Animal Science* 66, 669–680.

Smith, G.S., Middleton, K.R. and Edmonds, A.S. (1978) A classification of pasture and fodder plants according to their ability to translocate sodium from their roots into aerial parts. *New Zealand Journal of Experimental Agriculture* 6, 183–188.

Southern, P. (1979) The value of cereal crop residues. In: *Our Land*. (W. Australia), p. 3.

Spencer, J.D., Allee, G.L. and Sauber, T.E. (2000) Phosphorus availability and digestibility of normal and genetically modified low-phytate corn for pigs. *Journal of Animal Science* 78, 675–681.

Stock, R., Grant, R. and Klopfenstein, T. (1991) Average composition of feeds used in Nebraska. In: *Nebguide*. Cooperative Extension, University of Nebraska-Lincoln, Institute of Agriculture and Natural Resources.

Suleiman, A., Okine, E. and Goonewardne, L.A. (1997) Relevance of National Research Council feed composition tables in Alberta. *Canadian Journal of Animal Science* 77, 197–203.

Suttle, N.F. (1974) A technique for measuring the biological availability of copper to sheep using hypocupraemic ewes. *British Journal of Nutrition* 32, 395–401.

Suttle, N.F. (1983) Assessing the mineral and trace element status of feeds. In: Robards, G.E. and Packham, R.G. (eds) *Proceedings of the Second Symposium of the International Network of Feed Information Centres, Brisbane*. Commonwealth Agricultural Bureaux, Farnham Royal, UK, pp. 211–237.

Suttle, N.F. (1985) Estimation of requirements by factorial analysis: potential and limitations. In: Mills, C.F., Bremner, I. and Chesters, J.K. (eds) *Proceedings of the Fifth International Symposium on Trace Elements in Man and Animals*. Commonwealth Agricultural Bureaux, Farnham Royal, UK, pp. 881–883.

Suttle, N.F. (1991) Mineral supplementation of low quality roughages. In: *Proceedings of Symposium on Isotope and Related Techniques in Animal Production and Health*. International Atomic Energy Agency, Vienna, pp. 101–104.

Suttle, N.F. (2000) Minerals in livestock production. *Asian-Australasian Journal of Animal Science* 13, 1–9.

Suttle, N.F. and Jones, D. (1986) Copper and disease resistance in sheep: a rare natural confirmation of interaction between a specific nutrient and infection. *Proceedings of the Nutrition Society* 45, 317–325.

Suttle, N.F., Alloway, B.J. and Thornton, I. (1975) An effect of soil ingestion on the utilisation of dietary copper by sheep. *Journal of Agricultural Science, Cambridge* 84, 249–254.

Suttle, N.F., Abrahams, P. and Thornton, I. (1984) The role of a soil × dietary sulfur interaction in the impairment of copper absorption by ingested soil in sheep. *Journal of Agricultural Science, Cambridge* 103, 81–86.

Suttle, N.F., Bell, J., Thornton, I. and Agyriaki, A. (2003) Predicting the risk of cobalt deprivation in grazing livestock from soil composition data. *Environmental Geochemistry and Health* 25, 33–39.

Thornton, I. (1974) Biogeochemical and soil ingestion studies in relation to the trace-element nutrition of livestock. In: Hoekstra, W.G., Suttie, J.W., Ganther, H.E. and Mertz, W. (Eds) *Trace Element Metabolism in Animals – 2*. University Park Press, Baltimore, Maryland, pp. 451–454.

Turner, B.L., Paphazy, M.J., Haygarth, P.M. and McKelvie, I.D. (2002) Inositol phosphates in the environment. *Philosophical Transactions of the Royal Society of London, B* 357, 449–469.

Turnlund, J.R. (1989) The use of stable isotopes in mineral nutrition research. *Journal of Nutrition* 119, 7–14.

Underwood, E.J. (1977) *Trace Elements in Human and Animal Nutrition*, 4th edn. Academic Press, London.

Underwood, E.J. and Suttle, F. (1999) *The Mineral Nutrition of Livestock*, 3rd edn. CAB International, Wallingford, UK.

Vaithiyanathan, S. and Singh, M. (1994) Seasonal influences on soil ingestion by sheep in an arid region. *Small Ruminant Research* 14, 103–106.

White, C.L., Robson, A.D. and Fisher, H.M. (1981) Variation in nitrogen, sulfur, selenium, cobalt, manganese, copper and zinc contents of grain from wheat and two lupin species grown in a range of Mediterranean environments. *Australian Journal of Agricultural Research* 32, 47–59.

Widmer, M.R., McGinnis, L.M. and Stein, H.H. (2007) Energy, phosphorus, and amino acid digestibility of high-protein distillers dried grains and corn germ fed to growing pigs. *Journal of Animal Science* 85, 2994–3003.

Wilman, D. and Derrick, R.W. (1994) Concentration and availability to sheep of N, P, K, Ca, Mg and Na in chickweed, dandelion, dock, ribwort and spurrey compared with perennial ryegrass. *Journal of Agricultural Science, Cambridge* 122, 217–223.

Winks, L. (1990) Phosphorus and beef production in northern Australia. 2. Responses to phosphorus by ruminants – a review. *Tropical Grasslands* 24, 140–158.

Wright, D.E. Towers, N.R., Hamilton, P.B. and Sinclair, D.P. (1978) Intake of zinc sulphate in drinking water by grazing beef cattle. *New Zealand Journal of Agricultural Research* 21, 215–221.

3 Assessing and Controlling Mineral Status in Livestock

Chapter 2 should have helped erase belief that the mineral composition of the forage or ration alone can be relied upon in avoiding mineral deprivation in livestock. On a broad geographical basis, areas where mineral imbalances in livestock are more or less likely to occur can be predicted by soil-mapping techniques. Thus, areas of molybdenum-induced copper deficiency in cattle in the UK can be predicted from the results of stream sediment reconnaissance for molybdenum (Thornton and Alloway, 1974). The potential in applying such techniques in this (Suttle, 2008) and other spheres is large (Plant *et al.*, 1996; White and Zasoski, 1999), but they have rarely been successfully deployed to predict health problems in livestock (Fordyce *et al.*, 1996). Soil maps suggesting a widespread risk of zinc deficiency in crops (Alloway, 2003) bear no relationship to the small risk of zinc deprivation in livestock. Comprehensive maps of soil mineral status exist for the UK (McGrath and Loveland, 1992) and Europe (De Vos *et al.*, 2006), and multivariate analysis of the data may facilitate the prediction of mineral status in livestock (Suttle, 2000). However, many factors can interrupt the flow of mineral into the animal (Fig. 3.1), most of which are reviewed in Chapters 1 and 2. One not covered so far is the possibility that the feeds analysed may not be representative of the mixture of materials consumed by the animal.

Food Selection

Selective grazing

In foraging situations, the forage sample collected may not represent the material selected by the grazing animal. Animals show preferences for different types and parts of plants that can vary widely in mineral concentration. Selective grazing greatly influences the assessment of dietary phosphorus status and could equally influence mineral intakes where livestock have access to a mixture of pasture and browse material (Fordyce *et al.*, 1996). Plant material selected during grazing can be sampled via oesophageal fistulae, but the contamination of masticated food with saliva can complicate interpretation. For poorly absorbed trace minerals, faecal analysis can give reasonable predictions of mineral intake (Langlands, 1987). The washing of contaminating soil from a herbage sample by the analyst can also introduce errors because trace minerals consumed and partially utilized by the grazing animal are thereby deselected.

Diet selection

When offered mixtures of dietary ingredients differing in texture, moisture content and

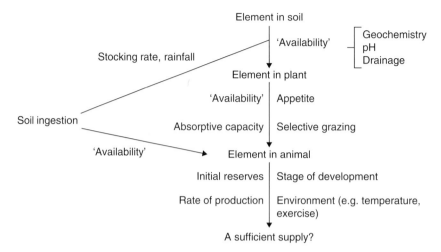

Fig. 3.1. Summary of the many and varied factors, in addition to plants, that can influence the flow of an element from the soil to the grazing animal and whether the supply will meet the animal's requirement.

palatability, animals may consume a blend that differs from the one provided. Thus, an experimental diet for ruminants containing oat hulls to provide adequate roughage may have to be pelleted to ensure that all hulls are consumed. With a loose feed, simply weighing feed refusals and relying on mineral analysis of the diet as fed to estimate unconsumed minerals can lead to false estimates of mineral intake. When offered limestone separately from the main ration, laying hens may vary their consumption of limestone according to the phosphorus supply from the ration (Barkley et al., 2004).

Clinical and Pathological Changes

The limited diagnostic value of measures of soil and plant mineral status means that focus must be placed on the animal. Unfortunately, the clinical and pathological abnormalities associated with most mineral imbalances are rarely specific (Table 3.1) and often bear a clinical similarity to other nutritional disorders and parasitic or microbial infections of the gut. The first requirement is for a scheme for differential diagnosis that should eliminate non-mineral causes of disorder (Appendix 1). Diagnosis of mild mineral imbalance is particularly difficult and the most reliable diagnostic criterion is often an improvement in health or productivity after the provision of a specific mineral supplement in a well-designed and well-executed trial (Phillippo, 1983), a subject that is returned to in Chapter 19. Few pictures of 'deficient animals' are presented in this book because they can be misleading. Clinical and pathological mineral disorders are often the final expressions of defects arising from a weakening of different links along the same chain of metabolic events. Lack of clinical specificity may also arise because anorexia, with resultant under-nutrition, is a common early expression of many mineral deficiencies. Impaired reproduction caused by deprivation of manganese, selenium and zinc may arise through different primary effects on the reproductive process. Poor lamb and calf crops in enzootic fluorosis are attributable to fluorine-induced anorexia in the mother and to deformities of the teeth and joints that restrict mastication and grazing. The nature of the clinical and pathological changes can vary greatly, even within species, depending on the age at onset of mineral deprivation. Goitre (an enlarged thyroid gland) or congenital swayback in newborn lambs are well known signs of iodine and copper deprivation, respectively, in late pregnancy, but deprivation at other times may impair health in flocks or herds unaffected by goitre or swayback.

Table 3.1. Examples of the sequence of pathophysiological events during the development of six mineral-responsive diseases.

Mineral	Depletion	Deficiency	Dysfunction	Disorder
Ca	Young: bone ↓	Serum ↓	Chondrodystrophy	Rickets
	Old: bone ↓	Serum ↓	Irritability ↓	Recumbency
Mg	Young: bone ↓	Serum ↓	Irritability ↑	Convulsion
	Old: bone ?	Serum ↓	Irritability ↑	Convulsion
Cu	Liver ↓	Serum ↓	Disulfide bonds ↓	Wool crimp ↓
Co	Liver vitamin B_{12} ↓	Serum vitamin B_{12} ↓	MMA ↑	Appetite ↓
	Serum vitamin B_{12} ↓			
I	T_4 ↓	T_3 ↓	BMR ↓	Goitre
	Thyroid colloid ↓		Thyroid hypertrophy	
	Thyroid iodine ↓			
Se	GPx in blood and liver ↓	Serum Se ↓	Serum creatine kinase ↑	Myopathy

BMR, basal metabolic rate; GPx, glutathione peroxidase; MMA, methylmalonic acid; T3, triiodothyronine; T4, thyroxine.

Biochemical Indicators of Mineral Deprivation

The non-specificity of the clinical and pathological consequences of mineral deprivation shifts the focus to detecting the biochemical changes in the tissues and fluids of animals that accompany mineral deprivation (Appendix 1, step 3c). These changes are also valuable in the pre-clinical detection of mineral imbalances (Mills, 1987).

A general macro-model

A sequence of events is presented in Fig. 3.2 to provide a rational basis for the biochemical assessment of mineral deprivation. The sequence consists of four phases: (i) 'depletion', during which storage pools of the mineral are reduced; (ii) 'deficiency', during which transport pools of the mineral are reduced; (iii) 'dysfunction', when mineral-dependent functions become rate-limited; and (iv) 'disorder', during which clinical abnormalities become apparent to the naked eye.

The model and terminology from the last edition of this book, modified slightly by Lee et al. (1999), is retained here to avoid the confusion that continues to arise from the use of the same word 'deficient' to describe diets low in minerals, healthy animals with subnormal mineral status and clinically sick animals. A diet is only 'mineral-deficient' if it is likely to be fed long enough to induce subnormal mineral status and – if fed – may not necessarily result in clinical disease.

Variations on a theme

The precise sequence of events during the development of a clinical disorder varies widely from mineral to mineral and with the rate of mineral deprivation. There are limited body stores of magnesium and zinc, and reductions in plasma magnesium and zinc (i.e. the transport pool) therefore occur with the onset of deprivation. With calcium, the mobilization of skeletal reserves can preserve serum calcium concentrations during long periods of depletion, but the rapid increase in demand for calcium with the onset of lactation can lead to acute hypocalcaemia and clinical disease (milk fever) while large reserves remain. A similar overlap between phases can occur in iron deprivation, with biochemical evidence of deficiency occurring before iron reserves are entirely exhausted (Chesters and Arthur, 1988). The point of transition from deficiency to dysfunction and disorder is also variable and depends partly upon the demand being placed

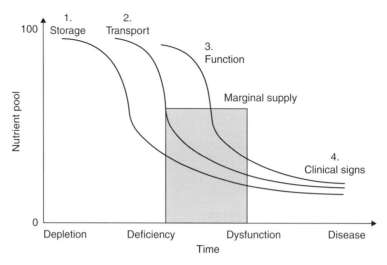

Fig. 3.2. The sequence of pathophysiological changes that can occur in mineral-deprived livestock. There are four possible phases, beginning with depletion and ending with overt disease, related to changes in body pools of mineral that serve storage (e.g. liver), transport (e.g. plasma) or functional (e.g. muscle enzyme) purposes. For some elements (e.g. calcium and phosphorus), storage and function are combined in bone; for cobalt (as vitamin B_{12}) there is role-reversal, with the plasma rather than the liver serving as the store; for elements with small or slowly mobilized stores (e.g. zinc), curves 1 and 2 are superimposed or interposed; for all elements there is a marginal area (shaded) where stores are all but exhausted and mineral-dependent functions begin to fail, but the animal remains outwardly healthy. The upper limit (100) on the mineral pool axis represents the maximum or normal attainable pool size.

on the critical pathway, particularly where cobalt is concerned. A low selenium status may be tolerable until changes in diet and increased freedom of movement (i.e. turnout to spring pasture) present a twofold oxidative stress; a low iodine status may be tolerable until a cold stress induces a thermogenic response; and simultaneous deficiencies of selenium and iodine increase the likelihood that deprivation of either element will cause the animal to enter the dysfunctional phase (Arthur, 1992).

A general 'nano'-model

With the 'snapshot' of macro-mineral metabolism presented in Fig. 1.1, a small-scale ('nano') model is inserted to indicate that the large model represents the summation of a myriad of balances struck in individual cells – the 'pixels' that make up the snapshot, as it were. Similarly, the sequence shown in Fig. 3.2 has its 'nano' equivalent. According to O'Dell (2000), the following sequence of events occurs in vulnerable cells during zinc deprivation:

- Depletion – of zinc stores in the plasma membrane.
- Deficiency – of alkaline phosphatase in the plasma membrane, accompanied by changes in protein and lipid content.
- Dysfunction – changes in intracellular calcium ions and dependent enzymes.
- Disorder – periparturient bleeding due to impaired platelet aggregation and increased erythrocyte fragility.

The biochemical markers used for diagnostic purposes act as remote sensors of rate-limiting events in the cells of vulnerable organs.

Differentiation of Mineral Deprivation

Few biochemical markers indicate the onset of dysfunction or disorder because sample sites and analytes have been selected for convenience rather than diagnostic insight. The popular biochemical criteria for all six elements listed in Table 3.1 convey information about depletion, deficiency, stores and transport forms. Subnormal analytical results at step 3d in Appendix 1 are not therefore synonymous with loss of health. By studying the relationships between markers of depletion or deficiency and dysfunction, the interpretation of each can be clarified (Mills, 1987).

Criteria of dysfunction

Biochemical markers of dysfunction, where available, are more useful than indices of depletion and deficiency because they are one or two phases nearer to clinical disease. Increases in methylmalonic acid (MMA) indicate that MMA coenzyme A isomerase activity becomes rate-limited in cobalt (vitamin B_{12}) deficiency, and increases in creatine kinase indicate that muscle membrane permeability increases in selenium deprivation. However, a rise in plasma MMA may precede loss of appetite, an important clinical sign of cobalt deprivation. Animals have internal biochemical systems for recognizing low mineral status and correcting it: mRNA for storage proteins such as ferritin and cytosolic glutathione peroxidase are switched off in iron and selenium deprivation, respectively, while the mRNA for cholecystokinin is induced in zinc deprivation. The use of mRNA for routine diagnosis is likely to remain prohibitively expensive, but its experimental use could resolve some long-standing arguments by letting the animal express the trace element function that it most needs to preserve.

Delineation of marginal values

Throughout this book, all biochemical criteria will be interpreted using a three-tier system (Table 3.2) with 'marginal bands' to separate the normal from the dysfunctional or disordered (Fig. 3.2), rather than a diagnostically unhelpful 'reference range' that abruptly separates the biochemically normal from the abnormal. Individual variation within the 'at-risk' flock or herd is also informative, with high coefficients of variation indicating possible dysfunction in some of its members.

Detailed temporal studies of the pathophysiological changes during exposure to mineral imbalances are required to delineate marginal bands for all biochemical criteria of mineral status. For example, White (1996) used pooled data for zinc-deprived lambs to fit an equation (Fig. 3.3) to the relationship between mean values for plasma zinc and growth rate and define a plasma zinc threshold value (6.7 μmol l^{-1}) – 95% of the critical value or asymptote – below which growth impairment was likely to occur. A more pragmatic approach is to relate a measure of mineral status to the likelihood of livestock responding positively to mineral supplementation in field trials. Thus, Clark et al. (1985) suggested that lambs in New Zealand would benefit from cobalt supplementation if their plasma vitamin B_{12} concentrations fell below 300 pmol l^{-1}. A similar study in New Zealand (Grace and Knowles, 2002) suggested that lamb growth was unlikely to be retarded by selenium deprivation until blood selenium fell below 130 nmol l^{-1}, a value far below the existing reference range limit of 500 nmol l^{-1}. However, the critical value depends on the index of adequacy and, with optimal wool growth as the goal, the lower limit of marginality for plasma zinc rose to 7.7 μmol l^{-1} (White et al., 1994). Similar results would probably be found for blood or plasma selenium during selenium deprivation.

Table 3.2. Interpretation of biochemical and other indices of mineral deprivation (or excess) in livestock is enhanced by the use of a marginal band between values consistent with health and ill-health.

Result	Phase[a]	Response to supplementation
Normal	Equilibrium or depletion	None
Marginal	Deficiency (or accretion)	Unlikely
	Dysfunction	Possible
Abnormal	Disorder	Probable

[a]See Fig. 3.2 for illustration of terminology.

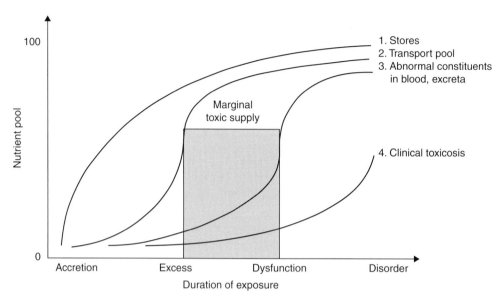

Fig. 3.3. The sequence of biochemical events during chronic exposure to excessive amounts of mineral. The sequence becomes telescoped during acute exposure. The rate of transition is also affected by dietary attributes (e.g. absorbability), animal attributes (e.g. mineral may be safely disposed of in products such as milk and eggs) and factors such as disease (e.g. liver toxins or pathogens).

Physiological and developmental effects

When using the mineral composition of animal tissues and fluids to indicate the quality of mineral nutrition, it is important to avoid the confounding effects of physiological and developmental changes. The normal newborn animal often has 'abnormal' tissue or blood mineral composition when judged inappropriately by adult standards, as in the case of liver and blood copper and selenium and serum iodine. Most blood mineral concentrations also change abruptly and briefly around parturition, and sampling at this time is only appropriate for acute disorders of calcium and magnesium metabolism.

Choice of Sample

The choice of tissue or fluid for analysis varies with the mineral under investigation and the purpose of the investigation. Blood, urine, saliva and hair have the advantage of accessibility by simple, minimally invasive procedures. Surveys of soil/plant/animal relationships indicate that soil and plant samples rarely give information that would be sufficient for diagnostic purposes.

Whole blood

Whole blood, serum and plasma are widely sampled and serum is usually chosen for analysis because it avoids the cost and possible analytical complications of adding an anticoagulant, and gives a more stable (i.e. haemolysis-free) form for transportation. Results for plasma and serum have been assumed to be the same, but serum invariably contains less copper than plasma (Laven and Livesey, 2006). Results for plasma samples generally reflect the mineral status of the transport pool of the element, and low values indicate onset of the 'deficiency' phase. The use of whole blood (or erythrocytes) brings both complications and new possibilities into the assessment of mineral status. Minerals are often incorporated as functional units in the immature erythrocyte prior to release into the bloodstream, and the mineral content of the younger erythrocytes alone accurately reflects recent mineral nutriture. In livestock deprived of a mineral, the decline in mineral concentration in the whole blood (or erythrocytes) is usually slower than that in plasma.

Erythrocytes

Once released, the mineral status of the erythrocyte may remain unchanged throughout its life, despite marked fluctuations in the mineral status of the diet. Thus, activities of erythrocyte enzymes indicative of copper and selenium status (copper:zinc superoxide dismutase and glutathione peroxidase, respectively) do not increase until about 3 weeks after a rise in copper or selenium provision – until there are sufficient enriched young cells to increase the mean values for the entire erythrocyte population. This may be an advantage if animals have been treated or have inadvertently gained access to a new mineral supply (e.g. during handling) prior to blood sampling, since erythrocyte enzyme activities in the subsequent blood sample would be unaffected. However, it could be disadvantageous in masking a recent downturn in copper or selenium supply. In extremely deficient animals, erythrocytes may be released in which the enzyme lacks its functional mineral (apo-enzyme). Assay of enzyme protein by ELISA or the extent of *in vitro* reconstitution may offer new insights into the severity of depletion (e.g. of zinc) in such circumstances (see Chapter 16). Where the normal function of an element is expressed by the erythrocyte, such as stability of the membrane and/or resistance to oxidative stress, as in copper, phosphorus and selenium deprivation, functional status may be indicated by *in vitro* haemolysis tests or staining for Heinz bodies; however, such tests would obviously be non-specific. In zinc deprivation, it is arguable that we should measure both zinc and alkaline phosphatase in the erythrocyte membrane rather than in plasma (O'Dell, 2000).

Secretions

Analysis of saliva provides diagnostic information on several minerals and samples are readily procured using an empty 20 ml syringe fitted with a semi-rigid polypropylene probe and using a 'trombone slide' gag to open the mouth. The composition of the saliva is particularly sensitive to sodium deprivation. As salivary sodium declines, potassium rises and the sodium to potassium ratio in parotid saliva can decrease 10-fold or more, providing a means of detecting sodium deficiency in the animal (Morris, 1980). Preparatory mouth washes can reduce contamination from feed and digesta without disturbing the sodium to potassium ratio. The analysis of sodium to potassium in muzzle secretions can also indicate sodium status (Kumar and Singh, 1981). Salivary zinc has been suggested as a sensitive indicator of zinc status in man, but it has yet to be evaluated in domestic livestock.

Urine and faeces

Mineral excretion by both urinary and faecal routes may correlate with either the mineral intake or mineral status of the animal, depending on the mineral. Subnormal urinary magnesium and sodium outputs point clearly to low intakes of the respective elements and a heightened risk of disorder, while raised urine fluorine can indicate fluorine toxicity. However, high urinary fluorine may reflect mobilization of skeletal fluorine that has built up during previous periods of exposure. Furthermore, concentrations of minerals in urine and faeces are influenced by dietary and animal factors such as water intake and the digestibility of the diet. Correction factors, such as mineral to creatinine ratios in urine and mineral to alkane or titanium ratios in the faeces of grazing animals, can be applied to minimize the affect of 'non-mineral' factors. After prolonged exposure to very low intakes, faecal mineral concentrations can contribute to a diagnosis even without correction (e.g. faecal phosphorus).

Tissues

Of the body tissues, liver and bone are the most commonly sampled because they serve as storage organs for several minerals and because they can be sampled by simple techniques such as aspiration biopsy of the liver and trephine for rib bone. For instance, subnormal concentrations of iron, copper and cobalt (or vitamin B_{12}) in the liver indicate early dietary deprivation and 'depletion' of these elements. Subnormal concentrations of calcium and phosphorus in bone can suggest deficiencies of calcium, phosphorus or vitamin D, and high fluorine levels in bone indicate an excess fluorine intakes. However, results may vary widely from bone to bone, with the age of animal and with mode of expression (fresh, dried, defatted or ashed bone basis), making interpretation difficult (see Chapter 6, in particular). With heterogeneous organs such as the kidneys, results can be affected by the area sampled (e.g. selenium concentrations are two- to fivefold higher in the cortex than in the medulla) (Millar and Meads, 1988), making a precise description of and adherence to sampling protocols essential. Relationships between mineral concentrations in storage organs and ill-health can, however, be poor because stores may be exhausted long before any disorder arises (Fig. 3.2).

Appendages

The analysis of minerals in body appendages such as hair, hoof, fleece or feathers for diagnostic purposes has had a chequered history (Combs, 1987). Relationships to other indices of status, both biochemical and clinical, have often been poor. There are many possible reasons for this, including exogenous surface contamination; the presence of skin secretions (e.g. suint in wool); failure to standardize on a dry weight basis; and variable time periods over which the sample has accumulated. Samples of body appendages should always be rigorously washed and dried. Sampling of newly grown hair or fleece from a recently shaved site (e.g. a liver biopsy site) can give a measure of contemporary mineral status, and historical profiles can be obtained by dividing hair, fleece and core samples of hoof into proximal, medial and distal portions. The colour of the sampled hair has no effect on mineral concentrations because minerals are not contained in the pigments.

Functional forms and indices

Advances in assay procedures for enzymes and hormones have greatly increased the range and sensitivity of the diagnostic techniques now available. Serum assays for vitamin B_{12} rather than cobalt, for triiodothyronine rather than protein-bound iodine, for caeruloplasmin rather than copper and for glutathione peroxidase rather than selenium have provided alternative indicators of dietary and body status of these elements. However, enzyme assays are associated with problems of standardization and their use still does not invariably improve the accuracy of diagnosis, as is the case with caeruloplasmin (Suttle, 2008). The diagnostic strengths and limitations of these and similar estimations are considered in the chapters dealing with individual elements and their interrelations.

Balance

The balance between the amounts of minerals absorbed from the diet can be important. Balance between macro-elements is important because cations such as Na^+, P^+, Ca^{2+} and Mg^{2+} are alkali-producing, whereas anions such as Cl^- and SO_4^{2-} are acid-producing. The preservation of the acid–base balance is important for all species from poultry to dairy cows, and cation–anion imbalance can result in clinical disorders (see Chapter 8). Highly digestible energy- and protein-rich diets sustain high levels of production and may increase the risk of mineral imbalance when compared with low-quality feeds, but they also lower pH in the gut and body to an extent that affects acid–base balance.

Biochemical Indicators of Mineral Excess

Chronic exposure to a mineral excess generally leads to a sequence of biochemical changes that is a mirror image of events during deprivation (compare Fig. 3.3 with 3.2).

1. There is *accretion* of the mineral at any storage sites.
2. Mineral concentrations in *transport* pools may rise.
3. *Dysfunction* may be manifested by the accumulation of abnormal metabolites or constituents in the blood, tissues or excreta.
4. Clinical signs of *disorder* become visible.

For example, marginal increases in plasma copper and major increases in serum aspartate aminotransferase (AAT) precede the haemolytic crisis in chronic copper poisoning in sheep: the rise in AAT is indicative of hepatic dysfunction. However, increases in AAT also occur during the development of white muscle disease caused by selenium deficiency because the enzyme can leak from damaged muscle as well as liver. Assays of glutamate dehydrogenase (found only in liver) and creatine kinase (found only in muscle) are of greater value than AAT in distinguishing the underlying site of dysfunction. The whole sequence in Fig. 3.3 can become telescoped during an acute toxicity. Marginal bands must again be used in conjunction with biochemical criteria of mineral excess to allow for variability in tolerance and in the rate of transition between phases.

Correction of Mineral Imbalance

Holistic approach

Ideally, mineral supplements should be used only when requirements cannot be met reliably by the judicious combination of natural feeds. The addition of protein concentrates to a grain mixture raises the content of such minerals as calcium, phosphorus, zinc, copper and iodine. While the use of plant protein sources can lower the availability of some minerals for pigs and poultry – notably of zinc – because of antagonism from phytate, natural or synthetic phytases can be used to counter the antagonism. Bran is freely available on many small, mixed farms and is an excellent source of phosphorus and iron for ruminants (Suttle, 1991). However, reliance on natural mineral sources requires accurate information on the mineral composition of local feeds and this is rarely available (Arosemena *et al.*, 1995).

Mineral supplementation

Inorganic mineral supplements are added routinely to home-mixed and purchased rations as an insurance against the inclusion of natural feeds that deviate from the norm of mineral or antagonist composition. In some localities, supplements of specific minerals are always necessary because the pastures or feeds are abnormal in their mineral composition as a consequence of local soil and climatic effects. The use of industrial nitrogen sources to provide protein for ruminants increases the need for minerals from other dietary components. Urea supplementation is increasingly used to improve the nutritive value of low-quality roughages and crop by-products in developing countries, but may only do so if accompanied by additional phosphorus and sulfur (Suttle, 1991). Successful procedures have been developed for the prevention and control of all disorders caused by mineral deprivation and for many of those caused by mineral toxicities. Methods can be divided into three categories: indirect methods that increase the mineral composition of pastures and feeds while they are growing; and continuous or discontinuous direct methods, which involve the administration of minerals to animals. The procedure of choice varies greatly with different elements, climatic environments, conditions of husbandry and economic circumstances.

Indirect methods

In favourable environments, treatment of the soil can often improve the mineral composition of herbage and is a logical first step when crop or pasture growth has been limited by mineral deprivation. By incorporating small proportions of cobalt or selenium into fertilizers such as superphosphate, the costs of application are minimized and the whole herd or flock is supplemented. Problems may persist if deprivation is caused by poor availability to the plant rather than poor mineral content of the soil. For example, cobalt-containing fertilizers are ineffective on calcareous or highly alkaline soils because the availability of cobalt to plants is low in such conditions. Soils that are low in available phosphorus are likely to trap added fertilizer phosphorus in unavailable forms. The application of sulfur fertilizers to seleniferous soils can prevent selenium toxicity in the grazing animal, but may be ineffective if the soil is already high in sulfate. Treatment of pastures very high in molybdenum with copper fertilizer may not raise herbage copper sufficiently to prevent induced copper deficiency. Indirect methods may be impracticable or uneconomic under extensive range conditions and ineffective under adverse conditions for plant growth, as is the case with phosphorus fertilizers applied during a dry season. The residual effects of all soil treatments and therefore the frequency with which they must be applied can vary widely from one location to another, and it is often impossible to treat or prevent a mineral imbalance in livestock indirectly.

Discontinuous direct methods

The oral dosing, drenching or injection of animals with mineral solutions, suspensions or pastes has the advantage that all animals receive known amounts of the required mineral at known intervals. This type of treatment is unsatisfactory where labour costs are high and animals have to be driven long distances and handled frequently and specifically for treatment. With minerals such as copper that are readily stored in the liver to provide reserves against periods of depletion, large doses given several months apart can be satisfactory. Oral dosing with selenium salts works well in selenium-deficient areas, particularly where it can be conveniently combined with oral dosing of anthelmintics. By contrast, cobalt deficiency cannot always be prevented fully if the oral doses of cobalt salts are weeks apart. With iron, iodine and copper, some of the disadvantages of oral dosing can be overcome by the use of injectable organic complexes of the minerals. Such complexes are more expensive, but when injected subcutaneously or intramuscularly they are translocated slowly to other tissues and prevent deprivation for lengthy periods. For instance, a single intramuscular injection of iron-dextran supplying 100 mg iron at 2–4 days of age can control piglet anaemia (Chapter 13).

SLOW-RELEASE ORAL METHODS. The efficacy of administering minerals to individual animals can be improved by using large oral doses in relatively inert, slowly solubilized forms. Heavy pellets (e.g. selenium, see Chapter 15), glass boluses (multi-element) (Miller et al., 1988) or particles of the mineral (e.g. copper oxide, see Chapter 11) are retained in the gastrointestinal tract and slowly release absorbable mineral. In early Australian studies, cobalt deprivation was solved by use of the heavy cobalt pellets that lodge in the reticulo-rumen and yield a steady supply of cobalt for many months, unless they are regurgitated or became coated (see Chapter 10). Retention of the mineral is dependent on specific gravity and particle size (Judson et al., 1991), but, by ingenious capsule design, mineral sources of low density can be retained in the rumen (Cardinal, 1997). Minerals can also be given with vitamins when compressed into a polymer capsule which erodes slowly from its 'open' end (Allan et al., 1993).

Continuous direct methods

Hand-fed or mechanically fed stock are best supplemented by mixing the required minerals with the food offered. When adequate trough-time or trough-space is allowed, variation in individual consumption is not pronounced and the amounts provided should not exceed average mineral requirements. Where animal productivity and farm labour costs are high, it is usual to buy feeds that contain a 'broad-spectrum' mineral supplement. Where production is less intensive and farm labour less costly, home-mixing is often practised, adding only those minerals known to be needed in amounts appropriate for the locality. Home-grown roughage rather than expensive concentrates can be used as a carrier for the mineral supplement. In sparse grazing conditions where there is no hand-feeding, treatment of the water supply may be practicable but cannot be relied on if the water supply is saline and may be unsatisfactory with an element like cobalt, which is required continuously by the animal. Trace minerals can be add to a piped water supply via a metering device (Farmer et al., 1982), but limitations remain because voluntary water intake is affected by the weather and season and may be uncontrollable under range conditions.

FREE-ACCESS MINERAL MIXTURES. The most widely used method of supplementation is the provision of mineral mixtures in loose ('lick') or block form. Common salt (NaCl) is the main ingredient (30–40%) because it is palatable and acts as a 'carrier' of other less palatable minerals. Individual animals are expected to seek out dispensers and consume enough mineral from them to meet their needs (McDowell et al., 1993). The siting of dispensers can be arranged to attract stock to areas that might otherwise be undergrazed. However, individual variations in consumption from licks or blocks can be marked (Bowman and Sowell, 1997). In one study, 19% of the ewe flock commonly consumed nothing, while others consumed a feed block at 0.4–1.4 kg day^{-1} (Ducker et al., 1981). In a study of salt supplementation (presented in protected, trailer-drawn blocks) of a beef suckler herd grazing Californian rangeland, the mean daily consumption per head (cow ± calf) ranged from 0 to 129 g day^{-1} about a mean of 27 g day^{-1} over consecutive 7-day periods (Fig. 3.4) (Morris et al., 1980). Provision is therefore 'semi-continuous' rather than 'continuous'.

The provision of several widely scattered mineral sources can minimize competition and hence individual variations in the uptake of minerals. Seasonal variation in mineral consumption can also occur, but intakes are more uniform from granulated than block sources of minerals (Rocks et al., 1982). Minerals can be used pharmacologically, and molybdate-containing salt licks have been highly effective in preventing chronic copper poisoning in sheep and cattle in certain areas.

Mineral Sources

The choice of components for a feed or mineral supplement is determined by the following:

- the purity and cost per unit of the element or elements required;
- the chemical form, which may determine the stability and availability of the required element;
- particle size, which can affect the ease and safety of mixing and availability to the animal;
- freedom from harmful impurities, such as cadmium and fluorine;
- compatibility with other ingredients, including vitamins; and
- the risk of toxicity, often the converse of availability and affecting man as well as animal.

With phosphorus supplements, the first four factors can each greatly influence the choice of material. With the trace elements, cost is of minor significance and availability and safety can be the major concerns.

Inorganic versus 'organic' sources

There is a popular misconception, sustained and encouraged particularly by aggressive marketing, that inorganic trace mineral supplements are intrinsically of less value to livestock (and man) than the natural forms in which elements are present in feeds, and are less available than synthetic organic complexes containing those elements. Trace metal–amino acid complexes can allegedly mimic the process by which trace elements are absorbed (as metal-peptides) and might thus be more available to livestock than inorganic mineral

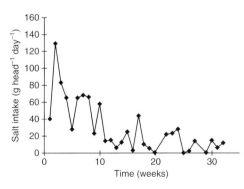

Fig. 3.4. Variation in average salt intake by a group of beef suckler cows given free access to salt from covered trailers on Californian rangeland. Observations began with calving in November (data from Morris et al., 1980).

sources. However, such complexes must be strong enough to resist natural dietary antagonists, yet deliver the complexed element to the tissues in a usable form. An early series of *in vitro* tests showed that a wide variety of amino acid–trace mineral complexes were not even strong enough to resist simulated acid digestion in the stomach (Fig. 2.4) and this has been confirmed in the specific case of manganese 'chelates' (Lu *et al.*, 2003). There is no evidence that such complexes with copper are superior to copper sulfate for ruminants (Suttle, 1994). While chelated minerals might be expected to be of benefit in non-ruminant diets where phytate is such an important antagonist, these expectations have yet to be fulfilled and a more sound approach is to attack the organic antagonist with phytase. Evidence that chelation alters the form or distribution of the complexed element at an early stage of digestion (e.g. in the rumen) is not proof of nutritional benefit (Wright *et al.*, 2008). Evidence that chelation increases tissue concentrations for elements whose uptake is normally regulated by need (e.g. selenium) may indicate a potentially dangerous by-passing of the homeostatic mechanism, and again provides no proof of nutritional benefit. Chelation may be useful for one element, chromium, in stressed animals for reasons that are not fully understood. Where the provision of an element in organic form influences metabolism, as it undoubtedly does with selenium given as selenomethionine, the assessment of mineral status is affected without necessarily conferring any nutritional benefit. Evidence of consistently superior value in a robust test of availability for organic complexes has yet to appear in a reputable, peer-reviewed scientific journal, with the singular exception of iron in non-ruminants.

Over-correction of Mineral Deficits

The provision of minerals beyond the needs of livestock increases costs with no benefit – other than ill-founded 'peace of mind' – and can be harmful. Excess phosphorus and magnesium can cause death from urinary calculi (Hay and Suttle, 1993), while an excess of copper is quickly toxic in sheep. Iron is sometimes added as iron oxide for cosmetic rather than nutritional purposes, to the possible detriment of copper status. The excess provision of iron may be partly responsible for the low availability of manganese in many pig rations (Finley *et al.*, 1997). Proprietary free-access mineral mixtures for the same class of stock can vary widely in composition, due largely to the use of different safety factors in seeking to deliver a sufficient supply of all essential minerals to all individuals, whatever their circumstances (Suttle, 1983). Concentrations of trace elements are also driven upward by competition between products and exploitation of a widespread misconception amongst farmers that 'more means better'. In Australia, government scientists have tried to reverse this trend by formulating a mineral supplement solely to meet local mineral requirements (Siromin®) (White *et al.*, 1992). The use of over-generous 'shot-gun' mixtures of all essential minerals should be periodically questioned because of the risk of causing long-term mineral imbalances. Surplus minerals are largely excreted and some (notably phosphorus and copper) are causing increasing concern as pollutants of the environment (see Chapter 20).

Self-correction of Mineral Imbalance

Contrary to popular belief, appetite for a mineral, particularly trace elements, is not a reliable measure of an animal's need, nor can it be relied upon to correct any dietary imbalance (Pamp *et al.*, 1976). Grazing sheep dosed orally with minerals to overcome deficiencies have been found to eat as much mineral lick as undosed animals (Cunningham, 1949), and mineral-deprived livestock often reject supplements containing the mineral they lack. Physiological and behavioural explanations probably lie in the absence of cognitive distress and/or the inability to associate immediate alleviation of that distress with consumption of a recognizable mineral source. The voluntary consumption of mineral mixtures or 'licks' is determined more by palatability than physiological need. The few examples of 'mineral wisdom' being expressed by livestock all involve macro-elements. Sodium-deficient ruminants develop an appetite for salt and will even consume sodium sources to which they are normally averse, such as bicarbonate of soda. However, if salt is provided in a free-access mixture that is also rich in magnesium, excessive consumption of magnesium may cause diarrhoea (Suttle et al., 1986). When offered choices between feeds containing excess or normal concentrations of salt and high- or low-energy and protein concentrations, sheep select combinations that improve the diet (Thomas et al., 2006). The relish with which livestock consume mineral deposits around receding water sources may have unintended nutritional benefits. Pullets approaching lay will self-select diets rich enough in calcium to meet the combined needs of bone deposition and shell calcification. The laying hen continues to show preferences for calcium-rich diets on egg-laying days, but a hen deprived of phosphorus will adapt to a decreased need for calcium by reducing its voluntary consumption of calcium carbonate (Barkley et al., 2004).

References

Allan, C.L., Hemingway, R.G. and Parkin, J.J. (1993) Improved reproductive performance in cattle dosed with trace element/vitamin boluses. *Veterinary Record* 132, 463–464.
Alloway, B.J. (2004) *Zinc in Soils and Crop Nutrition*. IZA Publications, International Zinc Association, Brussels, pp. 1–116.
Arosemena, A., DePeters, E.J. and Fadel, J.G. (1995) Extent of variability in the nutrient composition within selected by-product feedstuffs. *Animal Feed Science and Technology* 54, 103–120.
Arthur, J.R. (1992) Selenium metabolism and function. *Proceedings of the Nutrition Society of Australia* 17, 91–98.
Barkley, G.R., Miller, H.M. and Forbes, J.M. (2004) The ability of laying hens to regulate phosphorus intake when offered two feeds containing different levels of phosphorus. *British Journal of Nutrition* 92, 233–240.
Bowman, J.G.P. and Sowell, B.F. (1997) Delivery method and supplement consumption by grazing ruminants: a review. *Journal of Animal Science* 75, 543–550.
Cardinal, J.R. (1997) Intrarumunal devices. *Advanced Drug Delivery Systems* 28, 303–322.
Chesters, J.P. and Arthur, J.R. (1988) Early biochemical defects caused by dietary trace element deficiencies. *Nutrition Research Reviews* 1, 39–56.
Clark, R.G., Wright, D.F. and Millar, P.R. (1985) A proposed new approach and protocol to defining mineral deficiencies using reference curves. Cobalt deficiency in young sheep used as a model. *New Zealand Veterinary Journal* 33, 1–5.
Combs, D.P. (1987) Hair analysis as an indicator of the mineral status of livestock. *Journal of Animal Science* 65, 1753–1758.
Cunningham, I.J. (1949) Sheep dosed with minerals eat as much lick as undosed animals. *New Zealand Journal of Agriculture* 8, 583–586.
De Vos, W., Travainen, T. and Reeder, S. (eds) (2006) *Geochemical Atlas of Europe. Part 2: Interpretation of Geochemical Maps, Additional Tables, Figures, Maps, and Related Publications*. Geological Survey of Finland.
Ducker, M.J., Kendall, P.T., Hemingway, R.G. and McClelland, T.H. (1981) An evaluation of feed blocks as a means of providing supplementary nutrients to ewes grazing upland/hill pastures. *Animal Production* 33, 51–58.
Farmer, P.E., Adams, T.E. and Humphries, W.R. (1982) Copper supplementation of drinking water for cattle grazing molybdenum-rich pastures. *Veterinary Record* 111, 193–195.
Finley, J.W., Caton, J.S., Zhou, Z. and Davison, K.L. (1997) A surgical model for determination of true absorption and biliary excretion of manganese in conscious swine fed commercial diets. *Journal of Nutrition* 127, 2334–2341.
Fordyce, F.M., Masara, D. and Appleton, J.D. (1996) Stream sediment, soil and forage chemistry as predictors of cattle mineral status in northeast Zimbabwe. In: Appleton, J.D., Fuge, R. and McCall, G.H.J. (eds) *Environmental Geochemistry and Health*. Geological Society Special Publication No. 113, London, pp. 23–37.

Grace, G.D. and Knowles, S.O. (2002) A reference curve using blood selenium concentration to diagnose selenium deficiency and predict growth responses in lambs. *New Zealand Veterinary Journal* 50, 163–165.

Hay, L. and Suttle, N.F. (1993) Urolithiasis. In: *Diseases of Sheep*, 2nd edn. Blackwell Scientific, London, pp. 250–253.

Judson, G.J., Babidge, P.J., Brown, T.H., Frensham, A.B., Langlands, J.P. and Donald, G.E. (1991) Copper oxide particles as an oral supplement: the retention of wire particles in the alimentary tract of sheep. In: Momcilovic, B. (ed.) *Proceedings of the Seventh International Symposium on Trace Elements in Man and Animals, Dubrovnik*. IMI, Zagreb, pp. 15-10–15-11.

Kumar, S. and Singh, S.P. (1981) Muzzle secretion electrolytes as a possible indicator of sodium status in buffalo (*Bubalus bubalis*) calves: effects of sodium depletion and aldosterone administration. *Australian Journal of Biological Sciences* 34, 561–568.

Langlands, J.P. (1987) Assessing the nutrient status of herbivores. In: Hacker, J.B. and Ternouth, J.H. (eds) *Proceedings of the Second International Symposium on the Nutrition of Herbivores*. Academic Press, Sydney, Australia, pp. 363–385.

Laven, R.A. and Livesey, C.T. (2006) An evaluation of the effect of clotting and processing on the recovery of copper from bovine blood. *Veterinary Journal* 171, 295–300.

Lee, J., Masters, D.G., White, C.L., Grace, N.D. and Judson, G.J. (1999). Current issues in trace element nutrition of grazing livestock in Australia and New Zealand. *Australian Journal of Agricultural Research* 50 1341–1364.

Lu, L., Luo, X.G., Ji, C., Liu, B. and Yu, S.X. (2007) Effect of manganese supplementation on carcass traits, meat quality and lipid oxidation in broilers. *Journal of Animal Science* 85, 812–822.

McDowell, L.R., Conrad, J.H. and Hembry, F.G. (1993) *Minerals for Grazing Ruminants in Tropical Regions*, 2nd edn. Animal Science Department, University of Florida, Gainsville, Florida.

McGrath, S. and Loveland, P.J. (1992) *The Soil Geochemical Atlas of England and Wales*. Blackie Academic and Professional, Glasgow, UK.

Millar, P.R. and Meads, W.J. (1988) Selenium levels in the blood, liver, kidney and muscle of sheep after the administration of iron/selenium pellets or soluble glass boluses. *New Zealand Veterinary Journal* 36, 8–10.

Millar, P.R., Meads, W.J., Albyt, A.T., Scahill, B.G. and Sheppard, A.D. (1988) The retention and efficacy of soluble glass boluses for providing selenium, cobalt and copper to sheep. *New Zealand Veterinary Journal* 36, 11–14.

Mills, C.F. (1987) Biochemical and physiological indicators of mineral status in animals: copper, cobalt and zinc. *Journal of Animal Science* 65, 1702–1711.

Morris, J.G. (1980) Assessment of the sodium requirements of grazing beef cattle: a review. *Journal of Animal Science* 50, 145–152.

Morris, J.G., Dalmas, R.E. and Hall, J.L. (1980) Salt (sodium) supplementation of range beef cows in California. *Journal of Animal Science* 51, 722–731.

O'Dell, B.L. (2000) Role of zinc in plasma membrane function. *Journal of Nutrition* 130, 1432S–1436S.

Pamp, D.E., Goodrich, R.D. and Meiske, J.C. (1976) A review of the practice of feeding minerals free choice. *World Review of Animal Production* 12, 13–17.

Phillippo, M. (1983) The role of dose–response trials in predicting trace element disorders. In: Suttle, N.F., Gunn, R.G., Allen, W.M., Linklater, K.A. and Wiener, G. (eds) *Trace Elements in Animal Production and Veterinary Practice*. British Society of Animal Production Special Publication No. 7, Edinburgh, pp. 51–60.

Plant, J.A., Baldock, J.W. and Smith, B.I. (1996) The role of geochemistry in environmental and epidemiological studies in developing countries: a review. In: Appleton, D., Fuge, R. and McCall, G.H.J. (eds) *Environmental Geochemistry and Health*. Geological Society Special Publication No. 113, London, pp. 7–22.

Rocks, R.L., Wheeler, J.L. and Hedges, D.A. (1982) Labelled waters of crystallization in gypsum to measure the intake of loose and compressed mineral supplements. *Australian Journal of Experimental Agriculture and Animal Husbandry* 22, 35–42.

Suttle, N.F. (1983) Meeting the mineral requirements of sheep. In: Haresign, E. (ed.) *Sheep Production*. Butterworths, London, pp. 167–183.

Suttle, N.F. (1991) Mineral supplementation of low quality roughages. In: *Proceedings of Symposium on Isotope and Related Techniques in Animal Production and Health*. International Atomic Energy Agency, Vienna, pp. 101–104.

Suttle, N.F. (1994) Meeting the copper requirements of ruminants. In: Garnsworthy, P.C. and Cole, D.J.A. (eds) *Recent Advances in Animal Nutrition*. Nottingham University Press, Nottingham, UK, pp. 173–188.

Suttle, N.F. (2000) Minerals in livestock production. *Asian-Australasian Journal of Animal Science* 13, 1–9.

Suttle, N.F. (2008) Relationships between the concentrations of tri-chloroacetic acid-soluble copper and caeruloplasmin in the serum of cattle from areas with different soil concentrations of molybdenum. *Veterinary Record* 162, 237–240.

Suttle, N.F., Brebner, J., McLean, K. and Hoeggel, F.U. (1996) Failure of mineral supplementation to avert apparent sodium deficiency in lambs with abomasal parasitism. *Animal Science* 63, 103–109.

Thomas, D.T., Rintoul, A.J. and Masters, D.G. (2006) Sheep select combinations of high and low sodium chloride, energy and protein feed that improve their diet. *Applied Animal Behaviour Science* 105, 140–153.

Thornton, I. and Alloway, B.J. (1974) Geochemical aspects of the soil–plant–animal relationship in the development of trace element deficiency and excess. *Proceedings of the Nutrition Society* 33, 257–266.

White, C.L. (1996) Understanding the mineral requirements of sheep. In: Masters, D.G. and White, C.L. (eds) *Detection and Treatment of Mineral Nutrition Problems in Grazing Sheep*. ACIAR Monograph 37, Canberra, pp. 15–30.

White, C.L., Masters, D.G., Peter, D.W., Purser, D.B., Roe, S.P. and Barnes, M.J. (1992) A multi-element supplement for grazing sheep 1. Intake, mineral status and production responses. *Australian Journal of Agricultural Research* 43, 795–808.

White, C.L., Martin, G.B., Hynd, P.T. and Chapman, R.E. (1994) The effect of zinc deficiency on wool growth and skin and wool histology of male Merino lambs. *British Journal of Nutrition* 71, 425–435.

White, J.G. and Zasoski, R.J. (1999) Mapping soil micronutrients. *Field Crops Research* 60, 11–26.

Wright, C.L., Spears, J.W. and Webb, K.E. (2008) Uptake of zinc from zinc sulfate and zinc proteinate by ovine ruminal and omasal epithelia. *Journal of Animal Science* 86, 1357–1363.

4 Calcium

Introduction

The essentiality of calcium for growth and milk production was one of the first consequences of mineral deprivation to be unequivocally demonstrated (see Chapter 1) but the pullet provided a convenient experimental model for early studies of calcium deprivation (Common, 1938). As animal production intensified, energy-rich, grain-based diets were increasingly fed to poultry and other livestock in protected environments and the incidence of bone disorders multiplied. Unwittingly, animals were being fed diets naturally deficient in calcium while being simultaneously 'starved' of vitamin D_3 (essential for the efficient utilization of calcium) by the absence of sunlight. Animal breeders ensured that interest in calcium nutrition was maintained by selecting for traits with high calcium requirements: growth, milk yield, litter size and egg production. For example, between 1984 and 2004 the efficiency of broiler production greatly improved and broilers reached market weight (2.9 kg) in 6 weeks instead of 9 weeks, while consuming less feed and 11% less calcium in total (Driver et al., 2005a). Requirements for calcium have had to be re-examined, but bone disorders in broilers can persist even when calcium nutrition is optimal and the only solution may be to restrict growth by limiting feed intake (Angel, 2007). In the high-yielding dairy cow, an acute calcium-responsive disorder, milk fever, still strikes many animals with no sign of bone disorders at calving and the best means of prevention is a matter of dispute.

Functions of Calcium

Calcium is the most abundant mineral in the body and 99% is found in the skeleton. The skeleton not only provides a strong framework for supporting muscles and protects delicate organs and tissues, including the bone marrow, but is also jointed to allow movement, and is malleable to allow growth. Furthermore, the skeletal reserve of calcium actively supports calcium homeostasis in what Bain and Watkins (1993) described as an 'elegant compromise'.

Non-skeletal functions

The small proportion (1%) of body calcium that lies outside the skeleton is important to survival. It is found as the free ion, bound to serum proteins and complexed to organic and inorganic acids. Ionized calcium – 50–60% of the total plasma calcium – is essential for nerve conduction, muscle contraction and cell signalling (Carafoli, 1991). Changes in calcium ion concentrations within and between cells (Breitwieser, 2008) are modulated by vitamin D_3 and the calcium-binding proteins, calmodulin and osteopontin: among other things, they can trigger the immune response. Calcium can activate or stabilize some enzymes and is required for normal blood clotting (Hurwitz, 1996), facilitating the conversion of prothrombin to thrombin, which reacts with fibrinogen to form the blood clot, fibrin.

Bone growth and mineralization

Bones grow in length by the proliferation of cartilaginous plates at the ends of bones in response to growth hormone and other growth factors (Loveridge, 1999). Cells towards the end of the regular columns of chondrocytes that constitute the growth or epiphyseal plate become progressively hypertrophic and degenerative. They concentrate calcium and phosphorus at their peripheries and exfoliate vesicles rich in amorphous calcium phosphate ($Ca_3(PO_4)_2$) (Wuthier, 1993). Thus, a calcium-rich milieu is provided to impregnate the osteoid (organic matrix) laid down by osteoblasts. The crystalline bone mineral hydroxyapatite ($Ca_{10}(PO_4)_6(OH)_2$) accumulates in a zone of provisional calcification around decaying chondrocytes, replacing them with an apparently disorganized and largely inorganic matrix of trabecular bone. Bones grow in width through the inward deposition of concentric shells or lamellae of osteoid by osteoblasts, behind a cutting cone of osteoclasts and beneath the surface of the bone shaft or periosteum (Loveridge, 1999).

Eggshell formation

In avian species, calcium protects the egg through the deposition of an eggshell during passage down the oviduct. Unique changes in bone morphology coincide with the onset of sexual maturity, when the release of oestrogen halts cancellous and structural bone formation. New medullary bone is woven on to the surface of existing cortical bone and bone marrow to provide a labile calcium reserve that maintains plasma calcium concentrations during shell formation (Gilbert, 1983; Whitehead, 1995). The shell matrix becomes heavily impregnated with calcium carbonate ($CaCO_3$) and the need to furnish about 2g Ca for every egg produced dominates calcium metabolism in the laying hen (Gilbert, 1983; Bar and Hurwitz, 1984).

Metabolism of Calcium

Certain features of calcium metabolism must be described before the assessment of the calcium value of feeds can be addressed. Calcium is absorbed from the diet according to need by a hormonally regulated process in the small intestine (Schneider et al., 1985; Bronner, 1987) up to limits set by the diet and by the net movement of calcium into or out of the skeleton.

Control of absorption

Calcium is absorbed by an active process in the small intestine under the control of two hormones: parathyroid hormone (PTH) and the physiologically active form of vitamin D_3, dihydroxycholecalciferol (1,25-$(OH)_2D_3$, also known as calcitriol) (Schneider et al., 1985; Bronner, 1987). The parathyroid gland responds to small reductions in ionic calcium in the extracellular fluid by secreting PTH (Brown, 1991). This stimulates the double hydroxylation of vitamin D_3, first to 25-OHD_3 in the liver and then to 24,25-$(OH)_2D_3$ or 1,25-$(OH)_2D_3$ primarily in the kidneys (Omdahl and DeLuca, 1973; Borle, 1974), but also in the bone marrow, skin and intestinal mucosa (Norman and Hurwitz, 1993). Activated 1,25-$(OH)_2D_3$ opens calcium channels in the intestinal mucosa to facilitate calcium uptake and transfer with the help of a calcium-binding protein, calbindin (Hurwitz, 1996; Shirley et al., 2003). The role of calbindin in facilitating the absorption of calcium according to supply and demand is illustrated by data for fast- and slow-growing chicks given diets varying in calcium (Fig. 4.1) (Hurwitz et al., 1995). The full potential of a feed as a provider of absorbable calcium ($^{max}A_{Ca}$) can only be established under conditions where requirements (R) are barely met by intake (I), a condition rarely met in balance studies (AFRC, 1991).

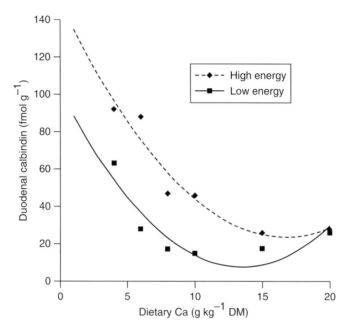

Fig. 4.1. Mechanism for absorbing calcium according to need. As dietary Ca concentrations decrease, the synthesis of a potentiator of absorption (calbindin) increases. If high-energy diets are fed, demand for Ca rises and the broiler chick synthesizes yet more duodenal calbindin (from Hurwitz et al., 1995).

Plots of calcium absorption against R/I in cattle (ARC, 1980) and sheep (Fig. 4.2) show that both species raise their calcium absorption as the value of R/I approaches unity, up to a maximum absorption of approximately 0.68 for all feeds, including forages. However, this rule does not apply in sheep when imbalanced diets are fed. When the calcium supply is excessive and concentrations in the gut exceed 1 mmol l^{-1}, calcium is passively absorbed and homeostatic mechanisms are reversed by the secretion of calcitonin, which inhibits the intestinal hydroxylation of vitamin D$_3$ (Beckman et al., 1994). In such circumstances, calcium may be absorbed from the rumen (Yano et al., 1991; Khorasani et al., 1997). The vast quantities of calcium required by the laying hen are mostly absorbed by passive mechanisms (Gilbert, 1983).

Movements of calcium to and from the skeleton

The net flow of calcium into or out of the skeleton plays an important role in regulating circulating concentrations of ionic calcium and thus calcium absorption. Mammalian young are born with poorly mineralized bones and while they suckle they do not receive enough calcium to fully mineralize all the bone growth that energy-rich milk can sustain (AFRC, 1991). After weaning there is normally a progressive increase in bone mineralization (for data on lambs, see Field et al., 1975; Spence et al., 1985), stimulated by increased exercise and load-bearing. Ash concentration in the broiler tibia can increase by 10% between 16 and 42 days of age (see Driver et al., 2005a), and both exercise and provision of a perch accelerate bone mineralization in poultry (Bond et al., 1991).

Differences in calcium concentrations in the immature bodies of livestock species (see Table 1.2 for a comparison of the horse, deer, and sheep) probably reflect the proportion of compact bone present and may influence calcium absorption. Around one-fifth of calcium in the skeleton is mobilized in sheep (Braithwaite, 1983b) and dairy cows (Ramberg et al., 1984) in association with parturition and the onset of lactation. A peri-parturient reduction in bone

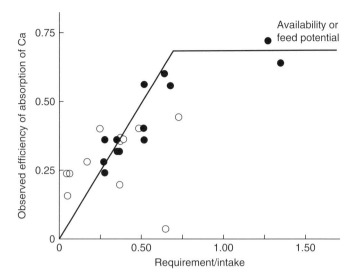

Fig. 4.2. Evidence that calcium is absorbed according to need. Pooled data for the efficiency of Ca absorption in lambs plotted against the adequacy of the dietary supply (ratio of net Ca requirement to Ca intake) show a curvilinear rise to a plateau representing the full potential of the dietary Ca source (from AFRC, 1991). ●, nutritionally balanced diets; ○, nutritionally imbalanced diets.

density can be detected in both sheep and goats (Liesegang, 2008). The resorption process is probably obligatory since it is not abated by supplying more dietary calcium (AFRC, 1991; Kamiya et al., 2005) or protein (Chrisp et al., 1989b) and is not necessarily accompanied by hypocalcaemia (Braithwaite, 1983a; Liesegang et al., 2000).

Hormonal control of skeletal calcium

The hormonal partnerships that facilitate absorption also regulate calcium fluxes to and from bone, but the mechanisms are complex and multicentred. Nuclear receptors for 1,25-$(OH)_2D_3$ on the chondrocytes and osteoblasts facilitate calcium accretion. Receptors for PTH are found linked to regulatory (G) proteins in organs such as the kidneys and brain, as well as in the parathyroid gland (Hory et al., 1998). Calcium accretion and resorption are coupled during growth (Eklou-Kalonji. et al., 1999) but uncoupled in late pregnancy and early lactation, partly through reductions in the activity of osteocalcin – a hormone that promotes bone growth. Serum osteocalcin concentrations decrease prior to calving (Peterson et al., 2005) and remain low for at least 14 days (Liesegang et al., 2000), halting bone growth to reduce the demand for calcium. Mammary tissue can up-regulate the expression and release of a PTH-like hormone, PTH-related protein (PTHrP) (Goff et al., 1991). PTHrP then enters the maternal circulation and mimics PTH. Bone-resorbing cells (osteoclasts) respond to 1,25-$(OH)_2D_3$ via cytokines released by osteoblasts (Norman and Hurwitz, 1993) and to oestradiol, increasing in sensitivity to PTHrP (Goff, 2007). Calcitonin is now believed to have an important role in limiting bone resorption during lactation (Miller, 2006), but this does not prevent the down-regulation of calcium absorption.

Excretion

The modulation of excretion by the faecal and urinary routes probably involves G-linked calcium ion receptors (Hory et al., 1998), but plays only a minor role in calcium homeostasis. The faecal endogenous excretion of calcium (FE_{Ca}) is generally unaffected by changes in dietary calcium supply, and the difference between intake and faecal excretion of calcium (apparent absorption, AA_{Ca}) in balance studies can be adjusted to give values for true calcium absorption (A_{Ca}) in ruminants (Fig. 4.2) (ARC, 1980). Increases in dry matter (DM) intake (DMI) are associated with proportional increases in FE_{Ca} in sheep on dry diets (AFRC, 1991) and a similar relationship applies to grass diets (Chrisp et al., 1989a,b). Losses of 0.6–1.0 g FE_{Ca} kg^{-1} DMI have been reported for lactating cows on corn silage (Martz et al., 1999) and for lambs on mixed dry diets (Roque et al., 2006), but much higher losses (up to 50 mg Ca kg^{-1} live weight (LW)) have been reported in lambs (Chrisp et al., 1989a). Little attention has been given to FE_{Ca} in non-ruminants: the Agricultural Research Council (ARC, 1981) assumed that it was 32 mg kg^{-1} LW in estimating the maintenance requirement (M) of pigs, but enormous and unlikely losses are predicted by the fattening stage (3.2 g Ca day^{-1} at 100 kg LW) and it may be more accurate to assume that FE_{Ca} is related to DMI, as in ruminants. In Fig. 4.3a, an alternative estimate of irreducible FE_{Ca} in pigs on a corn/soybean meal (SBM) diet is derived. This translates to a calcium loss of 1 g kg^{-1} DMI, similar to that found in ruminants but nearly threefold that found in pigs on semi-purified diets very low in calcium (Traylor et al., 2001).

Although FE_{Ca} is not regulated, it makes a significant contribution to faecal excretion in pigs and measurements of AA_{Ca} significantly underestimate calcium absorption in balance studies (Fig. 4.3b). Urinary calcium excretion tends to remain low and constant regardless of calcium status, although it may rise significantly in the acidotic dairy cow (Goff and Horst, 1998) and in the laying hen on days when no shell is formed (Gilbert, 1983). The retention of calcium is therefore roughly equivalent to AA_{Ca}, and at marginal calcium intakes reflects the full 'calcium value' of the calcium supplied.

Natural Sources of Calcium

Milk and milk replacers

Milks and milk replacers derived from bovine milk are rich in calcium (Table 2.5) that is well absorbed by the young – A_{Ca} 0.83 (Challa and Braithwaite, 1989) and AA_{Ca} 0.95 (Yuangklang et al., 2004) – but calcium is marginally less absorbable from replacers based on soy protein (AA_{Ca} 0.87) (Yuangklang et al., 2004). The effect of protein source was attributed to the formation of insoluble calcium

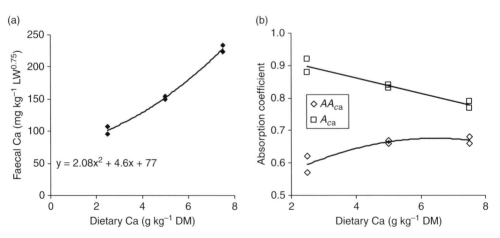

Fig. 4.3. (a) The relationship between faecal calcium excretion and dietary calcium in pigs for the two higher phosphorus groups of three studied by Vipperman et al. (1974). The intercept indicates an irreducible faecal endogenous loss (*IFE*) or maintenance requirement of 77 mg Ca kg^{-1} LW$^{0.75}$. With the pigs weighing 25 kg and consuming 0.85 kg of food daily, the intercept translates to an IFE_{Ca} of approximately 1 g Ca kg^{-1} DMI. (b) The IFE_{Ca} value can be used to derive true absorption coefficients (A_{Ca}) from faecal Ca excretion, and these show a tendency for Ca to be absorbed according to need.

phosphate, but may also involve the formation of insoluble complexes with phytate from the soy product. Phytate ceases to affect calcium absorption in weaned ruminants because it is degraded by the rumen microflora.

Forages

Forages are generally satisfactory sources of calcium for grazing livestock, particularly when they contain leguminous species. Minson (1990) gives global average calcium values of 14.2 and 10.1 g kg^{-1} DM for temperate and tropical legumes, and of 3.7 and 3.8 kg^{-1} DM for the corresponding grasses. In the UK, swards contain an average calcium level of 5.4 g kg^{-1} DM (MAFF, 1990) and should usually meet the calcium requirements of sheep, but a contribution from legumes may be needed for the dairy cow. Cultivar differences in calcium content can be marked, but the maturity of the sward and seasonal changes in the legume contribution have more widespread influences. The leaf generally contains twice as much calcium as the stem, and pasture calcium is therefore increased by applying nitrogenous fertilizer and decreases with advancing maturity. In New Zealand, pasture calcium increased from 4.4 to 5.7 g kg^{-1} DM between spring and summer due to a seasonal increase in clover content (Stevenson et al., 2003). The slowing of pasture growth by flooding or seasonal reductions in soil temperature increase herbage calcium concentrations. The calcium content of the soil (Jumba et al., 1995) and applications of lime, limestone or calcium chloride to soils has surprisingly little effect on forage calcium (e.g. Goff et al., 2007). Selective grazing is likely to result in higher calcium concentrations in ingested forage than in hand-plucked samples. Grass conserved as silage contains slightly more calcium than the more mature hay (6.4 versus 5.6 g kg^{-1} DM), while alfalfa silage is much richer in calcium and more variable (13.5–24.0 g kg^{-1} DM) (MAFF, 1990). Maize silages are generally low (1.4–3.0 g kg^{-1} DM) (see Table 2.1 and McNeil et al., 2002), root crops and straws marginal (3–4 g kg^{-1} DM) and leafy brassicas rich in calcium (10–20 g kg^{-1} DM).

Concentrates

Cereals are low in calcium, with maize typically containing only 0.2 g kg^{-1} DM, wheat 0.6 g kg^{-1} DM, oats and barley 0.6–0.9 g kg^{-1} DM and cereal by-products rarely more than 1.5 g kg^{-1} DM (see Table 2.1). Although richer in calcium than cereals, vegetable protein sources usually contain no more than 2–4 g kg^{-1} DM (MAFF, 1990) and most will not give adequate calcium levels when blended with cereals. Rapeseed meal, with 8.4 g Ca kg^{-1} DM, is exceptional and, like sugarbeet pulp (6.0–7.5 g kg^{-1} DM), is an excellent supplement for phosphorus-rich, grain-based diets. Fish meal and meat and bone meals are also good sources of calcium, with levels of 50–100 g kg^{-1} DM being commonplace.

Absorbability of Forage Calcium for Ruminants

The NRC (2001) accorded forage calcium an A_{Ca} coefficient of 0.30, largely because of the low AA_{Ca} values that are frequently reported for alfalfa and a low A_{Ca} value of 0.42 reported for corn silage in lactating cows in negative calcium balance (Martz et al., 1999). These and other low A_{Ca} values require critical evaluation because ($^{max}A_{Ca}$) can only be established if: (i) calcium requirements are barely met by calcium intake; (ii) calcium is not being removed from the skeleton; and (iii) the diet provides an adequate amount of other nutrients, notably phosphorus and protein (Fig. 4.2).

Alfalfa

Low A_{Ca} and AA_{Ca} coefficients are to be expected, since alfalfa contains three to four times more calcium than grasses and invariably provides more calcium than ruminants need if fed as the main dietary component, but low A_{Ca} values are widely attributed to the inhibitory effects of oxalate. Crystals of calcium oxalate were once seen in the faeces of cattle fed on alfalfa hay (Ward et al., 1979), but they contain only one-third by weight of calcium.

Since alfalfa hays contain around four times more calcium than oxalate (Hintz et al., 1984), only a small fraction of the calcium supplied by alfalfa can be excreted as unabsorbable oxalates. Furthermore, oxalate is partially broken down in the rumen (Duncan et al., 1997). A high A_{Ca} of 0.56 was reported for alfalfa harvested in the vegetative state and given as hay to non-lactating goats fed to calcium requirement, but alfalfa harvested at the full-bloom stage had a lower calcium absorption of 0.47 (Freeden, 1989). In a comparison of alfalfa and cereal silages given to high-yielding cows, alfalfa had the highest AA_{Ca} (0.34) (Khorasani et al., 1997) but a large proportion (45%) disappeared from the rumen, possibly by passive absorption at high rumen calcium concentrations (calcium-rich alfalfa, containing 17 g kg^{-1} DM, was given with calcium-rich concentrate, containing 8.5 g kg^{-1} DM). The A_{Ca} 0.21 for alfalfa hay given to lambs appeared to be far lower than that for three mineral sources (Roque et al., 2006), but the hay comprised less than 50% of the ration (Fig. 2.6) and the results for the mineral sources were atypical. Furthermore, the application of a new kinetic model gave results that were so similar for the different sources that the data were pooled, with A_{Ca} averaging 0.52 (Dias et al., 2006)! In tropical grass species relatively low in calcium (1.3–3.5 g kg^{-1} DM), A_{Ca} coefficients were 0.51 and 0.52 for two 'high oxalate' species and only marginally higher in two 'low oxalate' species (0.57 and 0.64, respectively) (Blayney et al., 1982) when fed to Brahman steers. There is unlikely to be an obstacle to post-ruminal absorption of calcium, judging from the high availability of calcium from alfalfa meal for non-ruminants (Soares, 1995), unless calcium forms unabsorbable complexes with phytate in the rumen in some circumstances.

Other forage species

The corn silage studied by Martz et al. (1999) was deficient in phosphorus, sufficiently so to have possibly reduced food intake, and this may have lowered the ability to retain calcium. Low A_{Ca} coefficients of 0.17 and 0.19 were reported for ryegrass/white clover and oat/ryegrass mixtures given to lactating ewes with high needs for calcium (Chrisp et al., 1989b), but milk yield was increased by 18% and A_{Ca} increased to 0.30 when a protected casein supplement was fed, indicating that the diet was imbalanced with respect to its rumen-degradable protein supply. The protected casein was given in expectation that bone resorption might be reduced in early lactation, but it was not. In a second experiment, A_{Ca} for unsupplemented ryegrass/clover increased from 0.24 to 0.32 between weeks 2 and 5 of lactation, indicating marked differences between animals and differences between experiments in calcium absorption. A decrease in the calcium removed from the skeleton (1.7 g day^{-1}) was thought to have increased the need to absorb calcium as lactation progressed. Low A_{Ca} coefficients of 0.24–0.47 have been reported for green-feed oats and ryegrass/white clover given to stags with high calcium requirements for antler growth (Muir et al., 1987), but values may have been suppressed by seasonal increases in bone resorption. Estimates of A_{Ca} ranging from 0.41 to 0.64 have been obtained for different grass species given to growing lambs (Thompson et al., 1988). The higher values, obtained with the most nutritious species, were considered to indicate how much calcium was absorbable if needed. Timothy hay supplemented with limestone has a similar A_{Ca} to that of alfalfa (Freeden, 1989). Citrus pulp had a lower A_{Ca} coefficient (0.12) than alfalfa hay (0.21) in Roque et al.'s (2006) study, but basal diets differed in composition (Fig. 2.6).

There are no unequivocal measurements of calcium absorption in any forage that support the low A_{Ca} coefficient used by the NRC (2001) in calculating calcium requirements for dairy cows.

Absorbability of Forage Calcium for Horses

Horses were thought to be particularly vulnerable to the calcium-depleting effects of oxalate-rich roughages, but similar high AA_{Ca} coefficients of 0.80 and 0.76 were reported for ponies given alfalfa meals that were low or high in oxalate, respectively. The valuation of the alfalfas was valid because the ponies were fed, as the main source of calcium, with oats (1:2) in diets low in calcium (1.5g kg^{-1} DM) to maximize calcium absorption (Hintz et al., 1984). In a recent study of grass-fed weanling Thoroughbred foals AA_{Ca} was relatively high (0.60), but did not decrease when calcium supplements raised dietary calcium from 3.5 to 12.2g kg^{-1} DM (Grace et al., 2002a). This suggests that horses might not absorb calcium according to need, but there may be another explanation. The body of the foal is relatively rich in calcium (Table 1.2), probably reflecting the relatively high proportion of bone in the equine carcass. The weanling foal may have an enhanced capacity to deposit calcium in under-mineralized bone, and the presence of this physiological 'sponge', soaking up the apparent excess calcium absorbed, may have pre-empted the down-regulation of calcium absorption. A high AA_{Ca} of 0.75 was found in Thoroughbred mares given a similar pasture to that fed to foals (Grace et al., 2002b), probably reflecting the higher calcium requirements associated with lactation. In calves reared on milk replacers and likely to have relatively 'soft' bones, a doubling of dietary calcium from 6.4 to 11.5g kg^{-1} DM produced a relatively small decrease in AA_{Ca} from 0.90 to 0.75 (Yuangklang et al., 2004).

Absorbability of Calcium in Poultry and Pig Feeds

Cereals and vegetable proteins contain calcium that is highly absorbable, but the phytate they contain can lower the absorbability of inorganic calcium (necessarily added to meet requirements) through the formation of unabsorbable complexes in the gut. Increases in dietary calcium in the young chick (Driver et al., 2005b) and pig (Vipperman et al., 1974) can have a far smaller effect on calcium absorption than in the ruminant. This may indicate a considerable capacity of rapidly growing 'soft' bones to accumulate calcium, as in the horse.

Poultry

The complex effects of interactions between calcium and phytate on calcium absorption are illustrated by the wide range in fractional retention (or AA_{Ca}) found in broilers fed diets with various calcium, phytase and vitamin D supplements (Fig. 4.4) (Qian et al., 1997). Whether or not phytase and vitamin D_3 were added to alleviate a marginal phosphorus deficiency, retention fell as dietary calcium increased, the absorptive mechanism was down-regulated and calcium was precipitated as insoluble phytate complexes. The principal difference between groups lies in the AA_{Ca} attained at low calcium intakes. With the basal diet, the first increment in dietary calcium had a small effect on AA_{Ca}, indicating that the chicks were absorbing almost as much calcium as the corn/SBM diet allowed and giving maximum attainable retention equivalent to a $^{max}AA_{Ca}$ of 0.58 when vitamin D and available phosphorus were lacking. When vitamin D_3 was added to the diet (at 660µg kg^{-1}), values increased to $^{max}AA_{Ca}$ 0.62; removing the second

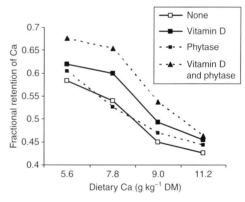

Fig. 4.4. Fractional retention[a] of dietary calcium in broiler chicks declines as dietary calcium exceeds the need for calcium. However, need depends upon the supply of available phosphorus and vitamin D from the diet. Supplementation with vitamin D increases the retention of Ca. Supplementation with phytase lifts the restriction imposed by a marginal P deficiency, but only raises the retention of Ca when vitamin D is also given (data from Qian et al., 1997, using the highest of four levels of phytase and vitamin supplementation). [a]Retention is roughly equivalent to apparent absorption in all species because little calcium is excreted via the urine.

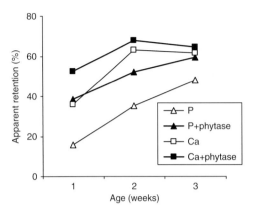

Fig. 4.5. The apparent (ileal) retention of calcium and phosphorus by broiler chicks increases markedly during the first 3 weeks of life. The addition of phytase to the diet increases both Ca and P retention, suggesting that – unaided – the newly hatched chick has a low capacity to degrade phytate and that Ca and phytate P are co-precipitated (data from Olukosi et al., 2007).

constraint with phytase (900 U kg^{-1}) increased that value to 0.67. The range in $^{max}AA_{Ca}$ is much less than that in AA_{Ca}, although still sufficiently wide for vitamin D and phytase to lower the calcium requirement by 13%. Similar results have been reported by Driver et al. (2005b). The ability of the newly hatched chick to extract calcium (and phosphorus) from a corn/SBM diet can be poor (Fig. 4.5), with only 36% (and 16%) being apparently retained. However, retention increased to 63% (and 35%) during the second week. The early constraint was partially alleviated by adding phytase to the diet and was presumed to be caused by the co-precipitation of calcium (largely from limestone) and phytate. By week 3 phytase had no effect on calcium retention. The impact of these interactions on calcium requirements and the efficacy of phytase supplements will be discussed later. Saturated fat supplements probably lower the retention of calcium from limestone in newly hatched broilers (Atteh and Leeson, 1983).

Pigs

The effect of phytate on AA_{Ca} is also evident in the literature for pigs. The main objective has been to improve the utilization of phytate phosphorus in corn/SBM diets, but there are associated improvements in A_{Ca} (see Figs 6.3 (p. 130) and 6.7 (p. 134)).

- Phytase: improvements in AA_{Ca} have been reported in 10 out of 11 studies in which phytase was added to rations for growing pigs. The mean improvement was 0.06 ± 0.02 on a basal AA_{Ca} coefficient of 0.61 ± 0.02 with dietary calcium of 5–6 g kg^{-1} DM.
- Dietary calcium: increasing concentrations from 3.2 to 4.1 and 5.0 g kg^{-1} DM reduced AA_{Ca} from 0.73 to 0.62 and 0.61, respectively, in finisher pigs when all diets contained added phytase (Liu et al., 1998). Corrected A_{Ca} coefficients (Fig. 4.3) were 1.04, 0.86 and 0.81, respectively.
- Phytate: in three comparisons of low phytate and normal (high phytate) corn (Spencer et al., 2000) or barley (Veum et al., 2001, 2007) the average improvement in the AA_{Ca} coefficient was 0.10 on a mean value of 0.57 for normal corn. In a ration virtually free of calcium, AA_{Ca} was 0.70 in low-phytate and 0.47 in high-phytate corn or SBM (Bohlke et al., 2005).

The enhanced utilization of phytate phosphorus in corn/SBM rations fed as gruels is likely to be matched by an increase in AA_{Ca}, but it may not be detected if the diet provides excess calcium or endogenous phytase. In a study by Larsen et al. (1999), the corrected A_{Ca} coefficient for a low-calcium ration in which all the calcium came from organic sources (barley, peas and rapeseed cake) was close to unity, but was decreased to 0.77 by pelleting the ration.

Calcium Requirements

The principal factors affecting the mineral requirements of farm animals, including calcium, are described in Chapter 1. Several developments in livestock husbandry have combined to steadily raise dietary calcium requirements, including: (i) genetic improvements, resulting in faster growth and higher yields; (ii) increasing use of energy-dense diets; (iii) early weaning on to solid diets, from which calcium is relatively poorly absorbed; (iv) breeding immature animals while they are still growing; and (v) the use of antibiotics, hormones and other feed additives as growth stimulants.

Although improved performance obtained through higher feed consumption does not increase the required dietary calcium concentration, improved feed conversion efficiency increases such requirements proportionately and the above factors probably contributed to the 50% increase in estimated calcium requirements of growing pigs between 1964 and 1976 (ARC, 1981). The net calcium requirement for growth is almost entirely for bone growth, but is particularly hard to define. Liberal supplies of vitamin D_3 are required to make the best use of dietary calcium in livestock housed for long periods.

Sheep and cattle

There have been no rigorous attempts to define the calcium requirements of sheep or cattle by means of feeding trials and it is still necessary to rely on factorial estimates of requirement. Unfortunately, these have varied considerably between authorities, largely due to disagreements on a realistic calcium absorption coefficient. The Agriculture and Food Research Council (AFRC, 1991) found further evidence to support the use of a maximal calcium absorption of 0.68 (Fig. 4.2) and their estimates for sheep and cattle are presented in Tables 4.1 and 4.2, respectively. The requirements for lambs (Table 4.1) were tested by feeding at 75%, 100% and 125% of the recommended level, and there is every indication that they still err on the generous side (Wan Zahari et al., 1990). The important features of these estimated requirements are as follows:

Table 4.1. Dietary calcium requirements of sheep at the given DMI (AFRC, 1991).

	Live weight (kg)	Production level/stage	Diet quality[a]	DMI (kg day^{-1})	Ca (g kg^{-1} DM)
Growing lambs	20	100 g day^{-1}	L	0.67	3.7
			H	0.40	5.7
		200 g day^{-1}	L	–	U
			H	0.57	7.0
	40	100 g day^{-1}	L	1.11	2.4
			H	0.66	3.4
		200 g day^{-1}	L	1.77	2.6
			H	0.93	4.0
Pregnant ewe carrying twins[b]	75	9 weeks	L	1.10	1.4[c]
			H	0.71	1.6[c]
		13 weeks	L	1.28	2.0
			H	0.85	2.6
		17 weeks	L	1.68	2.9
			H	1.13	3.9
		Term	L	2.37	3.2
			H	1.62	4.3
Lactating ewe nursing twins	75	2–3 kg milk day^{-1}	L	2.8–3.7	2.8 (m)[d]
			L	2.3–3.2	3.1 (bm)
			H	1.8–2.4	3.8 (m)
			H	1.5–2.1	4.3 (bm)

[a]L, a poorly digestible diet with a 'q' value of 0.5; H, a highly digestible diet with a 'q' value of 0.7.; U, unattainable performance.
[b]The requirements for small ewes carrying single lambs are similar, assuming they will eat proportionately less DM.
[c]Sufficient for a dry ewe.
[d]Requirements are influenced by the ability of the diet to meet energy needs and prevent loss of body weight (m); diets that allow loss of body weight at 0.1 kg day^{-1} (bm) are associated with higher requirements.

Table 4.2. Dietary calcium requirements of cattle at the given DMI (from AFRC, 1991).

	Live weight (kg)	Production level/stage	Diet quality	DMI (kg day^{-1})	Ca (g kg^{-1} DM)
Growing cattle	100	0.5 kg day^{-1}	L	2.8	5.2
			H	1.7	8.0
		1.0 kg day^{-1}	L	4.5	6.3
			H	2.4	10.8
	300	0.5 kg day^{-1}	L	5.7	3.0
			H	3.4	4.4
		1.0 kg day^{-1}	L	8.3	3.5
			H	4.7	5.5
	500	0.5 kg day^{-1}	L	10.9	2.6
			H	6.1	3.6
		1.0 kg day^{-1}	L	11.6	2.8
			H	6.5	4.3
Pregnant cow calf weight 40 kg at birth	600	23 weeks	L	6.3	2.1
			H	4.0	2.7
		31 weeks	L	7.2	2.3
			H	4.7	3.0
		39 weeks	L	9.1	2.7
			H	6.1	3.5
		Term	L	11.2	2.8
			H	7.5	3.6
Lactating cow	600	10 kg day^{-1}	L	12.0	2.9 (m)
			L	9.9	3.3 (bm)
		20 kg day^{-1}	H	11.4	4.6 (m)
			H	10.1	5.1 (bm)
		30 kg day^{-1}	H	15.3	4.8 (m)
			H	13.8	5.2 (bm)

L, low and H, high ratio of metabolizable to gross energy; m, fed to maintain body weight; bm, fed below maintenance.

- requirements for lambs and calves decrease with age, but increase with growth rate;
- requirements for ewes rise rapidly in late pregnancy to a level equal to that for lactation;
- requirements for cows increase slowly during pregnancy and abruptly at calving;
- when cows 'milk off their backs' (i.e. lose body weight to sustain production), requirements in concentration terms increase (by 8.0% in the example given); and
- no figures need be met on a day-to-day basis.

Sheep performance is unlikely to suffer if the diet provides an average of 3 g Ca kg^{-1} DM throughout the year. Because there is less time for cows to replenish lost calcium reserves between lactations, average annual dietary calcium concentrations should be about 25% higher than for ewes and 15% higher in late pregnancy for dairy heifers preparing to deliver a first calf than for subsequent calvings. Higher requirements were derived for dairy cows by the NRC (2001) on the questionable assumption that calcium absorption is much lower. If adhered to, their requirements might paradoxically increase the risk of milk fever.

Horses

The mineral requirements of horses have been reviewed by NRC (2007) and there are new factorial estimates of the calcium requirements of grazing foals (Grace et al., 1999a) and lactating mares (Grace et al., 1999b) from New Zealand, based on new data for the mineral compositions of the carcass (Table 1.2) and milk (Table 2.3). Assuming faecal endogenous losses (FE_{Ca}) the same as

Table 4.3. New factorial estimates of calcium requirements for growing pigs on corn/soybean meal diets. These are lower than previous published estimates (see Fig. 4.6 and Underwood and Suttle, 1999) and will lower the need for added inorganic phosphorus. Diets based on other cereals, including fibrous by-products, can provide 10% less calcium.

Live weight (kg)	DMI (kg day^{-1})	LWG (kg day^{-1})	Net need[a] (g day^{-1})	Gross need[b] (g kg^{-1} DM)
20	1.0	0.4	5.0	6.3
40	1.6	0.6	6.8	5.3
80	2.5	0.8	9.9	4.2
100	3.0	1.0	10.5	4.2

[a]Using ARC (1981) requirements for growth: from birth to 40 kg LW, 12.5–0.1 g Ca kg^{-1} LWG; for 80–100 kg LW, 7.5 g Ca day^{-1}; and 1 g Ca kg^{-1} DMI for maintenance (see Fig. 4.3).
[b]Absorption coefficients: 0.80.

those estimated for sheep (16 mg kg^{-1} LW) (ARC, 1980), a milk yield of 20 kg day^{-1}, and A_{Ca} coefficients of 0.7 or 0.5 from grass in the weaned Thoroughbred foal or lactating mare, the calcium needs for the weanling and mare were thus placed at 4.6 and 4.9 g kg^{-1} DM, respectively, but these may be overestimates. Young Thoroughbreds on grass containing 3.5 Ca kg^{-1} DM have shown no benefit from calcium supplementation (Grace et al., 2003) and a similar pasture was considered adequate for mares (Grace et al., 2002b). There is no evidence that mares absorb calcium less efficiently than weaned foals.

Pigs

Empirical feeding trials have been extensively used to define the calcium requirements of growing pigs (ARC, 1981), but may have overestimated need (Underwood and Suttle, 1999). Factorially derived calcium requirements depend greatly on the choice of calcium absorption coefficient: ARC (1981) assumed that A_{Ca} declined from 0.67 to 0.47 as pigs grew from 25 to 90 kg, but the calcium requirement declines markedly over that weight range and the decline in A_{Ca} probably reflected an increasingly generous calcium supply. Heavier pigs can absorb most of the required calcium from their diet (e.g. Liu et al., 1998). Pigs probably maintain a high absorptive efficiency for predominantly inorganic calcium if fed to requirements for calcium and phosphorus.

New requirements have been derived (Table 4.3) using an A_{Ca} coefficient of 0.8 and the maintenance (M) value derived from Fig. 4.3. The comparison in Fig. 4.6 shows that the new estimates are lower than current national standards, but they are supported by the results of feeding trials. Many studies have shown that the availability of phytate phosphorus is improved by reducing the amount of calcium added to pig rations due to the reduced formation of calcium–phytate complexes, and it is becoming increasingly

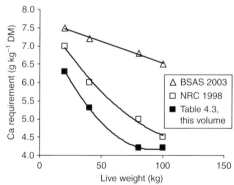

Fig. 4.6. The current recommended provision of calcium to growing pigs (Suttle, 2009; data from Table 4.3) is lower than that advocated in Europe by the British Society of Animal Science (Whittemore et al., 2003) and in the USA by the NRC (1998). In several studies (e.g. Reinhard and Mahan, 1986; Adeola et al., 2006), the performance of young pigs improved if dietary Ca was lowered to 5–6 g kg^{-1} DM.

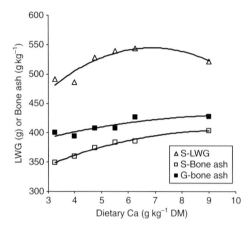

Fig. 4.7. Calcium requirements for live-weight gain (LWG) and bone mineralization (tibia ash) in broiler chicks at starter (S, 0–16 day) and grower (G, 19–42 day) stages (from Driver et al., 2005a). The penultimate dietary concentration (6.25 g Ca kg^{-1} DM) was optimal for growth in starters and for bone mineralization in growers; supraoptimal for growth in growers (average LWG was unaffected by Ca and is not shown); and suboptimal for bone mineralization in starters. (A basal corn/soybean meal diet provided 4.5 g 'available' P kg^{-1} DM and food intake was 40 g and 113 g day^{-1} in starters and growers, respectively).

important to define and feed to minimum calcium requirements to make the most efficient use of phytate phosphorus (see Chapter 6). Requirements should be lower for pigs on diets containing plant phytases from sources such as wheat, rye and bran than for those on corn/SBM diets.

Poultry: meat production

Recent research suggests that requirements of growing birds for both calcium and phosphorus are much lower than the latest NRC (1994) standards (Driver et al., 2005a; Fritts and Waldroup, 2006). A change of emphasis to how *little* inorganic phosphorus might be needed for growth has drawn attention to the advantage of lowering dietary calcium, whereupon the availability of calcium simultaneously improves, as in the pig. Newly hatched broiler chicks can grow unconstrained for 16–21 days with dietary calcium at 5–7 g kg^{-1} DM with 'available phosphorus' at (Henry and Pesti, 2002) or below (Driver et al., 2005b) the NRC standard: for growers, 3.5 g Ca kg^{-1} DM can suffice (Fig. 4.7), reflecting the more efficient absorption by older chicks (Fig. 4.5), but requirements for maximal bone ash were slightly higher. The widespread use of the 0–16 or 18-day chick growth model may have exaggerated the overall need for calcium. New standards based on the above experiments are given in Table 4.4. Feeding excess calcium leads to the formation of insoluble soaps in fat-supplemented diets. Inclusion of plant phytases will lower calcium requirements.

Poultry: egg production

Estimates of the calcium requirements for egg production and quality have ranged widely

Table 4.4. Dietary requirements (g kg^{-1} DM) of calcium for optimal growth and bone mineralization in broilers, indicated by recent research, are far lower than the latest NRC (1994) standards and have been extrapolated to give needs for other birds.

	Broiler[a] stage (weeks)		
	Early (0–3)	Middle (3–6)	Late (6–8)
Broilers – new proposal	7.0	5.0	4.0
Broilers – NRC (1994)	10.0	9.0	8.0
Leghorn chick[b]	6.0	4.0	20.0[c]
Turkey poult[b]	8.0	6.0	3.0

[a]Corresponding growth stages are 0–6, 6–18 and >18 weeks for Leghorns and 0–4, 8–12 and 20–24 weeks for poults.
[b]Provisional recommendations, based on the degree to which requirements in broilers fell with concurrent reductions in dietary calcium and phosphorus.
[c]This value is high in preparation for lay.

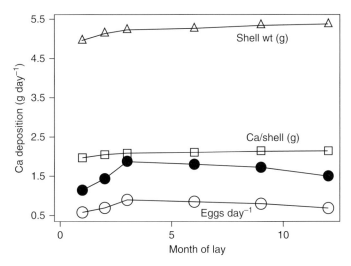

Fig. 4.8. Daily calcium deposition in eggs by laying hens (solid circle) peaks after about 3 months of lay due to the combined effects of increases in rate of egg production, the amount of shell per egg and the calcium concentration in the shell (data from Roland, 1986).

from 2.9 to 6.2 g day^{-1} and from 28 to 48 g kg^{-1} DM (Roland, 1986). The variations stem from differences in: phosphorus, vitamin D and plant phytases in basal diets; stage of lay; intake and energy density of food; type and strain of bird; egg size; temperature; contribution of calcium from the skeleton; source of supplemental calcium; and safety margins.

The basis for dietary effects was discussed earlier and the effect of stage of lay is illustrated in Fig. 4.8. Roland (1986) proposed a gradual increase in dietary calcium provision from 3.75 to 4.25 g kg^{-1} DM throughout lay and subsequent studies support this: one indicated a need for 3.6 g kg^{-1} DM for the first 4 months of lay in birds consuming 113 g food day^{-1} (Fig. 4.9); in another, reduction of calcium provision from 4.0 to 3.7 g kg^{-1} DM for the entire lay in hens consuming a wheat-based diet at 100 g day^{-1} had no adverse effects on performance, but there was a slight reduction in egg weight at the end of lay (Scott et al., 2000). NRC (1994) allowed for the effects of food intake (i.e. decreasing energy density) over the range of 80 to 120 g day^{-1} for white-egg-laying hens about a mean calcium provision of 3.25 g kg^{-1} DM; their recommendations were some 10% lower for brown hens, which consume more food. High environmental temperatures (>20°C) tend to reduce feed intake and increase the required dietary calcium concentration. Calcium requirements increase from 3.8 to 4.5 g kg^{-1} DM as egg size increases from 50 to 60 g (Simons, 1986). For pullets entering lay, the optimum transitional feeding regimen for calcium is to increase levels from 10 to 30 g kg^{-1} DM one week before the first egg is anticipated (Roland, 1986). Requirements for peak egg

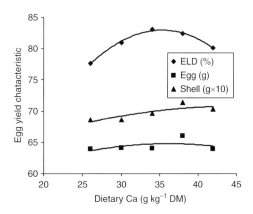

Fig. 4.9. Responses of Leghorn hens to increasing dietary calcium provision: egg laying days (ELD (%)) and both egg and shell weight were optimal with about 38 g Ca kg^{-1} DM, but bone strength (not shown) increased linearly as dietary Ca increased (from Narvaez-Solarte et al., 2006). (A basal corn/soybean meal diet provided 3.5 g available P kg^{-1} DM.)

production must sometimes be met on a daily and even an hourly basis if shell strength is to be maintained. Indeed, the laying hen will voluntarily consume more of a calcium supplement on laying than on non-laying days, and more of a low- than a high-calcium ration on any other day (Gilbert, 1983). At the high dietary calcium level needed for laying hens, interactions with saturated fat have not been not found (Atteh and Leeson, 1985).

Biochemical Consequences of Calcium Deprivation

Calcium-responsive disorders can arise in one of two ways: as a result of an acute increase in demand ('metabolic deprivation') or of chronic failure to meet dietary requirements. The biochemical changes that accompany such disorders are illustrated in Fig. 4.10. The model for chronic deprivation is based on data for pigs (Eklou-Kalonji et al., 1999).

Blood changes in chronic deprivation

The stimulus to PTH secretion in the marginally hypocalcaemic individual was described earlier and has been confirmed in growing pigs that are chronically deprived of calcium (Fig. 4.11a). The fall in plasma calcium is barely perceptible, while the increase in PTH gains momentum over a period of weeks and is accompanied by a steady rise in calcitriol $((OH)_2D_3)$. Bone metabolism is altered by the increased activities of PTH and calcitriol and the effects can be tracked by markers of bone growth (osteocalcin) and resorption (hydroxyproline), both of which show slight but significant increases in chronically deprived pigs (Fig. 4.11b). There is a tendency for plasma inorganic phosphorus to rise in chronically deprived lambs (see Chapter 6) and for plasma calcium to rise in chronic phosphorus deprivation (Fig. 4.12) in sheep, reflecting the simultaneous release of both elements from resorbed bone to meet a deficit of only one element. In lactating ewes confined to a cereal grain ration, serum calcium declines to <50% of normal values (2.2–2.9 mmol l^{-1}) within a few weeks, but the rates of fall in lambs, weaners and pregnant ewes are progressively slower due to lower demands for calcium (Franklin et al., 1948).

Blood changes in acute ('metabolic') deprivation

When a cow calves she can lose 23 g of calcium in 10 kg of colostrum within 24 h, yet there is only 3 g of calcium in her entire bloodstream and she is unlikely to have eaten for several hours. In response to these sudden changes, several biochemical and physiological responses can be detected and three are illustrated in Fig. 4.13. Plasma hydoxyproline (Goff et al., 2004), type 1 collagen metabolites and urinary deoxypyridinoline (Liesegang et al., 1998) and osteocalcin (Liesegang et al., 2000) concentrations also increase at parturition. Multiparous cows have lower plasma osteocalcin and PTH concentrations than primiparous cows after calving (Kamiya et al., 2005). Even heifers experience 'metabolic deprivation' and show a limited reduction in plasma calcium and a temporary increase in PTH after their first calving. The increase in calcitriol after parturition is relatively small and slow (Fig. 4.13) and all changes can last for 14 days (Liesegang et al., 2000). Serum inorganic phosphorus can increase at parturition in the healthy dairy cow (e.g. Peterson et al., 2005). Post-parturient hypocalcaemia tends to be episodic and may recur at roughly 9-day intervals as healthy cows readjust to the increasing demands of lactation until it peaks (Hove, 1986). The secretion of PTHrP enhances bone resorption and may protect the fetus and neonate from hypocalcaemia (Goff et al., 1991).

In milk fever, these homeostatic mechanisms fail, principally because of target-organ insensitivity to PTH and calcitriol (Goff et al., 1989b) and plasma calcium declines to an intolerable degree (<1.25 mmol l^{-1}). Serum inorganic phosphorus *declines* in milk fever to about one-third of normal (0.5 mmol l^{-1}), and serum magnesium may also be subnormal (<0.8–1.2 mmol l^{-1}; Fenwick, 1988). In ewes, the corresponding changes are less acute and begin before lambing – particularly in twin-bearing ewes – because

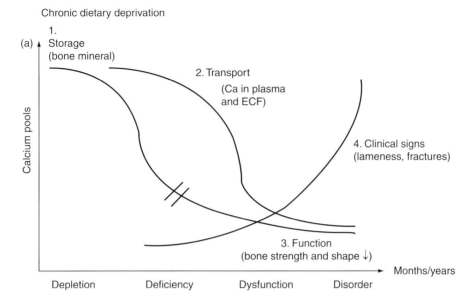
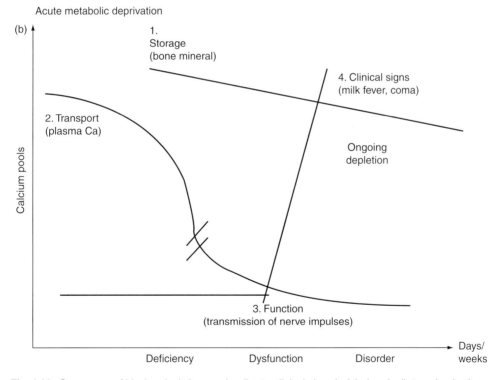

Fig. 4.10. Sequences of biochemical changes leading to clinical signs in (a) chronic dietary deprivation (i.e. skeletal disorders) and (b) acute metabolic deprivation of calcium (e.g. milk fever). The 'acute' model also applies to hypomagnesaemic tetany. ECF, extracellular fluid.

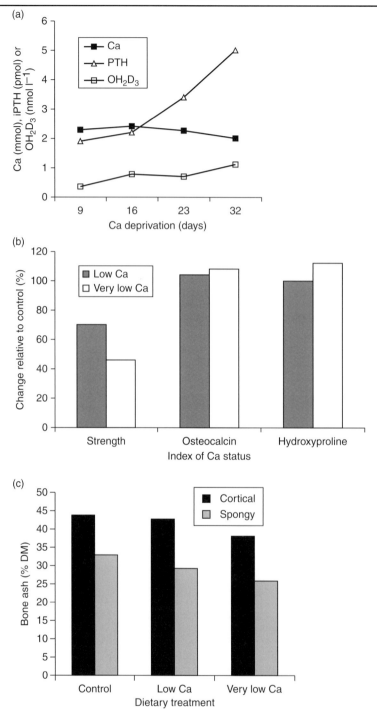

Fig. 4.11. Sequential (a) and terminal (b and c) changes in biochemical or physical measures of calcium status from a study in which weanling pigs were given diets containing 9.0 (control), 3.0 (low Ca) or 1.1 (very low Ca) g Ca kg^{-1} DM for 32 days (Eklou-Kalonji et al., 1999). In (a), note that the invocation of hormonal control (increases in plasma parathyroid hormone (PTH) and calcitriol OH$_2$D$_3$) delays the fall in plasma Ca. In (b), note that plasma markers indicate increased bone growth (osteocalcin) and resorptive (hydroxyproline) activities that probably cause a reduction in the failure load of a cortical bone (strength of metatarsal). In (c), note the greater effects of Ca deprivation on the ash content of spongy than of cortical bone.

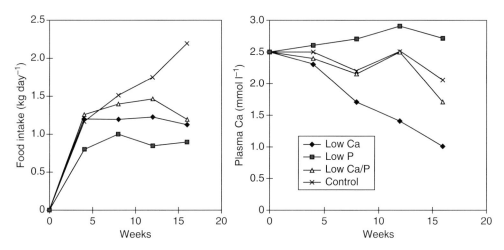

Fig. 4.12. Effects of single or combined deficiencies of calcium and phosphorus on food intake and plasma calcium in lambs. Note that Ca deprivation has lesser effects on both parameters when P is also in short supply (from Field et al., 1975).

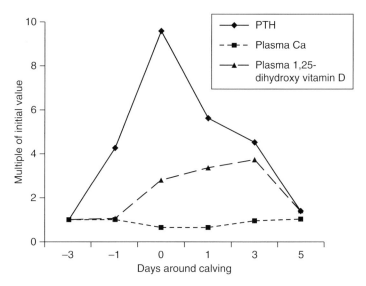

Fig. 4.13. Precipitous increases in the demand for calcium at calving prompt an early rise in plasma parathyroid hormone (PTH) concentrations and a slower rise in dihydroxylated vitamin D, in an attempt to curtail life-threatening reductions in plasma Ca (values are given relative to those 3 days before calving, which were approximately 60 pg ml^{-1} for PTH; 30 pg ml^{-1} for 1,25-(OH)$_2$D and 2.4 mmol l^{-1} for Ca (data for non-relapsing cows; Goff et al., 1989b).

the demand for calcium for the fetus in late pregnancy becomes greater than that for lactation (Field and Suttle, 1967; Sansom et al., 1982). Low plasma osteocalcin concentrations prior to parturition indicate suppression of osteoblast activity and bone growth in the ewe to 'spare' calcium (Farrugia et al., 1989).

Bone changes in chronic deprivation of the young

Chronic calcium deprivation reduces the degree of bone mineralization, with consequent biochemical, physical, histological and radiological changes. Some changes seen in baby pigs and

Table 4.5. Effects of dietary calcium concentrations on selected performance, skeletal and blood indices of calcium status in the baby pig given a synthetic milk diet from 0 to 6 weeks of age (Trial 2) (Miller et al., 1962).

Criterion	Dietary Ca (g kg^{-1} DM)			
	4	6	8	10
Performance				
Live weight gain (g day^{-1})	170	210	180	200
Dry matter intake (g)	230	240	230	250
Feed gain (g g^{-1})	1.33	1.16	1.31	1.27
Serum				
Inorganic P (mmol l^{-1})	3.29	3.54	3.43	3.52
Ca (mmol l^{-1})	2.60	2.60	2.72	3.00
Alkaline phosphatase (U l^{-1})	9.90	8.58	8.50	9.75
Bone				
Humerus ash (%)[a]	42.2	43.3	46.1	47.1
Femur specific gravity	1.14	1.15	1.16	1.18
Femur weight (g)[b]	47.9	–	47.2	–
Eighth rib weight (g)[b]	3.62	–	3.44	–
Femur breaking load (kg)	57.7	65.9	65.9	84.5

[a]Dry fat-free basis.
[b]Trial 1 data.

Table 4.6. Effects of feeding diets low in calcium and/or phosphorus for 18 weeks on the growth and mineral content of the lamb skeleton (Field et al., 1975).

Diet	Changes in fat-free skeleton				Total mineral ratios	
	Weight (kg)	Ca (g)	P (g)	Mg (g)	Ca:P	Ca:Mg
Control	+1.44	+197	+67	4.7	2.14	40.5
Low Ca	+1.00	+15	−18	2.4	2.13	25.7
Low P	+0.43	+57	−10	0.4	2.44	49.0
Low Ca/P	+0.92	+35	−14	0.8	2.30	39.4

lambs are given in Tables 4.5 and 4.6, respectively, and are similar to those seen when phosphorus is lacking (compare Tables 4.5 and 6.4) since mineralization requires both minerals. The average calcium to phosphorus ratio of 2.1:1 deviates only marginally during calcium or phosphorus deprivation (Table 4.6) but *changes in the calcium to magnesium ratio may distinguish calcium from phosphorus insufficiencies*. A marked reduction in the calcium to magnesium ratio from 40 to 26 has been noted in lambs depleted of calcium, whereas phosphorus depletion tends to raise the ratio in the entire skeleton to 49 (Table 4.6).

The 'elegant compromise' is illustrated by the substantial increase in skeletal size and weight that can occur, with little increase in bone calcium, in lambs on a low-calcium diet (Table 4.6). Bone growth and remodelling proceed, allowing the initial mineral reserve to be redistributed in a larger bone volume and enabling growth to take place without loss of function. The responses seen in rapidly growing species can be sharper: studies have seen a marked loss of bone strength in calcium-deprived pigs (Fig. 4.11b). In growing chicks, 25% of bone minerals can be lost, chiefly from the axial skeleton (Common, 1938), but the effects may be less spectacular in the much-studied tibia. Reducing dietary calcium from 10.4 to 4.4 g kg^{-1} DM reduced tibia ash in 16-day-old broiler chicks from 39.6% to 35.1%

on diets high in phosphorus (Driver et al., 2005b). In lambs, calcium deprivation did not reduce bone growth to the extent that phosphorus deprivation does (Table 4.6) because the formation of bone matrix is less impaired. Poor mineralization of bone matrix, leading to a widened and weakened growth plate, is more prominent when dietary calcium is lacking than when phosphorus is lacking in the growing animal. Normally the biochemical, histological, radiological and physical measures of bone quality agree well in both young (Eklou-Kalonji et al., 1999) and adult animals (Keene et al., 2005). More details are presented in Chapter 6, since they are most widely used in diagnosing phosphorus deprivation.

Bone changes in chronic deprivation of the adult

Since the epiphyseal plate has 'closed', changes are confined to the shafts and ends of bones where calcium accretion slows and resorption increases. Net withdrawal of minerals varies from bone to bone. Those low in ash ('spongy'), such as rib, vertebra, sternum and the cancellous end of long bones, are the first to be affected, while those high in ash ('compact'), such as the shafts of long bones and small bones of the extremities, are the last. A similar ranking of bones applies in the calcium-deprived young animal (Fig. 4.11c). Caudal vertebrae and metacarpal show no reduction in mineralization during the anticipated calcium deprivation of the dairy cow during lactation (Keene et al., 2005. In the sexually mature bird, non-physical measures can overestimate bone strength because medullary bone lacks the strength of structural bone (Whitehead, 1995). In demineralized adult bone, there is generally little alteration in the proportions of minerals present. Periods of bone resorption can be tolerated if the bones are fully calcified at the outset; depletion is restricted to some 10–20% of the skeletal reserve and repletion takes place before the next period of intensive demand.

Parasitism and bone mineralization

Studies with lambs infected experimentally with the larvae of nematodes that parasitize the gastrointestinal tract have revealed species- and dose-dependent demineralization of the skeleton of a severity that matches anything achieved with calcium- (or phosphorus-) deficient diets (Fig. 4.14) (Coop and Sykes, 2002).

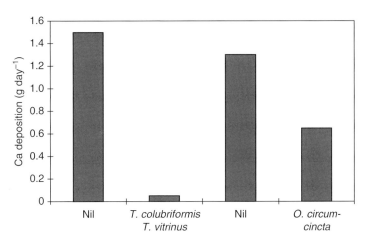

Fig. 4.14. Deposition of calcium in the skeleton of lambs is greatly reduced by parasitic infections of the abomasum (by *Ostertagia circumcincta*) or small intestine (by *Trichostrongylus vitrinus* and *Trichostrongylus colubriformis*). Trickle infections of the nematode larvae were used and controls were pair-fed (from Sykes, 1987). Nil, non-infected

Table 4.7. Effect of a low-calcium diet, supplemented (+) and unsupplemented (−) with calcium carbonate (CaCO$_3$), on blood calcium and egg production in hens (from Buckner et al., 1930).

Weeks on Ca-deficient diet	Serum Ca (mg l^{-1})		No. of eggs per hen		Weight of shell (g)		Weight of contents (g)	
	+Ca	−Ca	+Ca	−Ca	+Ca	−Ca	+Ca	−Ca
−2–0	213	209	6.3	6.9	5.0	4.8	52.3	53.2
0–2	248	152	6.6	4.7	5.4	3.6	53.9	53.5
2–4	215	161	9.0	1.5	5.7	3.8	56.7	47.4

Since natural nematode infections are commonplace in young grazing livestock, they may affect susceptibility to dietary calcium (and phosphorus) deprivation in the field. The demineralization associated with abomasal parasitism (by *Ostertagia circumcincta*) occurs in the presence of normal plasma calcium and phosphorus concentrations and normal absorption. Loss of appetite contributes to demineralization, but comparisons with pair-fed, uninfected controls have shown an additional loss of bone minerals. Parasitism of the small intestine has adverse effects on phosphorus metabolism (see Chapter 6) and the resultant hypophosphataemia may be partly responsible for the loss of calcium from the skeleton. The systemic effects of gut parasites on bone demineralization may be triggered by similar factors to those that mobilize bone matrix during early lactation to furnish amino acids for the mammary gland and milk protein synthesis. Parasites of the liver have far lesser effects on bone mineralization. It would be surprising if debilitating infections of the gut by other pathogens (e.g. *Coccidia*) or in other hosts (calves, kids, young deer, piglets and chicks) did not sometimes cause demineralization of the skeleton.

Changes in blood and bone in laying hens

The entire skeleton of the laying hen only contains enough calcium to supply 10 eggs and her daily intake of calcium from unsupplemented rations would be less than 0.03 g Ca day^{-1}. During a normal laying year, up to 30 times the hen's total-body calcium is deposited in shell (Gilbert, 1983). Skeletal resorption of calcium cannot make a significant contribution over a 12-month period of lay and is stretched to its limits every day that an egg is prepared for lay. To promote calcium deposition, laying hens normally have higher and more variable serum calcium values (5.0–7.5 mmol l^{-1}) than non-laying hens or chickens (2.2–3.0 mmol l^{-1}) due to the presence of a calcium phospholipoprotein, vitellogenin, a precursor of egg-yolk proteins (Hurwitz, 1996). During the period of shell calcification serum calcium values decline, particularly if a low-calcium diet is given, but mobilization of bone mineral may prevent serum calcium from falling further (Table 4.7).

Clinical and Pathological Features of Acute Calcium Deprivation

Milk fever

Milk fever is principally seen in high-yielding dairy cows. It is associated with parturition and the initiation of lactation in older cows (5–10 years old). Within 48 h of calving, the cow becomes listless, shows muscular weakness, circulatory failure, muscular twitching, anorexia and rumen stasis (i.e. a parturient paresis). The condition develops through a second stage of drowsiness, staring, dry eyes with dilated pupils and sternal recumbency to a final stage of lateral recumbency and coma. In affected cows, serum calcium is <1.0–1.25 mmol l^{-1} and the severe hypocalcaemia can be associated with loss of appetite (Huber et al., 1981), decreased blood flow to peripheral tissues (Barzanji and

Daniel, 1987), hypoxia (Barzanji and Daniel, 1988), hypothermia (Fenwick, 1994), hyponatraemia (Fenwick, 1988), hypomotility of the rumen, displacement of the abomasum (Oetzel, 1996) and lesions in the myocardium (Yamagishi et al., 1999). These secondary consequences may underlie the failure of some cows to respond to calcium alone. Serum magnesium is also often low and convulsions or tetany, as described in Chapter 5, may then accompany the usual signs of milk fever. Lactating females of all species, including the nanny, sow and mare, occasionally develop milk fever.

Hypocalcaemia in ewes

The biochemical term 'hypocalcaemia' is used to describe a clinical condition that develops with the approach of lambing. It is characterized by restlessness, apparent blindness, rumen stasis and, in the worst cases (55%) (Mosdol and Wange, 1981), recumbency, tetany and death. Symptomatically and to some extent aetiologically, hypocalcaemia resembles toxaemia of pregnancy (a consequence of energy deficiency), being most likely to occur in older, twin-bearing ewes exposed to a change in or shortage of feed and/or stressed by transportation (Hughes and Kershaw, 1958) or adverse weather. In a large UK flock, 43% of lambing ewes presenting with ataxia or recumbency were found not to be hypocalcaemic (Cockcroft and Whitely, 1999). Hypocalcaemia and pregnancy toxaemia can be distinguished biochemically by the presence of low plasma calcium (the former) or raised β-hydroxybutyrate concentrations (the latter), but each is likely to develop as a consequence of the other because loss of appetite is a feature of both conditions and calcium and energy supplies are simultaneously decreased. To complicate matters further, hypomagnesaemia often accompanies hypocalcaemia and may explain the wide range of clinical signs associated with hypocalcaemia, which include convulsions.

The disease rarely affects ewes in their first pregnancy (Mosdol and Wange, 1981) and is physiologically similar to milk fever in cattle in that it is caused by failure of supply to meet increasing demand for calcium. However, onset is pre- rather than post-parturient. There is no clear association with low dietary calcium, indicating the importance of mobilizing calcium from the skeleton. Suggestions that excess dietary phosphorus predisposes to hypocalcaemia (Jonson et al., 1971, 1973) might be explained by retarded bone resorption. There is disagreement on the role of excess magnesium (Pickard et al., 1988), but excess potassium may predispose to the disease (see Chapter 8). Vaccination with cortisol-inducing vaccines in late pregnancy may be a risk factor and vaccination should be initiated as early as possible before lambing (Suttle and Wadsworth, 1991). Hypocalcaemia may also occur at the end of a drought when a sudden luxuriant growth of grass becomes available following a period of feeding grain low in calcium (Larsen et al., 1986).

Transport tetany

Clinical disease in the form of paresis associated with hypocalcaemia is a risk factor associated with the transfer of weaned ruminants to feedlots. Lucas et al. (1982) reported such an incident involving lambs, and the transportation of horses can give rise to a similar condition. The problem is probably caused by a combination of factors including lack of food and therefore dietary calcium, lack of mobilization from a poorly mineralized skeleton and increased soft-tissue uptake of calcium.

Occurrence of Acute Disorders

Milk fever in grazing herds

Milk fever affects 3–10% of dairy herds in developed countries and is a common reason for veterinary call-outs (Heringstad et al., 2005). A similar proportion of affected animals fails to respond to treatment (Littledike et al., 1981). Incidence within herds can be as high as 25% and Allen and Sansom (1985) estimated that the disease costs the UK dairy industry £10 million per annum. Breeds vary in susceptibility to milk fever: in collated data from 37 experiments, mostly from the USA, the Jersey was 2.25 times more likely to develop the disease than the Holstein–Friesian (Lean et al., 2006). A meta-analysis identified risk factors for the Ruakura herd at year-round pasture in the North Island of New Zealand between 1970 and 2000 (Roche and Berry, 2006). The average incidence of milk fever was 13.1% up to 1979, but it declined to 4.5% in the next decade and to 3.2% in the last. The change was attributed to cross-breeding with the Friesian, begun in the 1970s, and possibly to the introduction of routine magnesium supplementation (12 g cow^{-1} day^{-1}) around 1978. Disease incidence was repeatable and a cow that had developed milk fever in the previous lactation was 2.2 times more likely to succumb in a current lactation than one that had not. Residual effects of weather conditions were found, with incidence being increased after days of high evaporation, large diurnal variation in air temperature, high rainfall and low grass minimum temperature, similar conditions to those associated with an increased risk of hypomagnesaemic tetany (see Chapter 5). Older cows and those needing assistance at calving had an increased risk of developing milk fever, as did cows of extremely low or high body-condition score. The susceptibility of older cows may be due to an age-related decline in receptors for calcitriol in the intestinal mucosa (Horst et al., 1990).

An association with body fatness was also indicated in a Canadian study (Oikawa and Kato, 2002), in which the same anomalies in serum lipid profile were shown by cows with milk fever and those with fatty liver syndrome. In Norwegian Red dairy cattle, the heritability of milk fever has been estimated to increase from 0.09 to 0.13 from the first to third lactation (Heringstad et al., 2005). High 'alkalinity' or dietary cation–anion difference (DCAD, >300 milli-equivalents (me) kg^{-1} DM, see p. 175 et seq.), due largely to high pasture potassium, may increase the risk of milk fever (Pehrson et al., 1999; McNeill et al., 2002; Roche et al., 2003a,b) by reducing the amount of calcium that can be mobilized from the skeleton (Goff, 2007). There is a tendency for hypocalcaemia to be associated with high urinary pH (Fig. 4.15).

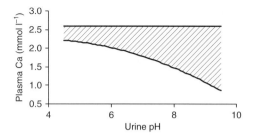

Fig. 4.15. Relationship between plasma calcium, measured within 24 h of calving, and urine pH, measured within 2 weeks before calving, in 172 dairy cows from 23 commercial herds in eastern Australia. Individual values were broadly scattered in the shaded area between the two lines, indicating a proneness to hypocalcaemia in some but not all alkalotic individuals (from McNeill et al., 2002).

Milk fever in housed herds

High potassium concentrations in conserved forage lead to raised DCAD in rations fed to housed herds, although natural values are generally less extreme than those found in pasture (i.e. <300 me kg^{-1} DM)). The effects of manipulating DCAD in the pre-partum diet on the incidence of milk fever have been widely studied under experimental conditions, but there is controversy over the relative importance of DCAD and the separate effects of ions. One study reported a greatly increased incidence of milk fever when dietary potassium was raised from 10 to 30 g kg^{-1} DM (see p. 178) (Goff and Horst, 1997), but the use of vulnerable cows (Jersey breed, >4 years old) and potassium bicarbonate (KHCO$_3$) as the potassium source may have exaggerated the risk presented by high pasture potassium. The addition of sodium bicarbonate (NaHCO$_3$) was just as potent. Lean et al. (2006), one of several groups to conduct meta-analyses of the DCAD literature, found that the prediction of milk-fever incidence in 37 studies was poorer with a model including the most popular DCAD calculation ((sodium + potassium) − (chlorine + sulfur)) (model 1) than when potassium, sulfur and other minerals were included as independent variables (model 2). A quadratic relationship with dietary calcium was evident, with the lowest risk of milk fever with diets relatively low (<5 g) or very high (>20 kg^{-1} DM) in calcium, confirming the earlier conclusion of Oertzel (1991). The influences of magnesium and phosphorus were also evident in model 2, which predicted a 62% *decrease* in the risk of milk fever as magnesium increased from 3 to 4g kg^{-1} DM and an 18% *increase* as phosphorus increased by the same amount. Using a restricted data set (22 studies) containing the same experiments, Charbonneau et al. (2006) found that a slightly modified equation (+0.65 sulfur) accounted for 44% of the variation in milk fever, with the incidence decreasing from 16.4% to 3.2% as DCAD fell from +300 to 0 me kg^{-1} DM. The high incidence of milk fever in housed dairy herds in the USA is associated with the widespread use of conserved alfalfa, rich in calcium and potassium, and the excessive use of phosphorus supplements (for review, see Goff and Horst, 2003).

Clinical and Pathological Features of Chronic Calcium Deprivation

Poor growth and survival

Feeding marginally inadequate calcium levels (5.1 versus 7.8g Ca kg^{-1} DM) during the rearing of chicks for egg laying has been seen to increase mortality from 1.9% to 10.6% and reduce body weight at 86 days from 1013 to 962g (Hamilton and Cipera, 1981). It is noteworthy that skeletal abnormalities were not seen. Most deaths were attributed to omphalitis, an infection of the umbilical stump. The first abnormality noted when lambs were weaned on to a diet very low in calcium (0.68g kg^{-1} DM) was loss of appetite after 4 weeks and a subsequent proportional retardation of growth (Field et al., 1975). However, severe demineralization can occur in young pigs deprived of calcium without effects on growth rate or feed conversion efficiency (Eklou-Kalonji et al., 1999)

Abnormalities of bones

Prolonged calcium deprivation can eventually manifest itself in a variety of ways depending on the species and stage of growth. Examples are as follows: lameness and abnormal 'peg-leg' gait; enlarged and painful joints; bending, twisting, deformation or fractures of the pelvis and long bones; arching of the back, with posterior paralysis in pigs (Miller et al., 1962); facial enlargements, involving particularly the submaxillary bones in horses; and malformations of the teeth and jaws, especially in young sheep. In prolonged calcium deprivation in growing chicks the long bones may fracture (Common, 1938). However, most of these abnormalities can also arise from causes other than calcium deprivation. Problems with movement and prehension can hamper the ability of grazing livestock to secure feed and water and indirectly result in death. The type of bone disorder that develops is partly dependent on the age at which calcium deprivation develops.

Rickets is a non-specific term used to denote the skeletal changes and deformities in young animals. It is characterized by a uniform widening of the epiphyseal–diaphyseal cartilage and an excessive amount of osteoid or uncalcified tissue, causing enlargement of the ends of the long bones and of the rib costochondral junction. Rickets is caused by calcium and phosphorus deprivation in young broilers (e.g. Driver et al., 2005b), and the two conditions can be distinguished histologically (see Chapter 6).

Osteomalacia describes the excessive mobilization of minerals leaving a surfeit of matrix in bones in which the growth plate has 'closed' (i.e. the calcium-deprived adult).

Osteoporosis indicates that bones contain less mineral than normal but proportionately less matrix, so that the degree of mineralization of matrix remains normal. Matrix osteoporosis is characteristic of protein deficiency rather than simple calcium or phosphorus deficiency (Sykes and Field, 1972; Sykes and Geenty, 1986) and can occur in young as well as mature animals. In adult birds, demineralization of the skeleton takes place normally in response to the demands of egg-laying, and most hens end lay in an osteoporotic state (Whitehead, 1995).

Tibial dyschondroplasia is inducible by feeding broilers a diet low in calcium (Driver et al., 2005b), but has other causes in commercial flocks. It is histologically distinguishable from rickets by the presence of a plug of avascular cartilage that forms beneath the growth plate (Edwards and Veltmann, 1983).

Teeth

Remarkably little attention has been given to the effects of calcium nutrition on tooth development. The chemical composition and histology of the outer regions of the tooth (i.e. cementum) bear a close similarity with those of bone. However, calcium in the tooth is not a metabolically active compartment. Duckworth et al. (1962) reported marked effects of calcium (and phosphorus) deprivation on mineralization of the mandible in sheep, but made no comment on tooth abnormalities. The rate of cementum deposition is probably reduced by calcium (or phosphorus) deprivation during tooth development, but not necessarily out of proportion with skeletal development. Disorders of the dentition of grazing animals, such as the premature shedding of incisor teeth in sheep (broken mouth), have often been linked to inadequate mineral nutrition, but careful experimentation has failed to confirm a causal link (Spence et al., 1985). Other factors, such as soil ingestion, can cause excessive tooth wear.

Depression of milk yield

When lactating animals have been fed on roughage and grains low in calcium for long periods after reaching peak lactation, the skeletal reserves can be depleted to the point where milk yield is impaired throughout lactation, particularly in the dairy cow (Becker et al., 1953). An almost complete failure of milk production has been observed in sows fed on a calcium-deficient diet during the previous pregnancy. This reduction in milk yield may result from the loss of appetite that can occur after severe calcium depletion.

Reduction in egg yield and quality

Laying hens are often unable to satisfy their high calcium requirements by demineralizing bone. Egg yield, egg weight, hatchability and eggshell thickness are consequently reduced, with shell strength being the most sensitive index of dysfunction (Hamilton and Cipera, 1981). In early experiments, egg production had virtually ceased by the 12th day after removal of supplementary calcium, and the ash content of eggshells from some birds was less than 25% of normal (Deobald et al., 1936). Gilbert (1983) found that egg production increased in direct proportion to dietary calcium over the range of 1.9–30.9 g kg^{-1}, rising from 10% to 93%.

Occurrence of Chronic Calcium Deprivation

Calcium responses are predominantly associated with the feeding of grain-based diets and are inevitable in pigs and poultry fed predominantly on low-calcium cereals and in aged, high-yielding dairy cows given concentrate diets, unless they are appropriately supplemented. Within a population, it is the highest-yielding individuals that are most at risk because they have the highest demands for calcium.

Growing birds

The incidence of calcium-responsive rickets in broiler chicks is increased by feeding excess phosphorus in low-calcium diets, but was hard to eliminate in one study (Driver et al., 2005b). Similarly, the leg weakness or tibial dyschondroplasia that commonly affect young broilers and older turkey poults can be induced by diets low in calcium (7.5 g kg^{-1} DM), yet also occur when supplies are apparently adequate. Control can be obtained by dietary supplementation with calcitriol (Table 4.8), but not with unhydroxylated D_3 (i.e. normal dietary sources). Modern hybrids may lack the ability to produce enough endogenous vitamin at the growth plate in what is the fastest growing bone in the broiler's body and calcium deprivation exposes this more basic disturbance (Whitehead, 1995). Inconsistent relationships between the incidences of bone disorders and dietary calcium may be partly related to concomitant responses

Table 4.8. Effects of dietary calcium and dihydroxycholecalciferol (1,25-(OH)$_2$D$_3$) on the incidence of tibial dyschondroplasia (TD) in broiler chicks (Whitehead, 1995).

	Ca (g kg^{-1})	1,25-(OH)$_2$D$_3$ (μg kg^{-1})			
		0	2.0	3.5	5.0
TD incidence (%)	7.5	50	15	5	0
	10	10	20[a]	5	5
	12.5	15	15	0	0

[a]The complex interactions between Ca and 1,25-(OH)$_2$D$_3$ are due mainly to: (i) useful enhancement of Ca uptake by the vitamin at the lowest dietary Ca level; (ii) a growth-inhibiting but TD-protective hypercalcaemia at high vitamin levels; and (iii) opposing influences of high Ca and 1,25-(OH)$_2$D$_3$ levels on P status.

in growth rate, with alleviation of growth restraint exacerbating the bone disorder.

Laying hens

Shell quality cannot be maintained for a single day without calcium supplementation and poor shell quality is a major source of loss to the poultry industry (Roland, 1986). The transitional calcium nutrition around the onset of lay is also crucial. If low-calcium diets are fed, the pullet over-consumes in an effort to meet her burgeoning calcium requirement, increasing the risk of fatty liver haemorrhagic syndrome (Roland, 1986). Inadequate calcium nutrition during lay will contribute to 'cage layer fatigue', but this osteoporotic condition also occurs when calcium nutrition is adequate and it is not a specific consequence of calcium dysfunction (Whitehead, 1995). Bone fractures commonly occur during the transportation and slaughter of broilers and 'spent' hens. As with cage layer fatigue and tibial dyschondroplasia, these cannot be attributed specifically to inadequate calcium nutrition (Wilson, 1991), but it may play a part.

Grain-feeding in other species

Without supplementation, grain-based diets will cause rickets in growing pigs and leg weakness in boars, caused by a similar growth plate lesion to that seen in tibial dyschondroplasia. Calcium deprivation may occur also in horses and sheep fed largely on grain diets designed as drought rations when little or no grazing is available (Larsen et al., 1986). Severe stunting of growth, gross dental abnormalities and some deaths have been observed in lambs and young weaned sheep fed a wheat-grain-based diet without additional calcium (Franklin et al., 1948). Young horses have been seen to develop 'osteofibrosis' and to lose weight when fed on cereals (Groenewald, 1937).

Grazing livestock

Calcium deprivation rarely affects grazing livestock (McDonald, 1968) and that situation has not changed. However, it can occur in grazing, high-producing dairy cows and lead to osteodystrophic diseases and low productivity in other livestock on acid, sandy or peaty soils in humid areas where the pastures are composed mainly of quick-growing grasses that contain <2g Ca kg^{-1} DM, as reported from parts of India, the Philippines and Guyana. The ability of an animal to absorb and use calcium depends on its vitamin D$_3$ status, and grazing livestock are dependent on UV irradiation of the skin to convert dietary precursors such as 7-dehydrocholesterol to D$_3$. Where animals are housed during the winter, or graze in areas of high latitude, few of the UV rays reach the skin and clinical abnormalities can develop at relatively high calcium intakes because metabolism is impaired. A marked seasonal decline in vitamin D status can occur in sheep kept outdoors during winter (Fig. 4.16) and is believed to have contributed to the development of rickets in

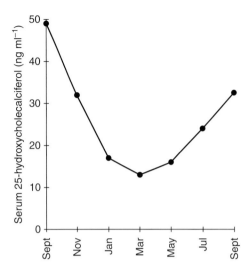

Fig. 4.16. Concentrations of vitamin D metabolite, 25-hydroxycholecalciferol, in the serum of pregnant ewes outwintered in Scotland, decline markedly due to lack of solar irradiation (data from Smith and Wright, 1981).

young sheep in central Scotland (Bonniwell et al., 1988).

Diagnosis of Calcium Disorders

Acute disorders

Diagnosis of milk fever is presumptive, being based heavily on clinical assessment and confirmed by a response to urgent treatment. Subsequent analysis of blood samples taken prior to treatment for serum calcium will confirm diagnosis, and appropriate diagnostic ranges are given in Table 4.9. Plasma calcium levels appear to remain equally low (approximately 1.0 mmol l^{-1}) at successive stages in the clinical progression of milk fever (Fenwick, 1988). If blood samples are unobtainable, analysis of calcium in vitreous humour, withdrawn from the eye of the dead animal, can be informative because there is a reasonable correlation with concentrations in plasma. The analysis of ionized calcium is useful as an experimental tool (Phillippo et al., 1994) but unhelpful in the diagnostic context. Acute calcium-responsive disorders such as hypocalcaemia in ewes and transport tetany are diagnosed on the same presumptive basis as milk fever.

Chronic disorders

Chronic bone disorders in all species of livestock can have various causes, but those specifically caused by calcium deprivation are confirmed by the presence of hypocalcaemia. Assessments of bone quality alone can never provide a specific diagnosis because calcium is only one of many factors that influence bone mineralization. Guidelines for the assessment of bone quality are given in Chapter 6. Diagnosis of health problems such as weak eggshells in the laying hen are complicated by

Table 4.9. Marginal bands[a] for mean total serum calcium in assessing low calcium status in various classes of livestock.

Livestock	Mean total serum Ca	
	mmol l^{-1}	mg l^{-1}
Peri-parturient dairy cow	1.3–2.0	50–80
Peri-parturient ewe	1.5–2.25	60–90
Young ruminants	1.8–2.0	70–80
Pigs	1.75–2.0	70–90
Poultry[b] laying	3.8–5.0[c]	180–200
Poultry non-laying	2.0–2.5	70–100
Horses	2.25–2.5	60–100

[a]Mean values within bands indicate possible dysfunction and values below bands probable dysfunction in some individuals. Individual values below bands indicate a need for wider sampling and the possibility of dysfunction.
[b]Turkeys have slightly higher ranges than those quoted, which are for fowl.
[c]Hypercalcaemia and risk of tissue damage, particularly to the kidneys, is indicated by values >3.2 mmol l^{-1} in most stock, but >6.0 mmol l^{-1} in laying hens.

diurnal fluctuations in serum calcium. The initial assumption in the case of eggshell problems should be that calcium nutrition is inadequate with respect to either the forms (lack of coarse particles) or daily quantities of calcium supplied. Diagnosis is again confirmed by positive responses to remedial measures.

Treatment and Prevention of Acute Disorders

Distinctive approaches are required to control acute disorders, and those used in the case of milk fever have been reviewed since the last edition of this book (Thilsing-Hansen et al., 2002; Goff, 2007; DeGaris and Lean, 2008). The treatment of milk fever requires the rapid reversal of hypocalcaemia and prevention of the post-parturient decline in plasma calcium to a level of 1.75 mmol (70 mg) l^{-1}. This is achieved by maximizing the supply of calcium absorbed from the gut and that mobilized from the skeleton.

Parenteral treatment of milk fever

A single intravenous calcium infusion, usually 600 ml of a 40% solution of calcium borogluconate, together with magnesium hypophosphite, will bring about 50% of affected animals to their feet within a few hours. A further 40% will recover after one or two further treatments. The need for repeat treatments increases with the delay in applying the first treatment and with the severity of the initial hypocalcaemia. Efficacy is not improved by extending the infusion period from the customary 15 min to 6 h (Braun et al., 2004). If a veterinary practitioner is not immediately available then the herdsman should intervene, warming the infusate before administration (Fenwick, 1994). About one in ten affected cows will remain recumbent, even after rolling the cow from side-to-side to lessen the risk of abomasal displacement ('downer' cows). Prevention is better than cure – the cow that recovers from milk fever is more prone to limb injuries and diseases such as mastitis and ketosis than unaffected cows, and milk yield can remain depressed by 1.1–1.7 kg day^{-1} for 4–6 weeks in Finnish Ayrshire cows (Rajala-Schulz et al., 1999).

Prevention of milk fever by parenteral methods

A massive dosage of vitamin D_3 prior to calving can be effective, but the prediction of calving date needed for the timing of treatment is critical. Parenteral administration of OHD_3 overrides the normal absorptive control mechanisms, enhancing calcium absorption from diets providing plentiful calcium (Braithwaite, 1983b). Doses are far lower and timing is less crucial when the 1α analogue of vitamin D_3 is used (Sachs et al., 1987a,b), but this may merely delay the onset of hypocalcaemia (Hove, 1986) and decrease the efficacy of subsequent calcium treatments (Horst, 1986). Intramuscular injections of PTH are effective but impractical (Goff et al., 1989a).

Prevention of milk fever in housed herds

Increases in dietary calcium supply prior to calving can be counterproductive (Ender et al., 1956; Goff and Horst, 1997; Lean et al., 2006) and the efficacy of very high calcium levels (>20 g kg^{-1} DM) may be partly due to a decrease in demand following appetite suppression and decreased milk production. One early radical approach was to feed diets low in calcium prior to calving, thus priming the homeostatic pathways, but the undoubted effectiveness of this method may have been at least partially attributable to associated changes in other dietary constituents, notably potassium (Horst et al., 1997). Low-calcium diets are not easily formulated, unless whole crop silage is available, and at some point the skeletal calcium deficit has to be made good. An alternative measure is to feed 'acidic' diets (DCAD < 100 me kg^{-1} DM) around calving (Block, 1994; Lean et al., 2006). This line of treatment was instigated by Ender et al. (1971), who noted that milk fever rarely occurred in cows given silage preserved with acids. The feeding of silage rather than hay as roughage would be expected to decrease the incidence of milk fever. Dietary acidity can easily be monitored via the pH of urine, which should be below 7.0. Sources of anions vary in their acidifying capacity, with HCl > NH$_4$Cl > CaCl$_2$ > CaSO$_4$ > MgSO$_4$ > elemental sulfur (Goff et al., 2004) when assessed via the effects on urine pH. The efficacy of acidic diets has been attributed exclusively to the effects of acid–base balance on the sensitivity of bone and kidneys to PTH and consequently on calcium resorption from bone (Phillippo et al., 1994; Goff and Horst, 1997; Goff, 2000). Such diets raise calcium ion concentrations in plasma prior to calving (Charbonneau et al., 2006). However, the role of DCAD per se in the aetiology of milk fever has probably been exaggerated because anion and cation supplements have many relevant effects that are not related to DCAD (see Chapter 7). In particular, anion supplements raise the availability of magnesium (Roche et al., 2003a) while potassium supplements reduce it, and supplementary magnesium is protective against milk fever. Chloride increases the flux of calcium across the ovine rumen epithelium, a response that may explain the efficacy of calcium chloride (CaCl$_2$) in preventing milk fever (Leonhard-Marek et al., 2007). Acidic diets have been found to reduce DMI by 11.3% on average (Charbonneau et al., 2006) and should be gradually discontinued after 21 days because they may depress milk yield (Roche et al., 2003b).

Prevention of milk fever in grazing herds

Large anion inputs are needed to offset the high potassium levels in the herbage available to spring calving herds (Pehrson et al., 1999), but they can be achieved by mixing anionic salts with supplementary food (Roche et al., 2003a,b). Alternatively, large CaCl$_2$ supplements (providing 50–75 g calcium) can be given by bolus or drench (Goff and Horst, 1993, 1994), but not without hazard. Administration in a large bolus carries a risk of damaging the throat. In addition, risks of acidosis, aspiration pneumonia and abomasal inflammation attend drenching. Hazards can be decreased by giving CaCl$_2$ in gel or emulsion form (in polyethylene glycol rather than oil) and/or as calcium propionate, but there may still be some loss of appetite.

Attention has switched to the use of fertilizers to raise pasture chloride (Perhson et al., 1999). Responses of pasture chloride to fertilizer chloride in pot trials with alfalfa followed the law of diminishing returns and the effects on DCAD were relatively small. Ammonium chloride (NH$_4$Cl) and CaCl$_2$ were equally effective, raising levels from 5.2 to 8.7 g Cl kg^{-1} DM at application rates of 112 kg Cl ha^{-1} (Goff et al., 2007). In field trials, the application of 150 kg Cl ha^{-1} to timothy pasture lowered DCAD in the harvested hays. When fed to cattle, low-DCAD hays (10 me kg^{-1} DM) have been found to significantly reduce urine pH and increase the amount of calcium mobilized from the skeleton by an infusion of EDTA; efficacy in preventing milk fever was implied (Heron et al., 2009).

- In practice, only pasture-based diets (as grass, hay or silage) intended for spring calvers need be treated to lower DCAD.
- Since the primary source of excess cations is potassium-rich forage, steps should be taken to minimize the potassium content of pastures (see Chapter 7).

Prevention of milk fever by selection

Heritabilities for milk fever within breeds can be as low as 2.6% (Holstein–Friesian) (Pryce et al., 1997) and it is therefore not possible to reduce herd incidence by sire selection within breeds. Crossing Jerseys with a less susceptible breed will reduce milk fever (Roche and Berry, 2006) as will culling cows that have had milk fever, but such a policy will tend to remove the highest yielders (Pryce et al., 1997; Rajala-Schulz et al., 1999).

Control of hypocalcaemia in ewe and goats

Hypocalcaemia in pregnant ewes and nannies is also treated with calcium borogluconate and response to treatment confirms the diagnosis. The administered dose is 100 ml of a 20% solution of calcium borogluconate with added magnesium. Preventive measures have not been studied in the peri-parturient animal, but acidic diets produce similar changes in bone metabolism in dry dairy goats to those seen in dairy cows and may prevent milk fever, provided they are introduced well before birth. The dairy sheep, however, may be less amenable to such preventive measures (Liesegang, 2008). The use of acid-preserved silage, 1α OHD_3 injections and oral supplementation with $CaCl_2$ and/or calcium propionate have been suggested, with the last having particular potential where there is a parallel risk of pregnancy toxaemia (Leonhard-Marek et al., 2007). The same treatments are indicated for transport tetany. Sustained protection requires adequate synthesis of $1,25\text{-}(OH)_2D_3$ from the unhydroxylated vitamin, and this may be in short supply in outdoor flocks in early spring (Smith and Wright, 1981) and housed dairy flocks. Both may require simultaneous supplementation with vitamin D_3.

Control and Prevention of Chronic Calcium Deprivation

Prevention in newly hatched chicks

Instead of feeding high dietary calcium levels to counteract the poor capacity of the hatchling to digest phytate, why not feed diets with less phytate and/or more phytase? If starter diets use less corn and more wheat, calcium deprivation should be less of a problem.

Prevention in non-ruminants

The skeletal defects caused by calcium deprivation are mostly irreversible and weak eggshells are irreparable. Preventive measures must therefore be adopted early in development. There is little to choose between the common mineral sources of calcium (Soares, 1995). The exceptions are dolomitic limestone and soft-rock phosphate, which have *relatively* low availabilities (0.65 and 0.70, respectively) compared with dibasic calcium phosphate ($CaHPO_4$) in pigs or poultry (Dilworth et al., 1964; Reid and Weber, 1976). All calcium sources are likely to have a lower availability when fed in high-phytate, low-phytase diets. Claims that the calcium from $CaCO_3$ is far more available to pigs than that from $CaHPO_4$ (Eeckhout et al., 1995) may reflect saturation of the inhibitory effect of phytate at high calcium levels, achieved only with $CaCO_3$, thus confounding the effect of source and level.

Particle size can be important in the choice of calcium supplements. The high calcium needs of laying hens are commonly met by supplementation with 4% calcium as limestone, a relatively cheap and plentiful source. However, replacing one-half to two-thirds with granular sources such as oyster shell has a sparing effect so that 3% calcium can be sufficient with respect to shell quality (Roland et al., 1974). Large particles of calcium may remain in the gizzard longer and provide more retainable calcium than finely ground sources (Scott et al., 1971), possibly by achieving synchrony with diurnal fluctuations in the calcium requirement for shell calcification (Whitehead, 1995). Particle size of $CaCO_3$ influences the phosphorus requirement of broilers. Monobasic calcium phosphate (CaH_2PO_4) is unsuitable for laying hens because it results in acidosis and poor shell quality. In the poultry industry, calcium formate is advocated for reducing the contamination of carcasses and eggs with pathogenic bacteria. A soluble calcium source, calcium citrate malate, was found to be of similar value to $CaCO_3$ in broiler chicks (Henry and Pesti, 2002).

Prevention in housed ruminants

With housed ruminants receiving regular concentrate supplements, additional calcium is either incorporated into the whole mixed diet or into the concentrate portion of the ration, usually as $CaCO_3$ or $CaHPO_4$. The same salts are routinely added to free-access mineral mixtures for housed and grazing livestock, although they are rarely needed continuously as far as the grazing animal is concerned. There is again little to choose between sources of calcium, with dicalcium phosphate (DCP) being marginally superior to $CaCO_3$ when averaged over five studies in cattle (Soares, 1995). According to Roque et al. (2007), calcium absorption from DCP was only two-thirds that of $CaCO_3$ and indistinguishable from oyster meal. However, the lambs were retaining large amounts of calcium (around $2g\ day^{-1}$) from DCP yet not gaining weight, suggesting differential retention of dense particles of mineral in the gut.

The acidotic salts, $CaCl_2$ or calcium sulfate ($CaSO_4$), may be useful for manipulating acid–base balance in dairy cows to avoid milk fever. When calcium and other minerals are mixed with the feed it is important to ensure that they are evenly distributed. Inadequate mixing of limestone with cereals may contribute to hypocalcaemia after drought feeding (Larsen et al., 1986) and can be avoided by using molasses to stick such supplements to the grain (see Chapter 9 for details). Phosphorus-free calcium sources are best in grain-based diets for ruminants. Increased use is being made of calcium complexes with organic acids ('calcium soaps') as energy sources in ruminant nutrition, although not as improved sources of calcium.

Calcium Toxicity

Calcium is not generally regarded as a toxic element because homeostatic mechanisms ensure that excess dietary calcium is extensively excreted in faeces. However, doubling the dietary calcium concentration of chicks to $20.5g\ kg^{-1}$ DM causes hypercalcaemia and growth retardation, with fast-growing strains being more vulnerable than slow-growing strains (Hurwitz et al., 1995). The adverse nutritional consequences of feeding excess calcium are generally indirect and arise from impairments in the absorption of other elements; thus, deficiencies of phosphorus and zinc are readily induced in non-ruminants. Dietary provision of calcium soaps of fatty acids (CSFA) represents a convenient way of increasing the fat (i.e. energy) content of the diet without depressing fibre digestibility in the rumen; significant improvements in milk yield can be obtained in dairy cows and ewes, but not without feeding gross excesses of calcium (about $10g\ kg^{-1}$ DM) and sometimes matching excesses of phosphorus ($6g\ kg^{-1}$ DM) (Underwood and Suttle, 1999). Whether these excesses are related to the adverse effects of CSFA on conception rate in primiparous cows remains to be investigated.

Hypercalcaemia can cause life-threatening tissue calcification, but usually occurs as a secondary consequence of phosphorus deprivation or over-exposure to vitamin D_3 (Payne and Manston, 1967) and its analogues (Whitehead, 1995), some of which occur naturally. Ingestion of $1\alpha\ OHD_3$ in Solanum malacoxylon can cause tissue calcification in cattle. The first sign of trouble is the development of osteophagia, and there may be no alternative but to temporarily destock pastures infested with the water-loving plant. Factors that safely maximize the deposition of calcium in bone (heavily boned breeds: ample dietary phosphorus; alkaline mineral supplements) or lower calcium intake (grain feeding) are worth exploring as preventive measures where S. malacoxylon is a problem. Trisetum flavescens also contains $1\alpha\ OHD_3$ and can cause calcification of the cardiovascular system and kidneys when fed as silage to cattle or sheep (Franz et al., 2007). Cattle were the more severely affected, moving unnaturally and exhibiting low body temperatures and erect hair. In addition, calcification of the aorta was detectable by ultrasonography. In exposed sheep, cardiac arrhythmia and kidney calcification were found.

References

Adeola, O., Olukosi, O.A., Jendza, J.A., Dilger, R.N. and Bedford, M.R. (2006) Response of growing pigs to Peniophora lycii- and Escherichia coli-derived phytases or varying ratios of total calcium to total phosphorus. Animal Science 82, 637–644.

AFRC (1991) Technical Committee on Responses to Nutrients Report No. 6. A reappraisal of the calcium and phosphorus requirements of sheep and cattle. Nutrition Abstracts and Reviews 61, 573–612.

Allen, W.M. and Sansom, B.F. (1985) Milk fever and calcium metabolism. Journal of Veterinary Pharmacology and Therapeutics 8, 19–21.

Angel, R. (2007) Metabolic disorders: limitations to growth of and mineral deposition into the broiler skeleton after hatch and potential implications for leg problems. *Journal of Applied Poultry Research* 16, 138–149.
ARC (1980) *The Nutrient Requirements of Ruminants*. Commonwealth Agricultural Bureaux, Farnham Royal, UK, pp. 184–185.
ARC (1981) *The Nutrient Requirements of Pigs*. Commonwealth Agricultural Bureaux, Farnham Royal, UK, pp. 215–248.
Atteh, J.O. and Leeson, S. (1983) Effects of dietary fatty acids and calcium levels on performance and mineral metabolism of broiler chickens. *Poultry Science* 62, 2412–2419.
Atteh, J.O. and Leeson, S. (1985) Responses of laying hens to dietary saturated and unsaturated fatty acids in the presence of varying dietary calcium levels. *Poultry Science* 62, 2412–2419.
Bain, S.D. and Watkins, B.A. (1993) Local modulation of skeletal growth and bone modelling in poultry. *Journal of Nutrition* 123, 317–322.
Bar, A. and Hurwitz, S. (1984) Egg shell quality, medullary bone ash, intestinal calcium and phosphorus absorption and Ca-binding protein in phosphorus-deficient hens. *Poultry Science* 63, 1975–1979.
Barzanji, A.A.H. and Daniel, R.C.W. (1987) Effect of hypocalcaemia on blood flow distribution in sheep. *Research in Veterinary Science* 42, 92–95.
Barzanji, A.A.H. and Daniel, R.C.W. (1988) The effects of hypocalcaemia on blood gas and acid–base parameters. *British Veterinary Journal* 144, 93–97.
Becker, R.B., Arnold, P.T.D., Kirk, W.G., Davis, G.K. and Kidder, R.W. (1953) *Minerals for Beef and Dairy Cattle*. Bulletin 153, Florida Agricultural Experiment Station.
Beckman, M.J., Goff, J.P., Reinhardt, T.A., Beitz, D.C. and Horst, R.L. (1994) In vivo regulation of rat intestinal 24-hydroxylase: potential new role of calcium. *Endocrinology* 135, 1951–1955.
Blaney, B.J., Gartner, R.J.W. and Head, T.A. (1982) The effects of oxalate in tropical grasses on calcium, phosphorus and magnesium availability to cattle. *Journal of Agricultural Science, Cambridge* 99, 533–539.
Block, E. (1994) Manipulation of dietary cation–anion balance difference on nutritionally related production diseases, productivity and metabolic responses of dairy cows. *Journal of Dairy Science* 77, 1437–1450.
Bohkle, R.A., Thaler, R.C. and Stein, H.H. (2005) Calcium, phosphorus and amino acid digestibility in low-phytate corn, normal corn and soybean meal by growing pigs. *Journal of Animal Science* 83, 2396–2403.
Bond, P.L., Sullivan, T.W., Douglas, J.H. and Robeson, L.G. (1991) Influence of age, sex and method of rearing on tibia length and mineral deposition in broilers. *Poultry Science* 70, 1936–1942.
Bonniwell, M.A., Smith, B.S.W., Spence, J.A., Wright, H. and Ferguson, D.A.M. (1988) Rickets associated with vitamin D deficiency in young sheep. *Veterinary Record* 122, 386–388.
Borle, A.B. (1974) Calcium and phosphate metabolism. *Annual Review of Physiology* 36, 361–390.
Braithwaite, G.D. (1983a) Calcium and phosphorus requirements of the ewe during pregnancy and lactation. 1. Calcium. *Journal of Agricultural Science, Cambridge* 50, 711–722.
Braithwaite, G.D. (1983b) Effect of 1α-hydroxy cholecalciferol on calcium and phosphorus metabolism in sheep given high or low calcium diets. *Journal of Agricultural Science, Cambridge* 96, 291–299.
Braun, U., Salis, F., Siegwart, N. amd Hassig, M. (2004) Slow intravenous infusion of calcium in cows with parturient paresis. *Veterinary Record* 154, 336–338.
Breitwieser, G. (2008) Extracellular calcium as an integrator of tissue function. *International Journal of Biochemistry and Cell Biology* 40, 1467–1480.
Bronner, F. (1987) Intestinal calcium absorption: mechanisms and applications. *Journal of Nutrition* 117, 1347–1352.
Brown, E.M. (1991) Extracellular Ca^{2+} sensing, regulation of parathyroid cell function and role of Ca^{2+} and other ions as extracellular (first) messengers. *Physiological Reviews* 71, 371–411.
Buckner, G.D., Martin, J.H. and Insko, W.M. Jr (1930) The blood calcium of laying hens varied by the calcium intake. *American Journal of Physiology* 94, 692–695.
Carafoli, E. (1991) Calcium pump of the plasma membrane. *Physiological Reviews* 71, 129–149.
Challa, J. and Braithwaite, G.D. (1989) Phosphorus and calcium metabolism in young calves with special emphasis on phosphorus homeostasis. 4. Studies on milk-fed calves given different amounts of phosphorus but a constant intake of calcium. *Journal of Agricultural Science, Cambridge* 113, 283–289.
Charbonneau, E., Pellerin, D. and Oetzel, G.R. (2006) Impact of lowering dietary cation–anion-balance in non-lactating cows: a meta-analysis. *Journal of Dairy Science* 89, 537–548.

Chrisp, J.S., Sykes, A.R. and Grace, N.D. (1989a) Faecal endogenous loss of calcium in young sheep. *British Journal of Nutrition* 61, 59–65.

Chrisp, J.S., Sykes, A.R. and Grace, N.D. (1989b) Kinetic aspects of calcium metabolism in lactating sheep offered herbages with different calcium concentrations and the effect of protein supplementation. *British Journal of Nutrition* 61, 45–58.

Cockcroft, P.D. and Whiteley, P. (1999) Hypocalcaemia in 23 ataxic/recumbent ewes: clinical signs and likelihood ratios. *Veterinary Record* 144, 529–532.

Common, R.H. (1938) Observations on mineral metabolism in pullets 3. *Journal of Agricultural Science, Cambridge* 28, 347–366.

Coop, R.L. and Sykes, A.R. (2002) Interactions between gastrointestinal parasites and nutrients. In: Freer M. and Dove H. (eds) *Sheep Nutrition*. Commonwealth Agricultural Bureaux, Farnham Royal, UK, pp. 318–319.

DeGaris, P.J. and Lean, I.J. (2008) Milk fever in dairy cows: a review of pathophysiology and control principles. *The Veterinary Journal* 176, 58–69.

Deobald, H.J., Lease, E.J., Hart, E.B. and Halpin, J.G. (1936) Studies on the calcium metabolism of laying hens. *Poultry Science* 15, 179–185.

Dias, R.S., Kebreab, E., Vitti, D.M.S.S., Roque, A.P., Bueno, I.C.S. and France, J. (2006) A revised model for studying phosphorus and calcium kinetics in growing sheep. *Journal of Animal Science* 84, 2787–2794.

Dilworth, B.C., Day, E.J. and Hill, J.E. (1964) Availability of calcium in feed grade phosphates to the chick. *Poultry Science* 43, 1121–1134.

Duckworth, J., Benzie, D., Cresswell, E., Hill, R. and Boyne, A.W. (1962) Studies of the dentition of sheep. III. A study of the effects of vitamin D and phosphorus deficiencies in the young animal on the productivity, dentition and skeleton of Scottish Blackface ewes. *Research in Veterinary Science* 3, 408–415.

Duncan, A.J., Frutos, P. and Young, S.A. (1997) Rates of oxalic acid degradation in the rumen of sheep and goats in response to different levels of oxalic acid administration. *Animal Science* 65, 451–455.

Driver, J.P., Pesti, G.M., Bakali, R.I. and Edwards, H.M. Jr (2005a) Calcium requirements of the modern broiler chicken as influenced by dietary protein and age. *Poultry Science* 84, 1629–1639.

Driver, J.P., Pesti, G.M., Bakali, R.I. and Edwards, H.M. Jr (2005b) Effects of calcium and nonphytate phosphorus concentrations on phytase efficacy in broiler chicks. *Poultry Science* 84, 1406–1417.

Edwards, H.M. and Veltmann, J.R. (1983) The role of calcium and phosphorus in the aetiology of tibial dyschondroplasia in young chicks. *Journal of Nutrition* 113, 1568–1575.

Eeckhout, W., de Poepe, M., Warnants, N. and Bekaert, H. (1995) An estimation of the minimal P requirements of growing-finishing pigs, as influenced by calcium level of the diet. *Animal Feed Science and Technology* 52, 29–40.

Eklou-Kalonji, E., Zerath, E.C., Lacroix, C.C., Holy, X., Denis, I. and Pointillart, A. (1999) Calcium-regulating hormones, bone mineral content, breaking load and trabecular remodeling are altered in growing pigs fed calcium-deficient diets. *Journal of Nutrition* 129, 188–193.

Ender, F., Dishington, I.W. and Helgebostad, A. (1956) Parturient paresis and related forms of hypocalcemic disorders induced experimentally in dairy cows. *Nordisk Veterinaermedicin* 8, 507–513.

Ender, F., Dishington, I.W. and Helgebostad, A. (1971) Calcium balance studies in dairy cows under experimental induction and prevention of hypocalcaemic parturient paresis. *Zeitschrift fur Tierphysiologie Tierernahrung und Fultermittelkunde* 28, 233–256.

Farrugia, W., Fortune, C.L., Heath, J., Caple, I.W. and Wark, J.D. (1989) Osteocalcin as an index of osteoblast function during and after ovine pregnancy. *Endocrinology* 125, 1705–1710.

Fenwick, D.C. (1988) The relationship between certain blood calcium ions in cows with milk fever and both the state of consciousness and the position of the cows when attended. *Australian Veterinary Journal* 65, 374–375.

Fenwick, D.C. (1994) Limitations to the effectiveness of subcutaneous calcium solutions as a treatment for cows with milk fever. *Veterinary Record* 134, 446–448.

Field, A.C. and Suttle, N.F. (1967) Retention of calcium, phosphorus, magnesium, sodium and potassium by the developing sheep foetus. *Journal of Agricultural Science, Cambridge* 69, 417–423.

Field, A.C., Suttle, N.F. and Nisbet, D.I. (1975) Effects of diets low in calcium and phosphorus on the development of growing lambs. *Journal of Agricultural Science, Cambridge* 85, 435–442.

Franklin, M.C., Reid, R.L. and Johnstone, I.L. (1948) *Studies on Dietary and Other Factors Affecting the Serum-calcium Levels of Sheep*. Bulletin, Council for Scientific and Industrial Research, Australia, No. 240.

Franz, S., Gasteiner, F., Schilcher, F. and Baumgartner, W. (2007) Use of ultrasonography for detecting calcification in cattle and sheep given calcinogenic *Trisetum flavescens* silage. *Veterinary Record* 161, 751–754.

Freeden, A.H. (1989) Effect of maturity of alfalfa (*Medicago sativa*) at harvest on calcium absorption in goats. *Canadian Journal of Animal Science* 69, 365–371.

Fritts, C.A. and Waldroup, P.W. (2006) Modified phosphorus program for broilers based on commercial feeding intervals to sustain live performance and reduce total water-soluble phosphorus in litter. *Journal of Applied Poultry Research* 15, 207–218.

Gilbert, A.B. (1983) Calcium and reproductive function in the hen. *Proceedings of the Nutrition Society* 42, 195–212.

Goff, J.P. (2007) The monitoring, prevention and treatment of milk fever and subclinical hypocalcaemia in dairy cows. *The Veterinary Journal* 176, 50–57.

Goff, J.P. and Horst, R.L. (1993) Oral administration of calcium salts for treatment of hypocalcaemia in cattle. *Journal of Dairy Science* 763, 101–108.

Goff, J.P. and Horst, R.L. (1994) Calcium salts for treating hypocalcaemia: carrier effects, acid–base balance and oral versus rectal administration. *Journal of Dairy Science* 77, 1451–1456.

Goff, J.P. and Horst, R.L. (1997) The effect of dietary potassium and sodium but not calcium on the incidence of milk fever in dairy cows. *Journal of Dairy Science* 80, 176–186.

Goff, J.P. and Horst, R.L. (1998) Use of hydrochloric acid as a source of anions for prevention of milk fever. *Journal of Dairy Science* 81, 2874–2880

Goff, J.P. and Horst, R.L (2003) Milk fever control in the United States. *Acta Veterinaria Scandinavica, Supplement* 97, 145–147.

Goff, J.P., Kerli, M.E. Jr and Horst, R.L. (1989a) Periparturient hypocalcaemia in cows: prevention using intramuscular parathyroid hormone. *Journal of Dairy Science* 72, 1182–1187.

Goff, J.P., Reinhardt, T.A. and Horst, R.L. (1989b) Recurring hypocalcaemia of bovine parturient paresis is associated with failure to produce 1,25 hydroxyvitamin D. *Endocrinology* 125, 49–53.

Goff, J.P., Reinhardt, T.A., Lee, S. and Hollis, B.W. (1991) Parathyroid hormone-related peptide content of bovine milk and calf blood assessed by radioimmunoassay and bioassay. *Endocrinology* 129, 2815–2819.

Goff, J.P., Riuz, R. and Horst, R.L. (2004) Relative acidifying activity of anionic salts commonly used to prevent milk fever. *Journal of Dairy Science* 87, 1245–1255.

Goff, J.P., Brummer, E.C., Henning, S.J., Doorenbos, R.K. and Horst, R.L. (2007) Effect of application of ammonium chloride and calcium chloride on alfalfa cation–anion content and yield. *Journal of Dairy Science* 90, 5159–5164.

Grace, N.D., Pearce, S.G., Firth, E.C. and Fennessy, P.F. (1999a) Content and distribution of macro- and trace-elements in the body of young pasture-fed horses. *Australian Veterinary Journal* 77, 172–176.

Grace, N.D., Pearce, S.G., Firth, E.C. and Fennessy, P.F. (1999b) Concentrations of macro- and trace-elements in the milk of pasture-fed Thoroughbred mares. *Australian Veterinary Journal* 77, 177–180.

Grace, N.D., Gee, E.K., Firth, E.C. and Shaw, H.L. (2002a) Determination of digestible energy intake, dry matter digestibility and mineral status of grazing New Zealand Thoroughbred yearlings. *New Zealand Veterinary Journal* 50, 182–185.

Grace, N.D., Shaw, H.L., Firth, E.C. and Gee, E.K. (2002b) Determination of digestible energy intake, and apparent absorption of macroelements of grazing, lactating Thoroughbred mares. *New Zealand Veterinary Journal* 50, 182–185.

Grace, N.D., Rogers, C.W., Firth, E.C., Faram, T.L. and Shaw, H.L. (2003) Digestible energy intake, dry matter digestibility and effect of calcium intake on bone parameters of Thoroughbred weanlings in New Zealand. *New Zealand Veterinary Journal* 51, 165–173.

Groenewald, J.W. (1937) Osteofibrosis in equines. *Onderstepoort Journal of Veterinary Science and Animal Industry* 9, 601–620.

Hamilton, R.M.G. and Cipera, J.D. (1981) Effects of dietary calcium levels during brooding, rearing and early egg laying period on feed intake, egg production and shell quality in white leghorn hens. *Poultry Science* 60, 349–357.

Henry, M.H. and Pesti, G.M. (2002) An investigation of calcium citrate-malate as a calcium source for broiler chicks. *Poultry Science* 81, 1149–1155.

Heringstad, B., Chang, Y.M., Gianola, D. and Klemetsdal, G. (2005) Genetic analysis of clinical mastitis, milk fever, ketosis and retained placenta in three lactations of Norwegian Red cows. *Journal of Dairy Science* 88, 3273–3281.

Heron, V.S., Tremblay, G.F. and Oba, M. (2009) Timothy hays differing in dietary cation–anion difference affect the ability of dairy cows to maintain their calcium homeostasis. *Journal of Dairy Science* 92, 238–246.

Hintz, H.F., Schryver, H.F., Doty, J., Lakin, C. and Zimmerman, R.A. (1984) Oxalic acid content of alfalfa hays and its influence on the availability of calcium, phosphorus and magnesium to ponies. *Journal of Animal Science* 58, 939–942.

Horst, R.L. (1986) Regulation of calcium and phosphorus homeostasis in the dairy cow. *Journal of Dairy Science* 69, 604–616.

Horst, R.L., Goff, J.P. and Reinhardt, T.A. (1990) Advancing age results in reduction of intestinal and bone 1,25 dihydroxyvitamin D receptor. *Endocrinology* 126, 1053–1057.

Hory, B., Rousanne, M.-C., Droke, T.B. and Bordeau, A. (1998) The calcium receptor in health and disease. *Experimental Nephrology* 6, 171–179.

Hove, K. (1986) Cyclic changes in plasma calcium and the calcium homeostatic endocrine system of the postparturient dairy cow. *Journal of Dairy Science* 69, 2072–2082.

Huber, T.L., Wilson, R.C., Stattelman, A.J. and Goetsch, D.D. (1981) Effect of hypocalcaemia on motility of the ruminant stomach. *American Journal of Veterinary Research* 42, 1488–1490.

Hughes, L.E. and Kershaw, G.F. (1958) Metabolic disorders associated with movement of hill sheep. *Veterinary Record* 70, 77–78.

Hurwitz, S. (1996) Homeostatic control of plasma calcium concentration. *Critical Reviews in Biochemistry and Molecular Biology* 31, 41–100.

Hurwitz, S., Plavnik, I., Shapiro, A., Wax, E., Talpaz, H. and Bar, A. (1995) Calcium metabolism and requirements of chickens are affected by growth. *Journal of Nutrition* 125, 2679–2686.

Jonson, G., Luthman, J., Mollerberg, L. and Persson, J. (1971) Hypocalcaemia in pregnant ewes. *Nordisk Veterinaermedicin* 23, 620–627.

Jonson, G., Luthman, J., Mollerberg, L. and Persson, J. (1973) Mineral feeding flocks of sheep with cases of clinical hypocalcaemia. *Nordisk Veterinaermedicin* 25, 97–103.

Jumba, I.O., Suttle, N.F., Hunter, E.A. and Wandiga, S.O. (1995) Effects of soil origin and mineral composition and herbage species on the mineral composition of forages in the Mount Elgon region of Kenya. 1. Calcium, phosphorus, magnesium and sulfur. *Tropical Grasslands* 29, 40–46.

Kamiya, Y., Kamiya, M., Tanaka, M. and Shioya, S. (2005) Effects of calcium intake and parity on plasma minerals and bone turnover around parturition. *Animal Science Journal* 76, 325–330.

Khorasani, G.R., Janzen, R.A., McGill, W.B. and Kennelly, J.J. (1997) Site and extent of mineral absorption in lactating cows fed whole-crop cereal grain silage or alfalfa silage. *Journal of Animal Science* 75, 239–248

Larsen, J.W.A., Constable, P.D. and Napthine, D.V. (1986) Hypocalcaemia in ewes after a drought. *Australian Veterinary Journal* 63, 25–26.

Larsen, T., Skoglund, E., Sandberg, A-S. and Engberg, E.R.M. (1999) Soaking and pelleting of pig diets alters the apparent absorption and retention of minerals. *Canadian Journal of Animal Science* 79, 477–483.

Lean, I.J., DeGaris, P.J., McNeill, D.M. and Block, E. (2006) Hypocalcaemia in dairy cows: meta-analysis and dietary cation–anion difference theory revisited. *Journal of Dairy Science* 89, 669–684.

Leonhard-Marek, S., Becker, G., Breves, G. and Schroder, B. (2007) Chloride, gluconate, sulphate and short-chain fatty acids affect calcium flux rates across the sheep forestomach epithelium. *Journal of Dairy Science* 90, 1516–1526.

Liesegang, A. (2008) Influence of anionic salts on bone metabolism in periparturient dairy goats and sheep. *Journal of Dairy Science* 91, 2449–2460.

Liesegang, A., Sassi, M.-L., Risteli, J., Eicher, R., Wanner, M. and Riond, J.-L. (1998) Comparison of bone resorption markers during hypocalcaemia in dairy cows. *Journal of Dairy Science* 81, 2614–2622.

Liesegang A., Eicher, R., Sassi, M.-L., Risteli J., Kraenzlin, M. and Riond, J.-L. (2000) Biochemical markers of bone formation and resorption around parturition and during lactation in dairy cows with high and low milk yields. *Journal of Dairy Science* 83, 1773–1781.

Littledike, E.T., Young, J.W. and Beitz, D.C. (1981) Common metabolic diseases in cattle: ketosis, milk-fever and downer cow complex. *Journal of Dairy Science* 64, 1465–1482.

Liu, J., Bollinger, D.W., Ledoux, D.R. and Veum, T.L. (1998) Lowering the dietary calcium to total phosphorus ratio increases phosphorus utilisation in low-phosphorus corn–soybean meal diets supplemented with microbial phytase for growing-finishing pigs. *Journal of Animal Science* 76, 808–813.

Loveridge, N. (1999) Bone: more than a stick. *Journal of Animal Science* 77 (Suppl. 2), 190–196.

Lucas, M.J., Huffman, E.M. and Johnson, L.W. (1982) Clinical and clinicopathological features of transport tetany of feedlot lambs. *Journal of American Veterinary Medical Association* 181, 381–383.

MAFF (1990) *UK Tables of the Nutritive Value and Chemical Composition of Foodstuffs.* In: Givens, D. I. (ed.) Rowett Research Services, Aberdeen, UK.

Martz, F.A., Belo, A.T., Weiss, M.F. and Belyca, R.L. (1999) True absorption of calcium and phosphorus from alfalfa and corn silage when fed to lactating cows. *Journal of Dairy Science* 82, 618–622.

McDonald, I.W. (1968) The nutrition of the grazing ruminant. *Nutrition Abstracts and Reviews* 38, 381–400.

McNeill, D.M., Roche, J.R., McLachlan, B.P. and Stockdale, C.R. (2002) Nutritional strategies for the prevention of hypocalcaemia at calving for dairy cows in pasture-based systems. *Australian Journal of Agricultural Research* 53, 755–770.

Miller, S. (2006) Calcitonin – guardian of the mammalian skeleton or is it just a 'fish story'? *Endocrinology* 147, 4007–4009.

Miller, E.R., Ullrey, D.C., Zutaut, G.L., Baltzer, B.V., Schmidt, D.A., Hoefer, J.A. and Luecke, R.W. (1962) Calcium requirement of the baby pig. *Journal of Nutrition* 77, 7–17.

Minson, D.J. (1990) *Forages in Ruminant Nutrition.* Academic Press, San Diego, California, pp. 208–229.

Mosdol, G. and Waage, S. (1981) Hypocalcaemia in the ewe. *Nordisk Veterinaemedicin* 33, 310–326.

Muir, P.D., Sykes, A.R. and Barrell, G.K. (1987) Calcium metabolism in red deer (*Cervus elaphus*) offered herbages during anterogenesis: kinetic and stable balance studies. *Journal of Agricultural Science, Cambridge* 109, 357–364.

Narvaez-Solarte, W., Rostagno, H.S., Soares, P.R., Uribe-Valasquez, L.L. and Silva, M.A. (2006) Nutritional requirement of calcium in white laying hens from 46–62 wk of age. *International Journal of Poultry Science* 5, 181–184.

Norman, A.W. and Horwitz, S. (1993) The role of vitamin D endocrine system in avian bone biology. *Journal of Nutrition* 123, 310–316.

NRC (1985) *Nutrient Requirements of Sheep*, 6th edn. National Academy of Sciences, Washington, DC.

NRC (1994) *Nutrient Requirements of Poultry*, 9th edn. National Academy of Sciences, Washington, DC.

NRC (1998) *Nutrient Requirements of Swine*, 10th edn. National Academy of Sciences, Washington, DC.

NRC (2000) *Nutrient Requirements of Beef Cattle*, 7th edn. National Academy of Sciences, Washington, DC.

NRC (2001) *Nutrient Requirements of Dairy Cows*, 7th edn. National Academy of Sciences, Washington, DC.

NRC (2007) *Nutrient Requirements of Horses*, 6th edn. National Academy of Sciences, Washington, DC.

Oetzel, G.R. (1991) Meta-analysis of nutritional risk factors for milk fever in dairy cattle. *Journal of Dairy Science* 74, 3900–3912.

Oikawa, S. and Kato, N. (2002) Decreases in serum apolipoprotein B-100 and A-1 concentrations in cows with milk fever and downer cows. *Canadian Journal of Veterinary Research* 66, 31–34.

Olukosi, O.A., Cowieson, A.J. and Adeola, O. (2007) Age-related influence of a cocktail of xylanase, amylase and protease or phytase individually or in combination in broilers. *Poultry Science* 86, 77–86.

Omdahl, J.L. and DeLuca, H.F. (1973) Regulation of vitamin D metabolism and functions. *Physiological Reviews* 53, 327–372.

Payne, J.M. and Manston, R. (1967) The safety of massive doses of vitamin D_3 in the prevention of milk fever. *Veterinary Record* 81, 214–216.

Perhson, B., Svensson, C., Gruvaeus, I. and Virkki, M. (1999) The influence of acidic diets on the acid–base balance of dry cows and the effect of fertilisation on the mineral content of grass. *Journal of Dairy Science* 82, 1310–1316.

Peterson, A.B., Orth, M.W., Goff, J.P. and Beede, D.K. (2005) Periparturient responses of multiparous Holstein cows fed different dietary phosphorus concentrations prepartum. *Journal of Dairy Science* 88, 3582–3594.

Phillippo, M., Reid, G.W. and Nevison, I.M. (1994) Parturient hypocalcaemia in dairy cows: effects of dietary acidity on plasma minerals and calciotropic hormones. *Research in Veterinary Science* 56, 303–309.

Pickard, D.W., Field, B.G. and Kenworthy, E.B. (1988) Effect of magnesium content of the diet on the susceptibility of ewes to hypocalcaemia in pregnancy. *Veterinary Record* 123, 422.

Pryce, J.E., Veerkamp, R.F., Thompson, R., Hill, W.G. and Simm, G. (1997) Genetic aspects of common health disorders and measures of fertility in Holstein Friesian herds. *Animal Science* 65, 353–360.

Qian, H., Kornegay, E.T. and Denbow, D.M. (1997) Utilisation of phytate phosphorus and calcium as influenced by microbial phytase, cholecalciferol and the calcium:total phosphorus ratio in broiler diets. *Poultry Science* 76, 37–46.

Rajala-Schulz, P.J., Grohn, Y.T. and McCulloch, C.E. (1999) Effects of milk fever, ketosis and lameness on milk yield in dairy cows. *Journal of Dairy Science* 82, 288–294.

Ramberg, C.F., Johson, E.K., Fargo, R.D. and Kronfeld, D.S. (1984) Calcium homeostasis in dairy cows with special reference to parturient hypocalcaemia. *American Journal of Physiology* 246, 698–704.

Reid, B.L. and Weber, C.W. (1976) Calcium availability and trace mineral composition of feed grade calcium supplements. *Poultry Science* 55, 600–605.

Reinhart, G.A. and Mahan, D.C. (1986) Effects of various calcium:phosphorus ratios at low and high dietary phosphorus for starter, grower and finishing swine. *Journal of Animal Science* 63, 457–466.

Roche, J.R. and Berry, D.P. (2006) Periparturient climatic, animal, and management factors influencing incidence of milk fever in grazing systems. *Journal of Dairy Science* 89, 2775–2783.

Roche, J.R., Dalley, D., Moate, P., Grainger, C., Rath, M. and O'Mara, F. (2003a) Dietary cation–anion difference and the health and production of pasture-fed dairy cows. 1. Dairy cows in early lactation. *Journal of Dairy Science* 86, 970–978.

Roche, J.R., Dalley, D., Moate, P., Grainger, C., Rath, M. and O'Mara, F. (2003b) Dietary cation–anion difference and the health and production of pasture-fed dairy cows. 2. Nonlactating periparturient cows. *Journal of Dairy Science* 86, 979–987.

Roland, D.A. (1986) Calcium and phosphorus requirements of commercial Leghorns. *World Poultry Science Journal* 42, 154–165.

Roland, D.A. Sr, Sloan D.R. and Harms, R.H. (1974) Effect of various levels of calcium with and without pullet-sized limestone on shell quality. *Poultry Science* 53, 662–666.

Roque, A.P., Dias, R.S., Vitti, D.M.S.S., Bueno, I.C.S., Cunha, E.A., Santos, L.E. and Bueno, M.S. (2007) True digestibility of calcium from sources used in lamb finishing diets. *Small Ruminant Research* 71, 243–249.

Sachs, M., Bar, A., Nir, O., Ochovsky, D., Machnai, B., Meir, E., Weiner, B.Z. and Mazor, Z. (1987a) Efficacy of 1α hydroxyvitamin D_3 in the prevention of bovine parturient paresis. *Veterinary Record* 120, 39–42.

Sachs, M., Perlman, R. and Bar, A. (1987b) Use of 1α hydroxyvitamin D_3 in the prevention of bovine parturient paresis. IX. Early and late effects of a single injection. *Journal of Dairy Science* 70, 1671–1675.

Sansom, B.F., Bunch, K.J. and Dew, S.M. (1982) Changes in plasma calcium, magnesium, phosphorus and hydroxyproline concentrations in ewes from twelve weeks before until three weeks after lambing. *British Veterinary Journal* 138, 393–401.

Schneider, K.M., Ternouth, J.H., Sevilla, C.C. and Boston, R.C. (1985) A short-term study of calcium and phosphorus absorption in sheep fed on diets high and low in calcium and phosphorus. *Australian Journal of Agricultural Research* 36, 91–105.

Scott, M.I., Hull, S.J. and Mullenhoff, P.A. (1971) The calcium requirements of laying hens and effects of dietary oyster shell upon egg quality. *Poultry Science* 50, 1055–1063.

Scott, T.A., Kampen, R. and Silversides, F.G. (2000) The effect of phosphorus, phytase enzyme and calcium on the performance of layers fed wheat-based diets. *Canadian Journal of Animal Science* 80, 183–190.

Shappell, N.W., Herbein, J.H., Deftos, L.J. and Aiello, R.J. (1987) Effects of dietary calcium and age on parathyroid hormone, calcitonin and serum and milk proteins in the periparturient dairy cow. *Journal of Nutrition* 117, 201–207.

Shirley, R.B., Davis, A.J., Compton, M.M. and Berry, W.D. (2003) The expression of calbindin in chicks that are divergently selected for low to high incidence of tibial dyschondroplasia. *Poultry Science* 82, 1965–1973.

Simons, P.C.M. (1986) Major minerals in the nutrition of poultry. In: Fisher, C. and Boorman, K.N. (eds) *Nutrient Requirements of Poultry and Nutritional Research*. Butterworths, London, pp. 141–145.

Smith, B.S.W. and Wright, H. (1981) Seasonal variation in serum 25-hydroxyvitamin D concentration in sheep. *Veterinary Record* 109, 139–141.

Soares, J.H. (1995) Calcium bioavailability. In: Ammerman, C.B., Baker, D.H. and Lewis A.J. (eds) *Bioavailability of Nutrients for Animals*. Academic Press, New York, pp. 1195–1198.

Spence, J.A., Sykes, A.R., Atkinson, P.J. and Aitchison, G.U. (1985) Skeletal and blood biochemical characteristics of sheep during growth and breeding: a comparison of flocks with and without broken mouth. *Journal of Comparative Pathology* 95, 505–522.

Spencer, J.D., Allee, G.L. and Sauber, T.E. (2000) Phosphorus bioavailability and digestibility of normal and genetically modified low-phytate corn for pigs. *Journal of Animal Science* 78, 675–681.

Stevenson, M.A., Williamson, N.B. and Russell, D.J. (2003) Nutrient balance in the diet of spring-calving, pasture-fed dairy cows. *New Zealand Veterinary Journal* 51, 81–88.

Suttle, N.F. and Wadsworth, I.R. (1991) Physiological responses to vaccination in sheep. *Proceedings of the Sheep Veterinary Society* 15, 113–116.

Sykes, A.R. (1987) Endoparasites and herbivore nutrition. In: Hacker, J.B. and Ternouth, J.H. (eds) *The Nutrition of Herbivores.* Academic Press, Canberra, pp. 211–232.

Sykes, A.R. and Field, A.C. (1972) Effects of dietary deficiencies of energy, protein and calcium on the pregnant ewe. I. Body composition and mineral content of the ewes. *Journal of Agricultural Science, Cambridge* 78, 109–117.

Sykes, A.R. and Geenty, K.C. (1986) Calcium and phosphorus balances of lactating ewes at pasture. *Journal of Agricultural Science, Cambridge* 106, 369–375.

Thilsing-Hansen, T., Jorgensen, R.J. and Ostergaard, S. (2002) Milk fever control principles: a review. *Acta Veterinaria Scandinavica* 43, 1–19.

Thompson, J.K., Gelman, A. and Weddell, J.R. (1988) Mineral retention and body composition of grazing lambs. *Animal Production* 46, 53–62.

Traylor, S.L., Cromwell, G.L., Lindemann, M.D. and Knabe, D.A. (2001) Bioavailability of phosphorus in meat and bone meal for swine. *Journal of Animal Science* 79, 2634–2642.

Underwood, E.J. and Suttle, F. (1999) *The Mineral Nutrition of Livestock*, 3rd edn. CAB International, Wallingford, UK p. 93.

Veum, T.L., Ledoux, D.R., Bollinger, D.W., Raboy, V. and Ertl, D.S. (2001) Low-phytic acid corn improves nutrient utilisation for growing pigs. *Journal of Animal Science* 79, 2873–2880.

Veum, T.L., Ledoux, D.R. and Raboy, V. (2007) Low-phytate barley cultivars improve the utilisation of phosphorus, calcium, nitrogen, energy and dry matter in diets fed to young swine. *Journal of Animal Science* 85, 961–971.

Vipperman, P.E. Jr, Peo, E.R. and Cunningham, P.J. (1974) Effect of dietary calcium and phosphorus level upon calcium, phosphorus and nitrogen balance in pigs. *Journal of Animal Science* 38, 758–765.

Wan Zahari, M., Thompson, J.K., Scott, D. and Buchan, W. (1990) The dietary requirements of calcium and phosphorus for growing lambs. *Animal Production* 50, 301–308.

Ward, G., Harbers, L.H. and Blaha, J.J. (1979) Calcium-containing crystals in alfalfa: their fate in cattle. *Journal of Dairy Science* 62, 715–722.

Whitehead, C.C. (1995) Nutrition and skeletal disorders in broilers and layers. *Poultry International* 34, 40–48.

Whittemore, C.T., Hazzledine, M.J. and Close, W.H. (2003) *Nutrient Requirement Standards for Pigs.* British Society of Animal Science, Penicuik, UK.

Wilson, J.H. (1991) Bone strength of caged layers as affected by dietary Ca and phosphorus concentrations, reconditioning and ash content. *British Poultry Science* 32, 501–508.

Wuthier, R.E. (1993) Involvement of cellular metabolism of calcium and phosphate in calcification of avian growth plate. *Journal of Nutrition* 123, 301–309.

Yamagishi, N., Ogawa, K. and Naito, Y. (1999) Pathological changes in the myocardium of hypocalcaemic parturient cows. *Veterinary Record* 144, 67–72.

Yano, F., Yano, H. and Breves, G. (1991) Calcium and phosphorus metabolism in ruminants. In: *Proceedings of the Seventh International Symposium on Ruminant Physiology.* Academic Press, New York, pp. 277–295.

Yuangklang, C., Wensing, T., Van den Broek, L., Jittakhot, S. and Beynen, A.C. (2004) Fat digestion in veal calves fed milk replacers low or high in calcium and containing either casein or soy protein isolate. *Journal of Dairy Science* 87, 1051–1056.

5 Magnesium

Introduction

Around 1930, lack of magnesium in the diet was found to cause hyperirritability and convulsions in laboratory rats, while similar abnormalities in grazing cows were being associated with subnormal serum magnesium and termed 'grass tetany' (Sjollema and Seekles, 1930). Studies of magnesium metabolism in ruminants were keenly pursued, and interest was sustained by the fact that intensification of grassland production through the increased use of fertilizers seemed to increase the incidence of a disorder that remains a problem to this day. By contrast, there has, until recently, been little research on the magnesium metabolism of pigs and poultry because common feedstuffs contain sufficient magnesium to meet their requirements and natural deficiencies are unheard of. The discovery that magnesium supplementation prior to slaughter can improve meat quality in stressed pigs has led to an upsurge of research activity, but the focus of this chapter remains with the magnesium nutrition of ruminants.

Functions

Although most of the body's magnesium (60–70%) is present in the skeleton (Underwood and Suttle, 1999), magnesium is second only to potassium in abundance in the soft tissues, which contain 0.1–$0.2\,g$ Mg kg^{-1} fresh weight. Unlike potassium, however, magnesium is largely (80%) protein bound. Magnesium is associated predominantly with the microsomes, where it functions as a catalyst of a wide array of enzymes, facilitating the union of substrate and enzyme by first binding to one or the other (Ebel and Günther, 1980). Magnesium is thus required for oxidative phosphorylation leading to ATP formation, sustaining processes such as the sodium ion/potassium ion pump; pyruvate oxidation and conversion of α-oxoglutarate to succinyl coenzyme A; phosphate transfers, including those effected by alkaline phosphatase, hexokinase and deoxyribonuclease; the β oxidation of fatty acids; and the transketolase reaction of the pentose monophosphate shunt (Shils, 1997).

Magnesium also performs non-enzymic functions: the binding of magnesium to phosphate groups on ribonucleotide chains influences their folding; exchanges with calcium influence muscle contraction (Ebel and Günther, 1980; Shils, 1997); and cell membrane integrity is partly dependent on the binding of magnesium to phospholipids. Magnesium is present in erythrocytes and deprivation changes the fluidity of red cell membranes (Tongyai et al., 1989). Rumen microorganisms require magnesium to catalyse many of the enzymes essential to cellular function in mammals, and feeding sheep on a semi-purified diet virtually devoid of magnesium rapidly impairs cellulolytic activity by the rumen microflora. Magnesium also occurs in relatively low but life-sustaining concentrations in the extracellular fluids, including the cerebrospinal fluid (CSF), where it governs the neuromuscular transmission of nerve impulses. Magnesium also antagonizes calcium-modulated transmitter release at the synapses and affects autonomic control in the heart.

Sources of Magnesium

Milk and milk replacers

Milks are relatively low in magnesium (0.1–0.2 g l^{-1}), but colostrums are three times richer (Table 2.4). Magnesium is absorbed very efficiently from milk by the young lamb and calf, with absorption from the large intestine making a large contribution to that high efficiency (Dillon and Scott, 1979). However, magnesium absorption (A_{Mg}) coefficients fall in calves from 0.87 at 2–3 weeks to 0.32 at 7–8 weeks of age (ARC, 1980). Prolonged rearing of calves on a simulated milk diet will induce hypomagnesaemic tetany (Blaxter et al., 1954), as will the artificial maintenance of adult ruminants on a similar diet (Baker et al., 1979). The magnesium concentration of milk cannot be raised by dietary magnesium supplementation.

Forages

The magnesium concentration in forages depends chiefly on the plant species and the seasonal and climatic conditions during plant growth, with soil origin being of little importance (Jumba et al., 1995). Grasses in temperate and tropical pastures contain on average 1.8 and 3.6 g Mg kg^{-1} dry matter (DM), respectively, and legumes 2.6 and 2.8 g kg^{-1} DM, respectively (results for 930 forage samples) (Minson, 1990). In temperate grasses 65% of samples contain <2 g Mg kg^{-1} DM, a statistic that has a bearing on the prevalence of magnesium disorders of livestock. There are species and varietal differences amongst grasses, with Ajax and S24 perennial ryegrass containing 30% less magnesium than Augusta (hybrid ryegrass), Melle (perennial ryegrass) and the average UK pasture (Underwood and Suttle, 1999). Seasonal variations within species are relatively small (Minson, 1990), but are larger in mixed swards and can assume nutritional significance when concurrent changes in potassium, an antagonist of magnesium, are considered (Fig. 5.1). Magnesium concentrations are low in spring when potassium concentrations are high, legume contribution is small and risk of grass tetany is maximal. The relatively high magnesium concentrations shown in Fig. 5.1, from a New Zealand study, compared with those from UK pastures (Table 5.1), probably reflect a high legume contribution (about 20%) in New Zealand pastures. The UK data show higher magnesium

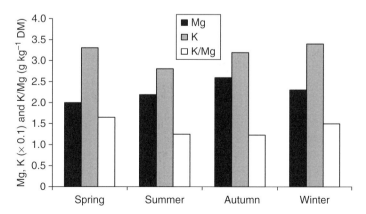

Fig. 5.1. Effect of season on mean magnesium and potassium concentrations in mixed (*Lolium perenne* and *Trifolium repens*) pastures on seven farms in Manawatu, New Zealand (from Stevenson et al., 2003). Fertilizer was applied in autumn.

Table 5.1. Mean magnesium (± standard deviation) concentrations (mg kg^{-1} DM) in livestock feeds commonly used in the UK (MAFF, 1990).

Roughages	Mg	Concentrates	Mg	By-products	Mg
Barley straw	0.7 (0.31)	Barley	1.2 (0.20)	Wheat bran	6.2 (2.70)
Oat straw	0.9 (0.31)	Cassava	1.1 (0.57)	Brewers' grains	1.7 (0.36)
Grass	1.6 (0.56)	Maize	1.3 (0.13)	Distillers' grains[a]	3.3 (0.34)
Kale	1.6 (0.18)	Oats, winter	1.0 (0.06)	Citrus pulp	1.7 (0.50)
White clover	2.2 (0.50)	Wheat	1.1 (0.13)	Sugarbeet pulp	1.1 (0.19)
Grass silage	1.7 (0.54)	Maize gluten	4.1 (0.70)		
Grass hay	1.4 (0.52)	Cottonseed meal	5.8 (0.43)		
Clover silage	2.3 (0.75)	Fish meal, white	2.3 (0.31)		
Lucerne hay	1.7 (0.27)	Groundnut meal	3.5 (0.21)		
Lucerne silage	1.8 (0.23)	Linseed meal	5.4 (0.09)		
Maize silage	2.2 (0.69)	Maize-germ meal	2.1 (0.65)		
Fodder beet	1.6 (0.30)	Palm-kernel meal	3.0 (0.41)		
Swedes	1.1 (0.07)	Rapeseed meal	4.4 (0.53)		
		Soybean meal	3.0 (0.23)		
		Sunflower meal	5.8 (0.49)		

[a]Barley-based.

concentrations in legume (alfalfa) than grass hays and in legume (clover) than grass silages, while conserved grass generally contains more magnesium than fresh grass (Table 5.1). As far as other roughages are concerned, straws are low in magnesium, while root and leafy brassica crops are similar to grasses (Table 5.1).

Concentrates

The common components of concentrates used in animal feeding vary widely in magnesium content. Generally, cereal grains are low (1.1–1.3g kg^{-1} DM), oilseed meals high (3.0–5.8g kg^{-1} DM) and fish meals intermediate (1.7–2.5g kg^{-1} DM) in magnesium concentration (MAFF, 1990). Animal products vary greatly in magnesium content, with the level increasing in proportion to bone content from 0.4g kg^{-1} DM in pure meat meal to 2g kg^{-1} DM in the average meat-and-bone meal. However, these are no longer used in Europe as protein supplements.

Absorption of Magnesium from the Rumen

The rumen is an important site for A_{Mg} in sheep (Tomas and Potter, 1976b; Field and Munro, 1977) and cattle (Greene et al., 1983b; Khorasani et al., 1997), and metabolic events in the rumen largely determine the 'magnesium value' of feeds. Magnesium is absorbed by a combination of passive and active transport processes. The primary process is normally passive and begins at the apical membrane of the rumen mucosa, where magnesium uptake is driven by the negative potential difference (PD_a) and inhibited by high lumenal potassium concentrations. An active carrier-mediated process – possibly involving a magnesium and hydrogen ion exchanger and insensitive to potassium – becomes the dominant process at high luminal magnesium concentrations (Martens and Schweigel, 2000). A_{Mg} is completed by a secondary active process, located in the basolateral membrane, that is saturable and controls efflux to the bloodstream (Dua and Care, 1995). Within a given species, the main influences on A_{Mg} are factors that affect the soluble magnesium concentration in the rumen and the PD across the whole rumen mucosa (PD_t).

Soluble magnesium and pH

In sheep, magnesium uptake from the rumen increases as rumen magnesium concentrations increase. At high luminal concentrations (>3–4 mmol l⁻¹) (Martens and Schweigel, 2000) magnesium is passively absorbed and does not affect PD_t (Jittakhot et al., 2004b). Rumen pH markedly influences the solubility of magnesium. Ultrafilterable magnesium in rumen liquor decreases from around 6.0 to <0.5 mmol l⁻¹ in cows as rumen pH increase from 5.6 to 7.2 (Johnson et al., 1988), with values 1.0–1.5 mmol l⁻¹ lower on grass than on hay with concentrate diets for a given pH. In sheep, rumen solubility of magnesium decreases from 80% to 20%, approximately, when pH is increased in vitro from 5 to 7 (Dalley et al., 1997b). However, rumen pH can fall to remarkably low levels (about pH 5) in animals grazing spring pasture (A. Sykes, personal communication, 2008).

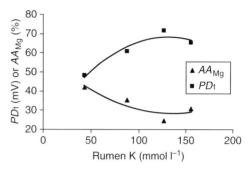

Fig. 5.2. Increasing the potassium concentration in rumen liquor by adding potassium bicarbonate (KHCO₃) to the concentrate diet of sheep causes the potential difference across the rumen mucosa (PD_t) to increase and the apparent absorption of magnesium (AA_{Mg}) to decrease. The curvilinear relationships are mirror images of each other and reflect the negative relationship between PD_t and A_{Mg} from the rumen (data from Jittakhot et al., 2004b).

Antagonism from potassium

Infusions of potassium into the rumen or perfusions of the isolated rumen mucosa decrease PD_a (i.e. depolarize the surface) but increase PD_t. These relationships between lumenal potassium and PD are logarithmic, and uptake of magnesium from the rumen also decreases logarithmically as potassium concentrations increase (Care et al., 1984; Martens and Schweigel, 2000). Dietary potassium supplements thus decrease apparent A_{Mg} (AA_{Mg}) (Fig. 5.2) unless magnesium intakes are very high and the potassium-independent pathway becomes dominant (Fig. 5.3). Potassium-rich diets may also reduce A_{Mg} by increasing the rumen volume and rate of outflow of digesta from the rumen (Dalley et al., 1997a; Ram et al., 1998). Increasing the proportion of roughage in the diet has similar effects that are not wholly attributable to the associated change in dietary potassium (Schonewille et al., 2002). Significant amounts of potassium normally enter the rumen in saliva (see Chapter 7).

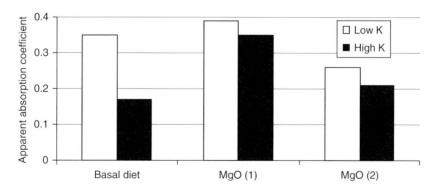

Fig. 5.3. Contrasting effects of potassium on magnesium absorption from the basal diet and an inorganic magnesium source in sheep (from Ram et al., 1998). The coefficient is calculated from the increments in Mg absorption expressed as a fraction of the increments in Mg intake from magnesium oxide (MgO) at each of two levels: (1) approximately 2.6 g day⁻¹ and (2) 3.4 g day⁻¹.

Effects of dietary sodium and salivary flow

Salivary and rumen potassium concentrations increase in sheep deprived of sodium in an aldosterone-mediated response (see Chapter 7), and magnesium absorption is decreased by inhibition of the potassium-sensitive passive absorption pathway (Martens et al., 1987). An independent effect of sodium on magnesium uptake has been ruled out by in vitro studies that showed that the substitution of lithium for sodium did not affect magnesium transport (Martens and Blume, 1986). The AA_{Mg} from dried grass can be markedly increased (from 0.02 to 0.35) by continuously infusing water into the ovine rumen (10 l day^{-1}) (Suttle and Field, 1966) and there are parallel increases in the urinary excretion of potassium and magnesium. It is tempting to conclude that the copious salivation normally induced during the consumption of dried grass is reduced by water infusion, lowering the amounts of potassium recycled to the rumen and enhancing AA_{Mg}. Succulent forages induce post-prandial salivary flow and the effect of feed type on the diurnal pattern of salivary potassium flow might influence magnesium absorption.

Effects of anions

Adding potassium as the acetate rather than the chloride to the diet of sheep appears to decrease AA_{Mg} (Suttle and Field, 1967) and the addition of anions to the diet of cows increases AA_{Mg} (Roche et al., 2003). In vitro, sulfate is a more potent inhibitor of magnesium uptake across the rumen epithelium than chloride, due possibly to a rise in lumenal pH (Martens and Blume, 1986). Potassium given as the bicarbonate can raise the rumen pH by 15% (Wylie et al., 1985; Ram et al., 1998) and inhibits AA_{Mg} more than potassium given as potassium chloride (KCl) (Schonewille et al., 1997a), but to the same extent as potassium in dried grass (Schonewille et al., 1999a) or given as citrate (Schonewille et al., 1999b). Studies of the antagonism between potassium and magnesium that used KCl may have underestimated the antagonism from herbage potassium.

Effects of dietary protein and higher fatty acids

Early Dutch studies suggested specific inhibitory effects of high crude protein and fatty acid concentrations in spring pasture on magnesium absorption. Transfer of ruminants from diets of hay plus concentrates to grass or grass products can raise rumen ammonia concentrations as well as pH (Johnson et al., 1988; Johnson and Aubrey Jones, 1989) and infusion of ammonium chloride (NH$_4$Cl) into the rumen can lower A_{Mg} without affecting PD_t (Care et al., 1984). However, the effects are inconsistent (Gabel and Martens, 1986) and transient (Martens and Schweigel, 2000). High levels of fatty acids were thought to lower the A_{Mg} by forming insoluble magnesium soaps in the rumen, which were excreted in the faeces, and A_{Mg} from the rumen can be decreased by adding fat to the diet (Kemp et al., 1966; Rahnema et al., 1994), but it is calcium rather than magnesium soaps that form at high mineral intakes (Ferlay and Doreau, 1995). Soaps are partially broken down in the intestine, which would release some of the magnesium that might have bypassed the rumen as magnesium soaps. Crude protein and higher levels of fatty acids are but two of many correlated abnormalities in spring pasture, the effects of which are impossible to separate from those of potassium.

Secretion and Absorption of Magnesium Beyond the Rumen

Fine contrasts in magnesium fluxes across the hind gut mucosa are found between cattle and sheep, adding to the weight of evidence that results for one species can no longer be assumed to apply to others. There is usually some net A_{Mg} from the hind gut of sheep (Tomas and Potter, 1976a; Dalley et al., 1997a, 1989; Bell et al., 2008), but the contribution to total A_{Mg} is small when the AA_{Mg} coefficient is high (0.83) (Field and Munro, 1977). If potassium has decreased A_{Mg} from the rumen, there can be a substantial compensatory increase in A_{Mg} from the hind gut on magnesium-rich diets (2.5 or 3.4 g kg^{-1} DM) (Dalley et al., 1997a), but not on diets with less magnesium (1 or 2 g kg^{-1} DM) (Greene et al.,

1983a). Post-ruminal A_{Mg} is higher from forages fed once daily rather than continuously (Grace and MacRae, 1972) and this may reflect a constraining effect of the rate of A_{Mg} from the rumen. Hind-gut A_{Mg} allows a source of low solubility in the rumen (magnesium-mica) to equal rumen-soluble sources such as magnesium hydroxide $(Mg(OH)_2)$ in terms of total tract A_{Mg} in sheep (Hurley et al., 1990), but the latter may be influenced by the pH of the caecal digesta (Dalley et al., 1997b). By contrast, there is a net *loss* of magnesium from the hind gut in forage-fed cattle (Bell et al., 2008). Losses via secretions into the hind gut may be greater and A_{Mg} may be lower in cattle than in sheep because concentrations of soluble magnesium in the hind gut are lower, with a higher proportion of ingested magnesium having been absorbed from the rumen. Infusion of potassium into the abomasum or ileum has no effect on A_{Mg} in sheep (Wylie et al., 1985) or cattle (Greene et al., 1983b).

The low AA_{Mg} coefficients from forages high in potassium and from artificially dried grass in both sheep and cattle (0.02–0.12) indicate either limited scope for compensatory hind gut absorption from forage or exceptionally high faecal endogenous losses (FE_{Mg}). However, models of magnesium metabolism that allow for hindgut absorption improve the prediction of results from balance studies in potassium-supplemented sheep (Robson et al., 1997) and the rate of development of hypomagnesaemia in cattle on an artificial low-magnesium diet (Robson et al., 2004). Absorption can occur from the bovine hind gut because the rectal perfusion of magnesium prevents hypomagnesaemia in cattle (Bacon et al., 1990).

Techniques for Measuring Magnesium Availability

The nutritive value of magnesium in feeds can be assessed in various ways.

Apparent and true absorption

AA_{Mg} from feeds varies widely (ARC, 1980; Henry and Benz, 1995), but is an attribute of the feed rather than the animal and is conventionally expressed as a fraction of intake. That convention has been questioned following a study in which AA_{Mg} varied with supplementary magnesium intake (Fig. 5.3), but intakes were taken to levels never found in natural feeds. Although rarely measured, A_{Mg} can be estimated indirectly from balance data by correcting for a constant FE_{Mg} because the faeces is not a major route of excretion for magnesium absorbed in excess of need.

Urinary magnesium excretion

Magnesium is less readily filtered at the glomerulus than most macro-minerals, but sufficient amounts are filtered and escape tubular reabsorption – once the renal threshold of 0.92 mmol l^{-1} is exceeded – to allow urine to be the major route of excretion (Ebel and Günther, 1980). The renal processes have been reviewed by Martens and Schweigel (2000), who confirmed the linear relationship between magnesium intake above requirement and urinary excretion in sheep, a relationship that has been taken to indicate A_{Mg} in mature ruminants (Fig. 5.4).

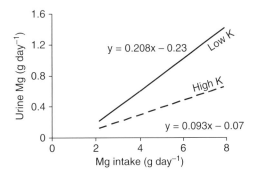

Fig. 5.4. The slope of relationships between the intake and urinary excretion of magnesium gives an approximate measure of the efficiency with which magnesium is absorbed (A_{Mg}): in this instance involving identical twin dairy heifers given diets high (broken line) or low in potassium (K; solid line; 20 or 40 g K kg^{-1} DM), addition of K reduces A_{Mg} from 20.8 to 9.3% (data from Field and Suttle, 1978).

Individual variation

Wide, repeatable individual variation in magnesium metabolism has often been reported within small groups and must be minimized by appropriate experimental designs (e.g. Latin square) (Schonewille et al., 1999) to obtain significant treatment effects. Rates of magnesium transport across the isolated rumen epithelium differ between breeds (Martens and Schweigel, 2000), but individuality is seen in sheep fed by abomasal infusion (Baker et al., 1979). Mean estimates of 'magnesium value' obtained with small groups of animals are unlikely to be representative of those for wider populations.

Availability of Magnesium in Feeds for Ruminants

Effects of potassium

Important effects of feed type and ruminant species on the antagonism between potassium and magnesium have been revealed by adhering to convention and comparing the linear relationships between fractional AA_{Mg} and dietary potassium using pooled data from the literature (Fig. 5.5) (Adediji and Suttle, 1999). As potassium concentrations increase, AA_{Mg} decreases in cattle and sheep, whether forage or mixed diets are fed. However, the magnitude of the effect varied: for each increase of 1% in forage potassium, the decrease in AA_{Mg} was maximal in sheep given forage (−0.112) and minimal in cattle given forage (−0.016). Subsequent studies have generally supported the strength of antagonism shown in Fig. 5.5. When dried grasses relatively low or high in potassium (26 or 43 g kg^{-1} DM) were fed to cattle, AA_{Mg} was 0.11 and 0.02, respectively, and the addition of extrinsic potassium (as bicarbonate (HCO_3)) to the low-potassium forage mimicked the effect of intrinsic potassium (Schonewille et al., 1999). Although AA_{Mg} was not correlated with potassium in six silages given to dry cows, all were relatively high in potassium (31–47 g kg^{-1} DM) (Schonewille et al., 1997a) and the range of AA_{Mg} (0.07–0.19) did not deviate markedly from the relationship shown for the forage-fed bovine (Fig. 5.5d). The low range of AA_{Mg} values recorded for dried grasses given to sheep (0.02–0.12) (Suttle and Field, 1966; Adam et al., 1996; Schonewille et al., 2002) is similar to that found in cattle. In Fig. 5.3, potassium (as HCO_3) reduced AA_{Mg} in the mixed, dry basal diet by 50%. Although dietary potassium is not the only determinant, AA_{Mg} is low in all types of forage at potassium concentrations commonly encountered by ruminants.

Interactions of potassium with animal species

When potassium is low (10 g kg^{-1} DM), sheep generally absorb forage magnesium three times more efficiently than cattle (compare Fig. 5.5a with 5.5c). When concentrations are high (50 g kg^{-1} DM), however, the average AA_{Mg} for sheep and cattle is equally low (about 0.10). The contrasting slopes suggest that AA_{Mg} is more sensitive to antagonism from potassium in sheep than in cattle. However, the lower intercept for the cattle regression might be attributable to a strong antagonism at low potassium concentrations.

The differences in AA_{Mg} between cattle and sheep, evident in Fig. 5.5, probably emanate from the rumen. The magnesium concentration required to saturate absorption from the rumen is three times greater in cattle than in sheep (12.5 versus 4 mmol l^{-1}) (Martens, 1983) and this may affect the relative importance of the rumen and hence the antagonism between potassium and magnesium in the two species. The decrease in AA_{Mg} from 0.30 to 0.22 recorded in goats when dietary potassium was increased from 7.8 to 34 g kg^{-1} DM (Schonewille et al., 1997b) sits more comfortably among data for sheep than for cattle data on mixed diets (Fig. 5.5b versus 5.5d).

Interactions of potassium and diet type

When mixed, dry (R/C) rations low in potassium were fed to sheep in the study shown in Fig. 5.5, AA_{Mg} was lower than in sheep fed forages that were equally low in potassium. In cattle, however AA_{Mg} was higher from R/C

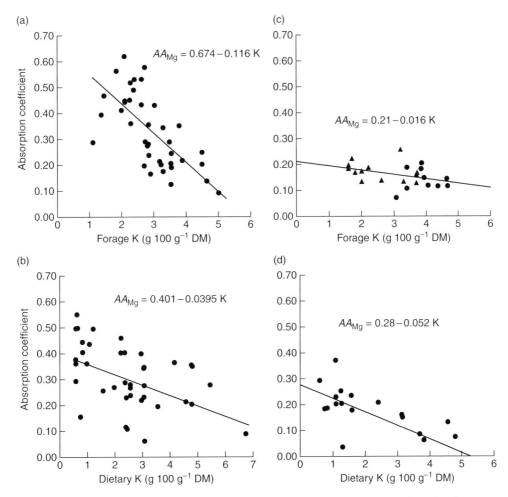

Fig. 5.5. Magnesium absorbability is inversely related to dietary potassium concentrations, but the interaction is species- and diet-dependent (from Adediji and Suttle, 1999). (a) Sheep given fresh grasses; (b) sheep given roughage plus concentrates; (c) cattle given hays or grasses; and (d) cattle given hay plus concentrates.

diets. The inclusion of concentrates attenuated the potassium antagonism in sheep but exacerbated it in cattle, leaving AA_{Mg} equally sensitive to increases in potassium in the two species on R/C diets (compare Figs 5.5b and d). Two similar regressions for cattle have since been reported, using pooled data wholly (Weiss 2004) and largely (Schonewille et al., 2008) for animals fed R/C diets at a single laboratory. The implied interactions between feed type and potassium are confirmed by other studies. When the potassium concentration in a cattle ration based on dried grass was reduced by increasing the proportion of concentrate from 0% to 60% (Schonewille et al., 2002), the change in AA_{Mg} (per 10 g K kg^{-1} DM) was 0.0695, 30% higher than in Fig. 5.5d. In goats, the substitution of starch-rich concentrate for roughage enhanced AA_{Mg} independently, to some degree, of its potassium-lowering effect (Schonewille et al., 1997b). The AA_{Mg} of whole-crop cereal silages (i.e. a mixed although succulent diet) tended to be lower (0.17–0.21) than that of alfalfa silage (0.24) in 50:50 mixtures with a concentrate that was relatively low in potassium (<16 g kg^{-1} DM)

(Khorasani et al., 1997). Variability in Fig. 5.5b and Fig. 5.5d may therefore be partly due to differences in the proportion of concentrate as well as magnesium concentration in the mixed ration (Ram et al., 1998).

Other dietary influences on 'magnesium value'

High intakes of calcium and phosphorus may suppress A_{Mg} (Littledike and Goff, 1987) and a daily supplement of 150 g calcium carbonate ($CaCO_3$) for 4 days post-calving has been found to have a hypomagnesaemic effect in cows (Roche et al., 2002), but Weiss (2004) found no evidence for antagonism from calcium in his cattle studies. In sheep, the effects of both calcium and phosphorus additions on AA_{Mg} were small (Chicco et al., 1973). The addition of the ionophore monensin to concentrate-based rations for sheep and cattle usually increases A_{Mg}, chiefly from the rumen, by about 10% (Kirk et al., 1994).

Genetic variation

Studies with identical twin cows have shown that variation in A_{Mg} is partly heritable (Field and Suttle, 1979). Repeatable differences in milk magnesium concentrations have been reported between individual cows and these may reflect genetic differences and affect individual susceptibility to grass tetany (Thielen et al., 2001). In a comparison of beef breeds on a common diet, Aberdeen Angus were found to have relatively low serum magnesium concentrations (Littledike et al., 1995). Marked bovine breed differences in magnesium metabolism have been linked to differences in susceptibility to grass tetany (Greene et al., 1986), but the finding requires confirmation in view of the small numbers representing each breed or cross ($n = 5$) and inconsistent rankings for crosses. Rates of magnesium transport across the isolated ovine rumen epithelium differ between breeds (Martens and Schweiger, 2000), but individuality is also seen in sheep fed by abomasal infusion (Baker et al., 1979). The effects of sire on urinary magnesium excretion in lambs provide confirmation of the genetic control of magnesium metabolism (Field et al., 1986).

Magnesium Availability in Other Species

Knowledge of A_{Mg} is meagre for pigs and poultry, but values are much higher than those for ruminants. Employing a combination of comparative balance and isotope dilution techniques, Güenter and Sell (1974) estimated that A_{Mg} in chickens was 0.55–0.58 from barley, maize and wheat, 0.83 from oats, 0.60 from soybean meal, 0.62 from dried skimmed milk and 0.43 from rice. In monogastric species magnesium is absorbed mostly from the small and large intestine (Güenter and Sell, 1973; Partridge, 1978). Guinea pigs can absorb magnesium from grass and alfalfa hays more than twice as efficiently as lambs given the same hays (Reid et al., 1978). The horse, another hind-gut fermenter, also absorbs magnesium well from grass. The mean (standard error) AA_{Mg} values from cut grass given to Thoroughbred yearlings and lactating mares were 0.50 (0.035) and 0.63 (0.04), respectively, three to four times higher than in cattle. These values were significantly reduced in yearlings to 0.38 (0.04) by large $CaCO_3$ supplements (Grace et al., 2003).

Magnesium Requirements

Net requirement for maintenance

The faeces is not a major route of excretion for magnesium absorbed in excess of need, but the average FE_{Mg} of 3 mg kg^{-1} live weight (LW) estimated by ARC (1980) is relatively high in proportion to the total body content of magnesium when compared to the respective numbers for calcium or minimum phosphorus loss. Reanalysis of published data has revealed a linear relationship between FE_{Mg} and plasma magnesium concentration, with values declining to approximately 1 mg kg^{-1} LW in the severely hypomagnesaemic individual (Bell et al., 2008). Allowing for the reduction in FE_{Mg} during magnesium deprivation improved a model for predicting the final outcome of experimental magnesium depletion in cattle (Bell et al., 2008). The amounts of magnesium secreted in saliva in sheep given fresh ryegrass or lucerne hay (0.85 and 0.30 mg kg^{-1} LW, respectively) (Grace et al., 1985) are too small to allow potassium to increase FE_{Mg} by inhibiting reabsorption from the rumen. Urinary magnesium losses are negligible at marginal magnesium intakes and relationships between urinary magnesium excretion and magnesium intake in mature cattle give no indication that more magnesium is required to initiate urinary magnesium excretion in diets high in potassium (Fig. 5.4). Indeed, FE_{Mg} at maintenance is estimated to be 2.8 and 3.6 mg kg^{-1} LW, respectively, and lower in the high-potassium group that became slightly hypomagnesaemic.

Net requirement for milk production

The magnesium concentration in milk is low (4 mmol l^{-1}) and only one-tenth that of calcium, but it is maintained during depletion and represents a continuous drain on maternal reserves. Furthermore, colostrum contains two to three times more magnesium than main milk and increases the maternal demand for magnesium around calving (Goff and Horst, 1997). Recent studies from New Zealand suggest that milk magnesium may be lower than the historic values used in factorial models (Theilin et al., 2001).

Gross requirements for sheep and cattle

The daily amounts of magnesium required vary with species, age and rate of growth or production, but the relatively low and variable efficiency with which magnesium is absorbed from foods makes A_{Mg} the primary determinant of gross need. Appropriate mean A_{Mg} values derived from Fig. 5.5 have been used to generate requirements as dietary magnesium concentrations for two levels of potassium and two different types of diet: forage or a roughage/concentrate mixture (1:1) for sheep and cattle (Tables 5.2 and 5.3, respectively). Noteworthy features of the estimates of magnesium requirement are as follows:

- cattle have higher requirements than sheep;
- cows relying on grass (e.g. in New Zealand) as opposed to housed herds (e.g. in the USA) have higher requirements;
- lactating animals have higher requirements than non-lactating animals;
- there is no increase in the required dietary magnesium concentration with increased levels of performance;
- sheep given mixed rations as opposed to forage need 50% less magnesium; and
- the average magnesium levels in UK (Table 5.1) and New Zealand pastures (Fig. 5.1) are marginal in spring.

Support for the cattle requirements is given by the fact that hypomagnesaemia can be induced with mixed diets containing 1.7 g Mg and 40 g K kg^{-1} DM (van de Braak et al., 1987). An additional 18 g Mg day^{-1} for every increase of 10 g K kg^{-1} DM has been advocated for dairy cows on mixed diets (Weiss, 2004). The requirements for the average grazing bovine in Table 5.3 cover the worst-case scenario predicted by the Dutch guidelines for potassium-rich pastures (Underwood and Suttle, 1999). It has, however, been argued that an allowance is needed to protect the most vulnerable individuals in a herd or flock (see ARC, 1980). The requirement for cellulose digestion and food intake for a semi-purified diet (30% cellulose) is 8–10 mg Mg kg^{-1} LW (Ammerman et al., 1971), equivalent to 0.5–0.8 g Mg kg^{-1} DM and high enough not to be met by some low-quality roughages.

Gross requirements for goats

The strength of a potassium–magnesium antagonism study in goats given a semi-purified diet suggests that requirements of goats may be closer to those given for sheep than those for cattle (Schonewille et al., 1997b).

Gross requirements for horses

A 200-kg horse gaining 1 kg day^{-1} has been estimated by factorial modelling to require only 0.7 g Mg kg^{-1} DM in its pasture (Grace et al., 2002b), while lactating mares require 1.7 g Mg kg^{-1} DM (Grace et al., 2002a). The models used an FE_{Mg}

Table 5.2. Factorially derived estimates of the magnesium requirements of grazing sheep.[a]

	Live weight (kg)	Growth/production (kg day^{-1})	Dietary requirement (g kg^{-1} DM)		DMI[b] (kg day^{-1})
			High K[c]	Low K[c]	
Growth	20	0.1	1.0	0.50	0.50
		0.2	0.9	0.45	0.76
	30	0.1	1.0	0.50	0.67
		0.2	0.9	0.45	1.00
		0.3	0.8	0.40	UA
	40	0.1	1.0	0.50	0.83
		0.2	0.9	0.45	1.23
		0.3	0.7	0.35	1.80
		Fetuses (−12 weeks to term)			
Pregnancy	40	1	0.9	0.45	0.64–0.96
		2	0.9	0.45	0.74–1.25
	75	1	1.1	0.55	1.03–1.51
		2	0.9	0.45	1.17–1.93
		Milk (kg day^{-1})			
Lactation	40	1	1.2	0.60	1.18
		2	1.2	0.60	1.90
	75	1	1.4	0.70	1.48
		2	1.3	0.65	2.18
		3	1.3	0.65	2.90

UA, unattainable performance on such diets.
[a] For housed sheep on mixed rations, requirements fall between those shown for high- and low-K diets.
[b] Dry matter intakes for diets of intermediate nutritive value, q = 0.6 (see ARC, 1980, which also gives the net requirements for maintenance (3 mg kg^{-1} LW), growth (450 mg kg^{-1} LWG) and milk production (170 mg kg^{-1})).
[c] Absorption coefficients were 0.20 for sheep grazing pasture high in K (40 g kg^{-1} DM) and 0.40 for pastures low in K (20 g kg^{-1} DM).

Table 5.3. Factorially derived estimates of the magnesium requirements of grazing cattle.

	Live weight (kg)	Growth/production (kg day^{-1})	Dietary requirement (g kg^{-1} DM)		DMI[a] (kg day^{-1})
			High K[b]	Low K[b]	
Growth	100	0.5	1.5	1.1	2.1
		1.0	1.4	1.1	3.2
	200	0.5	1.6	1.2	3.3
		1.0	1.4	1.1	4.7
	400	0.5	1.7	1.3	5.2
		1.0	1.4	1.1	7.3
		Fetuses (weeks to term)			
Pregnancy	600	12	2.1	1.6	5.8
		4	1.8	1.4	7.3
		Milk (kg day^{-1})			
Lactation	600	10	2.4	1.8	9.4
		20	2.1	1.7	14.0
		30	2.0	1.6	18.8

[a] Assuming a diet of moderate digestibility, with a 'q' value of 0.6 (see ARC, 1980, which also gives the net requirements for maintenance (3 mg kg^{-1} LW), growth (450 mg kg^{-1} LWG) and milk production (125 mg kg^{-1})).
[b] Based on absorption coefficients of 0.14 and 0.18 for grass high (40 g kg^{-1} DM) and low (20 g kg^{-1} DM) in K. Dutch authorities make a similar distinction between grass products and corn silage (Schonewille et al., 2008).

of 6 mg kg^{-1} LW – twice that of the ruminant – and an A_{Mg} of 0.4, *less* than the AA_{Mg} observed in the group's balance studies (0.5–0.6), and are likely to have overestimated needs that are closer to those of pigs and poultry (reported next). Natural diets should invariably meet the low magnesium needs of the weanling or mature equine. If there is a problem it might lie with the suckling foal, as mare's milk is relatively low in magnesium (2 mmol l^{-1}) (Table 2.3).

Gross requirements for pigs

The minimum magnesium requirement of baby pigs receiving a purified diet is 325 mg kg^{-1} DM (Miller *et al.*, 1965) and the NRC (1998) recommends 400 mg Mg kg^{-1} DM. Higher requirements have been reported for optimum growth in weaner pigs (400–500 mg kg^{-1} DM) (Mayo *et al.*, 1959). Little is known of the precise requirements of sows during pregnancy and lactation, except that the recommended level for baby pigs is adequate for pregnancy and 130 mg kg^{-1} DM is seriously inadequate for lactation (Harmon *et al.*, 1976). These empirical findings can be compared with factorial estimates derived by the ARC (1981). Assuming an endogenous loss of 0.4 mg Mg kg^{-1} LW, tissue accretion rates declining from 460 to 380 mg Mg kg^{-1} LW gain (LWG) and an A_{Mg} of 0.8 derived from studies with baby pigs, dietary requirement fell from 410 to 160 mg Mg kg^{-1} DM as pigs grew from 5 to 90 kg LW. The figure for older, fattening pigs that probably absorb magnesium with lower efficiency (perhaps A_{Mg} 0.6) may be nearer to 210 mg kg^{-1} DM.

Gross requirements for poultry

In growing birds, the growth rate probably affects requirements. In one study, 0.2 g Mg kg^{-1} DM sufficed for early growth in White Leghorn chicks, whereas faster growing broiler chicks needed 0.5 g kg^{-1} DM (McGillivray and Smidt, 1975). Calcium and phosphorus have small effects on the requirements of chicks (Nugara and Edwards, 1963): increasing dietary phosphorus from 3 to 9 g kg^{-1} DM and dietary calcium from 6 to 12 g kg^{-1} DM raised the requirement from 0.46 to 0.59 g Mg kg^{-1} DM, but 0.25 g Mg kg^{-1} DM was sufficient for survival and maximum growth with 6 g P kg^{-1} DM (Wu and Britton, 1974). A significant increase in weight gain to 3 weeks of age, but not to market weight, has been observed when magnesium is added to high-energy diets (McGillivray and Smidt, 1977).

The requirements of laying hens depend on the criteria of adequacy employed and the type of diet (Sell, 1979). The hierarchy of magnesium requirements is as follows: to maintain normal serum magnesium, 0.16 g kg^{-1} DM; to support maximal egg production, 0.26 g kg^{-1} DM; to obtain a normal magnesium concentration in the egg, egg weight and hatchability of chicks, 0.36 g kg^{-1} DM (Hajj and Sell, 1969). Supplemental magnesium sulfate (MgSO$_4$) at 0.05% has increased shell weight and the thickness of eggs produced by hens on a practical ration (Bastien *et al.*, 1978), but no such benefit has been observed by others (Underwood and Suttle, 1999). The recommendations of the NRC (1994) for uniformly high concentrations of 0.5 g Mg kg^{-1} DM for broilers, turkey poults and laying hens (with a food intake of 100 g day^{-1}) contain a generous safety margin. It would be unusual for poultry fed on commercial diets to need continuous supplementation with magnesium.

Biochemical Changes Associated with Magnesium Deprivation

It is helpful to consider the factors that govern plasma magnesium in the normal animal before considering the consequences of magnesium deprivation.

Plasma magnesium and its control

Plasma normally contains about 1 mmol (23 mg) Mg l^{-1} partitioned between two forms: one is tightly bound to protein (32%), the other free and ionized (68%). The concentrations are influenced principally by the A_{Mg}, unavoidable losses in secreted milk (and digestive juices) and equilibria attained with intracellular soft tissue and skeletal pools of magnesium (Martens and Schweigel, 2000). Most of the magnesium in bone is present in a 1:50 ratio with calcium in the hydration shell of hydroxyapatite crystals, is partly exchangeable and mobilized during experimental haemodialysis

(Gardner, 1973) or intravenous infusion of the calcium chelator EDTA (Robson et al., 2004).

There is pronounced seasonal cycling of bone magnesium in association with bovine lactation (Fig. 5.9), but this probably reflects bone turnover rather than magnesium depletion since plasma magnesium remains normal (Engels, 1981). The feeding of gross excesses of magnesium (24 g kg^{-1} DM) to lambs and steers has been found to increase rib magnesium from 3.6–4.1 to 5.8–6.4 g kg^{-1} DM, and there were linear relationships between plasma magnesium and magnesium intakes (Chester-Jones et al., 1989, 1990). Plasma magnesium is clearly not rigidly controlled and fasting causes a rapid fall in serum magnesium while cessation of milking causes abrupt increases (Littledike and Goff, 1987). Plasma magnesium is briefly increased by about 25% after parturition (e.g. Goff and Horst, 1997), the pattern being the mirror image of that seen in plasma calcium (Fig. 4.13), but the ionized proportion remains unchanged (Riond et al., 1995). Plasma magnesium may be influenced indirectly by the calcium-regulating hormones calcitonin and PTH, by the sodium/potassium-regulating hormone aldosterone (Littledike and Goff, 1987; Charlton and Armstrong, 1989) and by thyroxine and insulin (Ebel and Günther, 1980). The general model for biochemical responses to mineral deprivation, presented in Chapter 3, requires modification to fit the acute course of events in magnesium deprivation, but the picture is similar to that described for acute calcium deprivation (see Fig. 4.10, p. 69).

Depletion

In chronic magnesium deprivation in calves, calcium to magnesium ratios in bone increase from the normal 50:1 to 150:1, but this may be due mainly to decreased magnesium deposition in new bone (Underwood and Suttle, 1999). Modelling studies suggest that mobilization of magnesium from the bovine skeleton is so slow in the adult that it cannot significantly affect the rate at which hypomagnesaemia develops (Robson et al., 2004) and bone magnesium is not low in cases of grass tetany (Rook and Storry, 1962). Of the soft tissues examined, only kidney magnesium was reduced in sheep experimentally deprived of magnesium (McCoy et al., 2001a). However, skeletal magnesium can be mobilized by experimental procedures and inhibition of bone resorption lowers plasma magnesium (Matsui et al., 1994). In one field study, the oldest cows in a herd had the lowest plasma magnesium values (Fig. 5.6) and it was postulated that slower bone (and hence magnesium) turnover in older animals might be partly responsible. Further modelling studies have shown that small reductions in magnesium 'outflow', reducing FE_{Mg} from 3 to 1 mg kg LW^{-1}, improve the fit to experimental data, but still leave the model cow in a more hypomagnesaemic state than that observed (Bell et al., 2006). Equally small increases in magnesium 'inflow' from the skeleton may influence the onset of hypomagnesaemia and tetany.

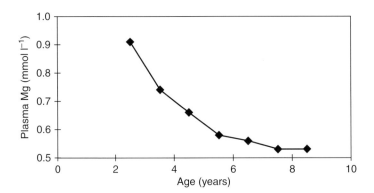

Fig. 5.6. Plasma magnesium concentrations can decrease markedly with age in a beef suckler herd, but cumulative nutritional deficits may contribute to the apparent 'age' effect (from Suttle et al., 1980).

Deficiency

Adult sheep and cattle, whether non-lactating or lactating (Suttle and Field, 1969; McCoy et al., 2001a,b), show a rapid fall in plasma magnesium when they are fed low diets low in magnesium, with the rate of fall reflecting the dietary magnesium concentration (Fig. 5.7). Urinary magnesium losses fall close to zero and losses via secretions into the gut (FE_{Mg}) follow the decline in plasma magnesium (Bell et al., 2008). Magnesium concentrations in the CSF also decline, but at a much slower rate (Pauli and Allsop, 1974), suggesting an active transport mechanism at the blood–brain barrier. There are probably gradual reductions in magnesium concentrations in other body fluids, such as vitreous humour in sheep (McCoy et al., 2001a) and cattle (McCoy et al., 2001b), but not in milk (Rook and Storry, 1962; McCoy et al., 2001c). In the early stages of magnesium deprivation, hypomagnesaemia is not accompanied by hypocalcaemia in sheep, whether dry or lactating ewes (Suttle and Field, 1969; McCoy et al., 2001a).

Dysfunction

Mild behavioural changes in the severely hypomagnesaemic individual indicate that neuromuscular activity begins to increase when magnesium concentrations in the CSF fall below 0.6 mmol l^{-1}, but the transition from deficiency to disorder is often so rapid in the adult that a period of dysfunction is never seen. Severe hypomagnesaemia can delay rumen contractions in response to feeding and impair motility in the lower intestinal region (Bueno et al., 1980), and this may explain the loss of appetite seen in some ewes (Suttle and Field, 1969) and cows (van de Braak et al., 1987; McCoy et al., 2001c) during magnesium deprivation. Loss of appetite lowers magnesium intake further, thus exacerbating hypomagnesaemia (Herd, 1967). Adult sheep given a grossly deficient diet showed an immediate, marked loss of appetite due to rumen dysfunction, but the effect was delayed and less pronounced in young lambs (Chicco et al., 1972). In parallel studies with non-lactating and lactating ewes, those individuals that subsequently developed tetany when deprived of magnesium showed a more rapid fall in plasma magnesium than those that did not succumb, and they became hypocalcaemic (McCoy et al., 2001a). Tetanic individuals were distinguished more by the rapid development of hypocalcaemia, 48 h before the onset of disorder, rather than the severity of hypomagnesaemia. Severe hypomagnesaemia inhibits the customary PTH response to mild hypocalcaemia, but the inappetence associated with severe hypomagnesaemia may also reduce responsiveness to PTH secretion (Littledike and Goff, 1987). In experimentally depleted cattle, haematocrit, white cell and platelet counts increased between 11 and 26 days, but red cell fragility was not compromised (McCoy et al., 2002). Decreases in plasma triglycerides and increases in tissue iron (in bovine heart and ovine liver) were also reported, but dietary effects – other than magnesium deprivation – and inappetence may have been partly responsible.

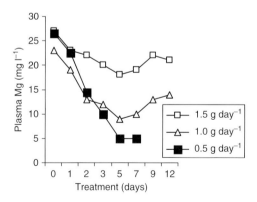

Fig. 5.7. Effect of magnesium intake (providing 0.5, 1.0 or 1.5 g Mg day^{-1}) on plasma Mg in dry ewes given a semi-purified diet, high in potassium (providing 47 g K day^{-1}; Suttle and Field, 1967): note the tendency for plasma Mg to increase after day 5, due possibly to mobilization of skeletal Mg reserves.

Changes in non-ruminants

Magnesium deprivation causes a number of biochemical changes in laboratory animals, including a decline in hepatic oxidative phosphorylation, increased prostanoid synthesis, reduction in blood pressure and body temperature and decreased thiamine concentrations in the tissues (Shils, 1997). Severe, prolonged magnesium deprivation causes an acute inflammatory response, hypertriglyceridaemia, hypercholesterolaemia and lipid peroxidation in rats, none of which has been evident in severely deprived cows (McCoy et al., 2002).

Clinical Signs of Magnesium Deprivation

Non-ruminants chronically deprived of magnesium eventually develop a wide range of clinical disorders, including retarded growth, peripheral vasodilatation, calcification of the kidneys and anorexia, that reflect the diverse functions of the element. These abnormalities are rarely seen in ruminants because they are pre-empted by an acute disorder.

Tetany in cattle and sheep

The initial signs of hypomagnesaemic tetany in cattle are those of nervous apprehension with the head held high, ears pricked and flicked, eyes staring and a readiness to kick out. Movements are stiff and stilted. Animals stagger when walking and there is a twitching of the muscles, especially of the face and ears. Within a few hours or days, extreme excitement and violent convulsions may develop, leaving the animal flat on its side and 'pedalling' its forelegs. Death usually occurs during one of the convulsions or after the animal has passed into a coma. The presence of a scuffed arc of pasture by the feet of a dead animal is an important diagnostic feature. Treatment with magnesium results in prompt recovery and confirms the diagnosis, but mortality among untreated clinical cases is high (30%). Pre-convulsive clinical signs in sheep are less clearly defined than in cattle and can be confounded with those of hypocalcaemia or pregnancy toxaemia. Affected ewes breathe rapidly and their facial muscles tremble. Some ewes cannot move, while others move with a stiff, awkward gait. Eventually they collapse and show repeated tetanic spasms with the legs rigidly extended.

Subclinical disorder

Magnesium deprivation may cause loss of production and signs of 'irritability' in a herd without causing overt clinical disease (Whittaker and Kelly, 1982). Moderate hypomagnesaemia is present in affected herds and the annual cost to the small dairy industry in Northern Ireland was once estimated to be £4 million (McCoy et al., 1993). Small improvements in milk-fat yield occurred in some grazing herds given magnesium supplements of $10 g day^{-1}$. Improvements were most noticeable in early lactation in underfed herds (Young et al., 1981) and may have been partly attributable to improved digestibility of low-quality forage (Wilson, 1980). No comparable subclinical disorder has been reported in sheep. Dairy cows on grain-based diets often produce milk with unacceptably low fat concentrations and supplements of calcined (c) magnesium oxide (cMgO) can remedy the problem (Xin et al., 1989). However, other minerals – some with buffering capacity (e.g. sodium bicarbonate, $NaHCO_3$) and others with none (e.g. sodium chloride, NaCl) – can be equally effective and 'low milk fat' syndrome is not a specific consequence of magnesium deprivation.

Occurrence of Hypomagnesaemia and Tetany

The overall incidence of hypomagnesaemic tetany in dairy cows in New Zealand, the USA and Europe is 1–4% (Cairney, 1964; Grunes et al., 1970; Whittaker and Kelly, 1982), but the incidence can be as high as 20–30% in individual herds and the mortality rate amongst clinical cases can be high (25% in beef cows, 55% in dairy cows) (Harris et al., 1983). The incidence of both clinical and subclinical disorder varies between breeds (Greene et al., 1985), seasons (Fig. 5.8) (McCoy et al., 1993, 1996) and years (Whittaker and Kelly, 1982), and may have declined in the 1990s in the UK (Fig. 5.8). Incidence of hypomagnesaemia was far higher in beef than dairy herds in North ern Ireland (McCoy et al., 1993) and in non-lactating than lactating cows in the UK

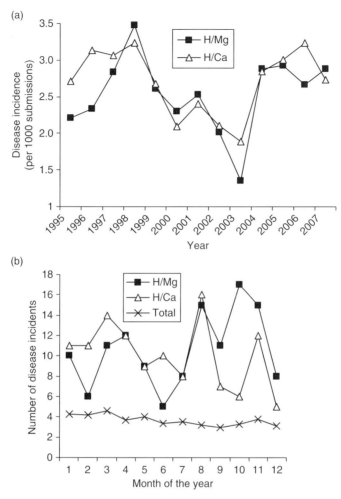

Fig. 5.8. Annual relative (a) or actual monthly incidence (b) of 'hypomagnesaemia' (H/Mg) and 'hypocalcaemia' (H/Ca) in UK cattle, between 1995 and 2007 (a) or in 2007 (b). In (a), results are expressed relative the total number of submissions for diseases of all types in an attempt to minimize confounded effects such as major disease outbreaks (e.g. foot-and-mouth disease in 1998) and structural changes in national disease surveillance; in (b), total monthly submissions (x) are given in thousands and show hypomagnesemia to reach a peak in autumn and early winter.

(Whittaker and Kelly, 1982) because beef cows and dry cows receive less concentrates and magnesium supplements. In New Zealand, where cows are pastured throughout the year, the disease occurs most frequently in late winter and early spring when almost half the herds in some areas can be hypomagnesaemic (Young et al., 1981). Outwintered beef cattle, often given low-magnesium diets containing a large proportion of straw, can develop hypomagnesaemic tetany (Menzies et al., 1994). The disease can occur when early growths of cereals are grazed (e.g. 'wheat' staggers; Littledike and Goff, 1987), and can occasionally affect stall-fed animals. In sheep, the disorder is most likely to develop following the onset of lactation, ewes bearing twins being particularly vulnerable (Sansom et al., 1982) and currently 20–40 outbreaks are reported annually in the UK. The following factors that predispose cattle to clinical disease also apply to sheep and to subclinical disorder in cattle.

Potassium fertilizers

Numerous studies have associated the incidence of tetany with heavy use of fertilizers. In an early large Dutch study, the incidence of tetany was 0.5% on pastures low in potassium and 4.3–6.5% on pastures treated with excess potassium, nitrogen (50 kg ha^{-1}) or both ('t Hart and Kemp, 1956). Such treatments may decrease herbage magnesium, especially if the legume content of the sward is suppressed, and will increase pasture potassium, thus lowering A_{Mg} (Fig. 5.5). Dutch workers predicted the risk of hypomagnesaemia from a relationship between plasma magnesium and the product of potassium and crude protein in pasture, but the basis for the relationship was unclear (Underwood and Suttle, 1999). High dietary potassium and nitrogen have also been implicated in 'wheat staggers' (Mayland et al., 1976). However, the two factors are usually confounded and potassium is probably the more influential. There is no evidence that nitrogen-rich diets per se are a risk factor, whereas hypomagnesaemia and tetany have been induced by simply increasing potassium intakes in sheep (Suttle and Field, 1969) and cattle (Bohman et al., 1969). High rates of potassium fertilizer application (150 or 225 kg ha^{-1} as KCl) have decreased urinary magnesium excretion in cows (Roche et al., 2002). Furthermore, potassium lowers the responsiveness of bone to PTH and hence the rate of release of magnesium from the skeleton.

Weather

A sudden rise in grass temperature in spring is a risk factor for grass tetany because pasture potassium can rise rapidly as soil temperature rises (Dijkshoorn and 't Hart, 1957; cited by Kemp et al., 1961) and allows rapid grass growth. Outbreaks can be precipitated in outwintered beef cattle when adverse weather conditions disrupt grazing and feeding (Menzies et al., 1994; see Fig. 5.8). In Australia, outbreaks in breeding ewes have been associated with periods of rapid winter growth of pastures (Blumer et al., 1939). In the UK, extreme spring weather conditions – either mild and wet or cold, with snowfall – have caused outbreaks of hypomagnesaemic tetany.

Dietary change to pasturage

There is a tendency for seasonal increase in the incidence of grass tetany herds in spring (Fig. 5.8) associated with the transition from winter housing to spring pastures. The change of diet is accompanied by many marked changes in dietary composition and rumen characteristics, of which potassium is but one, and most of the factors that change have been proposed as risk factors at some time (Underwood and Suttle, 1999). For example, pH and ammonia ion (NH_4^+) concentrations rise and the solubility of magnesium decreases in the rumen (Martens and Schweigel, 2000). Rumen composition gradually returns towards normal (Johnson et al., 1988, 1989) as both the rumen microflora and host adjust to the new diet, and there is usually gradual recovery in plasma magnesium in the weeks following turnout. However, there are several points to note:

- Changes in composition between winter and spring diets are usually confounded with abrupt increases in potassium intake and its manifold effects, including those on sodium (see Chapter 8).
- Higher intakes of digestible organic matter can increase milk yield and hence the demand for magnesium.
- Spring pasture may be carrying newly hatched nematode larvae that invade the gut of the immunosuppressed lactating ewe, causing sudden increases in salivary potassium.

Thus, many factors may contribute to the risk associated with dietary change, but specific contributions are hard to define in the face of the large influence of potassium.

Dietary change at parturition

The housed dairy cow experiences a dietary change at calving and transition diets (i.e. covering the pre- to post-calving period) with a low DCAD, fed for the primary purpose of

Table 5.4. Close similarity between the causative or associated factors in hypomagnesaemic tetany (HMT) and hypocalcaemia (HC) in cows and ewes suggests a common aetiology with poor mobilization of calcium *and* magnesium from the skeleton playing a unifying role. Proximity to parturition promotes HC in cows but not in twin-bearing ewes, whose demand for both minerals is already high. Preventive measures for one disease should work for both.

	Plasma Ca	Plasma Mg	Dietary K	Air temperature	Parturition
HMT	↓	↓	↑	↓	↓
HC	↓	↓	↑	↓	↑

preventing milk fever, raise plasma magnesium in early lactation (Roche et al., 2005). However, the effect is once again confounded with potassium, the major contributor of cations. Conditions that increase the risk of milk fever in herds at pasture (Roche and Berry, 2006) also increase the risk of grass tetany and the incidences of the two disorders may be related (Table 5.4; Fig. 5.8a).

Transport

Outbreaks occur when cattle and sheep are subjected to intense excitement, as in transit or sale-yard conditions, giving rise to the term 'transit tetany'. Lack of food and the effect of stressors on the distribution of magnesium between extracellular and intracellular pools (Martens and Schweigel, 2000) may compound the problem.

Diagnosis

Diagnoses of hypomagnesaemic tetany are made instantaneously on the basis of presenting signs and are confirmed initially by response to magnesium treatment and later by hypomagnesaemia in samples taken prior to treatment. Although tetany can occur when serum magnesium falls below 0.5 mmol (12 mg) l^{-1}, individual ewes and cows can survive with values <0.2 mmol l^{-1} (McCoy et al., 2001a,b). Subclinical disease in cattle has been associated with mean serum magnesium <0.6 mmol l^{-1} and values of 0.6–0.8 mmol l^{-1} are regarded as marginal (Sutherland et al., 1986; McCoy et al., 1996); slightly lower marginal ranges are used for sheep (Table 5.2). In CSF, diagnostic thresholds of 0.5 and 0.6 mmol Mg l^{-1} are indicated for cattle (McCoy et al., 2001b) and sheep (McCoy et al., 2001a), respectively; correlations with plasma magnesium were relatively weak (McCoy et al., 2001a, 2001b). In the dead animal, magnesium in the fresh vitreous humour from the eye correlates well with that in plasma in sheep and cattle; values of <0.6 and <0.5 mmol Mg l^{-1}, respectively, in fresh samples are indicative of hypomagnesaemic tetany. Tetanic and non-tetanic cows were most clearly differentiated by lower calcium concentrations in aqueous humour (mean 0.89 versus 1.33 mmol l^{-1}) (McCoy et al., 2001c), but the distinction was less evident in ewes (McCoy et al., 2001a). Values in vitreous humour increase within 24 h of death, but the addition of a small volume (3% v/v) of 4% formaldehyde can reduce anomalies caused by bacterial growth during sample storage for 72 h (McCoy et al., 2001b). The presence of calcium to magnesium ratios of >50:1 in the rib bone indicate a probability of hypomagnesaemic tetany in sucking or milk-fed calves and lambs, but are unreliable in the lactating animal because values fluctuate in synchrony with other minerals, including phosphorus (Fig. 5.9).

Treatment

Subcutaneous injection of a single dose of 200–300 ml of a 20% solution of $MgSO_4$, or intravenous injection of a similar dose of magnesium lactate, restores the serum magnesium of an affected cow to near normal within about 10 min and is almost always followed by improvement or disappearance of the signs of tetany. Equally rapid responses have been found in artificially depleted calves given 51 mg

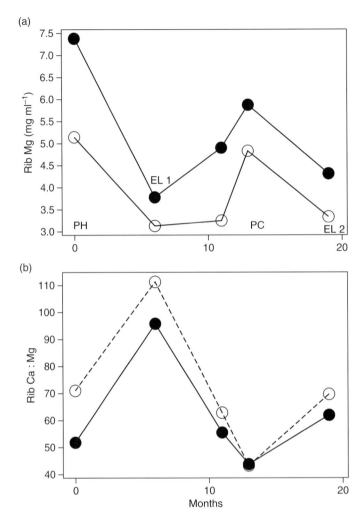

Fig. 5.9. Effects of lactation and phosphorus status on (a) rib magnesium and (b) rib calcium to magnesium ratio. Samples taken from two groups of beef cows, one receiving supplementary phosphorus (●) and the other not (○), grazing phosphorus-deficient pasture (data from Engels, 1981). PH or PC, pregnant heifer or cow; EL, early lactation.

Mg kg^{-1} LW per rectum as magnesium chloride (MgCl$_2$) (Bacon et al., 1990). Serum magnesium is likely to fall again unless the diet is immediately changed to hay and magnesium-rich concentrates, or the animal is given oral doses of magnesium. The affected ewe will respond quickly to intravenous magnesium but it is invariably given with calcium, usually as 50 ml of a 250 g l^{-1} solution of calcium borogluconate containing 25 g of magnesium hypophosphite. Calcium should also be given to the recovering cow using a dose of 500 ml to correct any hypocalcaemia.

Prevention by Direct Intervention

Prevention is far better than cure for an acute disorder that requires immediate skilful treatment. Magnesium supplies can be improved via the pasture or feed.

Magnesium fertilizers

The use of fertilizers is popular on large dairy farms in Northern Ireland (McCoy et al.,

1996), but efficacy is dependent upon soil type. When cMgO, kieserite ($MgSO_4.H_2O$) and dolomitic limestone ($CaCO_3.MgCO_3$) were compared in early studies, the last of these was found to be the least effective. The two other sources increased pasture magnesium from 2 to about 3 g kg^{-1} DM and recommended application rates were much lower than for dolomitic limestone (1125 and 5600 kg ha^{-1}, respectively) (Underwood and Suttle, 1999). Foliar dusting of pastures with fine cMgO before or during tetany-prone periods increases pasture magnesium, but the effects are short-lived regardless of the amount applied (Fig. 5.10) (Kemp, 1971). Dustings with at least 17 kg ha^{-1}, repeated after heavy rainfall and at 10-day intervals, have been recommended (Rogers, 1979). Adhesion to the leaf can be improved by applying cMgO in a slurry with bentonite. Pasture dusting is more reliable when integrated with a 'strip-grazing' system and is now the most popular preventive method in New Zealand. Spray application of $Mg(OH)_2$ in a fine suspension can produce short-term increases in herbage and serum magnesium from low application rates (5 kg Mg ha^{-1}) (Parkins and Fishwick, 1998). Foliar dusting and spraying use the pasture as a short-term carrier for the dietary magnesium supplement.

Dietary magnesium supplements

Supplementation with magnesium via concentrates is preferred by many small dairy farmers, but the historic prophylactic doses of cMgO (50–60 g (25–30 g Mg) day^{-1} for adult dairy cattle, 7–15 g day^{-1} for calves and 7 g day^{-1} for lactating ewes) were over-generous (Underwood and Suttle, 1999), being sufficient to raise urine pH (Xin et al., 1989) and possibly increase the risk of milk fever. Smaller doses (12 g Mg day^{-1} for cows) are routinely used in New Zealand and are believed to have reduced the incidence of milk fever (Roche and Berry, 2006), but chloride or sulfate salts are now preferred. Commercial sources of cMgO vary widely in origin, particle size and processing (e.g. calcination temperature) and hence in nutritive value (Xin et al., 1989; Adams et al., 1996). Coarse particle sizes (>500 μm) and low calcination temperatures (<800°C) are associated with low AA_{Mg}.

Magnesium compounds have also been incorporated into mineral mixes, drenches, salt- and molasses-based free-choice licks, and sprinkled on to feeds such as cereals, chopped roots or silage, to varying effect. The provision of magnesium-containing blocks was common amongst beef suckler herds in Northern Ireland (McCoy et al., 1996), but did not reduce the incidence of hypomagnesaemia (McCoy et al., 1996). The addition of a soluble magnesium salt to the drinking water is generally effective (see Rogers and Poole, 1976), although water intake varies widely between individuals and with weather and pasture conditions. Magnesium salts are metered into the water supply during milking in New Zealand (Dosatron, Florida, USA) during the tetany season, but this is a time of low drinking activity because of the high water intake from spring pasture. Another practice that ensures uniform individual treatment is to add magnesium (10–20 g as chloride or sulfate) to bloat control drenches, given daily from the milking platform. This method is costly in labour terms if magnesium is given to the whole herd (or flock) continuously during the tetany season.

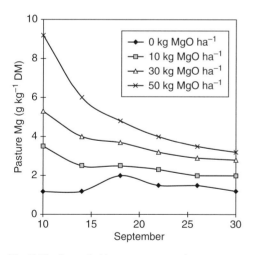

Fig. 5.10. Control of hypomagnesaemic tetany by dusting pastures with magnesite (MgO) is short-lived because the method relies on surface adherence (Kemp, 1971). Repeated applications are necessary and a single treatment with a suspension of magnesium hydroxide may be more effective.

Magnesium alloy bullets that lodge in the rumen provide insufficient daily magnesium to prevent hypomagnesaemia in dairy cows but may be more effective in the beef cow, which requires less magnesium (Stuedemann et al., 1984).

Merits of different mineral sources

Many comparisons have been made of the *relative* availability of different magnesium sources in ruminants. The oxide and hydroxide have been given low average values of 0.75 (n = 18) and 0.60 (n = 3) relative to $MgSO_4$ for sheep (Henry and Benz, 1995), but the hydroxide can be superior to MgO (Hurley et al., 1990). Rankings in individual studies may depend on the rumen solubility of a particular magnesium source. Absolute A_{Mg} values are influenced by attributes of the diet being supplemented. For example, the A_{Mg} for $MgCl_2$ is four times higher when infused into the rumen of sheep on a pelleted semi-purified diet than when chopped hay is offered (0.55 versus 0.13) (Field and Munro, 1977), and cMgO is a consistently better magnesium source for sheep when added to a maize-based rather than a dried-grass diet (Adams et al., 1996). The AA_{Mg} from a supplement such as cMgO can exceed that of a basal diet, although the advantage as well as the antagonism from potassium can be masked at high magnesium intakes in sheep (Fig. 5.3). In cattle, cMgO halves in 'magnesium value' when potassium is added to the diet (Jittakhot et al., 2004a). A source such as magnesium-mica that releases its magnesium post-ruminally may escape antagonism from potassium (Hurley et al., 1990). Magnesium phosphates can be effective, despite being rumoured as a possible source of antagonism between magnesium and phosphorus through the formation of insoluble phosphates in the gut (Underwood and Suttle, 1999). The oxide, carbonate and sulfate are all sources of highly available magnesium for broiler chickens (McGillivray and Smidt, 1975). The A_{Mg} value for reagent grade $MgSO_4$ was 0.57 in a semi-purified diet for chickens (Güenter and Sell, 1974), confirming the contrast with ruminants.

Prevention by Indirect Methods

The risk of disorder can be reduced by pre-empting the risk factors listed earlier.

Predictors

Reliable predictors of hypomagnesaemic tetany are needed and focus was initially placed on the chemical composition of pasture. Kemp and 't Hart (1957) used the ratio of potassium to calcium plus magnesium in forage as a prognostic threshold: when the ratio was >2.2, the risk of grass tetany in the Netherlands was high at 6.7%. The general predictive value of this ratio has been confirmed in many countries, but the relationship with disease incidence is linear rather than threshold in nature, imprecise and rarely used. The relationship between plasma magnesium, pasture potassium and crude protein, devised in the Netherlands (Underwood and Suttle, 1999), has never been routinely used elsewhere and emphasis has shifted to the indices of magnesium status of the animal. There is a curvilinear relationship between plasma and urinary magnesium (Fig. 5.11), and individual urinary magnesium to creatinine ratios (molar) of <0.4 are considered to be 'critical' by some European workers (van de Braak et al., 1987). In New Zealand, herds with mean values of <1.0 may benefit from supplementary magnesium (Sutherland et al., 1986). The marginal bands given in Table 5.5 are based on the above findings. A modelling approach to the prediction of hypomagnesaemia is being developed in New Zealand using, among other things, easily recorded variables such as an individual cow's milk magnesium output (Bell et al., 2008). This model has yet to be tested.

Restricting use of potassium fertilizers

Conservative use of potassium-containing fertilizers, including dairy slurry (Soder and Stout, 2003), is recommended, but there is a limit to the extent to which herbage potassium can be

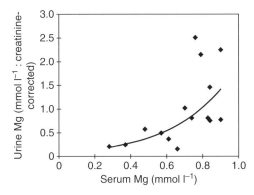

Fig. 5.11. Curvilinear relationships between two diagnostic parameters – in this case urine and plasma magnesium – will mean that each will have a different contribution to make. Urine Mg reflects the dietary supply of Mg in excess of need, while plasma Mg reflects the degree to which need is not met (data from Sutherland et al., 1986).

reduced without restricting pasture yield. The need to dispose of slurry is a constraint, but harmful effects can be restricted by limiting dispersal to fields intended for conservation or avoiding slurry-treated pastures at times of high risk. Potassic fertilizers can be similarly restricted, and potassium-rich feeds such as molasses and molassed beet pulp should not be used at vulnerable times. The requirements presented (Tables 5.2 and 5.3) suggest that the ratio of potassium to magnesium in herbage should be kept below 16 for cattle and 30 for sheep. It is interesting to note that with fertilizer applied in autumn, the potassium to magnesium ratio in the average Manitoa (New Zealand) pasture stays just above the cattle threshold (Fig. 5.2).

Managing turnout

Management of dietary change at turnout is important. The policy of gradually extending the daily grazing period to save money on supplementary feeds should reduce the rate of dietary change and hence the risk of hypomagnesaemia. If weather forecasts indicate an imminent abrupt change in temperature, wind chill or heavy rainfall, turnout should be delayed. Carbohydrate supplements lower rumen pH and improve magnesium absorption (Schonewille et al., 1997b) and may contribute to the efficacy of liquid mixtures of molasses and magnesium. The feeding of grain as a preventive measure for milk fever should also improve magnesium status, and a short period of grain feeding at times of risk in an otherwise pasture-based system may save lives and inconvenient treatments. Reductions in sodium intake at turnout should be avoided.

Transport stress

The introduction of stricter regulations governing animal transportation and lairage on welfare grounds may reduce the incidence of transport tetany in sheep and cattle. Protection may also be afforded by allowing stock access to magnesium-supplemented drinking water

Table 5.5. Marginal bands for assessing the magnesium status of ruminant livestock.

	Serum[a] (mmol l^{-1})[b]	CSF[c] (mmol l^{-1})[b]	Vitreous humour (fresh)[c] (mmol l^{-1})[b]	Urine[a] (mg creatinine^{-1}, molar ratio)	Rib (mg g^{-1} ash)[b]
Cattle	0.5–0.75	0.4–0.6	0.5–0.75	0.4–0.8	<12 (6–12 months) <7 (>36 months)
Sheep	0.6–0.75	0.6–0.8	0.4–0.8	0.4–0.8	–

CSF, cerebrospinal fluid.
[a]When mean values for sampled populations fall below these ranges, the flock or herd will probably benefit from supplementation. Values within bands indicate a possibility of current or future responses in some individuals that increases with proximity to the lower limit.
[b]Multiply values by 24.3 to convert units from mmol to mg and divide rib data by 24.3 to convert mg to mmol.
[c]When the mean post-mortem value is below the ranges given, hypomagnesaemic tetany is a probable cause of death.

for 24 h prior to transportation. This suggestion is made on the basis of evidence that pigs given access to magnesium prior to transportation exhibit attenuated stress responses, but, to the author's knowledge, comparable responses in ruminants have yet to be tested.

Genetic selection

Since magnesium metabolism and susceptibility to grass tetany are subject to genetic variation, the possibility of selecting for resistance to the disease via indices that indicate high efficiencies of magnesium utilization should be explored. (Note that a selection strategy targets the vulnerable minority of any population, whereas preventive supplementation methods involve the treatment of the whole herd or flock in perpetuity and ensures the survival of the vulnerable.) Sire effects on urinary magnesium have been found in sheep (Field et al., 1986) and ratios of magnesium to creatinine in the urine of bulls held at artificial insemination centres on a standard ration might allow the selection of resistant sires. Progress can also be made by selecting for plants rich in magnesium, and hypomagnesaemia may be controlled in sheep by grazing them on a grass cultivar high in magnesium (Moseley and Baker, 1991). The quest for low potassium genotypes that decrease the risk of milk fever should yield cultivars with a lower likelihood of causing hypomagnesaemic tetany.

Magnesium and Meat Quality

Dietary magnesium supplementation prior to slaughter can moderate some biochemical changes in pigs associated with transport stress (Apple et al., 2005) and reduce post-mortem changes in pork, associated with muscle glycolysis, that cause the meat to become pale, soft and exudative (PSE pork) (D'Souza et al., 1999). The incidence of PSE pork is a major concern amongst meat retailers and research on the efficacy of different sources of magnesium for controlling PSE has received massive support from manufacturers of 'organic' mineral supplements. A bewildering array of responses to magnesium, given in one of ten or more forms, has been reported and citation lists to papers get ever longer (e.g. Apple et al., 2005). The variability of magnesium responses is partly due to other factors that increase the incidence of PSE, such as genotype (halothane sensitivity), lengthy transportation, lack of lairage and goading. When stress between departure from the piggery and the moment of slaughter is minimal, a positive response to magnesium is unlikely.

Experimental protocols have measured so many indices of quality (or putative determinants of quality) that at least one was likely to register as 'improved', allowing a claim of efficacy (or potential efficacy) to be made for the particular magnesium source used. There is no physiological evidence to indicate that the form or route of magnesium supplementation has any effect on the glycolytic changes in muscle that cause PSE. Magnesium supplementation is only needed for a short period prior to slaughter (with 1.6g for 1 day being as effective as 3.2g for 5 days) regardless of source (sulfate, propionate or proteinate) (Hamilton et al., 2003) and may be most conveniently provided via the drinking water (Frederick et al., 2004). Supplementation with mica improved feed conversion efficiency during the early finishing phase in one pig study (Apple et al., 2005) but effects on performance have been rarely seen in other studies.

Magnesium and Heat Stress

Magnesium supplementation has decreased the effect of heat stress on 10-day-old quail, whether it was given as MgO or magnesium proteinate (Sahin et al., 2005). The authors claimed superior protection for the organic source, but there was no verification that the intended supplementation rates (1 or 2g Mg kg^{-1} DM) were attained or that the MgO was of high purity and fine particle size.

Magnesium Toxicity

The lack of homeostatic mechanisms for the control of metabolism for a given mineral is usually associated with tolerance of extreme intakes, but this is barely true of magnesium. The NRC (2005) has placed the tolerable limits of magnesium for pigs, poultry, ruminants and horses at 2.4, 5.0–7.5, 6 and 8 g kg^{-1} DM, respectively. In laying hens, 8–12 g kg^{-1} DM has reduced egg production and eggshell strength (Hess and Britton, 1997), and chicks develop diarrhoea with >30 g Mg (as MgO) kg^{-1} DM (Lee and Britton, 1987).

Ruminants

Tolerance is of most practical importance in forage-fed ruminants because they are often given large magnesium supplements to prevent deprivation. In sheep and steers given roughage/concentrate mixtures, 14 g Mg kg^{-1} DM has been found to be mildly toxic (Chester-Jones et al., 1989, 1990). The first abnormalities were loosening of the surface (stratum corneum) of rumen papillae. Higher levels (25 and 47 g kg^{-1} DM) caused severe diarrhoea and drowsiness and severe degeneration of the stratified squamous epithelium of the rumen in the steers. Early loss of appetite can occur with >13 g Mg kg^{-1} DM (Gentry et al., 1978), but this may be partly attributable to the unpalatability of the MgO used to raise magnesium intakes. Linear increases in serum magnesium with increasing magnesium intakes can take serum Mg to >3 mmol l^{-1} after 71 days' supplementation at 47 g kg^{-1} DM in steers (Chester-Jones et al., 1990). At such levels there is likely to be loss of muscle tone, deep tendon reflexes and muscle paralysis (Ebel and Günther, 1980). Increases in soft tissue magnesium were, however, small and mostly non-significant. The significance of such studies relates to the practice of providing grazing animals with free access to magnesium-rich mineral mixtures. In one instance, a mixture containing 130 g Mg kg^{-1} DM exacerbated diarrhoea in lambs (Suttle and Brebner, 1995). Presumably, high intakes of potassium will increase the laxative effects of magnesium in ruminants by increasing the concentrations of magnesium in digesta reaching the large intestine. Given the variable individual and day-to-day intakes of such mixtures, some members of a flock or herd may ingest enough magnesium-rich mineral to disturb gut function. Excessive magnesium intakes increase the risks of urolithiasis in goats (James and Chandron, 1975), sheep (Chester-Jones et al., 1989) and calves (Chester-Jones et al., 1990), but incidence may partly depend on the phosphorus concentration in the urine (see Chapter 6). The addition of magnesium-mica to feeds as a pelleting agent may increase the risk of urolithiasis.

References

Adams, C.L., Hemingway, R.G. and Ritchie, N.S. (1996) Influence of manufacturing conditions on the bioavailability of magnesium in calcined magnesites measured in vivo and in vitro. *Journal of Agricultural Science, Cambridge* 127, 377–386.

Adediji, O. and Suttle, N.F. (1999) Influence of diet type, K and animal species on the absorption of magnesium by ruminants. *Proceedings of the Nutrition Society* 58, 31A.

Ammerman, C.B., Chicco, C.F., Moore, J.E., van Walleghem, P.A. and Arrington, L.R. (1971) Effect of dietary magnesium on voluntary food intake and rumen fermentations. *Journal of Dairy Science* 54, 1288–1293.

Apple, J.K., Kegley, E.B., Maxwell, C.C., Rakes, L.K., Galloway, D. and Wistuba, T.J. (2005) Effects of magnesium and short-duration transportation on stress response, post-mortem muscle metabolism and meat quality in finishing swine. *Journal of Animal Science* 83, 1633–45.

ARC (1980) *The Nutrient Requirements of Ruminants*. Commonwealth Agricultural Bureaux, Farnham Royal, UK, pp. 184–185.

ARC (1981) *The Nutrient Requirements of Pigs*. Commonwealth Agricultural Bureaux, Farnham Royal, UK.

Bacon, J.A., Bell, M.C., Miller, J.K., Ramsey, N. and Mueller, F.J. (1990) Effect of magnesium administration route on plasma minerals in Holstein calves receiving either adequate or insufficient magnesium in their diets. *Journal of Dairy Science* 73, 470–473.

Baker, R.M., Boston, R.C., Boyes, T.E. and Leever, D.D. (1979) Variations in the response of sheep to experimental magnesium deficiency. *Research in Veterinary Science* 26, 129–133.

Bastien, R.W., Bradley, J.W., Pennington, B.L. and Ferguson, T.M. (1978) Bone strength and egg characteristics as affected by dietary minerals. *Poultry Science* 57, 1117.

Bell, S.T., McKinnon, A.E. and Sykes, A.R. (2006) Estimating the risk of hypomagnesaemic tetany in dairy herds. In: Kebreab, E., Dijkstra, J., Bannink, A., Gerrits, W.J.J. and France, J. (eds) *Nutrient Digestion and Utilization in Farm Animals: Modelling Approaches*. CAB International, Wallingford, UK, pp. 211–228.

Blaxter, K.L., Rook, J.A.F. and MacDonald, A.M. (1954) Experimental magnesium deficiency in calves. 1. Clinical and pathological observations. *Journal of Comparative Pathology* 64, 157–175.

Blumer, C.C., Madden, F.J. and Walker, D.J. (1939) Hypocalcaemia, grass tetany or grass staggers in sheep. *Australian Veterinary Journal* 15, 2–27.

Bohman, V.R., Lesperance, A.L., Harding, G.D. and Grunes, D.L. (1969) Induction of experimental tetany in cattle. *Journal of Animal Science* 29, 99–102.

Bueno, L., Fioramonti, J., Geux, E. and Raissiguer, Y. (1980) Gastrointestinal hypomotility in magnesium deficient sheep. *Canadian Journal of Animal Science* 60, 293–301.

Cairney, I.M. (1964) Grass staggers in beef cattle: results of survey of disease in Hawke's Bay. *New Zealand Journal of Agriculture* 109, 45–49.

Care, A.D., Brown, R.C., Farrar, A.R. and Pickard, D.W. (1984) Magnesium absorption from the digestive tract of sheep. *Quarterly Journal of Experimental Physiology* 69, 577–587.

Charlton, J.A. and Armstrong, D.G. (1989) The effect of an intravenous infusion of aldosterone upon magnesium metabolism in the sheep. *Quarterly Journal of Experimental Physiology* 74, 329–337.

Chester-Jones, H., Fontenot, J.P., Veit, H.P. and Webb, K.E. (1989) Physiological effects of feeding high levels of magnesium to sheep. *Journal of Animal Science* 67, 1070–1081.

Chester-Jones, H., Fontenot, J.P., Veit, H.P. and Webb, K.E. (1990) Physiological effects of feeding high levels of magnesium to steers. *Journal of Animal Science* 68, 4400–4413.

Chicco, C.F., Ammerman, C.B., Feaster, J.P. and Dunavant, B.G. (1973) Nutritional interrelationships of dietary calcium, phosphorus and magnesium in sheep. *Journal of Animal Science* 36, 986–993.

Dalley, D.E., Isherwood, P., Sykes, A.R. and Robson, A.B. (1997a) Effect of intraruminal infusion of potassium on the site of magnesium absorption within the digestive tract of sheep. *Journal of Agricultural Science, Cambridge* 129, 99–106.

Dalley, D.E., Isherwood, P., Sykes, A.R. and Robson, A.B. (1997b) Effect of *in vitro* manipulation of pH on magnesium solubility in ruminal and caecal digesta in sheep. *Journal of Agricultural Science, Cambridge* 129, 107–112.

Dillon, J. and Scott, D. (1979) Digesta flow and mineral absorption in lambs before and after weaning. *Journal of Agricultural Science, Cambridge* 92, 289–297.

D'Souza, D.N., Warner, R.D., Dunshea, F.R. and Leury, B.J. (1999) Comparison of different dietary magnesium supplements on pork quality. *Meat Science* 51, 221–225.

Dua, K. and Care, A.D. (1995) Impaired absorption of magnesium in the aetiology of grass tetany. *British Veterinary Journal* 151, 413–426.

Ebel, H. and Günther, T. (1980) Magnesium metabolism: a review. *Journal of Clinical Chemistry and Clinical Biochemistry* 18, 257–270.

Engels, E.A.N. (1981) Mineral status and profiles (blood, bone and milk) of the grazing ruminant with special reference to calcium, phosphorus and magnesium. *South African Journal of Animal Science* 11, 171–182.

Ferlay, A. and Doreau, M. (1995) Influence of method of administration of rapeseed oil in dairy cows. 2. Status of divalent cations. *Journal of Dairy Science* 78, 2239–2246.

Field, A.C. and Munro, C.S.M. (1977) The effect of site and quantity on the extent of absorption of Mg infused into the gastro-intestinal tract of sheep. *Journal of Agricultural Science, Cambridge* 89, 365–371.

Field, A.C. and Suttle, N.F. (1979) Effect of high and low potassium intakes on the mineral metabolism of monozygotic twin cows. *Journal of Comparative Pathology* 89, 431–439.

Field, A.C., Woolliams, J.A. and Woolliams, C. (1986) The effect of breed of sire on the urinary excretion of phosphorus and magnesium in lambs. *Animal Production* 42, 349–354.

Frederick, B.R., van Heugten, E. and See, M.T. (2004) Timing of magnesium supplementation administered through drinking water to improve fresh and stored pork quality. *Journal of Animal Science* 82, 1454–1460.

Gabel, G. and Martens, H. (1986) The effect of ammonia on the magnesium metabolism of sheep. *Journal of Animal Physiology and Animal Nutrition* 55, 268–277.

Gardner, J.A.A. (1973) Control of serum magnesium levels in sheep. *Research in Veterinary Science* 15, 149–157.
Gentry, R.P., Miller, W.J., Pugh, D.G., Neathery, M.W. and Bynum, J.B. (1978) Effects of feeding high magnesium to young dairy calves. *Journal of Dairy Science* 61, 1750–1754.
Goff, J.P. and Horst, R.L. (1997) Physiological changes at parturition and their relationship to metabolic disorders. *Journal of Dairy Science* 80, 1260–1268.
Grace, N.D. and MacRae, J.C. (1972) Influence of feeding regimen and protein supplementation on the sites of net absorption of magnesium in sheep. *British Journal of Nutrition* 27, 51–55.
Grace, N.D., Carr, D.H. and Reid, C.S.W. (1985) Secretion of sodium, potassium, phosphorus, calcium and magnesium via the parotid and mandibular saliva in sheep offered chaffed lucerne hay or fresh 'Grasslands Ruanui' ryegrass. *New Zealand Journal of Agricultural Research* 28, 449–455.
Grace, N.D., Shaw, H.L., Firth, E.C. and Gee, E.K. (2002a) Determination of digestible energy intake, and apparent absorption of macroelements of grazing, lactating Thoroughbred mares. *New Zealand Veterinary Journal* 50, 182–185.
Grace, N.D., Shaw, H.L., Firth, E.C. and Gee, E.K. (2002b) Determination of digestible energy intake, dry matter digestibility and mineral status of grazing, New Zealand Thoroughbred yearlings. *New Zealand Veterinary Journal* 50, 63–69.
Grace, N.D., Rogers, C.W., Firth, E.C., Faram, T.L. and Shaw, H.L. (2003) Digestible energy intake, dry matter digestibility and effect of calcium intake on bone parameters of Thoroughbred weanlings in New Zealand. *New Zealand Veterinary Journal* 51, 165–173.
Greene, L.W., Fontenot, J.P. and Webb, K.E. Jr (1983a) Site of magnesium and other macro-mineral absorption in steers fed high levels of potassium. *Journal of Animal Science* 57, 503–513.
Greene, L.W., Webb, K.E. and Fontenot, J.P. (1983b) Effect of potassium level on site of absorption of magnesium and other macroelements in sheep. *Journal of Animal Science* 56, 1214–1221.
Greene, L.W., Baker, J.F., Byers, F.M. and Schelling, G.T. (1985) Incidence of grass tetany in a cow herd of a five-breed diallel during four consecutive years. *Journal of Animal Science* 61 (Suppl. 1), 60.
Greene, L.W., Solis, J.C., Byers, F.M. and Schelling, G.T. (1986) Apparent and true digestibility of magnesium in mature cows of five breeds and their crosses. *Journal of Animal Science* 63, 189–196.
Grunes, D.L., Stout, P.R. and Brownell, J.R. (1970) Grass tetany of ruminants. *Advances in Agronomy* 22, 331.
Güenter, W. and Sell. J.L. (1973) Magnesium absorption and secretion along the gastrointestinal tract of the chicken. *Journal of Nutrition* 103, 875–881.
Güenter, W. and Sell, J.L. (1974) A method for determining 'true' availability of magnesium from foodstuffs using chickens. *Journal of Nutrition* 104, 1446–1457.
Hajj, R.N. and Sell, J.L. (1969) Magnesium requirement of the laying hen for reproduction. *Journal of Nutrition* 97, 441–448.
Hamilton, D.N., Elis, M., McKeith, F.K. and Eggert, J.M. (2003) Effect of level, source and time of feeding prior to slaughter of supplementary dietary magnesium on pork quality. *Meat Science* 65, 853–857.
Harmon, B.G., Liu, C.T., Jensen, A.H. and Baker, D.H. (1976) Dietary magnesium levels for sows during gestation and lactation. *Journal of Animal Science* 42, 860–865.
Harris, D.J., Lambell, R.G. and Oliver, C.J. (1983) Factors predisposing beef and dairy cows to grass tetany. *Australian Veterinary Journal* 60, 230–234.
Henry, P.R. and Benz, S.A. (1995) Magnesium bioavailability. In: Ammerman, C.B., Baker, D.H. and Lewis, A.J. (eds) *Bioavailability of Nutrients for Animals*. Academic Press, New York, pp. 239–256.
Herd, R.P. (1966) Fasting in relation to hypocalcaemia and hypomagnesaemia in lactating cows and ewes. *Australian Veterinary Journal* 42, 269–272.
Hess, J.B. and Britton, W.M. (1997) Effect of dietary magnesium excess in white Leghorn hens. *Poultry Science* 76, 703–710.
Hurley, L.A., Greene, L.W., Byers, F.M. and Carstens, G.E. (1990) Site and extent of apparent magnesium absorption by lambs fed different sources of magnesium. *Journal of Animal Science* 68, 2181–2187.
James, C.S. and Chandran, K. (1975) Enquiry into the role of minerals in experimental urolithiasis in goats. *Indian Veterinary Journal* 52, 251–258.
Jittakhot, S., Schonewille, J.Th., Wouterse, H.S., Yuangklang, C. and Beynen, A.C. (2004a) Apparent magnesium absorption in dry cows fed at 3 levels of potassium and two levels of magnesium intake. *Journal of Dairy Science* 87, 379–385.
Jittakhot, S., Schonewille, J. Th., Wouterse, H.S., Yuangklang, C. and Beynen, A.C. (2004b) The relationships between potassium intakes, transmural potential difference of the rumen epithelium and magnesium absorption in wethers. *British Journal of Nutrition* 91, 183–189.

Johnson, C.L. and Aubrey Jones, D.A. (1989) Effect of change of diet on the mineral concentration of rumen fluid, on magnesium metabolism and on water balance in sheep. *British Journal of Nutrition* 61, 583–594.

Johnson, C.L., Helliwells, S.H. and Aubrey Jones, D.A. (1988) Magnesium metabolism in the rumens of lactating dairy cows fed on spring grass. *Quarterly Journal of Experimental Physiology* 73, 23–31.

Jumba, I.O., Suttle, N.F., Hunter, E.A. and Wandiga, S.O. (1995) Effects of soil origin and composition and herbage species on the mineral composition of forages in the Mount Elgan region of Kenya. 1. Calcium, phosphorus, magnesium and sulfur. *Tropical Grasslands* 29, 40–46.

Kemp, A. (1971) The effects of K and N dressings on the mineral supply of grazing animals. In: *Proceedings of the 1st Colloquium of the Potassium Institute*. IBS, Wageningen, The Netherlands, pp. 1–14.

Kemp, A. and 't Hart, M.L. (1957) Grass tetany in grazing milking cows. *Netherlands Journal of Agricultural Science* 5, 4–17.

Kemp, A., Deijs, W.B., Hemkes, O.J. and Van Es, A.J.H. (1961) Hypomagnesaemia in milking cows: intake and utilization of magnesium from herbage by lactating cows. *Netherlands Journal of Agricultural Science* 9, 134–149.

Kemp, A., Deijs, W.B. and Kluvers, E. (1966) Influence of higher fatty acids on availability of magnesium in milking cows. *Netherlands Journal of Agricultural Science* 14, 290–295.

Khorasani, G.R., Janzen, R.A., McGill, W.B. and Kenelly, J.J. (1997) Site and extent of mineral absorption in lactating cows fed whole-crop cereal grain silage or alfalfa silage. *Journal of Animal Science* 75, 239–248.

Kirk, D.J., Fontenot, J.P. and Rahnema, S. (1994) Effects of lasalocid and monensin on digestive tract flow and partial absorption on minerals in sheep. *Journal of Animal Science* 72, 1029–1103.

Lee, S.R. and Britton, W.M. (1987) Magnesium-induced catharsis in chicks. *Journal of Nutrition* 117, 1907–1912.

Littledike, E.T. and Goff, J.P. (1987) Interactions of calcium, phosphorus and magnesium and vitamin D that influence their status in domestic meat animals. *Journal of Animal Science* 65, 1727–1743.

Littledike, E.T., Wittam, J.E. and Jenkins, T.G. (1995) Effect of breed, intake and carcass composition on the status of several macro and trace minerals of adult beef cattle. *Journal of Animal Science* 73, 2113–2119.

MAFF (1990) *UK Tables of the Nutritive Value and Chemical Composition of Foodstuffs*. In: Givens, D.I. (ed). Rowett Research Services, Aberdeen, UK.

Martens, H. (1983) Saturation kinetics of magnesium efflux across the rumen wall in heifers. *British Journal of Nutrition* 49, 153–158.

Martens, H. and Blume, I (1986) Effect of intraruminal sodium and potassium concentrations and of the transmural potential difference on magnesium absorption from the temporarily isolated rumen of sheep. *Quarterly Journal of Experimental Physiology* 71, 409–415.

Martens, H. and Schweigel, M. (2000) Pathophysiology of grass tetany and other hypomagnesemias. Implications for clinical management. *Veterinary Clinics of North America: Food Animal Practice* 16, 339–368.

Martens, H., Kubel, O.W., Gabel, G. and Honig, H. (1987) Effects of low sodium intake on magnesium metabolism of sheep. *Journal of Agricultural Science, Cambridge* 108, 237–243.

Matsui, T., Yans, H., Kawabata, T. and Harumoto, T. (1994) The effect of suppressing bone resorption on magnesium metabolism in sheep (*Ovis aries*). *Comparative Biochemistry and Physiology* 107A, 233–236.

Mayland, H.F., Grunes, D.L. and Lazar, V.A. (1976) Grass tetany hazard of cereal forages based upon chemical composition. *Agronomy Journal* 68, 665–667.

Mayo, R.H., Plumlee, M.P. and Beeson, W.M. (1959) Magnesium requirement of the pig. *Journal of Animal Science* 18, 264–274.

McCoy, M.A., Goodall, E.A. and Kennedy, D.G. (1993) Incidence of hypomagnesaemia in dairy and suckler cows in Northern Ireland. *Veterinary Record* 132, 537.

McCoy, M.A., Goodall, E.A. and Kennedy, D.G. (1996) Incidence of bovine hypomagnesaemia in Northern Ireland and methods of supplementation. *Veterinary Record* 138, 41–43.

McCoy, M.A., Bingham, V., Hudson, J., Cantley, L., Hutchinson, T., Davison, G., Fitzpatrick, D.A. and Kennedy, D.G. (2001a) Postmortem biochemical markers of experimentally- induced hypomagnesaemic tetany in sheep. *Veterinary Record* 148, 233–237.

McCoy, M.A., Hudson, J., Hutchinson, T., Davison, G., Fitzpatrick, D.A. and Kennedy, D.G. (2001b) Postsampling stability of eye fluid magnesium concentrations in cattle. *Veterinary Record* 148, 312–313.

McCoy, M.A., Hutchinson, T., Davison, G., Fitzpatrick, D.A., Rice, D.A. and Kennedy, D.G. (2001c) Postmortem biochemical markers of experimentally induced hypomagnesaemic tetany in cattle. *Veterinary Record* 148, 268–273.

McCoy, M.A., Young, P.B., Edgar, E.M., McCarville, G., Fitzpatrick, D.A and Kennedy, D.G. (2002) Biochemical changes induced by hypomagnesaemia in lactating cows and ewes. *Veterinary Record* 150, 176–181.

McGillivray, J.J. and Smidt, M.J. (1975) Biological evaluation of magnesium sources. *Poultry Science* 54, 1792–1793.

McGillivray, J.J. and Smidt, M.J. (1977) Energy level, potassium, magnesium and sulfate interaction in broiler diets. *Poultry Science* 56, 1736–1737.

Menzies, F.D., Bryson, D.G. and McCallion, T. (1994) Bovine mortality survey. *Department of Agriculture for Northern Ireland Publication*, p. 11.

Miller, E.R., Ullrey, D.E., Zutaut, C.L., Baltzer, B.V., Schmidt, D.A., Hoefer, J.A. and Luecke, R.W. (1965) Magnesium requirement of the baby pig. *Journal of Nutrition* 85, 13–20.

Minson, D.J. (1990) *Forages in Ruminant Nutrition*. Academic Press, San Diego, California, pp. 208–229.

Moseley, G. and Baker, D.H. (1991) The efficacy of a high magnesium grass cultivar in controlling hypomagnesaemia in grazing animals. *Grass and Forage Science* 46, 375–380.

NRC (1985) *Nutrient Requirements of Sheep*, 6th edn. National Academy of Sciences, Washington, DC.

NRC (1994) *Nutrient Requirements of Poultry*, 9th edn. National Academy of Sciences, Washington, DC.

NRC (1998) *Nutrient Requirements of Swine*, 10th edn. National Academy of Sciences, Washington, DC.

NRC (2000) *Nutrient Requirements of Beef Cattle*, 7th edn. National Academy of Sciences, Washington, DC.

NRC (2001) *Nutrient Requirements of Dairy Cows*, 5th edn. National Academy of Sciences, Washington, DC.

Nugara, D. and Edwards, H.M. Jr (1963) Influence of dietary Ca and P levels on the Mg requirements of the chick. *Journal of Nutrition* 80, 181–184.

Parkins, J.J. and Fishwick, G. (1998) Magnesium hydroxide as a novel herbage spray supplement for lactating cows and ewes. *Animal Science* 66, 93 (abstract).

Partridge, I.G. (1978) Studies on digestion and absorption in the intestines of growing pigs. 3. Net movements of mineral nutrients in the digestive tract. *British Journal of Nutrition* 39, 527–537.

Pauli, J.V. and Allsop, T.P. (1974) Plasma and cerebrospinal fluid magnesium, calcium and potassium in dairy cows with hypomagnesaemic tetany. *New Zealand Veterinary Journal* 22, 227–231.

Rahnema, S., Wu, Z., Ohajuruka, O.A., Weiss, W.P. and Palmquist, D.L. (1994) Site of mineral absorption in lactating cows fed high-fat diets *Journal of Animal Science* 72, 229–235.

Ram, L., Schonewille, J.Th., Martens, H., Van't Klooster, A.Th., Wouterse, H. and Beynen, A.C. (1998) Magnesium absorption by wethers fed potassium bicarbonate in combination with different magnesium concentrations *Journal of Dairy Science* 81, 2485–2492.

Reid, R.L., Jung, G.A., Roemig, I.J. and Kocher, R.E. (1978) Mineral utilisation in lambs and guinea pigs fed Mg-fertilized grass and legume hays. *Agronomy Journal* 70, 9–14.

Riond, J.-L., Kocabagli, N., Spichiger, V.E. and Wanner, M. (1995) The concentration of ionized magnesium in serum during the peri-parturient period of non-paretic cows. *Veterinary Research Communications* 19, 195–203.

Robson, M.R., Field, A.C. and Sykes, A.R. (1997) A model for magnesium metabolism in sheep. *British Journal of Nutrition* 78, 975–992.

Robson, M.R., Sykes, A.R., McKinnon, A.E. and Bell, S.T. (2004) A model for magnesium metabolism in young sheep. *British Journal of Nutrition* 91, 73–79.

Roche, J.R. and Berry, D.P. (2006) Periparturient climatic, animal, and management factors influencing incidence of milk fever in grazing systems. *Journal of Dairy Science* 89, 2775–2783.

Roche, J.R., Morton, J. and Kolver, E.S. (2002) Sulfur and chlorine play a non-acid base role in periparturient calcium homeostasis. *Journal of Dairy Science* 85, 3444–3453.

Roche, J.R., Dalley, D., Moate, P., Grainger, C., Rath, M. and O'Mara, F.O. (2003) Dietary cation–anion difference and the health and production of pasture-fed dairy cows 2. Non-lactating peri-parturient cows. *Journal of Dairy Science* 86, 979–987.

Roche, J.R., Petch, S. and Kay, J.K. (2005) Manipulating the dietary cation–anion difference via drenching to early-lactation dairy cows grazing pasture. *Journal of Dairy Science* 88, 264–276.

Rogers, P.A.M. (1979) Hypomagnesaemia and its clinical syndromes in cattle: a review. *Irish Veterinary Journal* 33, 115–124.

Rogers, P.A.M. and Poole, D.B.R. (1976) Control of hypomagnesaemia in cows: a comparison of magnesium acetate in the water supply with magnesium oxide in the feed. *Irish Veterinary Journal* 30, 129–136.

Rook, J.A.F. and Storry, J.E. (1962) Magnesium in the nutrition of ruminants. *Nutrition Abstracts and Reviews* 32, 1055–1076.

Sahin, N., Onderci, M., Sahin, K., Cikim, G. and Kucuk, O. (2005) Magnesium proteinate is more protective than magnesium oxide in heat-stressed quail. *Journal of Nutrition* 135, 1732–1737.

Sansom, B.F., Bunch, K.G. and Dew, S.M. (1982) Changes in plasma calcium, magnesium, phosphorus and hydroxyproline concentrations in ewes from twelve weeks before until three weeks after lambing. *British Veterinary Journal* 138, 393–401.

Schonewille, J.Th., Ram, L., Van Klooster, A.Th., Wouterse, H. and Beynen, A.C. (1997a) Intrinsic potassium in grass silage and magnesium absorption in dry cows. *Livestock Production Science* 48, 99–110.

Schonewille, J.Th., Ram, L., Van Klooster, A.Th., Wouterse, H. and Beynen, A.C. (1997b) Native corn starch versus either cellulose or glucose in the diet and the effects on apparent magnesium absorption in goats. *Journal of Dairy Science* 80, 1738–1743.

Schonewille, J. Th., Beynen, A.C., Van Klooster, A. Th. and Wouterse, H. (1999a) Dietary potassium bicarbonate and potassium citrate have a greater inhibitory effect than does potassium chloride on magnesium absorption in wethers. *Journal of Nutrition* 129, 2043–2047.

Schonewille, J.Th., Van Klooster, A.Th., Wouterse, H. and Beynen, A.C. (1999b) The effects of intrinsic potassium in artificially dried grass and supplemental potassium bicarbonate on apparent magnesium absorption in dry cows. *Journal of Dairy Science* 82, 1824–1830.

Schonewille, J. Th., Wouterse, H. and Beynen, A.C. (2002) The effect of iso-energetic replacement of artificially dried grass by pelleted concentrate on apparent magnesium absorption in dry cows. *Livestock Production Science* 76, 59–69.

Schonewille, J. Th., Everts, H., Jittakhot, S. and Beynen, A.C. (2008) Quantitative prediction of magnesium absorption in dairy cows. *Journal of Dairy Science* 91, 271–278.

Sell, J.L. (1979) Magnesium nutrition of poultry and swine. In: *Proceedings of the Second Annual International Minerals Conference, St Petersburg Beach, Florida, Illinois.* International Minerals and Chemical Corporation.

Shils, M.E. (1997) Magnesium. In: O'Dell, B.L. and Sunde, R.A. (eds) *Handbook of Nutritionally Essential Mineral Elements.* Marcel Dekker, New York, pp. 117–152.

Sjollema, B. and Seekles, L. (1930) Uber Storungen des mineralen Regulationsmechanismus bei Krankheiten des Rindes. (Ein Beitrag zur Tetaniefrage.) *Biochemische Zeitschrift* 229, 338–380.

Soder, K.J. and Stout, W.L. (2003) The effect of soil type and fertilisation level on mineral concentration in pasture: potential relationships to ruminant performance and health. *Journal of Animal Science* 81, 1603–1610.

Stevenson, M.A., Williamson, N.B. and Russell, D.J. (2003) Nutrient balance in the diet of spring-calving, pasture-fed dairy cows. *New Zealand Veterinary Journal* 51, 81–88.

Stuedemann, J.A., Wilkinson, S.R. and Lowrey, R.S. (1984) Efficacy of a large magnesium alloy rumen bolus in the prevention of hypomagnesaemic tetany in cows. *American Journal of Veterinary Research* 45, 698–702.

Sutherland, R.J., Bell, K.C., McSporran, K.D. and Carthew, G.W. (1986) A comparative study of diagnostic tests for the assessment of herd magnesium status in cattle. *New Zealand Veterinary Journal* 34, 133–135.

Suttle, N.F. and Brebner, J. (1995) A putative role for larval nematode infection in diarrhoeas which did not respond to anthelmintic drenches. *Veterinary Record* 137, 311–316.

Suttle. N.F. and Field, A.C. (1966) Studies on magnesium in ruminant nutrition. 6. Effect of intake of water on the metabolism of magnesium, calcium, sodium, potassium and phosphorus in sheep. *British Journal of Nutrition* 20, 609–619.

Suttle. N.F. and Field, A.C. (1967) Studies on magnesium in ruminant nutrition. 8. Effect of increased intakes of potassium and water on the metabolism of magnesium, phosphorus, sodium, potassium and calcium in sheep. *British Journal of Nutrition* 21, 819–831.

Suttle. N.F. and Field, A.C. (1969) Studies on magnesium in ruminant nutrition. 9. Effect of potassium and magnesium intakes on development of hypomagnesaemia in sheep. *British Journal of Nutrition* 23, 81–90.

't Hart, M.L. and Kemp, A. (1956) De invloed van de weersomstandigheden op het optreden van kopziekte bij rundvee. *Tijdschrift voor Diergeneeskunde* 81, 84–95.

Theilin, M., Sedcole, J.R. and Sykes, A.R. (2001) Changes in plasma and milk magnesium concentrations in pasture-fed dairy cows in early lactation. *Proceedings of the New Zealand Society of Animal Production* 61, 152–155.

Tomas, F.M. and Potter, B.J. (1976a) Interaction between site of magnesium absorption in the digestive tract of the sheep. *Australian Journal of Agricultural Research* 27, 437–446.

Tomas, F.M. and Potter, B.J. (1976b) The site of magnesium absorption from the ruminant stomach. *British Journal of Nutrition* 36, 37–45.

Tongyai, S., Raysigguer, Y., Molta, C., Gueux, E., Maurois, P. and Heaton, F.W. (1989) Mechanism of increased erythrocyte membrane fluidity during magnesium deficiency in rats. *American Journal of Physiology* 257, 270–276.

Underwood, E.J. and Suttle, F. (1999) *The Mineral Nutrition of Livestock*, 3rd edn. CAB International, Wallingford, UK.

van de Braak, A.E., van't Klooster, A.T. and Malestein, A. (1987) Influence of a deficient supply of magnesium during the dry period on the rate of calcium mobilisation by dairy cows at parturition. *Research in Veterinary Science* 42, 101–108.

Weiss, W.P. (2004) Macromineral digestion by lactating dry cows: factors affecting digestibility of magnesium. *Journal of Dairy Science*, 87, 2167–2171.

Whittaker, D.A. and Kelly, J.M. (1982) Incidence of clinical and subclinical hypomagnesaemia in dairy cows in England and Wales. *Veterinary Record* 110, 450–451.

Wilson, G.F. (1980) Effects of magnesium supplements on the digestion of forages and milk production of cows with hypomagnesaemia. *Animal Production* 31, 153–157.

Wu, C.L. and Britton, W.M. (1974) The influence of phosphorus on magnesium metabolism in the chick. *Poultry Science* 53, 1645 (abstract).

Wylie, M.J., Fontenot, J.P. and Greene, L.W. (1985) Absorption of magnesium and other macrominerals in sheep infused with potassium in different parts of the digestive tract. *Journal of Animal Science* 61, 1219–1229.

Xin, Z., Tucker, W.B. and Hemken, R.W. (1989) Effect of reactivity rate and particle size of magnesium oxide on magnesium availability, acid–base balance, mineral metabolism and milking performance of dairy cows. *Journal of Dairy Science* 72, 462–470.

Young, P.W., Rys, G. and O'Connor, M.B. (1981) Hypomagnesaemia and dairy production. *Proceedings of the New Zealand Society of Animal Production* 41, 61–67.

6 Phosphorus

An early indication that a shortage of phosphorus could seriously impair livestock production came from the pioneering work of Sir Arnold Theiler (1912) as he investigated two debilitating diseases ('styfziekte' and 'lamziekte') of cattle and sheep grazing veldt pastures at Armoedsvlakte in South Africa's Northern Cape. The diseases were characterized by high mortality, poor growth and fertility, fragile bones and periodic osteophagia. Phosphorus deprivation emerged as a possible cause and details of successful preventive measures were eventually published (Theiler and Green, 1932; DuToit et al., 1940). Other pastoral areas lacking in phosphorus were identified on many continents, including Australia and North America. In Florida, precise requirements for phosphorus were soon being estimated for lambs (Beeson et al., 1944). However, remarkable improvements in production obtained in South Africa with supplements of blood meal, providing both phosphorus and protein, exaggerated the impact of phosphorus deprivation per se and the optimal phosphorus supplementation for cattle is still under investigation on veldt pastures (De Waal et al., 1996; de Brouwer et al., 2000) and elsewhere (Karn, 2001). It is now apparent that phosphorus has equally important roles to play in soft and hard tissues and that exchanges of phosphorus between them can protect the grazing animal from phosphorus deprivation.

Ruminants differ markedly from non-ruminants, in that the presence of microbial phytase in the rumen allows them to utilize much of the abundant phosphorus present as phytates (Py) in grains. Early success in supplementing pig and poultry rations with mixtures of calcium and phosphorus (e.g. bone meal) masked an underlying dietary antagonism whereby calcium supplements reduce the availability of Py phosphorus (PyP). The discovery that phosphorus could be liberated from PyP by adding phytase to the chick diet (Nelson et al., 1971) began a fruitful line of research that continues to this day. Nevertheless, profligate use of inorganic phosphorus (P_i) supplements in feeds and fertilizers for all classes of livestock has so enriched soils and watercourses with phosphorus in intensively farmed areas that it has contributed to the eutrophication of surface waters and led to mandatory restrictions on farm 'inputs' of phosphorus (Jongbloed and Hitchens, 1996). This problem will be partly solved as a better understanding of phosphorus metabolism lowers estimates of phosphorus requirement. The application of biotechnology that allows better utilization of plant sources of phosphorus by pigs and poultry has been hailed as a breakthrough, but does not overcome the basic problem – cereals and vegetable proteins contain more phosphorus than livestock need.

Functions of Phosphorus

Phosphorus is the second most abundant mineral in the animal body and about 80% is found in the bones and teeth. The formation and maintenance of bone are quantitatively the most important functions of phosphorus, and the changes in bone structure and composition that result from phosphorus deprivation are in most respects the same as those described for calcium deprivation (see Chapter 4). Phosphorus is, however, required for the formation of the organic bone matrix as well as the mineralization of that matrix. The remaining 20% of body phosphorus is widely distributed in the fluids and soft tissues of the body where it serves a range of essential functions. Phosphorus is a component of deoxy- and ribonucleic acids, which are essential for cell growth and differentiation. As phospholipid it contributes to cell membrane fluidity and integrity and to the myelination of nerves; and as phosphate (PO_4) helps to maintain osmotic and acid–base balance. Phosphorus also plays a vital role in a host of metabolic functions, including energy utilization and transfer via AMP, ADP and ATP, with implications for gluconeogenesis, fatty acid transport, amino acid and protein synthesis and activity of the sodium/potassium ion pump.

Disturbances of glycolytic metabolism have been noted in the erythrocytes from phosphorus-deficient cattle (Wang et al., 1985). In ruminants, the requirements of the rumen and caeco-colonic microflora are also important, and microbial protein synthesis may be impaired on low-phosphorus diets (Petri et al., 1989; Ternouth and Sevilla, 1990b). Phosphorus is further involved in the control of appetite (in a manner not yet fully understood) and in the efficiency of feed utilization (Ternouth, 1990). In a fascinating twist, it now appears that Py is an important constituent of mammalian tissues as well as a dietary antagonist of phosphorus absorption (A_p), being present in the basolateral membrane of cells where it 'anchors' alkaline phosphatases (Vucenik et al., 2004). Synthesis of Py can take place in the tissues, where its strong affinity for ferric ions provides antioxidant capacity and affinity for calcium ions allows participation in gene regulation, RNA export, DNA repair and cell signalling (Bohn et al., 2008).

Dietary Sources of Phosphorus

The concentrations of phosphorus in feeds vary more widely than those of most macro-minerals as the data for the UK show (Table 6.1).

Table 6.1. Mean (standard deviation) phosphorus concentrations (g kg^{-1} DM) in livestock feeds commonly used in the UK (MAFF, 1990).

Roughages	P	Concentrates	P	By-products	P
Barley straw	1.1 (1.10)	Barley	4.0 (0.46)	Wheat bran	12.6 (3.00)
Oat straw	0.9 (0.24)	Maize	3.0 (0.32)	Rice bran	17.4 (1.70)
Grass	3.0 (0.68)	Oats	3.4 (0.46)	Brewers' grains	5.1 (1.00)
Kale	4.2 (0.71)	Wheat	3.3 (0.42)	Distillers' grains[a]	9.6 (0.80)
White clover	3.8 (0.28)	Maize gluten	2.8 (1.20)	Citrus pulp	1.1 (0.09)
Grass silage	3.2 (0.62)	Cottonseed meal	8.9 (1.1)	Sugarbeet pulp	0.8 (0.15)
Clover silage	2.3 (0.75)	Fish meal, white	38.0 (14.5)	Cassava meal	0.9 (0.14)
Lucerne silage	3.3 (0.35)	Groundnut meal	6.0 (0.20)	Feather meal	3.1 (0.60)
Maize silage	2.6 (1.20)	Linseed meal	8.7 (0.80)		
Fodder beet	1.8 (0.27)	Maize-gluten meal	2.8 (1.20)		
Swedes	2.6 (0.28)	Palm-kernel meal	6.2 (0.61)		
Grass hay	2.6 (0.77)	Rapeseed meal	11.3 (1.50)		
		Soybean meal (ext.)	7.9 (0.15)		
		Sunflower-seed meal	10.8 (1.90)		

[a]Barley-based.

Milk and animal by-products

Milk is an important source of phosphorus for the young of all species and contains 1.0–1.6g P l^{-1}. Concentrations are more uniform on a dry weight basis at around 8g kg^{-1} dry matter (DM), but main milk from mares is relatively low in phosphorus (0.54g l^{-1}) (see Table 2.1). In all species, colostrum is much richer in phosphorus than main milk. Milk phosphorus is mostly present in casein, and products derived from bovine milk vary in phosphorus according to the manufacturing process: by-products of cheese-making such as whey are much lower in phosphorus than those from butter-making such as dried skim-milk, but liquid whey can still provide useful amounts of phosphorus for pigs. Rendered animal by-products such as meat and bone meal are valuable as sources of phosphorus and must not contain <4% phosphorus in the USA. Although their use in the European Community is currently prohibited, these constraints may be lifted.

Cereals

Cereals are rich in phosphorus (Table 6.1), most of which is stored in the seed coat in the form of *Py* complexes with calcium, magnesium and potassium, with *Py* taking the form of inositol tetra-, penta- and hexaphosphates, respectively. The extensive literature on feed *Py* has been reviewed by Reddy and Sathe (2001). The polydentate *Py* molecule has been likened to a turtle to help memorize the nomenclature: the inositol ring is the 'body' with four esterified PO_4 at the corners, represented by four 'paws', and one at either end pointing up ('head') or down ('tail') (Fig. 6.1). Each PO_4 is capable of binding minerals with varying affinity *in vitro* (copper > zinc > nickel > cobalt > manganese > iron > calcium) (Bohn et al., 2008). Of the total phosphorus in cereals, *PyP* usually contributes 70–80% (Selle et al., 2003; Jendreville et al., 2007).

Plant protein sources and by-products

Vegetable protein sources are richer in phosphorus than cereals (Table 6.1) but the proportion reported to be present as *PyP*, 50–60%, is generally lower (e.g. 55% in 25 samples of soybean meal (SBM) obtained from across the USA) (Manangi and Coon, 2006). However, a comparison across legume and oilseed types indicates that they contain a higher proportion of *PyP* than cereals (Viveros et al., 2000). In the Manangi and Coon (2006) study, mean (standard deviation) *PyP* and non-*PyP* concentrations were 3.87 (0.23) and 3.16 (0.64) g kg^{-1} DM, respectively. Attention has focused on the more variable *Py* component of grains, but non-*PyP* in SBM may be predictable from the total phosphorus (Fig. 6.2) (Gao et al., 2007). The widely used sugarbeet pulp, whether molassed or not, is notoriously low in phosphorus (Table 6.1), but cereal-based by-products such as brewers' and distillers' grains

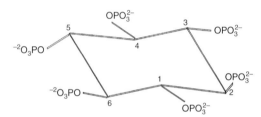

Fig. 6.1. Structure of the fully deprotonated form of phytic acid (*myo*-inositol hexakis phosphate) (from Bedford and Partridge, 2001).

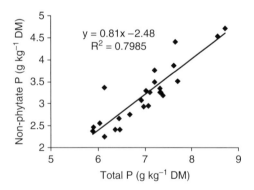

Fig. 6.2. Soybean meals vary widely in total phosphorus and non-phytate P concentration, but non-phytate P is a predictable fraction of total P, increasing from 0.40 to 0.53 across the range of total P values (from Manangi and Coon, 2006).

retain most of the phosphorus present in the grain and are high in phosphorus (Table 6.1). Little attention has been given to the concentration of P_yP in other feeds, but analytical technique is important.

The 'standard' ferric chloride ($FeCl_3$) extraction method for measuring P_y does not extract all of the P_yP from certain feeds, including the seeds of lupins and field beans (Selle et al., 2003). Recent modifications of the assay methods for P_y have been reviewed by Gao et al. (2007). Nevertheless, substantial concentrations of P_yP have been detected in alfalfa hay, bagasse and citrus pulp (0.82, 0.32 and 0.54 g kg^{-1} DM, respectively, equivalent to 21%, 64% and 45% of total phosphorus) (Dias et al., 2006). The use of genetic engineering to produce crop varieties with less P_yP (e.g. Bretziger and Roboy, 2006) is discussed in more detail in Chapter 20.

Plant sources of phytase and phosphatase

Cereals contain phytases and acid phosphatases to release PO_4 from stored P_y during germination. These enzymes can perform a similar service in the gut and are an important determinant of 'phosphorus value'. Wheat, triticale, rye and barley are relatively rich in both enzymes, while maize, millet and sorghum contain little (Selle et al., 2003; Bohn et al., 2008). Phytase activity is associated with outer layer of the grain and brans from rye and wheat are much richer in phytase than the respective grains (7339 versus 5147 U kg^{-1} and 4624 versus 1637 U kg^{-1}). Acid phosphatase is similarly partitioned (Viveros et al., 2000). Legumes and oilseeds contain negligible phytase but moderate acid phosphatase activity (Viveros et al., 2000). Phytases vary in where they begin to hydrolyse the P_y molecule, generally preferring to hydrolyse 'equatorial ('paw') rather than axial ('head' or 'tail') PO_4 groups: they also vary in pH optima and stability in the stomach (Bohn et al., 2008) and therefore in their ability to improve phosphorus utilization.

Herbage

On a worldwide scale, herbage phosphorus varies more widely than is shown by the UK data in Table 6.1, being influenced primarily by the phosphorus status of the soil, stage of plant maturity and climate. On average, herbage phosphorus increases by 0.03–0.05 g kg^{-1} DM mg^{-1} extractable soil phosphorus (Minson, 1990; Jumba et al., 1995). Temperate forages generally contain more phosphorus than tropical forages (3.5 versus 2.3 g kg^{-1} DM), but this may be due to the greater use of fertilizer phosphorus. Alfalfa is generally richer in phosphorus than grasses and tropical legumes generally contain more phosphorus than tropical grasses (3.2 versus 2.7 g kg^{-1} DM) (Minson, 1990), but there are exceptions.

Table 6.2 presents hitherto unpublished data for grasses and legumes from Costa Rica. These data show that phosphorus is consistently low in four contrasting legume species. Some *Stylosanthes* species are renowned for a capacity to grow vigorously on soils that are low in phosphorus, but their phosphorus status remains low (often <1.0 g kg^{-1} DM). This tolerance of low soil phosphorus may be shared by other legumes. Note that legumes are much richer in calcium than grasses, giving calcium to phosphorus ratios that are almost fivefold higher than those in the grasses sampled. There is a marked reduction in whole-plant phosphorus concentrations as most forages mature, particularly during any dry season.

Selective grazing

Grazing animals prefer phosphorus-fertilized pasture when given a choice (Jones and Betteridge, 1994) and select pasture components that are richer in phosphorus than the whole sward. Workers at Armoedsvlakte noted two- to threefold higher phosphorus levels in herbage selected by cattle (i.e. in samples from oesophageal fistulae) than in hand-plucked samples (Engels, 1981) and ceased using herbage samples as an index of phosphorus supply (Read et al., 1986a,b; De Waal et al., 1996), as have others (Karn, 1997). Selectivity is most important when the sward contains a mixture

Table 6.2. Mean (standard deviation) concentrations of phosphorus and calcium (g kg^{-1} DM) and Ca:P ratios in some tropical grasses and legumes from Costa Rica.[a]

	n	P	Ca	Ca:P
Grass species				
Kikuyu (*Pennisetum clandestinum*)	5	3.1 (0.41)	2.5 (0.25)	0.8
Ratana (*Ischaemum ciliare*)	5	1.8 (0.70)	3.0 (0.57)	1.7
Signal grass (*Brachiaria decumbens*)	5	1.4 (0.27)	3.8 (0.72)	2.7
Star grass (*Cynodon nlemfuensis*)	4	1.5 (0.72)	2.4 (0.09)	1.6
Guinea (*Panicum maximum*)	3	2.4 (1.35)	3.1 (0.40)	1.3
Jaragua (*Hyparrhenia rufa*)	4	1.2 (0.55)	2.4 (0.20)	2.0
Legume species				
Stylo (*Stylosanthes guianensis*)	4	1.4 (0.19)	9.0 (0.39)	6.4
Centro (*Centrosema pubescens*)	3	1.3 (0.19)	9.1 (2.96)	7.0
Arachis (*Arachis glabrata*)	4	1.5 (0.16)	11.2 (0.77)	7.5
Naked Indian (*Bursera simaruba*)	5	1.1 (0.30)	10.4 (2.68)	9.5

[a]O. Adedeji and N.F. Suttle, unpublished data.

of young, mature and senescent material and/or a mixture of species that vary in phosphorus concentration and palatability (McClean et al., 1990). The selection of particular parts of a plant at a given stage of maturity is unimportant because distribution within a plant is relatively uniform. Cohort, oesophageally fistulated animals yield more reliable measures of phosphorus intake when parenteral doses of ^{32}P allow correction for contamination by salivary phosphorus (Coates et al., 1987; Langlands, 1987; Coates and Ternouth, 1992). If the entire stand is virtually removed during a grazing period, the early selection of phosphorus-rich herbage may have little effect on cumulative phosphorus intake. Hand-plucked samples can be standardized to give samples of known maturity for particular species (Kerridge et al., 1990) and herbage phosphorus concentrations should thus be measured whenever possible to record changes in the available phosphorus supply from the soil from year to year.

Roughages and forage crops

The higher phosphorus levels reported for grass silages than grass hays in Table 6.1 probably reflect the earlier growth stage at which silage is harvested. Immature, whole-crop cereal silages are now widely used as feeds for ruminants in autumn and winter, but often have relatively low phosphorus concentrations. In the UK, corn silage usually contains 30–50% less phosphorus than grass silage. Although ensilage of the whole crop raises phosphorus concentrations slightly, the immature grain lacks a phosphorus-rich seedcoat. Brans are rich in phosphorus, straws low in phosphorus (1–2 g kg^{-1} DM) and root crops generally of intermediate content (Table 6.1). Selection between constituents that vary widely in phosphorus content can affect phosphorus intakes from 'complete' diets offered to cows.

Assessing the Value of Phosphorus Sources

Certain aspects of phosphorus metabolism must be covered before proceeding to assess the 'phosphorus value' of feeds.

Absorption

The absorption of phosphorus (A_P) takes place principally in the proximal small intestine (Bown et al., 1989; Yano et al., 1991) and is largely unregulated in all species. Specific, sodium-dependent transporters for phosphate (PO_4) have been identified in the jejunal mucosa and kidneys of poultry, but phosphorus deprivation only affects transporter expression in the kidneys (Huber et al., 2006). There is measurable absorption of phosphorus from the large intestine in pigs (Liu et al., 2000).

Absorption in ruminants

In weaned sheep (Braithwaite, 1986; Portilho et al., 2006), calves (Challa et al., 1989) and goats (Vitti et al., 2000), A_P is linearly related to P_i intake over wide ranges, with high coefficients of 0.68–0.80. Dephosphorylation and hydrolysis of Py in ingested grains and seeds by microbial phytases and phosphatases releases PO_4 in the rumen (Reid et al., 1947; Nelson et al., 1976; Morse et al., 1992) and this is largely incorporated into microbial protein but the degradation of PyP can be far from complete. Microbial phosphorus is marginally less well absorbed than PO_4, which is – like other anions – readily absorbed.

Absorption in non-ruminants

In young pigs A_{P_i} is even higher than in ruminants, with absorption coefficients of 0.89–0.97 being reported (Peterson and Stein, 2006; Pettey et al., 2006), but there are wide differences between species in the ability to absorb organic phosphorus. While non-ruminants can free most of the Py from organic complexes in the feed during its digestion (dephosphorylation), they remain heavily dependent on the limited presence of phytases in grains to free phosphorus from Py (conversion) (Zyla et al., 2004). Phytase activity is very low in the intestinal mucosa of pigs and has not been detected in the lumen (Pointillart et al., 1987), but the enzyme is expressed in poultry (Cho et al., 2006). Conversion produces a family of partially degraded Py molecules whose spectrum changes during passage through acid and alkaline regions of the gut, allowing different phytases to act synergistically (Zyla et al., 2004).

Recycling and excretion in ruminants

The measurement of A_P in ruminants is complicated by the copious secretion of phosphorus in saliva, which adds greatly to the flow of phosphorus into the rumen (Tomas et al., 1973). Salivary phosphorus is absorbed as efficiently as P_i (Challa et al., 1989; AFRC, 1991), allowing considerable recycling to take place, but some escapes and the faeces provides the primary route for excreting surplus absorbed phosphorus on forage-based diets. The faeces thus contains unabsorbed phosphorus from saliva, as well as the diet and rumen microbes, and the widely measured apparent absorption of phosphorus (AA_P) underestimates 'phosphorus value'.

Salivary phosphorus secretion is largely determined by the concentration of phosphorus in saliva, which is highly correlated with plasma phosphorus (Challa et al., 1989; Valk et al., 2002). Low phosphorus intakes cause plasma and salivary phosphorus to fall, but irreducible faecal endogenous losses (IFE_P) remain the predominant source of faecal phosphorus (e.g. 70–90%) (Breves et al., 1985).

Excretion in pigs and poultry

Comparatively little saliva is secreted by pigs and the urine is normally the major route for excreting excess absorbed phosphorus. The difference between intake and faecal excretion (or AA_P) is a valid indicator of 'phosphorus value' in pigs. However, estimates of IFE_P in pigs vary 10-fold between studies (Petersen and Stein, 2006), probably due to artefacts of the techniques employed, and the conversion of AA_P to A_P by deducting IFE_P is unreliable in pigs (see p. 29). The urine is also the route for excreting surplus absorbed phosphorus in poultry, but the measurement of AA_P is complicated by the entry of urinary phosphorus into the gut, distal to the ileum, so that uncorrected AA_P provides no useful information on the phosphorus value of feeds for poultry. A combination of indigestible markers and ileal cannulae must be used to determine phosphorus flows leaving the ileum, but they have shown that P_i is again absorbed almost completely (Rodehutscord et al., 2002).

Excretion in the horse

Mature trotters given a diet of hay and concentrate in equal measure and containing 2.9 g

P kg^{-1} DM had a low AA_P coefficient of 0.024 and excreted no phosphorus in their urine (van Doorn et al., 2004). The addition of monocalcium phosphate or bran to the diet raised phosphorus intake by 23–25 g day^{-1}, urinary phosphorus by only 2 g day^{-1} and AA_P to 0.15 or 0.11. These findings indicate that the horse excretes excess absorbed phosphorus via the faecal rather than the urinary route, probably by virtue of a high salivary output. Studies with rabbits indicate that hind-gut fermenters can degrade Py in the caecum (Marounek et al., 2003), but this will not be detected by measuring AA_P until the capacity to excrete phosphorus via the saliva and faeces has been saturated. The 'phosphorus value' of feeds for equines is therefore largely unknown because the widely measured AA_P reflects animal rather than feed attributes.

If lactating mares are given similar forage to foals, they appear to absorb phosphorus far more efficiently (AA_P coefficient 0.47 versus 0.07–0.14) (Grace et al., 2002, 2003), but this will merely reflect the diversion of absorbed phosphorus from salivary to mammary secretion. The lack of effect of calcium (Grace et al., 2003) or phytase (van Doorn et al., 2004) supplements on AA_P does not rule out possible changes in A_P. Similarly the lack of effect of replacing a P_i supplement with bran, with or without phytase (1000 U kg^{-1} DM), on urinary phosphorus excretion or AA_P does not rule out the occurrence of an antagonism between calcium and Py in the gut.

Retention and relative availability

Most attempts to assess the 'phosphorus value' of feeds for pigs and poultry have tried to avoid the difficulties of measuring A_P by comparing repletion rates on a low-phosphorus basal diet supplemented with 'unknown' (test) sources of phosphorus or an inorganic source of known, high availability phosphorus such as monosodium phosphate (MSP). Such comparisons yields values for relative absorption (RA_P), but the technique presents further problems, notably the effect of performance index on the RA_P value (see Chapter 2) and the fact that it is relative retention that is measured, thus ignoring the urinary excretion of excess absorbed phosphorus. In a recent attempt to measure the RA_P of bone meal (Traylor et al., 2005), the slopes ratio test gave a value of 87% or 95%, depending on whether a linear regression was forced through the origin. Markedly curvilinear responses in live-weight gain (LWG) with bone meal appear identical to those for MSP.

Phosphorus Availability

Milk and milk replacers

All animals that are suckled or reared on milk substitutes (e.g. Challa and Braithwaite, 1989) absorb phosphorus with almost complete efficiency. Data from Petty et al., (2006) indicate that the phosphorus in casein is absorbed by pigs as efficiently as P_i. With the introduction of solid feeds into the diet, wide differences in 'phosphorus value' emerge in different species.

Feeds for ruminants

A high proportion of the phosphorus in fresh or dry forages is absorbed by both sheep and cattle (AFRC, 1991; NRC, 2001) regardless of forage type, with a mean A_P coefficient of 0.74±0.09 (standard deviation) (AFRC, 1991). The phosphorus in wheat straw (Ternouth, 1989) and brans (Knowlton et al., 2001) is also relatively well absorbed, although only one A_P coefficient has been published (0.63 for rice bran) (Field et al., 1984). Feeds may vary in A_P according to the extent to which they promote microbial protein synthesis (Bravo et al., 2003b), but well-formulated diets will vary little in this respect and uniformly high A_P values of 0.78–0.81 have been reported for two cereals (wheat and barley), three vegetable protein sources (rapeseed meal, maize gluten and SBM) and a fish meal for sheep (Field et al., 1984).

Three recent observations suggest that the long-held view that PyP is always completely degraded in the rumen may no longer be valid. First, dairy cows given diets containing 38% grain in processed forms showed increases in plasma P_i when phytase was added to the diet (Kincaid et al., 2005). Second, a survey of the form of phosphorus in faeces from US dairy cows found that 18% was present as PyP (Toor et al., 2005). Third, specific measurement of the digestibility of PyP (mostly from corn and SBM) showed that 53% escaped degradation and was excreted in the faeces of lambs (Dias et al., 2006). The treatment of vegetable proteins with formaldehyde to reduce the degradability of protein in the rumen also restricts the degradability of Py (Park et al., 1999), and the provision of high-yielding lactating ewes and cows with bypass protein may reduce the A_p of vegetable proteins such as SBM that are otherwise rapidly degraded. With the more slowly degradable rapeseed meal, formaldehyde treatment did not affect the repletion rate in phosphorus-depleted sheep (Bravo et al., 2003a).

High dietary calcium (>8 g kg^{-1} DM) exacerbates phosphorus deprivation in sheep (Young et al., 1966; Field et al., 1975; Wan Zahari et al., 1990). Increases in dietary calcium (from 3.4 to 5.4 g kg^{-1} DM) decreased A_{Pi} in sheep by 18%, but had no such effect with the low-phosphorus basal diet (1.5 g P kg^{-1} DM) (Field et al., 1983). In a survey of faecal phosphorus in commercial dairy herds, a positive correlation with dietary calcium was noted (Chapuis-Hardy et al., 2004). The possibility that excess calcium can sometimes form insoluble complexes with PyP in the rumen, partially protecting it from degradation, requires urgent investigation.

Feeds for pigs

Feeds are conventionally given values relative to MSP (RA_p) and values for cereals and vegetable proteins are generally low, with large differences between sources, particularly cereals (Soares, 1995). Coefficients for AA_p can be a more useful statistic, although still variable, and the published data for common feeds and blends are summarized in Table 6.3. The low AA_p values for maize compared with wheat (Jongbloed and Kemme, 1990) have been confirmed by Rapp et al. (2001). Studies with low-Py cereals show that Py is a major determinant of AA_p in a given cereal and that AA_p is marginally higher from normal barley than normal maize.

Table 6.3. Coefficients for the apparent absorption of phosphorus (AA_p) from common feeds and blends fed to pigs in 11 studies, selecting data for treatments in which no inorganic P supplements were used. The 'P availability' values, which are commonly used in USA to predict the need for P_i supplementation (NRC, 1998), are also shown. These underestimate AA_p, particularly for feed blends, the single exception being a pelleted maize/soybean meal blend.

Food	AA_p	NRC (1998) available P[b]	References
Barley, normal	0.39, 0.40	0.28	Jongloed and Kemme (1990); Veum et al. (2002)
Maize, normal	0.17, 0.37	0.13	Jongloed and Kemme (1990); Veum et al. (2001)
Wheat	0.41	0.45	Jongloed and Kemme (1990)
SBM, normal	0.38	0.23	Jongbloed and Henkens (1996)
Blends			
Maize/SBM[a] (74:20 on average)	0.32, 0.43, 0.37, 0.16	0.15, 0.15	Spencer et al. (2000); Adeola et al. (2004); Brana et al. (2006)
Maize/barley/SBM (32:30:11)	0.34	0.15	Kies et al. (2006)
Barley/SBM (54:23)	0.54	0.20	Veum et al. (2007)

[a]SBM, soybean meal.
[b]Using measures for the relative availability of the feed when compared to monosodium phosphate, normally given a value of 1.0 (or 100%), adjusted to a probable AA_p of 0.9 (90%).

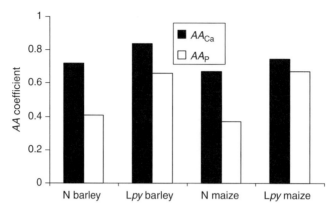

Fig. 6.3. Genetic modification of barley and maize to produce varieties low in phytate (Lpy) greatly increases the apparent absorption (AA) of phosphorus; there is a corresponding though smaller increase in apparent absorption of calcium, and barley is marginally superior to maize as a source of both minerals (Veum et al., 2001, 2002). N, normal.

Blending maize with phytase-rich cereals such as triticale (Pointillart et al., 1987) or by-products such as rice bran (Pointillart, 1991) and wheat middlings (Han et al., 1998) significantly increases AA_P, partly because of increased degradation of PyP from maize. The AA_P values for blends (e.g. Kies et al., 2006) are nearly twice those calculated by summating standard RA_P values for individual feeds, as is the common practice (Table 6.3). This convention is impractical for blends other than corn/SBM because it does not allow for interactions with phytase-rich feeds. Indeed, it may not be sufficiently precise even for corn/SBM blends because the NRC (1998) relies heavily on briefly reported RA_P values (e.g. 0.14 in corn and 0.31 in extracted SBM obtained in single trials) (Cromwell, 1992).

As dietary calcium increases, AA_P decreases in low-phosphorus diets, even those supplemented with phytase (Fig. 6.4). Soaking the diet can increase AA_P (Liu et al., 1997), while pelleting reduces it (Table 6.3) (Larsen et al., 1998). High-moisture corn contains phosphorus of high RA_P (approximately 0.5) (Soares, 1995). Recent evaluations of meat and bone meal (Traylor et al., 2005) placed RA_P above the top of the range in the literature (50–80%), and a high protein content (around 50% crude protein) allows the partial replacement of sources of low RA_P such as SBM.

Feeds for poultry

It is still common practice to consider PyP to be totally unavailable to poultry, following NRC (1994), but this assumption may only be close to the truth for laying hens and newly hatched chicks on corn-based diets. The chick's ability to retain PyP increases dramatically in the second week of life (see Fig. 4.5, p. 62) and many studies have shown significant pre-caecal disappearance of Py

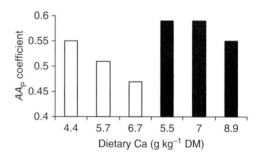

Fig. 6.4. Raising dietary calcium concentrations lowers the apparent absorption of phosphorus (AA_P) from phytase-supplemented diets in weanling pigs given diets of low (3.6 g P kg^{-1} DM, open columns) or normal (4.5 g P kg^{-1} DM, solid columns) phosphorus concentration. The effect is attenuated by the increase in inorganic P supplementation (from 0.7 to 1.6 g P kg^{-1} DM) (data from Qian et al., 1996, pooling the results from two levels of phytase supplementation).

from the gut in older chicks. With corn/SBM diets, 0.19–0.25 of Py is degraded by 2–3-week-old broiler chicks, whether the diet is high (Camden et al., 2001; Tamin et al., 2004) or low (Mangani and Coon, 2006) in calcium. In diets with no added P_i, laying hens absorb less phosphorus from maize than barley (AA_p 0.25 vs 0.34 at the ileum), reflecting the higher Py content of maize (Francesch et al., 2005). However, maize still provides substantial amounts of AA_P. With exogenous phytase added, AA_P rises to 0.51 and 0.58, respectively, clearly indicating the constraint imposed upon absorption by undegraded Py. In hens given a diet relatively low in calcium (30 g kg^{-1} DM), only 10–21% of PyP in a corn/SBM diet is degraded (Van der Klis et al., 1997).

Cereal type and blend is again important. Blending triticale rather than corn with SBM (45:30) in phosphorus-deficient diets for chicks is equivalent to adding up to 0.81 g P_i kg^{-1} DM to the diet (Jondreville et al., 2007). Turkeys may utilize Py much more efficiently: in an early study, an RA_P of 80% was found for maize and cottonseed meal (Andrews et al., 1972). Phosphorus retention was relatively high (50%) in newly hatched poults given a corn/SBM diet that provided only one-fifth of the total phosphorus as P_i (Applegate et al., 2003).

Effects of calcium in poultry

Dietary calcium has a major effect on the utilization of PyP in poultry. In one study with chicks, reducing dietary calcium from 10 to 5 g kg^{-1} DM by removing limestone supplements increased the fraction of Py degraded from 0.51 to 0.65 (Mohammed et al., 1991). In a similar study, the fraction of Py degraded was increased from 0.25 to 0.69 (Tamin et al., 2004). Lowering both calcium and P_i in the diet of chicks increased Py degradation from 0.12 to 0.37 at 18 days (Yan et al., 2005). The impact of excess dietary calcium on phosphorus retention in broiler chicks given a corn/SBM diet is demonstrated in Fig. 6.5. Similar results have been reported by Driver et al. (2005), who stressed that they

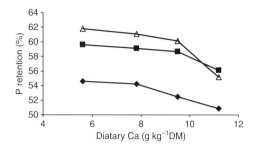

Fig. 6.5. The effects of dietary calcium alone (♦) and with added phytase; without (■) or with vitamin D (△) on phosphorus retention in chicks (data from Qian et al., 1997).

are effects of calcium per se and not of the calcium to phosphorus ratio. In both studies, the effect of calcium excess was far less severe when exogenous phytase was added. Maximal phosphorus retention was achieved when phytase and vitamin D was added to diets 'low' in calcium, giving three distinct 'plateau' values representing the maximal AA_P or potential 'phosphorus value' of the corn/SBM diet: 0.54 for the basal diet; 0.60 with phytase added to the low-calcium diet; and 0.62 with vitamin D also added. Supplementary calcium reduced all measures of phosphorus status in phytase-supplemented broiler chicks and reduced the amount of dialysable phosphorus released in vitro by each of three phytases by up to 10% (Zyla et al., 2004). At the much higher dietary calcium levels fed to laying hens (e.g. 40 g kg^{-1} DM), a 25% reduction in dietary calcium increased Py degradation from 36 to 48% (Van der Klis et al., 1997). However, shell strength would probably be weakened by the end of lay by such a sustained reduction in calcium provision.

Effects of calcium in pigs

Excess dietary calcium can exacerbate phosphorus deprivation in pigs (Lei et al., 1994; Eekhout et al., 1995; Liu et al., 1998). The dominant effect of calcium over calcium to phosphorus ratio on phosphorus metabolism is as marked in pigs as it is in poultry. Effects on AA_P (Fig. 6.4) translate into effects on performance, as Fig. 6.6 shows with data for

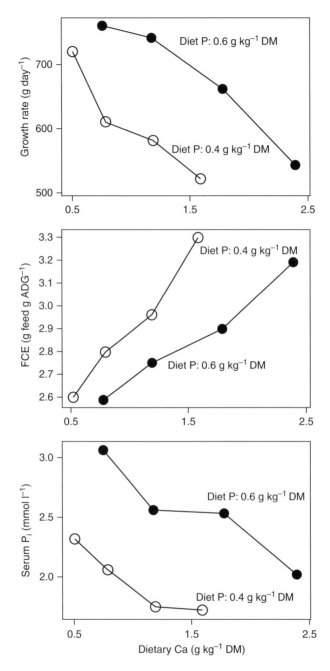

Fig. 6.6. The effects of dietary phosphorus and calcium concentrations on growth, feed conversion efficiency (FCE) and serum inorganic phosphorus (P_i) in 'grower' pigs (data from Reinhart and Mahan, 1986).

newly weaned pigs that were given diets 5% below or 10% above the 1979 NRC requirements for phosphorus and with stepped increases in dietary calcium. The steep gradients for each criterion of phosphorus status, plotted against dietary calcium, indicate large calcium effects, but the similarity of gradients for the two dietary phosphorus levels indicates small calcium–phosphorus interactions. The lower phosphorus level failed to meet requirements, but the smallest addition of calcium increased the deficit.

Phosphorus availability as an attribute of the experiment

From the foregoing review, it is clear that single experiments with pigs and poultry generate coefficients for A_P, AA_P or RA_P that are determined by the following:

- the chosen dietary calcium, P_i and vitamin D_3 levels;
- any inclusion of plant phytases in the chosen cereals or by-products;
- the effects of processing (pelleting) and presentation (wet or dry) on phytase activity; and
- the age of the animal.

Estimates of phosphorus availability are thus characteristic of the experiment as much as the feed, yet have profound effects on estimates of phosphorus requirements and the efficacy of phytase supplements, as described later. High estimated availabilities for phosphorus in corn and SBM in pigs and chicks on calcium-deficient diets overestimate the 'phosphorus value' and are of no practical use (see p. 29). One possible explanation for the low 'NRC available P' values for corn/SBM blends, shown in Table 6.3, is that relatively high dietary calcium levels were used in the relevant trials (Cromwell, 2002). Since calcium is also rendered unavailable by the interaction with P_y, the same criticisms can be made of measurements of A_{Ca} and derived estimates of calcium requirements (see Chapter 4).

New approaches to assessing 'phosphorus value'

Assessments must be based on a meta-analysis to be representative and take into account all known variables. If a simple linear regression technique is applied to data for pigs on diets varying in PyP contribution to total phosphorus (Fig. 6.7), several interesting facts emerge:

- non-PyP in barley/SBM blends has a similar AA to that reported for P_i (0.82);
- PyP has an average AA of 0.29; and
- Applying both coefficients to the data in Fig. 6.2 suggests that AA_p in SBM increases from 0.49 to 0.58 as their total P content increases.

While these results are still characteristic of the experiments, notably the chosen dietary calcium level (5.0–5.5 g kg^{-1} DM), extension of the approach to a larger database could generate reliable AA_p coefficients for factorial estimates of phosphorus requirements.

Requirements for Pigs and Poultry

The phosphorus requirements of pigs and chicks decline as they grow, as do those for calcium, but vary according to the chosen measure of phosphorus status. An early feeding trial generated phosphorus requirements for baby pigs of 4.0 g kg^{-1} DM for normal growth and feed utilization; 5.0 g kg^{-1} DM to maintain normal serum P_i and alkaline phosphatase values and for bone development; and 6.0 g kg^{-1} DM for maximum bone density and breaking strength (Table 6.4). Similar results were obtained by Combs et al. (1991a,c), but such requirements only apply under the particular trial conditions. Responses to added phosphorus in a more recent trial (Traylor et al., 2005) suggest that the corn/SBM diet requires phosphorus additions (as MSP) of about 0.7 g kg^{-1} DM for optimum growth, 1.3 g kg^{-1} DM for optimum feed conversion efficiency (FCE) and at least 2.5 g kg^{-1} DM for optimum femur strength. Traditionally, bone density has set the standard, but environmental considerations demand that it should only be used on welfare or survival (e.g. poultry) grounds, as discussed in Chapter 20. The view that calcium to phosphorus ratios have little independent influence on phosphorus requirements (Underwood and Suttle, 1999) has been endorsed by recent experiments with broiler chicks (Driver et al., 2005) and the foregoing review. Emphasis should be placed on maintaining minimal calcium to Py ratios while meeting calcium and phosphorus requirements, since there are no 'ideal' calcium to phosphorus ratios.

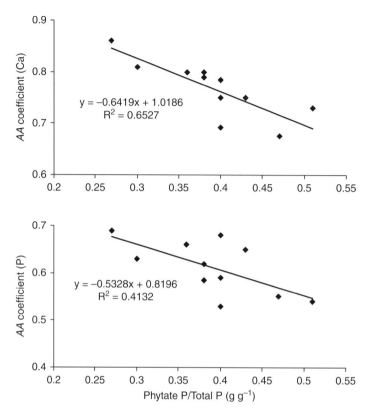

Fig. 6.7. The effect of phytate (*Py*) content of barley/soybean meal diets on apparent absorption (*AA*) coefficients for phosphorus and calcium in young pigs (data from Veum *et al.*, 2002, 2007). *Py* contribution to total P was varied by either genetic manipulation or dehulling the barley and adding inorganic P to some blends. The intercepts denote a fractional *AA* of 0.82 for non-*PyP* and virtually complete absorption of Ca from diets low in *Py*. The downward slopes indicate: (i) inhibitory effects of *Py* on the availability of total P and Ca, the effect on Ca being slightly the greater; (ii) an *AA* for *PyP* of 0.29; and (iii) a value relative to non-*PyP* of 0.32.

Table 6.4. Effects of dietary phosphorus concentrations on selected performance, skeletal and blood indices of P status in the baby pig given a synthetic milk diet from 0–6 weeks of age (trial 1) (Miller *et al.*, 1964).

	Dietary P (g kg^{-1} DM)			
Criterion	2	4	6	Index of P status
Live-weight gain (g day^{-1})	130	290	290	
Dry matter intake (g)	240	380	370	Performance
Feed/gain (g g^{-1})	1.85	1.31	1.28	
Serum inorganic P (mmol l^{-1})	1.03	2.71	3.03	
Serum Ca (mmol l^{-1})	3.20	3.05	2.58	Blood
Serum alkaline phosphatase (U l^{-1})	20.5	8.3	4.5	
Humerus ash (%)	33.4	44.1	47.5	
Femur (specific gravity)	1.11	1.15	1.17	
Eighth rib weight (g)	3.30	4.49	4.67	Bone
Femur breaking load (kg)	24	61	81	

Gross requirements for growing pigs

Factorial estimates of phosphorus requirement have a broader application than estimates from feeding trials and criterion of requirement may be of lesser importance in the weaned animal (Fig. 6.8). However, the choice of values for maintenance requirements (M) and A in the factorial model (see Chapter 1) is crucial and the use of a single value to cover diets varying in Py and calcium is unrealistic. Table 6.5 gives phosphorus requirements for diets with AA_P values of 0.33 or 0.50, ignoring M_P as a separate model component because of the difficulties

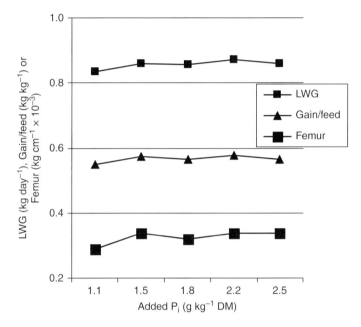

Fig. 6.8. A test of new phosphorus requirements of growing pigs (Table 6.5) by feeding trial. Pigs of 34 kg LW were given a corn/soybean basal diet containing 4.5 g P and 5.3 g Ca kg^{-1} DM. The addition of 1.5 g inorganic P (P_i), as monocalcium phosphate, was optimal for LW gain (LWG), feed conversion efficiency (FCE) and bending moment of the femur after 26 days (Hastad et al., 2004). Thus, providing 6 g total P kg^{-1} DM satisfied all needs: the 7.5 g recommended in Table 6.5 for such high PyP diets appears to be generous.

Table 6.5. Factorial estimates[a] of the dietary phosphorus requirements of growing pigs, illustrating the impact of dietary phytate (Py) and stage of growth on the need for P. The estimates should be used in conjunction with those for calcium given in Table 4.3 for optimum use of PyP.

Live weight (kg)	Growth rate (kg day^{-1})	Assumed DMI (kg day^{-1})	P needed (g kg^{-1} DM)	
			High PyP	Low PyP
5	0.1	0.3	9.2	9.2
25	0.3	1.05	9.5	6.3
45	0.6	1.8	7.5	5
90	0.9	2.8	4.5	3.0

[a]The simplified factorial model assumes net P requirements for growth of 10.0 g kg^{-1} LWG up to 25 kg, 7.5 g kg^{-1} LWG up to 50 kg and 5.0 g kg^{-1} LWG up to 90 kg LW (ARC, 1981) and the following apparent absorption coefficients for P AA_P: baby pigs, 0.8; weaned pigs, 0.33 for high-phytate corn/SBM diets and 0.5 for low-phytate diets that provide phytin P AA_P 0.2) and non-phytin P (AA_P 0.8) in equal proportions.

mentioned earlier (p.127), and with the minimum necessary calcium provision (see Chapter 4). Requirements derived with the higher AA_P value apply to diets with cereals other than corn, animal protein components or added phytase, while the higher requirements derived with the lower AA_P apply to corn/SBM diets with no added phytase. The validity of these estimates is hard to assess from feeding trials based on the old practice of measuring responses to P_i added to a diet (e.g. Fig. 6.8) containing an unreliably determined concentration of 'NRC available phosphorus', but are supported by feeding trials with corn/SBM (Fig. 6.6) and wheat/SBM diets (Eeckhout et al., 1995). The withdrawal of supplementary phosphorus from rations for finishing pigs may not impair performance (O'Quinn et al., 1997). While some studies appear to disagree with this finding (e.g. Hastad et al., 2004; Shelton et al., 2004; Jendza et al., 2005), the advantage associated with the addition of P_i may have been attributable to the unnecessarily high or low calcium levels used.

Requirements for boars, gilts and sows

Although the supply of sufficient phosphorus to obtain maximum bone strength may appear extravagant, bone strength is important in boars (ARC, 1990) and they should ideally be fed phosphorus (and calcium) at the higher level of 7.5 g P kg^{-1} DM from Table 6.5. Gilts should be fed similar generous amounts of phosphorus (and calcium) during the first reproductive cycle. Nimmo et al. (1981a) found that a high proportion of gilts (30%) became lame when fed 5.5 g P (and 7.2 g Ca) kg^{-1} DM during gestation, whereas those given 8.3 g P (and 10.7 g Ca) kg^{-1} DM did not. There were associated differences in bone strength, but no residual effects on offspring at the lower levels of supplementation (Nimmo et al., 1981b).

Gross requirements for poultry: growth

NRC recommendations for growing chicks and hens have declined over the years, but are still too generous. Keshavarz (2000a) slashed the NRC (1994) sliding scale for Leghorn chicks by 50% or more and found no reduction in performance or bone quality (Table 6.6). Response curves to added P_i in broiler chicks (Augsburger and Baker, 2004) indicate that 2 g P_i kg^{-1} DM is close to an optimal dietary supply. The NRC provision for growing turkeys has increased over the years, and the

Table 6.6. Dietary requirements of non-phytate phosphorus (N*Py*P) for growth in Leghorn or broiler chicks and turkey poults, as recommended by the NRC (1994). In the case of Leghorn chicks, far less N*Py*P can be fed without impairing performance or bone quality, according to the results of Keshavarz (2000a), given in parentheses, and the lower, bold estimates should be used in current practice.

		Growth stage*		
		Early	Middle	Late
N*Py*P (g kg^{-1} DM)	Leghorn chick	4.0 (**2.0**)	3.5 (**1.5**)	3.2 (**1.0**)
	Broiler chick	4.5	3.5	3.0
	Poult	6.0	4.2	2.8

	Age (weeks)			Dietary energy density (kcal me kg^{-1} DM)		
	Leghorn	Broiler	Poult	Leghorn	Broiler	Poult
*Early	0–6	0–3	0–4	2850	3200	2800
Middle	6–18	3–6	8–12	2850	3200	3000
Late	>18	6–8	20–24	2900	3200	3300

contrast with chicks cannot be linked to changes in the type of bird, its rate of growth or efficiency of feed use because similar changes have occurred in the different species. The sliding scale (i.e. phase feeding) for turkey poults given in Table 6.6 has been supported by subsequent studies with turkey poults up to 21 days of age (Applegate et al., 2003), in which 5.5 g non-PyP kg^{-1} DM was sufficient for growth, but did not allow maximal bone mineralization. Temporary underfeeding may not be a major concern because the 18-day-old broiler chick can compensate for previous phosphorus (and calcium) deprivation (Yan et al., 2005).

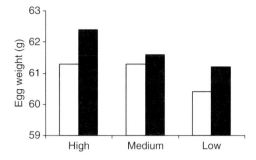

Fig. 6.9. The effects of phosphorus nutrition (high, medium or low) and the addition of phytase (solid columns) on mean egg weight in the final phase (age 54–66 weeks) of a study with laying hens (Keshavarz, 2000b). Inorganic phosphorus (P_i) was added to a corn/soybean meal diet to provide 3 (high), 1.5 (medium) or 1.0 (low) g P_i kg^{-1} DM, having been stepped down by 1.0 in each of two preceding 2-week phases for each group. Only the low level reduced egg weight, whereas the benefit from adding phytase was evident at all phosphorus levels, suggesting that this benefit comes partly from factors other than enhanced phosphorus availability.

Gross requirements for poultry: egg production

The onset of lay increases the need for phosphorus more than would be expected from the small amount of phosphorus in the egg (about 120 mg) because the increase in phosphorus catabolism associated with egg production increases endogenous losses. Early studies suggested that 50% more P_i was needed to prevent 'cage layer fatigue' than to maintain egg production (equivalent to 3 versus 2 g P_i kg^{-1} DM) (Owings et al., 1977) but the NRC (1994) reduced its recommendation to the equivalent of 2.5 g P_i kg^{-1} DM for all stages of lay. Assuming that phosphorus requirements decrease as lay progresses, Keshavarz (2000b) reduced the phased provision of P_i in a corn/SBM diet from 3.0–2.5–2.0 to 1.5–1.0–0.5 g kg^{-1} DM for weeks 30–42, 42–54 and 54–66 of age, respectively, without detriment to health or performance. A further decrement (−0.5 g at each stage) impaired production (Fig. 6.9). Adding a uniform 0.5 g P_i kg^{-1} DM throughout lay to a similar diet sustained egg production and quality, although LW and tibia ash were suboptimal by the end of lay (Boling et al., 2000a). Calcium was provided in both studies at a constant level (3.8 g kg^{-1} DM) throughout lay rather than allowing for a steady increase in requirement from 3.5 to 4.2 g kg^{-1} DM (see Chapter 4).

With phased increases in calcium provision, the uniform addition of 1 g P_i kg^{-1} DM is advocated, far less than the latest NRC (1994) guideline of 2.5 g P_i kg^{-1} DM.

Requirements are affected by environmental conditions. High temperatures reduce feed consumption and hence raise the required dietary phosphorus concentration (NRC, 1994). Floor-housed birds recycle phosphorus through coprophagy and their requirements are therefore lower than those of caged hens (Singsen et al., 1962).

An alternative framework for estimating phosphorus requirements

More precise allowances should be made for the available phosphorus from cereal/protein concentrate blends than the widely used convention of calculating 'available phosphorus' from average NRC (1994) feed composition values. For poultry, the common assumption is that corn/SBM diets yield 1 g non-PyP kg^{-1} DM. However, the analysed non-PyP can be 30% greater than the calculated value (e.g. Applegate et al., 2003) and the variation in non-PyP in SBM is so wide (Fig. 6.2) that, at a

25% inclusion rate, the richest samples should meet the total requirement for non-PyP. Furthermore, roughly one-third of the PyP consumed was not excreted in Kesharvarz's (2000b) study, and presumably contributed about 0.7 g A_P kg^{-1} DM. Corn/SBM diets for laying hens may not always need additions of P_i.

Phosphorus Requirements for Ruminants

The literature has been reviewed for all cattle by Underwood and Suttle (1999) and separately for beef and dairy cows by NRC (1996 and 2001, respectively). Estimates of phosphorus requirements from different authorities vary, even when the same method (i.e. factorial modelling) is used, because different values for M and A are used in the models (AFRC, 1991).

Net requirement for maintenance: sheep

A marginal improvement in A_P reduces IFE_P slightly when the phosphorus supply is inadequate (Ternouth, 1989; AFRC, 1991; Coates and Ternouth, 1992), but IFE_P at a marginal phosphorus intake should determine M_P. The AFRC (1991) used a 'roughage factor' to increase M_P by 60% for diets consisting principally of roughage on the assumption that roughages would stimulate salivary phosphorus output and cause more absorbed phosphorus to be lost in faeces. The precise value of 1.6 was based on the difference in FE_P between sheep fed entirely on loose dry roughage and those given a pelleted concentrate, providing similar 'loads' of absorbed phosphorus at least 30% above their basal requirement. This was probably an overestimate.

A single new estimate of M_P has been published: 0.61 g P day^{-1} for growing goats on a hay/concentrate (60/40) diet (Vitti et al., 2000). This is equivalent to only 0.43 g P kg^{-1} DMI, less than the AFRC's (1991) calculation for highly digestible diets (0.6 g P kg^{-1} DM). Since A_P is subject to genetic variation in sheep, marked hereditable differences in IFE_p are found both within and between breeds (Field et al., 1984, 1986) and some appropriate allowance might be needed in providing adequate phosphorus to all sheep. On pelleted, energy-dense diets that stimulate little salivary secretion, significant urinary phosphorus excretion may occur at relatively lower phosphorus intakes that fail to allow complete mineralization of bone, despite normal plasma P_i, and the amounts should be added to M_P (Braithwaite, 1984a). However, the majority of sheep and cattle fed roughage diets excrete very little phosphorus in their urine, even at plasma P_i levels as high as 2 mmol l^{-1} (AFRC, 1991).

Net requirement for maintenance: cattle

In the absence of data, the AFRC (1991) used its 'roughage adjusted' value for M_P (1 g P kg DMI^{-1}) for cattle fed on diets consisting mostly of roughage, whereas the NRC (2001) used the roughage factor for all dietary circumstances. There is now tentative indirect evidence for a 'long, dry roughage effect' in cattle. A low AA_P coefficient (0.08) recorded during the dry period with cows given a low-phosphorus diet consisting mostly of wheat straw (Valk et al., 2002) is compatible with an M_P of 1 g kg^{-1} DMI, but the higher AA_P (0.49–0.63) during lactation on corn silage suggests a much lower value for M_P. Replacing unchopped alfalfa hay (3%) with ground soy hulls (10%) in a dairy ration consisting mainly of silage and providing 3.3 g P kg^{-1} DM, increased AA_P from 0.34 to 0.44 (Wu, 2005), despite the small increase in total dietary fibre intake. The effect may be smaller in animals fed closer to their phosphorus requirement. The 'roughage effect' may not apply to grazing beef cattle, in which far lower estimates of M_P have been recorded (Ternouth et al., 1996; Ternouth and Coates, 1997). Although increases in the alfalfa silage content of a diet (from 48 to 58%) of lactating cows tend to reduce the mean AA_P (from 37 to 31%) (Wu et al., 2003), this would partly reflect the associated reduction in milk yield and loss of phosphorus in milk rather than a 'succulent forage' effect, surplus phosphorus having been lost via the faeces. The AFRC (1991) roughage factor (1.6) should only be used for sheep and cattle fed exclusively on dry roughages.

Absorption coefficients

The NRC (2001) modified the AFRC (1991) model by using higher A_P coefficients of 0.7 for concentrates and 0.64 for roughages instead of 0.58 for all feeds, but gave no supporting evidence for these modifications. Underwood and Suttle (1999) used an A_P coefficient of 0.74 for all feeds, giving lower estimates of phosphorus requirement than the NRC. The literature reviewed above would support a small differentiation between feeds with an A value of 0.74 for dry roughages and 0.80 for grass, succulent forages and concentrates.

New estimates of requirements for ruminants

Two different models are proposed for diets known or likely to stimulate low (grass, silage or roughage-free diets) or high (dry roughages) salivation rates with respective values for M_P of 0.6 and 1.0 g P kg^{-1} DMI and slight differences in the respective A_P of 0.8 and 0.7. The resultant requirements for cattle are given in Table 6.7. The requirements estimated with the 'low roughage' model for beef cattle match findings in grazing animals in Queensland (Hendricksen et al., 1994; Ternouth et al., 1996; Ternouth and Coates, 1997) and 'finishing' steers, gaining 1.7 kg day^{-1} on a roughage-free diet (Erickson et al., 1999). The latter showed no response in either growth rate or bone quality to phosphorus additions to a diet containing only 1.4 g P kg^{-1} DM, even when dietary calcium was raised from 3.5 to 7.0 g kg^{-1} DM. Feeding trials with dairy cows with a mixed diet (Valk and Sebek, 1999), probably falling between 'low' and 'high' in terms of the models, concluded that 2.4 g P kg^{-1} DM was inadequate for lactating dairy cows, while 2.8 g P kg^{-1} DM sustained both health and yield (30 kg day^{-1}). Several trials have merely shown no adverse effects when a generous level of phosphorus provision for the dairy cow was reduced. In the most demanding study, feeding 3.1 g P kg^{-1} DM for 2–3 years to cows yielding 40 kg milk day^{-1} (throughout two lactations for most cows) had no adverse effects on health or milk yield compared with 3.9 or 4.7 g P kg^{-1} DM (Wu et al., 2001). Rib bone ash fell from 56.2% to 53.9% DM, but without affecting bone strength.

The vast majority of mixed diets for high-yielding cows will not need phosphorus supplementation. Similar distinctions between diet types are made in the recommendations for sheep (Table 6.8) on the two different diet types, following the arguments made by Scott et al. (1984, 1985). Feeding to minimum requirements rather than exceeding generously the NRC recommendations, a common practice, will greatly reduce environmental phosphorus pollution (see Chapter 20).

Phased phosphorus feeding for ruminants

Feeding trials have usually employed constant dietary phosphorus concentrations, yet phosphorus requirements clearly decline during growth and lactation. In the study of Wu et al. (2001), faecal phosphorus increased steadily throughout lactation from approximately 5 to 6.5 g kg^{-1} faecal DM at the lowest phosphorus level (3.1 g kg^{-1} DM), indicating an increasing excess of supply over need. In the study by Valk et al. (2002), a 'deficient' phosphorus level (2.4 g kg^{-1} DM) allowed the retention of up to 3.2 g P day^{-1} as lactation progressed and milk yield declined. However, the answer is not to feed more phosphorus in early lactation because this does not correct a negative balance in either cows (Knowlton and Herbein, 2002) or ewes (Braithwaite, 1983) when phosphorus is being unavoidably mobilized from the skeleton. Requirements for phosphorus need *not* be met on a day-to-day basis. Instead, phosphorus should be briefly and moderately overfed, along with calcium, during late lactation and/or the early dry period to provide a sufficient skeletal reserve for late gestation and early lactation. In practice, decreases in the amount of concentrates fed to cows as lactation progresses will normally allow the pattern of phosphorus supply to follow need.

Table 6.7. Requirements of cattle for dietary phosphorus at the given DMI. Modified from the AFRC (1991)[a] by using two different factorial models for diets that promote low (L) or high (H) salivary flow rates (e.g. fresh pasture and hay, respectively).

	Live weight (kg)	Production level	DMI (kg day^{-1}) L–H	Need for P (g kg^{-1} DM) H	Need for P (g kg^{-1} DM) L
Growing cattle	100	0.5 kg day^{-1}	1.7–2.8	3.8	3.3
		1.0 kg day^{-1}	2.4–4.5	4.3	4.4
	300	0.5 kg day^{-1}	3.4–5.7	2.4	1.7
		1.0 kg day^{-1}	4.7–8.3	2.7	2.3
	500	0.5 kg day^{-1}	6.1–10.9	1.6	1.0
		1.0 kg day^{-1}	6.5–11.6	2.3	1.8
Pregnant cow (dairy/beef)	600	23 weeks	4.0–6.3	1.7	0.8
		31 weeks	4.7–7.2	1.9	1.0
		39 weeks	6.1–9.1	2.1	1.3
Calf 40 kg at birth		Term	7.5–11.2	2.2	1.5
Lactating cow	600	10 kg day^{-1}	12.0 (m)	2.6	–
			9.9 (bm)	2.8	–
		20 kg day^{-1}	11.4 (m)	n.a.	2.3
			10.1 (bm)	n.a.	2.5
		40 kg day^{-1}	19.3 (m)	n.a.	2.6
			17.8 (bm)	n.a.	2.8

bm, fed below maintenance; m, fed to maintain body weight; n.a., unattainable performance.
[a]In the AFRC (1991) factorial model (see Chapter 1), where need = $(M + P) / A$, the absorption coefficient (A) is increased to 0.8 for L and 0.7 for H diets.

Table 6.8. Requirements of sheep for dietary phosphorus at the given DMI. Modified from the AFRC (1991)[a] by using two different factorial models for diets that promote low (L) or high (H) salivary flow rates (e.g. fresh pasture and hay, respectively).

	Live weight (kg)	Production level/stage	DMI (kg day^{-1}) L–H	Need for P (g kg^{-1} DM) H	Need for P (g kg^{-1} DM) L
Growing lambs	20	100 g day^{-1}	0.40–0.67	2.4	3.0
		200 g day^{-1}	0.57	n.a	4.1
	40	100 g day^{-1}	0.66–1.11	1.9	2.1
		200 g day^{-1}	0.93–1.77	2.1	2.5
Pregnant ewe carrying twins[b]	75	9 weeks	0.71–1.10	1.5[d]	1.1[d]
		13 weeks	0.85–1.28	1.9	1.7
		17 weeks	1.13–1.68	2.1	2.1
		Term	1.62–2.37	2.0	1.9
Lactating ewe	75	2–3 kg milk day^{-1}	1.8 (m)[c]	n.a.	2.8
			1.5 (bm)	n.a.	3.2

bm, fed below maintenance; m, fed to maintain body weight; n.a., unattainable performance.
[a]Predictions from two models, as described in Table 6.7.
[b]Requirements for small ewes carrying single lambs are similar, assuming that they will eat proportionately less DM.
[c]Requirements are influenced by the ability of the diet to meet energy need and prevent loss in body weight (m); diets that allow loss of body weight at 0.1 kg day^{-1} (bm) are associated with higher requirements.
[d]Sufficient for dry ewes.

Phosphorus requirements for horses

Experiments with young growing ponies (Schryver et al., 1974) and with mature horses (Schryver et al., 1970, 1971) indicate relatively high requirements for the younger animals. The net requirement for growth has increased over the years because of an increased rate of growth (Underwood and Suttle, 1999). An estimated phosphorus need of 0.7 g kg^{-1} DM for grazing Thoroughbred foals (200 kg LW and gaining 1 kg day^{-1}) has been derived factorially from the mineral composition of the growth increment (8.3 g kg^{-1} LWG), an M_P of 10 mg kg^{-1} LW (ARC, 1980) and an A_P coefficient of 0.5 (Grace et al., 1999). Although use of the higher AFRC (1991) value for M_P would increase the estimated requirement from 6.0 to 8.7 g day^{-1} for foals consuming 6 kg DM day^{-1}, A_P may well be proportionately higher, giving a similar factorial estimate. Either way, growing horses are unlikely to receive less phosphorus than they need from natural feeds, including grains.

Biochemical Consequences of Phosphorus Deprivation

When phosphorus requirements are not met by the feeds and forages on offer, the sequence of events (Fig. 6.10) differs markedly from the general scheme for most minerals. Relative rates of change in different compartments vary with the circumstances, and this can be better understood if the norm is sketched out first.

Normal fluctuations in plasma phosphorus

It is arguable whether there is any specific hormonal regulation of plasma phosphorus (Underwood and Suttle, 1999). Concentrations can fluctuate widely in healthy animals since they simply reflect the equilibrium between inflows and outflows of phosphorus. Serum P_i rises after feeding and is lowered by handling stress and the approach of parturition. In dairy cows, large increases in phosphorus supply

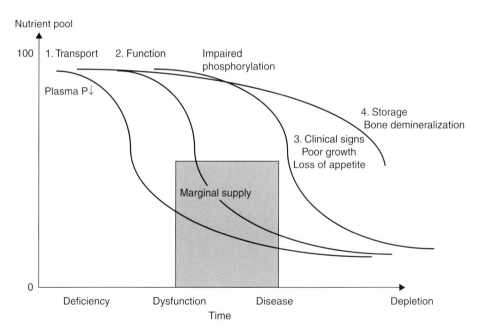

Fig. 6.10. The sequence of pathophysiological changes that occurs when livestock are given an inadequate dietary phosphorus supply. Unlike with most minerals, the transport pool shows an early decline and appetite is lost long before the skeleton becomes clinically affected.

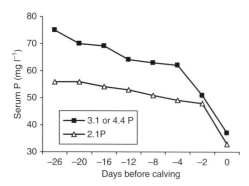

Fig. 6.11. Raising the dietary phosphorus concentration (from 2.1 to 3.1 or 4.4 g kg^{-1} DM) prior to calving does not prevent a rapid periparturient decline in serum P to 'subnormal' levels in the dairy cow (from Peterson et al., 2005).

prior to parturition cannot prevent a decline in serum phosphorus to levels that might be classed as 'subnormal' (Fig. 6.11). Serum phosphorus rose to about 70 mg l^{-1} in all groups on a common regimen (4.4 g P kg^{-1} DM) within 4 days of calving, presumably supplied partly from bone resorption (see Chapter 4), but then declined (Peterson et al., 2005).The supply of phosphorus relative to calcium for milk production from bone is imbalanced, with a calcium to phosphorus mass ratio in milk of only 1.3:1 as opposed to 2:1 in bone.

There are clear relationships between growth rate or bone ash and serum P_i in the pig and chick and the optimum serum P_i is around 2.5 mmol l^{-1} (Underwood and Suttle, 1999), nearly twice that considered optimal for the lamb. Plasma parathyroid hormone falls during phosphorus deprivation in the pig, presumably in response to a rise in plasma calcium (Somerville et al., 1985).

For ruminants in particular, different ranges are required for different stages of development and levels of production. The suckling lamb has a higher serum P_i than its dam until it is weaned and should be accorded a high normal range (>1.7 mmol l^{-1}) (Sharifi et al., 2005), as should the milk-fed calf. With little 'drain' of phosphorus into saliva, higher plasma P_i can probably be sustained at a given phosphorus intake, prior to weaning. If this is the case, the 'drain' of phosphorus into the milk might lower what is considered to be a 'normal' serum P_i for the cow and lower M_P by reducing losses via saliva.

Normal bone mineralization

Bone mineralization is also largely free from phosphorus-specific hormonal regulation, but varies between species and with the stage of development. In the pig, bone mineralization increases substantially after weaning even while growth is being retarded by low phosphorus intakes (Fig. 6.12). In broiler chicks, the ash content of the tibia can increase by 25% between 16 and 32 days in the face of phosphorus deprivation (Driver et al., 2005). In cows, the phosphorus concentration in rib biopsy samples can change markedly in early lactation (Fig. 6.13) without necessarily indicating the onset of phosphorus deprivation. The effects of phosphorus deprivation on bone density will depend on the stage of development or reproduction at the outset.

Depletion

In ruminants, the pool of phosphorus in the rumen constitutes a store. Rumen phosphorus concentrations decline on diets low in phosphorus partly because of a reduction in the flow of salivary phosphorus into the rumen (Wright et al., 1984; Challa et al., 1989; Valk et al., 2001). However, the dietary and physiological factors that influence salivary flow and plasma P_i affect the rumen phosphorus concentration attained on a given phosphorus intake. For example, rumen P_i concentrations in lambs are 66% lower if their diet, marginal in phosphorus (1.9 g kg^{-1} DM), contains 6.8 rather than 3.5 g Ca kg^{-1} DM (Wan Zahari et al., 1990). All growing animals adapt to phosphorus deprivation through changes in bone growth and mineralization, with the adaptation occurring sooner in non-ruminants than ruminants. Net demineralization (Table 6.9) involves decreases in the amount of bone matrix in the limb bones and in the degree of matrix mineralization in cancellous bones (Koch and Mahan, 1985). Weanling lambs deprived of phosphorus can increase their skeletal size without requiring more phosphorus (Table 4.6, p. 72), but the ash and organic

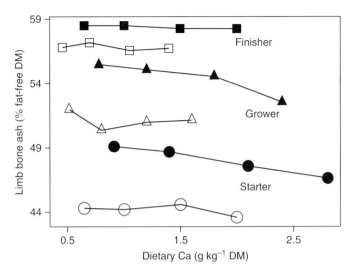

Fig. 6.12. The effects of dietary phosphorus and age on the mineralization of limb bones in young pigs. Note the large increases in bone ash from the starter to the finisher stage, regardless of whether diets low (open symbols: 0.5, 0.4 and 0.3 g P kg^{-1} DM) or high in phosphorus (closed symbols: 0.7, 0.6 and 0.5 g P kg^{-1} DM for successive stages) were fed (data from Reinhart and Mahan, 1986).

matter content of bones is again reduced (Field et al., 1975). In ruminants, most reports of demineralization are for final slaughter groups, with the first coming from a study of prolonged phosphorus deprivation in sheep (Stewart, 1934–5). These studies show that long bones are depleted more rapidly than the more widely studied metacarpals or metatarsals (Little, 1984). In young calves depleted of phosphorus, the ends of the long bones show greater reductions than the shaft, while the ribs show no change after short-term (6 week) depletion (Fig. 6.14) (Miller et al., 1987).

Deficiency

In young lambs given diets low in phosphorus (1.2 g kg^{-1} DM), onset of deficiency (hypophosphataemia) is delayed from 5 to 10 weeks when the diet is also low in calcium (0.66 versus 4.3 g kg^{-1} DM) (Fig. 6.15). This may be partly due to a decrease in phosphorus retention by the skeleton in animals also deprived of calcium (Braithwaite, 1984b; Field et al., 1985; Rajaratne et al., 1994). Other abnormalities in serum composition can occur at this stage, including a rise in alkaline phosphatase (Table 6.4), a small rise in serum calcium (Fig. 4.12, p. 71) due to bone resorption and a fall in osteocalcin reflecting impaired bone formation. The last may be an effect rather than a cause and can occur without an increase in bone resorption (Scott et al., 1994). Increases in renal 1α hydroxylase activity, plasma 1,25-dihydroxyvitamin D$_3$ (1,25-(OH)$_2$D$_3$) and intestinal calbindin have been reported in the phosphorus-depleted pig or chick (for review, see Littledike and Goff, 1987; Pointillart et al., 1989), but there is no such response in sheep (Breves et al., 1985). There may be teleological explanations for species contrasts because the risk of phosphorus deprivation is endemic in seed-eating monogastric animals that are left without phosphorus supplements, but not in ruminants. However, species differences may be secondary consequences of aberrations in calcium metabolism, such as other hormonal changes and the responses to vitamin D in pigs (Pointillart et al., 1986; Lei et al., 1994) and poultry (Mohammed et al., 1991; Biehl et al., 1995). The demineralization of bone continues during the deficiency phase (e.g. in dairy cows) (Wu et al., 2001).

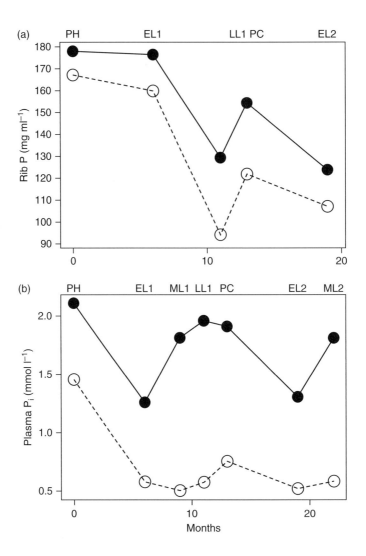

Fig. 6.13. The effects of lactation and phosphorus status on (a) P concentrations in rib biopsy; (b) plasma inorganic P(P$_i$). Samples were taken from two groups of beef cows, one receiving supplementary phosphorus (●) and one not (○), grazing phosphorus-deficient pasture. Note that consistently low plasma P$_i$ from month 6 did not reflect the skeletal changes (data from Engels, 1981). PH or PC, pregnant heifer or cow; EL, ML and LL, early, mid- or late lactations.

Dysfunction

There are limits to the extent to which growth and bone strength and shape can be sustained by a demineralizing skeleton, particularly in species with high relative growth rates such as the pig. Growth is probably limited by a reduction in appetite, which then becomes a sign of dysfunction (Fig. 6.10). Eventually the breaking and shear strength of bone weakens (Combs et al., 1991c). Histological examination of the epiphyseal growth plate in young animals reveals other signs of dysfunction. In comparisons of the rachitic long bones from phosphorus- and calcium-deprived chicks, the former could be distinguished by elongated metaphyseal vessels within the physis and by hypertrophic chondrocytes (Lacey and Huffer, 1982) and a less widened growth plate (Edwards and Veltman, 1983). Similar changes

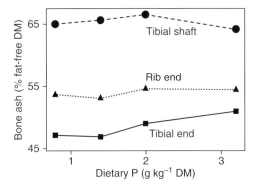

Fig. 6.14. The effects of sample site on the degree of bone demineralization recorded after short-term phosphorus depletion in calves (Miller et al., 1987).

are seen in lambs deprived of phosphorus, particularly in the rib (Field et al., 1975).

Clinical Manifestations of Phosphorus Deprivation

Poor appetite

Loss of appetite can cause weight loss, an early characteristic of phosphorus deprivation in both growing and mature animals of all species. This is illustrated by data for yearling cattle in Table 6.9 (Little, 1968) and for young pigs in Table 6.4.

Anorexia usually takes several weeks or months to develop in lambs (Wan Zahari et al., 1990), beef cattle (Call et al., 1978, 1986; Gartner et al., 1982) and dairy cows (Call et al., 1986). Appetite can be closely related to serum P_i, but may be governed by gut-based as well as systemic mechanisms (Ternouth and Sevilla, 1990a,b), including changes in rumen microbial activity. In one study (Wan Zahari et al., 1990), rumen phosphorus fell to <3 mmol l^{-1}, the minimum required by rumen microbes. Early reductions in appetite in ruminants have been reported when semi-purified diets very low in phosphorus have been fed (Field et al., 1975; Miller et al., 1987; Ternouth and Sevilla, 1990a). These diets stimulate little or no rumination – depriving the rumen microflora of salivary as well as dietary phosphorus – and cause relatively severe anorexia. Reduced egg production in phosphorus-deprived hens has been attributed to a reduction in food intake (Bar and Hurwitz, 1984). However, growth can be retarded before appetite is impaired in newly weaned pigs deprived of phosphorus (Fig. 6.12) (Underwood and Suttle, 1999).

Pica

Loss of appetite is often paralleled by a craving for and a consumption of abnormal materials such as soil, wood, flesh and bones. The

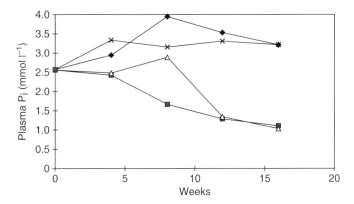

Fig. 6.15. The effects of diets low in phosphorus and/or calcium on plasma inorganic phosphorus (P_i) in lambs. Note that feeding a diet low in both minerals (△) delays the fall in P_i compared to a diet low only in phosphorus (■); feeding a diet low in calcium but adequate in phosphorus (◆) raises P_i when compared with a diet adequate in both minerals (×) (from Field et al., 1975).

Table 6.9. The effect of phosphorus supplementation on growth and feed consumption of yearling cattle fed on Townsville lucerne (*Stylosanthes humilis*) containing 10% crude protein and 0.07% phosphorus for 8 weeks (mean values ± standard error) (from Little, 1968).

Rate of supplementation with P	P intake (g day^{-1})	Weight gain (kg week^{-1})	DMI (kg day^{-1})	Blood P (mg l^{-1})[a]
Nil	3.3 ± 0.1	0.9 ± 0.2	5.2 ± 0.1	33 ± 1.0
Low	7.4 ± 0.1	1.7 ± 0.2	5.8 ± 0.4	60 ± 1.0
Medium	11.4 ± 0.1	1.7 ± 0.2	6.0 ± 0.2	63 ± 3.0
High	15.5 ± 0.3	2.7 ± 0.7	6.0 ± 0.5	67 ± 0.4

[a]Divide by 31 to give mmol l^{-1}.

depraved appetite (pica) may be general (allotriophagia) or specific (osteophagia, a craving for bones; sarcophagia, a craving for flesh). Pica is not a specific sign of phosphorus deprivation and has been observed in animals suffering from lack of sodium, potassium, energy and protein (Underwood *et al.*, 1940). However, bone-chewing in cows has been induced by exteriorizing the parotid salivary duct and feeding a low-phosphorus diet, thus preventing the recycling of phosphorus, and stopped within an hour of intravenously infusing sodium phosphate (Na$_2$PO$_4$) (D. Denton, personal communication, 1980). Pica has been disastrous in areas such as South Africa, where carcasses were infected with *Clostridium botulinum* because toxin formation during the process of putrefaction caused botulism (toxic paralysis) and death. This stimulated investigations that led to the discovery of phosphorus deprivation (Theiler, 1912). Losses of sheep and cattle from botulism have been reported from several parts of the world due to the widespread occurrence of *C. botulinum* and of phosphorus deprivation (Underwood *et al.*, 1939), but animals can now be vaccinated against the disease.

Reproductive disturbances

Poor lamb and calf crops were recurring features of early studies on the severely phosphorus-deficient veldt, but spectacular responses to supplementary phosphorus given as bone meal may have been attributable to the additional protein thus given. Subnormal fertility, depressed or irregular oestrus and delayed conception, responsive to phosphorus alone, have subsequently been reported in cattle in the same area, but sheep have not been affected (Read *et al.*, 1986a,b). In long-term experiments, no adverse effects on age of puberty or pregnancy have occurred in Hereford heifers fed on a low-phosphorus (1.4 g kg^{-1} DM) diet for 2 years (Call *et al.*, 1978), or on conception rate during rebreeding (Call *et al.*, 1986). Infertility did not develop until the seventh year and by that time other symptoms of deprivation had developed. In dairy cows, appetite and milk yield can be reduced by phosphorus deprivation while fertility remains normal (Call *et al.*, 1987). Improvements in reproduction rate have rarely been recorded under Australian conditions, and where they occur they are probably secondary to improved body weight and condition at conception (Winks, 1990). However, delayed conception causing temporary infertility has been observed in dairy cows in Australia mated during full lactation with marginal hypophosphataemia (Snook, 1958). Improved fertility in dairy cows, also with slightly subnormal serum P$_i$, has been reported after defluorinated superphosphate was added to the drinking water (Scharp, 1979). The lucerne (alfalfa) pasture being grazed was apparently adequate in phosphorus (3.9 g kg^{-1} DM) but high in calcium (15.5 g kg^{-1} DM). Following phosphorus supplementation, the first-service pregnancy rate increased from 36.5% to 63.2%, the mean calving to conception interval decreased from 109 to 85 days and the number of cows culled each year for infertility decreased from 15 to five. Similar 'before and after' responses to phosphorus supplements have been reported in cows on red-clover-dominant pastures containing 2.9 g P kg^{-1} DM, but again rich in calcium (7.0–14.5 g kg^{-1} DM) (Brooks et al., 1984).

Abnormalities of bones

The skeletal abnormalities associated with phosphorus deprivation are almost identical to those described for calcium deprivation (see Chapter 4) because bone mineral cannot be produced if either mineral is lacking. This has been illustrated by parallel studies of phosphorus and calcium requirements in baby pigs (Table 6.4 versus Table 4.5) (Miller et al., 1964), which revealed similar rachitic lesions whichever element was lacking. In broiler chicks, it is possible to distinguish 'phosphorus rickets' from 'calcium rickets' by the abnormalities in and immediately distal to the growth plate. In 'phosphorus rickets', the growth plate is normal but the primary spongiosa is elongated – the opposite of the description for 'calcium rickets' (Edwards and Veltman, 1983).

Hens given insufficient phosphorus can become lethargic, unwilling to stand and eventually die. The symptoms occur 5 weeks earlier in hens that are relatively old at the onset of phosphorus deprivation (70–76 versus 20–70 weeks) (Boling et al., 2000a). The condition fits the description of 'cage layer fatigue', but phosphorus deprivation is probably not the only cause.

Subclinical Manifestations of Phosphorus Deprivation

Impaired feed utilization

FCE is invariably poor in phosphorus-deprived pigs (Fig. 6.6) (Koch and Mahan, 1985; Reinhart and Mahan, 1986) and poultry (Qian et al., 1997; Driver et al., 2005), but in the absence of pair-fed controls it is impossible to say to what extent this is due to a reduction in food intake. Reductions in FCE in ruminants are rare (Ternouth, 1990), but were evident in the calves deprived of phosphorus by Challa and Braithwaite (1988). Their study was unusual in that all of the groups were continuously offered a restricted amount of food each day that was totally consumed irrespective of its phosphorus content, showing that FCE can be reduced independently of food intake in the phosphorus-deprived animal. Reductions in FCE have been reported in phosphorus-deprived steers (Long et al., 1957), lactating beef cows (Fishwick et al., 1977) and goats (Muschen et al., 1988), but in all cases there was an accompanying reduction in appetite.

Depression of milk yield

A lactating animal can respond to phosphorus deprivation by reducing its food intake and milk yield without changing the concentrations of phosphorus in the milk produced (Muschen et al., 1988; Valk et al., 2002). Severe experimental phosphorus deprivation imposed during late pregnancy and early lactation lowered milk yield in beef cows (Fishwick et al., 1977). In lactating goats, transfer to a low-phosphorus diet after 6 weeks' lactation was sufficient to suppress milk yield (Muschen et al., 1988). Dairy cows, pregnant with their second calf and given diets low in phosphorus (2.4 g kg^{-1} DM) but high in calcium (7.2 g kg^{-1} DM) from the seventh month of gestation, showed maximum depression of milk yield (32–35%) between 18 and 34 weeks of lactation and proportionate reductions in food intake (Call et al., 1987). A similar low phosphorus supply throughout 1.3 lactations reduced milk yield, body weight and appetite during the second lactation (Valk and Sebek 1999). In both studies, wheat straw was fed during the dry period to exacerbate phosphorus deprivation. In the study by Valk and Sebek (1999), the transitional diet consisted mostly (63%) of wheat straw, containing only 1.6 g P kg^{-1} DM and 1.3 g Ca kg^{-1} DM, and it allowed only slight retention of phosphorus (0.8 g day^{-1}) despite recovery of plasma P_i to 2.0 mmol l^{-1}. Had that diet provided the dry cow with sufficient calcium to replenish its skeletal reserves it is possible that milk yield would have been sustained in the second lactation. After several successive lactations on phosphorus-deficient veldt pastures, improvements of up to 27.5% in weaning weight were obtained in beef herds given phosphorus supplements (Read et al., 1986a,b), and these probably reflected improved milk yields.

Improvements in milk yield or weaning weight are rare consequences of phosphorus deprivation in either dairy or beef herds and are most likely to occur during the first lactation when animals are still growing (Winks, 1990; Karn, 1997, 2001). Reductions in milk yield may be entirely due to loss of appetite and/or reduced synthesis of rumen microbial protein.

Bone demineralization in nematodiasis

The remarkable demineralization of the skeleton induced by parasitic nematode infections of the sheep's gut was outlined in Chapter 4; the specific effects on phosphorus metabolism are dealt with here. Infections of the small intestine (e.g. by *Trichostrongylus colubriformis*) can reduce the absorption of dietary and salivary phosphorus by about 40% and induce hypophosphataemia, but infections of the abomasum (e.g. by *Ostertagia circumcincta*) have little or no effect (Wilson and Field, 1983). Combined infections have similar phosphorus-depleting effects – which last up to 17 weeks, depending on the acquisition of immunity – that are not merely secondary consequences of reduced food intake (Bown et al., 1989). While the sustained dual-infectious challenge in Bown et al.'s study was severe (generally 12,000 infectious larvae kg^{-1} herbage DM) by comparison with intensively grazed temperate pastures, seasonal nematode infections may increase the severity of phosphorus deprivation (Suttle, 1994).

Reduction in egg yield and quality

In hens, phosphorus deprivation is manifested by a decline in egg yield, hatchability and shell thickness, but this is much less likely to occur than calcium deprivation because of the far smaller requirement for phosphorus. A study found that the egg production and the hatchability of fertile eggs from caged hens decreased rapidly on diets containing 3.4–5.4 g phosphorus, all of plant origin, until 0.9 g P_i kg^{-1} DM was added, but the same amount of organic phosphorus (as hominy) was ineffective (Waldroup et al., 1967). In later studies, the addition of 1.0 g P_i kg^{-1} DM was insufficient to sustain performance (Boling et al., 2000a). Egg weight and body weight were improved by adding phytase to a diet providing a similar level of organic phosphorus (Barkley et al., 2004). Supplementation of commercial, grain-based diets with P_i or a Py-degrading additive is necessary for maximum egg yield and quality, but the benefits can be exaggerated by a prior failure to build up the skeletal reserves of calcium needed to sustain egg production. Calcium intakes may decline as a result of reduced voluntary intake of limestone on a diet low in available phosphorus, where the calcium source is offered separately and *ad libitum* (Barkley et al., 2004).

Occurrence of Phosphorus Deprivation

Natural shortages of phosphorus in livestock usually develop in different circumstances from those of calcium. Situations in which the two minerals are both limiting are rare.

Grazing cattle

Phosphorus deprivation is predominantly a chronic condition of grazing cattle, arising from a combination of soil and climatic factors that reduce herbage phosphorus. Soils low in plant-available phosphorus (<10 mg kg^{-1} DM) yield herbage low in phosphorus (Kerridge et al., 1990). As pasture productivity and milk yield rise with high-nitrogen fertilizer use, the critical soil phosphorus value can rise to 30 mg kg^{-1} DM (Davison et al., 1997). Acid, iron-rich soils are particularly likely to provide insufficient phosphorus. The occurrence of a dry period when plants mature and seed is shed accentuates any soil effect, unless the foraging animal can recover the seed (as it does from subterranean clover). In the South African veldt, herbage phosphorus concentrations typically fall from 1.3–1.8 to 0.5–0.7 g kg^{-1} DM between the wet summer and the dry winter and remain low for 6–8 months (Underwood, 1981). Similarly, in the Mitchell grass (*Astrebla* species) pastures of Northern Australia, phosphorus values fall from 2.5 to as low as 0.5 g kg^{-1} DM (Davies et al., 1938). Forages from the northern Great Plains of the USA are marginal in phosphorus (Karn, 1997).

Protein, digestible energy and sulfur concentrations in herbage also fall with maturity (for sulfur, see Chapter 9) (Grant et al., 1996; Karn, 1997) so that other deficiencies often contribute to the malnutrition of livestock in phosphorus-deficient areas. This is apparent from early studies of Zebu cattle in East Africa (Lampkin et al., 1961), of the savannah pastures during the dry winter season in South Africa (Van Niekerk and Jacobs, 1985) and in subtropical Australia (Cohen, 1979). Responses to phosphorus supplements are unlikely to be obtained while other nutritional deficits exist (Miller et al., 1990). In beef cattle, improvements in food intake and growth rate usually occur if dietary nitrogen is >15 g kg^{-1} DM and phosphorus <1.5 g kg^{-1} DM (Winks, 1990). The introduction of tropical legumes (e.g. *Stylosanthes* species), tolerant of low available soil phosphorus, lifted the constraint on herbage DM production in northern Queensland (Jones, 1990), but merely exacerbated deficiencies in livestock by producing highly digestible forage that was exceedingly low in phosphorus (often <1 g kg^{-1} DM) (Miller et al., 1990).

The treatment of low-quality roughages with urea to improve their digestibility will often produce a phosphorus-deficient diet and additional phosphorus should be included in the treatment.

Dairy cattle are more vulnerable than beef cattle to phosphorus deprivation because of their longer lactation and earlier conception. The higher prevalence of aphosphorosis in beef cattle in South Africa than in Australian has been attributed to an adverse influence of calcareous soils in South Africa (Cohen, 1979). It is worth noting that tropical legumes often have extremely high calcium concentrations (Table 6.2).

Other grazing livestock

Phosphorus deprivation is less common and usually less severe in grazing sheep than in grazing cattle. There are several possible explanations for this. Sheep (and also goats) have a higher feed consumption per unit LW, but the maintenance requirements for phosphorus increases with DMI (AFRC, 1991), putting smaller ruminants at a disadvantage. They also have a smaller proportion of bone to LW than cattle and therefore smaller growth requirements. In addition, because of their different methods of prehension, sheep are probably more able to select plants that are less phosphorus deficient from mixed herbage. The most important species difference is, however, the much shorter period of the annual reproductive cycle during which sheep are lactating (Read et al., 1986a,b). This allows time for depleted skeletal reserves to be replenished. Early Australian studies found no benefit from phosphorus supplements for sheep in areas where aphosphorosis existed in cattle (Underwood et al., 1940), and this has been confirmed in South Africa (Read et al., 1986a,b).

Poultry and pigs

Phosphorus deprivation would be endemic in poultry and pigs (due to the over-feeding of calcium and consequently poor availability of PyP) if P_i were not routinely added to commercial rations. For example, mortality is high in broiler chicks given a corn/mixed-vegetable protein diet containing 4.5 g P kg^{-1} DM unless it is supplemented with phosphorus (Simons et al., 1990). Phosphorus-responsive bone deformities (rickets) and fragilities will eventually develop, but they are usually preceded by life-threatening reductions in food intake. Nevertheless, a histological study of the proximal tibiotarsus from lame 35-day-old broiler chicks from Holland, where there is great pressure to reduce phosphorus levels in feeds, revealed a high (44%) incidence of hypophosphataemic rickets, which may increase susceptibility to tibial dyschrondroplasia and bacterial chondronecrosis (Thorp and Waddington, 1997). Growing pigs are less vulnerable than poultry because they are normally fed rations containing less calcium and need less phosphorus.

Diagnosis of Phosphorus Deprivation in Grazing Animals

The only definite evidence that lack of phosphorus has critically restricted livestock performance comes from improvements, particularly a recovery of appetite, seen when specific phosphorus supplements have been given. The following biochemical indices provide only a *rough* assessment of phosphorus status.

Plasma inorganic phosphorus

Although plasma (or serum) P_i can fall spectacularly in animals deprived of phosphorus (Fig. 6.15) and is used in 'metabolic profiles' to indicate phosphorus status, it is of questionable diagnostic value because of the many factors that affect it (p. 141). Inconsistent responses to supplementation in beef herds with serum P_i in the range of 1.2–1.5 mmol l^{-1} (e.g. Wadsworth *et al.*, 1990; Coates and Ternouth, 1992) suggest this is an appropriate band between the normal and deprived state in such herds and may be applicable to dairy herds, judging from the findings of Valk *et al.* (2001). Marginal serum P_i concentrations are most likely to precede clinical aphosphorosis in young, rapidly growing stock or high-yielding lactating animals. Older beef cattle on dry-season pastures of low nutritive value can have mean serum P_i values around 1.0 mmol l^{-1} and not benefit from phosphorus supplementation (Engels, 1981; Wadsworth *et al.*, 1990), but this may indicate other nutritional constraints such as protein or sulfur deficiency.

The equivalent marginal range for young pigs is higher (Fig. 6.7) and probably narrower than for ruminants: 2.2–2.8 mmol l^{-1} is suggested because growth retardation can occur with mean serum P_i as high as 1.9 mmol l^{-1} (Koch and Mahan, 1985). Sampling and analytical procedures must be standardized because plasma P_i values vary with the site of sampling (coccygeal and mammary vein > jugular vein) and are increased by delays in removing serum after formation of the blood clot (Teleni *et al.*, 1976; Forar *et al.*, 1982). The use of buffered trichloroacetic acid as a protein precipitant can reduce post-sampling differences, but plasma P_i cannot realistically be interpreted using a single threshold and marginal bands are necessary (Table 6.10).

Bone criteria in general

Deprivation of phosphorus and calcium is reflected by demineralization of the skeleton in sheep (Table 4.6), cattle (Little, 1984; Williams *et al.*, 1991a), pigs (Table 6.4 and Fig. 6.12) and turkeys (Fig. 1.4), but one element cannot be removed without the other and abnormalities are only indicative of a phosphorus deficit if calcium is non-limiting (compare Table 6.4 with Table 4.2). Protein depletion can also lead to poor mineralization of bones through the resorption of bone matrix or failure to form that matrix. Furthermore, phosphorus deprivation can cause loss of production without significant reductions in bone quality (Wan Zahari *et al.*, 1990; Coates and Ternouth, 1992). The inconsistency is to be expected given that the primary determinant of health and performance is appetite rather than bone strength and that the bone sample is a minute portion of a large heterogeneous reserve. Having noted the limitations of bone indices, it is important to choose the most informative from a bewildering array of possibilities.

In vivo bone assessment methods

Rib biopsy was first used in cattle (Little and Minson, 1977) and later applied to pigs (Combs *et al.*, 1991b), and is particularly useful in experimental studies of serial changes in skeletal mineralization, provided that the site is consistent (Beighle *et al.*,

Table 6.10. Marginal bands[a] for interpreting diagnostically useful indices of phosphorus or calcium status in the skeleton and blood (P only). Where bone growth is restricted by mineral deprivation, ash and other indicators of bone mineral density underestimate the degree of deprivation.

		Bone ash[b] (% dry weight)	Plasma inorganic P[e] (mmol l^{-1})
Cattle	Calf – milk-fed	60 (R,d)[c]	2.0–2.5
	Calf – weaned	48–56 (R,c)	1.3–1.9
	Cow	50–60 (R,c)	1.0–1.5
Pigs	Starter (10–25 kg)	45–48 (d)	2.6–3.2
	Grower (25–65 kg)	52–55 (d)	2.3–2.6
	Finisher (65–80 kg)	56–58 (d,M)	2.3–2.6
Sheep	Lamb – suckling	–	–
	Lamb – weaned	20–30 (V)	1.3–1.9
	Ewe – lactating	30–36 (T)	1.0–1.5
Broiler	Chick – <4 weeks	40–45 (T,d)[d]	2.2–2.4
	Chick – >6 weeks	45–50 (T,d)	2.0–2.2
		50–55 (T,c,d)	
Hen	Laying	40–50 (T,d)	1.0–1.6
Turkey	Poult	38–40 (T)	–

[a]Mean values within bands or individual values below lower limits indicate the possibility of impaired health or performance if P (or Ca) supplies are not improved.
[b]Bone ash can be converted approximately to specific mineral concentrations on the assumption that ash contains 0.18 g P, 0.36 g Ca and 0.09 g Mg kg^{-1}. Ash unit^{-1} volume is similar to ash unit^{-1} dry fat-free weight.
[c]Source and pre-treatment of bone sample: c, cortical bone; d, defatted bone; M, metacarpal/tarsal; R, whole rib; T, tibia; V, lumbar vertebra.
[d]In broiler chicks, the marginal range for the fatty tibia is 39–42% ash. Toes can be sampled *in vivo* and toe ash <13% is indicative of P (or Ca) deprivation.
[e]Multiply by 31 to obtain units in mg l^{-1}.

1993). A disc of bone is obtained by trephine (1.5 cm diameter in cattle) and repeated biopsies of the same rib are possible, but they should removed dorsally to earlier biopsies after a reasonable interval (>3 months) (Little and Radcliff, 1979). The analytical options available are given in the next section. Caudal vertebra biopsies can be used, but samples contain less ash and are probably less responsive to changes in phosphorus status than rib samples (Keene *et al.*, 2004). Toe ash concentrations are significantly reduced in clippings from the phosphorus-deprived fowl (e.g. Cheng *et al.*, 2004). Cortical bone thickness can be monitored by X-ray, dual-photon absorptiometry and computer-assisted tomography (Grace *et al.*, 2003). Radiographic photometry and ultrasound have also been used to obtain *in vivo* measurements of bone mineralization (Ternouth, 1990; Williams *et al.*, 1991b; Keene *et al.*, 2004). Biochemical tests on serum or urine can also indicate the extent to which bone is being grown or resorbed. These include assays of osteocalcin, hydroxyproline, deoxypyridinoline and type 1 collagen (McLean *et al.*, 1990; Leisegang *et al.*, 2000; Peterson *et al.*, 2005). These biochemical tests are subject to large fluctuations at parturition and are under the control of calcium-regulating hormones (see Chapter 4). They are not specific indicators of phosphorus deprivation and have more relevance to racehorses than commercial livestock since avoidance of stress injury is crucial to racing performance (Price *et al.*, 1995).

Post-mortem assessment of bone mineralization

In dead animals, the X-ray, morphometric or mineral analysis of bones may be the only diagnostic option, but interpretation is complicated by considerable variations between species, within and between particular bones at a given age and by age per se, since values for most indices increase with age irrespective of mineral nutrition (Fig. 6.12) (Beighle et al., 1994). Mineral to volume or ash to volume ratios or specific gravity in the entire bone are the simplest and best indices of bone mineralization in all species (e.g. lambs, Field et al., 1975; cattle, Williams et al., 1991a; chicks, Cheng et al., 2004). Guideline values are given in Table 6.10, which summarizes the contribution each parameter can make towards a diagnosis. In the young rapidly growing chick or piglet, where growth of long bones is restricted by phosphorus (or calcium) deprivation, total bone weight, ash or mineral content is required to give the full measure of deprivation. The mineral composition of bone mass is not greatly changed by phosphorus (or calcium) deprivation in growing animals (Williams et al., 1991a; Qian et al., 1996a), but is more variable in the laying hen. Fluctuations in the mineralization of rib biopsy samples in lactating animals can be marked (Fig. 6.13) (Read et al., 1986c) and give a different assessment of phosphorus deprivation from that given by plasma P_i. Phosphorus depletion tends to increase calcium to magnesium ratios in the skeleton of cattle as well as in lambs (Table 4.6, p. 72).

Faecal and urinary phosphorus

Although faecal phosphorus reflects endogenous loss as well as intake of phosphorus, concentrations have been used to monitor the response of grazing cattle to phosphorus supplementation on pastures deficient in both nitrogen and phosphorus (Holechek et al., 1985; Grant et al., 1996; Karn, 1997). A 'critical threshold' value of <2g P kg^{-1} faecal DM has been advocated and was a significant but weak predictor of milk yield in rangeland goats when tested recently (Mellado et al., 2006). If a simple qualitative 'stick' test for the presence of phosphorus in urine could be devised, a 'negative' result for ruminants at pasture would raise the possibility of phosphorus deprivation since little or no phosphorus is excreted in urine at marginal intakes of phosphorus.

The Prevention and Control of Phosphorus Deprivation in Ruminants

Phosphorus deprivation can be prevented or overcome by direct treatment of the animal through supplementing the diet or the water supply and indirectly by appropriate fertilizer treatment of the soils. The choice of procedure depends on the conditions of husbandry.

Use of fertilizers

In climatically favoured and intensively farmed areas with sown pastures, PO_4 applications designed primarily to increase herbage yields also increase herbage phosphorus (Falade, 1973). Minson (1990) calculated that, on average, pasture phosphorus was increased from 1.7 to 2.4 g kg^{-1} DM with 47 kg fertilizer P ha^{-1} applied, but the range of increases was wide (0.2–3.5 to 0.5–3.9 g P kg^{-1} DM from 9–86 kg fertilizer phosphorus). Phosphorus can be applied as rock phosphate (RP), single or triple superphosphate or combined with nitrogen and potassium in complete fertilizers, but it is impossible to generalize about recommended rates (Jones, 1990). For example, milk yields were increased in dairy cows on tropical grass pasture when 22.5 and 45.0 kg P ha^{-1} was applied with 300 kg N ha^{-1}, but not with 100 kg N ha^{-1} or none (Davison et al., 1997). The responses were attributed to increases in the amount of green leaf on offer rather than the twofold increase in herbage phosphorus (from 0.6 to 1.3 g kg^{-1} DM). No benefit was gained from providing additional dietary phosphorus (Walker et al., 1997), although fertilizer treatments that maximize herbage yield do not necessarily meet the requirements of grazing animals at all times. Fertilization with phosphorus is inefficient on acid, iron-rich soils with a high sorption capacity. On sparse, extensive, phosphorus-deficient native pastures, other methods are necessary because transport and application costs are high and herbage productivity is usually limited by climatic disadvantages. The continuous use of superphosphate fertilizers in New Zealand has led to the substantial accumulation of cadmium in soils due to its presence as a contaminant of RP (see Chapter 18). The use of superphosphate was popular in New Zealand because it provides a vehicle for the simultaneous distribution of copper and cobalt for preventing deficiencies of those elements.

Dietary supplementation

The easiest and cheapest procedure is to provide a phosphatic lick in troughs or boxes, protected from rain and situated near watering places. Typical levels of phosphorus provision for beef cattle are at least 5 g head^{-1} day^{-1} for growing cattle and twice as much for breeding stock (Miller et al., 1990), but intakes >8 g day^{-1} in growing cattle can retard growth (Grant et al., 1996). Phosphorus can also be provided directly in salt blocks and a simple 1:1 mixture of dicalcium phosphate (DCP, $CaHPO_4$) and salt with a small proportion of molasses is well consumed by most animals. However, consumption can vary greatly between individual animals and at different times of the year (see Chapter 3). Supplementation through the drinking water is possible where the access to water is controlled (i.e. there are no natural water sources), but requires water-soluble phosphorus sources such as sodium phosphate or ammonium polyphosphate (Hemingway and Fishwick, 1975), which are more expensive than the relatively insoluble DCP. Furthermore the amounts ingested fluctuate with water consumption and treatment may limit water consumption, cause toxicity and corrode dispensers (Miller et al., 1990). Superphosphate has long been used as a cheap source of water-soluble PO_4 (Du Toit et al., 1940; Snook, 1949; Scharp, 1979). Recent research found a that soluble phosphorus source, monocalcium phosphate (MCP, $Ca(H_2PO_4)_2$), stimulated liquid-associated rumen microbes and increased intake of a low phosphorous diet when compared with DCP (Ramirez-Perez et al., 2008). Where MCP and DCP are used in free-access mixtures, supplementary calcium is automatically supplied although it is rarely needed by grazing livestock and may already be present at levels sufficiently high to have contributed to any aphosphorosis. In low-cost production systems, where reliance is placed on low-quality roughages, phosphorus is often lacking and can be cheaply provided by including 20–30% wheat or rice bran in a molasses–urea block or loose supplement.

Combined methods

Where legume species tolerant of phosphorus-deficient soils have been sown, the most efficient procedure is to combine direct and indirect methods of supplementation. Phosphorus given as fertilizer is restricted in amount to that required to maximize herbage yield (up to 10 kg P ha^{-1} year^{-1}) and any additional phosphorus required is given directly (Miller et al., 1990).

Mineral sources

Essentially similar findings with respect to a variety of mineral PO_4 supplements have been reported for sheep, growing beef cattle and horses (Soares, 1995). Miller et al. (1987) found that defluorinated phosphate (DFP) and DCP provided phosphorus of equal availability to calves. A wider range of mineral supplements, including urea phosphate, monoammonium phosphate (Fishwick and Hemingway, 1973), two magnesium phosphates and tricalcium phosphate (Fishwick, 1978), have given similar responses in growth and phosphorus retention when tested with growing, phosphorus-deficient sheep. In sheep given coarsely ground RP, particles can be trapped in the alimentary tract giving a falsely high estimate (0.66) of A_P (and presumably A_{Ca}) (Suttle et al., 1989): a value two-thirds that of DCP (0.55 versus 0.80) is considered to be more appropriate and a similar low value for RP has been reported by Dayrell and Ivan (1989). Ground RP is relatively unpalatable and can contain enough fluoride to cause fluorine toxicity (see Chapter 18). In vitro solubility in citric acid differentiated two phosphorus sources, RP and DCP, as well as an in vivo measurement of A_P in sheep (Suttle et al., 1989), and neutral ammonium citrate can also be used as an extractant.

The Prevention and Control of Phosphorus Deprivation in Non-ruminants

Inorganic phosphorus supplementation

Historically, the problem presented by the low A_P from cereals and vegetable protein sources was overcome by liberal supplementation with P_i. The choice of supplement depended on chemical composition, biological availability, cost, accessibility and freedom from toxic impurities and dust hazards of the P_i source. The RA_P from different compounds has been extensively evaluated against MSP (Soares, 1995) and values of 0.90–0.95 have been accorded to bone meal, DFP, diammonium phosphate and DCP given to poultry (Soares, 1995). For pigs, a similarly high RA_P was given to DCP, but bone meal (0.8) and DFP (0.85) were generally less available. DFP was not inferior to MSP in a separate study with pigs (Coffey et al., 1994). Contrasts between species may depend on the fineness of grinding of the phosphorus sources used because poultry have the ability to trap and grind down coarse particles in the gizzard. The average AA_P 0.82 derived for P_i in trials with pigs (Fig. 6.7) refers to DCP, the predominant supplement used, and is comparable with a value of 0.82 found for DCP by Peterson and Stein (2006), who also found that MSP sources of different purity can vary in AA_P from 88% to 98% A_P, affecting the assessment of RA_P (Peterson and Stein, 2006).

Phytase supplementation

A wide range of exogenous phytases of microbial or fungal origin have been tested for their P_i-sparing effect, particularly in pigs (e.g. Kornegay and Qian, 1996; Murry et al., 1997). Phytases can be divided into three categories according to preferred sites for initiation of dephosphorylation or their pH optimum (acid, neutral and alkaline), as summarized in Table 6.11. Once the preferred site has been dephosphorylated, the enzyme becomes progressively less effective at removing further PO_4 groups. Fungal phytases are less efficacious than microbial phytases (Augspurger and Baker, 2004; Paditz et al., 2004), but responses to Escherichia coli 6-phytase in pigs (Brana et al., 2006) and broiler chicks (Fig. 6.16) given corn/SBM diets followed the law of diminishing returns and left almost half of the ingested Py in the excreta of the birds. The precise amount of P_i spared by 'first-generation' phytase supplements varied widely, even after adjusting for variations in AA_P in the basal diet (Knowlton et al., 2004) and probably for the same reasons that assessments of 'phosphorus value' of feeds vary (see p. 133). The efficacy of phytase in chicks decreases sharply with age in the first 3 weeks of life (Fig. 4.5, p. 62) and with the lowering of dietary calcium in chicks (Fritts and Waldroup, 2006), hens (Keshavarz, 2000b) and pigs (Lei et al., 1994). Many commercially biased experiments have exaggerated the efficacy of phytase by performing experiments with either very young chicks (<3 weeks old) or excess dietary calcium.

Table 6.11. Important characteristics of some commercially available phytase supplements, including the species of origin, vehicle for expression, specific locus of substrate, optimum pH, and degradability by pepsin.

Origin	Vehicle	Specificity[a]	Optimum pH	Degradability	Brand names
Aspergillus niger	Fungus	3-phytase	2.5–5.5	High	Finase P Natuphos
Peniophora lycii	Fungus	6-phytase	2.5–3.5	High	Ronozyme
Escherichia coli	Yeast	6-phytase	2.5–3.5	Low	Quantum OptiPhos Ecophos

[a]See Fig. 6.1 for the numbering of target PO_4 groups on the phytate molecule.

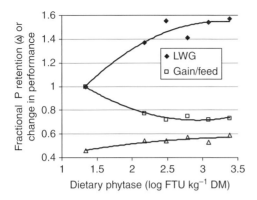

Fig. 6.16. The effects of adding an *Escherichia coli* 6-phytase on the fractional retention of phytate phosphorus, relative growth rate and feed conversion rate of broiler chicks in the first 14 days of life (data from Cowieson et al., 2006). LWG, live-weight gain.

Adding 500 U phytase allows the removal of 0.84 g P_i kg^{-1} DM from the average ration (Knowlton et al., 2004), but represents degradation of as little as 20–29% of the *Py*P present in corn/SBM diets (e.g. Cromwell et al., 1995). The 'P_i equivalence' of phytase supplements in poultry rations was broadly similar (e.g. Augspurger et al., 2003). Myo-inositol, the end product of complete *Py*P digestion, also stimulates growth (Zyla et al., 2004) and could partly explain variations in 'phosphorus equivalence' and why benefits in performance and bone mineralization from phytase supplementation including those in Fig. 6.9, sometimes exceed those attributable to improved A_P alone. Phytase supplements can increase nutrient digestibility in pigs (Johnston et al., 2004) and increase transit rate through the gut, and thus increase food intake in chicks (Watson et al., 2006).

A second generation of genetically modified microbial phytases has appeared with improved heat stability and resistance to digestion (e.g. Augspurger et al., 2003; Gentile et al., 2003): one derived from an *E. coli* that colonizes the porcine colon can obviate the need to add P_i to diets for finishing pigs and laying hens (Augspurger et al., 2007); others remain undestroyed by pelleting (Timmons et al., 2008). Mixtures of phytases with acid and alkaline pH optima might increase efficacy (Zyla et al., 2004), but the strategy adds to an already significant feed cost. The role of exogenous and endogenous phytases in reducing pollution of watercourses with phosphorus is discussed in Chapter 20.

Citrate and vitamin D supplements

Additions of citrate or sodium citrate (up to 6% of the diet) to diets marginally deficient in phosphorus have produced linear increases in chick growth rate and tibia ash. This response was attributed to chelation of calcium and the consequent increased conversion of *Py*P. An additional response was found when phytase was also added (Boling et al., 2000b). Pigs have shown much smaller responses to citrate, but their basal diet was lower in calcium (5 versus 6.2 g kg^{-1} DM). Supplementation with vitamin D_3 (Qian et al., 1997) and synthetic analogues of the vitamin (Biehl and Baker, 1997a,b; Snow et al., 2001; Cheng et al., 2004) enhances *Py*P utilization, probably by stimulating secretion of endogenous alkaline phosphatases (Edwards, 1993). The use of *Solanum malacoxylon* (or glaucophyllum SG) as a natural source of vitamin D_3 analogue has been advocated to reduce the cost of supplementation. This improves the utilization of both phosphorus and calcium in broiler chicks (Cheng et al., 2004). The combination of supplements that (probably) act at different stages of *Py*P degradation (e.g. citrate, phytase and vitamin D_3) can optimize the utilization of *Py*P and there is possible synergism between citrate and $1\alpha(OH)D_3$ (Fig. 6.17). A combination of citrate (15 g), wheat phytase (461 U as middlings) and microbial phytase (300 U kg^{-1} DM) can replace P_i in rations for growing pigs (Han et al., 1998). Inconsistent additive effects, such as the lack of benefit of adding *S. malacoxylon* with phytase (Cheng et al., 2004), are once again probably due to characteristics of the experiment, such as the high dose of phytase (1200 FTU as fungal 3-phytase) or a calcium-deficient basal diet. When *S. malacoxylon* was used without reducing dietary calcium, its calcinogenic activity caused anorexia in broiler chicks.

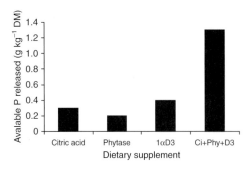

Fig. 6.17. Dietary supplements of citrate (Ci; 30–40 g), 3-phytase (300 FTU) and 1α(OH)D$_3$ (5 μg kg^{-1} DM) can each release significant amounts of available phosphorus from a P-deficient, corn/soybean meal for use by broiler chicks. Combining the supplements can produce synergistic responses when each is used at a suboptimal level (data from Snow et al., 2004).

Phosphorus Toxicity

The PO$_4$ ion is intrinsically well-tolerated and animals can allow circulating levels of PO$_4$ to fluctuate widely. This, coupled with ready excretion of excess PO$_4$ via the urine, means that livestock generally tolerate excessive phosphorus intakes. Problems arise through adverse interactions or cumulative effects with other minerals. High phosphorus intakes predispose animals to urinary calculi, but in ruminants these are only likely to form on diets consisting predominantly of concentrates. Diets rich in magnesium predispose to phosphatic calculi because magnesium phosphates are integral to growth of the calculus (Suttle and Hay, 1986). Sheep with a high inherited efficiency for phosphorus absorption appear particularly vulnerable (Field et al., 1984). Dietary excesses of phosphorus predispose broilers to tibial dyschrondroplasia (see Chapter 4) unless excess calcium is also fed (Edwards and Veltmann, 1983). Linear relationships between dietary phosphorus and the incidence of tibial dyschrondroplasia and 'calcium rickets' have been reported, with relatively small increments of phosphorus (from 4.8 to 5.9 g kg^{-1} DM) sufficient to increase incidences unless accompanied by an increase in calcium (from 4.4 to 6.4 g kg^{-1} DM) (Driver et al., 2005). Excesses of acidogenic sources such as monobasic phosphate disturb the acid–base balance in laying hens (Keshavarz, 1994). In the horse, feeding excess phosphorus in diets low in calcium can cause secondary hyperparathyroidism.

References

Adeola, O., Sands. J.S., Simmins, P.H. and Schulze, H. (2004) The efficacy of an *Escherichia coli*-derived phytase preparation. *Journal of Animal Science* 82, 2657–2666.

AFRC (1991) Technical Committee on Responses to Nutrients Report No. 6. A reappraisal of the calcium and phosphorus requirements of sheep and cattle. *Nutrition Abstracts and Reviews* 61, 573–612.

Andrews, T.L., Damron, B.L. and Harms, R.H. (1972) Utilisation of various sources of plant phosphorus by the turkey poult. *Nutrition Reports International* 6, 251–257.

Applegate, T.J., Webel, D.M. and Lei, X.G. (2003) Efficacy of phytase derived from *Escherichia coli* and expressed in yeast on phosphorus utilisation and bone mineralisation in turkey poults. *Poultry Science* 82, 1726–1732.

ARC (1980) *The Nutrient Requirements of Ruminants*. Commonwealth Agricultural Bureaux, Farnham Royal, UK, pp. 184–185.

Augspurger, N.R. and Baker, D.H. (2004) High dietary phytase levels maximise phytate-phosphorus utilisation but do not affect protein utilisation in chicks fed phosphorus- or amino acid deficient diets. *Journal of Animal Science* 82, 1100–1107.

Augspurger, N.R., Webel, D.M., Lei, X.G. and Baker, D.H. (2003) Efficacy of an *E. coli* phytase expressed in yeast for releasing phytate-bound phosphorus in young chicks and pigs. *Journal of Animal Science* 81, 474–483.

Augspurger, N.R., Webel, D.M. and Baker, D.H. (2007) An *Escherichia coli* phytase expressed in yeast effectively replaces inorganic phosphorus for finishing pigs and laying hens. *Journal of Animal Science* 85, 1192–1198.

Bar, A. and Hurwitz, S. (1984) Egg shell quality, medullary bone ash, intestinal calcium and phosphorus absorption and calcium-binding protein in phosphorus-deficient hens. *Poultry Science* 63, 1975–1980.

Barkley, G.R., Miller, H.M. and Forbes, J.M. (2004) The ability of laying hens to regulate phosphorus intake when offered two feeds containing different levels of phosphorus. *British Journal of Nutrition* 92, 233–240.

Bedford, M.R. and Partridge, G.G. (2001) *Enzymes in Farm Animal Nutrition*. CAB International, Wallingford, UK, pp. 62.

Beeson, W.M., Johnson, R.F., Bolin, D.W and Hickman, C.W. (1944) The phosphorus requirement for fattening lambs. *Journal of Animal Science* 3, 63–70.

Beighle, D.E., Boyazoglu, P.A. and Hemken, R.W. (1993) Use of bovine rib bone in serial sampling for mineral analysis. *Journal of Dairy Science* 76, 1047–1072.

Beighle, D.E., Boyazoglu, P.A., Hemken, R.W. and Serumaga-Zake, P.A. (1994) Determination of calcium, phosphorus and magnesium values in rib bones from clinically normal cattle. *American Journal of Veterinary Research* 55, 85–89.

Biehl, R.R. and Baker, D.H. (1997a) 1α-Hydroxy cholecalciferol does not increase the specific activity of intestinal phytase but does improve phosphorus utilisation in both caectomised and sham-operated chicks fed cholecalciferol-adequate diets. *Journal of Nutrition* 127, 2054–2059.

Biehl, R.R. and Baker, D.H. (1997b) Utilisation of phytate and non-phytate phosphorus in chicks as affected by source and amount of vitamin D_3. *Journal of Animal Science* 75, 2986–2993.

Biehl, R.R., Baker, D.H. and DeLuca, H.F. (1995) 1α-Hydroxylated cholecalciferol compounds act additively with microbial phytase to improve phosphorus, zinc and manganese utilization in chicks fed soy-based diets. *Journal of Nutrition* 125, 2407–2416.

Bohn, L., Meyer, A.S. and Rasmussen, S.K. (2008) Phytate: impact on environment and human nutrition. A challenge for molecular breeding. *Journal of Zhejiang University Science B* 9, 165–191.

Boling, S.D., Douglas, M.W., Johnson, M.L., Wang, X., Parsons, C.M., Koelkebeck, K.W. and Zimmerman, R.A. (2000a) The effects of available phosphorus levels and phytase on performance of young and older laying hens. *Poultry Science* 79, 224–230.

Boling, S.D., Webel, D.M., Mavromichalis, I., Parsons, C.M. and Baker, D.H. (2000b) The effects of citric acid on phytate-phosphorus utilisation in young chicks and pigs *Journal of Animal Science* 78, 682–689.

Bown, M.D., Poppi, D.P. and Sykes, A.R. (1989) The effect of a concurrent infection of *Trichostrongylus colubriformis* and *Ostertagia circumcincta* on calcium, phosphorus and magnesium transactions along the digestive tract of lambs. *Journal of Comparative Pathology* 101, 11–20.

Braithwaite, G.D. (1983) Calcium and phosphorus requirements of the ewe during pregnancy and lactation. 2. Phosphorus. *Journal of Agricultural Science, Cambridge* 50, 723–730.

Braithwaite, G.D. (1984a) Some observations in phosphorus homeostasis and requirements. *Journal of Agricultural Science, Cambridge* 102, 295–306.

Braithwaite, G.D. (1984b) Changes in phosphorus metabolism in sheep in response to the increased demands for phosphorus associated with an intravenous infusion of calcium. *Journal of Agricultural Science, Cambridge* 102, 135–139.

Braithwaite, G.D. (1986) Phosphorus requirements of ewes in pregnancy and lactation. *Journal of Agricultural Science, Cambridge* 106, 271–278.

Brana, D.V., Elis, M., Casteneda, E.O., Sands, J.S. and Baker, D.H. (2006) Effect of novel phytase on growth performance, bone ash, and mineral availability in nursery and grower-finisher pigs. *Journal of Animal Science* 84, 1839–1849.

Bravo, D., Bogaert, C., Meschy, F. and Sauvant, D. (2003a) Plasma phosphorus content and dietary phosphorus availability in adult sheep. *Animal Research* 52, 427–435.

Bravo, D., Sauvant, D., Bogaert, C. and Meschy, F. (2003b) Quantitative aspects of phosphorus excretion in ruminants. *Reproductive and Nutritional Development* 43, 285–300.

Bretziger, P. and Raboy, V. (2006) Effect of four independent low-phytate mutations on barley agronomic performance. *Crop Science* 46, 1318–1322.

Breves, G., Ross, R. and Holler, H. (1985) Dietary phosphorus depletion in sheep: effects on plasma inorganic phosphorus, calcium, 1,25-$(OH)_2$-Vit. D and alkaline phosphatase on gastrointestinal P and Ca balances. *Journal of Agricultural Science, Cambridge* 105, 623–629.

Brooks, H.V., Cook, T.G., Mansell, G.P. and Walker, G.A. (1984) Phosphorus deficiency in a dairy herd. *New Zealand Veterinary Journal* 32, 174–176.

Call, J.W., Butcher, J.E., Blake, J.T., Smart, R.A. and Shupe, J.L. (1978) Phosphorus influence on growth and reproduction of beef cattle. *Journal of Animal Science* 47, 216–225.

Call, J.W., Butcher, J.E., Shupe, J.L., Blake, J.K. and Olsen, A.E. (1986) Dietary phosphorus for beef cows. *American Journal of Veterinary Research* 47, 475–481.

Call, J.W., Butcher, J.E., Shupe, J.L., Lamb, R.C., Boman, R.L. and Olsen, A.E. (1987) Clinical effects of low dietary phosphorus concentrations in feed given to lactating cows. *American Journal of Veterinary Research* 48, 133–136.

Camden, B.J., Morel, P.C.H., Thomas, D.V., Ravindran, V. and Bedford, M.R. (2001) Effectiveness of exogenous microbial phytase in improving the bioavailabilites of phosphorus and other nutrients in maize–soya-bean meal diets for broilers. *Animal Science* 73, 289–297.

Challa, J. and Braithwaite, G.D. (1988) Phosphorus and calcium metabolism in growing calves with special emphasis on phosphorus homeostasis. 2. Studies of the effect of different levels of phosphorus, infused abomasally, on phosphorus metabolism. *Journal of Agricultural Science, Cambridge* 110, 583–589.

Challa, J. and Braithwaite, G.D. (1989) Phosphorus and calcium metabolism in growing calves with special emphasis on phosphorus homeostasis. 4. Studies on milk-fed calves given different amounts of dietary phosphorus but a constant intake of calcium. *Journal of Agricultural Science, Cambridge* 113, 285–289.

Challa, J., Braithwaite, G.D. and Dhanoa, M.S. (1989) Phosphorus homeostasis in growing calves. *Journal of Agricultural Science, Cambridge* 112, 217–226.

Chapuis-Lardy, L., Fiorini, J., Toth, J. and Dou, Z. (2004) Phosphorus concentration and solubility in dairy faeces: variability and affecting factors. *Journal of Dairy Science* 87, 4334–4341.

Cheng, Y.H., Goff, J.P., Sell, J.L., Dalloroso, M.E., Gil, S., Pawlak, S.E. and Horst, R.L. (2004) Utilizing *Solanum glaucophyllum* alone or with phytase to improve phosphorus utilisation in broilers. *Poultry Science* 83, 406–413.

Cho, J., Choi, K., Darden, T., Reynolds, P.R., Petitte, J.N. and Shears, S.B. (2006) Avian multiple inositol polyphosphate phosphatase is an active phytase that can be engineered to help ameliorate the planet's 'phosphate crisis'. *Journal of Biotechnology* 126, 248–259.

Coates, D.B. and Ternouth, J.H. (1992) Phosphorus kinetics of cattle grazing tropical pastures and implications for the estimation of their phosphorus requirements. *Journal of Agricultural Science, Cambridge* 119, 401–409.

Coates, D.B., Schachenmann, P. and Jones, R.J. (1987) Reliability of extrusa samples collected from steers fistulated at the oesophagus to estimate the diet of resident animals in grazing experiments. *Australian Journal of Experimental Agriculture* 27, 739–745.

Coffey, R.D., Mooney, K.W., Cromwell, G.L. and Aaron, D.K. (1994) Biological availability of phosphorus in defluorinated phosphates with different phosphorus solubilities in neutral ammonium citrate for chicks and pigs. *Journal of Animal Science* 72, 2653–2660.

Cohen, R.D.H. (1979) The calcium–phosphorus interaction and its relation to grazing ruminants. In: *Symposium on phosphorus. Nutrition Society of Australia (Queensland Branch)*, 8 May 1979.

Combs, N.R., Kornegay, E.T., Lindemann, M.D. and Notter, D.R. (1991a) Calcium and phosphorus requirements of swine from weaning to market weight. I. Development of response curves for performance. *Journal of Animal Science* 69, 673–681.

Combs, N.R., Kornegay, E.T., Lindemann, M.D., Notter, D.R. and Welker, F.H. (1991b) Evaluation of a bone biopsy technique for determining the calcium and phosphorus status of swine from weaning to market weight. *Journal of Animal Science* 69, 664–673.

Combs, N.R., Kornegay, E.T., Lindemann, M.D., Notter, D.R., Wilson, J.H. and Mason, J.P. (1991c) Calcium and phosphorus requirements of swine from weaning to market weight. II. Development of response curves for bone criteria and comparison of bending and shear bone testing. *Journal of Animal Science* 69, 682–693.

Cowieson, A.J., Acamovic, T. and Bedford, M.R. (2006) Supplementation of corn–soy-based diets with an *Escherichia coli*-derived phytase: effects on broiler chick performance and the digestibility of amino acids and metabolizability of minerals and energy. *Poultry Science* 85, 1389–1397.

Cromwell, G.L. (1992) The biological availability of phosphorus in feedstuffs for pigs. *Pig News Information* 13, 75.

Cromwell, G.L., Coffey, R.D., Parker, G.R., Monegue, H.J. and Randolph, J.H. (1995) Efficacy of a recombinant-derived phytase in improving the bioavailability of phosphorus in corn–soybean meal diets for pigs. *Journal of Animal Science* 73, 2000–2008.

Davies, J.G., Scott, A.E. and Kennedy, J.F. (1938) The yield and composition of a Mitchell grass pasture for a period of 12 months. *Journal of the Council for Scientific and Industrial Research, Australia* 11, 127–139.

Davison, T.M., Orr, W.N., Silver, B.A., Walker, R.G. and Duncalfe, F. (1997) Phosphorus fertilizer for nitrogen-fertilized dairy pastures. 1. Long-term effects on pasture, soil and diet. *Journal of Agricultural Science, Cambridge* 129, 205–215.

Dayrell, M. de S. and Ivan, M. (1989) True absorption of phosphorus in sheep fed corn silage and corn silage supplemented with dicalcium or rock phosphate. *Canadian Journal of Animal Science* 69, 181–186.

de Brouwer, C.H.M., Cilliers, J.W., Vermaak, L.M., van der Merwe, H.J. and Groenewald, P.C.N. (2000) Phosphorus supplementation to natural pasture grazing for beef cows in the Western Highveld region of South Africa. *South African Journal of Animal Science,* 30, 43–51.

De Waal, H.O., Randall, J.H. and Keokemoer, G.J. (1996) The effect of phosphorus supplementation on body mass and reproduction of grazing beef cows supplemented with different levels of phosphorus at Armoedsvlakte. *South African Journal of Animal Science* 26, 29–36.

Dias, R.S., Kebreab, E., Vitti, D.M.S.S., Roque, A.P., Bueno, I.C.S. and France, J. (2006) A revised model for studying phosphorus and calcium kinetics in growing sheep. *Journal of Animal Science* 84, 2787–2794.

DuToit, P.J., Malan, A.I., Van der Merwe, P.K. and Louw, J.G. (1940) Mineral supplements for stock: the composition of licks. *Farming in South Africa* 15, 233–248.

Driver, J.P., Pesti, G.M., Bakalli, R.L. and Edwards, H.M. Jr (2005) Effects of calcium and nonphytate phosphorus on phytase efficacy in broiler chicks. *Poultry Science* 84, 1406–1417.

Edwards, H.M. (1993) Dietary 1,25-dihydroxycholecalciferol supplementation increases natural phytate phosphorus utilisation in chickens. *Journal of Nutrition* 123, 567–577.

Edwards, H.M. and Veltmann, J.R. (1983) The role of calcium and phosphorus in the etiology of tibial dyschondroplasia in young chicks. *Journal of Nutrition* 113, 1568–1575.

Eeckhout, W., de Paepe, M., Warnants, N. and Bekaert, H. (1995) An estimation of the minimal P requirements for growing-finishing pigs, as influenced by the Ca level of the diet. *Animal Feed Science and Technology* 52, 29–40.

Engels, E.A.N. (1981) Mineral status and profiles (blood, bone and milk) of the grazing ruminant with special reference to calcium, phosphorus and magnesium. *South African Journal of Animal Science* 11, 171–182.

Erickson, G.E., Klopfenstein, T.J., Milton, C.T., Hanson, D. and Calkins, C. (1999) Effect of dietary phosphorus on finishing steer performance, bone status, and carcass maturity. *Journal of Animal Science* 77, 2832–2836.

Falade, J.A. (1973) Effect of phosphorus on the growth and mineral composition of four tropical forage legumes. *Journal of the Science of Food and Agriculture* 24, 795–802.

Field, A.C., Suttle, N.F. and Nisbet, D.I. (1975) Effects of diets low in calcium and phosphorus on the development of growing lambs. *Journal of Agricultural Science, Cambridge* 85, 435–442.

Field, A.C., Kamphues, J. and Woolliams, J.A. (1983) The effect of dietary intake of calcium and phosphorus on the absorption and excretion of phosphorus in Chimaera-derived sheep. *Journal of Agricultural Science, Cambridge* 101, 597–602.

Field, A.C., Woolliams, J.A., Dingwall, R.A. and Munro, C.S. (1984) Animal and dietary variation in the absorption and metabolism of phosphorus by sheep. *Journal of Agricultural Science, Cambridge* 103, 283–291.

Field, A.C., Woolliams, J.A. and Dingwall, R.A. (1985) The effect of dietary intake of calcium and dry matter on the absorption and excretion of calcium and phosphorus by growing lambs. *Journal of Agricultural Science, Cambridge* 105, 237–243.

Field, A.C., Woolliams, J.A. and Woolliams, C. (1986) The effect of breed of sire on the urinary excretion of phosphorus and magnesium in lambs. *Animal Production* 42, 349–354.

Fishwick, G. (1978) Utilisation of the phosphorus and magnesium in some calcium and magnesium phosphates by growing sheep. *New Zealand Journal of Agricultural Research* 21, 571–575.

Fishwick, G. and Hemingway, R.G. (1973) Magnesium phosphates as dietary supplements for growing sheep. *Journal of Agricultural Science, Cambridge* 81, 441–444.

Fishwick, G., Fraser, J., Hemingway, R.G., Parkins, J.J. and Ritchie, N.S. (1977) The effects of dietary phosphorus inadequacy during pregnancy and lactation on the voluntary intake and digestibility of oat straw by beef cows and the performance of their calves. *Journal of Agricultural Science, Cambridge* 88, 143–150.

Forar, F.L., Kincaid, R.L., Preston, R.L. and Hillers, J.K. (1982) Variation of inorganic phosphorus in blood plasma and milk of lactating cows. *Journal of Dairy Science* 65, 760–763.

Francesch, M., Broz, J. and Brufau, J. (2005) Effects of an experimental phytase on performance, egg quality, tibia ash content and phosphorus availability in laying hens fed on maize- or barley-based diets. *British Poultry Science* 46, 340–348.

Fritts, C.A. and Waldroup, P.W. (2006) Modified phosphorus program for broilers based on commercial feeding intervals to sustain live performance and reduce total and water-soluble phosphorus in litter. *Journal of Applied Poultry Research* 15, 207–218.

Gao, Y., Shang, C., Saghai Maroof, M.A., Biyashev, R.M., Grabau, E.A., Kwanyuen, P., Burton, J.W. and Buss, G.R. (2007) A modified colorimetric method for phytic acid analysis. *Crop Science* 47, 1797–1803.

Gartner, R.J.W., Murphy, G.M. and Hoey, W.A. (1982) Effects of induced, subclinical phosphorus deficiency on feed intake and growth in heifers. *Journal of Agricultural Science, Cambridge* 98, 23–29.

Gentile, J.M., Roneker, K.R., Crowe, S.E., Pond, W.G. and Lei, X.G. (2003) Effectiveness of an experimental consensus phytase in improving phytate-phosphorus utilisation by weanling pigs. *Journal of Animal Science* 81, 2751–2757.

Grace, N.D., Pearce, S.G., Firth, E.C. and Fennessy, P.F. (1999) Content and distribution of macro- and trace-elements in the body of young pasture-fed horses. *Australian Veterinary Journal* 77, 172–176.

Grace, N.D., Shaw, H.L., Firth, E.C. and Gee, E.K. (2002) Determination of digestible energy intake, and apparent absorption of macroelements of grazing, lactating Thoroughbred mares. *New Zealand Veterinary Journal* 50, 182–185.

Grace, N.D., Rogers, C.W., Firth, E.C., Faram, T.L. and Shaw, H.L. (2003) Digestible energy intake, dry matter digestibility and effect of calcium intake on bone parameters of Thoroughbred weanlings in New Zealand. *New Zealand Veterinary Journal* 51, 165–173.

Grant, C.C., Biggs, H.C., Meisnner, H.H. and Basson, P.A. (1996) The usefulness of faecal phosphorus and nitrogen in interpreting differences in live-mass gain and the response to phosphorus supplementation in grazing cattle in arid regions. *Onderstepoort Journal of Veterinary Research* 63, 121–126.

Han, Y.N., Roneker, K.R., Pond, W.G. and Lei, X.G. (1998) Adding wheat middlings, microbial phytase and citric acid to corn–soybean meal diets for growing pigs may replace inorganic phosphorus supplementation. *Journal of Animal Science* 76, 2649–2656.

Hastad, C.W., Dritz, S.S., Tokach, M.D., Goodband, R.D., Nelssen, J.L., DeRouchey, J.M., Boyd, R.D. and Johnston, M.E. (2004) Phosphorus requirements of growing-finishing pigs reared in a commercial environment. *Journal of Animal Science* 82, 2945–2952.

Hemingway, R.G. and Fishwick, G. (1975) Ammonium polyphosphate in the drinking-water as a source of phosphorus for growing sheep. *Proceedings of the Nutrition Society* 34, 78A–79A.

Hendricksen, R.E., Ternouth, J.D. and Punter, L.D. (1994) Seasonal nutrient intake and phosphorus kinetics of grazing steers in northern Australia. *Australian Journal of Agricultural Research* 45, 1817–1829.

Holechek, J.L., Galyean, M.L., Wallace, J.D. and Wofford, H. (1985) Evaluation of faecal indices for predicting phosphorus status of cattle. *Grass and Forage Science* 40, 489–492.

Huber, K., Hempel, R. and Rodehutscord, M. (2006) Adaptation of epithelial sodium-dependent phosphate transport in jejunum and kidney of hens to variations in dietary phosphorus intake. *Poultry Science* 85, 1980–1986.

Jendreville, C., Genthon, C., Bouguennec, A., Carre, B. and Nys, Y. (2007) Characterisation of European varieties of triticale with special emphasis on the ability of plant phytases to improve phytate phosphorus utilisation. *British Poultry Science* 48, 678–689.

Jendza, J.A., Dilger, R.N., Adedokun, S.A., Sands, J.S. and Adeola, O. (2005) *Escherichia coli* phytase improves growth performance of starter, grower and finisher pigs fed phosphorus-deficient diets. *Journal of Animal Science* 83, 1882–1899.

Johnston, S.L., Williams, S.B., Southern, L.L., Bidner, T.D., Bunting, L.D., Matthews, J.O. and Olcott, B.M. (2004) Effect of phytase addition and dietary calcium and phosphorus levels on plasma metabolites and ileal and total-tract digestibility in pigs. *Journal of Animal Science* 82, 705–714.

Jones, R.J. (1990) Phosphorus and beef production in northern Australia. 1. Phosphorus and pasture productivity – a review. *Tropical Grasslands* 24, 131–139.

Jones, R.J. and Betteridge, K. (1994) Effect of superphosphate on its component elements (phosphorus, sulfur and calcium), on the grazing preference of steers on a tropical grass–legume pasture grown on a low phosphorus soil. *Australian Journal of Experimental Agriculture* 34, 349–353.

Jongbloed, A.W. and Henkens, C.H. (1996) Environmental concerns of using animal manure – the Dutch case. In: Kornegay, E.T. (ed.) *Nutrient Management of Food Animals to Enhance and Protect the Environment.* CRC Press, Boca Raton, Florida, pp 315–333.

Jongbloed, A.W. and Kemme, P.A. (1990) Apparent digestible phosphorus in the feeding of pigs in relation to availability, requirement and environment. 1. Digestible phosphorus in feedstuffs from plant and animal origin. *Netherlands Journal of Agricultural Science* 38, 567–575.

Jumba, I.O., Suttle, N.F., Hunter, E.A. and Wandiga, S.O. (1995) Effects of soil origin and composition and herbage species on the mineral composition of forages in the Mount Elgan region of Kenya. 1. Calcium, phosphorus, magnesium and sulfur. *Tropical Grasslands* 29, 40–46.

Karn, J.F. (1997) Phosphorus supplementation of range cows in the Northern Great Plains. *Journal of Range Management* 50, 2–9.

Karn, J.F. (2001) Phosphorus nutrition of grazing cattle: a review. *Animal Feed Science and Technology* 89, 133–153.

Keene, B.E., Knowlton, K.F., McGilliard, M.L., Lawrence, L.A., Nickols-Richardson, S.M., Wilson, J.H., Rutledge, A.M., McDowell, L.R. and Van Amburgh, M.E. (2004) Measures of bone mineral content in mature dairy cows. *Journal of Dairy Science* 87, 3816–3825.

Kerridge, P.L., Gilbert, M.A. and Coates, D.B. (1990) Phosphorus and beef production in northern Australia. 8. The status and management of soil phosphorus in relation to beef production. *Tropical Grasslands* 24, 221–230.

Keshavarz, K. (1994) Laying hens respond differently to high dietary levels of phosphorus in monobasic and dibasic calcium phosphate. *Poultry Science* 73, 687–703.

Keshavarz, K. (2000a) A reevaluation of the non-phytate phosphorus requirement of growing pullets with and without phytase. *Poultry Science* 79, 748–763.

Keshavarz, K. (2000b) Non-phytate phosphorus requirement of laying hens with and without phytase on a phase feeding program. *Poultry Science* 79, 748–763.

Kies, A.K., Kemme, P.A., Sebek, L.B.J., van Diepen, J.Th.M. and Jongbloed, A.W. (2006) Effect of graded doses and a high dose of microbial phytase on the digestibility of various minerals in weaner pigs. *Journal of Animal Science* 84, 1169–1175.

Kincaid, R.L., Garikipati, D.K., Nennich, T.D. and Harrison, J.H. (2005) Effect of grain source and exogenous phytase on phosphorus digestibility in dairy cows. *Journal of Dairy Science* 88, 2893–2902.

Knowlton, K.F. and Herbein, J.H. (2002) Phosphorus partitioning during early lactation in dairy cows fed diets varying in phosphorus content. *Journal of Dairy Science* 85, 1227–1236.

Knowlton, K.F., Herbein, J.H., Meister-Werisbarth, M.A. and Wark, W.A. (2001) Nitrogen and phosphorus partitioning in lactating Holstein cows fed different sources of dietary protein and phosphorus. *Journal of Dairy Science* 85, 1210–1217.

Knowlton, K.F., Radcliffe, J.S., Novak, C.L. and Emmerson, D.A. (2004) Animal waste management to reduce phosphorus losses to the environment. *Journal of Animal Science* 82 (E. Suppl.), E173–E195.

Koch, M.E. and Mahan, D.C. (1985) Biological characteristics for assessing low phosphorus intake in growing swine. *Journal of Animal Science* 60, 699–708.

Kornegay, E.T. and Qian, H. (1996) Replacement of inorganic phosphorus by microbial phytase for young pigs fed a maize–soya bean meal diet. *British Journal of Nutrition* 76, 563–578.

Lacey, D.L. and Huffer, W.E. (1982) Studies on the pathogenesis of avian rickets. 1. Changes in epiphyseal and metaphyseal vessels in hypocalcaemic and hypophosphataemic rickets. *American Journal of Pathology* 109, 288–301.

Lampkin, G.H., Howard, D.A. and Burdin, M.L. (1961) Studies on the production of beef from zebu cattle in East Africa. 3. The value of feeding a phosphatic supplement. *Journal of Agricultural Science, Cambridge* 57, 39–47.

Langlands, J.P. (1987) Assessing the nutrient status of herbivores. In: Hacker, J.B. and Ternouth, J.H. (eds) *The Nutrition of Herbivores*. Academic Press, Sydney, Australia, pp. 363–390.

Lei, X.G., Ku, P.K., Miller, E.R., Yokoyama, M.T. and Ullrey, D.E. (1994) Calcium level effects the efficacy of supplemental microbial phytase in corn–soya bean meal diets for pigs. *Journal of Animal Science* 72, 139–143.

Liesegang A., Eicher, R., Sassi, M.-L., Risteli, J., Kraenzlin, M. and Riond, J.-L. (2000) Biochemical markers of bone formation and resorption around parturition and during lactation in dairy cows with high and low milk yields. *Journal of Dairy Science* 83, 1773–1781.

Little, D.A. (1968) Effect of dietary phosphate on the voluntary consumption of Townsville lucerne (*Stylosanthes humilis*) by cattle. *Proceedings of the Australian Society of Animal Production* 7, 376–380.

Little, D.A. (1984) Definition of objective criterion of body phosphorus reserves in cattle and its evaluation *in vivo*. *Canadian Journal of Animal Science* 64 (Suppl.), 229–231.

Little, D.A. and Minson, D.J. (1977) Variation in the phosphorus content of bone samples obtained from the last three ribs of cattle. *Research in Veterinary Science* 23, 393–394.

Little, D.A. and Radcliff, D. (1979) Phosphorus content of bovine rib: influence of earlier biopsy of the same rib. *Research in Veterinary Science* 27, 239–241.

Littledike, E.T and Goff, J.P. (1987) Interactions of calcium and phosphorus, magnesium and vitamin D that influence the status of domestic meat animals. *Journal of Dairy Science* 70, 1727–1743.

Liu, J., Bollinger, D.W., Ledoux, D.R., Ellersieck, M.R. and Veum, T.L. (1997) Soaking increases the efficacy of supplemental microbial phytase in a low-phosphorus corn–soybean meal diet for growing pigs. *Journal of Animal Science* 75, 1292–1298.

Liu, J., Bollinger, D.W., Ledoux, D.R. and Veum, T.L. (1998) Lowering the dietary calcium to total phosphorus ratio increases phosphorus utilisation in low-phosphorus corn–soybean meal diets supplemented with microbial phytase for growing-finishing pigs. *Journal of Animal Science* 76, 808–813.

Liu, J., Bollinger, D.W., Ledoux, D.R. and Veum, T.L. (2000) Effects of calcium:phosphorus ratios on apparent absorption of calcium and phosphorus in the small intestine, caecum and colon of pigs *Journal of Animal Science* 78, 106–109.

Long, T.A., Tillman, A.D., Nelson, A.B., Gallup, W.D. and Davis, W. (1957) Availability of phosphorus in mineral supplements. *Journal of Animal Science* 16, 444–450.

MAFF (1990) *UK Tables of the Nutritive Value and Chemical Composition of Foodstuffs.* In: Givens, D.I. (ed.) Rowett Research Services, Aberdeen, UK.

McLean, R.W., Hendricksen, R.E., Coates, D.B. and Winter, W.H. (1990) Phosphorus and beef production in northern Australia. 6. Dietary attributes and their relation to cattle growth. *Tropical Grasslands* 24, 197–208.

Manangi, M.K. and Coon, C.N. (2006) Evaluation of phytase enzyme with chicks fed basal diets containing different soybean samples. *Journal of Applied Poultry Research* 15, 292–306.

Marounek, M., Duskova, D. and Skivanova, V. (2003) Hydrolysis of phytic acid and its availability to rabbits. *British Journal of Nutrition* 89, 287–294.

Mellado, M., Rodriguez, S., Lopez, R. and Rodriguez, A. (2006) Relation among milk production and composition and blood profiles and faecal P and nitrogen in goats on rangeland. *Small Ruminant Research* 65, 230–236.

Miller, C.P., Winter, W.H., Coates, D.B. and Kerridge, P.C. (1990) Phosphorus and beef production in northern Australia. 10. Strategies for phosphorus use. *Tropical Grasslands* 24, 239–249.

Miller, E.R., Ullrey, D.C., Zutaut, C.L., Baltzer, B.V., Schmidt, D.A., Hoefer, J.A. and Luecke, R.W. (1964) Phosphorus requirement of the baby pig. *Journal of Nutrition* 82, 34–40.

Miller, W.J., Neathery, M.W., Gentry, R.P., Blackmon, D.M., Crowe, C.I., Ware, G.C. and Fielding, A.J. (1987) Bioavailability of phosphorus from defluorinated and dicalcium phosphates and phosphorus requirement of calves. *Journal of Dairy Science* 70, 1885–1892.

Minson, D.J. (1990) *Forages in Ruminant Nutrition.* Academic Press, San Diego, California, pp. 208–229.

Mohammed, A., Gibney, M. and Taylor, T.G. (1991) The effects of levels of inorganic phosphorus, calcium and cholecalciferol on the digestibility of phytate-P by the chick. *British Journal of Nutrition* 66, 251–259.

Morse, D., Head, H.H. and Wilcox, C.J. (1992) Disappearance of phosphorus in phytase from concentrates *in vitro* and from rations fed to lactating dairy cows. *Journal of Dairy Science* 75, 1979–1986.

Murry, A.C., Lewis, R.D. and Amos, H.E. (1997) The effect of microbial phytase in a pearl millet–soybean meal diet on apparent digestibility and retention of nutrients, serum mineral concentration and bone mineral density of nursery pigs. *Journal of Animal Science* 75, 1284–1291.

Muschen, H., Petri, A., Breves, G. and Pfeffer, E. (1988) Response of lactating goats to low phosphorus intake. 1. Milk yield and faecal excretion of P and Ca. *Journal of Agricultural Science, Cambridge* 111, 255–263.

Nelson, T.S., Shieh, T.R., Wodzinski, R.J. and Ware, J.H. (1971) Effect of supplemental phytase on the utilisation of phytate phosphorus by chicks. *Journal of Nutrition* 101, 1289–1293.

Nelson, T.S., Daniels, L.B., Hall, J.R. and Shields, L.G. (1976) Hydrolysis of natural phytate phosphorus in the digestive tract of calves. *Journal of Animal Science* 42, 1509–1512.

Nimmo, R.D., Peo, E.R., Crenshaw, D.D., Maser, B.D. and Lewis, A.J. (1981a) Effect of level of dietary calcium and phosphorus during growth and gestation on calcium–phosphorus balance and reproductive performance of first litter sows. *Journal of Animal Science* 52, 1343–1349.

Nimmo, R.D., Peo, E.R., Maser, B.D. and Lewis, A.J. (1981b) Effect of level of dietary calcium and phosphorus during growth and gestation on performance, blood and bone parameters of swine. *Journal of Animal Science* 52, 1330–1342.

NRC (1994) *Nutrient Requirements of Poultry*, 9th edn. National Academy of Sciences, Washington, DC.
NRC (1998) *Nutrient Requirements of Swine*, 10th edn. National Academy of Sciences, Washington, DC.
NRC (2000) *Nutrient Requirements of Beef Cattle*, 7th edn. National Academy of Sciences, Washington, DC.
NRC (2001) *Nutrient Requirements of Dairy Cows*, 5th edn. National Academy of Sciences, Washington, DC.
Olukosi, O.A., Cowieson, A.J. and Adeola, O. (2007) Age-related influence of a cocktail of xylanase, amylase and protease or phytase individually or in combination in broilers. *Poultry Science* 86, 77–86.
O'Quinn, P.R., Knabe, D.A. and Gregg, E.J. (1997) Efficacy of Natuphos® in sorghum-based diets for finishing swine. *Journal of Animal Science* 75, 1299–1307.
Owings, W.J., Sell, J.L. and Balloun, S.L. (1977) Dietary phosphorus needs of laying hens. *Poultry Science* 56, 2056–2060.
Paditz, K., Kluth, H. and Rodehutscord, M. (2004) Relationship between graded doses of three microbial phytases and digestible phosphorus in pigs. *Animal Science* 78, 429–438.
Park, W.-Y., Matsui, T., Konishi, C., Sung-Won, K., Yano, F. and Yano, H. (1999) Formaldehyde treatment suppresses ruminal degradation of phytate in soyabean meal and rapeseed meal. *British Journal of Nutrition* 81, 467–471.
Petersen, G. and Stein, N.H. (2006) Novel procedure for estimating endogenous losses and measurement of apparent and true digestibility of phosphorus by growing pigs. *Journal of Animal Science* 84, 2126–2132.
Peterson, A.B., Orth, M.W., Goff, J.P. and Beede, D.K. (2005) Periparturient responses of multiparous Holstein cows fed different dietary phosphorus concentrations prepartum. *Journal of Dairy Science* 88, 3582–3594.
Petri, A., Muschen, H., Breves, G., Richter, O. and Pfeiffer, E. (1989) Response of lactating goats to low phosphorus intake. 2. Nitrogen transfer from rumen ammonia to rumen microbes and proportion of milk protein derived from microbial amino-acids. *Journal of Agricultural Science, Cambridge* 111, 265–271.
Pettey, L.A., Cromwell, G.L. and Lindemann, M.D. (2006) Estimation of endogenous phosphorus loss in growing-finishing pigs fed semi-purified diets. *Journal of Animal Science* 84, 618–626.
Pointillart, A.L. (1991) Enhancement of phosphorus utilisation in growing pigs fed phytate-rich diets by using rye bran. *Journal of Animal Science,* 69, 1109–1115.
Pointillart, A.L., Fontaine, N. and Thomasset, M. (1986) Effects of vitamin D on calcium regulation in vitamin D deficient pigs given a phytate-phosphorus diet. *British Journal of Nutrition* 56, 661–669.
Pointillart, A.L., Fourdin, A. and Fontaine, N. (1987) Importance of cereal phytase activity for phytate phosphorus utilisation by growing pigs fed triticale or corn. *Journal of Nutrition* 117, 907–913.
Pointillart, A.L., Fourdin, A., Bordeau, A. and Thomasset, M. (1989) Phosphorus utilisation and hormonal control of calcium metabolism in pigs fed phytic phosphorus diets containing normal or high calcium levels. *Nutrition Reports International* 40, 517–527.
Portilho, F.P., Vitti, D.M.S.S., Abdalla, A.L., McManus, C.M., Rezende, M.J.M. and Louvandini, H. (2006) Minimum phosphorus requirement for Santa Iles lambs reared under tropical conditions. *Small Ruminant Research* 63, 170–176.
Price, J.S., Jackson, B., Eastell, R., Goodship, A.E., Blumsohn, A., Wright, I., Stoneham, S., Lanyon, L.E. and Russell, R.G.G. (1995) Age-related changes in biochemical markers of bone metabolism in horses. *Equine Veterinary Journal* 27, 210–217.
Qian, H., Kornegay, E.T. and Connor, D.E. (1996) Adverse effects of wide calcium:phosphorus ratios on supplemental phytase efficacy for weanling pigs fed phosphorus at two levels. *Journal of Animal Science* 74, 1288–1297.
Qian, H., Kornegay, E.T. and Veit, H.P. (1996) Effects of supplemental phytase and phosphorus on histological, mechanical and chemical traits of tibia and performance of turkeys fed on soyabean meal-based semi-purified diets high in phytate phosphorus. *British Journal of Nutrition* 76, 263–272.
Qian, H., Kornegay, E.T. and Denbow, D.M. (1997) Utilisation of phytate phosphorus and calcium as influenced by microbial phytase, cholecalciferol and the calcium:total phosphorus ratio in broiler diets. *Poultry Science* 76, 37–46.
Rajaratne, A.A.J., Scott, D. and Buchan, W. (1994) Effects of a change in phosphorus requirement on phosphorus kinetics in the sheep. *Research in Veterinary Science* 56, 262–264.
Ramirez-Perez, A.H., Sauvant, D. and Meschy, F. (2008) Effect of phosphate solubility on phosphorus kinetics and rumenal fermentation activity in dairy goats. *Animal Feed Science and Technology* 149, 209–227.

Rapp, C., Lantzsch, H.J. and Drochner, W. (2001) Hydrolysis of phytic acid by intrinsic plant or supplemented microbial phytase (*Aspergillus niger*) in the stomach and small intestine of minipigs fitted with re-entrant cannulas. *Journal of Animal Physiology and Animal Nutrition* 85, 406–413.

Read, M.P., Engels, E.A.N. and Smith, W.A. (1986a) Phosphorus and the grazing ruminant: the effect of supplementary P on sheep at Armoedsvlakte. *South African Journal of Animal Science* 16, 1–6.

Read, M.P., Engels, E.A.N. and Smith, W.A. (1986b) Phosphorus and the grazing ruminant: the effect of supplementary P on cattle at Glen and Armoedsvlakte. *South African Journal of Animal Science* 16, 7–12.

Read, M.P., Engels, E.A.N. and Smith, W.A. (1986c) Phosphorus and the grazing ruminant: 3. Rib bone samples as an indicator of the P status of cattle. *South African Journal of Animal Science* 16, 13–27.

Reddy, N.R. and Sathe, S.K. (2001) *Feed Phytates*. CRC Press Ltd., Boca Raton, Florida.

Reid, R.L., Franklin, M.C. and Hallsworth, E.G. (1947) The utilization of phytate phosphorus by sheep. *Australian Veterinary Journal* 23, 136–140.

Reinhart, G.A. and Mahan, D.C. (1986) Effects of various calcium:phosphorus ratios at low and high dietary phosphorus for starter, grower and finishing swine. *Journal of Animal Science* 63, 457–466.

Rodehutscord, M., Haverkamp, R. and Pfeffer, E. (1998) Inevitable losses of phosphorus in pigs, estimated from balance data using diets deficient in phosphorus. *Archiv fur Tierernahrung* 51, 27–38.

Rodehutscord, M., Sanver, F. and Timmler, R. (2002) Comparative study on the effect of variable phosphorus intake at two different calcium levels on P excretion and P flow at the terminal ileum of laying hens *Archiv fur Tierernahrung* 56, 189–198.

Scharp, D.W. (1979) Effect of adding superphosphate to the drinking water on the fertility of dairy cows. *Australian Veterinary Journal* 55, 240–243.

Schryver, H.F., Craig, P.H. and Hintz, H.F. (1970) Calcium metabolism in ponies fed varying levels of calcium. *Journal of Nutrition* 100, 955–964.

Schryver, H.F., Hintz, H.F. and Craig, P.H. (1971) Phosphorus metabolism in ponies fed varying levels of phosphorus. *Journal of Nutrition* 101, 1257–1263.

Schryver, H.F., Hintz, H.F., Lowe, J.E., Hintz, R.L., Harper, R.B. and Reid, J.T. (1974) Mineral composition of the whole body, liver and bone of young horses. *Journal of Nutrition* 104, 126–132.

Scott, D., McLean, A.F. and Buchan, W. (1984) The effect of variation in phosphorus intake on net intestinal phosphorus absorption, salivary phosphorus secretion and pathway of excretion in sheep fed roughage diets. *Quarterly Journal of Experimental Physiology* 69, 439–452.

Scott, D., Whitelaw, F.C., Buchan, W. and Bruce, L.A. (1985) The effect of variation in phosphorus secretion, net intestinal phosphorus absorption and faecal endogenous phosphorus excretion in sheep. *Journal of Agricultural Science, Cambridge* 105, 271–277.

Scott, D., Robins, S.P., Nicol, P., Chen, X.B. and Buchan, W. (1994) Effects of low phosphorus diets on bone and mineral metabolism and microbial protein synthesis in lambs. *Experimental Physiology* 79, 183–187.

Scott, D., Rajarante, A.A.J. and Buchan, W. (1995) Factors affecting faecal endogenous phosphorus loss in the sheep. *Journal of Agricultural Science, Cambridge* 124, 145–151.

Selle, P.H., Walker, A.R. and Bryden, W.L. (2003) Total and phytate-phosphorus contents and phytase activity of Australian-sourced ingredients for pigs and poultry. *Australian Journal of Agricultural Research* 43, 475–479.

Sharifi, K., Mohri, M., Abedi, V., Shahinfar, R., Farzaneh, M. and Shalchi, M.H. (2005) Serum and blood inorganic phosphorus in lambs from birth to 400th day of life: effect of weaning as a cut-off point between neonatal and adult levels. *Comparative Clinical Pathology* 14, 160–167.

Shelton, J.L., Southern, L.L, LeMieux, F.M., Bidner, T.D. and Page, T.G. (2004) Effects of microbial phytase, low calcium and phosphorus, and removing the dietary trace mineral premix on carcass traits, pork quality, plasma metabolites and tissue mineral content in growing-finishing pigs. *Journal of Animal Science* 82, 2630–2639.

Simons, P.C.M., Versteegh, H.A.J., Jongbloed, A.W., Kemme, P.A., Slump, P., Bos, K.D., Walters, M.G.E., Beudeker, R.F. and Verschoar, G.J. (1990) Improvement of phosphorus availability by microbial phytase in broilers and pigs. *British Journal of Nutrition* 64, 525–540.

Singsen, E.P., Spandorf, A.H., Matterson. L.D., Serafin, J.A. and Tlustohowicz, J.J. (1962) Phosphorus in the nutrition of the adult hen. 1. Minimum phosphorus requirements. *Poultry Science* 41, 1401–1414.

Snook, L.C. (1949) Phosphorus deficiency in dairy cows: its prevalence in South-Western Australia and possible methods of correction. *Journal of the Department of Agriculture for Western Australia* 26, 169–177.

Snook, L.C. (1958) Phosphorus deficiency as a cause of bovine infertility. *Proceedings of the Australia and New Zealand Association for the Advancement of Science, Adelaide.*

Snow, J.L., Baker, D.H. and Parsons, C.M. (2004) Phytase, citric acid and 1α hydroxycholecalciferol improve phytate phosphorus utilization in chicks fed a corn–soybean meal diet. *Poultry Science* 83, 1187–1192.

Soares, J.H. (1995) Phosphorus bioavailability. In: Ammerman, C.B., Baker, D.H. and Lewis, A.J. (eds) *Bioavailability of Nutrients for Animals.* Academic Press, New York, pp. 257–294.

Somerville, B.A., Maunder, E., Ross, R., Care, A.D. and Brown, R.C. (1985) Effect of dietary calcium and phosphorus depletion on vitamin D metabolism and calcium binding protein in the growing pig. *Hormone and Metabolism Research* 17, 78–81.

Spencer, J.D., Allee, G.L. and Sauber, T.E. (2000) Phosphorus bioavailability and digestibility of normal and genetically modified low-phytate corn for pigs. *Journal of Animal Science* 78, 675–681.

Stewart, J. (1934–35) The effect of phosphorus deficient diets on the metabolism, blood and bones of sheep. *University of Cambridge Institute of Animal Pathology Fourth Report*, pp. 179–205.

Suttle, N.F. (1994) Seasonal infections and nutritional status. *Proceedings of the Nutrition Society* 53, 545–555.

Suttle, N.F. and Hay, L. (1986) Urolithiasis. In: Martin, W.B. and Aitken, I.D. (eds) *Diseases of Sheep*, 2nd edn. Blackwell Scientific Publications, Oxford, UK, pp. 250–254.

Suttle, N.F., Dingwall, R.A. and Munro, C.S. (1989) Assessing the availability of dietary phosphorus for sheep. In: Southgate, D.A.T., Johnson, I.T. and Fenwick, G.R. (eds) *Nutrient Availability: Chemical and Biological Aspects.* Royal Society of Chemistry Special Publication No. 72, Cambridge, UK, pp. 268–270.

Tamin, N.M., Angel, R. and Christman, M. (2004) Influence of dietary calcium and phytase on phytate phosphorus utilisation in broiler chickens. *Poultry Science* 83, 1358–1367.

Teleni, E., Dean, H. and Murray, R.M. (1976) Some factors affecting the measurement of blood inorganic phosphorus in cattle. *Australian Veterinary Journal* 52, 529–533.

Ternouth, J.H. (1989) Endogenous losses of phosphorus in sheep. *Journal of Agricultural Science, Cambridge* 113, 291–297.

Ternouth, J.H. (1990) Phosphorus and beef production in Northern Australia. 3. Phosphorus in cattle – a review. *Tropical Grasslands* 24, 159–169.

Ternouth, J.H. and Coates, D.B. (1997) Phosphorus homeostasis in grazing breeder cattle. *Journal of Agricultural Science, Cambridge* 128, 331–337.

Ternouth, J.H. and Sevilla, C.L. (1990a) The effect of low levels of dietary phosphorus upon dry matter intake and metabolism in lambs. *Australian Journal of Agricultural Research* 41, 175–184.

Ternouth, J.H. and Sevilla, C.L. (1990b) Dietary calcium and phosphorus repletion in lambs. *Australian Journal of Agricultural Research* 41, 413–420.

Ternouth, J.H., Bortolussi, G., Coates, D.B., Hendricksen, R.E. and McLean, R.W. (1996) The phosphorus requirements of growing cattle consuming forage diets. *Journal of Agricultural Science, Cambridge* 126, 503–510.

Theiler, A. (1912) Facts and theories about styfziekte and lamziekte. In: *Second Report of the Directorate of Veterinary Research.* Onderstepoort Veterinary Institute, Pretoria, South Africa, pp. 7–78.

Theiler, A. and Green, H.H. (1932) Aphosphorosis in ruminants. *Nutrition Abstracts and Reviews* 1, 359–385.

Thorp, B.H. and Waddington, D. (1997) Relationships between the bone pathologies, ash and mineral content of long bones in 35-day-old broiler chickens. *Research in Veterinary Science* 62, 67–73.

Timmons, J.R., Angel, R., Harter-Dennis, J.M., Saylor, W.W. and Ward, N.E. (2008) Evaluation of heat-stable phytases in pelleted diets fed to broilers from day zero to thirty-five during the summer months. *Journal of Applied Poultry Research* 17, 482–489.

Tomas, F.M., Moir, R.J. and Somers, M. (1973) Phosphorus turnover in sheep. *Australian Journal of Agricultural Research* 18, 635–645.

Toor, G.S., Cade-Menun, B.J. and Sims, J.T. (2005) Establishing a linkage between phosphorus forms in dairy diets, feces and manures. *Journal of Environmental Quality* 34, 1380–1391.

Traylor, S.L., Cromwell, G.L. and Lindemann, M.D. (2005) Bioavailability of phosphorus in meat and bone meal for swine. *Journal of Animal Science* 83, 1054–1061.

Underwood, E.J. (1981) *Mineral Nutrition of Livestock*, 2nd edn. Commonwealth Agricultural Bureaux, Farnham Royal, Slough, UK, p. 31.

Underwood, E.J. and Suttle, F. (1999) *The Mineral Nutrition of Livestock*, 3rd edn. CAB International, Wallingford, UK.

Underwood, E.J., Beck, A.B. and Shier, F.L. (1939) Further experiments on the incidence and control of pica in sheep in the botulism areas of Western Australia. *Australian Journal of Experimental Biology and Medical Science* 17, 249–255.

Underwood, E.J., Shier, F.L. and Beck, A.B. (1940) Experiments in the feeding of phosphorus supplements to sheep in Western Australia. *Journal of the Department of Agriculture for Western Australia* 17, 388–405.

Valk, H. and Sebek, L.B.J. (1999) Influence of long-term feeding of limited amounts of phosphorus on dry matter intake, milk production and bodyweight change of dairy cows. *Journal of Dairy Science* 82, 2157–2163.

Valk, H., Sebek, L.B.J. and Beynen, A.C. (2002) Influence of phosphorus intake on excretion and blood plasma in dairy cows. *Journal of Dairy Science* 85, 2642–2649.

Van der Klis, J.D., Versteegh, H.A., Simons, P.C. and Kies, A.K. (1997) The efficacy of phytase in corn–soybean-based diets for laying hens. *Poultry Science* 76, 1535–1542.

Van Doorn, D.A., Everts, H., Wouterse, H. and Beynen, A.C. (2004) The apparent digestibility of phytate phosphorus and the influence of supplemental phytase in horses. *Journal of Animal Science* 82, 1756–1763.

Van Niekerk, B.D.H. and Jacobs, G.A. (1985) Protein energy and phosphorus supplementation of cattle fed low-quality roughage. *South African Journal of Animal Science* 15, 133–136.

Veum, T.L., Ledoux, D.R., Bollinger, D.W., Raboy, V. and Ertl, D.S. (2001) Low-phytic acid corn improves nutrient utilisation for growing pigs. *Journal of Animal Science* 79, 2873–2880.

Veum, T.L., Ledoux, D.R., Bollinger, D.W., Raboy, V. and Cook, A. (2002) Low-phytic acid barley improves calcium and phosphorus utilisation and growth performance in growing pigs. *Journal of Animal Science* 80, 2663–2670.

Veum, T.L., Ledoux, D.R. and Raboy, V. (2007) Low-phytate barley cultivars improve the utilisation of phosphorus, calcium, nitrogen, energy and dry matter in diets fed to young swine. *Journal of Animal Science* 85, 961–971.

Vitti, D.M.S.S., Kebreab, E., Lopes, J.B., Abdalla, A.L., De Carvalho, F.F.R., De Resende, K.T., Crompton, L.A. and France, J. (2000) A kinetic model of phosphorus metabolism in growing goats. *Journal of Animal Science* 78, 2706–2712.

Viveros, A., Centeno, C., Brenes, A., Canales, R. and Lozano, A. (2000) Phytase and acid phosphatases in plant feedstuffs. *Journal of Agricultural and Food Chemistry* 48, 4009–4013.

Vucenik, I., Passaniti, A., Vitolo, M.I., Tantivejkul, K., Eggleton, P. and Shamsuddin, A.M. (2004) Anti-angionetic activity of inositol hexaphosphate (IP6). *Carcinogenesis* 25, 2115–2123.

Wadsworth, J.C., McClean, R.W., Coates, D.B. and Winter, J.H. (1990) Phosphorus and beef production in northern Australia. 5. Animal phosphorus status and diagnosis. *Tropical Grasslands* 24, 185–196.

Waldroup, P.W., Simpson, C.F., Damron, B.L. and Harms, R.H. (1967) The effectiveness of plant and inorganic phosphorus in supporting egg production in hens and hatchability and bone development in chick embryos. *Poultry Science* 46, 659–664.

Walker, R.G., Davison, T.M., Orr, W.N. and Silver, B.A. (1997) Phosphorus fertilizer for nitrogen-fertilized dairy pastures. 3. Milk responses to a dietary phosphorus supplement. *Journal of Agricultural Science, Cambridge* 129, 233–236.

Wang, X.-L., Gallagher, C.H., McClure, T.J., Reeve, V.E. and Canfield, P.J. (1985) Bovine post-parturient haemoglobinuria: effect of inorganic phosphate on red cell metabolism. *Research in Veterinary Science* 39, 333–339.

Wan Zahari, M., Thompson, J.K., Scott, D. and Buchan, W. (1990) The dietary requirements of calcium and phosphorus for growing lambs. *Animal Production* 50, 301–308.

Watson, B.C., Matthews, J.O., Southern, L.L. and Shelton, J.L. (2006) The interactive effects of *Eimeria acervulina* and phytase for broiler chicks. *Poultry Science* 84, 910–913.

Williams, S.W., Lawrence, L.A., McDowell, L.R., Wilkinson, N.S., Ferguson, P.W. and Warmick, A.C. (1991a) Criteria to evaluate bone mineralisation in cattle. I. Effect of dietary phosphorus on chemical, physical and mechanical properties. *Journal of Animal Science* 69, 1232–1242.

Williams, S.W., McDowell, L.R., Lawrence, L.A., Wilkinson, N.S., Ferguson, P.W and Warmick, A.C. (1991b) Criteria to evaluate bone mineralisation in cattle. II. Non-invasive techniques. *Journal of Animal Science* 69, 1243–1254.

Wilson, W.D. and Field, A.C. (1983) Absorption and secretion of calcium and phosphorus in the alimentary tract of lambs infected with daily doses *of Trichostrongylus colubriformis* or *Ostertagia circumcincta*. *Journal of Comparative Pathology* 93, 61–71.

Winks, L. (1990) Phosphorus and beef production in northern Australia. 2. Responses to phosphorus by ruminants – a review. *Tropical Grasslands* 24, 140–158.

Wright, R.D., Blair-West, J.R., Nelson, J.F. and Tregear, G.W. (1984) Handling of phosphate by the parotid gland (ovine). *American Journal of Physiology* 246, F916–F926.

Wu, Z. (2005) Utilisation of phosphorus in lactating cows fed varying amounts of phosphorus and sources of fibre. *Journal of Dairy Science* 88, 2850–2859.

Wu, Z., Satter, L.H., Blohowiak, A.J., Stauffacher, R.H. and Wilson, J.H. (2001) Milk production, estimated phosphorus excretion and bone characteristics of dairy cows fed different amounts of phosphorus for two or three years. *Journal of Dairy Science* 84, 1738–1748.

Wu, Z., Tallam, S.K., Ishler, V.A. and Archibald, D. D. (2003) Utilisation of phosphorus in lactating cows fed varying amounts of phosphorus and forage. *Journal of Dairy Science* 86, 3300–3308.

Yan, F., Angel, R., Ashwell, C., Mitchell, A. and Christman, M. (2005) Evaluation of a broiler's ability to adapt to an early deficiency of phosphorus and calcium. *Poultry Science* 84, 1232–1241.

Yano, F., Yano, H. and Breves, G. (1991) Calcium and phosphorus metabolism in ruminants. In: *Proceedings of the Seventh International Symposium of Ruminant Physiology*. Academic Press, New York, pp. 277–295.

Young, V.R., Richards, W.P.C., Lofgreen, G.P. and Luick, J.R. (1966) Phosphorus depletion in sheep and the ratio of calcium to phosphorus in the diet with reference to calcium and phosphorus absorption. *British Journal of Nutrition* 20, 783–794.

Zyla, K., Mika, M., Stodolak, B., Wikiera, A., Koreleski, J. and Swiatkiewicz, S. (2004) Towards complete dephosphorylation and total conversion of phytates in poultry feeds. *Poultry Science* 83, 1175–1186.

7 Potassium

The essentiality of potassium in animal diets was demonstrated experimentally early in the 19th century. The element has been found to sustain nerve and muscle excitability as well as water and acid–base balance (Ward, 1966). Naturally occurring potassium deficiency in livestock is, however, rare. The element is so abundant in common rations and pastures that 'nutritionists have generally regarded potassium as a useful but non-critical nutrient' (Thompson, 1972). That view has been challenged with respect to cattle (Preston and Linser, 1985), but unequivocal benefits from potassium supplementation have yet to be shown in the field. The nutritional problems presented by potassium continue to involve excesses rather than deficiencies, particularly in ruminants whose natural diet frequently provides a manifold surplus of potassium that perturbs sodium and calcium homeostasis.

Potassium inevitably contributes to the regulation of acid–base balance and participates in respiration via the chloride shift (see Chapter 8). All soft tissues are much richer in potassium than in sodium, making potassium the third most abundant mineral in the body at about $3.0\,g\,kg^{-1}$ live weight (LW), ahead of sodium at $1.2\,g\,kg^{-1}$ LW (Fig. 7.1). The highest potassium concentrations are found in muscle (approximately $4\,g\,kg^{-1}$) (ARC, 1980) and it is possible to estimate lean body mass by measuring the activity of the naturally occurring potassium radioisotope, ^{40}K, in a whole-body monitor (Ward, 1966). Many enzymes have specific or facilitative requirements for potassium and the element influences many intracellular reactions involving phosphate, with effects on enzyme activities and muscle contraction (Ussing, 1960; Thompson, 1972).

Functions of Potassium

Potassium is the major intracellular ion in tissues. It is usually present at concentrations of $100–160\,mmol\,l^{-1}$, which are 25–30 times greater than those in plasma (Ward, 1966). The established gradients create an electrical potential that is essential for the maintenance of responsiveness to stimuli and muscle tone.

Dietary Sources of Potassium

Unlike the macro-minerals discussed so far, calcium, magnesium and phosphorus, there is no difficulty in measuring 'potassium value' and little systematic variation in availability between feeds. By and large, the potassium that is detected by feed analysis is absorbed by the animal.

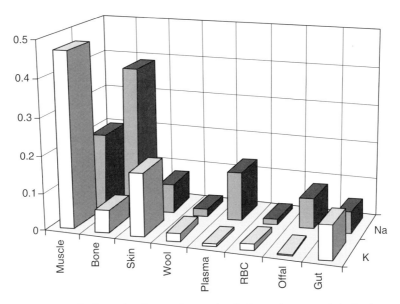

Fig. 7.1. Contrasting distributions of sodium and potassium in the unshorn, empty body of young sheep (data from Grace, 1983). RBC, red blood cell.

Forages

Potassium concentrations in forages are highly variable. In the UK, a potassium range of 10–45 g kg^{-1} dry matter (DM) about a mean of 24.3 g kg^{-1} DM has been reported for fresh herbage (Table 7.1) (MAFF, 1990). Similar variation has been found in Victoria and Queensland forages (McNeill et al., 2002) and in grass silage in the Netherlands (Schonewille et al., 1997). High pasture potassium is attributable to large farm inputs of potassium in organic and artificial fertilizers. Potassium is recycled via slurry application and the relationship between application rate and pasture potassium in one study is shown in Fig. 7.2. The effect varied slightly with soil type (Soder and Stout, 2003). At a given soil status, herbage potassium is influenced by the plant species, its state of maturity and the way the sward is managed. Cool-season grass species (e.g. *Lolium perenne*) maintain higher potassium concentrations than warm-season species (e.g. Robinson, 1985) and temperate legumes have higher levels than tropical legumes (Table 7.2) (Lanyon and Smith, 1985). For a given sward, pasture potassium can decrease markedly as the season progresses (Reid et al., 1984; Pelletier et al., 2006), although the effect varies between species (Fig. 7.3) (Perdomo et al., 1977; Cherney and Cherney, 2005).

Other feeds

Most feeds show less variation in potassium concentration than forage, but there is considerable overlap between types (Table 7.1). Carbohydrate sources are generally low in potassium (3–5 g kg^{-1} DM), but protein sources generally contain 10–20 g kg^{-1} DM and it is hard to compound a balanced ration from the ingredients given in Table 7.1 that does not meet the requirements of any class of livestock. Over-reliance on cereals and non-protein (also non-potassium) nitrogen sources might, however, create a potassium-deficient diet. If alkali-treated straw is used as the source of roughage, treatment with ammonia or sodium hydroxide appears to reduce potassium concentrations by about 25%. At the other extreme, root crops, molasses, molassed by-products (beet pulp, bagasse), cottonseed meal and soybean meal

Table 7.1. Mean (standard deviation) potassium concentrations (g kg^{-1} DM) in pastures and foodstuffs in the UK (data mostly from MAFF, 1990).

Forages	K	Concentrates	K	Roots and by-products	K
Herbage	24.3 (6.6)	Fish meal	10.2 (1.3)	Fodder beet	17.5 (4.8)
Grass silage	25.8 (6.8)	Soybean meal (ext.)	25.0 (1.0)	Swedes	28.6[a]
Clover silage (mixed)	27.4 (7.6)	Safflower meal	17.1 (1.8)	Turnips	37.8[a]
Lucerne silage	24.6 (4.0)	Palm-kernel cake	6.9 (1.2)	Cassava meal	8.1 (1.5)
Maize silage	12.3 (4.1)	Maize gluten	12.5 (2.7)	Meat and bone meal	5.2 (0.6)
Grass hay	20.7 (5.3)	Rapeseed meal	14.3 (2.2)	Molasses (beet)	49.1 (5.4)
Lucerne hay	27.5 (5.0)	Cottonseed meal	15.8 (0.8)	Molasses (cane)	38.6 (15.7)
Dried grass	26.0 (8.0)	Linseed meal	11.2 (0.13)	Molassed beet pulp	18.2 (1.9)
Dried lucerne	25.4 (8.3)	Maize	3.5 (0.22)	Beet pulp	11.7 (7.6)
Barley straw	16.0 (6.5)	Barley	5.0 (0.7)	Brewers' grains	0.6 (1.0)
Wheat straw	10.2 (3.7)	Wheat	4.6 (0.4)	Wheat feed	13.0 (2.2)
Alkali-treated barley straw	11.6 (5.3)	Oats	5.0 (0.9)	Distillers' dark grains: barley	10.2 (0.8)
		Sorghum	3.2 (0.1)		

[a]Data from New Zealand (Cornforth et al., 1978).

can add significant amounts of potassium to supplements for dairy cows. These should be avoided in pre-partum rations and where there is difficulty in restricting potassium accumulation in the pasture.

Absorbability

Dietary sources of potassium are highly soluble and almost completely absorbed, regardless of the amount ingested (Miller, 1995). In a study in sheep, the apparent absorption of potassium (AA_K) from four tropical grasses cut at three stages of regrowth varied little about a high mean of 0.86 (Perdomo et al., 1977). In cattle, AA_K increased marginally from 0.95 to 0.98 in steers as dietary potassium increased from 6 to 48 g kg^{-1} DM. Availability can be measured from urinary responses to dietary supplements, and one study thus estimated that 0.9 of potassium, given as the acetate or bicarbonate, is available to young pigs (Combs et al., 1985).

Potassium Requirements

Pigs and poultry

A reinvestigation of weanling cross-bred pigs placed the potassium requirement at 2.6–3.3 g kg^{-1} DM, using a purified diet supplemented with potassium acetate or bicarbonate (Combs et al., 1985). This figure is close to that originally proposed by Meyer et al. (1950). The potassium requirements for growth in poultry appear to vary according to dietary conditions, being increased by dietary phosphorus (Gillis, 1948, 1950), chloride

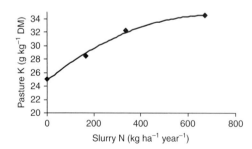

Fig 7.2. The dispersal of dairy slurry on pasture as a nitrogen fertilizer recycles potassium, greatly increasing the pasture K concentrations (from Soder and Stout, 2003). (*Dactylis glomerata* pasture; slurry K:N was 0.75.)

Table 7.2. Critical concentrations for potassium (g kg^{-1} DM) below which plant yield may be reduced in forage species (from Lanyon and Smith, 1985; Robinson, 1985).

Climate	Grass	K	Legume	K
Cool temperate	Poa pratensis	16–20	Trifolium repens	10
	Dactylis glomerata	23–35	Medicago sativa	12
	Lolium multiflorum	26–30	Trifolium fragiferum	10
	Lolium perenne	26–30		
	Festuca elatior	25–28		
Tropical	Digitaria decumbens	12–14	Stylosanthes humilis	6.0
	Sorghum sudanese	15–18	Centrosema brasilianum	11.2[a]
	Cynodon dactylon	15–17	Lotononis bainesii	9.0
			Lespedeza capitata	9.8–12.2[a]
			Capsicum pubescens	7.5[a]

[a]In the dry season critical values are <7 g kg^{-1} DM.

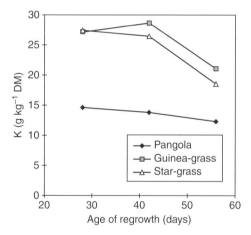

Fig. 7.3. Potassium concentrations in herbage decline as the sward matures (data for tropical grasses from Perdomo et al., 1977).

(Nesheim et al., 1964), carbohydrate (Leach et al., 1959) and protein supplies, but the effects of carbohydrate and protein probably reflect the higher growth rate achieved by feeding such supplements. The potassium requirement was once put at 4–6 g kg^{-1} DM for chickens (Thompson, 1972) and was similar to that of turkey poults (Supplee, 1965). Subsequent recommendations (Table 7.3) distinguished between different types and ages of bird, with particularly high potassium needs for the young turkey poult (7 g kg^{-1} DM) and a low requirement as pullets approach lay (2.5 g kg^{-1} DM). Optimum dietary concentrations will be influenced by interactions with sodium and chloride but 1 g K kg^{-1} DM may not support optimum shell thickness (Sauveur and Mongin, 1978).

Cattle

Precise data from feeding trials on the potassium requirements for growth or milk production are still not available. Thompson (1972) recommended 6–8 g kg^{-1} DM for growing beef cattle and 8–10 g kg^{-1} DM for lactating cows. NRC (2000, 2001) recommendations are similar. The increase in potassium need with the onset of lactation reflects the high potassium concentration in milk (1.5 g l^{-1}) and confirms the possibility that lack of potassium might limit milk production in some management systems. Factorial estimates have confirmed that while the potassium requirement increases with milk yield, needs for beef and dairy cattle are both lower than previously recommended, with the requirement for growth diminishing as the animals mature (Table 7.4) (ARC, 1980). Feeding trials with calves of 80 kg LW gaining 0.74 kg day^{-1} have suggested a potassium requirement of between 3.4 and 5.8 g kg^{-1} DM (Weil et al., 1988), while steers of 300 kg LW gaining 1.3 kg day^{-1} did not benefit when their dietary potassium was increased from 5.5 to 10 g kg^{-1} DM (Brink et al., 1984). Both of these studies support the lower ARC (1980) estimates (Table 7.4). Comparison of ARC

Table 7.3. Dietary potassium requirements (g kg^{-1} DM) of growing broiler and Leghorn chicks, turkey poults and laying birds (from NRC, 1994).

Bird type	Growth stage[a]			Lay stage[b]		
	Early	Middle	Late	High	Medium	Low
Broiler	3.0 (5.0)[c]	3.0	3.0	–	–	–
Leghorn	2.5 (4.0)[c]	2.5	2.5	1.9	1.5	1.3
Poults	7.0	5.0	4.0	–	6.0	–

[a]For broilers, Leghorn chicks and poults, respectively: early = 0–3, 0–6 or 0–8 weeks; middle = 3–6, 6–12, or 8–12 weeks; late = 6–18, >18 weeks to first lay and 20–24 weeks of age.
[b]High, medium and low represent declining energy densities and food intakes of 80, 100 and 120 g day^{-1}, respectively.
[c]A factorial model would predict higher potassium requirements for early growth in chicks and suggested values are given in parentheses, matching a corresponding increase in sodium requirement (Table 8.4).

Table 7.4. Examples of dietary requirements for potassium (g) of cattle and sheep (after ARC, 1980).

Animal	Weight gain or milk yield (kg day^{-1})	Assumed DM intake (kg day^{-1})	Faecal loss (g day^{-1})	Inevitable loss[c] (g day^{-1})	Production requirement (g day^{-1})	Total dietary requirement[a]	
						(g day^{-1})	(g kg^{-1} DM)
50 kg calf	0.5	0.5	1.3	1.9	1.0	4.2	8.4
250 kg bullock	0.5	5	13.0	9.5	1.0	23.5	4.7
600 kg cow, weeks 30–40 of pregnancy	0.5[b]	8	20.8	22.7	2.8	46.3	5.8
600 kg cow	10	10	26.0	22.7	15	63.7	6.4
600 kg cow	30	14	36.4	22.7	45	104.1	7.4
40 kg sheep	0.2	1	1.0	1.6	0.4	3.0	3.0

[a]Estimates did not incorporate an absorption coefficient, but with almost complete absorption (approximately 90%), this should not matter.
[b]Weight gain of fetus.
[c]In urine and through skin, saliva, etc.

(1980) maintenance requirements with those calculated by Karn and Clanton (1977) (3 g potassium 100 kg^{-1} LW) gives no indication of underestimation.

Sheep

Thompson (1972) gave 7–8 g K kg^{-1} DM as the requirement for sheep, based on two experiments with feedlot 'finishing lambs' given diets with graded but widely spaced concentrations of potassium as the carbonate (Telle et al., 1964; Campbell and Roberts, 1965). In both studies the best weight gains and feed efficiency were obtained at the highest levels employed (6.2 and 7 g K kg^{-1} DM, respectively). However, ARC (1980) concluded from the same data that growing sheep require only 3–5 g K kg^{-1} DM and their factorial estimates agreed with this lower estimate (Table 7.4).

Metabolism of Potassium

Absorption

Potassium absorption occurs principally in the small intestine in non-ruminants by unregulated

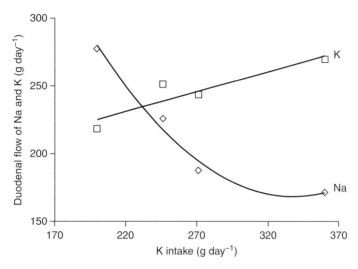

Fig 7.4. The effect of dietary potassium intake on duodenal flows of sodium and K in lactating cows given four different silages (containing 15–26 g K kg^{-1} DM), together with a Na-rich concentrate (data from Khorasani et al., 1997). Up to 100 g ingested K fails to reach the duodenum and, as K intake rises, the flow of Na to the duodenum decreases. A similar K versus Na antagonism is shown in Fig. 8.4, Chapter 8.

processes. In the ruminant, however, over 50% of the potassium entering the rumen is passively absorbed from it (Fig. 7.4), lowering the apical potential difference at the mucosal surface in the process (see Chapter 5). Potassium enters the bloodstream largely via conductance channels in the basolateral membrane of the gut mucosa.

Membrane transport

There are more mechanisms for transporting potassium across membranes than there are for any other element, reflecting the difficulty but essentiality of maintaining high intracellular concentrations of potassium. In addition to the familiar sodium/potassium ATPase pump and co-transporters (see Chapter 8), there are hydrogen/potassium ATPases and six types of potassium channel, each distinctively regulated (Peterson, 1997). Short-term adjustments to a fluctuating potassium supply can be made through changes in the net flux of potassium into cells, under the influence of insulin (Lindeman and Pederson, 1983). A further need for regulation lies in the cytotoxicity of high circulatory levels of potassium, with serum values >6 mmol l^{-1} being sufficient to cause heart failure. The use of ionophores in ruminant nutrition to manipulate rumen metabolism can have disturbing systemic effects (e.g. lowering erythrocyte potassium) (Spears and Harvey, 1987), which deserve wider consideration.

Excretion

The regulation of body potassium status is principally achieved by the kidneys, where tubular reabsorption is restricted during overload under the influence of aldosterone (see also Chapter 8) (Kem and Trachewsky, 1983). However, adaptation to potassium loading begins in the gut, where splanchnic sensors provide early warning of the ingestion of potentially lethal amounts (Rabinowitz, 1988). Response to the sensors involves an increase in sodium/potassium ion ATPase activity and an increase in the number of pumps in the basolateral membrane of both the distal renal tubule and colon, leading to increased potassium excretion by both the urinary and faecal routes (Hayslett and Binder, 1982). Aldosterone modulates the renal but not the colonic response. Michell (1995) has suggested that aldosterone may be more important to the grazing ruminant by controlling potassium excess than in the more widely studied alleviation of sodium deficiency. However, the mechanisms are not as effective as those for sodium, and plasma potassium rises in response to increases in dietary potassium in animals as diverse as pigs (Combs et al., 1985) and steers (Greene et al., 1983). Early suggestions that a rise in plasma potassium increases sodium appetite were later questioned (Michell, 1978).

Secretion

In ruminants, potassium is the major cation in sweat, possibly due to the high potassium to sodium ratio in ruminants' natural diet of grass (Bell, 1995). Losses increase with environmental temperature and are greater in *Bos indicus* than in *Bos taurus* (Johnson, 1970) at a given temperature, despite a lower sweating rate. Muzzle secretions in buffalo heifers can contain twice as much potassium as sodium (12.6 ± 0.5 versus 6.4 ± 0.4 mmol l^{-1}) (Singh and Rani, 1999). Large amounts of potassium are recycled to the rumen via the secretion of saliva, which normally contains around 7 mmol K l^{-1}. Regression analysis of data illustrated in the Chapter 8 indicates that, on diets low in potassium (6 g kg^{-1} DM), the amounts of potassium entering the abomasum of lambs (36 kg) and steers (261 kg) are 6 and 20 g day^{-1}, respectively, exceeding the amounts ingested. Potassium is also the major cation secreted in milk (36 mmol l^{-1} in sheep and cattle); concentrations are not increased at high dietary potassium intake, but they decline slightly during severe potassium deprivation (Pradhan and Hemken, 1969). The excretory losses of potassium in calves may be increased by transport stress as a result of increased aldosterone activity (Hutcheson and Cole, 1986). Simulation of stress by the intravenous administration of cortisol has been seen to lower plasma potassium in steers from 4.7 to 3.3 mmol l^{-1} (Parker et al., 2004).

Biochemical Consequences of Potassium Deprivation

In most species there are most reductions in serum potassium below the normal range of 5–6 mmol l^{-1} when dietary potassium supplies are inadequate (Table 7.5). However, in dairy calves marginally deprived of potassium, loss of appetite and reduced weight gain have been seen to occur with plasma potassium unaffected at 6.1–6.6 mmol l^{-1} (Weil et al., 1988). In beef cattle, urinary creatinine to potassium ratios (molar) below 2.1 indicate the likelihood of a negative potassium balance (Karn and Clanton, 1977). Depletion of intracellular potassium may be partially offset by the uptake of hydrogen ions, but only at the expense of intracellular acidosis.

Table 7.5. Suggested marginal bands[a] for potassium in serum, urine and diet as a guide to the diagnosis of dietary deprivation (D) or excess (E) in livestock.

		Serum (mmol l^{-1})	Urine (mmol l^{-1})	Diet (g kg^{-1} DM)
Cattle	D	<2.5	<19	5–8
	E	6–10	>120	30–40
Sheep	D	2.4–4.0	–	4–8
	E	5–10	–	30–40
Pigs	D	2.5–3.5	3–5	3–6
	E	5–10	–	>18
Poultry	D	3.0–5.5	–	1.0–3.5
	E	>10	–	>9

[a]Mean values below (for D) or above (for E) the given bands indicate probability of ill-health from deprivation or toxicity. Values within bands indicate a possibility of future disorder.

Clinical Consequences of Potassium Deprivation

Reduced appetite was one of the first signs shown when growing pigs (Hughes and Ittner, 1942) and feeder lambs (Campbell and Roberts, 1965) were given semi-purified, low-potassium diets. Poor growth and muscular weakness, stiffness and paralysis and intracellular acidosis have also been reported. The transition to a state of dysfunction is rapid because the animal body contains virtually no reserves of potassium. Poor growth derives primarily from inappetence and partly from an impairment of protein metabolism. When lactating dairy cows were given diets based on beet pulp, brewers' grains and maize containing only 0.6 or 1.5 g K kg^{-1} DM, severe anorexia developed within 4 weeks (Pradham and Hemken, 1969). Milk yield fell and pica, loss of coat condition and decreased pliability of the hide were evident. Potassium deprivation led to a greater reduction in milk potassium than in plasma potassium and there was a compensatory rise in milk sodium. Haematocrit increased from 35.3% to 38.1% on the low-potassium diets. In a study involving less severe depletion in dairy cows, improved recovery of body weight (+92.9 versus +15.7 kg) was observed after peak lactation when the diet contained potassium at 6.6 rather than 4.5 g kg^{-1} DM (Dennis et al., 1976). However, there was no accompanying improvement in milk yield, despite an improved appetite.

Natural Occurrence of Potassium Deprivation

Potassium concentrations decline in dry swards due to a combination of ageing and leaching effects so that potassium values of 0.9–5.7 g kg^{-1} DM can occur in winter-range pasture (Preston and Linser, 1985). Responses to potassium supplementation have been obtained under such rangeland conditions when urea was also given, but they varied from year to year (Karn and Clanton, 1977). Responses are more likely to occur if cereal/urea supplements are deployed to alleviate the more serious constraints of digestible energy and degradable nitrogen in such pastures. Potassium supplementation of corn-based diets has been claimed to have several benefits, including improved gains and mortality in finishing beef cattle, improved milk production in dairy cows and alleviated stiffness and excessive irritability in swine (Preston and Linser 1985). However, no published peer-reviewed papers support such claims. The inclusion of potassium in supplements fed during droughts may be beneficial. Unusual conditions such as acidosis, stress and diarrhoea may lead to potassium depletion (Preston and Linser, 1985). Raising the potassium concentration in the 'receiving' diet of recently transported calves from 7–9 to 13–14 g kg^{-1} DM improved weight gain in the first 4–7 weeks and reduced calf mortality (Hutcheson et al., 1984), but a response to the accompanying anion (chloride) cannot be ruled out.

Potassium and Acid–Base Balance

Optimal production requires the avoidance of acid–base imbalance (i.e. of acidosis and alkalosis) and potassium is the major determinant of equilibrium in forage-based diets (Tremblay et al., 2006). Imbalance has historically been measured by estimating dietary cation–anion differences (DCAD) in milli-equivalents (me) by a variety of formulae, including only fixed, monovalent ions ((Na^+ + K^+) – Cl^-) ($DCAD_1$) or divalent and metabolized ions ($DCAD_2$: e.g. (Na^+ + K^+ + Ca^{2+} + Mg^{2+}) – (Cl^- + S^{2-})). Differences in ion absorption or urine-acidifying capacity have sometimes been allowed for (for review, see Lean et al., 2006), but the chosen absorption coefficients for calcium (0.4) and sulfur (0.65) are questionable. The importance of DCAD has been considered sufficient for the NRC (2001) to quote characteristic values for feeds. Forages were accorded a high $DCAD_1$ (>100 me kg^{-1} DM), cereals a low $DCAD_1$ (<20 me kg^{-1} DM) and protein supplements a high (e.g. soybean meal) or low (e.g. fish meal) $DCAD_1$. Forages can vary widely in $DCAD_2$, with values for Swedish pastures ranging from −14 to +726 me kg^{-1} DM and those for Australian grasses, hays and silages from −17 to 480, −13 to 470 and −2 to 390 me kg^{-1} DM, respectively (McNeill et al., 2002). However, the effects of manipulating DCAD on production have been inconsistent, possibly due to the confounded effects of ions not incorporated in the chosen formula (e.g. Ca^{2+} from CaCl$_2$, NH$_4^+$ from NH$_4$Cl and HCO$_3^-$ or CO$_3^{2-}$ from sodium and potassium salts), local buffering effects in the rumen (Underwood and Suttle, 1999) and interactions among cations (magnesium, calcium and chlorine). Species differ in their response to acidification of the diet, with dairy goats and cow showing evidence of increased bone resorption, but dairy sheep showing no such changes (Liesegang, 2008).

Potassium, DCAD and milk fever

One of the main areas of recent research on DCAD in ruminant nutrition concerns the positive correlation between DCAD in transition diets and the incidence of milk fever (Table 7.6). The discovery that potassium and sodium supplementation as bicarbonate (HCO$_3$) prior to calving raises the incidence of milk fever showed that the effect was attributable to DCAD and not to potassium per se (Goff and Horst, 1997), but it may have been amplified by the use of HCO$_3$ salts. Studies with a grazing herd showed that the hypocalcaemic effect of potassium fertilization occurred without affecting DCAD$_2$ in the sampled pasture (Roche et al., 2002).

Furthermore, provision of magnesium as sulfate or chloride rather than magnesium oxide (MgO) alleviated the temporary hypocalcaemia with no accompanying difference in DCAD. It was therefore suggested that sulfate and chloride have independent effects on milk fever. Since subclinical hypocalcaemia can reduce subsequent milk yield, improved yields following the addition of these anions to the transition diet may not be wholly or even partly attributable to changes in DCAD. A comprehensive meta-analysis of past DCAD trials showed that potassium, sulfur, magnesium and calcium each had more significant effects than any of the four DCAD formulae evaluated (Lean et al., 2006) and potassium was the least influential of the minerals. Decreases in DCAD$_2$ may be less useful in preventing milk fever in grazing than housed herds (McNeill et al., 2002) and may reduce milk yield (Roche et al., 2003). Meta-analysis of a restricted data set gave stronger evidence for a positive relationship between DCAD$_1$ and milk fever incidence (Charbonneau et al., 2006). Further discussion of the effects of acidic diets on bone mineral metabolism can be found in Chapter 8.

Heat stress

Heat stress can induce respiratory alkalosis associated with panting and supplementation with potassium chloride (KCl) or sodium bicarbonate (NaHCO$_3$), but not potassium carbonate (KCO$_3$), can increase milk yield in the heat-stressed dairy cow (West et al., 1991). The specific role of potassium is questionable because renal conservation of potassium allows plasma potassium to remain stable during heat stress, even in the relatively susceptible B. taurus (Beatty et al., 2006). Studies with heat-stressed chicks indicate a protective role for the anion rather than the cation.

Acid–base balance in pigs and poultry

Acid–base balance is important in monogastric species, but approaches to regulation have been less complex than those described for ruminants. Mongin (1981) suggested the use of dietary

Table 7.6. The influence of dietary potassium on the incidence of milk fever in a dairy herd and on the success of treatment at two dietary calcium levels (from Goff and Horst, 1997). The table shows the incidence of milk fever cases (% C) and the number of treatments per case (T).

Dietary K (g kg^{-1} DM)		Dietary Ca (g kg^{-1} DM)	
		5	15
11	C	0	20
	T	0	1.0
21	C	36	66
	T	2.25	1.5
31	C	80	23
	T	2.0	1.25

electrolyte balance (DEB), defined as sodium plus potassium minus chloride, and diets formulated to contain 250 me DEB kg^{-1} DM were recommended for optimum growth in chicks. Similar recommendations were made for pigs (Austin and Calvert, 1981). In heat-stressed broiler chicks, the feeding of KCl or ammonium chloride (NH$_4$Cl) in the drinking water (1.5 and 2.0 g l^{-1}, respectively) with a diet adequate in potassium (7.3 g kg^{-1} DM) improved growth and feed conversion efficiency (Teeter and Smith, 1986). Higher levels or mixtures of the two salts were, however, detrimental. Later work confirmed the benefits of drinking water supplemented with 0.067 mol KCl or NaCl l^{-1} on growth and water consumption, but there was no restoration of lowered plasma sodium or potassium and aldosterone levels remained high (Deyhim and Teeter, 1994). It was concluded that the osmotic stress of the heat-stressed broiler had not been alleviated. Where there is evidence of production responses to sodium or effective replacement of sodium by potassium in diets apparently adequate in sodium (e.g. in poultry), these responses probably reflect optimization of the acid–base balance, rather than responses to potassium per se.

Potassium Toxicity: Primary

Potassium is poorly tolerated under acute challenge (Neathery et al., 1979). Calves discriminate against diets containing 20 g K as KCl kg^{-1} DM when given a choice, but appetite is not depressed when they are given no choice (Neathery et al., 1980). However, appetite and growth have been seen to decline with 60 g K kg^{-1} DM, a level sometimes encountered in herbage under intensive grassland management and found in molasses (Table 7.1). The major direct effects of excess potassium are a disturbance of acid–base balance, hyperkalaemia and cardiac arrest (Neathery et al., 1979).

Potassium Toxicity: Secondary

At exposure levels insufficient to cause toxicity, potassium can interact with the metabolism of other minerals in a 'domino' effect with potential or realized adverse effects on animal health.

Subclinical effects on sodium metabolism

As potassium intakes rise there is a marked reduction in the duodenal flow of sodium (Fig 7.4), which probably arises from a reduction in the flow of unabsorbed salivary sodium. There are four possible explanations: salivary flow is reduced; more sodium is absorbed from the rumen; or the salivary sodium to potassium ratio decreases under the influence of aldosterone, as shown in studies with sheep (Humphery et al., 1984). A fourth explanation that this a residual effect of earlier displacement of sodium by potassium begs the question of how such a displacement occurred. The range of dietary potassium over which the 'aldosterone-like' response occurs (10–46 g kg^{-1} DM) embraces concentrations found in most forages and occurs on diets relatively high in sodium (>4 g kg^{-1} DM). Furthermore, the same response is seen in sheep. Suggestions that a low salivary sodium to potassium ratio may predispose cattle to bloat have not been confirmed in New Zealand (Edmeades and O'Connor, 2003).

Adverse interactions with magnesium

Potassium inhibits the absorption of magnesium from the rumen. Expression of the antagonism varies with feed type, but often results in a doubling of magnesium requirement over the range of potassium levels commonly found in herbages (Table 5.2, p. 102). The potassium/magnesium antagonism is exacerbated by sodium deprivation and the induction of aldosterone, which raises rumen potassium concentrations (Charlton and Armstrong, 1989). High potassium intakes can thus increase the risk for hypomagnesaemic tetany through a triple interaction (potassium × sodium × magnesium).

Adverse interactions with calcium

High pre-partum intakes of potassium predispose cows to hypocalcaemia and milk fever (Table 7.6; also see Chapter 4). While the large effect shown in Table 7.6 is arguably a 'worst-case scenario' in that the study used highly vulnerable, aged Jersey cows, there are many reports of potassium supplements inducing hypocalcaemia. In a recent study in sheep, the hypocalcaemic effect was accompanied by increased erythrocyte calcium (Phillips et al., 2006). The underlying mechanism is a decreased mobilization of calcium from bone at a time of increased calcium demand for lactation, an effect exacerbated by hypomagnesaemia (see Chapter 5), giving scope for a quadruple interaction (potassium × sodium × magnesium × calcium). The dependence of aldosterone secretion on calcium-ion-signalling pathways (Spat and Hunyady, 2004) provides a possible pathway for interactions with parathyroid hormone.

Prevention

Reductions in forage potassium may have a threefold benefit in reducing incidence of metabolic diseases in ruminants. Although many pastoral systems require inputs of potassium in artificial fertilizers to maintain grass yields, extensive recycling of potassium via urination and slurry dispersal reduces the required input. The best practice is to use, whenever possible, mixtures of nitrogen and phosphorus *without* potassium for the first fertilizer application in spring, when herbage potassium is maximal, and to restrict subsequent applications to the minimum needed to reach the critical concentration for maximum plant yield (Cherney and Cherney, 2005). Advantage may also be taken of variation between plant species in potassium concentration. Replacing alfalfa with timothy hay in the transition diet will greatly reduce potassium intakes prior to calving and thus reduce the risk of milk fever. Agronomists have belatedly looked for pasture species (Cherney and Cherney, 2005; Tremblay et al., 2006) and husbandry practices (Pelletier et al., 2006; Swift et al., 2007) that give low DCAD forages as a strategy for improving herd health, and harvesting low DCAD species such as *Phleum pratense* at the flower-heading stage has been advocated. Selection for strains with a low potassium requirement would probably be more effective than selecting for low DCAD while lessening the risks of hypomagnesaemia and sodium deprivation as well as hypocalcaemia.

References

ARC (1980) *The Nutrient Requirements of Ruminants*. Commonwealth Agricultural Bureaux, Farnham Royal, UK, pp. 184–185.

Beatty, D.T., Barnes, A., Taylor, E., Pethick, D., McCarthy, M. and Maloney, S.K. (2006) Physiological responses of *Bos taurus* and *Bos indicus* to prolonged continuous heat and humidity. *Journal of Animal Science* 84, 972–985.

Bell, F.R. (1995) Perception of sodium and sodium appetite in farm animals. In: Phillips, C.J.C. and Chiy, P.C. (eds) *Sodium in Agriculture*. Chalcombe Publications, Canterbury, UK, pp. 82–90.

Brink, D.R., Turgeon, O.A. Jr, Harmon, D.L., Steele, R.T., Mader, T.L. and Britton, R.A. (1984) Effects of additional limestone of various types on feedlot performance of beef cattle fed high corn diets differing in processing method and potassium level. *Journal of Animal Science* 59, 791–798.

Campbell, L.D. and Roberts, W.K (1965) The requirements and role of potassium in ovine nutrition. *Canadian Journal of Animal Science* 45, 147–156.

Charlton, J.A. and Armstrong, D.G. (1989) The effect of an intravenous infusion of aldosterone upon magnesium metabolism in sheep. *Quarterly Journal of Experimental Physiology* 74, 329–337.

Cherney, J.H. and Cherney, D.J.R. (2005) Agronomic response of cool season grasses to low-intensity harvest management and low potassium fertility. *Agronomy Journal* 97, 1216–1221.

Charbonneau, E., Pellerin, D. and Oetzel, G.R. (2006) Impact of lowering dietary cation–anion balance in non-lactating cows: a meta-analysis. *Journal of Dairy Science* 89, 537–548.

Combs, N.R., Miller, E.R. and Ku, P.K. (1985) Development of an assay to determine the bioavailability of potassium in feedstuffs for the young pig. *Journal of Animal Science* 60, 709–714.

Cornforth, I.S., Stephen, R.C., Barry, T.N. and Baird, G.A. (1978) Mineral content of swedes, turnips and kale. *New Zealand Journal of Experimental Agriculture* 6, 151–156.

Dennis, R.J., Hemken, R.W. and Jacobson, D.R. (1976) Effect of dietary potassium percent for lactating cows. *Journal of Dairy Science* 59, 324–328.

Deyhim, F. and Teeter, R.G. (1994) Effect of heat stress and drinking water salt supplements on plasma electrolytes and aldosterone concentrations in broiler chickens. *International Journal of Biometeorology* 38, 219–217.

Edmeades, D.C. and O'Connor, M.B. (2003) Sodium requirements for temperate pastures in New Zealand. *New Zealand Journal of Agricultural Research* 46, 37–47.

Gillis, M.B. (1948) Potassium requirement of the chick. *Journal of Nutrition* 36, 351–357.

Gillis, M.B. (1950) Further studies on the role of potassium in growth and bone formation. *Journal of Nutrition* 42, 45–57.

Goff, J.P. and Horst, R.L. (1997) The effect of dietary potassium and sodium but not calcium on the incidence of milk fever in dairy cows. *Journal of Dairy Science* 80, 176–186.

Grace, N.D. (1983) Amounts and distribution of mineral elements associated with fleece-free empty body weight gains in the grazing sheep. *New Zealand Journal of Agricultural Research* 26, 59–70.

Greene, L.W., Webb, K.E. and Fontenot, J.P. (1983) Effect of potassium level on site of absorption of magnesium and other macroelements in sheep. *Journal of Animal Science* 56, 1214–1221.

Hayslett, J.P. and Binder, H.J. (1982) Mechanism of potassium adaptation. *American Journal of Physiology* 243, F103–F112.

Hughes, E.H. and Ittner, N.R. (1942) The potassium requirement of growing pigs. *Journal of Agricultural Research* 64, 189–192.

Humphery, T.J., Coghlan, J.P., Denton, D.A., Fan, J.S., Scoggins, B.B., Stewart, K.W. and Whitworth, J.A. (1984) Effect of potassium, angiotensin II on blood aldosterone and cortisol in sheep on different dietary potassium and sodium intakes. *Clinical and Experimental Pharmacology and Physiology* 11, 97–100.

Hutcheson, D.P. and Cole, A.J. (1986) Management of transit stress syndrome in cattle: nutritional and environmental factors *Journal of Animal Science* 58, 700–707.

Hutcheson, D.P., Cole, A.J. and McLaren, J.B. (1984) Effects of pretransit diets and post-transit potassium levels for feeder calves. *Journal of Animal Science* 58, 700–707.

Johnson, K.G. (1970) Sweating rate and the electrolyte content of skin secretions of *Bos taurus* and *Bos indicus* cross-bred cows. *Journal of Agricultural Science, Cambridge* 75, 397–402.

Karn, J.F. and Clanton, D.C. (1977) Potassium in range supplements. *Journal of Animal Science* 45, 1426–1434.

Kem, D.C. and Trachewsky, D. (1983) Potassium metabolism. In: Whang, R. (ed.) *Potassium: Its Biological Significance*. CRC Press, Boca Raton, Florida, pp. 25–35.

Khorasani, G.R., Janzen, R.A., McGill, W.B. and Kennelly, J.J. (1997) Site and extent of mineral absorption in lactating cows fed whole-crop cereal grain silage or alfalfa silage. *Journal of Animal Science* 75, 239–248.

Lanyon, L.E. and Smith, F.W. (1985) Potassium nutrition of alfalfa and other forage legumes: temperate and tropical. In: Munson, RD. (ed.) *Potassium in Agriculture*. American Society of Agronomy, Madison, Wisconsin, pp. 861–894.

Leach, R.M., Jr., Dam, R., Zeigler, T.R. and Norris, L.C. (1959) The effect of protein and energy on the potassium requirement of the chick. *Journal of Nutrition* 68, 89–100.

Lean, I.J., DeGaris, P.J., McNeill, D.M. and Block, E. (2006) Hypocalcaemia in dairy cows: meta-analysis and dietary cation–anion difference theory revisited. *Journal of Dairy Science* 89, 669–684.

Liesegang, A. (2008) Influence of anionic salts on bone metabolism in dairy goats and sheep. *Journal of Dairy Science* 91, 2449–2460.

Lindeman, R.D. and Pederson, J.A. (1983) Hypokalaemia. In: Whang, R. (ed.) *Potassium: Its Biological Significance*. CRC Press, Boca Raton, Florida, pp. 45–75.

MAFF (1990) *UK Tables of the Nutritive Value and Chemical Composition of Foodstuffs*. In: Givens, D. I. (ed.) Rowett Research Services, Aberdeen, UK.

McNeill, D.M., Roche, J.R., McLachlan, B.P. and Stockdale, C.R. (2002) Nutritional strategies for the prevention of hypocalcaemia at calving for dairy cows in pasture-based systems. *Australian Journal of Agricultural Research* 53, 755–770.

Meyer, J.H., Grummer, R.H., Phillips, R.H. and Bohstedt. G. (1950) Sodium, chlorine, and potassium requirements of growing pigs. *Journal of Animal Science* 9, 300–306.

Michell, A.R. (1978) Plasma potassium and sodium appetite: the effect of potassium infusion in sheep. *British Veterinary Journal* 134, 217–224.

Michell, A.R. (1995) Physiological roles for sodium in mammals. In: Phillips, C.J.C. and Chiy, P.C. (eds) *Sodium in Agriculture*. Chalcombe Publications, Canterbury, UK, pp. 91–106.

Miller, E.R. (1995) Potassium bioavailability. In: Ammerman, C.B., Baker, D.H. and Lewis, A.J. (eds) *Bioavailability of Nutrients for Animals*. Academic Press, New York, pp. 295–302.

Mongin, P. (1981) Recent advances in dietary anion–cation balance applications in poultry. *Proceedings of the Nutrition Society* 40, 285–294.

Neathery, M.W., Pugh, D.G., Miller, W.J., Whitlock, R.H., Gentry, R.F. and Allen, J.C. (1979) Potassium toxicity and acid:base balance from large oral doses of potassium to young calves. *Journal of Dairy Science* 62, 1758–1765.

Neathery, M.W., Pugh, D.G., Miller, W.J., Gentry, R.F. and Whitlock, R.H. (1980) Effects of sources and amounts of potassium on feed palatability and on potassium toxicity in dairy calves. *Journal of Dairy Science* 63, 82–85.

Nesheim, M.C., Leach, R.M. Jr, Zeigler, T.R. and Serafin, J.A. (1964) Interrelationships between dietary levels of sodium, chlorine and potassium. *Journal of Nutrition* 84, 361–366.

NRC (1994) *Nutrient Requirements of Poultry*, 9th edn. National Academy of Sciences, Washington, DC.

NRC (2000) *Nutrient Requirements of Beef Cattle*, 7th edn. National Academy of Sciences, Washington, DC.

NRC (2001) *Nutrient Requirements of Dairy Cows*, 7th edn. National Academy of Sciences, Washington, DC.

Parker, A.J., Hamlin, G.P., Colemen, C.J. and Fitzpatrick, L.A. (2004) Excess cortisol interferes with a principal mechanism of resistance to dehydration in *Bos indicus* steers. *Journal of Animal Science* 82, 1037–1045.

Pelletier, S., Belanger, G., Tremblay, G.F., Bregard, A. and Allard, G. (2006) Dietary cation–anion difference of Timothy as affected by development stage and nitrogen and phosphorus fertilisation. *Agronomy Journal* 98, 774–780.

Perdomo, J.T., Shirley, R.L. and Chicco, C.F. (1977) Availability of nutrient minerals in four tropical forages fed freshly chopped to sheep. *Journal of Animal Science* 45, 1114–1119.

Peterson, L.N. (1997) Potassium in nutrition. In: O'Dell, B.L. and Sunde, R.A. (eds) *Handbook of Nutritionally Essential Mineral Elements*. Marcel Dekker Inc., New York, pp. 153–183.

Phillips, C.J.C., Mohammed, M.O. and Chiy, P.C. (2006) The critical potassium concentration for induction of mineral disorders in non-lactating Welsh Mountain sheep. *Small Ruminant Research* 63, 32–38.

Pradhan, K. and Hemken, R.W. (1969) Potassium depletion in lactating dairy cows. *Journal of Dairy Science* 51, 1377–1381.

Preston, R.L. and Linser, J.R. (1985) Potassium in animal nutrition. In: Munson, R.D. (ed) *Potassium in Agriculture*. American Society of Agronomy, Madison, Wisconsin, pp. 595–617.

Rabinowitz, L. (1988) Model of homeostatic regulation of potassium excretion in sheep. *American Journal of Physiology* 254, R381–R388.

Reid, R.L., Baker, B.S. and Vona, L.C. (1984) Effects of magnesium sulphate supplementation and fertilization on quality and mineral utilisation of timothy hays by sheep. *Journal of Animal Science* 59, 1403–1410.

Robinson, D.L. (1985) Potassium nutrition of forage grasses. In: Munson, R.D. (ed) *Potassium in Agriculture*. American Society of Agronomy, Madison, Wisconsin, pp. 895–903.

Roche, J.R., Dalley, D., Moate, P., Grainger, C., Rath, M. and O'Mara, F. (2003) Dietary cation–anion difference and the health and production of pasture-fed dairy cows. 2. Nonlactating periparturient cows. *Journal of Dairy Science* 86, 979–987.

Roche, J.R., Morton, J. and Kolver, E.S. (2002) Sulfur and chlorine play a non-acid base role in periparturient calcium homeostasis. *Journal of Dairy Science* 85, 3444–3453.

Sauveur, B. and Mongin, P. (1978) Interrelationships between dietary concentrations of sodium, potassium and chloride in laying hens. *British Poultry Science* 19, 475–485.

Schonewille, J.Th., Ram, L., Van't Klooster, A.Th., Wonterse, H. and Beynen, A.C. (1997) Intrinsic potassium in grass silage and magnesium absorption in dry cows. *Livestock Production Science* 48, 99–110.

Singh, S.P. and Rani, D. (1999) Assessment of sodium status in large ruminants by measuring the sodium-to-potassium ratio in muzzle secretions. *American Journal of Veterinary Research* 60, 1074–1081.

Soder, K.J. and Stout, W.L. (2003) The effect of soil type and fertilisation level on mineral concentration in pasture: potential relationships to ruminant performance and health. *Journal of Animal Science* 81, 1603–1610.

Spat, A. and Hunyady, L. (2004) Control of aldosterone secretion: a model for convergence in cell-signalling pathways. *Physiological Reviews* 84, 489–539.

Spears, J.W. and Harvey, R.W. (1987) Lasalocid and dietary sodium and potassium effects on rumen mineral metabolism, ruminal volatile fatty acids and performance of finishing steers. *Journal of Animal Science* 65, 830–840.

Supplee, W.C. (1965) Observations on the requirement of young turkeys for dietary potassium. *Poultry Science* 44, 1142–1144.

Swift, M.L., Bittman, S., Hunt, D.E. and Kowalenko, C.G. (2007) The effect of formulation and amount of potassium fertiliser on macromineral concentration and cation–anion difference in tall fescue. *Journal of Dairy Science* 90, 1063–1072.

Teeter, R.G. and Smith, M.O. (1986) High chronic ambient temperature stress effects on acid–base balance and the response to supplemental ammonium chloride and potassium carbonate. *Poultry Science* 65, 1777–1781.

Telle, P.P., Preston, R.L., Kintner, L.D. and Pfander, W.H. (1964) Definition of the ovine potassium requirement. *Journal of Animal Science* 23, 59–66.

Thompson, D.J. (1972) *Potassium in Animal Nutrition*. International Minerals and Chemical Corporation, Libertyville, Illinois.

Tremblay, G.F., Brassard, H., Belanger, G., Seguin, P., Drapeau, R., Bregard, A., Michaud, R. and Allard, G. (2006) Dietary cation anion differences in five cool-season grasses. *Agronomy Journal* 98, 339–348.

Ussing, H.H. (1960) The biochemistry of potassium. In: *Proceedings of the 6th Congress*. International Potash Institute, Amsterdam.

Ward, G.M. (1966) Potassium metabolism of domestic ruminants: a review. *Journal of Dairy Science* 49, 268–276.

Weil, A.B., Tucker, W.B. and Hemken, R.W. (1988) Potassium requirement of dairy calves. *Journal of Dairy Science* 71, 1868–1872.

West, J.W., Mullinix, B.G. and Sandifer, T.G. (1991) Changing dietary electrolyte balance for dairy cows in cool and hot environments. *Journal of Dairy Science* 74, 1662–1674.

8 Sodium and Chloride

Sodium and chloride are considered together because of their related metabolism, functions and requirements in the animal and their interactions with each other. Although plants use sodium and potassium interchangeably, the distinctive metabolism of potassium in animals merits a specific chapter (see Chapter 7) and only certain interactions with sodium and chloride will be dealt with here. The need for salt was recognized long before its nature and extent were established. A craving for salt by grazing animals was recorded by early settlers and the provision of salt became synonymous with good stock husbandry. Indeed, the transition from a nomadic to an agricultural way of life, with dependence on cereals or vegetables rather than meat and milk, was only sustainable with dietary supplements of salt (Denton, 1982). Long after the first experimental induction of salt deprivation in livestock in the 19th century (see Chapter 1), Australian workers began to unravel the complex hormonal control of sodium metabolism by ingenious surgical interventions that prevented the recycling of sodium via salivary secretion in sheep, some of which had their adrenal glands removed (Denton, 1982). A need for supplemental sodium in pigs and poultry fed on cereal-based rations was also established (Aines and Smith, 1957). By contrast, chloride deprivation rarely occurs naturally (Summers et al., 1967), although it has been induced experimentally in poultry (Leach and Nesheim, 1963) and calves (Neathery et al., 1981).

Attention has now switched to the problems of salt excess. Approximately one-third of the Earth's land surface is affected by salinity, sodicity and aridity (Chiy and Phillips, 1995) and the exploitation of halophytic browse species and sea water for irrigation in arid coastal areas are only possible when sodium excesses are controlled (Pasternak et al., 1985; Masters et al., 2006). Recently, welfare problems associated with dehydration during prolonged transportation have focused attention once more on the hormonal control of thirst and sodium metabolism (Hogan et al., 2007) and the findings may have a bearing on the current controversy surrounding sodium requirements of dairy cows (Edmeades and O'Connor, 2003).

Functions

Sodium and chloride maintain osmotic pressure, regulate acid–base equilibrium and control water metabolism in the body, with Na^+ being the major cation in the extracellular fluid (ECF) and Cl^- the major anion, at concentrations of 140 and 105 mmol l^{-1}, respectively. Sodium plays a key role, providing an 'osmotic skeleton' that is 'clothed' with an appropriate volume of water (Michell, 1995). When ion intakes increase, water intakes also increase (Wilson, 1966; Suttle and Field, 1967) to protect the gut, facilitate excretion and to 'clothe'

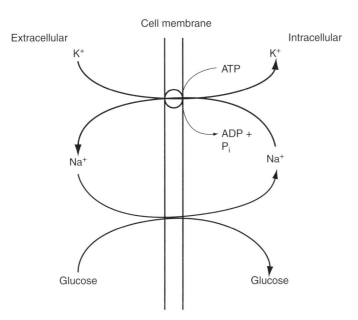

Fig. 8.1. The 'sodium pump' is vital to the maintenance of electrochemical differences across membranes and the cellular uptake of glucose through the action of a sodium/potassium ion-dependent ATPase, but activity is sustained at a significant energy cost.

the enlarged 'skeleton'. The 'osmotic skeleton' is sustained by the Na$^+$/K$^+$ ATPase pump in cell membranes, which actively transports sodium out of the cell, converting the energy of ATP into osmotic gradients along which water can flow (Fig. 8.1) and fuelling other cation-transporting mechanisms. Thus, transmembrane potential differences are established, and these influence the uptake of other cations and are essential for excitability (i.e. response to stimuli).

Amino acid and glucose uptake are dependent on sodium and sodium-dependent phosphate transporters have been isolated from the avian jejunum and kidneys (Huber et al., 2006). Exchange of Na$^+$ with H$^+$ influences pH regulation, while that with Ca^{2+} influences vascular tone. The contribution of sodium to acid–base balance is covered in more detail in Chapter 7. There are some marked contrasts between sodium and potassium (Fig. 7.1), with sodium making up over 90% of the bases of the serum, but little is present in the blood cells. There are significant stores of sodium in the skeleton, which may contain approximately 3–5 g kg^{-1} fresh weight in sheep. Chloride is found both within the cells, including the gastric goblet cells where it occurs as hydrochloride (HCl), and in the body fluids in the form of salts. Respiration is based on the 'chloride shift' whereby the potassium salt of oxyhaemoglobin exchanges oxygen for carbon dioxide via the bicarbonate ion (HCO$_3^-$) in the tissue and reverses that process in the lung, where reciprocal chloride exchanges maintain the anion balance (Block, 1994).

Clinical Consequences of Deprivation

Sodium

The first sign of sodium deprivation in milking cows is pica or a craving for salt, manifested by avid licking of wood, soil, urine or sweat. This can occur within 2–3 weeks of deprivation. Water consumption can become excessive (polydipsia) and urine output greatly increased. After several weeks, appetite and milk yield begin to decline and the animal loses weight and develops a haggard appearance because of a rough coat, while milk fat content decreases. In a high-producing cow, the breakdown can be sudden and death ensues. However, if supplementary salt is provided then a dramatic recovery occurs. An impaired conception rate has been linked to sodium deprivation in dairy cows in Victoria, where pastures were rich in potassium and urinary sodium outputs were low (Harris et al., 1986). In growing pigs, poultry, sheep and goats, sodium deprivation is also manifested within a few weeks by inappetence, growth retardation and inefficient feed use due to impairment of protein and energy metabolism, but digestibility is not affected. Laying hens on low-salt rations lose weight, are prone to cannibalism and reduce both egg production and egg size (Sherwood and Marion, 1975). The feeding of sodium at levels only marginally below requirement induces moult in laying hens, reducing egg production from 60% to 20% over 15 days (Ross and Herrick, 1981). In sodium-depleted, lactating sows, the interval between weaning and oestrus increases from 6.2 to 12.6 days, although the live weight of the sow and growth rate of her litter remain unaffected (Seynaeve et al., 1996). The reproductive consequences of sodium deprivation merit further consideration.

Chloride

When lactating cows were given a diet low in chloride they developed pica, lowered milk yield, constipation and cardiovascular depression (Fettman et al., 1984). Without the demand of lactation, however, extreme measures are needed to induce ill-health. Chloride deprivation has been produced experimentally by giving young calves a diet low in chloride while removing their chloride-rich abomasal contents twice daily (Neathery et al., 1981). The calves became anorexic and lethargic after 7 days, with mild polydipsia and polyuria. Severe eye defects developed after 24–46 days of depletion, but control calves given a diet with added chloride grew normally despite the removal of digesta. Chloride-deficient animals showed a craving for potassium chloride (KCl) as well as sodium chloride (NaCl). In chicks given a diet containing only $0.2\,g\ Cl^-\ kg^{-1}$ dry matter (DM), mortality increased and nervous symptoms were reported (Leach and Neisheim, 1963).

Dietary Sources of Sodium and Chloride

There are no problems of low or variable sodium and chloride availability in feeds, and sources are easily ranked by their sodium and chloride concentrations.

Sodium in forages

Forages are generally poor in sodium due to soil, plant and husbandry factors. Sodium is readily leached from soils with a low cation exchange capacity, such as the pumices of New Zealand (Edmeades and O'Connor, 2003), and such soils support pastures low in sodium. Soils and pastures within 25 km of the east coast of New Zealand are enriched with sodium by the deposition of sea spray, and coastal pastures can be three times richer in sodium (0.3 versus $0.1\,g\ kg^{-1}$ DM) than those growing inland (Ledgard and Upsdell, 1991; Edmeades and O'Connor, 2003). The distribution of pasture sodium concentrations worldwide is skewed towards low values, with 50% of samples containing $<1.5\,g\ kg^{-1}$ DM, and low values are more common in tropical than in temperate pastures: 50% of tropical legume samples contain $<0.4\,g\ Na\ kg^{-1}$ DM (Minson, 1990).

There are consistent differences amongst species and varieties in leaf sodium concentration. Among the more widely grazed native range species the order is *Holcus lanatus* > *Dactylis glomerata* > *Festuca pratensis* > *Phleum pratense*. Varietal differences are most marked in the richest (natrophilic) species such as *Lolium perenne* (Chiy and Phillips, 1995). *L. perenne* is a natrophilic species and *P. pratense* a natrophobic species. The former, whether they be grasses (e.g. *Phalaris* species) or legumes (e.g. white clover, subterranean clover and *Lotus*), commonly contain 3–4 g Na kg^{-1} DM, while natrophobic species, including *Pennisetum clandestinum* (kikuyu), lucerne and red clover, commonly contain <0.5 g kg^{-1} DM (Edmeades and O'Connor, 2003). Sodium concentrations in grasses (but not legumes) change as they mature, with values decreasing from 0.5 to 0.1–0.2 g kg^{-1} DM in Californian rangeland pasture between spring and September (Morris *et al.*, 1980). They are also affected by applications of potassium and nitrogen fertilizers: nitrogen increases pasture sodium in a dose-dependent manner, but the concurrent application of potassium limits that response to nitrogen, particularly at high application rates (Fig. 8.2) (Kemp, 1971). Maize silage, like some other cereal silages (Khorasani *et al.*, 1997), is a poor source of sodium.

Chloride in forages

Most pastures are appreciably richer in chloride than in sodium irrespective of species or state of maturity. For example, respective mean values for sodium and chloride of 0.8 and 4.0 g kg^{-1} DM in legumes and 1.4 and 5.0 g kg^{-1} DM in grasses have been reported (Thomas *et al.*, 1952). Pastures invariably provide sufficient chloride for grazing animals, but they can be enriched by fertilizer application (Chiy and Phillips, 1993a) and this may have beneficial consequences in terms of acid–base balance (Goff *et al.*, 2007), but decreases in DCAD are relatively small (Fig. 8.3).

Concentrates

A survey in the USA (Berger, 1990) showed that many foods contain less sodium than expected from the widely quoted data of NRC; the mean value for maize (0.07 g kg^{-1} DM) was only 23% of the listed value. More reliable data for sodium and chloride in common foodstuffs are given in Table 8.1. Most grain and plant protein sources (with the exception of sunflower meal) are poor in sodium, although values are not normally distributed, while root crops, root by-products and animal products are relatively rich in sodium. Cereal grains, like grasses, generally provide more chloride than sodium, with maize providing the least and barley the most (0.5 and 1.8 g Cl$^-$ kg^{-1} DM, respectively). Cereal straws contain three- to sixfold more chloride than the grains.

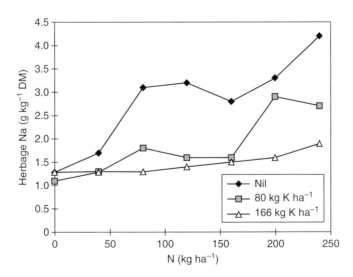

Fig. 8.2. Herbage sodium concentrations are increased by the application of nitrogenous fertilizers but if potassium is simultaneously applied the effect can be lost (data from Kemp, 1971).

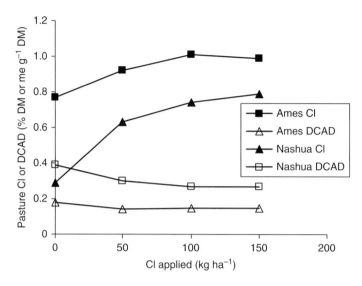

Fig. 8.3. Results of an attempt to manipulate the dietary cation–anion difference (DCAD) of pasture by applying chloride (Cl) as a fertilizer (from Goff *et al.*, 2007). While the increase in pasture Cl was significant, it varied between sites (Ames and Nashua) and had relatively small effects on DCAD. The source of Cl (ammonium chloride or calcium chloride) did not affect either Cl or calcium concentrations in pasture.

Vegetable protein supplements are consistently low in chloride (0.3–0.7 g kg^{-1} DM) and it is possible to produce blends with cereals that do not meet the chloride requirements of pigs and poultry. The inclusion of animal by-products and grass meal would correct any sodium deficit.

Drinking water

Water sources vary enormously in the concentrations of sodium, chloride and other minerals they contain, from virtually none in some streams and reticulation sources to highly saline

Table 8.1. Mean (standard deviation) sodium concentrations (g kg^{-1} DM) in pastures and foodstuffs in the UK (from MAFF, 1990) and chloride (Cl) concentrations for corresponding foodstuffs from the USA (NRC as quoted by McDowell, 2003).

Forages	Na	Cl	Concentrates	Na	Cl	Roots and by-products	Na	Cl
Herbage	2.5 (2.1)	4–6	Fish meal	11.2 (1.5)	–	Fodder beet	3.0 (1.6)	–
Grass silage	2.6 (1.6)	–	Soybean meal (ext)	0.2 (0.1)	0.4	Swedes	0.7[a]	–
Clover silage (mix)	0.8 (0.5)	–	Safflower meal	1.0 (1.2)	1.3	Turnips	2.0[a]	–
Lucerne silage	1.3 (1.0)	4.1	Palm-kernel cake	0.2 (0.1)	–	Cassava meal	0.6 (0.3)	–
Maize silage	0.3 (0.2)	1.8	Maize gluten	2.6 (1.4)	1.0	Meat and bone meal	8.0 (1.0)	8–12
Grass hay	2.1 (1.7)	4–5	Rapeseed meal	0.4 (0.3)	–	Molasses (beet)	25.0 (8.2)	16.4
Lucerne hay	0.6 (0.1)	3.0	Cottonseed meal	0.2 (0.2)	0.5	Molasses (cane)	1.2 (0.9)	31.0
Dried grass	2.8 (1.6)	–	Linseed meal	0.7 (0.0)	0.4	Molassed beet pulp	4.4 (0.3)	–
Dried lucerne	1.3 (0.8)	4.8	Maize	–	0.5	Beet pulp	3.2 (2.1)	0.4
Barley straw	1.3 (1.4)	6.7	Barley	0.3 (0.4)	1.8	Brewers' grains	0.3 (0.3)	1.7
Wheat straw	0.6 (1.0)	3.2	Wheat	0.1 (0.1)	0.5	Wheat feed	0.1 (0.1)	0.5
Alkali-treated barley straw	32.0 (10.5)	–	Oats	0.2 (0.1)	1.1	Distillers' grains Wheat	3.1 (3.9)	–
			Sorghum	0.5 (0.0)	1.0	Barley	0.3 (0.3)	–

[a]New Zealand data from Cornforth *et al.* (1978).

supplies, usually from deep wells or bores (Shirley, 1978). Drinking water can therefore constitute a valuable source of sodium (and of other minerals, such as sulfates and magnesium salts). Sodium-deprived animals show a marked preference for salty water (Blair-West et al., 1968) and they will even consume bicarbonate of soda (NaHCO$_3$) which is usually avoided (Bell, 1993). When the diet provides sufficient sodium, cattle will discriminate against water containing 12.5 g NaCl l^{-1} in favour of pure water; pygmy goats show a similar preference, while both sheep and normal goats remain indifferent (Goatcher and Church, 1970). Wide individual and seasonal variation in water consumption (see Chapter 2) affects the contribution of drinking water to total intakes of sodium. It has been calculated that cows yielding 20 kg milk and consuming 50 l water each day would meet their sodium requirement from drinking water alone if it contained 1 g NaCl l^{-1}. Supplementary NaCl can be given to poultry via drinking water (Ross, 1979).

Metabolism of Sodium and Chloride

Palatability

The presence of salt contributes to the palatability of a feed (Grovum and Chapman, 1988), but the addition of salt to a feed replete with sodium can lower feed intake (Wilson, 1966; Moseley, 1980) and may be used as a means of restricting the intake of supplementary foods in ruminants (De Waal et al., 1989). However, sodium appetite is relative rather than absolute and can easily be learned (i.e. affected by experience).

Absorption

Both sodium and chloride are readily absorbed, but each element can influence the absorption of the other (for a review, see Henry, 1995). Studies with the isolated rumen showed that the active transports of sodium and chloride are coupled, with the absorption of one ion requiring the presence of the other (Martens and Blume, 1987). Ammonium at >8 mmol l^{-1} can inhibit sodium transport while enhancing chloride transport across the isolated rumen epithelium via separate effects on uncoupled pathways (Abdoun et al., 2005), but the effects are dependent on pH and influenced by the dietary treatment of donor sheep (Abdoun et al., 2003). However, the rumen is not normally a site of net sodium absorption. In sheep and cattle the major site is the large intestine, particularly on diets high in potassium (Greene et al., 1983a,b; Khorasani et al., 1997). Sodium uptake from the intestinal lumen is achieved by coupling to glucose and amino acid uptake via co-transporters and also by exchange with hydrogen ions (H$^+$) (Harper et al., 1997). Absorption of chloride from dietary and endogenous (gastric secretions) sources is also achieved by exchange for another anion, HCO$_3^-$.

Membrane transport

Sodium and chloride are highly labile in the body. At the cellular level, the continuous exchange of Na$^+$ and K$^+$ via ATP-dependent Na$^+$–K$^+$ pumps provides the basis for glucose and amino acid uptake (by co-transport) (Fig. 8.1), maintaining high intracellular potassium concentrations but requiring about 50% of the cell's maintenance need for energy (Milligan and Summers, 1986). Production increases Na$^+$/K$^+$ transport and modelling studies suggest that the associated contribution to energy expenditure rises from 18% to 23% as lamb growth increases from 90 to 230 g day^{-1}, with increased metabolite transport (e.g. amino acids) across the gut mucosa being primarily responsible (Gill et al., 1989). Sodium transport across membranes is also achieved by a wide variety of complementary mechanisms: by a sodium–hydrogen exchanger; by electroneutral sodium–potassium–2 chloride co-transporters; by sodium–chloride and sodium–magnesium exchangers; and by voltage-gated Na$^+$ channels. Chloride transport is almost as complex, with both electrically and mechanically ('stretch') driven calcium-activated channels and chloride–HCO$_3^-$ exchangers contributing to fluxes (Harper et al., 1997;

Burnier, 2007). Dietary supplements of ionophores interact with dietary sodium to affect ruminal metabolism (Spears and Harvey, 1987), increase magnesium absorption and decrease sodium retention by increasing urinary sodium loss (Kirk et al., 1994).

Hormonal influences

Sodium metabolism is hormonally regulated as part of osmotic control. The activation of renin to form angiotensins (I, II and III), with vasopressin can modulate responses to aldosterone, changing ECF volume and blood pressure through appropriate adjustments in thirst and water balance (Bell, 1995; Michell, 1995; Burnier, 2007). The hormonal control axes are active in the fetus, which responds to variations in maternal sodium intake (Rouffet et al., 1990), and in the newborn calf (Safwate, 1985). Control mechanisms and outcomes vary between species: for example, the interaction between aldosterone and angiotensin does not affect centrally mediated sodium appetite in sheep (Shade et al., 2002). Plasma sodium can vary significantly through the oestrus cycle in lactating cows (from 138 to 160 mmol l^{-1}), reaching a peak between days 5–10 and a trough on day 20. The pattern is less striking in heifers and is presumed to reflect the activity of oestrogen, rather than aldosterone, which shows a slightly earlier peak during the cycle rather than a trough (Rouffet et al., 1983). The cyclic changes in plasma sodium are more pronounced than those induced by sodium loading in pregnant cows (Rouffet et al., 1990).

Secretion

Much of the sodium that enters the gastrointestinal tract comes from saliva. This is particularly so in ruminants, which secrete about 0.3 l kg^{-1} live weight (LW) day^{-1}, containing 150 mmol Na l^{-1} as the major cation. Experiments with sheep and cattle show that when potassium is added to a dry ration marginal in sodium the flow of sodium into the abomasum is reduced, probably by reducing salivary sodium secretion (Fig. 8.4). Similar changes may occur when potassium-rich silages are mixed with a sodium-rich concentrate (Fig. 7.4, p. 173). Sodium, chloride and potassium are all lost via skin secretions, but there are major differences between species. In non-ruminants, including the horse, sodium is the major cation in sweat and salt concentrations in sweat can reach 45 g l^{-1}. Horses, mules and donkeys sweat profusely when exercised (p. 6), but the high loss of sodium balances the loss of water and provides a defence against hypernatraemia. Milk normally contains 17 and 27 mmol Na l^{-1} in sheep and cattle, respectively, but in mares the figure is only 5 mmol l^{-1}, indicating a further possible line of defence against deprivation.

Excretion

Regulation of sodium status is principally achieved in the kidneys by control of reabsorption of sodium, secreted in the proximal tubule, by changes in active transport and membrane permeability. Reabsorption from the distal tubule can be enhanced by aldosterone so that urinary losses become negligible when sodium and potassium intakes are low (see Harper et al., 1997 for detail on regulatory processes). Potassium

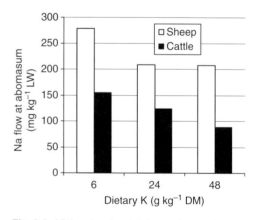

Fig. 8.4. High potassium intakes reduce sodium flow from the rumen in sheep and cattle. Since respective Na intakes were only 40 and 55 mg kg^{-1} live weight (LW), respectively, the flow of salivary Na from the rumen must have been reduced by K supplementation (data from Greene et al., 1983a,b).

supplementation increases the urinary and decreases the faecal excretion of sodium in sheep and cattle (Greene et al., 1983a,b). This change in partition is the expected consequence of the events just described in the rumen, but contains a paradox. Potassium (4g day^{-1}) enhances the ovine aldosterone response to sodium depletion (Humphery et al., 1984), but aldosterone normally reduces the urinary excretion of sodium. Somewhere along the line potassium must be affecting sodium metabolism by pathways that do not involve aldosterone.

Dietary excesses of chloride are predominantly excreted through the kidneys via a potassium/chloride co-transporter in the renal tubules. Although chloride can be reabsorbed, the process is passive.

Water deprivation

Water deprivation causes a rise in plasma sodium in steers. The risk of osmotic imbalance is countered by complex physiological control mechanisms, some mediated by the central nervous system and influenced by cortisol (Parker et al., 2004). In ruminants, the rate of outflow of digesta from the rumen is increased. This has adverse consequences for digestibility (Arieli et al., 1989), but possible benefits in terms of undegraded protein outflow (Hemsley et al., 1975). The effects of sodium and potassium/sodium imbalance on magnesium metabolism in ruminants are considered elsewhere (Chapters 5 and 7).

Heat stress

Sodium losses increase as temperature and humidity increase. In unacclimatized cattle, the dribbling of saliva can result in daily losses of sodium and chloride of up to 14 and 9mg kg^{-1} LW, respectively (Aitken, 1976). Plasma sodium and potassium decrease during exposure to heat and high humidity in steers that have free access to water, but changes and residual effects are less marked in species accustomed to such conditions (e.g. Bos indicus versus B. taurus) (Beatty et al., 2006)

Requirements for Sodium and Chloride

There is no need for a safety factor to allow for variable availability because absorption is usually almost complete, but one may be needed for livestock unsuited to hot, humid environments and for young, grazing livestock that have yet to acquire resistance to infectious nematode larvae on the pasture. However, the presence of large reserves of sodium in the rumen and skeleton of ruminants means that they can tolerate diets low in sodium for long periods.

Sheep

Merino wethers fed on high-grain diets need >0.6g sodium (McClymont et al., 1957) and a minimum requirement of 1.0g kg^{-1} DM has been derived for growing lambs in feeding trials (Hagsten et al., 1975). Field trials with grazing lambs and ewes give a similar range of sodium requirement (0.5–1.0g kg^{-1} DM) (Edmeades and O'Connor, 2003). Morris and Peterson (1975) used salivary sodium to potassium ratios to assess the sodium requirement for lactating ewes (Fig. 8.5) and concluded that the minimum sodium level needed to maximize sodium status was 0.80–0.87g

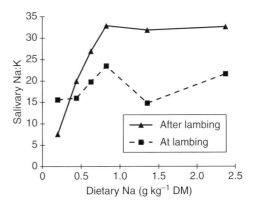

Fig. 8.5. The salivary sodium to potassium ratio reflects dietary Na supply after but not before lambing because of the higher Na requirement for lactation than gestation and possibly because of a change in the set point for Na:K around parturition (data from Morris and Peterson, 1975).

kg^{-1} DM, only 37.5% of the NRC requirement at the time. Even this provision was generous in terms of animal performance because the unsupplemented diet containing 0.2g kg^{-1} DM did not reduce appetite, ewe body weight or lamb growth rate. Tolerance of sodium depletion commencing in ovine pregnancy has also been demonstrated (Michell et al., 1988). Factorial estimates of need yielded high requirements of 0.8–2.7g Na kg^{-1} DM (ARC, 1980), but allowed for a high faecal endogenous loss (FE_{Na} 20mg Na kg^{-1} LW). Obligatory faecal losses on diets low in potassium are <6mg Na kg^{-1} LW (Michell et al., 1988) and give lower factorial estimates of sodium requirement (Underwood and Suttle, 1999). However, a moderate increase in dietary potassium (from 6 to 24 kg^{-1} DM) increased urinary sodium losses from 30 to 40mg kg^{-1} LW in wether lambs and induced a negative sodium balance (Greene et al., 1983a); dietary sodium, at 1.8g kg^{-1} DM, had clearly become marginal. Since most forage-based diets are relatively rich in potassium, a higher set of requirements than those given previously (Underwood and Suttle, 1999) is given in Table 8.2. Minimum chloride requirements have apparently not been studied experimentally.

Sodium in cattle

Field experiments indicate that 0.6–0.8g Na kg^{-1} DM is sufficient for growth under climatic conditions that are not extreme (Morris and Gartner, 1971; Morris and Murphy, 1972), but lactating cows need slightly more (1.0g kg^{-1} DM) (Morris, 1980; Edmeades and O'Connor, 2003). Using the salivary sodium to potassium ratio as the indicator of need, the requirement of housed buffalo heifers would appear to be between 0.6 and 1.0g kg^{-1} DM (Singh and Rani, 1999). Reductions in the salivary sodium to potassium ratio after parturition in grazing beef cows has confirmed the increased demand associated with lactation (Morris et al., 1980). However, the provision of supplemental NaCl in block form was not found to improve calf weaning weight on pasture containing only 0.15–0.50g Na kg^{-1} DM and the drinking water only 10–25 mg Na l^{-1}. Substantially higher sodium requirements of 1.8g Na kg^{-1} DM have been given for lactating dairy cows by the NRC (2001) than by the ARC (1980), whose factorial estimates make no allowance for obligatory losses in urine.

The requirements given in Table 8.2 have been increased to allow for a similar effect of potassium to that described in sheep: they confirm that requirement increases slightly with a rise in milk yield. Applications of NaCl to grazed pasture in Wales that increased herbage sodium from an apparently adequate level of 2g kg^{-1} DM to 5g kg^{-1} DM also improved milk fat yield, giving high estimates of the prevalence of sodium deficiency in dairy cows (Chiy and Phillips, 1993a), although the effect was confined to low-yielding cows in a subsequent study (Chiy and Phillips, 2000). The responses may be attributable to short-term effects on palatability or rumen physiology that increase grazing and ruminating time and bite rate rather than alleviation of sodium deprivation (Chiy and Phillips, 1993b, 2000). Such effects might not be seen in longer-term studies because of a lack of herbage to graze, and they have not been found in New Zealand (Edmeades and O'Connor, 2003). Increased requirements for sodium (and potassium) at the onset of lactation have been proposed (Silanikove et al., 1997) to avoid large negative balances, but massive amounts of sodium (and potassium) were being excreted in urine at the time (1.0–2.6g day^{-1}), suggesting either a failure of normal homeostatic mechanisms or – as seems more likely – unavoidable losses due to bone catabolism.

Chloride in cattle

The daily chloride requirement of beef cattle gaining 1.0kg day^{-1} estimated by ARC (1980) is equivalent to a dietary concentration of 0.7g Cl kg^{-1} DM, close to the need for sodium. Requirements for lactating dairy cows should be substantially higher because cow's milk contains more than twice as much chloride as sodium. In the experiment of Fettman et al. (1984), a supplement of 0.6g chloride was sufficient to

Table 8.2. Estimates of the minimum dietary sodium concentrations required by sheep[a] and cattle at the given DMI (after ARC, 1980).

	Live weight (kg)	Growth or product	DMI (kg day^{-1})	Gross[b] Na requirement	
				(g day^{-1})	(g kg^{-1} DM)
Lamb	20	0.1 kg day^{-1}	0.5	0.35	0.7
		0.2 kg day^{-1}	0.8	0.47	0.6
	40	0.1 kg day^{-1}	0.83	0.58	0.7
		0.2 kg day^{-1}	1.23	0.71	0.6
Ewe	75	0	0.8	1.0	1.25
Ewe, pregnant	75	1 fetus, week 12	1.03	1.0	1.0
		Term	1.51	1.0	0.7
Ewe, lactating	75	1 kg day^{-1}	1.48	1.2	0.81
		2 kg day^{-1}	2.18	1.8	0.83
		3 kg day^{-1}	2.90	2.2	0.76
Steer	200	0.5 kg day^{-1}	3.3	2.3	0.70
		1.0 kg day^{-1}	4.7	3.1	0.65
	400	0.5 kg day^{-1}	5.2	3.9	0.75
		1.0 kg day^{-1}	7.3	4.7	0.65
Cow, dry	600	0	5.0	4.5	0.9
Cow, pregnant	600	Week 12	5.8	4.3	0.9
		Term	9.0	7.2	0.8
Cow, milking	600	10 kg day^{-1}	9.4	10.3	1.1
		20 kg day^{-1}	14.0	16.8	1.2
		30 kg day^{-1}	18.8	22.6	1.2

[a]Values for sheep have been reduced by using a lower value for endogenous loss of 5 as opposed to 25 mg Na kg^{-1} live weight.
[b]Absorption coefficient = 0.91.

prevent the symptoms of deprivation produced by a basal diet providing 1 g Cl kg^{-1} DM, but the NRC (2001) suggest a generous provision of 2.0 g Cl kg^{-1} DM. Milk chloride, in contrast to most minerals, appears to rise with advancing lactation. A study in Holstein cows reported values of 1.16, 1.29 and 1.90 g Cl l^{-1} milk in months 1, 5 and 10 of lactation, respectively (Lengemann et al., 1952). However, milk may represent a route of excretion for chloride and losses of chloride in milk may not need to be wholly replaced.

Goats

From a study of growth, reproduction and lactation of goats at two dietary concentrations of sodium (0.3 and 1.7 g kg^{-1} DM) (Schellner, 1972/1973), the lowest level was severely inadequate and the higher level appeared to be sufficient. No report on the chloride requirements of goats is available.

Pigs

In experiments with weaner pigs on diets very low in sodium, potassium and chloride, 0.8–1.1 g Na and 1.2–1.3 g Cl kg^{-1} DM were required for optimum growth (Meyer et al., 1950). The need for sodium was later confirmed for conventional maize and soybean meal diets (Hagsten and Perry, 1976; Honeyfield et al., 1985). The ARC (1981) used a factorial model to define requirements (Table 8.3) and predicted that the need for sodium falls markedly with age to values well below the latest recommendation from NRC (1998) of 1 g Na kg^{-1} DM from 20 to 120 kg LW. Applying the model to the lactating sow also gives a far lower need than the 1.9 g Na

Table 8.3. Sodium requirements of pigs predicted from a factorial model[a] (after ARC, 1981).

Live weight (kg)	Growth rate (kg day^{-1})	Net requirement (g day^{-1})	Total requirement (g day^{-1})	DM intake (kg day^{-1})	Dietary Na concentration (g kg^{-1} DM)[b]
5	0.27	0.416	0.462	0.37	1.25
25	0.55	0.659	0.732	1.04	0.71
45	0.78	0.851	0.946	1.78	0.53
90	0.79	0.833	0.926	2.78	0.33
150	5.5[c]	2.08	2.14	5.00	0.43

[a]Model components were 1.0–1.5 g Na kg^{-1} live-weight (LW) gain for growth, 56 g Na total need for pregnancy, 0.3–0.4 g Na kg^{-1} for milk yield, 1 mg kg^{-1} LW for maintenance and an absorption coefficient of 0.9.
[b]Corresponding requirements for NaCl are 2.62-fold greater.
[c]Milk yield.

kg^{-1} DM recommended by NRC (1998). Generous provision during pregnancy (1.5 g Na kg^{-1} DM: NRC, 1998) would establish a store to meet any lactational deficit. The NRC (1998) requirements for chloride were doubled for the early stages of growth on the basis of studies with early weaned pigs given animal protein (Mahan et al., 1996, 1999). Significant growth responses were obtained with additions of HCl to diets containing 2.3 g Cl kg^{-1} DM and the requirement for the first 2 weeks after weaning at 22 days old was 3.8 g Cl kg^{-1} DM (Mahan et al., 1999). The NRC (1998) recommends a substantial reduction in chloride provision to 0.8 g kg^{-1} DM by the fattening stage, but feeding trials have suggested higher needs for finisher than grower pigs and improved food conversion efficiency with 1.7 g Cl kg^{-1} DM, particularly if diets are low in sodium (Honeyfield et al., 1985).

Poultry

The latest NRC requirements are summarized in Table 8.4 and are considerably greater than those indicated by early studies with less productive birds (Underwood and Suttle, 1999), but more sodium may be needed for very young birds. Broiler chicks can need 4–5 g kg^{-1} DM in the first week, declining by 30% at 3 weeks of age (Edwards, 1984; Britton, 1990). Chloride requirements are also much higher (4–5 g kg^{-1} DM) early in life. Requirements for sodium, potassium and chloride probably decline with stage of growth in all types of bird, as is the case with pigs. The average hen's egg contains 73 mg sodium, 88 mg chloride and 82 mg potassium, but the increases in daily requirement with the onset of lay are more than met by increases in food intake and requirements remain low when stated as dietary concentrations, although they increase with the use of high-energy diets (Table 8.4). The NRC standards suggest that supplementation of conventional poultry rations with sodium is essential, however, turkey poults have been reared successfully to 35 days of age on diets containing only 0.9 g Na kg^{-1} DM, well below the NRC (1994) requirement (Frame et al., 2001). The latest chloride requirements for poultry (Table 8.4) (NRC, 1994) are of a similar order to those for sodium.

Biochemical and Physiological Signs of Sodium Deprivation

Depletion

The existence and physiological significance of an initial depletion phase, as shown in the general model (Chapter 3), is questionable. The large pools of sodium present in the rumen and skeleton are not essential for body function and increase with sodium intake, and therefore have the characteristics of 'stores'. However, their presence may not greatly delay the onset of far-reaching adaptations in sodium metabolism when sodium deprivation is acute and severe. Depletion of the rumen pool of sodium occurs concurrently with the changes about to be described and the skeletal sodium pool may also decline, though at a much slower rate.

Table 8.4. Dietary requirements (g kg⁻¹ DM) of growing broiler (B) and Leghorn (L) chicks, turkey poults (P) and laying birds for sodium and chloride (modified from NRC, 1994).

	Bird type	Growth stage[a]			Lay stage[b]		
		Early	Middle	Late	High	Medium	Low
Na	B	3.0[c]	1.5	1.2	–	–	–
	L	3.0[c]	1.5	1.5	1.9	1.5	1.3
	P	1.7	1.2	1.2	–	1.2	–
Cl	B	3.0[c]	1.5	1.2	–	–	–
	L	3.0[c]	1.2	1.5	1.6	1.3	1.1
	P	1.5	1.4	1.2	–	1.2	–

[a] For broilers, Leghorn chicks and poults, respectively: early = 0–3, 0–6 or 0–4 weeks; middle = 3–6, 6–12, or 8–12 weeks; late = 6–8, 18 weeks to first lay and 20–24 weeks of age.
[b] High, medium and low represent declining energy densities and food intakes of 80, 100 and 120 g day⁻¹, respectively.
[c] For the first week of life, 8–10 g NaCl kg⁻¹ DM is indicated by some studies (see text).

Sodium concentrations in the equine body are almost twice as high as those found in lambs and almost half is found in the skeleton (Grace et al., 1999), representing yet another possible line of defence against deprivation.

Urinary sodium concentrations rapidly decline to extremely low values and urinary potassium excretion also declines (Singh and Rani, 1999). Faecal sodium is reduced (Jones et al., 1967; Michell et al., 1988) through reabsorption against a concentration gradient in the lower intestine (Bott et al., 1964). The mammary gland is unresponsive to aldosterone and milk

Deficiency

Sodium concentrations in plasma must be maintained to preserve osmotic balance and do not decline until the sodium-deprived animal is *in extremis* (e.g. Seynaeve et al., 1996). The principal mechanism for sodium conservation in ruminants involves the replacement of salivary sodium on a molar basis by potassium in an adaptation modulated by aldosterone (Blair-West et al., 1963). Sodium deprivation causes a greater aldosterone response in sheep on a high-potassium than on a potassium-free diet (Humphery et al., 1984). The relationship between plasma aldosterone and the salivary sodium to potassium ratio is curvilinear, with rapid increases in hormone concentrations occurring when the sodium to potassium ratio falls to <5.0 in sheep and goats (Fig. 8.6). In bullocks grazing sodium-deficient pasture, salivary sodium and potassium values were found to be 40 and 90 mmol l⁻¹, respectively, compared with 145 and 7 mmol l⁻¹ in normal animals (Murphy and Plasto, 1972). Qualitatively similar reciprocal changes occur in sweat, but they are less marked than those in saliva (Fig. 8.7).

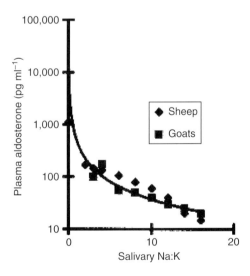

Fig. 8.6. Sodium is conserved through the action of the adrenal hormone aldosterone, which facilitates the replacement of Na by potassium in saliva and urine. Low Na:K ratios in saliva are therefore correlated with plasma aldosterone and are a useful indicator of sodium deprivation (data from McSweeney et al., 1988).

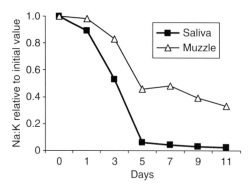

Fig. 8.7. Sodium deprivation of buffalo heifers decreases the sodium to potassium ratio in muzzle secretions as well as in saliva (data from Singh and Rani, 1999). The initial Na:K ratios were 30 and 0.46 for saliva and muzzle secretions, respectively.

sodium declines only slightly during sodium deficiency, with no compensatory changes in milk potassium in the cow (Schellner, 1972, 1973) or ewe (Morris and Petersen, 1975). Sodium deprivation of young pigs caused plasma potassium to increase from 5.6 to 7.4 ± 0.17 (standard error) mmol l^{-1} as dietary sodium decreased from 1.8 to 0.3 g kg^{-1} DM, and disturbances of amino acid metabolism were indicated by changes in the plasma lysine to arginine ratio (Honeyfield et al., 1985).

Dysfunction

Changes in plasma potassium in pigs are accompanied by evidence of haemoconcentration, with blood haemoglobin increasing from 124 to 159 ± 8 g l^{-1} in sodium deprivation (Honeyfield et al., 1985). Remarkable changes have been reported in the lower intestine of hens depleted of sodium (Elbrond et al., 1991). After 3–4 weeks on a low-sodium diet, the microvillus surface area more than doubled due to an increase in epithelial cell number and the surface area per cell. A greater density of open sodium channels and an increase in the short-circuit current contributed to greatly increased net sodium transport. Morphological changes are also seen in the adrenals of sodium-depleted cattle, where the width of the zona glomerulosa, as a proportion of the cortex, doubles (Morris, 1980). A dietary deficiency of sodium in laying hens, given 0.5 g Na kg^{-1} DM, was aggravated by restriction of chloride to 0.8 g kg^{-1} DM, but partly alleviated by increasing potassium from 7 to 12 g kg^{-1} DM (Sauveur and Mongin, 1978).

Diagnosis of Sodium Deprivation

The earliest and most obvious sign of sodium deprivation, a craving for salt, does not provide a definitive diagnosis because some healthy individuals eat salt avidly (Denton, 1982) and salt craving is a feature of other disorders such as phosphorus deprivation. Other manifestations of sodium deprivation such as polydipsia and polyuria (Whitlock et al., 1975) are also non-specific, and reliance must be placed on biochemical criteria such as the salivary sodium to potassium ratio. Early reductions in the salivary sodium to potassium ratio precede the appearance of disorder, and a band of marginal sodium to potassium ratios must be recognized (see Table 8.5). Although ratios can vary widely within groups and between frequent samplings from the same animal, reciprocal movements in plasma aldosterone indicate extreme physiological sensitivity of sodium to potassium ratios rather than random changes (N.F. Suttle, unpublished data).

The *median* salivary sodium to potassium ratio and the *proportion* of values <2.0 should be used to assess risk of sodium deprivation in a flock or herd, rather than the mean sodium to potassium ratio.

Parotid saliva can be easily drawn into a syringe from the hollow between cheek and teeth using a 'trombone slide' gag to open the mouth (Murphy and Connell, 1970). In bovines, the sodium to potassium ratio in muzzle secretions also decreases during sodium deprivation and droplets procured by syringe may constitute a more convenient sample than saliva, but the critical ratio is much lower (<0.2) than it is for saliva (Singh and Rani, 1999). Faecal sodium values <1 g kg^{-1} DM and a sodium to potassium ratio of <1.0 indicate deprivation (Little, 1987), but their dependability has been questioned (Morris, 1980) and would be influenced by factors such

Table 8.5. Suggested marginal bands[a] for sodium and chloride in serum, urine and diet as a guide to the diagnosis of dietary deprivation (D) or excess (E) of each element for livestock.

			Serum[b] (mmol l^{-1})	Urine (mmol l^{-1})	Diet (g kg^{-1} DM)
Na	Cattle[c]	D	130–140	<3	0.5–1.0
		E	140–145	40–60	30–60
	Sheep[c]	D	140–145	–	0.5–1.0
		E	150–160	40–60	30–60
	Pigs	D	137–140	1–3	0.3–1.0
		E	150–180	>11	20–40
	Poultry	D	130–145	–	0.8–1.5
		E	150–160	–	50–80
Cl	Cattle	D	70–85	2–5	1–3
		E	>150	–	–
	Sheep	D	70–85	–	1–3
		E	>150	>100	>50
	Pigs	D	88–100	–	0.9–2.4
		E	>110	–	20–49
	Poultry	D	90–115	–	0.8–2.0
		E	>174	>100	6.0–12.0

[a]Mean values below (for D) or above (for E) the given bands indicate probability of ill-health from deprivation or toxicity. Values within bands indicate the possibility of future disorder.
[b]If serum is unobtainable, similar interpretations can be applied to concentrations in vitreous humour.
[c]Salivary ratios of Na to K are more useful for diagnosing insufficiency in cattle and sheep. Median values of 4.0–10.0 are regarded as 'marginal'; the equivalent range for bovine muzzle secretions is 0.2–0.3.

as DM intake and digestibility. Urine analysis is more dependable, with sodium concentrations of <3 mmol l^{-1} indicating deprivation in *B. taurus* (Morris, 1980) and *B. indicus* (Singh and Rani, 1999). However, values are influenced by urine output and therefore differ between animals on lush or dry diets unless creatinine is used as a marker. Low plasma sodium concentrations indicate terminal sodium deprivation and the values given in Table 8.5 should be carefully interpreted. The surest indication of sodium deprivation is a positive response in appetite, appearance and productivity to salt supplementation.

Occurrence of Sodium Deprivation

Sodium deprivation can be precipitated by dietary, climatic and disease factors and occurs in the following conditions:

- rapidly growing animals given cereal-based diets that are inherently low in sodium (Table 8.1);
- animals grazing pastures on soils naturally low in sodium;
- animals grazing pastures heavily fertilized with potassium;
- lactating animals secreting large amounts of sodium in milk, particularly cows;
- tropical or hot, semi-arid climates, causing large losses of water and sodium in sweat;
- heavy or intense physical work that causes profuse sweating; and
- animals with gut infections that cause diarrhoea.

When one or more of these conditions applies continuously for long periods and extra salt is not provided, sodium deprivation is inevitable.

Dietary sodium

Extensive areas of sodium deprivation in livestock occur in the centres of many continents, including Saharan (Little, 1987), southern (Walker, 1957; De Waal *et al.*, 1989) and tropical Africa and Latin America (McDowell et al., 1993). In arid inland areas of Australia (Murphy and Plasto, 1972; Denton, 1982) and Africa (Howard et al., 1962) where the water table is low, grass pastures commonly contain 0.1–0.8 g Na kg^{-1} DM and even grains and seeds are low in sodium (Chamberlain, 1955; Morris and Gartner, 1971). Temperate continents (Chiy and Phillips, 1995) and islands such as New Zealand (Towers and Smith, 1983) can also be low in sodium. One-third of New Zealand pasture species have been found to contain <0.5 g Na kg^{-1} DM (Edmeades and O'Connor, 2003) and the introduction of legumes such as alfalfa, also low in sodium (Table 8.1), to New Zealand has been implicated in sodium deprivation in dairy cows (Joyce and Brunswick, 1975).

Pasture potassium

Potassic fertilizer applications may accentuate sodium deprivation in two ways: by reducing pasture sodium and by impairing sodium metabolism in the animal. Applications of KCl (80 kg ha^{-1}) reduce sodium concentrations to 0.4 g kg^{-1} DM in natrophilic pasture species such as *Dactylis glomerata* (McNaught, 1959) and a similar problem was identified in Dutch pastures (Lehr et al., 1963). A moderate pastures in dietary potassium (from 6 to 24 g kg^{-1} DM) can greatly reduce the flow of sodium from the rumen, particularly in cattle (Fig. 8.4). The reduction in flow may exceed dietary sodium intakes and can be attributable to either reduced salivary secretion of sodium or increased absorption of salivary sodium from the rumen. Exposure to excess potassium causes an immediate reduction in rumen sodium concentration (Scott, 1974) and an immediate large increase in urinary sodium excretion (Suttle and Field, 1967). Natriuresis is still evident 10–17 days after potassium exposure (Greene et al., 1983a,b) and may be attributable to the diversion of sodium from the rumen and a consequent increase in renal tubular load, possibly involving an aldosterone response and reduction in salivary sodium secretion (see Chapter 7).

Climatic factors

There are risks of sodium deprivation amongst working animals in hot, humid environments – particularly in equine species, which secrete sweat rich in sodium. The increased use of the lactating cow for draught purposes in developing countries will increase the risk of sodium deprivation. With mature, non-lactating, non-working animals or with younger ruminants whose diets permit little growth, sodium homeostasis is so effective that deprivation is likely only in hot environments with exceedingly low sodium intakes from herbage.

Transport

Supplementation with sea water has improved the health of sheep that became starved and diarrhoeic during prolonged transportation by sea (Black, 1997). Lack of sodium was believed to have hindered water absorption from the gut.

Disease factors

Infections of the small and large intestine perturb mechanisms that normally contribute to sodium homeostasis. Acute bacterial infections of the gut that cause severe diarrhoea in calves require the provision of sodium, together with glucose, bicarbonate and even potassium in a carefully balanced rehydration therapy (Michell, 1989). Parasitic infections of the abomasum in lambs cause sodium efflux into the lumen, but losses can be reabsorbed at uninfected distal sites since faecal sodium excretion is not increased, and the salivary sodium to potassium ratio is unaffected in infected lambs on a diet marginal in sodium (E. Ortolani, personal communication, 2000). When the small intestine is also parasitized, as is often the case in young lambs and calves at pasture, sodium deficiency and salt appetite may be induced on occasions despite mineral supplementation (Suttle et al., 1996) and adequate pasture sodium levels (Suttle and Brebner, 1995).

Prevention of Sodium Deprivation

Supplementary sodium is invariably supplied commercially as common salt because of its palatability and ready availability. In the past, plentiful, cheap supplies coupled with low toxicity encouraged the extravagant use of salt, but practices have changed as requirements have become more closely defined and costs to the environment, to the animal's energy metabolism (Arieli et al., 1989) and to small farmers in tropical countries become apparent. In pig and poultry feeding it was common to add 5 g NaCl kg^{-1} DM to rations, but 2.5 g should be sufficient even for the youngest pigs with the highest requirements (see Table 8.3), especially if the ration contains animal protein. For growing or fattening beef cattle, hand- or mechanically fed, salt supplements need not add >0.5 g Na kg^{-1} DM to the total ration, but the level should be doubled for housed lactating dairy cows.

With milking cows and ewes at pasture, it is customary to rely on the voluntary consumption of loose or block salt made continuously available: this is usually sufficient, although individual consumption can be highly variable and is sometimes greatly in excess of need (Fig. 3.4, p. 49) (Suttle and Brebner, 1995). The physical form of the salt affects voluntary consumption. Lactating cows and heifers at pasture consume significantly more loose than block salt, but intakes of block salt are still sufficient to meet the sodium needs of lactation (Smith et al., 1953). Sheep show a similar preference for loose salt. Applications of salt as a grassland fertilizer rarely influence pasture yields in New Zealand, even on soils low in sodium and high in potassium. The relationship between application rate and herbage sodium is linear, but 50 kg NaCl ha^{-1} is sufficient to raise herbage sodium in natrophobic species to the level required by the grazing animal (Edmeades and O'Connor, 2003). Where the problem is one of potassium-induced sodium deficiency, potassium inputs to the farm should be restricted and seasonal applications delayed (see Chapter 7). The use of a low-grade kainite (KCl), containing 22% subsidiary sodium, should maintain herbage sodium concentrations.

Biochemical Consequences of Chloride Deprivation

In experimental deprivation of calves, plasma chloride fell from 96 to 31 mmol l^{-1}, salivary chloride fell from 25 to 16 mmol l^{-1} and there was secondary alkalosis, with reductions in plasma sodium and potassium (Neathery et al., 1981). Although the technique used to induce deprivation in lactating cows was less severe (Fettman et al., 1984) the biochemical consequences were similar.

Abomasal parasitism inhibits the normal secretion of HCl into the gut lumen (Coop, 1971), but the consequences of this in terms of acid–base balance and the pathogenesis of infection have not been explored. Acute challenge infections with *Haemonchus contortus* raise salivary and presumably serum chloride and perturb salivary sodium to potassium ratios by increasing salivary potassium, indicating the possibility of acidosis and loss of intracellular potassium (N.F. Suttle, unpublished data). Guidelines for the assessment of chloride status are given in Table 8.5.

Sodium, Chloride and Acid–Base Balance

The role of sodium and chloride in determining acid–base balance has been overshadowed by that of potassium (Chapter 7), but was demonstrated in an experiment with lambs in which diets were supplemented with 2% $NaHCO_3$ or 1% NH_4Cl to produce alkaline and acidic diets, respectively. Retention of calcium and phosphorus was reduced on the acidic diet, and changes in acid and alkaline phosphatase activities were suggestive of increased osteoclast and decreased osteoblast activity in bone (Abu Damir et al., 1981). In a comparable study with pigs (Budde and Crenshaw, 2003), no effects of 'acidic' diets on bone mineralization were found despite apparent intracellular retention of chloride, but this may have been attributable to the use of $CaCl_2$ as a source of chloride.

Occurrence of Chloride Deprivation

While chloride deficiency is not believed to occur naturally, Coppock (1986) has questioned the assumption that sodium will always be the more limiting element and identified diets based on maize (as grain, gluten and silage) as presenting a risk of chloride deficiency. Chloride depletion might also occur in hot climates since cattle exposed to 40°C for 7 h are estimated to lose slightly more chloride than sodium in sweat (1 g day^{-1} for a 200-kg steer and 1.69 g^{-1} for a 500-kg cow) (ARC, 1980). Increases in muscle chloride have been reported in 'downer cows' (see Chapter 4) and may reflect a perturbed 'chloride shift'.

Spontaneous turkey cardiomyopathy (round heart)

Sudden deaths in turkeys, associated with swollen hearts, ascites and hydropericardium at post-mortem examination, have been linked to electrolyte imbalance (Frame et al., 2001). The disease occurs at a young age (11–21 days), particularly in females, with an increased frequency in birds kept at high altitude. It is thought that the relatively high potassium intakes provided by high-protein diets can raise an already high haematocrit and that increased blood viscosity increases cardiac workload. The incidence of spontaneous turkey cardiomyopathy (STC) is usually low (<3%), but has been increased by raising dietary sodium (from 1.4 to 2.4 g kg^{-1} DM) and decreased by raising chloride (from 1.6 to 4.0 g kg^{-1} DM) in experiments conducted at the high altitude of Utah (Frame et al., 2001). Chloride supplementation was thought to have increased potassium excretion. In two experiments in which mean 35-day body weights differed (950 versus 850 g), STC incidence was significantly lower in slower-growing poults (1.15 versus 0.50). The disease can be controlled by restricting growth rate.

Heat stress

Heat stress can induce respiratory alkalosis associated with panting, and supplementation with sodium or potassium salts can increase milk yield in heat-stressed cows (West et al., 1991). In heat-stressed broiler chicks, the feeding of KCl or ammonium chloride (NH_4Cl) in the drinking water (1.5 and 2 g l^{-1}, respectively) with a diet adequate in potassium improved growth rate and food conversion efficacy (Teeter and Smith, 1986); higher levels or mixtures of the two salts were, however, detrimental. Later work confirmed the benefits of supplementing drinking water with KCl or NaCl (0.067 mmol l^{-1}), but there was no restoration of lowered plasma sodium or potassium and aldosterone levels remained high (Dayhim and Teeter, 1994). It would appear that the critical disturbance in electrolyte metabolism involves chloride rather than sodium or potassium.

Sodium, Chloride and Salt Toxicities

Dietary excesses of osmotically active elements such as sodium and chloride can disturb body functions (e.g. induce oedema), are usually concurrent and occur under natural circumstances through the consumption of saline drinking water or plants growing on saline soils. Excesses can also arise from accidental or intentional human interventions (e.g. access to NaCl-containing fertilizers or mineral mixtures, failed irrigation schemes or alkali treatment of grain and roughage). In many respects chloride toxicity is synonymous with salt toxicity, but excess chloride has been specifically incriminated as a factor contributing to tibial dyschondroplasia (Shirley et al., 2003). Guidelines for the assessment of excesses in the diet and animal are given in Table 8.5.

Salt supplements

Tolerance of dietary salt is largely dependent on the availability of salt-free drinking water. Daily bolus doses of 10.5 g NaCl kg^{-1} LW given via rumen cannulae to grazing lambs depressed growth within 4 weeks and after 9 months reductions of 26% in weight gain and 14% in clean wool yield were recorded (De Waal et al., 1989). Protein supplementation reduced the adverse effects of NaCl dosage, raising the possibility that the mineral reduces microbial synthesis of protein in the rumen. With continuous dietary supplementation in calves, 1.83 g NaCl kg^{-1} LW day^{-1} has been tolerated with no disturbance of electrolyte status or growth (Wamberg et al., 1985). It seems unlikely that the individual intake of salt from free-access mixtures would ever be sufficient to impair production, provided there is access to drinking water.

Excess sodium in the diet increases water consumption in all species, including poultry (Mongin, 1981), increasing the volume and moisture content of excreta. Feeding to lower sodium requirements will reduce water and salt accumulation in excreta and associated environmental problems.

Saline drinking water

Water containing up to 5 g NaCl l^{-1} is safe for lactating cattle and up to 7 g l^{-1} is safe for non-lactating cattle and sheep (Shirley, 1978). Stock can adapt to concentrations considerably higher than these, at least in temperate climates. In a climate where the winters were cool to mild and the pasture lush, sheep tolerated water containing 13 g NaCl l^{-1} (Peirce, 1957, 1965). However, with 20 g NaCl l^{-1} feed consumption and body weight declined and some animals became weak and emaciated. In high environmental temperatures and dry grazing conditions, the tolerable salt concentration is reduced because of increased water consumption. Toxicity varies with the constituent salts in saline water: sodium chloride appears to be the least harmful component, while magnesium sulfates and carbonates are much more toxic than the corresponding sodium salts. Both young and old pigs have been found to tolerate drinking water containing 10 g NaCl l^{-1}, but 15 g l^{-1} was toxic (Heller, 1932). An increase in dietary sodium from 1 (low) to 4 (modest excess) g kg^{-1} DM increased the water intake of lactating sows from 12.4 to 13.9 l day^{-1} (Seynaeve et al., 1996). Laying hens are much less tolerant of saline water than pigs and adverse effects are of particular concern because of the speed with which egg-shell quality is reduced by relatively low concentrations (2 days exposure to 0.2 g NaCl l^{-1}) and the residual effects of prolonged exposure (at least 15 weeks) (Balnave and Yoselewitz, 1985; Balnave and Zhang, 1998). The percentage of damaged shells increased linearly to a maximum of 8.9% as sodium concentrations increased up to 0.6 g l^{-1}, a level found in some of the local (New South Wales, Australia) underground sources.

Salt-tolerant plants

Saline soils support an unusual flora made up of species that tolerate sodium by various means involving avoidance (deep roots), exclusion (salt glands) and dilution (bladder cells, succulence) (Gorham, 1995). There is therefore no simple correlation between soil salinity and plant sodium concentrations. Species such as the bladder saltbush (*Atriplex vesicaria*) are essential to livestock production in many arid regions, but ash concentrations (200–300 g kg^{-1} DM, mostly as salt) depress digestibility and lead to energy losses associated with low digestibility and greatly increased sodium turnover (Arieli et al., 1989). Tolerance of salt-rich plants is highly dependent on free access to water low in salt but high relative to the suggested limits for NaCl in the diet (15 g kg^{-1} DM for pigs and up to 25 g NaCl kg^{-1} DM for cattle) under favourable conditions (for review, see Marai et al., 1995). Toxicity is indicated clinically by anorexia and water retention and physiologically by intracellular dehydration due to hypertonicity of the ECF, with sodium and chloride both contributing to disturbances. Control is only likely to be achieved by a variety of approaches including the encouragement of deep-rooted (saline-soil-avoiding) species, irrigation systems that limit the upward migration of salt and the use of complementary feeds low in sodium (e.g. maize and lucerne) (Masters et al., 2006).

Sodium hydroxide-treated feeds

The treatment of grains, straws and mature or whole crop silages with sodium hydroxide (NaOH) can improve nutrient utilization by ruminants (e.g. Leaver and Hill, 1995; Chaudhry et al., 2001) and the amounts of sodium added are substantial. In the former study, wheat and wheat straw containing 18 g Na kg^{-1} were fed with silage (40:60) to dairy cows without mishap for 12 weeks from day 54 of lactation. Calves can tolerate 1.2 g NaOH kg^{-1} LW day^{-1} and show no benefits in terms of acid–base status from neutralization of the alkali with HCl (Wamberg et al., 1985). However, the preferred alkali is ammonium hydroxide (NH_4OH) because it provides nitrogen and may be less likely than NaOH to cause kidney lesions (nephritis) (D. Rice, personal communication, 1985). Alkali-treated feeds should not be used for transition cows because they may increase the incidence of milk fever (see Chapter 4).

References

Abdoun, K., Wolf, K., Arndt, G. and Martens, H. (2003) Effect of ammonia on Na^+ transport across isolated rumen epithelium of sheep is diet dependent. *British Journal of Nutrition* 90, 751–758.

Abdoun, K., Stumpff, F., Wolf, K. and Martens, H. (2005) Modulation of electroneutral Na transport in sheep rumen epithelium by luminal ammonia. *American Journal of Physiology – Gastrointestinal and Liver Physiology* 289, G508–G520.

Abu Damir, H., Scott, D., Loveridge, N., Buchan, W. and Milne, J. (1991) The effects of feeding diets containing either $NaHCO_3$ or NH_4Cl on indices of bone formation and resorption and on mineral balance in the lamb. *Journal of Experimental Physiology* 76, 725–732.

Aines, P.D. and Smith, S.E. (1957) Sodium versus chloride for the therapy of salt-deficient dairy cows. *Journal of Dairy Science* 40, 682–688.

Aitken, F.C. (1976) Technical communication of the Commonwealth Bureau of Nutrition No. 26.

ARC (1980) *The Nutrient Requirements of Ruminants*. Commonwealth Agricultural Bureaux, Farnham Royal, UK, pp. 213–216.

ARC (1981) *The Nutrient Requirements of Pigs*. Commonwealth Agricultural Bureaux, Farnham Royal, UK.

Arieli, A., Naim, E., Benjamin, R.W. and Pasternak, D. (1989) The effect of feeding saltbush and sodium chloride on energy metabolism in sheep. *Animal Production* 49, 451–457.

Balnave, D. and Yoselewitz, I. (1985) The relation between sodium chloride concentration in drinking water and egg-shell damage. *British Journal of Nutrition* 58, 503–509.

Balnave, D. and Zhang, D. (1998) Adverse responses in egg shell quality in late-lay resulting from short-term use of saline drinking water in early- or mid-lay. *Australian Journal of Agricultural Research* 49, 1161–1165.

Beatty, D.T., Barnes, A., Taylor, E., Pethick, D., McCarthy, M. and Maloney, S.K. (2006) Physiological responses of *Bos taurus* and *Bos indicus* to prolonged continuous heat and humidity. *Journal of Animal Science* 84, 972–985.

Bell, F.R. (1995) Perception of sodium and sodium appetite in farm animals. In: Phillips, C.J.C. and Chiy, P.C. (eds) *Sodium in Agriculture*. Chalcombe Publications, Canterbury, UK, pp. 82–90.

Berger, L.L. (1990) Comparison of National Research Council feedstuff mineral composition data with values from commercial laboratories. In: *Proceedings of Georgia Nutrition Conference, Atlanta, 1990*. University of Georgia, Atlanta, Georgia, pp. 54–62.

Black, H. (1997) Sea water in the treatment of inanition in sheep. *New Zealand Veterinary Journal* 45, 122.

Blair-West, J.R., Coghlan, J.P., Denton, D.A., Goding, J.R., Wintour, M. and Wright, R.D. (1963) The control of aldosterone secretion. In: Plincus, S. (ed.) *Recent Progress in Hormone Research*. Academic Press, New York. pp. 311–383.

Blair-West, J.R., Coghlan, J.P., Denton, D.A., Nelson, J.F., Orchard, E., Scoggins, B.A., Wright, R.D., Myers, K. and Junqueira, C.L. (1968) Physiological, morphological and behavioural adaptation to a sodium deficient environment by wild native Australian and introduced species of animals. *Nature UK* 217, 922–928.

Block, E. (1994) Manipulation of cation-difference on nutritionally related production diseases, productivity and metabolic responses in dairy cows. *Journal of Dairy Science* 77, 1437–1450.

Bott, E., Denton. D.A., Goding, J.R. and Sabine, J.R. (1964) Sodium deficiency and corticosteroid secretion in cattle. *Nature UK* 202, 461–463.

Britton, W.M. (1990) Dietary sodium and chlorine for maximum broiler growth. In: *Proceedings of Georgia Nutrition Conference, Atlanta*, 1990. University of Georgia, Atlanta, Georgia, pp. 152–157.

Budde, R.A. and Crenshaw, T.D. (2003) Chronic metabolic acid load induced by changes in dietary electrolyte balance increased chloride retention but did not compromise bone in growing swine. *Journal of Animal Science* 81, 197–208.

Burnier, M. (2007) *Sodium in Health and Disease*. CRC Press, Boca Raton, Florida.

Chamberlain, G.T. (1955) The major and trace element composition of some East African feeding stuffs. *East African Agricultural Journal* 21, 103–107.

Chaudhry, A.S., Cowan, R.T., Granzan, B.C. and Klieve, A.V. (2001) The nutritive value of Rhodes grass (*Chloris guyana*) when treated with CaO, NaOH or a microbial inoculant and offered to dairy heifers as big-bale silage. *Animal Science* 73, 329–340.

Chiy, P.C. and Phillips, C.J.C. (1993) Sodium fertilizer application to pasture. 1. Direct and residual effects on pasture production and composition. *Grass and Forage Science* 48, 189–202.

Chiy, P.C. and Phillips, C.J.C. (1995) Sodium in forage crops and Sodium fertilization. In: Phillips, C.J.C. and Chiy, P.C. (eds) *Sodium in Agriculture*. Chalcombe Publications, Canterbury, UK, pp. 43–81.

Chiy, P.C. and Phillips, C.J.C. (2000) Sodium fertilizer application to pasture. 10. A comparison of the responses of dairy cows with high and low milk yield potential. *Grass and Forage Science* 55, 343–350.

Chiy, P.C., Phillips, C.J.C. and Bello, M.R. (1993a) Sodium fertilizer application to pasture. 2. Effects on dairy cow production and behaviour. *Grass and Forage Science* 48, 203–212.

Chiy, P.C., Phillips, C.J.C. and Omed, H.M. (1993b) Sodium fertilizer application to pasture. 3. Rumen dynamics. *Grass and Forage Science* 48, 249–259.

Coop, R.L. (1971) The effect of large doses of *Haemonchus contortus* on the level of plasma pepsinogen and the concentration of electrolytes in the abomasal fluid of sheep. *Journal of Comparative Pathology* 81, 213–219.

Coppock, C.E. (1986) Mineral utilisation by the dairy cow – chlorine. *Journal of Dairy Science* 69, 595–603.

Cornforth, I.S., Stephen, R.C., Barry, T.N. and Baird, G.A. (1978) Mineral content of swedes, turnips and kale. *New Zealand Journal of Experimental Agriculture* 6, 151–156.

Dayhim, F. and Teeter, R.G. (1994) Effect of heat stress and drinking water salt supplements on plasma electrolytes and aldosterone concentrations in broiler chickens. *International Journal of Bacteriology* 38, 216–217.

Denton, D.A. (1982) *The Hunger For Salt: An Anthropological, Physiological and Medical Analysis*. Springer-Verlag, Berlin/Heidelberg.

De Waal, H.O., Baard, M.A. and Engels, E.A.N. (1989) Effects of sodium chloride on sheep. 1. Diet composition, body mass changes and wool production in young Merino wethers grazing mature pasture. *South African Journal of Animal Science* 19, 27–42.

Edmeades, D.C. and O'Connor, M.B. (2003) Sodium requirements for temperate pastures in New Zealand. *New Zealand Journal of Agricultural Research* 46, 37–47.

Edwards, H.M. (1984) Studies on the aetiology of tibial dyschondroplasia in chickens. *Journal of Nutrition* 114, 1001–1013.

Elbrond, V.S., Danzer, V., Mayhew, T.M. and Skadhouge, E. (1991) Avian lower intestine adapts to dietary salt (NaCl) depletion by increasing transepithelial sodium transport and microvillus membrane surface area. *Experimental Physiology* 76, 733–744.

Fettman, M.J., Chase, L.E., Bentinck-Smith, C., Coppock, E. and Zinn, S.A. (1984) Nutritional chloride deficiency in early lactation Holstein cows. *Journal of Dairy Science* 67, 2321–2335.

Frame, D.D., Hooge, D.M. and Cutler, R. (2001) Interactive effects of dietary sodium and chloride on the incidence of spontaneous cardiomyopathy (round heart) in turkeys. *Poultry Science* 80, 1572–1577.

Gill, M., France, J., Summers, M., McBride, B.W. and Milligan, L.P. (1989) Simulation of the energy costs associated with protein turnover and Na^+K^+-transport in growing lambs. *Journal of Nutrition* 119, 1287–1299.

Goatcher, W.D. and Church, D.C. (1970) Taste responses in ruminants. III. Reactions of pigmy goats, normal goats, sheep and cattle to sucrose and sodium chloride. *Journal of Animal Science* 31, 364–372.

Goff, J.P., Brummer, E.C., Henning, S.J., Doorenbos, R.K. and Horst, R.L. (2007) Effect of application of ammonium chloride and calcium chloride on alfalfa cation–anion content and yield. *Journal of Dairy Science* 90, 5159–5164.

Gorham, J. (1995) Sodium content of agricultural crops. In: Phillips, C.J.C. and Chiy, P.C. (eds) *Sodium in Agriculture*. Chalcombe Publications, Canterbury, UK, pp. 17–32.

Grace, N.D., Pearce, S. G., Firth, E.C. and Fennessy, P.F. (1999) Content and distribution of macro- and trace-elements in the body of young pasture-fed horses. *Australian Veterinary Journal* 77, 172–176.

Greene, L.W., Fontenot, J.P. and Webb, K.E. Jr (1983a) Site of magnesium and other macro-mineral absorption in steers fed high levels of potassium. *Journal of Animal Science* 57, 503–513.

Greene, L.W., Webb, K.E. and Fontenot, J.P. (1983b) Effect of potassium level on site of absorption of magnesium and other macroelements in sheep. *Journal of Animal Science* 56, 1214–1221.

Grovum, W.L. and Chapman, H.W. (1988) Factors affecting the voluntary intake of food by sheep. 4. The effect of additives representing the primary tastes on sham intakes by oesophageally fistulated sheep. *British Journal of Nutrition* 59, 63–72.

Hagsten, I. and Perry, T.W. (1976) Evaluation of dietary salt levels for swine. 1. Effect on gain, water consumption and efficiency of feed conversion. *Journal of Animal Science* 42, 1187–1190.

Hagsten, I., Perry, T.W. and Outhouse, J.B. (1975) Salt requirements of lambs. *Journal of Animal Science* 42, 1187–1190.

Harper, M.-E., Willis, J.S. and Patrick, J. (1997) Sodium and chloride in nutrition. In: O'Dell, B.L. and Sunde, R.A. (eds) *Handbook of Nutritionally Essential Mineral Elements*. Marcel Dekker Inc., New York, pp. 93–116.

Harris, J., Caple, I.W. and Moate, P.J. (1986) Relationships between mineral homeostasis and fertility of dairy cows grazing improved pastures. In: *Proceedings of the Sixth International Conference on Production Disease in Farm Animals, Belfast*. Veterinary Research Laboratory, Stormont, UK, pp. 315–318.

Heller, V.G. (1932) Saline and alkaline drinking waters. *Journal of Nutrition* 5, 421–429.

Hemsley, J.A., Hogan, J.P. and Weston, R.H. (1975) Effect of high intake of sodium chloride on the utilisation of a protein concentrate by sheep. *Australian Journal of Agricultural Research* 26, 715–727.

Henry, P.R. (1995) Sodium and chlorine bioavailability. In: Ammerman, C.B., Baker, D.H. and Lewis, A.J. (eds) *Bioavailability of Nutrients for Animals*. Academic Press, New York, pp. 337–348.

Hogan, J.P., Petherick, J.C. and Phillips, J.C. (2007) The physiological and metabolic impacts on sheep and cattle of feed and water deprivation before and during transport. *Nutrition Research Reviews* 20, 17–28.

Honeyfield, D.C., Frosseth, J.A. and Barke, R.J. (1985) Dietary sodium and chloride levels for growing-finishing pigs. *Journal of Animal Science* 60, 691–698.

Howard, D.A., Burdin, M.L. and Lampkin G.H. (1962) Variation in the mineral and crude-protein content of pastures at Muguga in the Kenya Highlands. *Journal of Agricultural Science, Cambridge* 59, 251–256.

Huber, K., Hempel, R. and Rodehutscord, M. (2006) Adaptation of epithelial sodium-dependent phosphate transport in jejunum and kidney of hens to variations in dietary phosphorus intake. *Poultry Science* 85, 1980–1986.

Humphery, T.J., Coghlan, J.P., Denton, D.A., Fan, J.S., Scoggins, B.B., Stewart, K.W. and Whitworth, J.A. (1984) Effect of potassium, angiotensin II on blood aldosterone and cortisol in sheep on different dietary potassium and sodium intakes. *Clinical and Experimental Pharmacology and Physiology* 11, 97–100.

Jones, D.I.H., Miles, D.G. and Sinclair, K.B. (1967) Some effects of feeding sheep on low-sodium hay with or without a sodium supplement. *British Journal of Nutrition* 21, 391–397.

Joyce, J.P. and Brunswick, I.C.F. (1975) Sodium supplementation of sheep and cattle. *New Zealand Journal of Experimental Agriculture* 3, 299–304.

Kemp, A. (1971) *The Effects of K and N Dressings on the Mineral Supply of Grazing Animals*. Proceedings of the 1st Colloquium of the Potassium Institute. IBS, Wageningen, The Netherlands, pp. 1–14.

Khorasani, G.R., Janzen, R.A., McGill, W.B. and Kenelly, J.J. (1997) Site and extent of mineral absorption in lactating cows fed whole-crop cereal grain silage or alfalfa silage. *Journal of Animal Science* 75, 239–248.

Kirk, D.J., Fontenot, J.P. and Rahnema, S. (1994) Effects of lasalocid and monensin on digestive tract flow and partial absorption on minerals in sheep. *Journal of Animal Science* 72, 1029–1037.

Leach, R.M. Jr (1974) Studies on the potassium requirement of the laying hen. *Journal of Nutrition* 104, 684–686.
Leach, R.M. and Nesheim, M.C. (1963) Studies on chloride deficiency in chicks. *Journal of Nutrition* 81, 193–199.
Leaver. J.D. and Hill, J. (1995) The performance of dairy cows offered ensiled whole-crop wheat, urea-treated whole-crop wheat or sodium hydroxide-treated wheat grain and wheat straw in a mixture with silage. *Animal Science* 61, 481–489.
Ledgard, S.F. and Upsdell, M.P. (1991) Sulfur inputs from rainfall throughout New Zealand. *New Zealand Journal of Agricultural Research* 34, 105–111.
Lehr, J.J., Grashuis, J. and Van Koetsveld, E.E. (1963) Effect of fertilization on mineral-element balance in grassland. *Netherlands Journal of Agricultural Science* 11, 23–37.
Lengemann, F.W., Aines, P.D. and Smith, S.E. (1952) The normal chloride concentration of blood plasma, milk and urine of dairy cattle. *Cornell Veterinarian* 42, 28–35.
Little, D.A. (1987) The influence of sodium supplementation on the voluntary intake and digestibility of low-sodium *Setaria sphacelatae* cv Nandi by cattle. *Journal of Agricultural Science, Cambridge* 108, 231–236.
MAFF (1990) *UK Tables of the Nutritive Value and Chemical Composition of Foodstuffs*. In: Givens, D.I. (ed.) Rowett Research Services, Aberdeen, UK.
Mahan, D.C., Newton, E.A. and Cera, K.R. (1996) Effect of supplemental sodium chloride, sodium phosphate or hydrochloric acid in starter pig diets containing dried whey. *Journal of Animal Science* 74, 1217–1222.
Mahan, D.C., Wiseman, T.D., Weaver, E. and Russell, L. (1999) Effect of supplemental sodium chloride and hydrochloric acid added to initial starter diets containing spray-dried blood plasma and lactose on resulting performance and nitrogen digestibility of 3-week-old weaned pigs. *Journal of Animal Science* 77, 3016–3021.
Marai, I.F.M., Habeeb, A.A. and Kamal, T.H. (1995) Response of livestock to excess sodium intake. In: Phillips, C.J.C. and Chiy, P.C. (eds) *Sodium in Agriculture*. Chalcombe Publications, Canterbury, UK, pp. 173–180.
Martens, H. and Blume, I. (1987) Studies on the absorption of sodium and chloride from the rumen of sheep. *Comparative Biochemistry and Physiology A – Comparative Physiology* 86, 653–656.
Masters, D.G., Benes, S.E. and Norman, C. (2006) Biosaline agriculture. *Agriculture, Ecosystems and Environment* 119, 234–248.
McClymont, G.L., Wynne, K.N., Briggs, P.K. and Franklin, M.C. (1957) Sodium chloride supplementation of high-grain diets for fattening Merino sheep. *Australian Journal of Agricultural Research* 8, 83–90.
McDowell, L.R. (2003) *Minerals in Animal and Human Nutrition*. Elsevier, Amsterdam, p. 688.
McDowell, L.R., Conrad, J.H. and Hembry, F.G. (1993) *Minerals for Grazing Ruminants in Tropical Regions*, 2nd edn. Animal Science Department, University of Florida, Gainsville, Florida.
McNaught, K.J. (1959) Effect of potassium fertiliser on sodium, calcium and magnesium in plant tissues. *New Zealand Journal of Agriculture* 99, 442–448.
McSweeney, C.S., Cross, R.B., Wholohan, B.T. and Murphy, M.R. (1988) Diagnosis of sodium status in small ruminants. *Australian Journal of Agricultural Research* 39, 935–942.
Meyer, J.H., Grummer, R.H., Phillips, R.H. and Bohstedt, G. (1950) Sodium, chlorine, and potassium requirements of growing pigs. *Journal of Animal Science* 9, 300–306.
Michell, A.R. (1989) Practice tip: oral and parenteral rehydration therapy. *In Practice* May, 96–99.
Michell, A.R. (1995) Physiological roles for sodium in mammals. In: Phillips, C.J.C. and Chiy, P.C. (eds) *Sodium in Agriculture*. Chalcombe Publications, Canterbury, UK, pp. 91–106.
Michell, A.R., Moss, P., Hill, R., Vincent, I.C. and Noakes, D.E. (1988) The effect of pregnancy and sodium intake on water and electrolyte balance in sheep. *British Veterinary Journal* 144, 147–157.
Milligan, L.P. and Summers, M. (1986) The biological basis of maintenance and its relevance to assessing responses to nutrients. *Proceedings of the Nutrition Society* 45, 185–193.
Minson, D. J. (1990) *Forages in Ruminant Nutrition*. Academic Press, San Diego, California, pp. 208–229.
Mongin, P. (1981) Recent advances in dietary anion–cation balance applications in poultry. *Proceedings of the Nutrition Society* 40, 285–294.
Morris, J.G. (1980) Assessment of sodium requirements of grazing cattle: a review. *Journal of Animal Science* 50, 145–151.

Morris, J.G. and Gartner, R.J.W. (1971) The sodium requirements of growing steers given an all-sorghum grain ration. *British Journal of Nutrition* 25, 191–205.

Morris, J.G. and Murphy, G.W. (1972) The sodium requirements of beef calves for growth. *Journal of Agricultural Science, Cambridge* 78, 105–108.

Morris, J.G. and Peterson, R.G. (1975) Sodium requirements of lactating ewes. *Journal of Nutrition* 105, 595–598.

Morris, J.G., Delmas, R.E. and Hull, J.L. (1980) Salt supplementation of range beef cows in California. *Journal of Animal Science* 51, 71–73.

Moseley, G. (1980) Effects of variation in herbage sodium levels and salt supplementation on the nutritive value of perennial ryegrass for sheep. *Grass and Forage Science* 35, 105–113.

Murphy, G.M. and Connell, J.A. (1970) A simple method of collecting saliva to determine the sodium status of cattle and sheep. *Australian Veterinary Journal* 46, 595–598.

Murphy, G.M. and Plasto, A.W. (1972) Sodium deficiency in a beef cattle herd. *Australian Veterinary Journal* 48, 129.

Neathery, M.W., Blackmon, D.M., Miller, W.J., Heinmiller, S., McGuire, S., Tarabula, J.M., Gentry, R.F. and Allen, J.C. (1981) Chloride deficiency in Holstein calves from a low chloride diet and removal of abomasal contents. *Journal of Dairy Science* 64, 2220–2233.

NRC (1994) *Nutrient Requirements of Poultry*, 9th edn. National Academy of Sciences, Washington, DC.

NRC (1998) *Nutrient Requirements of Swine*, 10th edn. National Academy of Sciences, Washington, DC.

NRC (2001) *Nutrient Requirements of Dairy Cows*, 5th edn. National Academy of Sciences, Washington, DC.

Parker, A.J., Hamlin, G.P., Coleman, C.J. and Fitzpatrick, L.A. (2004) Excess cortisol interferes with a principal mechanism of resistance to dehydration in *Bos indicus* steers. *Journal of Animal Science* 82, 1037–1045.

Pasternak, D., Danon, A., Arouson, J.A. and Benjamin, R.W. (1985) Developing the seawater agriculture concept. *Plant and Soil* 89, 337–348.

Peirce, A.W. (1957) Saline content of drinking water for livestock. *Veterinary Reviews and Annotations* 3, 37–43.

Peirce, A.W. (1965) Studies on salt tolerance of sheep. 5. The tolerance of sheep for mixtures of sodium chloride, sodium carbonate and sodium bicarbonate in the drinking water. *Australian Journal of Agricultural Research* 49, 815–823.

Ross, E. (1979) The effect of water sodium on the chick requirement for dietary sodium. *Poultry Science* 58, 626–630.

Ross, E. and Herrick, R.B. (1981) Forced rest induced by moult or low-salt diet and subsequent hen performance. *Poultry Science* 60, 63–67.

Rouffet, J.D., Clement, T.J. and Aranas, T.J. (1983) Changes of aldosterone in blood serum of dairy cattle during estrous cycle. *Journal of Dairy Science* 66, 1734–1737.

Rouffet, J.D., Dalle, M., Tournaire, C., Barlet, J.-P. and Delost, P. (1990) Sodium intake by pregnant cows and plasma aldosterone and cortisol concentrations in the foetus during late pregnancy. *Journal of Dairy Science* 73, 1762–1765.

Sauveur, B. and Mongin, P. (1978) Interrelationships between dietary concentrations of sodium, potassium and chloride in laying hens. *British Poultry Science* 19, 475–485.

Safwate, A. (1985) Urinary sodium excretion and the renin–aldosterone system in the new-born calf. *Journal of Physiology* 362, 261–271.

Schellner, G. (1972/1973) Die Wirkung von Natriummangel und Natriumbeifutterung auf Wachstum, Milch- und Milchfettleistung und Fruchtbarkeit bei Ziegen. *Jahrbuch fur Tierernahrung und Futterung* 8, 246–259.

Scott, D. (1974) Changes in water, mineral and acid–base balance associated with feeding and diet. In McDonald, I.W. and Warner, A.C. (eds) *Digestion and Metabolism in the Ruminant*. Proceedings of the VIth International Symposium on Ruminant Physiology. University of New England Publishing Unit, Armidale, Australia, pp. 205–215.

Seynaeve, M., de Widle, R., Janssens, G. and de Smet, B. (1996) The influence of dietary salt level on water consumption, farrowing and reproductive performance of lactating sows. *Journal of Animal Science* 74, 1047–1055.

Shade, R.E., Blair-West, J.R., Carey, K.D., Madden, L.J., Weisunger, R.S. and Denton, D.A. (2002) Synergy between angiotensin and aldosterone in evoking sodium appetite in baboons. *American Journal of Physiology – Regulatory Integrative and Comparative Physiology* 283, R1070–R1078.

Sherwood, D.H. and Marion, J.E. (1975) Salt levels in feed and water for laying chickens. *Poultry Science* 54, 1816 (abstract).

Shirley, R.L. (1978) Water as a source of minerals. In: Conrad, J.H. and McDowell, L.R. (eds) *Latin American Symposium on Mineral Nutrition Research with Grazing Ruminants*. Animal Science Department, University of Florida, Gainsville, Florida, pp. 40–47.

Shirley, R.B., Davis, A.J., Compton, M.M. and Berry, W.D. (2003) The expression of calbindin in chicks that are divergently selected for low or high incidence of tibial dyschondroplasia. *Poultry Science* 82, 1965–1973.

Silanikove, N., Malz, E., Halevi, A. and Shinder, D. (1997) Metabolism of water, sodium, potassium and chloride by high yielding dairy cows at the onset of lactation. *Journal of Dairy Science* 80, 949–956.

Singh, S.P. and Rani, D. (1999) Assessment of sodium status in large ruminants by measuring the sodium-to-potassium ratio in muzzle secretions. *American Journal of Veterinary Research* 60, 1074–1081.

Smith, S.E., Lengemann, F.W. and Reid, J.T. (1953) Block versus loose salt consumption by dairy cattle. *Journal of Dairy Science* 36, 762–765.

Spears, J.W. and Harvey, R.W. (1987) Lasalocid and dietary sodium and potassium effects on rumen mineral metabolism, ruminal volatile fatty acids and performance of finishing steers. *Journal of Animal Science* 65, 830–840.

Summers, J.D., Moran, E.T. and Pepper, W.F. (1967) A chloride deficiency in a practical diet encountered as a result of using common sulphate antibiotic potentiating procedure. *Poultry Science* 46, 1557–1560.

Suttle, N.F. and Brebner, J. (1995) A putative role for larval nematode infection in diarrhoeas which did not respond to anthelmintic drenches. *Veterinary Record* 137, 311–316.

Suttle, N.F. and Field, A.C. (1967) Studies on magnesium in ruminant nutrition. 8. Effect of increased intakes of potassium and water on the metabolism of magnesium, phosphorus, sodium, potassium and calcium in sheep. *British Journal of Nutrition* 21, 819–826.

Suttle, N.F., Brebner, J., McClean, K. and Hoeggel, U. (1996) Failure of mineral supplementation to avert apparent sodium deficiency in lambs with abomasal parasitism. *Animal Science* 63, 103–109.

Teeter, R.G. and Smith, M.O. (1986) High chronic ambient temperature stress effects on acid–base imbalance and their response to supplemental ammonium chloride, potassium chloride and potassium carbonate. *Poultry Science* 65, 1777–1781.

Thomas, B., Thompson, A., Oyenuga, V.A. and Armstrong, R.H. (1952) The ash constituents of some herbage plants at different stages of maturity. *Empire Journal of Experimental Agriculture* 20, 10–22.

Towers, N.R. and Smith, G.S. (1983) Sodium (Na). In: Grace, N.D. (ed) *The Mineral Requirements of Grazing Ruminants*. Occasional Publication No. 9, New Zealand Society of Animal Production, Palmerston North, New Zealand, pp. 115–124.

Underwood, E.J. and Suttle, F. (1999) *The Mineral Nutrition of Livestock*, 3rd edn. CAB International, Wallingford, UK, pp. 200–202.

Walker, C.A. (1957) Studies of the cattle of Northern Rhodesia. 1. The growth of steers under normal veld grazing and supplemented with salt and protein. *Journal of Agricultural Science, Cambridge* 49, 394–400.

Wamberg, S., Engel, D. and Stigsen, P. (1985) Acid–base balance in ruminating calves given sodium-hydroxide treated straw. *British Journal of Nutrition* 54, 655–657.

West, J.W., Mullinix, B.G. and Sandifer, T.G. (1991) Changing dietary electrolyte balance for dairy cows in cool and hot environments. *Journal of Dairy Science* 74, 1662–1674.

Whitlock, R.H., Kessler, M.J. and Tasker, J.B. (1975) Salt (sodium) deficiency in dairy cattle: polyuria and polydypsia as prominent clinical features. *Cornell Veterinarian* 65, 512–526.

Wilson, A.D. (1966) The tolerance of sheep to sodium chloride in food or drinking water. *Australian Journal of Agricultural Research* 17, 503–514.

9 Sulfur

Introduction

Recognition that sulfur is an important dietary component stems from the discovery in the 1930s that the essential amino acid, methionine, contains one atom of sulfur per molecule (McCollum, 1957). Subsequent investigations showed that sulfur is an integral part of the majority of tissue proteins, comprising 0.5–2.0% by weight. Other sulfur amino acids such as cystine, cysteine, cystathionine and taurine contribute to protein sulfur, but each can be derived from methionine and they are not essential constituents of the diet. Strictly speaking, sulfur is only an essential nutrient for plants and microbes because only they can synthesize sulfur amino acids and hence proteins from degradable inorganic sulfur sources. Mammals digest plant proteins and recombine the sulfur amino acids to form their own unique tissue proteins. Non-ruminants require particular balances amongst their dietary amino acids, and methionine is often the second most-limiting amino acid behind lysine. Since manipulation of dietary sulfur can make only a slender contribution to sulfur amino acid supply in the non-ruminant – either by sparing cystine (Lovett et al., 1986) or via microbial synthesis of sulfur amino acids in the large intestine – those interested in this aspect of monogastric nutrition should consult texts on amino acid nutrition (D'Mello, 2003) and availability (Henry and Ammerman, 1995). Ruminants are quite different: they possess a substantial microbial population in the forestomach or rumen that incorporates degradable inorganic sulfur sources into microbial proteins (Kandylis, 1984a). Following digestion in the small intestine, microbial protein provides a balanced amino acid supply for all purposes except the growth of wool, hair and mohair. Furthermore, since sulfur is much cheaper to provide than dietary protein, efficient production requires the pragmatic if not essential use of dietary sulfur in ruminant nutrition.

Functions

The functions of sulfur are as diverse as the proteins of which it is a part. Sulfur is frequently present as highly reactive sulfhydryl (SH) groups or disulfide bonds, maintaining the spatial configuration of elaborate polypeptide chains and providing the site of attachment for prosthetic groups and the binding to substrates that are essential to the activity of many enzymes. Hormones such as insulin and oxytocin contain sulfur, as do the vitamins thiamine and biotin. The conversion between homocysteine and methionine facilitates many reactions involving methylation. Cysteine-rich molecules such as metallothionein play a vital role in protecting animals from excesses of copper, cadmium and zinc, while others influence selenium transport and protect tissues from selenium toxicity. Glutathione may facilitate the uptake of copper by the liver (see Chapter 11). Glutathione also protects tissues from oxidants by interconverting between the reduced (GSH) and oxidized (GSSG) state, and may thus protect erythrocytes from lead toxicity. Sulfur is present as sulfate (SO_4) in the chondroitin SO_4 of connective tissue, in the natural anticoagulant heparin and is particularly abundant in the keratin-rich appendages (hoof, horn, hair, feathers, wool fibre and mohair). However, the primary rate-limiting function in sulfur-deprived ruminants is related not to the functions or associations listed above, but to rumen fermentation and microbial protein synthesis (MPS) (Durand and Komisarczuk, 1988).

Sulfur Sources for Ruminants

Sulfur in forages

The sulfur concentrations found in pasture and conserved forages range widely from 0.5 to >5.0 g kg^{-1} dry matter (DM) (Table 9.1), depending mainly on the plant species, availability of soil sulfur, nitrogen and phosphorus, and on maturity of the sward. Species vary in their sulfur requirement: legumes show optimal growth with around 2 g kg DM^{-1} in their tissues, while maize tolerates 1.4 g kg DM^{-1} and sugarcane <1 g kg DM^{-1}. Any area away from the coast and industrial activity, where the annual rainfall is moderate (>500 mm) and the soils derive from weathered volcanic material, is likely to have leached soils, low in sulfur and swards that respond to supplementation. In the UK, the range of sulfur concentrations in fresh grass is relatively narrow (1.2–4.0 g kg DM^{-1}) and the mean (2.2 g kg DM^{-1}) (MAFF, 1990) is adequate for plant growth. In many other countries, sulfur deficiency limits forage production. Areas of deficiency have been delineated in Australia and in the south-east, north-east, mid-west, west and north of the USA (Beaton et al., 1971; Underwood and Suttle, 1999). Many regions of China are also deficient in sulfur (Qi et al., 1994), and a need for regular inputs to maintain the productivity of clover-based pastures in Australasia (Goh and Nguyen, 1997) and maize in Nigeria (Ojeniyi and Kayode, 1993) has been confirmed.

There is little published information on seasonal variations in sulfur concentrations, but sulfur and protein are highly correlated and protein concentrations decline at a rate of 1–3 g kg^{-1} DM day^{-1}, depending on species and growth conditions (Minson, 1990) with the approach of maturity; this is equivalent to a decline in sulfur of 0.1–0.3 g kg^{-1} DM $week^{-1}$. Sulfur concentrations in a mixed sward can rise with the approach of summer due to a rise in the contribution from legumes such as white clover (see Chapter 2). In the Mt Elgon region of West Kenya, 25% of dry-season forages contain <1 g S kg^{-1} DM (Jumba et al., 1996). Sulfur deficiency is rare in arid regions (<250 mm rainfall per annum) because of the upward movement of SO_4 through the soil profile. Application of SO_4 fertilizer increases herbage sulfur even when baseline levels are high, mostly by increasing the SO_4 (i.e. non-protein) component (Spears et al., 1985). Pastures in coastal or industrial areas are enriched with sulfur from sea spray (3–5 kg ha^{-1}) (Ledgard and Upsdel, 1991) or emissions (approximately 8 kg ha^{-1} $year^{-1}$) (McClaren, 1975) and are unlikely to be deficient.

Sulfur in other foodstuffs

Cereal grains tend to be low in sulfur (approximately 1 g kg^{-1} DM) (Todd, 1972), cereal straws slightly richer (1.4–4.0 g kg^{-1} DM) (Suttle, 1991), protein supplements rich (2.2–4.9 g kg^{-1} DM) (Qi et al., 1994) and brassica crops too rich (4.8–9.0 g kg^{-1} DM) (Cornforth et al., 1978) in sulfur. Leafy brassicas such as kale and the leaves of root crops such as swedes and turnips contain the most sulfur. Among by-products, molasses contains 3–7 g S kg^{-1} DM, while animal by-products are moderate sources.

Sulfur in drinking water

Surface and ground water supplies can contain SO_4-sulfur at as much as 1.5 g l^{-1}. In regions

Table 9.1. Reported values for sulfur and total sulfur to nitrogen ratios in some common feedstuffs for ruminants.

	Replication	S (g kg^{-1} DM)	S:N[a]	Reference
Grasses	137	2.2	0.088	MAFF (1990)
Dried grass	8	3.5	0.12	MAFF (1990)
Stylo hay	2	1.25	0.071	Bird (1974)
Alfalfa hay	2	3.6	0.095	Qi et al. (1994)
Spear-grass hay	2	0.5, 0.8	0.055, 0.205	Kennedy and Siebert (1972a,b)
Corn silage	1	–	–	Buttrey et al. (1986)
Sorghum silage	1	1.0	0.067	Ahmad et al. (1995)
Barley straw	4	2.0	0.33	MAFF (1990)
Wheat straw	2	1.0	0.32	Kennedy (1974)
Barley	10	1.5	0.073	MAFF (1990)
Maize	5	1.6	0.10	MAFF (1990)
Distillers' dried grain	5	3.7	0.084	MAFF (1990)
Swedes	–	4.8	0.179	Cornforth et al. (1978)
Turnips	–	6.1	0.233	Cornforth et al. (1978)
Kales	–	9.0	0.313	Cornforth et al. (1978)
Feather meal	5	18.1	0.127	MAFF (1990)
Rapeseed meal	5	16.9	0.26	MAFF (1990)
Linseed meal	5	4.1	0.065	MAFF (1990)
Cottonseed meal	5	5.0	0.083	MAFF (1990)
Soybean meal	5	4.6	0.058	MAFF (1990)
Safflower meal	1	2.2	0.029	Qi et al. (1994)
Fish meal	1	4.9	0.046	Qi et al. (1994)
Molasses (cane)	1	7.3	0.811	Bogdanovic (1983)

[a]This ratio only improves the assessment of nutritive value for ruminants if the S and N have similar rates and extents of degradability in the rumen.

such as the state of Nevada, 23% of samples in one early survey contained >250 mg SO$_4$ l^{-1} (Miller et al., 1953; cited by Weeth and Hunter, 1971). In Saskatchewan, Canada, water from deep aquifers contained 500 mg S l^{-1} as SO$_4$ and was sufficient to induce hypocuprosis (see also Chapter 11) (Smart et al., 1986). Drinking water can therefore make a significant contribution to sulfur intake, but not necessarily enhance the value of the ruminant diet.

Nutritive Value of Sulfur in Feeds

The value of feeds as sources of sulfur for ruminants depends entirely on their availability for MPS. This in turn depends on the co-availability of other factors needed for MPS and therefore presents unique problems. Nutritive value is regarded by some to be more closely related to ratios of sulfur to nitrogen than to sulfur concentration per se, and sulfur and nitrogen concentrations are well correlated within feeds. However, there are wide differences in sulfur to nitrogen ratios between feeds (Table 9.1), with ratios being sixfold higher in brassicas at the high extreme (0.2–0.3) than in cereals and vegetable protein sources at the low extreme (0.03–0.05), and the range bears no relation to nutritive value. While plant sulfur is comprised mostly of protein sulfur, approximately a third is in the inorganic form and some in fibrous foods may be present in acid detergent-insoluble forms, unavailable to rumen microbes. The sulfur incorporated into microbial nucleic acids may not be available to the animal. Plant tannins have complex effects on the nutritive value of plant methionine and cysteine. With tannin-rich species such as *Lotus corniculatus*, plant sulfur amino acids are protected from degradation by complexation in the rumen, but also from the proteolysis necessary for absorption from the small intestine; the net effect is an increase the amounts methionine and cysteine absorbed (Wang et al., 1996).

Metabolism

Influence of rumen microbiota

All rumen microorganisms require sulfur, but different species procure sulfur by different pathways. Some degrade inorganic sources to sulfide and incorporate it into sulfur amino acids (dissimilatory) while others utilize only organic sulfur (assimilatory) (Kandylis, 1984a; Henry and Ammerman, 1995). Assimilatory microbes require rumen-degradable protein (RDP) sources of sulfur. Most of the degradable sulfur entering the rumen, whether in organic or inorganic form, enters the sulfide pool (Gawthorne and Nader, 1976); approximately 50% leaves it as absorbed sulfide when SO_4 is the major dietary source (Fig. 9.1) (Kandylis and Bray 1987), thus having little or no nutritional value but a potential for toxicity. The proportion of total sulfur-flow rumen that is 'captured' as rumen microbial protein varies widely and is determined by factors such as the sulfur source (organic versus inorganic; methionine versus other sulfur amino acids) (Bird, 1972a), dietary sulfur concentration and the co-availability of other substrates (chiefly degradable nitrogen). Approximately 50% of bacterial organic sulfur is derived from sulfide in sheep given grass or lucerne hay (Kennedy and Milligan, 1978). The dietary supply is supplemented slightly by sulfur secreted in saliva in both inorganic and organic forms in sheep, and more liberally in cattle (Kennedy and Siebert, 1972b; Bird, 1974), but the capacity for recycling is limited (Kandylis and Bray, 1987).

Optimal MPS and sulfur capture occur when fermentable energy, degradable sulfur, nitrogen and phosphorus are supplied at rates that match the capacity for MPS (i.e. by 'rumen synchrony') (Moir, 1970; Beever, 1996). Microbial protein is generally assumed to contain sulfur to nitrogen at a ratio of 0.067 (ARC, 1980), but lower ratios have been reported (0.054 and 0.046 for mixed rumen bacteria and protozoa, respectively) (Harrison and McAllan, 1980) and they can be significantly increased (from 0.075 to 0.11) by high SO_4 intakes (Kandylis and Bray, 1987). The rumen protozoa feed on rumen bacteria, returning sulfur to the sulfide pool as they do so (Kandylis, 1984a) and the efficiency of MPS can be increased by defaunating the rumen (i.e. removing the protozoa) (Bird and Leng, 1985), a process that should lower rumen sulfide concentrations (Hegarty et al.,

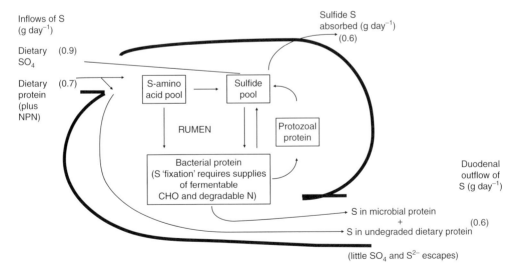

Fig. 9.1. Primary flows of sulfur in the rumen: metabolizable energy and nitrogen supplies as well as sulfur determine the most important product, microbial protein, entering the duodenum (amounts of S, g day^{-1} are from the 'low' S treatment of Kandylis and Bray, 1987).

1994). The considerable involvement of rumen sulfide in copper metabolism is discussed in Chapter 11. Anaerobic fungi may play a significant role in the structural degradation of fibre in the rumen: fungal activity is dependent on the dietary sulfur supply and contributes significantly to the synthesis of sulfur amino acids (Weston et al., 1988).

Post-rumenal metabolism

Sulfur in the digesta leaves the rumen in several forms, but principally as undegraded dietary, microbial and fungal protein (Fig. 9.1). The proportions are determined by dietary composition and the efficiency of microbial capture (Bray and Till, 1975). From that point, sulfur metabolism in the ruminant is similar to that of the monogastric animal. Sulfur amino acids have first to be liberated by proteolytic digestion, but 80–90% of sulfur amino acids entering the small intestine are normally absorbed by sheep (Wang et al., 1996). SO_4 is absorbed by active transport. Further degradation of SO_4^{2-} to S^{2-} and incorporation into bacterial protein can occur in the hind gut, but probably makes little contribution to sulfur utilization (Kandylis, 1984a). Methionine is demethylated in the tissues to provide methyl groups (CH_3) for the elaboration of carbon chains. This leaves homocysteine, from which the non-essential sulfur amino acids cystathionine and cysteine are formed by transsulfuration in the liver, viscera and skin (Liu and Masters, 2000). Tissue protein synthesis can now proceed and cysteine accumulates at sites such as muscle and the mammary gland in proteins such as myosin and casein, which contain sulfur to nitrogen in a ratio of about 0.067. Wool, hair and mohair are much richer in sulfur, containing 2.7–5.4% sulfur, mostly as cysteine, and with a sulfur to nitrogen ratio of 0.2 (Langlands and Sutherland, 1973).

Excretion

Sulfur is excreted principally via the urine as SO_4 derived from the oxidation of sulfide and sulfur-containing amino acids and catabolism of other organic molecules in the tissues. SO_4 is filtered at the glomerulus at rates that exceed the tubular reabsorption rate, and is therefore rapidly excreted in urine. SO_4 competes with molybdate for renal tubular reabsorption (Bishara and Bray, 1978); the macronutrient (sulfur) can greatly impede the micronutrient (molybdenum), but molybdate can only slightly influence the urinary excretion of SO_4. Dick (1956) demonstrated a remarkable enhancement of urinary molybdenum excretion and decreased retention in molybdenum-loaded sheep when SO_4 was subsequently added to the diet. When sulfur and molybdenum are simultaneously added to the diet, sulfur can *decrease* urinary molybdenum excretion by decreasing molybdate absorption and increasing retention (Grace and Suttle, 1979) (see also Chapter 17). The sulfur excreted via faeces is largely in undegraded organic forms, with the amount increasing as MPS increases (Kandylis and Bray 1987).

Homeostasis

Homeostatic regulation of SO_4 metabolism is of little significance in the monogastric animal because conserved SO_4 cannot be used to synthesize sulfur amino acids. In the ruminant, conservation might theoretically be effective because sulfur secreted in saliva can be incorporated into microbial protein. However, salivary sulfur secretion decreases as sulfur intakes and plasma SO_4 concentrations decline, and there is no evidence that the capacity of the salivary glands to extract and secrete sulfur increases during sulfur deficiency. During sulfur amino acid or protein deficiency, the catabolism of tissue protein will allow redistribution of sulfur amino acids from the more expendable protein pools (e.g. bone matrix in the pregnant or lactating female) to less expendable sites and processes (e.g. fetus and mammary gland). Wool growth is more sensitive to a reduced protein supply than fetal growth, maternal weight or milk production (Masters et al., 1996) and reductions in wool growth during sulfur deprivation might allow the synthesis of far more protein elsewhere because other proteins contain far fewer sulfur amino acids.

Sulfur Requirements of Ruminants

The definition of ruminant requirements for sulfur poses unique problems that need a unique solution. Like those of monogastric animals (D'Mello, 2003), net requirements are best stated as needs for protein and sulfur amino acids, but they have only recently been estimated in this way and then only in sheep (Fig. 9.2) (Liu and Masters, 2000). Unlike in non-ruminants, gross sulfur amino acid requirements cannot be derived by dividing net requirements by an absorption coefficient for sulfur amino acid absorption. The flow of sulfur amino acids from the rumen exceeds intake because of MPS (Fig. 9.1), thus reducing the dietary need for pre-formed sulfur amino acids. Since microbes themselves need sulfur for the pivotal role of MPS, their needs must come first. Traditional statements of requirements as dietary sulfur concentrations (McDowell, 2003) and novel estimates of net sulfur amino acid needs (Fig. 9.2) are unsatisfactory indices of gross dietary needs.

Needs for hair and wool production

Sheep and goats differ from cattle in that hair and wool growth adds significantly to their need for sulfur (Williams et al., 1972). Since hair and wool growth is inevitable and is related to body size and growth rate, net sulfur and sulfur amino acid needs increase with body size and growth rate (Fig. 9.2) and are considered part of the maintenance need (ARC, 1980; AFRC, 1993). In double-coated, cashmere-producing goats, there is an additional need during the cashmere-growing season in autumn (Souri et al., 1998). The closeness of sulfur to nitrogen ratios in major animal tissues and products means that the efficiencies of utilization of microbial protein sulfur for most types of production are similar, regardless of species. However, the efficiency of protein use for fleece growth is singularly low (only 0.26 compared to 0.68 for milk production and growth) (AFRC, 1993) because of the high sulfur amino acid content of wool protein. This amplifies the difference in sulfur requirements for maintenance between cattle and sheep.

A new system for estimating sulfur requirement

Sulfur requirements were first linked to MPS by harnessing estimates to the calculation of protein requirements (ARC, 1980). It was assumed that the sulfur and nitrogen naturally present feed proteins would be appropriately balanced for MPS, having a similar sulfur to nitrogen ratio of 0.067 whether degraded (RDP) or not. If additional non-protein nitrogen was required, sulfur should be supplied in same ratio. However, early studies with mature male

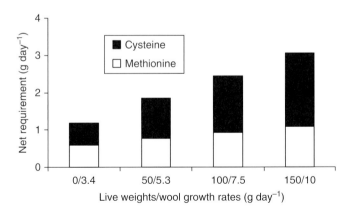

Fig. 9.2. The effects of live weight and wool growth rates on net methionine and cysteine requirements in a 40-kg sheep (from Liu and Masters, 2000).

Merino sheep indicated that when low-quality roughages were supplemented with urea and SO_4, higher total sulfur to nitrogen ratios were required (approximately 0.10) (Moir, 1970; Bird, 1974). Using a similar animal model, Kandylis and Bray (1987) showed that both the digestibility of the diet and MPS were improved by supplements that increased the dietary sulfur to nitrogen ratio from 0.11 to 0.18 (Fig. 9.3). A major factor contributing to this high sulfur to nitrogen requirement is the lower efficiency (50 versus 70%) with which sulfur was incorporated into MPS at both levels of supplementation, presumably due to the high proportion of ingested sulfur leaving the rumen as sulfide (Fig. 9.1). The 'metabolizable protein' system for estimating protein requirements (AFRC, 1993) divided dietary crude protein (nitrogen × 6.25) into three components – quickly degradable (forages), slowly degradable and non-rumen-degradable protein (QDP, SDP and UDP, respectively) – but made no provision for sulfur. That gap is filled by the estimates for different types of production shown in Table 9.2. These follow the ARC (1980) principle, following the needs for protein but use an S:N ratio of 0.1 rather than 0.067 to allow for inefficient capture of plant sulfur. The needs for sulfur are generally higher in sheep than in cattle at a given stage in development; higher in younger than older growing animals; and little affected by milk yield within species (because dry matter intake increases with yield). Further information on the degradability of feed and synthetic sources of sulfur is needed to make full use of the 'metabolizable protein system' (AFRC, 1993).

Application and tests of the new system

Where ration formulation models indicate scope for supplementation with quickly degradable (QD) protein, 'QD sulfur' (e.g. sodium sulfate, Na_2SO_4) should be added in a ratio of 0.1 to QD nitrogen to allow for the relatively inefficient use of SO_4. Feeding trials confirm that cattle need less sulfur than sheep to digest the same urea-supplemented, low-quality diet (Kennedy and Siebert, 1972a; Bird, 1974). Responses to sulfur supplementation obtained in sheep grazing pastures containing 0.95–1.18g S kg^{-1} DM (total S:N 0.08–0.09) support the new estimates (Mata et al., 1997). An abomasal infusion of methionine and casein increased wool growth by 28% in Merino wethers on a diet containing 1.1g S kg-1 DM (S:N 0.091), as oaten hay and oat grain, but a 23% increase in food allowance above maintenance did not (Revell et al., 1999), suggesting that MPS was being constrained by lack of sulfur. Studies with Angora goats, in which responses in production and metabolic parameters were used to define optimum sulfur intakes (Qi et al., 1992, 1993) indicated high optimum total sulfur to nitrogen ratios of 0.1–0.4 and it has been argued that other ruminants may also benefit from higher sulfur to nitrogen ratios than those recommended here (Qi et al., 1994). There is no doubt that protein and therefore sulfur requirements are high for rapid mohair growth. However, there is a limit to the extent to which the higher requirement can be met by increasing *degradable* sulfur to nitrogen ratios unless the sulfur amino acid concentration in microbial protein can be increased. In fact, the optimal supplements in the experiments of Qi et al. (1992, 1993) provided degradable sulfur to nitrogen ratios of between 0.047 and 0.088

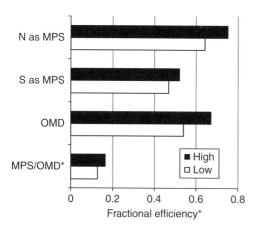

Fig. 9.3. The efficiency of organic matter digestion (OMD) and use of dietary sulfur and nitrogen for microbial protein synthesis (MPS) is improved by increasing the amounts (high versus low) of degradable S (as Na_2SO_4) and N (as urea) added to a poor-quality basal diet for sheep, but S is used less efficiently than N (data from Kandylis and Bray, 1987). *Results for MPS/OMD are g kg^{-1} organic matter digested.

Table 9.2. Theoretical minimum sulfur requirements for microbial protein synthesis (MPS) in ruminants given balanced diets adequate in metabolizable energy (ME) and metabolizable protein (MP)[a].

Ruminant type	Live weight (kg)	'Production' (kg day^{-1} or no.)	Required MPS (g day^{-1})	Required S[b] (g day^{-1})	DMI (kg day^{-1})	Dietary S (g kg^{-1} DM)
Lamb	20	0.1	61	0.97	0.6	1.6
		0.2	88	1.40	1.0	1.3
	40	0.1	72	1.15	1.1	1.0
		0.2	95	1.52	1.9	0.85
Ewe	40	1[c]	78	1.24	0.9	1.3
		2	90	1.43	1.1	1.3
	80	1	116	1.85	1.4	1.3
		2	137	2.19	1.8	1.2
	40	1	133	2.12	1.2	1.8
		2	209	3.34	1.9	1.8
	80	2	234	3.70	2.2	2.1
		3	309	4.94	2.9	1.6
Goat (Angora)	30	0.1	95	1.52	0.9	1.6
Steer	100	0.5	220	3.51	2.1	1.6
		1.0	348	5.55	3.2	1.8
	300	0.5	299	4.78	4.3	1.0
		1.0	409	6.54	6.1	1.0
	500	0.5	373	5.96	7.0	0.9
		1.0	477	7.60	10.1	0.8
Cow	600	1[c]	420	6.70	8.8	0.8
		10[d]	762	12.16	11.8	1.0
		20	1229	19.63	14.4	1.3
		40	2165	34.58	23.9	1.5

[a]Requirements of ME and MP for MPS are those given by AFRC (1993).
[b]The formula to obtain S requirements from the MPS required for a given level of production is (MPS / 6.25) × 0.1.
[c]The number of fetuses at week 20 of ovine pregnancy and week 36 of bovine pregnancy.
[d]Dry milk yield.

for mature goats and between 0.061 and 0.11 for growing kids. For all production purposes, sulfur requirements should be calculated on the basis of common needs for rumen MPS (Table 9.2), rather than on dietary S:N ratios.

Biochemical Abnormalities of Sulfur Deprivation

Rumen fluid changes in both biochemical (Whanger, 1972) and microbial (Weston et al., 1988; Morrison et al., 1990) composition, associated with a striking fall in rumen sulfide concentrations. Sulfur deprivation causes hypoalbuminaemia in the host, but blood urea will be high if the intake of degradable dietary nitrogen is high. SO_4 concentrations in plasma generally reflect sulfur intake if sulfur alone is limiting for MPS, but if another factor (such as nitrogen) is limiting, serum SO_4 values for a given sulfur intake are increased (Kennedy and Siebert, 1972b). Low plasma SO_4 levels of around 10 mg S l^{-1} are found on poor, dry-season pasture, but they rise to 20–40 mg l^{-1} with the onset of rain and the growth of green herbage (Kennedy and Siebert, 1972a; White et al., 1997). Concentrations of methionine in plasma and liver fall when calves are given a diet sufficiently low in sulfur (0.4 g kg^{-1} DM) to retard growth (Slyter et al., 1988). If dietary sulfur supply alone is limiting rumen microbial synthesis, the provision of additional sulfur should lower blood urea levels and urinary nitrogen excretion. Diets low in sulfur may enhance the tissue accumulation of cadmium by reducing rumen sulfide concentrations and limiting the formation of insoluble and unabsorbable cadmium sulfide (Smith and White, 1997).

Clinical Symptoms of Sulfur Deprivation

The symptoms of sulfur deprivation in ruminants are not specific, being shared by any nutritional deficiency or factor that depresses rumen MPS. On fibrous diets there is often an early depression of appetite and digestibility due to a failure to digest cellulose in sheep (Kennedy and Siebert, 1972a,b; Hegarty et al., 1994) and cattle (Kennedy, 1974). Sheep deprived of sulfur spend more time ruminating and have an increased rumen liquor volume (Weston et al., 1988). Eventually growth is retarded, wool or hair is shed, there is profuse salivation and lacrimation and the eyes become cloudy; eventually, the emaciated animal dies (Thomas et al., 1951; Kincaid, 1988).

Occurrence of Sulfur Deprivation

Plant factors

If crop husbandry is ideal, sulfur deficiency will not occur in livestock grazing legume-rich swards because rumen microbes need less sulfur than the plants for protein synthesis. However, livestock fed predominantly on sugarcane or maize by-products (including silage) might need supplementation. Some species of tropical grasses such as spear grass (*Heteropogon contortus*) are often low in sulfur (approximately $0.5\,g\,kg^{-1}$ DM) (Kennedy and Siebert, 1972a,b; Morrison et al., 1990). Sulfur fertilization of ryegrass pasture (Jones et al., 1982) or forage crops (Buttrey et al., 1986; Ahmad et al., 1995) or direct supplementation of diets based on sulfur-deficient forages can improve animal production. In all pastoral systems, risks of sulfur deprivation (but also nitrogen and phosphorus deficiencies) in livestock increase with distance from the coast and with increasing herbage maturity. The steppes of Inner Mongolia have recently been identified as a low-sulfur region on the basis of the sulfur to nitrogen ratio (50% of samples <0.07) (Wang et al., 1996). Improvements in wool production and live-weight gain are commonly found when sulfur supplements are provided for sheep grazing dry-season pastures in Western Australia (Mata et al., 1997; White et al., 1997). Mature forage is often harvested for winter use as a maintenance feed and attempts to improve its nutritive value by supplementation with rumen-degradable energy and nitrogen sources will only succeed if sulfur is also added (Leng, 1990; Suttle, 1991). The converse is also true: responses will only be obtained with sulfur if the energy, nitrogen and phosphorus requirements of microbes are met (Suttle, 1991). Marked improvements in production can be obtained by the balanced supplementation of low-quality roughages with molasses–urea mixtures, but they must contain added sulfur. Intensive animal production from cereals and non-protein nitrogen sources requires sulfur supplements. The nutritive value of all silages is low for reasons that are not understood (AFRC, 1993): since silages have unusually high concentrations of soluble nitrogen, it is possible that they lack the equally soluble sulfur necessary for the microbial capture of ammonia.

Animal factors

The incidence of sulfur deprivation will obviously depend on the type of stock being reared. With mixed stocking systems, pastures that are sufficient for cattle may provide insufficient sulfur for sheep. In temperate zones, lambing coincides with the spring grass growth, rich in protein at times of maximum need of the ewe and young lamb (Table 9.2). However, factors that increase protein requirements, such as infections by gut nematodes (Coop and Sykes, 2002), will increase sulfur requirements. Widespread deficiencies of sulfur have been predicted for cattle on the basis of recommended sulfur to nitrogen ratios of 0.1 for beef and 0.08 for dairy animals (Qi et al., 1994), but these have yet to be tested. In tropical areas of Queensland and Western Australia, it has been suggested that seasonal supplements of sulfur (and nitrogen) will be needed to obtain full production advantages from the use of defaunated sheep because of suboptimal rumen sulfide concentrations (Hegarty et al., 1994).

Diagnosis of Sulfur Deprivation

Since the problem of sulfur deprivation in ruminants relates to the well-being of the rumen microflora, the best diagnosis may be afforded by obtaining samples of rumen fluid by stomach tube and determining whether they contain sufficient sulfide for unrestricted, dissimilatory MPS. The critical sulfide sulfur level was thought to be as low as $1\,\text{mg}\,\text{l}^{-1}$ (Bray and Till, 1975; Hegarty et al., 1991), but other studies suggest that it may be between 1.6 and $3.8\,\text{mg}\,\text{l}^{-1}$ (Kandylis and Bray, 1987; Weston et al., 1988) and the critical value may depend on the relative importance of assimilatory and dissimilatory pathways. Unlike total sulfur, sulfide is easily determined by acid displacement of hydrogen sulfide and titration (Kennedy and Siebert, 1972a). Samples of rumen fluid can be obtained by means of a simple vacuum source (a 50 ml disposable syringe fitted with a two-way valve) and interchangeable tubes for receiving the samples. Ease of sampling depends on the consistency of the digesta and relevance of the result upon the continuity of feeding since rumen sulfide increases after a meal. Three samples should be taken, one pre-prandial, one post-prandial (1–2 h) and another after about 5 h in discontinuously fed livestock. Low serum SO_4 sulfur ($<10\,\text{mg}\,\text{l}^{-1}$) is also indicative of sulfur deprivation. Australian workers have recently examined blood and liver GSH concentrations as indicators of sulfur status (Mata et al., 1997; White et al., 1997). Blood GSH showed a similar marked decline to that of plasma SO_4 during the dry season (Fig. 9.4), reaching levels that would be regarded as subnormal ($340\,\text{mg}\,\text{l}^{-1}$; the norm is $>500\,\text{mg}\,\text{l}^{-1}$). However, the decline was not diminished by dietary SO_4 supplements (White et al., 1997) and only partially reduced by 'protected methionine', suggesting that GSH synthesis was restricted by lack of protein rather than sulfur (Mata et al., 1997). Blood GSH could be as a useful complement to serum SO_4 when nitrogen is also lacking. The surest diagnosis is provided by responses in appetite or growth (of wool or body) to continuous sulfur supplementation, provided that no other factors are limiting rumen microbial activity.

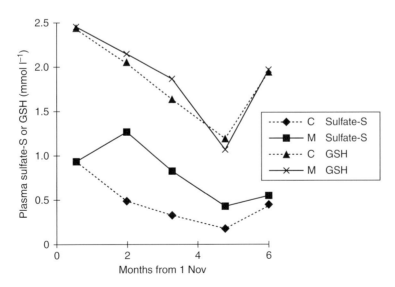

Fig. 9.4. Seasonal fluctuations in two biochemical indices of sulfur status, plasma sulfate (SO_4) and blood glutathione (GSH), in two groups of sheep grazing in Western Australia. One group (■,×) received a sulfur-rich mineral (M), while the other (♦,▲) did not (C) (data from White et al., 1997).

Prevention and Control of Sulfur Deprivation

Fertilizers

Many phosphorus fertilizers used to be relied upon to provide sufficient adventitious sulfur to avoid the need for specific sulfur sources, but the sulfur to phosphorus ratio in phosphorus fertilizers has steadily and markedly declined and so has the sulfur status of pastures and crops. As with phosphorus (see Chapter 6), the primary goal must be to ensure that shortage of sulfur does not limit pasture or crop growth. Sophisticated pasture growth models have been developed based on two soil tests that measure available SO_4^{2-} and the organic sulfur component from which SO_4^{2-} is 'mineralized' (Goh and Nguyen, 1997). Thus, it has been estimated that 6.7–13.5 kg S ha^{-1} year^{-1} is required to replace the net loss of sulfur from the soil SO_4^{2-} pool, stemming from removal of the element in animal products, leaching and the 'export' of excreta. Using a simpler yardstick (soil extractable SO_4 <12 mg kg^{-1}), biennial sulfur applications of 10 kg ha^{-1} were recommended for maize grown in a rainforest–savannah transition zone in Nigeria (Ojeniyi and Kayode, 1993). Sulfur in fertilizer form is most usually distributed as calcium sulfate ($CaSO_4$) or potassium sulfate (K_2SO_4) and the responses in herbage sulfur are dependent on application rate (Fig. 9.5) (Spears et al., 1995).

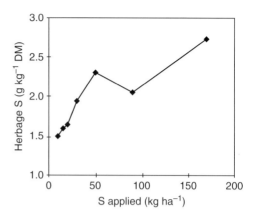

Fig. 9.5. Responses in herbage sulfur concentration to different application rates of fertilizer sulfur as calcium sulfate ($CaSO_4$) in south-east Scotland (McLaren et al., 1975).

Direct supplements

Lack of sulfur in the diet can readily be corrected by the inclusion of elemental sulfur, Na_2SO_4 or $CaSO_4$, but only if other constraints upon MPS are lifted (Kennedy and Siebert, 1972b; Bird, 1974). Poor-quality pasture or roughage is usually lacking in metabolizable energy, RDP and possibly phosphorus as well as sulfur. Inclusion of $CaSO_4$ in a molasses–urea–bran block yields an ideal supplement for such pastures, with the proportions of $CaSO_4$ and urea providing 0.1 rather than 0.067 S:N as previously proposed (Underwood and Suttle, 1998). One procedure is to heat the molasses to 90°C for 20 min, cool to 60°C, then thoroughly mix in (kg kg^{-1}): urea, 0.1; $CaSO_4$, 0.022; magnesium oxide (MgO), 0.01; calcium hydroxide ($Ca(OH)_2$), 0.1; sodium chloride (NaCl), 0.05; wheat bran, 0.23; and trace elements as needed. The mixture is poured into moulds and allowed to solidify ($CaSO_4$, modified from Habib et al., 1991). For stall-fed animals, a loose dry mixture can be spread over hand-fed fodder or feeds in the trough and molasses may not be necessary. In a trial, pastured sheep offered 175 g head^{-1} day^{-1} of a mineral mixture containing 46 g S kg^{-1} while grazing consumed enough supplement (167 g day^{-1}) to provide 1.15 g S day^{-1}. The additional sulfur probably contributed to the reported improvement in performance (White et al., 1997). Where cereals are used for drought feeding the supplement should provide 10 g urea and $CaSO_4$ kg^{-1} DM (Bogdanovic, 1983). The supply of both nitrogen and sulfur in slowly degradable forms (e.g. biuret and $CaSO_4$) may improve MPS by up to 25% compared with that attained with urea and Na_2SO_4 by improving microbial capture, thus minimizing losses as ammonia and sulfide.

Nutritive value of inorganic sources

Many comparative assessments of different sulfur sources have been reported (Henry and Ammerman, 1995), but those based on sulfur balance give little or no indication of ability to promote MPS. Techniques based on *in vitro* cultures of rumen microbes get closer to true nutritional value to the extent that they adequately simulate rumenal events *in vivo*, but they can give implausible results compared with animal responses to supplementation. For example, elemental sulfur is accorded low nutritive values of 0–36% by *in vitro* methods, 28–69% by balance methods and 73–102% by sheep growth tests when compared to Na_2SO_4 or methionine (Henry and Ammerman, 1995). There is clear evidence that elemental sulfur is a good source for ruminants (Slyter et al., 1988) and potentiates the same rumen-based antagonisms as other sulfur sources (see Chapter 11 for the copper-molybdenum antagonism). There is no evidence that inorganic sulfur sources differ in nutrient value for MPS, but they may do so. SO_4^{2-} ions have been accorded metabolic significance on account of their effects on acid–base balance and feature in some formulae for calculating dietary cation–anion difference (DCAD) and preventive strategies for milk fever (Chapter 8). However, giving magnesium as SO_4 rather than chloride or oxide in the prepartum period has alleviated hypocalcaemia by a greater extent than could be accounted for by effects on DCAD and significantly improved milk protein concentration on pastures of high sulfur content (3.2–3.7 g kg^{-1} DM) (see p. 176; Roche et al., 2002).

'Protected' sulfur sources

Where MPS cannot provide sufficient sulfur amino acids to meet the needs of the ruminant, it may be more economical to feed sulfur sources that are 'protected' (^{pr}S) from rumen degradation, such as the analogues malyl and methyl malyl methionine (Met) than to feed sulfur in expensive, SDP sources such as fish meal. It has been estimated that an additional 100 g lupin seeds day^{-1} would be needed to provide the improvements in wool and body growth afforded by 3 g ^{pr}Met day^{-1} (Mata et al., 1997). However, the first priority must be to maximize MPS using inorganic sulfur supplements if necessary. The second priority is to supplement with UDP if the metabolizable protein requirement exceeds the capacity for MPS (AFRC 1993). All claims for benefits from feeding costly ^{pr}S supplements should be evaluated against the background of what was or what might have been achieved with rumen-degradable sulfur and nitrogen sources and UDP supplements. If the supply of sulfur per se is limiting rumen microbial activity, the provision of 'protected' forms might be disadvantageous.

Methionine analogues have been mostly used for improving production with high sulfur amino acid requirements, such as merino wool (Coetzee et al., 1995) or angora and cashmere fibre (Souri et al., 1998), but in the last case the sulfur content of the diet was unspecified and responses may have been wholly or partially obtainable with SO_4. Furthermore, there is far more cysteine than methionine in wool (Fig. 9.2) and since transsulfuration is far from complete (Liu and Masters, 2000), it would be more logical to supply protected cysteine for such animals. Dietary ^{pr}Met supplements improve milk fat yield in high-yielding dairy cows (Overton et al., 1996), but not in low-yielding cows on ryegrass/white clover pasture (Pacheco-Rois et al., 1999) or in those on a diet in which methionine supply was thought to be inadequate (Casper et al., 1987). When fed at high levels of 12–24 g day^{-1} to sheep, ^{pr}Met reduced appetite (Kandylis, 1984a). However, ^{pr}Met may have a sparing effect on selenium by reducing the incorporation of the predominant selenium source in natural diets into body protein and increasing its availability for the synthesis of functional selenoproteins.

Sulfur Toxicity

Direct effects

The margin between the desirable and harmful concentration of sulfur in the ruminant diet is surprisingly small at two- to threefold. Reductions in appetite and growth rate are found in cattle and sheep given diets with 3–4 g S kg^{-1} DM (for review, see Kandylis, 1984b). This low tolerance is most commonly ascribed to sulfide toxicity (Bird, 1972b). The first site to be affected is the rumen, where sulfide is generated, and motility may be impaired. The variable tolerances to different amounts and sources of sulfur in the literature (Kandylis, 1984b) partly reflect differences in the rate of ingestion of degradable sulfur, the rate of sulfide absorption across the rumen wall (which is pH dependent) and the rate of sulfide capture by rumen microbes (see Fig. 9.1). The simultaneous addition of urea can lessen the depression of appetite and digestibility caused by sulfur alone (Bird, 1974). However, the performance of steers on a concentrate diet was adversely affected when $(NH_4)_2SO_4$ was used to raise dietary sulfur from 2.0 to 2.5 g kg^{-1} DM, a very modest increase (Zinn et al., 1997). There is evidence that the rumen microflora adapt to high SO_4 intakes and gradually increase their dissimilatory capacity, thus increasing rumen sulfide concentrations (Cummings et al., 1995b).

Absorbed sulfide may have harmful effects on the central nervous system, whether from drinking water (Smart et al., 1986) or dietary sulfur supplements (Raisbeck, 1982; Cumming et al., 1995a), contributing to the development of polioencephalomalacia (PEM). PEM developed in calves and lambs when a dietary acidifier, included to reduce risk of urolithiasis, was changed from ammonium bicarbonate to ammonium sulfate (Jeffrey et al., 1994). Losses from a clinically similar disease in lambs were associated with a high sulfur concentration (4.1 g kg^{-1} DM) in a concentrate fed ad libitum (Low et al., 1996). The disorders differed in fine pathological detail and responsiveness to vitamin B$_1$ from those associated with PEM caused by thiamine deficiency, the common cause of the disease. Cattle reduced both water and food consumption when offered drinking water containing 2.83 g SO_4 as Na$_2$SO$_4$ l^{-1} for 30 days (Weeth and Hunter, 1971); the treatment was equivalent to an increase in dietary sulfur to 8.4 g kg^{-1} DM. SO_4 can be removed from 'saline' water supplies by reverse osmosis (Smart et al., 1986). These adverse effects are in complete contrast to those found in non-ruminants, in which undegraded SO_4 entering the small and large intestine causes an osmotically driven diarrhoea.

Indirect effects

Chronic exposure to high sulfur intakes in the diet or drinking water can have adverse indirect effects on ruminants, particularly by inducing copper deficiency, with or without the assistance of molybdenum (see Chapter 11).

Toxic sulfur compounds

Some of the sulfur in crops and forages is present in potent organic forms. The most widely studied is S-methylcysteine sulfoxide (SMCO), a free amino acid resembling methionine, which commonly causes a haemolytic anaemia in ruminants consuming large quantities of brassicas such as kale (Barry et al., 1981; Barry and Manley, 1985) or rape (Suttle et al., 1987) (for review, see Whittle et al., 1976). Conversion of SMCO to dimethyl disulfoxide in the rumen plays a key role in pathogenesis (Smith, 1978). SMCO levels can be reduced by plant breeding and limiting the available sulfur supply from the soil (T. Barry, personal communication, 1978). Oral dosing with methionine provides partial protection from kale anaemia while stimulating live-weight gain and wool growth. It has been suggested that methionine reduces the normally rapid and complete degradation of SMCO in the rumen (Barry and Manley, 1985).

References

AFRC (1993) *Energy and Protein Requirements of Ruminants.* An advisory manual prepared by the AFRC Technical Committee on Responses to Nutrients. Commonwealth Agricultural Bureaux, Farnham Royal, UK.

ARC (1980) *The Nutrient Requirements of Ruminant Livestock.* Commonwealth Agricultural Bureaux, Farnham Royal, UK, pp. 166–168.

Ahmad, M.R., Allen, V.G., Fontenot, J.P. and Hawkins, G.W. (1995) Effect of sulfur fertilization on chemical composition, ensiling characteristics and utilisation by lambs on sorghum silage. *Journal of Animal Science* 73, 1803–1810.

Barry, T.N. and Manley, T.R. (1985) Responses to oral methionine supplementation in sheep fed on kale (*Brassica oleracea*) diets containing S-methyl-L-cysteine sulphoxide. *British Journal of Nutrition* 54, 753–761.

Barry, T.N., Reid, T.C., Miller, K.R. and Sadler, W.A. (1981) Nutritional evaluation of kale (*Brassica oleracea*) diets 2. *Journal of Agricultural Science, Cambridge* 96, 269–282.

Beaton, J.D., Tisdale, S.L. and Platou, J. (1971) Crop responses to sulfur in North America. *Technical Bulletin 18*. The Sulfur Institute, Washington, DC, pp. 1–10.

Beever, D.E. (1996) Meeting the protein requirements of ruminant livestock. *South African Journal of Animal Science* 26, 20–26.

Bird, P.R. (1972a) Sulfur metabolism and excretion studies in ruminants. V. Ruminal desulfuration of methionine and cysteine. *Australian Journal of Biological Sciences* 25, 185–193.

Bird, P.R. (1972b) Sulfur metabolism and excretion studies in ruminants. X. Sulfide toxicity in sheep. *Australian Journal of Biological Sciences* 25, 1087–1098.

Bird, P.R. (1974) Sulfur metabolism and excretion studies in ruminants. XIII. Intake and utilisation of wheat straw by sheep and cattle. *Australian Journal of Agricultural Sciences* 25, 631–642.

Bird, S.H. and Leng, R.A. (1985) Productivity responses to eliminating protozoa from the rumen of sheep. In: Leng, R.A., Barker, J.S.F., Adams, D. and Hutchinson, K. (eds) *Reviews in Rural Science* 6. University of New England Publishing Unit, Armidale, Australia, pp. 109–117.

Bishara, H.N. and Bray, A.C. (1978) Competition between molybdate and sulfate for renal tubular reabsorption in sheep. *Proceedings of the Australian Society of Animal Production* 12, 123.

Bogdanovic, B. (1983) A note on supplementing whole wheat grain with molasses, urea, minerals and vitamins. *Animal Production* 37, 459–460.

Bray, A.C. and Till, A.R. (1975) Metabolism of sulfur in the gastro-intestinal tract. In: McDonald, I.W. and Warner, A.C.I. (eds) *Digestion and Metabolism in the Ruminant*. The University of New England Publishing Unit, Armidale, Australia, pp. 243–260.

Buttrey, S.A., Allen, V.G., Fontenot, J.P. and Renau, R.B. (1986) Effect of sulfur fertilization on chemical composition, ensiling characteristics and utilisation of corn silage by lambs. *Journal of Animal Science* 63, 1236–1245.

Casper, D.P., Schingoethe, D.J., Yang, C.-M.J. and Mueller, C.R. (1987) Protected methionine supplementation with extruded blend of soybeans and soybean meal for dairy cows. *Journal of Dairy Science* 70, 321–330.

Coetzee, J., de Wet, P.J. and Burger, W.J. (1995) Effects of infused methionine, lysine and rumen-protected methionine derivatives on nitrogen retention and wool growth of Merino wethers. *South African Journal of Animal Science* 25, 87–94.

Coop, R.L. and Sykes, A.R. (2002) Interactions between gastrointestinal parasites and nutrients. In: Freer, M. and Dove, H. (eds) *Sheep Nutrition*. CAB International, Wallingford, UK, pp. 313–331.

Cornforth, I.S., Stephen, R.C., Barry, T.N. and Baird, G.A. (1978) Mineral content of swedes, turnips and kale. *New Zealand Journal of Experimental Agriculture* 6, 151–156.

Cummings, B.A., Caldwell, D.R., Gould, D.H. and Hawar, D.W. (1995a) Identity and interactions of rumen microbes associated with sulfur-induced polioencephalomalacia in cattle. *American Journal of Veterinary Research* 56, 1384–1389.

Cummings, B.A., Gould, D.H., Caldwell, D.R. and Hawar, D.W. (1995b) Ruminal microbial alterations associated with sulfide generation in steers with dietary sulfate-induced polioencephalomalacia. *American Journal of Veterinary Research* 56, 1390–1395.

Dick, H.T. (1956) Molybdenum in animal nutrition. *Soil Science* 81, 229–258.

D'Mello, J.P.F. (2003) *Amino Acids in Animal Nutrition*, 2nd ed. CAB International, Wallingford, UK.

Durand, M. and Komisarczuk, K. (1988) Influence of major minerals on rumen microbiota. *Journal of Nutrition* 118, 249–260.

Gawthorne, J.M. and Nadar, C.J. (1976) The effect of molybdenum on the conversion of sulfate to sulfide and microbial protein sulfur in the rumen of sheep. *British Journal of Nutrition* 35, 11–23.

Goh, K.M. and Nguyen, M.L. (1997) Estimating net annual soil sulfur mineralisation in New Zealand grazed pastures using mass balance models. *Australian Journal of Agricultural Research* 48, 477–484.

Grace, N.D. and Suttle, N.F. (1979) Some effects of sulfur intake on molybdenum metabolism in sheep. *British Journal of Nutrition* 41, 125–136.

Habib, C., Basit Ali Shah, S., Wahidullah, Jabbar, G. and Ghufranullah (1991) The importance of urea–molasses blocks and by-pass protein in animal production: the situation in Pakistan. In: *Proceedings of Symposium on Isotope and Related Techniques in Animal Production and Health*. International Atomic Energy Agency Vienna, pp. 133–144.

Harrison, D.G. and McAllan, A.B. (1980) Factors affecting microbial growth yields in the reticulo-rumen. In: Ruckebusch, Y. and Thivend, P. (eds) *Digestive Physiology and Metabolism in Ruminants*. MTP Press, Lancaster, UK, pp. 205–226.

Hegarty, R.S., Nolan, J.V. and Leng, R.A. (1991) Sulfur availability and microbial fermentation in the fauna free rumen. *Archive fur Animal Nutrition, Berlin* 41, 725–736.

Hegarty, R.S., Nolan, J.V. and Leng, R.A. (1994) The effects of protozoa and of supplementation with nitrogen and sulfur on digestion and microbial metabolism in the rumen of sheep. *Australian Journal of Agricultural Research* 45, 1215–1227.

Henry, P.R. and Ammerman, C.B. (1995) Sulfur bioavailability. In: Ammerman, C.B., Baker, D.H. and Lewis, A.J. (eds) *Bioavailability of Nutrients For Animals*. Academic Press, New York, pp. 349–366.

Jeffrey, M., Duff, J.P., Higgins, R.J., Simpson, V.R., Jackman, R., Jones, T.O., Machie, S.C. and Livesey, C.T. (1994) Polioencephalomalacia associated with the ingestion of ammonium sulfate by sheep and cattle. *Veterinary Record* 134, 343–348.

Jones, M.B., Rendig, V.V., Torell, D.T and Inouye, T.S. (1982) Forage quality for sheep and chemical composition associated with sulfur fertilization on a sulfur deficient site. *Agronomy Journal* 74, 775–780.

Jumba, I.O., Suttle, N.F. and Wandiga, S.O. (1996) Mineral composition of tropical forages in the Mount Elgon region of Kenya. *Tropical Agriculture* 73, 108–112.

Kamprath, E.J. and Jones, U.S. (1986) Plant response to sulfur in southeastern United States. In: Tabatabai, M.A. (ed.) *Sulfur in Agriculture*. American Society of Agronomy, Madison, Wisconsin, pp. 323–343.

Kandylis, K. (1984a) The role of sulfur in ruminant nutrition. A review. *Livestock Production Science* 11, 611–624.

Kandylis, K. (1984b) Toxicity of sulfur in ruminants: review. *Journal of Dairy Science* 67, 2179–2187.

Kandylis, K. and Bray, A.C. (1987) Effects of variation in dietary sulfur on movement of sulfur in the rumen. *Journal of Dairy Science* 70, 40–49.

Kennedy, P.M. (1974) The utilisation and excretion of sulfur in cattle fed on tropical roughages. *Australian Journal of Agricultural Research* 25, 1015–1022.

Kennedy, P.M. and Milligan, L.P. (1978) Quantitative aspects of the transformations of sulfur in sheep. *British Journal of Nutrition* 39, 65–84.

Kennedy, P.M. and Siebert, B.D. (1972a) The utilisation of spear grass (*Heteropogon contortus*). II. The influence of sulfur on energy intake and rumen and blood parameters in sheep and cattle. *Australian Journal of Agricultural Research* 23, 45–46.

Kennedy, P.M. and Siebert, B.D. (1972b) The utilisation of spear grass (*Heteropogon contortus*). III. The influence of the level of dietary sulfur on the utilisation of spear grass by sheep. *Australian Journal of Agricultural Research* 24, 143–152.

Kincaid, R. (1988) Macro elements for ruminants. In: Church, D.C. (ed.) *The Ruminant Animal – Digestive Physiology and Nutrition*. Prentice Hall, Englewood Cliffs, New Jersey, pp. 326–330.

Langlands, J.P. and Sutherland, H.A.M. (1973) Sulfur as a nutrient for Merino sheep. I. Storage of sulfur in tissues and wool and its secretion in milk. *British Journal of Nutrition* 30, 529–535.

Ledgard, S.F. and Upsdell, M.P. (1991) Sulfur inputs from rainfall throughout New Zealand. *New Zealand Journal of Agricultural Research* 34, 105–111.

Leng, R.A. (1990) Factors affecting the utilisation of 'poor-quality' forages by ruminants particularly under tropical conditions. *Nutrition Research Reviews* 3, 277–303.

Liu, S.M. and Masters, D.G. (2000) Quantitative analysis of methionine and cysteine requirements for wool production of sheep. *Animal Science* 71, 175–185.

Lovett, T.D., Coffey, M.T., Miles, R.D. and Combs, G.E. (1986) Methionine, choline and sulfate interrelationships in the diet of weanling pigs. *Journal of Animal Science* 63, 467–471.

Low, J.C., Scott, P.R., Howie, F., Lewis, M., Fitzsimons, J. and Spence, J.A. (1996) Sulfur-induced polioencephalomalacia in sheep. *Veterinary Record* 138, 327–329.

MAFF (1990) *UK Tables of the Nutritive Value and Chemical Composition of Foodstuffs*. In: Givens, D.I. (ed.) Rowett Research Services, Aberdeen, UK.

Masters, D.G., Stewart, C.A., Mata, G. and Adams, N.R. (1996) Responses in wool and liveweight when different sources of dietary protein are given to pregnant and lactating ewes. *Animal Science* 62, 497–506.

Mata, G., Masters, D.G., Chamberlain, N.L. and Young, P. (1997) Production and glutathione responses to rumen-protected methionine in young sheep grazing dry pastures over summer and autumn. *Australian Journal of Agricultural Research* 48, 1111–1120.

McClaren, R.G. (1975) Marginal sulfur supplies for grassland herbage in south east Scotland. *Journal of Agricultural Science, Cambridge* 85, 571–573.

McCollum, E.V. (1957) *A History of Nutrition*. Houghton Mifflin, Boston, Massachusetts.

McDowell, L.R. (2003) *Minerals in Animal and Human Nutrition*. Elsevier, Amsterdam.

Minson, D.J. (1990) *Forages in Ruminant Nutrition*. Academic Press, San Diego, California, pp. 208–229.

Moir, R.J. (1970) Implications of the N:S ratio and differential recycling. In: Muth, O.H. and Oldfield J.E. (eds) *Sulfur in Nutrition – Symposium Proceedings*. AVI Publishing Company, Westport, Connecticut.

Morrison, M., Murray, R.M. and Boniface, A.N. (1990) Nutrient metabolism and rumen micro-organisms in sheep fed a poor quality tropical grass hay supplemented with sulfate. *Journal of Agricultural Science* 115, 269–275.

Ojeniyi, S.O. and Kayode, G.O. (1993) Response of maize to copper and sulfur in tropical regions. *Journal of Agricultural Science, Cambridge* 120, 295–299.

Overton, T.R., LaCount, D.W., Cicela, T.M. and Clark, J.H. (1996) Evaluation of a ruminally protected methionine product for lactating dairy cows. *Journal of Dairy Science* 79, 631–638.

Pacheco-Rios, D., McNabb, W.C., Hill, J.P., Barry, T.N. and McKenzie, D.D.S. (1999) The effects of methionine supplementation upon milk composition and production of forage-fed cows. *Canadian Journal of Animal Science* 79, 235–241.

Qi, K., Lu, C.D., Owens, F.N. and Lupton, C.J. (1992) Sulfate supplementation of Angora goats: metabolic and mohair responses. *Journal of Animal Science* 70, 2828–2837.

Qi, K., Lu, C.D. and Owens, F.N. (1993) Sulfate supplementation of growing goats: effects on performance, acid–base balance and nutrient digestibilities. *Journal of Animal Science* 71, 1579–1587.

Qi, K., Owens, F.N. and Lu, C.D. (1994) Effects of sulfur deficiency on performance of fiber-producing sheep and goats: a review. *Small Ruminant Research* 14, 115–126.

Raisbeck, M.F. (1982) Is polioencephalomalacia associated with high sulfate diets? *Journal of the American Veterinary Medical Association* 180, 1303–1305.

Rasmussen, P.E. and Kresge, P.O. (1986) Plant responses to sulfur in the western United States. In: Tabatabai, M.A. (ed.) *Sulfur in Agriculture*. American Society of Agronomy, Madison, Wisconsin, pp. 357–374.

Revell, D.K., Baker, S.K. and Purser, D.B. (1999). Nitrogen and sulfur mobilised from body tissue can be used for wool growth. *Australian Journal of Agricultural Research* 50, 101–108.

Roche, J.R., Morton, J. and Kolver, E.S. (2002) Sulfur and chlorine play a non-acid base role in periparturient calcium homeostasis. *Journal of Dairy Science* 85, 3444–3453.

Slyter, L.L., Chalupa, W. and Oltjen, R.R. (1988) Response to elemental sulfur by calves and sheep fed purified diets. *Journal of Animal Science* 66, 1016–1027.

Smart, M.E., Cohen, R., Christensen, D.A. and Williams, C.M. (1986) The effects of sulfate removal from the drinking water on the plasma and liver copper and zinc concentrations of beef cows and their calves. *Canadian Journal of Animal Science* 66, 669–680.

Smith, G.M. and White, C.L. (1997) A molybdenum–sulfur–cadmium interaction in sheep. *Australian Journal of Agricultural Research* 48, 147–154.

Smith, R.H. (1978) S-Methyl cysteine sulphoxide: the kale anaemia factor (?) *Veterinary Science Communications* 2, 47–61.

Souri, M., Galbraith, H. and Scaife, J.R. (1998) Comparisons of the effects of genotype and protected methionine supplementation on growth, digestive characteristics and fibre yield in cashmere-yielding and angora goats. *Animal Science* 66, 217–223.

Spears, J.W., Burns, J.C. and Hatch, P.A. (1985) Sulfur fertilisation of cool season grasses and effect on utilisation of minerals, nitrogen and fibre by steers. *Journal of Dairy Science* 68, 347–355.

Suttle, N.F. (1991) Mineral supplementation of low quality roughages. In: *Proceedings of Symposium on Isotope and Related Techniques in Animal Production and Health*. International Atomic Energy Agency, Vienna, pp. 101–114.

Suttle, N.F., Jones, D.G., Woolliams, C. and Woolliams, J.A. (1987) Heinz body anaemia in lambs with deficiencies of copper and selenium. *British Journal of Nutrition* 58, 539–548.

Thomas, W.E., Loosli, J.K., Williams, H.H. and Maynard, L.A. (1951) The utilisation of urea and inorganic sulfates by lambs. *Journal of Nutrition* 43, 515–523.

Todd, J.R. (1972) Copper, molybdenum and sulfur contents of oats and barley in relation to chronic copper poisoning in housed sheep. *Journal of Agricultural Science, Cambridge* 79, 191–195.

Underwood, E.J. and Suttle, F. (1999) *The Mineral Nutrition of Livestock*, 3rd edn. CAB International, Wallingford, UK, p. 232.

Wang, Y., Waghorn, G.C., McNabb, W.C., Barry, T.N., Hedley, M.J. and Shelton, I.D. (1996) Effect of condensed tannins upon the digestion of methionine and cysteine in the small intestine of sheep. *Journal of Agricultural Science, Cambridge* 127, 413–421.

Weeth, H.J. and Hunter, J.E. (1971) Drinking of sulfate water by cattle. *Journal of Animal Science* 32, 277–281.

Weston, R.H., Lindsay, J.R., Purser, D.B., Gordon, G.L.R. and Davis, P. (1988) Feed intake and digestion responses in sheep to the addition of inorganic sulfur to a herbage diet of low sulfur content. *Australian Journal of Agricultural Research* 39, 1107–1119.

Whanger, P.D. (1972) Sulfur in ruminant nutrition. *World Review of Nutrition and Dietetics* 15, 225–237.

White, C.L., Kumagai, H. and Barnes, M.J. (1997) The sulfur and selenium status of pregnant ewes grazing Mediterranean pastures. *Australian Journal of Agricultural Research* 48, 1081–1087.

Whittle, P.J., Smith, R.H. and McIntosh, A. (1976) Estimation of S-methyl sulphoxide (kale anaemia factor) and its distribution among brassica and forage crops. *Journal of Science in Food and Agriculture* 27, 633–642.

Williams, A.J., Robards, G.E. and Saville, D.G. (1972) Metabolism of cystine by Merino sheep genetically different in wool production. II. The responses in wool growth to abomasal infusion of L-cystine and DL-methionine. *Australian Journal of Biological Science* 25, 1269–1276.

Zinn, R.A., Aloarez, E., Mendez, M., Montano, M., Ramirez, E. and Shen, Y. (1997) Influence of dietary sulfur level on growth, performance and digestive function in feedlot cattle. *Journal of Animal Science* 75, 1723–1728.

10 Cobalt

Introduction

The essentiality of cobalt in the forages of sheep and cattle emerged from investigations of two naturally occurring disorders: 'coast disease' of sheep in South Australia (Lines, 1935; Marston, 1935) and 'wasting disease' of cattle in Western Australia (Underwood and Filmer, 1935). In both situations the pastures had underlying calcareous soils of Aeolian origin. A condition called 'cobalt pine' was being concurrently recognized in sheep on similar soils in north-west Scotland (Suttle, 1988). Before long, similar severe disorders were recognized in other areas: 'bush-sickness' in New Zealand, 'salt sickness' in Florida, 'Nakuruitis' in Kenya, 'lecksucht' in the Netherlands and Germany and 'Grand Traverse disease' in Michigan. All described a similar wasting disease that was preventable by cobalt supplementation, although the geological associations sometimes differed.

The mode of action of cobalt was not recognized until 1948, when the anti-pernicious anaemia factor, subsequently designated vitamin B_{12}, was found to contain cobalt (Smith, 1948). To be absorbed, vitamin B_{12} must be bound to intrinsic factor, secreted by the gastric mucosa, and partial gastrectomy increased susceptibility to pernicious anaemia. Cobalt deprivation in ruminants is primarily a vitamin B_{12} deficiency brought about by the inability of the rumen microorganisms to synthesize sufficient vitamin B_{12} from their cobalt supply. Remission of all signs of cobalt deprivation in lambs has been secured within 3 years by parenteral injections of vitamin B_{12} (Smith et al., 1951). Hindgut fermenters such as the horse do not succumb to cobalt deprivation and may have the capacity to absorb vitamin synthesized in the caecum with the help of undegraded intrinsic factor (Davies, 1979; Lewis et al., 1995). Pigs and poultry obtain their vitamin either preformed in food or indirectly by ingesting faeces. There is no evidence that any species requires cobalt other than as vitamin B_{12} and this chapter will again deal almost exclusively with ruminants, whose symbiotic rumen microbial population also needs vitamin B_{12} to produce propionate, a metabolite that is a major determinant of the host's cobalt responsiveness.

Structure and Function of Vitamin B_{12}

Vitamin B_{12} has the general formula $C_{63}H_{88}N_{14}O_{14}PCo$, a molecular weight of 1357 daltons and contains 4.4% cobalt. Rumen microbes incorporate cobalt into a corrin (tetrapyrrole) ring to form vitamin B_{12} (cobalamin, Cbl) and similar, highly complex molecules (cobinamides). The essentiality of cobalt for mammals and microbes is linked to two distinct forms of vitamin B_{12} formed by attachment of

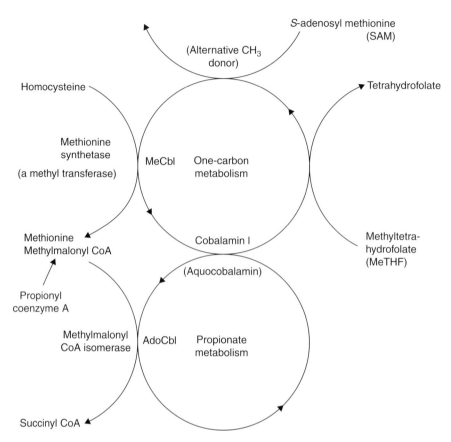

Fig. 10.1. Vitamin B_{12} (cobalamin, Cbl) has two distinct yet linked coenzyme functions: facilitating the build up of carbon skeletons via methyl-cobalamin (MeCbl) (one-carbon pathway) and gluconeogenesis from propionate via adenosyl-cobalamin (AdoCbl) (propionate pathway). Failure of these respective pathways is indicated by the accumulation of the abnormal metabolites methylmalonyl acid and formiminoglutamic acid. Rumen microbes use AdoCbl to convert succinate to propionate and failure is indicated by the accumulation of succinate.

α ligands that project above the plane of the ring to produce adenosyl-cobalamin (AdoCbl) and methyl-cobalamin (MeCbl). A β ligand, dimethylbenzimidazole, projects below the plane and distinguishes the molecule from cobinamides, which have other β ligands. Further information on the biochemistry of vitamin B_{12} is given by Bannerjee (1999).

coenzyme A (CoA) mutase to form succinate from propionate, thus facilitating gluconeogenesis chiefly in the liver (Fig. 10.1). Microbes are equally dependent on AdoCbl, but for the reverse purpose: in the *Propioni* bacteria, formation of propionate from succinate is catalysed by the same mutase. Several aminomutases, including leucine 2,3 mutase are also AdoCbl-dependent.

Succinate–propionate exchange

As AdoCbl, cobalt influences intermediary energy metabolism, keeping the tricarboxylic acid cycle turning by assisting methylmalonyl-

Methylation

As MeCbl, vitamin B_{12} assists methionine synthase to transfer methyl groups from 5-methyltetrahydrofolate to homocysteine via *S*-adenosyl methionine for the regeneration of methionine (and tetrahydrofolate) (Fig. 10.1). MeCbl thus plays a pivotal role in one-carbon metabolism (i.e. the build up of carbon 'skeletons' and nucleotide synthesis) (Bassler, 1997), facilitating the synthesis of a wide range of molecules including formate, noradrenaline, myelin and phosphatidyl ethanolamine. While interest has focused on the hepatic activity of methionine synthase, enzyme activity is three times higher in the intestinal mucosa and spleen than in the liver (Lambert et al., 2002). MeCbl is also important for microbes and is needed for methane, acetate and methionine synthesis by rumen bacteria (Poston and Stadman, 1975).

Clinical Consequences of Cobalt Deprivation

A number of distinct clinical conditions that respond to treatment with cobalt or vitamin B_{12} have now been described in ruminants (Table 10.1). These occur principally in sheep and to a lesser extent in goats. The earliest visible abnormality may be seen below the eye, where excessive lachrymation mats the fleece: its precise cause is unknown. The whole fleece or hair coat usually appears dull, limp and scruffy. Loss of hair colour has been reported in the bovine (Judson et al., 1982). In chronic disorders, progressive inappetence leads to marked weight loss, listlessness, muscular wasting (marasmus) and pica (depraved appetite). The overall appearance is indistinguishable from that caused by starvation. It is still possible to find mortality rates of 30% in groups of unsupplemented lambs on pastures low in cobalt (Grace and Sinclair, 1999). Anaemia eventually develops, causing the visible mucous membranes to appear blanched and the skin pale and fragile. Without intervention, death ensues. An internal examination will reveal a body with little or no visible fat, heavy haemosiderin deposits in the spleen and hypoplasia of erythrogenic tissue in the bone marrow. The anaemia is normocytic and normochromic in lambs, unlike the megaloblastic anaemia seen in humans. Neurological damage may occur (Fell, 1981).

Fatty liver diseases

White-liver disease (WLD), OWLD in ovines, was first reported in grazing lambs at times of prolific pasture growth (Sutherland et al., 1979) and is often accompanied by clinical signs of photosensitization and biochemical signs of liver damage. The disease has been produced experimentally (Kennedy et al., 1994b) and progresses from mild fatty infiltration of hepatocytes to a point where bile ducts proliferate, enzymes leak into the bloodstream from damaged liver cells and plasma bilirubin levels rise (Kennedy et al., 1997). Affected lambs are anorexic, anaemic and ill-thriven in appearance. The disease is sometimes accompanied by polioencephalomalacia and is preventable

Table 10.1. Summary of the principal cobalt-responsive disorders found in sheep and goats, their causes and distinguishing anomalies. All are associated with but not necessarily distinguished by low plasma and liver vitamin B_{12} concentrations.

Disorder	Time course	Principal feature	Underlying dysfunction	Distinguishing anomaly	Vulnerable stock
Ill-thrift	Semi-chronic	Slow growth at pasture	AdoCbl	MMA ↑	Young lambs
'Pine'	Chronic	Weight loss	MeCbl	Homocysteine ↑ Haemoglobin ↓	Lambs and ewes
White-liver disease	Semi-acute	Fatty liver	AdoCbl and MeCbl	TG ↑ Liver enzymes ↑	Lambs and Angora kids
'Hepatic lipidosis'	Semi-acute	Fatty liver	MeCbl	MCV ↓ MCHC ↓	Kids on low-protein diets

AdoCbl, adenosyl-cobalamin; MeCbl, methyl-cobalamin; MCV, mean (red) cell volume; MCHC, mean (red) cell haemoglobin concentration; MMA, methylmalonic acid; TG, triglycerides.

by cobalt or vitamin B_{12} supplementation (Sutherland et al., 1979; McLoughlin et al., 1986). A few cases of WLD have been described in Angora goats in New Zealand (Black et al., 1988) but caprine WLD is seen more widely in other breeds in Oman, where an incidence of 5.3% was reported from an abattoir survey in the Muscat area (Johnson et al., 1999). The disorder has been reproduced experimentally on a diet consisting mostly of Rhodes grass hay, 'low' in cobalt (<0.1 mg kg^{-1} dry matter (DM)), but a vitamin B_{12}-responsive anaemia that was microcytic and hypochromic developed at an early stage (Johnson et al., 2004). These unusual features may be attributable to inadequate microbial protein synthesis.

Infertility and perinatal mortality

While infertility is always likely to arise as a secondary consequence of debilitating conditions such as severe cobalt deprivation, it may also occur in the absence of ill-thrift. In a study of beef cows, five out of nine unsupplemented cows failed to conceive compared to only two or three in each of three cobalt-supplemented groups of nine to 11 cows (Judson et al., 1997). Ewes deprived of cobalt during early pregnancy may give birth to fewer lambs than normal (Duncan et al., 1981; Quirk and Norton, 1987) and those that are born may be stillborn or slow to suck and have diminished prospects of survival (Fisher and MacPherson, 1991). Poor reproductive performance is not attributable to anorexia because ewe live weight and body condition is not affected but may be caused by impaired gluconeogenesis at a time of high glucose need. Depression of milk yield can reduce the growth rate of surviving lambs (Quirk and Norton, 1987). Newborn calves are less affected by cobalt deprivation of their dams, although their subsequent growth can be depressed (Quirk and Norton, 1988).

Disease susceptibility

Increased susceptibility to infection by the abomasal parasite *Ostertagia circumcincta* has been reported in lambs (Ferguson et al., 1989) and the immune responses of cattle against a similar parasite (*O. ostertagi*) are compromised by cobalt deprivation (MacPherson et al., 1987). Increased faecal egg counts have been reported in a natural, mixed nematode infection in grazing Texel lambs of low plasma B_{12} status (<100 pmol l^{-1}) (Vellema et al., 1996). Adverse interactions with infection by abomasal parasites may be particularly pronounced because these nematodes also suppress appetite and destroy the glands that produce intrinsic factor. However, infection with the intestinal parasites *Trichostrongylus vitrinus* and *O. circumcincta* did not accelerate a decline in plasma vitamin B_{12} in grazing lambs (Suttle et al., 1986). There is no evidence that susceptibility to microbial infection is increased by cobalt deprivation. Antibody responses to *Mycobacterium avium* subspecies *paratuberculosis* and bovine herpes virus type 1 were not affected in the Vellema et al. (1996) study. Although *in vitro* lymphoproliferative responses were increased by cobalt supplementation, cultured lymphocytes can show a 10-fold increase in vitamin B_{12} turnover under such challenges (Quadros et al., 1976), calling into question their physiological significance.

Sources of Cobalt

A regular dietary cobalt supply is clearly essential for survival, health and efficient production.

Forages

Cobalt concentrations in pastures and forages vary widely between plant species and with soil conditions (Minson, 1990). In early Scottish studies, cobalt concentrations in mixed pasture ranged from 0.02 to 0.22 mg kg^{-1} DM and legumes were richer in cobalt than grasses grown in the same conditions, with red clover and ryegrass containing 0.35 and 0.18 mg kg^{-1} DM, respectively (Underwood and Suttle, 1999). Lower values were reported for Japanese-grown legumes and grasses, but the species difference was proportionately greater (respective means, 0.12 and 0.03 mg kg^{-1} DM) (Hayakawa, 1962). The advantage of legumes can be lost if soils are cobalt deficient. Twofold variation amongst grass species grown under similar conditions has been reported, with *Phleum pratense* and *Dactylis glomerata* poor sources compared to *Lolium perenne*. Pasture cobalt is reduced as the soil pH rises (see Chapter 2) and increased by poor soil drainage, but is largely unaffected by advancing plant maturity.

Ingested soil

Soil contamination can greatly increase herbage cobalt, particularly in unwashed spring and autumn samples. Washing will often reduce herbage cobalt concentrations without necessarily improving the assessment of cobalt supply since ingested soil cobalt is partially available for ruminal synthesis of vitamin B_{12} (Brebner et al., 1987), raises serum vitamin B_{12} (Grace, 2006) and probably alleviates deficiency at times (Macpherson, 1983; Clark et al., 1989). Grass species can differ in the degree to which they are contaminated with soil as well as in their ability to absorb cobalt from the soil (Jumba et al., 1995).

Concentrates

Data on cobalt in grains and other concentrates are meagre, partly because of the difficulty in measuring low concentrations in silica-rich samples. Cereal grains are poor cobalt sources with concentrations usually within the range 0.01–0.06 mg kg^{-1} DM, and are used to produce cobalt-deficient diets (Field et al., 1988; Kennedy et al., 1991b, 1992). Like most trace elements, cobalt is highly concentrated in the outer fibrous layers, so that bran and pollard are much richer in cobalt than the whole grain. Feeds of animal origin, other than liver meal, are mostly poor sources of cobalt.

Milk and milk products

Cow's milk normally contains 0.5–0.9 μg Co l^{-1}. However, only the cobalt that is present as vitamin B_{12} – <10% judging from the range of values found in fat-free milk (0.3–3.0 nmol l^{-1}) (Judson et al., 1997) – is of nutritional value. Both milk cobalt and vitamin B_{12} can be increased several-fold by large dietary cobalt supplements. Milk is an important source of vitamin B_{12} for the pre-ruminant, although concentrations are low in all species (typically <5 μg l^{-1} in the ewe) (Grace et al., 2006).

Cobalt Availability and the Synthesis of Vitamin B_{12}

Dietary cobalt is only of use to ruminants if incorporated into vitamin B_{12} by rumen microbes, but they incorporate most of the supply into physiologically inactive corrinoids (Gawthorne, 1970). The efficiency of conversion to Cbl is thought to be primarily affected by dietary cobalt supply, but may also be influenced by the source of cobalt, diet type and animal species.

Effect of cobalt intake

The efficiency of Cbl synthesis appears to decrease as cobalt intake increases, with 15% captured as 'Cbl' by sheep given hay low in cobalt and only 3% captured from a 1 mg daily supplement given by drench (Smith and Marston, 1970a). With <0.07 mg Co kg^{-1} DM in a hay diet, cobalt levels in rumen fluid were <0.5 μg l^{-1} and microbial vitamin B_{12} output was suboptimal (Smith and Marston, 1970b). The apparent effect of cobalt intake was recently confirmed in sheep given a similar range of cobalt intakes by continuous intraruminal infusion (Fig. 10.2). Ruminal synthesis of vitamin B_{12} increases within hours of cobalt supplementation. In one study that used an area under the plasma vitamin B_{12} versus time after dosing curve as a measure of vitamin synthesis by depleted lambs, synthesis increased in proportion to the square root of cobalt dose over the range 1–32 mg (Suttle et al., 1989). With cobalt given daily, 0.1 mg day^{-1} increased rumen vitamin B_{12} to about 80% of the response attained with 4.0 mg day^{-1} (Wiese et al., 2007).

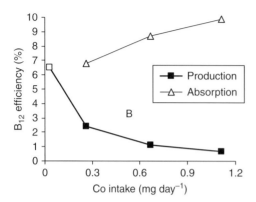

Fig. 10.2. The efficiency of rumen production of vitamin B_{12} and subsequent absorption in sheep continuously fed a mixed, low-cobalt diet of hay and concentrates (□) or given cobalt by intra-ruminal infusion (■,△). Cannulation of the abomasum allowed the flow of vitamin B_{12} from the rumen to be measured. Absorption was measured by calibrating responses in plasma vitamin B_{12} against those obtained with continuous intravenous infusions of vitamin B_{12} (Ludemann, 2007).

Effect of cobalt source

The high efficiencies of vitamin B_{12} production on the basal diets used by Smith and Marston (1970a) and Ludemann (2009) might be partly attributable to an intrinsically higher utilization of cobalt inherent in the feed than of extrinsic added cobalt. The rationale for such an effect is that those rumen microbes that colonize ingested feed particles (protozoa and fibre digesters) may make better use of the local supply of inherent cobalt than of a soluble source distributed throughout the rumen fluid. In Fig. 10.2, the efficiencies of use of added cobalt increase from 0.6% to 1.9% as less cobalt is added, but there may 'step-changes' to the 6.5% efficiency measured for the basal diet and the 15% reported for unsupplemented hay (Smith and Marston, 1970a).

Effect of energy and protein substrates

The first indication of an effect of diet type on rumen vitamin B_{12} synthesis was a decrease in vitamin B_{12} production in sheep following the addition of cereal to the diet (Sutton and Elliot, 1972; Hedrich et al., 1973). In the study described in Fig 10.2, sheep were given a cereal-supplemented diet (0.6 parts hay to 0.4 barley), low in cobalt, and the efficiencies of vitamin B_{12} synthesis were <0.5 of those found previously with a hay diet at all cobalt intakes (Smith and Marston, 1970a). Although the manner of supplementation differed, it seems unlikely that continuous cobalt infusion of continuously fed animals would decrease the efficiency of cobalt capture. In a recent study of vitamin B_{12} production in dairy cows, the efficiency of incorporation of cobalt into vitamin B_{12} (Schwab et al., 2006) (Table 10.2) was reduced from 13% to 7%, on average, by the inclusion of 23% grain (mostly maize) in the ration. Rumen production of vitamin B_{12} is linearly related to food intake in sheep (Hedrich et al., 1973) and cattle (Zinn et al., 1987), but the efficiency of capture was unaffected by the maize-based diet used in the latter study (trial 3: Table 10.2). In the Schwap et al. (2006) study, increasing the forage (predominantly maize silage) content of the diet from 35% to 60% reduced the efficiency of rumen vitamin B_{12} production from 12% to 8% on average. Maize and maize silage are characterized by relatively low rates of energy and protein degradability (AFRC, 1993) when compared to barley and grass silage and may therefore sustain relatively low rates of microbial protein and vitamin B_{12} synthesis.

Effect of species

Vitamin B_{12} production has been estimated in three recent cattle studies involving a wide range of diets but similar generous supplies of cobalt, of which about 70% came from inorganic sources. The derived efficiencies of cobalt capture ranged from 6.3% to 16.2% (Table 10.2), much higher than those found in sheep given plentiful supplies of dietary cobalt. This may explain the lower cobalt requirement of cattle.

In vitro synthesis of vitamin B_{12}

The study of vitamin B_{12} (and volatile fatty acid) production in continuous cultures of

Table 10.2. Efficiency of 'capture' of dietary cobalt as microbial vitamin B_{12} is affected by the nature of both the roughage and carbohydrate constituents of cattle diets. Low efficiencies are associated with high intakes of maize silage (MS) and ground maize (GM) or ground barley (GB).

Capture of Co as microbial vitamin $B_{12}{}^a$ (%)	Roughage (%)	Forage (%)	Carbohydrate (%)	Reference
14.0	45	45 AH	45 FM	Zinn et al. (1987)
11.0	20	0	64 DRM	
15.4	59	44 GLS, 15 MS	34 HMM	Sanschi et al. (2005)
16.2	86[b]	18 MS, 12 AH, 6 GH	0	Schwab et al. (2006)
8.2	61[b]	18 MS, 12 AH, 6 GH	16 GM, 8 GB	
9.5	88[b]	30 MS, 20 AH, 19 GH	0	
6.3	69	30 MS, 20 AH, 19 GH	15 GM, 7 GB	

AH, alfalfa hay; DRM, dried, rolled maize; FM, flaked maize; GH, grass hay; GLS, grass-legume silage; HMM, high-moisture maize.
[a]Efficiency calculated from the daily flow of vitamin B_{12} at the duodenum, converted to Co (×0.044) and expressed as a percentage of daily Co intake, which came predominantly from the mineral supplement (0.7 mg added Co kg^{-1} DM in each study). Only Schwab et al. (2006) analysed diets for Co and a total Co of 1 mg kg^{-1} DM was assumed for the other studies.
[b]The differences between roughage and total forage percentage were made up by a mixture of soybean hulls and sugarbeet pulp.

rumen microbes is yielding useful information on bioavailability (Fig. 10.3) (Kawashima et al., 1997b; Tiffany et al., 2006) after initial disappointments (McDonald and Suttle, 1986), but it has only been used to evaluate sources of supplementary cobalt. Accessibility of cobalt to the rumen microbes probably varies between culture systems, but the relatively high vitamin B_{12} output from unsupplemented cultures in the above studies suggests that the cobalt in oat hulls (Brebner, 1987), hay (McDonald and Suttle, 1986) and corn (Tiffany et al., 2006) is of relatively high availability. Cobalt can form strong complexes with methionine (NRC, 2005) and it is conceivable that some cobalt in feeds is or becomes complexed and escapes rumen degradation. Availability for vitamin B_{12} synthesis probably correlates with solubility or extractability of the cobalt source (Kawashima et al., 1997a,b) and preliminary work suggests that rumen vitamin B_{12} concentrations in sheep correlate with the acetate-extractable cobalt in ingested soil in vivo (Brebner et al., 1987) as well as in vitro (Fig. 10.3). Acetate-extractability of cobalt can vary from 3% to 13% between soil types (Suttle et al., 2003).

The extractability of cobalt in feeds could indicate its bioavailability to ruminants.

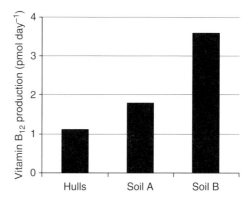

Fig. 10.3. Mean vitamin B_{12} production from three pairs of continuous cultures of rumen microbes 'fed' on oat hulls and liquid substrates. Soils low (soil A) or high (soil B) in acetate-extractable cobalt (0.09 and 0.58 mg kg DM^{-1}, respectively) were added to two pairs (Brebner, 1987).

Metabolism of Vitamin B_{12}

The absorption, transport and cellular utilization of vitamin B_{12} is facilitated by a trio of structurally related, Cbl-transporting proteins: haptocorrin, intrinsic factor (IF) and transcobalamin (TC) II (Wuerges et al., 2006).

Absorption

Cobalamins leave the rumen as dead microbes or bound to salivary haptocorrin, but are released during subsequent digestion and selectively bound to IF produced by parietal cells in the abomasum (McKay and McLery, 1981). In sheep, some 10,000–24,000 IU of IF is secreted in the gastric juice each day. Bile salts facilitate binding of the Cbl:IF complex to a multi-purpose receptor, 'cubilin', in the ileal mucosal brush border (Smith, 1997; Seetheram et al., 1999) and endocytosis of the complex allows selective discrimination against other corrinoids. Only 3–5% of the vitamin B_{12} that leaves the rumen is thought to be absorbed by cobalt-deprived sheep (Kercher and Smith, 1955; Marston, 1970). Higher apparent absorption has been reported at higher dietary cobalt intakes (Fig. 10.2) (Hedrich et al., 1973) and absorptive efficiencies of 11% have been reported in cattle on generous cobalt intakes (Zinn et al., 1987; Schwab et al., 2006). Features of vitamin B_{12} absorption that are found in non-ruminant species, such as an increase in absorptive capacity during pregnancy (Nexo and Olesen, 1982) and in the colostrum-fed neonate (Ford et al., 1975), may also occur in ruminants.

Mucosal transfer, transport and cellular uptake

The Cbl:IF complex is bound to TCII in the mucosal villi and exported into the bloodstream (Seetheram et al., 1999). Specific receptors on cell surfaces allow the TCII:Cbl complex to be taken up and converted to either MeCbl or AdoCbl. The two pathways coordinate (Riedel et al., 1999) and recycling of Cbl occurs in the tissues (Fig. 10.1) and via biliary secretion. Other Cbl-binding proteins are found in plasma, and mammalian species show important differences in the distribution and binding properties of three proteins, TC0, TCI and TCII. Bovine plasma contains high concentrations of the strong, high-molecular-weight binders TC0 and TC1 (Polak et al., 1979; Schulz, 1987a,b; Price, 1991), which play havoc with some assays for vitamin B_{12}, as described later. There is usually excess vitamin B_{12}-binding capacity in plasma. TCII is particularly strongly expressed in the kidneys, for reasons unknown.

Storage

Although it is customary to refer to excess supplies of vitamin B_{12} as being 'stored' in the liver, the early work of Marston (1970) gave little evidence of this. There was no significant increase in liver vitamin B_{12} concentrations after 36 weeks of daily oral treatment with cobalt at ten and 100 times a level regarded as 'marginal' (0.1 mg day^{-1}). Plasma vitamin B_{12} was slightly more responsive (Fig. 10.4), and Marston concluded that the poor capacity of sheep to store vitamin B_{12} in liver was 'a limiting factor' in the aetiology of cobalt-responsive disorders. Similar contrasts in patterns of vitamin B_{12} accretion in plasma and liver are evident in grain-fed cattle (Tiffany et al., 2003). Vitamin B_{12} resembles the other water-soluble vitamins in being poorly stored. Placental transfer of vitamin B_{12} is not marked. Liver vitamin B_{12} concentrations in the ovine fetus are less than half of those found in the mother (Grace et al., 1986) and may not be increased by supplementing the pregnant ewe with vitamin B_{12} (Grace et al., 2006).

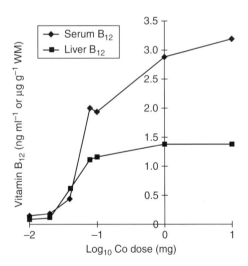

Fig. 10.4. When cobalt-deprived sheep are continuously repleted with diets providing 0.1 to 10 mg Co day^{-1}, vitamin B_{12} concentrations reach an earlier plateau in liver than in serum, suggesting limits to hepatic storage capacity (data from Marston, 1970).

Secretion

Vitamin B_{12} is secreted in milk and administering cobalt bullets to the depleted pregnant ewe can increase concentrations in the milk (Hart and Andrews, 1959), with values increasing from approximately 0.5 nmol l^{-1} by more than twofold in a recent study (Gruner et al., 2004a). Survey data in cows indicate that milk vitamin B_{12} concentrations decline exponentially as serum vitamin B_{12} falls to subnormal levels (Judson et al., 1997). However, the transfer of parenterally administered vitamin B_{12} is inefficient in dairy cows that are given ample dietary cobalt. Data for lactating cows given 10 mg of the vitamin by weekly injection (Girard and Matte, 2005) indicate that only 5.6% of the dose was secreted in milk. Raising dietary cobalt from 0.14 to 0.27 mg kg^{-1} DM during pregnancy had no significant effect on milk vitamin B_{12} or serum B_{12} in the newborn calf: colostrum levels were five times higher than those in main milk (19 versus 3.8–4.8 µg l^{-1}), but the ingestion of colostrum did not affect serum vitamin B_{12} in the calf (Stemme et al., 2006).

Excretion

Little is known about the excretion of vitamin B_{12} in ruminants. In humans, the primary route can be via biliary secretion to the faeces with negligible catabolism or urinary loss. In dairy cows, high cobalt intakes (17.8–31.0 mg day^{-1}) increase urinary vitamin B_{12} to peak values of around 12 µg l^{-1} during lactation, higher than those found in milk and serum (Walker and Elliot, 1972). The predominance of urinary excretion is a characteristic of all water-soluble vitamins. Urinary losses probably fall to negligible levels during cobalt deprivation and may be useful in assessing vitamin B_{12} status.

Biochemical Consequences of Cobalt Deprivation

Inadequate cobalt intakes cause a sequence of biochemical changes in the tissues and fluids of the body that is influenced by the nature of the dietary carbohydrate (Fig. 10.5).

Depletion

As soon as cobalt intake declines, concentrations of vitamin B_{12} in rumen fluid decline due to depletion of rumen microbes and total serum vitamin B_{12} declines because it reflects the amount of synthesized vitamin being absorbed (Somers and Gawthorne, 1969). The decline in liver vitamin B_{12} occurs *after* the decline in serum vitamin B_{12}, indicating that any liver store is slowly mobilized for extra-hepatic purposes.

Deficiency

At the cellular or 'nano' level, uptake of TCII by cells is up-regulated and the specific activities of NADH/NADPH-linked aquaCbl reductases (aquaCbl is an intermediary transfer protein, see Fig. 10.1) in the liver increase (Watanabe et al., 1991). These adaptations help to explain the lag before vitamin B_{12} concentrations in liver decline. The relationship between vitamin B_{12} concentrations in plasma and liver is generally curvilinear, with concentrations in liver showing little decline, relative to plasma, in the early stages of cobalt deprivation (Marston, 1970; Field et al., 1988). This pattern contrasts with the relationship between plasma and liver copper (see Chapter 11).

Dysfunctions in propionate metabolism – rumen microbes

Cobalt deprivation of the rumen microbes causes massive increases in succinate levels in the rumen (to 100 mmol l^{-1}) within days of transfer of lambs to a barley-based diet, low in cobalt (Fig. 10.6). Rumen succinate also increases in cobalt-deprived lambs at pasture and the inhibition of propionate-producing microbial species such as *Selenomonas ruminantium* has been implicated (Kennedy et al., 1991b, 1996). Rumen succinate was increased to a lesser degree (approximately 10 mmol l^{-1}) in sheep given a ground 3:1 blend of an oaten hay/lupin seed meal diet (Wiese et al., 2007) containing a similar cobalt concentration to that featured in Fig. 10.6, and in cattle given low-cobalt barley as their main feed (Tiffany and Spears, 2005).

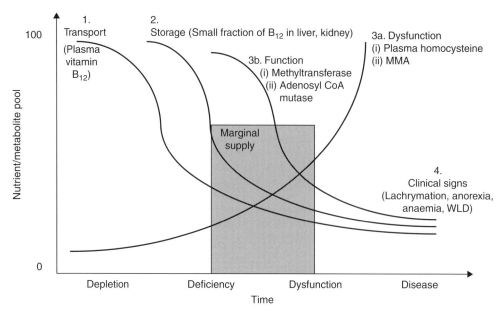

Fig. 10.5. The sequence of biochemical changes preceding the appearance of clinical symptoms in ruminants deprived of cobalt (and hence vitamin B_{12}) differs from the general model (see Fig. 3.1), in that the earliest change occurs in the transport pool and signs of metabolic dysfunction (3a, below) precede critical losses of enzyme function (3b). The affected pathway (i or ii) is determined by the load placed upon it by diet (i, propionate) or reproductive cycle (ii, lactation) and differs between sheep and cattle (see also Fig. 10.9; MMA, methylmalonic acid; WLD, white-liver disease).

Dysfunctions in propionate metabolism – ruminants

The main source of energy in ruminants is volatile fatty acids, including propionic acid. These are produced by rumen fermentation, and a breakdown in AdoCbl-dependent conversion of propionate to succinate occurs in the tissues of the cobalt-deprived ruminant. The rate of clearance of propionate from the blood falls and the intermediary metabolite methylmalonic acid (MMA) accumulates to levels of 10–100 µmol l^{-1} (Marston et al., 1961, 1972; Gawthorne, 1968). There is a correlation between appetite and propionate clearance rate (Marston et al., 1972); raised propionate concentrations in the portal blood flow induce powerful satiety signals in sheep (Farningham et al., 1993). By lowering the intake of fermentable carbohydrate, satiety will reduce the propionate load and equilibrium may be attained. Changes in dietary cobalt and/or food substrate that shift fermentation from propionate to succinate will weaken the satiety signal and improve appetite and growth. Thus, growth was retarded from the outset in lambs on a barley diet (Fig. 10.7), but took 77 days to appear on a roughage-based diet that was equally low in cobalt (Table 10.3) (Wiese et al., 2007) but that would yield more acetate and less propionate in the rumen. In this study, plasma MMA increased before growth was retarded and, in sheep given low-cobalt hay, MMA increased before a marker of MeCbl dysfunction (Fig. 10.6). In the grazing flock, increases in plasma MMA provide an early warning of disorder, but show little change in the cobalt-deprived steer (<6 µmol l^{-1}), whether the diet is based on roughage (Stangl et al., 2000c) or cereal (Kennedy et al., 1995; Tiffany et al., 2003; Tiffany and Spears, 2005).

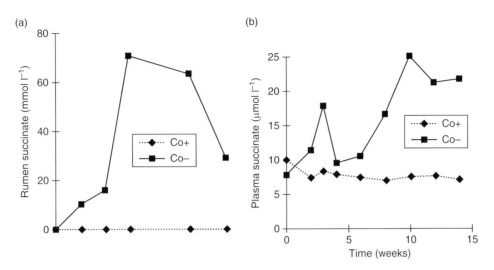

Fig. 10.6. Lambs given a barley-based diet very low in cobalt show marked early rises in rumen succinate (a) followed by increases in plasma succinate (b), a glucose precursor. This may partially overcome any constraint upon glucogenesis due to impaired propionate metabolism in the liver (from Kennedy et al., 1991).

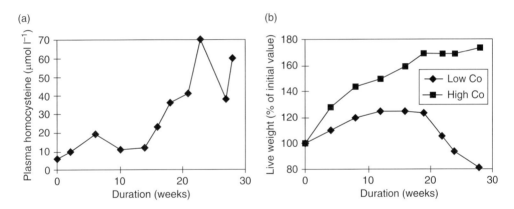

Fig. 10.7. The live-weight gain in lambs deprived of cobalt on a barley diet is a mirror image of concurrent changes in plasma homocysteine. The early slowing of growth is accompanied by little or no change, but homocysteine sharply increases as lambs start losing weight (from Kennedy et al., 1992).

Dysfunction in methylation

Lack of MeCbl has not been reported to restrict rumen methionine synthesis on low-cobalt diets, but it eventually impairs methylation in the tissues, causing a rise in plasma homocysteine (Fig. 10.7) and final reductions in methionine synthase activity in certain tissues in cobalt-deprived lambs (Table 10.4). Similar changes occur in cobalt-deprived cattle (Stangl et al., 2000c). Concurrent increases in S-adenosyl methionine (SAM) and an increase in the SAM to S-adenosyl homocysteine ratio in the liver confirm the reduction in methylation (see Fig. 10.1). MeCbl is an important methyl donor in folate metabolism (Scott, 1992) and liver folate concentrations decrease in cobalt-deprived sheep (Gawthorne and Smith, 1974) and steers (Stangl et al., 2000b,c). In earlier studies, a rise in formiminoglutamic acid (FIGLU) excretion in urine in cobalt-deprived sheep was attributed to reduced methylation of folate (Gawthorne, 1968). The accumulation

Table 10.3. In a study by Wiese et al. (2007), prolonged cobalt deprivation was required to reduce live weight, appetite and wool growth, but eventually the efficiency of converting feed to wool was decreased in lambs on a ground oaten hay/lupinseed meal diet (<0.026 mg Co kg^{-1} DM; 'low Co') compared to those given a Co-supplemented diet (0.1 mg day^{-1} orally by drench; 'plus Co') from day 56.

	Live-weight gain[a]		Food intake[b]		Wool growth rate	
	Low Co	Plus Co	Low Co	Plus Co	Low Co	Plus Co
Depletion (days)	(kg)		(g day^{-1})		(mg 100 cm^{-2})	
56–77	1	1.5	780^{AL}	780^{AL}	58 ± 2.5	59 ± 2.3
78–105	−2	−1	550^{AL}	550^{PF}	53 ± 2.3	74 ± 3.0

[a]Approximate figures estimated from graph.
[b]Both groups were fed *ad libitum* (AL) initially, but the food intake of the 'plus Co' group was restricted by pair feeding (PF) to match the AL intake of the 'low Co' group after 78 days.

of phosphatidyl choline in the liver and brain indicates a functional disturbance of phosphatidyl ethanolamine methyltransferase (Table 10.4), confirming earlier observations in sheep (Gawthorne and Smith, 1974), but these changes were not seen in cattle given a similar barley diet (Kennedy et al., 1995). Methionine supplementation via intravenous (Gawthorne and Smith, 1974) or abomasal infusion (Lambert et al., 2002) lowers hepatic methionine synthase activity, and it is possible that a negative balance between methionine supply and demand increases activity and the risk of MeCbl dysfunction. The discovery of another methionine synthase that is zinc dependent and uses betaine-homocysteine as a substrate (Millian and Garrow, 1998) raises the possibility of compensatory increases that attenuate the effects of cobalt deprivation.

Abnormal lipid metabolism

Major defects in lipid metabolism involving both Cbl-dependent pathways may explain the pathogenesis of OWLD on barley diets, given the biochemical abnormalities recorded in Table 10.4 (Kennedy et al., 1994b). Accumulating MMA CoA may inhibit the β-oxidation of free fatty acids, which the cobalt-deprived lamb mobilizes from fat depots to offset loss of appetite, and MMA becomes misincorporated into branched-chain fatty acids (Duncan et al., 1981; Kennedy et al., 1991a). The latter anomaly is associated with diets that generate high propionate concentrations in the rumen. It affects sheep and goats but not cattle or deer, which may lack the capacity to activate MMA CoA or metabolize propionate in their adipose tissue (Wahle et al., 1979). Such differences may explain why sheep and goats develop WLD but cattle and deer do not appear to. The intravenous infusion of propionate increases the expression of most lipogenic genes in ovine adipose tissue (Lee and Hossner, 2002). Accumulation of free fatty acids normally increases the hepatic synthesis of triglycerides and their export from the liver as very-low-density lipoprotein. However, very-low-density lipoprotein assembly requires MeCbl and methionine synthase, the activity of which is reduced by cobalt deprivation (Table 10.4). Accumulating triglycerides are peroxidizable and high circulating levels of homocysteine could initiate a chain of lipid peroxidation, explaining the presence of the oxidation product lipofuscin, depletion of the antioxidant vitamin E and damage to mitochondrial structure in the liver of lambs with OWLD (Kennedy et al., 1997). However, homocysteine-induced oxidant stress has not been seen to affect steers deprived of cobalt on a corn-silage diet (Stangl et al., 2000b).

Table 10.4. Biochemical markers of cobalt deprivation and abnormal lipid and vitamin E metabolism in the livers of lambs with ovine white-liver disease (OWLD) and in cobalt-sufficient controls (Kennedy et al., 1994b). Data are mean ± standard error of the mean for three (OWLD) or four (control) lambs.

Marker	Control	OWLD
Vitamin B_{12} (pmol g^{-1})	396±9.7	15±4.1
Vitamin E (nmol g^{-1})	13.1±0.94	5.9±0.56[a]
MMCoA mutase (U g^{-1})	30.8±2.2	4.4±0.6
Methionine synthase (U g^{-1})	869±102	105±26
SAM:SAH	5.4±0.76	2.5±0.68
PC:PE	1.5±0.04	1.1±0.09
Total lipid (g kg^{-1} FW)	50.8±8.0	165±13.9
Triglyceride fatty acids (mg kg^{-1} FW)	5.2±1.00	36.8±14.9
Free fatty acids (mg g^{-1} FW)	2.9±0.80	10.5±1.80

MMCoA, methylmalonyl coenzyme A; PC, phosphatidyl choline; PE, phosphatidyl ethanolamine; SAH, S-adenosyl homocysteine; SAM, S-adenosyl methionine.
[a]Dilution effect of the accumulated fat.

Rate-limiting function in sheep

The critical 'biochemical lesion' in cobalt-deprived lambs on a roughage-based diet (or grazing) is probably located systemically rather than in the rumen because appetite can be rapidly restored by the parenteral administration of vitamin B_{12} without repletion of rumen microbes or a reduction in rumen succinate production in lambs (Wiese et al., 2007). In Wiese et al.'s study, early inadequacy along the systemic AdoCbl pathway, revealed by increases in plasma MMA (to 30 mmol l^{-1}), did not limit appetite or growth. Growth cessation coincided with later rapid increases in plasma homocysteine, indicating failure along the MeCbl pathway on a roughage diet. A similar sequence of dysfunction is shown in Fig. 10.8. A different pattern was seen in lambs depleted on a barley diet, with growth retardation occurring early in association with increased plasma MMA, but before a significant increase in plasma homocysteine (Fig. 10.7). It appears that propionate load upon the liver can cause early growth retardation in cobalt-deprived lambs. The association between weight loss and inefficient wool growth in the study of Wiese et al. (2007) (Table 10.3) supports the notion that dysfunction along the MeCbl pathway eventually became limiting to performance.

Rate-limiting function in cattle

Fourfold increases in plasma homocysteine can occur after prolonged (>40 weeks) cobalt depletion on a barley diet without evidence of growth retardation (Kennedy et al., 1995). The time course of the hyperhomocysteinaemia has not been reported, but plasma MMA is only marginally increased, suggesting that cattle give a lower priority to the MeCbl than to the AdoCbl pathway. Cereal type affects the metabolic responses of cattle to cobalt deprivation, with a low-cobalt barley diet producing higher rumen succinate and lower propionate concentrations than a low-cobalt corn diet, but no growth retardation (Tiffany and Spears, 2005). Growth retardation and low liver folate were found in cattle given a maize silage diet low in cobalt (Schwarz et al., 2000; Stangl et al., 2000b,c). Anomalies in lipid metabolism are barely evident in cobalt-deprived cattle given barley (Kennedy et al., 1995). Cattle may divert less vitamin B_{12} to the MeCbl pathway than sheep

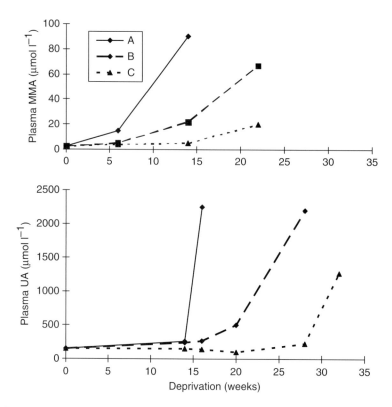

Fig. 10.8. The onset of biochemical dysfunction can be seen to vary in three cobalt-deprived sheep. The adenosyl cobalamin pathway is affected first, as shown by earlier rises in plasma methylmalonyl acid (MMA) than urocanate (UA), indicators of methyl cobalamin dysfunction (Price, 1991).

because they do not require large amounts of sulfur amino acids for wool synthesis (see Chapter 9). These contrasts between species and influence of energy substrates on the biochemical response to cobalt deprivation are illustrated in Fig. 10.9. The concept of linked variations in the two Cbl-dependent enzymes has been advanced from studies of cultured human glioma cells (Riedel et al., 1999).

Cobalt Requirements

The foregoing differences in vitamin B_{12}, energy and sulfur amino acid metabolism probably cause cobalt requirements to vary between species, with the type and amount of cereal in the ration (Table 10.5), stage of the reproductive cycle and the criterion of adequacy used.

Production from grass

Precise estimates of need are difficult to gauge from field studies because of seasonal changes in herbage cobalt and soil contamination. Cobalt responsiveness is poorly correlated with pasture cobalt even in washed samples of pasture in sheep (Gruner et al., 2004a,b). However, it is clear that grazing cattle have lower needs than sheep and cattle may not benefit from cobalt or vitamin B_{12} supplementation on pastures containing 0.06 mg Co kg^{-1} DM that cause clinical deprivation in lambs (Clark et al., 1986). Variation in animal need is also important. Pasture cobalt concentrations of 0.03–0.05 mg kg^{-1} DM may be tolerated by mature ewes, but requirements increase with the onset of lactation (Quirk and Norton, 1987, 1988; Gruner et al., 2004a), possibly because the lactating ewe's liver has to process more propionate and body fat is mobilized to meet energy needs but releases

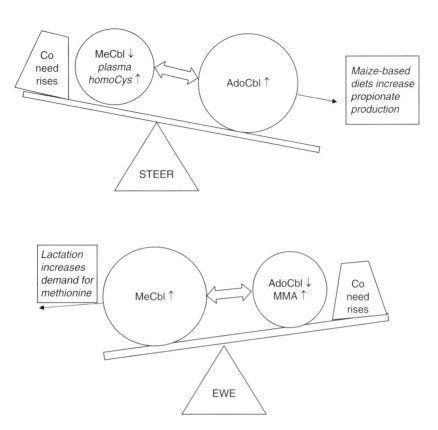

Fig. 10.9. A 'propionate load hypothesis' to explain the contrasting metabolic responses of sheep and cattle to cobalt deprivation. This hypothesis is based on Fig. 10.1 and makes two assumptions: in sheep, the balance struck between the methyl-cobalamin (MeCbl) and adenosyl-cobalamin (AdoCbl) pathways may be set to favour MeCbl, whereas in cattle AdoCbl is favoured; dietary or physiological factors that increase activity along the favoured pathway cause dysfunction along the unfavoured pathway. Thus, a diet that favours propionate production in the rumen of cattle will raise plasma homocysteine rather than methylmalonic acid (MMA), while the increased demand for methionine associated with the onset of lactation in sheep will cause an increase in plasma MMA rather than homocysteine.

Table 10.5. Provisional maximum[a] cobalt requirements of weaned ruminants at pasture or housed and fed mixed diets based on grass–legume silage plus barley or maize silage plus maize.

		Mixed diets	
	Grass	Grass silage and barley	Maize silage and maize
Cattle and deer[b]	0.06	0.04	0.15
Sheep and goats	0.12	0.08	0.30[b]

[a]All classes can perform well with less Co in situations where the quality of the diet and pasture can only sustain moderate levels of performance.
[b]The classification of deer with cattle and the allocation of a high Co need to sheep on maize-based diets are both speculative.

no additional vitamin B_{12}. There may be times when high soluble carbohydrate concentrations in fresh pasture raise cobalt requirements for growth and freedom from OWLD to ≥ 0.1 mg kg^{-1} DM; under normal circumstances, such high levels may only be needed to maintain maximal liver vitamin B_{12} (ARC, 1980). Fresh or conserved herbage cobalt concentrations of 0.05–0.12 mg kg^{-1} DM are therefore marginal for sheep and concentrations of 0.03–0.06 mg kg^{-1} DM are marginal for cattle, deer and goats.

Production from cereals

Studies with housed cattle have suggested that the cobalt requirement for optimum growth might be as high as 0.12–0.15 mg kg^{-1} DM on maize-based diets (Schwarz et al., 2000; Tiffany et al., 2003; Tiffany and Spears, 2005). Schwarz et al. (2000) fed corn silage with concentrate variously supplemented with cobalt. Cobalt requirements for 'optimal' haemoglobin, plasma homocysteine, liver folate, liver vitamin B_{12} and plasma vitamin B_{12} were estimated to be 0.1, 0.16, 0.19, 0.24 and 0.26 mg kg^{-1} DM, respectively. Higher requirements might reflect the inefficient rumen production of vitamin B_{12} from corn and corn silage (Table 10.2). However, with complete diets consisting mostly (85%) of ground corn or barley with monensin, an additive that normally favours propionate production growth requirements were met with only 0.02 mg Co kg^{-1} DM from the barley diet (Tiffany and Spears, 2005). When corn comprised only 49% of the ration during the growth phase of another trial (Tiffany et al., 2003), raising dietary cobalt from 0.04 to $\geq$ 0.09 mg kg^{-1} DM did not improve growth, but with 85% corn in the finishing ration the cobalt supplements did increase growth. Plasma vitamin B_{12} in unsupplemented steers in the two phases was equally low (approximately 50 pmol l^{-1}), but rumen propionate was 50% higher (46 versus 31 mmol l^{-1}) on the finishing diet.

Cobalt requirements for diets consisting mostly of maize silage or grain are uniquely high because such diets enhance propionate production in the rumen (Fig. 10.9). The use of monensin to manipulate rumen metabolism in grazing livestock may increase the cobalt requirement. Cobalt requirements of lambs on corn and barley-based diets have not been accurately defined, but are likely to exceed those of growing cattle and 0.2–0.3 mg kg^{-1} DM should be allowed until appropriate studies have been conducted (Table 10.5).

Production of milk

A hypothesis has been advanced that cows in early lactation have high systemic requirements for vitamin B_{12} to optimize methylation if the diet provides insufficient methionine (Girard, 2002). This hypothesis appeared to gain support when yields of milk and milk protein were increased by weekly intramuscular injections of 10 mg vitamin B_{12} in cows given diets supplemented with rumen-protected methionine and excess cobalt (0.66 mg added Co kg^{-1} DM) (Girard and Matte 2005). The vitamin B_{12} injections were believed to have increased the methylation of folate, but milk yield improved in the presence of normal concentrations of plasma MMA and liver vitamin B_{12}, apparently confounding existing ideas about assessing cobalt status as well as cobalt requirements. When a diet without protected methionine was supplemented with folic acid (2.6 g day^{-1}) and/or vitamin B_{12} (0.5 g day^{-1}), only folic acid increased milk yield (Graulet et al., 2007). It would appear that 'supranutritional' supplementation with vitamin B_{12} is only necessary when microbial synthesis of methionine is inadequate. The folic acid supplement doubled the total liver lipid content, 2 weeks after calving, unless vitamin B_{12} was also given, raising the possibility of involvement of these B vitamins in the aetiology of the fatty liver syndrome. The large amounts of propionate that high-yielding cows have to process may increase the priority given to the AdoCbl pathway and thus expose the MeCbl pathway. However, milk-yield responses to parenteral vitamin B_{12} may be less pronounced on diets providing adequate sulfur and optimum conditions for microbial protein and vitamin B_{12} synthesis. Feeding supranutritional dietary concentrations of cobalt (up to 1.71 mg kg^{-1} DM) does not increase milk yield (Kincaid and Socha, 2007), but a requirement similar to that of grain-fed cattle (i.e. 0.12 mg kg^{-1} DM) should be allowed for.

Subclinical Consequences of Cobalt Deprivation

The primary effect of cobalt deprivation has always been regarded as one on appetite, with secondary effects on feed conversion efficiency, but it has been claimed that cobalt deprivation can lower feed digestibility in goats (Kadim et al., 2003). This digestibility trial was unusual in that individuals were given a fixed intake of concentrate, allowed to express appetite for roughage (given *ad libitum*) and fed only once daily. The finding has yet to be confirmed under 'steady-state' conditions of ration composition and intake. Wool growth rate declines after prolonged cobalt deprivation (Table 10.4); when compared with pair-fed controls, wool growth in the cobalt-deprived lamb was eventually reduced by 28%. In the low-cobalt group, plasma vitamin B_{12} was <0.3μg l^{-1} throughout, appetite fell steadily after 77 days and body weight was lost but wool growth rate maintained. In the pair-fed plus cobalt group, plasma B_{12} was >1μg l^{-1} throughout, but an associated advantage in wool growth only emerged when food intake was restricted. Wool growth is more sensitive to cobalt deprivation than liveweight gain (Masters and Peter, 1990).

Occurrence of Cobalt Deprivation in Ruminants

Geographical distribution

Severe cobalt deprivation initially became recognized in localized areas within many countries, but much larger areas exist in which deprivation is mild or marginal and the only manifestation is unsatisfactory growth in 'tail-end' lambs. Cobalt deprivation arises on well-drained soils of diverse geological origin, including coarse volcanic pumice soils of New Zealand (Suttle, 1988), leached podzolized sands of Jutland, sandy loams of Goatland in Sweden (Schwan et al., 1987), soils derived from granites and ironstone gravels, as well as on the calcareous, wind-blown shell sands that were a feature of the earliest studies (p. 223). High levels of soil manganese depress the availability of soil cobalt to plants and the risk of cobalt deprivation may be predicted from equations that relate soil available cobalt to soil manganese (Suttle et al., 2003; Li et al., 2004). Cobalt deprivation may be induced by heavy liming, which also reduces the amount of plant-available (soil acetate-extractable) cobalt, but extremely acid soils (e.g. pH 5.0) can also be low in available cobalt (Suttle et al., 2003).

Geochemical mapping of areas of risk has been advocated (Suttle, 2000). The particular symptoms of WLD have been reported less frequently than 'ill-thrift', but the disease has been identified in lambs in Northern Ireland, Holland and Norway (for review see Suttle, 1988) and a variant occurs in Omani goats (Johnson et al., 1999, 2004).

Effects of species, breed, physiological state and husbandry

Young deer (Clark et al., 1986) and goat kids (Clark et al., 1987) – like calves and steers – are less likely to be retarded in growth than lambs confined to the same areas in New Zealand, and probably fit the cattle model in Fig. 10.8. In lambs, the incidence can still be high and in two recent studies 68–72% of untreated lambs had to be withdrawn from trials because of substantial weight loss (Grace et al., 2003; Gruner et al., 2004c). The high vulnerability of sheep and angora goats may be attributable to the high requirements of sulfur amino acids for fleece growth (see Chapter 9). A remarkable feature the 'low-cobalt' lambs in Table 10.4 is that wool growth was sustained while body weight was lost: wool growth rates were probably 7–8g day^{-1} and Fig. 9.2 suggests that >2g sulfur amino acids day^{-1} was still being deposited in the fleece. A different set of rules may apply in lactation when increased energy needs are met partly from endogenous sources (fat), while the increased need for methionine must be met largely from the diet. The model prediction that plasma MMA will be more responsive than plasma homocysteine to cobalt deprivation in the lactating dairy cow (Fig. 10.9) is supported by the responses to vitamin B_{12} injections reported by Girard and Matte (2005): only MMA was significantly reduced.

The incidence of cobalt deprivation in lambs can vary markedly from year to year on the same property (Lee, 1951) and, within a flock, young male lambs can be particularly vulnerable (Shallow et al., 1989). Variations in the degree of soil ingestion with weather conditions and grazing pressure may also be important (Grace, 2006), and disorders occur less often in sheep grazing long than short pasture because less soil is ingested (Andrews et al., 1958).

Effects of energy substrate

In Norway, the incidence of OWLD has been reported to vary between groups on adjacent fields on the same farm and was more closely related to high pasture concentrations of fructosan than of cobalt, which was relatively high (0.10–0.12 mg kg^{-1} DM) (Ulvund, 1995). Pasture fructosan concentrations are increased by sudden decreases in air temperature and may raise rumen propionate levels, precipitating disorder, or escape rumen degradation, thus limiting rumen synthesis of vitamin B_{12}. Other diets rich in readily or poorly fermented carbohydrate (e.g. molasses or maize and maize silage) may be associated with a relatively high risk of cobalt deprivation, particularly in sheep (according to Fig. 10.9).

Diagnosis of Cobalt-responsive Disorders

Mild cobalt deprivation is impossible to diagnose clinically because the unthrifty appearance is indistinguishable from the effects of nematodiasis or lack of dietary protein or digestible organic matter. A number of biochemical criteria are used to confirm diagnosis, but the most commonly used criterion, serum or plasma vitamin B_{12}, remains hard to interpret, particularly in cattle. Cobalt deprivation is confirmed beyond doubt by a response in appetite and weight to cobalt or vitamin B_{12} supplements. A flexible framework for the assessment of vitamin B_{12} status is presented in Table 10.6, based on the following assessments of individual criteria.

Table 10.6. Marginal bands[a] for the most common biochemical indices used to assess the mean cobalt and vitamin B_{12} status of groups of ruminants when grazing or intensively fed on maize-based diets indoors.

	Ruminant species	Grazing	Housed	Interpretive problems
Serum B_{12}[b] (pmol l^{-1})	Bovine: S	30–60[c]	Unknown	Prone to gross underestimation and laboratory variation
	Bovine: W	40–80[c,d]	70–150[d]	
	Ovine and others:[e] S	100–200	100–200	
	Ovine and others: W	200–300	Unknown	Propionate load
Liver B_{12}[b] (nmol kg^{-1} FW)	Bovine: S	70–100	70–100	
	Bovine: W	110–220	336–500	Propionate load
	Ovine and others: S	70–100	Unknown	
	Ovine and others: W	110–220		Propionate load
Serum homocysteine (μmol l^{-1})	All species:[e] S	8–12	8–12	Affected by methionine deficiency and possibly rate of transsulfuration in the gut during nematodeinfection
	All species:[e] W	Unreliable[f]		
Serum MMA[b] (μmol l^{-1})	All species: S	Unreliable	Unreliable	No propionate load
	Ovine and others: W	5–10	Unknown	Propionate load

MMA, methylmalonic acid; S, suckling; W, weaned.
[a]Mean values within a band denote the possibility of sufficient individuals benefiting to justify supplementation for all.
[b]Individual values close to below the **bold** limit are suggestive of production-limiting dysfunction. The higher the proportion in a sampled population, the stronger the case for supplementing with Co or vitamin B_{12}.
[c]Values may also be applied to cervines (Wilson and Grace, 2001).
[d]The apparent contrast between grazing and housed stock may be wholly or partly an artefact, attributable to different analytical methods.
[e]Lack of discrimination between species reflects lack of species-specific data rather than commonality of vitamin B_{12} status.
[f]Furlong et al. (2009).

Dietary and liver cobalt

Although dietary cobalt concentrations have a major influence on the rate of microbial vitamin B_{12} synthesis and vitamin B_{12} status of the host reflects microbial vitamin B_{12} output, the onset of disorder is affected by other factors, including the degree and duration of cobalt depletion and propionate load. Severity of disorder in sheep varies between years on the same low-cobalt pasture (Lee, 1951) and, in the case of OWLD, between pastures with equally adequate cobalt concentrations (>0.1 mg kg^{-1} DM) (Ulvund, 1995). With the added complication of contaminant soil cobalt, dietary cobalt concentration is of little value in diagnosis. Liver cobalt concentrations broadly reflect cobalt intake provided that analytical interference from high liver iron concentrations is avoided (Kawashima et al., 1997a; Underwood and Suttle, 1999). Early suggestions were that a liver cobalt level <0.04–0.06 indicated cobalt deprivation and that values >0.08–0.12 mg kg^{-1} DM ruled it out in sheep and cattle. This has been confirmed in cattle (Mitsioulis et al., 1995). The relationship between liver cobalt and liver vitamin B_{12} is curvilinear with wide scatter at high liver cobalt concentrations: the two parameters correlate linearly and well over the lower ranges (Suttle, 1995), but how reliable is the liver vitamin B_{12} as an index of dysfunction?

Liver vitamin B12

Low liver vitamin B_{12} concentrations reflect suboptimal cobalt intakes (Fig. 10.3). Biopsy samples can be obtained from the live animal, but considerations of welfare and cost have limited their use. Early guidelines were that values of 81–140 nmol kg^{-1} fresh weight (FW) were 'borderline' and <74 nmol kg^{-1} 'low' in sheep (Andrews et al., 1959). These guideline values are broadly supported by results from more recent dose–response trials in grazing sheep (Grace et al., 2003; Gruner et al., 2004b). On an individual basis, appetite can be reduced when liver vitamin B_{12} is <74 nmol kg^{-1} FW (Marston, 1970; Millar and Lorentz, 1979). Assessment of mean growth responses to cobalt supplements in weaned lambs in New Zealand indicated that, on a flock basis, a much higher mean threshold value of 375 nmol kg^{-1} FW was required to avoid growth retardation in the summer (Clark et al., 1989) and a similar relationship has been found in housed cattle (Stangl et al., 2000c). The apparent contrast has a simple explanation: within all flocks there is variation in vitamin B_{12} status and cobalt responsiveness, and individuals with below-average vitamin B_{12} status will contribute the most to the mean growth response. It is important to distinguish between biochemical criteria for individual and flock diagnosis of all mineral disorders (see Table 10.3). The current range of 'marginal' liver status for grazing cattle in New Zealand (75–250 nmol kg^{-1} FW) may be too low for cattle on grain-based diets, but cobalt-responsive groups cannot be distinguished by reference to liver vitamin B_{12} concentrations in the range 250–500 nmol kg^{-1} FW on such diets (see Tiffany et al., 2003; Tiffany and Spears, 2005). Any fatty infiltration will lead to an underestimation of effective vitamin B_{12} concentrations in the liver.

Total serum vitamin B_{12} in sheep

Correlations between the size of the mean growth response and mean serum or liver vitamin B_{12} in lambs in mid-trial were equally good and indicated that as serum values fell below 300 pmol l^{-1} there was an exponential increase in the size of the response (Fig. 19.5, p. 535). However, when the growth of individual lambs was plotted against their average serum vitamin B_{12} in one recent trial, values >220 pmol l^{-1} indicated a 95% probability of normal growth (Grace et al., 2003). A need to lower existing diagnostic guidelines for New Zealand was inferred. That view was apparently supported by findings in other trials in which lambs with minimum mean plasma vitamin B_{12} values <200 pmol l^{-1} showed no benefit from cobalt supplementation (Gruner et al., 2004c). However, plasma vitamin B_{12} values increased during those trials and final liver vitamin B_{12} values were normal and the earlier recommendations have been defended (Clark, 2005). The accuracy of prediction from a single sampling will depend on the previous and subsequent changes in plasma vitamin B_{12}. In Scotland, values generally decline between birth and weaning (Suttle et al., 1986). Since a small proportion of lambs benefiting from cobalt supplementation can cover the expense of treating the whole flock, large numbers of individuals should be sampled at random (at least ten and preferably one-tenth of the flock) and the sampling repeated, if possible (Grace et al., 2003). A note of caution should be attached to extrapolations from cobalt-responsive flocks in the above trials (Grace et al., 2003; Gruner et al., 2004c). In these trials, two-thirds of untreated lambs were removed, leaving highly selected populations of tolerant lambs. Suckling lambs with a mean plasma vitamin B_{12} of 140 pmol l^{-1} do not respond to supplementation (Grace and Sinclair, 1999; Gruner et al., 2004b) and higher diagnostic levels for the very young lamb and calf are therefore presented (Table 10.6).

Serum vitamin B_{12} in cattle

Assay results for cattle vary widely between laboratories and both diagnostic thresholds and responses to cobalt supplementation are much lower than those pertaining to sheep, whether microbiological or radioisotope dilution (RID) methods are employed (e.g. Givens and Simpson, 1983; Judson et al., 1997; Underwood and Suttle, 1999). Incomplete release of vitamin from binders can cause large underestimates of vitamin B_{12} (Price et al., 1993), but extraction at 100°C at a consistently high pH (9–12) overcomes both non-specific and residual binding problems (Babidge and Babidge, 1996). Contrasts between grazing sheep and cattle have since been encountered, as illustrated in Fig. 10.10 by the responses of grazing lambs and calves to the same dose (0.23 mg kg^{-1} live weight (LW)) of a slow-release, injectable formulation of vitamin B_{12}, albeit in separate trials. Pasture cobalt was adequate for lambs (0.07 mg kg^{-1} DM) (Grace and Lewis, 1999) and generous for calves (0.16 mg kg^{-1} DM) (Grace and West, 2000), yet untreated lambs had serum vitamin B_{12} values that were fourfold higher and the injection produced sustained increases in plasma vitamin B_{12} that were not seen in calves. Relationships between liver and serum vitamin B_{12} levels in lambs and calves can clearly be quite different, with the high liver vitamin B_{12} status of calves not being reflected by serum vitamin B_{12}, as measured by RID methods.

A variety of immunoassays developed for human blood samples (e.g. Vogeser and Lorenzl, 2007) have been used with bovine samples, but with conflicting results. Using a radioimmuno assay with alkaline extraction and an 'organic enhancer', Stangl et al. (2000c) claimed results similar to those of Price et al. (1993) when an increase in dietary cobalt from 0.08 to 0.2 mg kg^{-1} DM increased serum vitamin B_{12} in steers from 218 to 905 pmol l^{-1} after 43 weeks. With the same assay kit but no mention of an 'enhancer', it took between 112 and 132 days for an increment in dietary cobalt from 0.04 to 0.14 mg kg^{-1} DM to increase serum vitamin B_{12}, and the response was then small (from 48.7 to 132 pmol l^{-1}) (Tiffany et al., 2003). Results in this study were more characteristic of earlier, pre-1990 analyses. Unsupplemented cattle in the cobalt responsive phase of that study (fattening) could not be distinguished from those in the non-responsive phase (growing) in terms of plasma vitamin B_{12}, and neither could the cobalt responsiveness of steers given corn or barley diets (Tiffany and Spears, 2005). Chemiluminescent methods that do not rely on IF-binding may offer a way forward (Sato et al., 2002).

'Functional B_{12}' in serum

The analytical wheel may be turning full circle. In human clinical medicine, the current focus is on the 'metabolically active' component of plasma vitamin B_{12} bound to TCII, holotranscobalamin (Refsum et al., 2006; Brady et al., 2008). It could be that the earlier RID methods with acid extractants were fortuitously measuring only (or largely) the functional (TCII) B_{12} component in bovine samples, now regarded as the 'gold standard' for assessing vitamin B_{12} status in humans! The outstanding problem may be with ovine samples, where the use of assays that detect only TCII may improve specificity. Residual growth responses were found in lambs weaned off ewes treated with cobalt bullets during pregnancy, but these were not accompanied by a response in

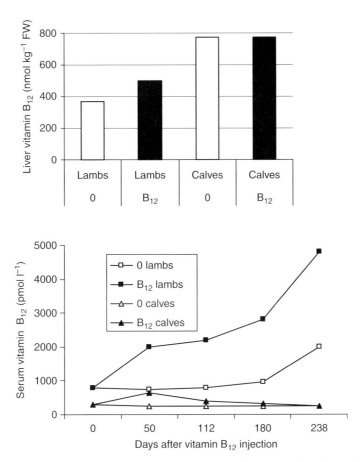

Fig. 10.10. Lambs and calves show contrasting responses in serial serum and final liver vitamin B_{12} concentrations to similar parenteral injections of vitamin B_{12} (0.23 mg kg^{-1} live weight), given in slow-release form (lamb data from Grace and Lewis, 1999; calf data from Grace and West, 2000). Pastures were of adequate cobalt content and the same radioisotope dilution method was used for vitamin B_{12} assays in both trials.

total plasma vitamin B_{12} (Gruner et al., 2004b), emphasizing the diagnostic limitations of total serum B_{12}. RIA methods that are specific for MeCbl and AdoCbl (i.e. the 'really functional' Cbl) may prove even more helpful than TCII.

values correlate well with liver vitamin B_{12} concentrations. Milk B_{12} levels <300 pmol l^{-1} indicate low vitamin status (Judson et al., 1997) and can be measured by optical biosensor methods (Indyk et al., 2002).

Vitamin B_{12} in milk

Analysis of vitamin B_{12} in milk has a particular advantage in assessing the status of cattle because there appear to be fewer analytical problems than for serum. Concentrations of vitamin B_{12} in milk can show large responses to supplementation when none is apparent in the plasma, and

Indicators of impaired propionate metabolism

Indicators of AdoCbl-related dysfunction such as raised urine and plasma or serum MMA provide a more reliable indication of cobalt responsiveness than vitamin B_{12} assays in weaned or adult sheep. However, plasma

MMA rises before appetite declines (Fisher and MacPherson, 1991) and can be raised in unresponsive flocks (Gruner et al., 2004c). For diagnostic purposes, emphasis should be placed on *individual* high MMA values rather than population means. If some weaned lambs in a grazing flock have MMA values above a marginal band of 7–14 µmol l^{-1} then growth retardation is likely (Gruner et al., 2004c). Relationships between serum MMA and serum vitamin B_{12} are curvilinear (Fig. 10.11), but the point of maximum curvature is influenced by diet. In a study of barley-fed lambs, increases in serum MMA occurred in most individuals with serum vitamin B_{12} levels of <220 pmol l^{-1} (Rice et al., 1987), but a higher threshold (375 pmol l^{-1}) was indicated in grazing sheep (Gruner et al., 2004c). A similar progression is seen as lambs age (Underwood and Suttle, 1999; Gruner et al., 2004c), reflecting the gradual replacement of milk lactose by propionate as the major energy substrate. Plasma MMA is less useful in suckled lambs and calves, since growth can be retarded before MMA increases (Quirk and Norton, 1987, 1988).

The pregnant and lactating ewe may show no abnormal MMA metabolism yet provide her offspring with insufficient vitamin B_{12} for optimal growth (Quirk and Norton, 1987). MMA can increase with the onset of lactation without accompanying changes in serum or liver vitamin B_{12}, possibly induced by higher propionate turnover (Fig. 10.9) (Gruner et al., 2004c). MMA is an insensitive marker in weaned cattle, showing only small increases in the cobalt-responsive animal (Tiffany and Spears, 2005). The suggestion that raised rumen succinate is a useful diagnostic aid (Wiese et al., 2007) is hard to fathom in view of the earliness of the increase (Fig. 10.6) and reciprocal relationship with rumen propionate. The diagnostic merits of plasma propionate have not been studied.

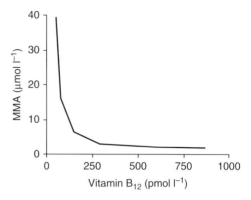

Fig. 10.11. The relationship between vitamin B_{12} and methylmalonic acid (MMA) concentrations in lambs grazing low-cobalt pasture is exponential, with vitamin B_{12} having to reach very low levels before MMA accumulates (Rice et al., 1987). A similar relationship has since been published (Gruner et al., 2004c).

and in the young suckled lamb. Early work focused on the increased excretion of another abnormal metabolite, FIGLU (Russel et al., 1975). Suckling calves and lambs both show increases in FIGLU excretion when growth is retarded by cobalt deprivation, but MeCbl dysfunction can become less important as they grow (Quirk and Norton, 1987, 1988). Ewes and heifers not excreting FIGLU in their urine can rear lambs that do so (Quirk and Norton, 1987, 1988), indicating that a different balance is struck between MeCbl- and AdoCbl-dependent pathways in dam and offspring (Fig. 10.9). Plasma homocysteine is used as a convenient marker of impaired methylation and is far more responsive than MMA in the weaned, cobalt-deprived calf. However, it may be affected relatively late in the weaned lamb that is deprived of cobalt on a barley diet (Fig. 10.7) and completely unaffected in lambs deprived while grazing (Furlong et al., 2010). Plasma concentrations of metabolites such as homocysteine may be influenced by factors other than vitamin B_{12} status, including the rate of conversion to methionine by transsulfuration in the gut mucosa, and transsulfuration rate may be influenced by larval nematode infection of that mucosa (Furlong et al., 2009). Reductions in liver folate concentrations are secondary consequences of impaired methionine synthase activity (e.g. Stangl et al., 2000c).

Indicators of abnormal methylation

Indicators of MeCbl dysfunction are the preferred diagnostic criteria in cattle of all ages

Other indicators of dysfunction

Other biochemical abnormalities have been reported in cobalt-deprived sheep, including low plasma glucose, ascorbic acid and thiamine concentrations and high 'liver enzymes', bile acids and pyruvate values, but these are not consistent or specific features (Table 10.1) (Ulvund, 1995; Underwood and Suttle, 1999). No biochemical marker in a suckling lamb or calf will indicate the full impact of a dietary cobalt deprivation that is severe enough to reduce milk yield in the mother.

Prevention and Control of Cobalt Deprivation

Vitamin B_{12} deprivation can be cured or prevented by a variety of methods. The choice of method is dictated by local circumstances and cost–benefit analysis.

Pasture fertilizers

The inclusion of cobalt salts or oxides in pasture fertilizers increases pasture cobalt and may also stimulate legume growth. Application rates and frequencies vary with the soil type, terrain and husbandry practice. In New Zealand hill country, a single application of 1.5 kg $CoSO_4.7H_2O$ ha^{-1} every 3–4 years has been found to be effective, while on more accessible sandy soils where phosphatic fertilizers are applied every year or two, less $CoSO_4$ suffices (Sherrell, 1984). Fertilizer treatment is unreliable on highly alkaline soils, such as calcareous shell sands and heavily limed soils, and on soils high in manganese oxides that fix cobalt in an unavailable form (Adams et al., 1969). Cobalt fertilizers are expensive and one way of using them sparingly is to treat strips of pasture only (MacPherson, 1983), and avoiding recently limed pastures altogether.

Dietary supplements

Cobalt salts can be added to loose mineral mixtures, licks and blocks, and to rations or drinking water for grazing or stall-fed animals. The former is the simplest and most economical method of cobalt supplementation and is commonly practised, regardless of need. A mineral mix containing 40 mg Co kg^{-1}, included at 25 kg t^{-1} in a concentrate that constitutes at least one-tenth of the total ration will provide sufficient cobalt. The provision of salt licks containing about 0.1% cobalt is generally effective under range conditions, but uptake may be inadequate in some individuals at vital times of the year. Continuous oral supplementation is recommended for ewes in late pregnancy to avoid loss of appetite and risk of pregnancy toxaemia, while increasing vitamin B_{12} status of the lambs. Cobalt supplementation of the pregnant heifer can produce massive increases in milk vitamin B_{12} (Quirk and Norton, 1988).

Parenteral or oral dosing

Early studies showed that injections of vitamin B_{12} rapidly alleviated signs of deprivation, but the commonly used dose in lambs (0.1 mg kg^{-1} LW) maintains normal serum vitamin B_{12} for <6 weeks (Grace et al., 1998; Underwood and Suttle, 1999) and repeated dosages become prohibitively expensive. Oral drenching is similarly effective, but the required frequency of dosing (every 2–4 weeks) makes it impractical unless combined with other therapies such as de-worming. Manufacturers of anthelmintic drenches commonly add cobalt to their products, but the amounts provided range widely and give short-lived increases in plasma vitamin B_{12} (e.g. Field et al., 1988). Responses have increased by 40% when doses for lambs are increased from 8 to 32 mg cobalt (Suttle et al., 1989). Conservative use of anthelmintics to avoid the build up of 'anthelmintic-resistant' nematodes on pasture restricts the efficacy of combining cobalt and anthelmintic doses.

Slow-release oral supplementation

Supplementary cobalt should be supplied continuously to maintain rumen liquor cobalt above the critical level of 5 μg l^{-1}, and this can be achieved by the use of cobalt-containing pellets or bullets. Early types consisted of cobalt oxide (60%) and their high specific gravity (4.5–5.0) facilitated retention in the rumino-reticulum for months or years, while slow dissolution supplied cobalt to the rumen bacteria (Underwood and Suttle, 1999). The efficacy of such pellets has been established under a wide range of conditions, including changes of formulation that halved the cobalt content (Judson et al., 1997) and combined it with zinc oxide for the prevention of facial eczema (Grace et al., 1997). In young lambs and calves, in which the rumen is not fully developed, and in lactating cows, in which feed intake is large and rumination frequent, a few animals regurgitate the pellets. A more serious problem, the development of an impervious coat of calcium phosphate on the surface of the pellet, was minimized by introducing a small steel screw with the pellet or by administering two pellets together to abrade the coating. A single treatment can supply sufficient cobalt to sheep for more than 5 years (Lee and Marston, 1969) and in beef cows for up to 19 months (Judson et al., 1997). Soluble glass boluses are not susceptible to coating and also yield a steady supply of cobalt for many months (Judson et al., 1988; Zervas, 1988; Masters and Peter, 1990; Kendall et al., 2001).

Slow-release parenteral supplementation

The slow release of soluble vitamin B_{12} can be contrived by incorporating it into biodegradable microspheres of lactide/glycolide polymer (Fig. 10.10) (Grace and Lewis, 1999). Single subcutaneous doses of 'microencapsulated' vitamin B_{12}, providing 0.14–0.24 mg kg^{-1} LW, have raised serum vitamin B_{12} concentrations for ≤ 187 days in lambs (Grace et al., 2003), 176 days in ewes (Grace et al., 2006) and 110 days in calves (Grace and West, 2000). The capacity of parenteral vitamin B_{12} to prevent cobalt deprivation in the grazing lamb (Grace et al., 2003) confirms that any changes in rumen microbial metabolism caused by low cobalt intakes are not in themselves growth-limiting.

Cobalt sources

Inorganic cobalt sources must be readily soluble in the rumen to be of nutritional value to ruminants when used as food supplements. While successful as a constituent of heavy pellets because of the large doses given, the relatively insoluble cobaltous and cobaltic oxides have lower nutritive value than more soluble inorganic salts ($CoCO_3$ and $CoSO_4$) when assessed by comparative increases in liver cobalt or vitamin B_{12} synthesis by cultured rumen microbes (Kawashima et al., 1997a,b). In an early study, Co_3O_4 and $CoSO_4$ were equally effective in raising serum and liver vitamin B_{12} (Andrews et al., 1966), but the amounts given (300 mg) far exceed the capacity for ruminal vitamin B_{12} synthesis (Fig. 10.4), thus putting the more soluble $CoSO_4$ at a disadvantage. Chelates have been no more effective in stimulating microbial vitamin B_{12} synthesis than simple inorganic salts whether the comparison was conducted in vitro between $CoCl_2$ and CoEDTA (N.F. Suttle, unpublished data) or glucoheptonate (Kawashima et al., 1997b), in vivo between CoEDTA and $CoSO_4$ in grazing lambs (Millar and Albyt, 1984) or in vivo between cobalt propionate and $CoCO_3$ in steers (Tiffany et al., 2003).

Protection from toxins

An abnormal substance that appears to affect cobalt requirements occurs in the perennial grass Phalaris tuberosa. Consumption of this grass is responsible, in restricted areas, for the disease in sheep and cattle known as 'Phalaris staggers' or 'Ranpha staggers'. This disease can be prevented but not cured by regular oral dosing with cobalt salts (50 μg Co twice daily or 28 mg Co week^{-1}) (Lee et al., 1957), by the use of cobalt pellets (Lee and Kuchel, 1953) or by treatment of pastures with a cobalt-containing fertilizer. P. tuberosa contains a neurotoxin that is responsible for the staggers syndrome, but the neurotoxin is inactivated or its absorption is reduced at high rumen cobalt concentrations. The action of cobalt in this respect differs from its action in controlling deficiency because administration of vitamin B_{12} is ineffective against Phalaris staggers. The amount of cobalt required for protection rises with increasing toxic potential of the pastures. In some areas, staggers does not develop because the cobalt intake from soil and herbage is sufficient to detoxify the neurotoxin. Cobalt supplements have also provided protection against annual ryegrass toxicity in sheep. The pathogenic agents are coryne toxins produced by a bacterium, *Clavibacter toxicus*, which is carried into the seed head by an invading nematode, *Anguina agrostis*. Cobalt supplements delay but do not prevent the onset of neurological signs, irrespective of the amount of toxin (0.15 or 0.30 mg kg^{-1} LW day^{-1}) or cobalt (4 or 16 mg day^{-1}) given (Davies et al., 1995).

Non-ruminants

Vitamin B_{12} performs the same functions in non-ruminants and ruminants. In pigs, administration of a methionine synthase inhibitor (nitrous oxide) induces symptoms of vitamin B_{12} deficiency (Kennedy et al., 1991a) and prolonged feeding of a vitamin B_{12}-free diet induces mild hyperhomocysteinaemia and macrocytic anaemia (Stangl et al., 2000a). Reduced activity of leucine 2,3 aminomutase has been implicated in pernicious anaemia in humans (Poston, 1980) and prolonged vitamin B_{12} deprivation of hens impairs the synthesis of leucine, but does not induce anaemia or signs of neurological impairment (Ward et al., 1988). Non-ruminants have low dietary needs for vitamin B_{12} and requirements are said to decline from 20 to 5 μg kg^{-1} DM as pigs grow from 5 to 50 kg LW (NRC, 1998), and from 9 to 3 μg kg^{-1} DM as Leghorn chicks progress between 0–6 and 6–12 weeks of age (NRC, 1984). However, there is no evidence to justify such scaling to body weight or age. Pigs and poultry rations that are derived entirely from plant sources contain little or no vitamin B_{12}, but they can be fed for long periods of time without greatly affecting the performance of laying hens (e.g. Ward et al., 1988) or growing pigs (Stangl et al., 2000a). Vitamin B_{12} is synthesized in the colon, but little or none is absorbed there. Faeces therefore contain vitamin B_{12} and useful amounts of the vitamin may be obtained via the consumption of faeces (Stangl et al., 2000a). If access to non-dietary vitamin B_{12} is denied, growth responses can be obtained from supplementary vitamin B_{12} or from the provision of feeds of animal or microbial origin to pigs and poultry. There are no reports of vitamin B_{12} deficiency in horses (Lewis et al., 1995). Clinical signs of cobalt (as opposed to vitamin B_{12}) deprivation have not been demonstrated in any species.

Cobalt Toxicity

Cobalt is commonly stated to be of low toxicity to all species, but this is only true if toxicity is assessed as a multiple of minimum requirement when there is a 100-fold margin of safety. In terms of dietary concentrations, cobalt is surpassed only by copper, selenium and iodine among the trace elements as a threat to health. Field cases of suspected cobalt toxicity in ruminants have been reported. There was no distinctive pathology, but liver cobalt was very high (20–69 mg kg^{-1} DM) (Dickson and Bond, 1974). The ARC (1980) has summarized the available evidence and – noting that >4 mg kg^{-1} and >1 mg kg^{-1} LW are toxic to sheep and young cattle, respectively – recommends that dietary cobalt for ruminants should not exceed 30 mg kg^{-1} DM. The NRC (2005) has reduced that figure to 25 mg kg^{-1} DM. Tolerance may be higher when cobalt is given continuously in the diet, rather than as a drench. Pigs have been found to tolerate 200 mg kg^{-1} DM, but higher levels (400 and 600 mg kg^{-1} DM) caused anorexia, stiffness, incoordination and muscular tremor (Huck and Clawson, 1976). Symptoms were alleviated by supplements of methionine or a combination of iron, manganese and zinc. Anaemia and decreases in tissue iron indicated a cobalt × iron antagonism. In day-old chicks given 125, 250 or 500 mg Co kg^{-1} DM for 14 days, the lowest level reduced feed intake, weight gain and gain to feed ratios, while the two higher levels caused pancreatic fibrosis, hepatic necrosis and muscle lesions (Diaz et al., 1994). In the chick embryo, cobalt was the third most toxic of eight elements examined, surpassed only by cadmium and arsenic (Gilani and Alibhai, 1990). The mechanism of cobalt toxicity may involve free-radical generation and consequent tissue damage (NRC, 2005).

References

Adams, S.N., Honysett, J.L., Tiller, K.G. and Norrish, K. (1969) Factors controlling the increase of cobalt in plants following the addition of a cobalt fertilizer. *Australian Journal of Soil Research* 7, 29–36.

AFRC (1993) *Energy and Protein Requirements of Ruminants.* An advisory manual prepared by the AFRC Technical Committee on Responses to Nutrients. Commonwealth Agricultural Bureaux, Farnham Royal, UK.

Andrews, E.D., Hart, L.I. and Stephenson, B.J. (1959) A comparison of the vitamin B_{12} and cobalt contents of livers from normal lambs, cobalt-dosed lambs and others with a recent history of mild cobalt deficiency disease. *New Zealand Journal of Agricultural Research* 2, 274–282.

Andrews, E.D., Stephenson, B.J., Anderson, J.P. and Faithful, W.C. (1958) The effect of length of pastures on cobalt-deficiency disease in lambs. *New Zealand Journal of Agricultural Research* 31, 125–139.

Andrews, E.D., Stephenson, B.J., Isaccs, C.E. and Register, R.H. (1966) The effects of large doses of insoluble and soluble forms of cobalt given at monthly intervals on cobalt-deficiency disease in lambs. *New Zealand Veterinary Journal* 14, 191–195.

ARC (1980) *The Nutrient Requirements of Ruminants.* Commonwealth Agricultural Bureaux, Farnham Royal, UK, pp. 184–185.

Bannerjee, R. (1999) *Chemistry and Biochemistry of B_{12}.* John Wiley & Co., New York, pp. 921.

Bassler, K.H. (1997) Enzymatic effects of folic acid and vitamin B_{12}. *International Journal of Vitamin and Nutrition Research* 67, 385–388.

Black, H., Hulton, J.B., Sutherland, R.J. and James, M.P. (1988) White liver disease in goats. *New Zealand Veterinary Journal* 36, 15–17.

Brady, J., Wilson, L., McGregor, L., Valente, E. and Orning, L. (2008) Active B_{12}: a rapid, automated assay for holotranscobalamin on the Abbot AxSYM analyser. *Clinical Chemistry* 54, 567–573.

Brebner, J. (1987) The role of soil ingestion in the trace element nutrition of grazing livestock. PhD Thesis, Imperial College, University of London, London.

Brebner, J., Suttle, N.F. and Thornton, I. (1987) Assessing the availability of ingested soil cobalt for the synthesis of vitamin B_{12} in the ovine rumen. *Proceedings of the Nutrition Society* 46, 766A.

Clark, R.G. (2005) Cobalt deficiency in sheep and diagnostic reference ranges. *New Zealand Veterinary Journal* 53, 265–266.

Clark, R.G., Burbage, J., Marshall, J. McD., Valler, T. and Wallace, D. (1986) Absence of vitamin B_{12} weight gain response in two trials with growing red deer (*Cervus elaphus*). *New Zealand Veterinary Journal* 34, 199–201.

Clark, R.G., Mantelman, L. and Verkerk, G.A. (1987) Failure to obtain weight gain response to vitamin B_{12} treatment in young goats grazing pasture that was cobalt-deficient for sheep. *New Zealand Veterinary Journal* 35, 38–39.

Clark, R.G., Wright, D.F., Millar, K.R. and Rowland, J.D. (1989) Reference curves to diagnose cobalt deficiency in sheep using liver and serum vitamin B_{12} levels. *New Zealand Veterinary Journal* 37, 1–11.

Davies, M.E. (1979) Studies on the microflora of the large intestine of the horse by continuous culture in an artificial colon. *Veterinary Research Communications* 3, 39–44.

Davies, S.C., White, C.L. and Williams, I.H. (1995) Increased tolerance to annual ryegrass toxicity in sheep given a supplement of cobalt. *Australian Veterinary Journal* 72, 221–224.

Diaz, G.J., Julian, R.J. and Squires, E.J. (1994) Lesions in broiler chickens following experimental intoxication with cobalt. *Avian Diseases* 38, 308–316.

Dickson, J. and Bond, M.P. (1974) Cobalt toxicity in cattle. *Australian Veterinary Journal* 50, 236.

Duncan, W.R.H., Morrison, E.R. and Garton, G.A. (1981) Effects of cobalt deficiency in pregnant and postparturient ewes and their lambs. *British Journal of Nutrition* 46, 337–343.

Farningham, D.A.H., Mercer, J.G. and Lawrence, C.B. (1993) Satiety signals in sheep: involvement of CCK, propionate and vagal CCK binding sites. *Physiology and Behaviour* 54, 437–442.

Fell, B.F. (1981) Pathological consequences of copper deficiency and cobalt deficiency. *Philosophical Transactions of the Royal Society of London, Series B* 294, 153–169.

Ferguson, E.G.W., Mitchell, G.B.B. and MacPherson, A. (1989) Cobalt deficiency and *Ostertagia circumcincta* infection in lambs. *Veterinary Record* 124, 20.

Field, A.C., Suttle, N.F., Brebner, J. and Gunn, G.W. (1988) An assessment of the efficacy and safety of selenium and cobalt included in an anthelmintic for sheep. *Veterinary Record* 123, 97–100.

Fisher, G.E.J. and MacPherson, A. (1991) Effect of cobalt deficiency in the pregnant ewe on reproductive performance and lamb viability. *Research in Veterinary Science* 50, 319–327.

Ford, J.E., Scott, K.J., Sansom, B.F. and Taylor, P.J. (1975) Some observations on the possible nutritional significance of vitamin B_{12}- and folate-binding proteins in milk. Absorption of [^{58}Co] cyano-cobalamin by suckling piglets. *British Journal of Nutrition* 34, 469–492.

Furlong, J.M., Sedcole, J.R. and Sykes, A.R. (2010) An evaluation of plasma homocysteine in the assessment of vitamin B_{12} status of pasture-fed sheep. *New Zealand Veterinary Journal* 58, 11–16.

Gawthorne, J.M. (1968) The excretion of methylmalonic and formiminoglutamic acids during the induction and remission of vitamin B_{12} deficiency in sheep. *Australian Journal of Biological Sciences* 21, 789–794.

Gawthorne, J.M. (1970) The effect of cobalt intake on the cobamide and cobinamide composition of the rumen contents and blood plasma of sheep. *Australian Journal of Experimental Biology and Medical Science* 48, 285–292.

Gawthorne, J.M. and Smith, R.M. (1974) Folic acid metabolism in vitamin B_{12}-deficient sheep. Effects of injected methionine on methotrexate transport and the activity of enzymes associated with folate metabolism in liver. *Biochemical Journal* 142, 119–126.

Gilani, S.H. and Alibhai, Y. (1990) Teratogenicity of elements to chick embryos. *Journal of Toxicology and Environmental Health* 30, 23–31.

Girard, C.L. (2002) A new look at the requirements of high-producing dairy cows for B-complex vitamins. In: Wiseman, J. and Garnsworthy, P.C. (eds) *Recent Developments in Ruminant Nutrition*. Nottingham University Press, Nottingham, UK, pp. 237–254.

Girard, C.L. and Matte, J.J. (2005) Effects of intramuscular injections of vitamin B_{12} on lactation performance of dairy cows fed dietary supplements of folic acid and rumen protected methionine. *Journal of Dairy Science* 88, 671–676.

Givens, D.I. and Simpson, V.R. (1983) Serum vitamin B_{12} concentrations in growing cattle and their relationship with growth rate and cobalt bullet therapy. *Occasional Publication No. 7, British Journal of Animal Production*, pp. 145–146.

Grace, N.D. (2006) Effect of ingestion of soil on the iodine, copper, cobalt (vitamin B_{12}) and selenium status in grazing sheep. *New Zealand Veterinary Journal* 54, 44–46.

Grace, N.D. and Lewis, D.H. (1999) An evaluation of the efficacy of microencapsulated vitamin B_{12} in increasing and maintaining serum and liver vitamin B_{12} concentrations in lambs. *New Zealand Veterinary Journal* 47, 3–7.

Grace, N.D. and Sinclair, G.R. (1999) Growth responses in lambs injected with a long-acting microencapsulated vitamin B_{12}. *New Zealand Veterinary Journal* 47, 213–214.

Grace, N.D. and West, D. (2000) Effect of an injectable microencapsulated vitamin B_{12} in increasing serum and liver vitamin B_{12} concentrations in calves. *New Zealand Veterinary Journal* 48, 70–73.

Grace, N.D., Clark, R.G. and Mortleman, L. (1986) Hepatic storage of vitamin B_{12} by the pregnant ewe and foetus during the third trimester. *New Zealand Journal of Agricultural Research* 29, 231–232.

Grace, N.D., Munday, R., Thompson, A.M., Towers, N.R., O'Donnell, K., McDonald, R.M., Stirnemann, M. and Ford, A.J. (1997) Evaluation of intraruminal devices for combined facial eczema control and trace element supplementation in sheep. *New Zealand Veterinary Journal* 45, 236–238.

Grace, N.D., West, D. and Sargison, N.R. (1998) The efficacy of a subcutaneous injection of soluble vitamin B_{12} in lambs. *New Zealand Veterinary Journal* 46, 194–196.

Grace, N.D., Knowles, S.O., Sinclair, G.R. and Lee, J. (2003) Growth responses to increasing doses of microencapsulated vitamin B_{12} and related changes in tissue B_{12} concentrations in cobalt-deficient lambs. *New Zealand Veterinary Journal* 51, 89–92.

Grace, N.D., Knowles, S.O. and West, D. (2006) Dose–response effects of long-acting injectable vitamin B_{12} plus selenium (Se) on the vitamin B_{12} and Se status of ewes and their lambs. *New Zealand Veterinary Journal* 54, 67–72.

Graulet, B., Matte, J.J., Desrochers, A., Doepel, L., Palin, M.-F. and Girard, C.L. (2007) Effects of dietary supplements of folic acid and vitamin B_{12} on metabolism of dairy cows in early lactation. *Journal of Dairy Science* 90, 3442–3455.

Gruner, T.M., Sedcole, J.R., Furlong, J.M., Grace, N.D., Williams, S.D., Sinclair, G.R., Hicks, J.D. and Sykes, A.R. (2004a) Concurrent changes in serum vitamin B_{12} and methylmalonic acid during cobalt or vitamin B_{12} supplementation of lambs while suckling and after weaning on properties in the South Island of New Zealand considered to be cobalt-deficient. *New Zealand Veterinary Journal* 52, 17–28.

Gruner, T.M., Sedcole, J.R., Furlong, J.M., Grace, N.D., Williams, S.D., Sinclair, G.R. and Sykes, A.R. (2004b) Changes in serum concentrations of MMA and vitamin B_{12} in cobalt-supplemented ewes and their lambs on two cobalt-deficient properties. *New Zealand Veterinary Journal* 52, 17–128.

Gruner, T.M., Sedcole, J.R., Furlong, J.M. and Sykes, A.R. (2004c) A critical evaluation of serum methylmalonic acid and vitamin B_{12} for the assessment of cobalt deficiency of growing lambs in New Zealand. *New Zealand Veterinary Journal* 52, 137–144.

Hart, L.I. and Andrews, E.D. (1959) Effect of cobaltic oxide pellets on the vitamin B_{12} content of ewes milk. *Nature (London)* 184, 1242–1243.

Hayakawa, T. (1962) Amounts of trace elements contained in grass produced in Japan. *National Institute of Animal Health Quarterly* 2, 172–181.

Hedrich, M.F., Elliot, J.M. and Lowe, J.E. (1973) Response in vitamin B_{12} production and absorption to increasing cobalt intake in the sheep. *Journal of Nutrition* 103, 1646–1651.

Huck, D.W. and Clawson, A.J. (1976) Excess dietary cobalt in pigs. *Journal of Animal Science* 43, 1231–1246.

Indyk, H.E., Persson, B.S., Caselunghe, M.C.B., Moberg, A., Filonzi, E.L. and Woolard, D.C. (2002) Determination of vitamin B_{12} in milk products and selected foods by optical biosensor protein-binding assay: method comparison. *Journal of AOAC International* 85, 72–81.

Johnson, E.H., Muirhead, D.E., Annamalai, K., King, G.J., Al-Busaidy, R. and Hameed, M.S. (1999) Hepatic lipidosis associated with cobalt deficiency in Omani goats. *Veterinary Research Communications* 23, 215–221.

Johnson, E.H., Al-Habsi, K., Kaplan, E., Srikandakumar, A., Kadim, I.T., Annamalai, K., Al-Busaidy, R. and Mahgoub, O. (2004) Caprine hepatic lipidosis induced through the intake of low levels of dietary cobalt. *The Veterinary Journal* 168, 174–179.

Judson, G.J., Brown, T.H., Kempe, B.R. and Turnbull, R.K. (1988) Trace element and vitamin B_{12} status of sheep given an oral dose of one, two or four soluble glass pellets containing copper, selenium and cobalt. *Australian Journal of Experimental Agriculture* 28, 299–305.

Judson, G.J., McFarlane, J.D., Mitsioulis, A. and Zviedrans, P. (1997) Vitamin B_{12} responses to cobalt pellets in beef cows. *Australian Veterinary Journal* 75, 660–662.

Judson, G.J., McFarlane, J.D., Riley, M.J., Milne, M.L. and Horne, A.C. (1982) Vitamin B_{12} and copper supplementation in beef calves. *Australian Veterinary Journal* 58, 249–252.

Jumba, I.O., Suttle, N.F., Hunter, E.A. and Wandiga, S.O. (1995) Effects of soil origin and mineral composition and herbage species on the mineral composition of forages in the Mount Elgon region of Kenya. 2. Trace elements. *Tropical Grasslands* 29, 47–52.

Kadim, I.T., Johnson, E.H., Mahgoub, O., Srikandakumar, A., Al-Ajmi, D., Ritchie, A., Annamalai, K. and Al-Halhali, A.S. (2003) Effect of low levels of dietary cobalt on apparent digestibility in Omani goats. *Animal Feed Science and Technology* 109, 209–216.

Kawashima, T., Henry, P.R., Bates, D.G., Ammerman, C.B., Littell, R.C. and Price, J. (1997a) Bioavailability of cobalt sources for ruminants. 2. Estimation of the relative value of reagent grade and feed grade cobalt sources from tissue cobalt accumulation and vitamin B_{12} concentrations. *Nutrition Research* 17, 957–974.

Kawashima, T., Henry, P.R., Bates, D.G., Ammerman, C.B., Littell, R.C. and Price, J. (1997b) Bioavailability of cobalt sources for ruminants. 3. *In vitro* ruminal production of vitamin B_{12} and total corrinoids in response to different cobalt sources and concentrations. *Nutrition Research* 17, 975–987.

Kendall, N.R., Jackson, D.W., Mackenzie, A.M., Illingworth, D.V., Gill, I.M. and Telfer, S.B. (2001) The effect of a zinc, cobalt and selenium soluble glass bolus on the trace element status of extensively grazed sheep over winter. *Animal Science* 73, 163–169.

Kennedy, D.G., Molloy, A.M., Kennedy, S., Scott, J.M., Blanchflower, W.J. and Weir, D.G. (1991a) Biochemical changes induced by nitrous oxide in the pig. In: Momcilovic, B. (ed.) *Proceedings of the Seventh International Symposium of Trace Elements in Man and Animals, Dubrovnik.* IMI, Zagreb, pp. 17-17–17-18.

Kennedy, D.G., Young, P.B., McCaughey, W.J., Kennedy, S. and Blanchflower, W.J (1991b) Rumen succinate production may ameliorate the effects of cobalt-vitamin B_{12} deficiency on methylmalonyl CoA mutase in sheep. *Journal of Nutrition* 121, 1236–1242.

Kennedy, D.G., Blanchflower, W.J., Scott, J.M., Weir, D.G., Molloy, A.M., Kennedy, S. and Young, P.B. (1992) Cobalt–vitamin B_{12} deficiency decreases methionine synthase activity and phospholipid methylation in sheep. *Journal of Nutrition* 122, 1384–1390.

Kennedy, D.G., Kennedy, S., Blanchflower, W.J., Scott, J.M., Weir, D.G., Molloy, A.M. and Young, P.B. (1994a) Cobalt–vitamin B_{12} deficiency causes accumulation of odd-numbered, branched chain fatty acids in the tissues of sheep. *British Journal of Nutrition* 71, 67–76.

Kennedy, D.G., Young, P.B., Blanchflower, W.J., Scott, J.M., Weir, D.G., Molloy, A.M. and Kennedy, S. (1994b) Cobalt–vitamin B_{12} deficiency causes lipid accumulation, lipid peroxidation and decreased alpha tocopherol concentrations in the liver of sheep. *International Journal of Vitamin Nutrition Research* 64, 270–276.

Kennedy, D.G., Young, P.B., Kennedy, S., Scott, J.M., Molloy, A.M., Weir, D.G. and Price, J. (1995) Cobalt–vitamin B_{12} deficiency and the activity of methyl malonyl CoA mutase and methionine synthase in cattle. *International Journal for Vitamin and Nutrition Research* 65, 241–247.

Kennedy, D.G., Kennedy, S. and Young, P.B. (1996) Effects of low concentrations of dietary cobalt on rumen succinate concentrations in sheep. *International Journal for Vitamin and Nutrition Research* 66, 86–92.

Kennedy, S., McConnell, S., Anderson, D.G., Kennedy, D.G., Young, P.B. and Blanchflower, W.J. (1997) Histopathologic and ultrastructural alterations of white liver disease in sheep experimentally depleted of cobalt. *Veterinary Pathology* 34, 575–584.

Kercher, C.J. and Smith, S.E. (1955) The response of cobalt-deficient lambs to orally administered vitamin B_{12}. *Journal of Animal Science* 14, 458–464.

Kincaid, R.L. and Socha, M.T. (2007) Effect of cobalt supplementation during late gestation and early lactation on milk and serum measures. *Journal of Dairy Science* 90, 1880–1886.

Lambert, B.D., Titgemeyer, E.C., Stokka, G.L., DeBey, B.M. and Loest, C.A. (2002) Methionine supply to growing steers affects hepatic activities of methionine synthase and betaine-homocysteine methyl transferase but not cystathione synthase. *Journal of Nutrition* 132, 2004–2009.

Lee, H.J. (1951) Cobalt and copper deficiencies effecting sheep in South Australia. Part 1. Symptoms and distribution. *Journal of Agricultural Science, South Australia* 54, 475–490.

Lee, H.J. and Kuchel, R.E. (1953) The aetiology of Phalaris staggers in sheep. 1. Preliminary observations on the preventive role of cobalt. *Australian Journal of Agricultural Research* 4, 88–99.

Lee, H.J. and Marston, H.R. (1969) The requirement for cobalt of sheep grazed on cobalt-deficient pastures. *Australian Journal of Agricultural Research* 20, 905–918.

Lee, H.J., Kuchel, R.E., Good, B.F. and Trowbridge, R.F. (1957) The aetiology of Phalaris staggers in sheep III. The preventive effect of various oral dose rates of cobalt. *Australian Journal of Agricultural Research* 8, 494–501.

Lee, S.H. and Hossner, K.L. (2002) Coordination of ovine adipose tissue gene expression by propionate *Journal of Animal Science* 80, 2840–2849.

Lewis, L.D., Knight, A., Lewis, B. and Lewis, C. (1995) In: Lewis, L.D. (ed.) *Equine Clinical Nutrition: Feeding and Care*. Wiley-Blackwell, New York, pp. 85–86.

Li, Z., McLaren, R.G. and Metherell, R.G. (2004) The availability of native and applied soil cobalt to ryegrass in relation to soil cobalt and manganese status and other soil properties. *New Zealand Journal of Agricultural Research* 47, 33–43.

Lines, E.W. (1935) The effect of the ingestion of minute quantities of cobalt by sheep affected with 'coast disease': a preliminary note. *Journal of the Council for Scientific and Industrial Research, Australia* 8, 117–119.

Ludemann, M. (2009) Development of an experimental approach to measure vitamin B_{12} production and absorption in sheep. PhD Thesis, Lincoln University, Christchurch, New Zealand, Chapter 5.

MacPherson, A. (1983) Oral treatment of trace element deficiencies in ruminant livestock. *Occasional Publication No. 7*. British Society of Animal Production, Edinburgh, pp. 93–103.

MacPherson, A., Gray, D., Mitchell, G.B.B. and Taylor, C.N. (1987) Ostertagia infection and neutrophil function in cobalt-deficient and cobalt-supplemented cattle. *British Veterinary Journal* 143, 348–355.

Marston, H.R. (1935) Problems associated with 'coast disease' in South Australia. *Journal of the Council for Scientific and Industrial and Industrial Research, Australia* 8, 111–116.

Marston, H.R. (1970) The requirement of sheep for cobalt or vitamin B_{12}. *British Journal of Nutrition* 24, 615–633.

Marston, H.R., Allen, S.H. and Smith, R.M. (1961) Primary metabolic defect supervening on vitamin B_{12} deficiency in the sheep. *Nature, UK* 190, 1085–1091.

Marston, H.R., Allen, S.H. and Smith, R.M. (1972) Production within the rumen and removal from the bloodstream of volatile fatty acids in sheep given a diet deficient in cobalt. *British Journal of Nutrition* 27, 147–157.

Masters, D.G. and Peter, D.W. (1990) Marginal deficiencies of cobalt and selenium in weaner sheep: response to supplementation. *Australian Journal of Experimental Agriculture* 30, 337–341.

McDonald, P. and Suttle, N.F. (1986) Abnormal fermentation in continuous cultures of rumen microorganisms given cobalt deficient hay or barley as the food substrate. *British Journal of Nutrition* 56, 369–378.

McKay, E.J. and McLery, L.M. (1981) Location and secretion of gastric intrinsic factor in the sheep. *Research in Veterinary Science* 30, 261–265.

McLoughlin, M.F., Rice, D.A. and McMurray, C.H. (1986) Hepatic lesions associated with vitamin B_{12} deficiency. In: *Proceedings of Sixth International Conference on Production Disease in Farm Animals, Belfast.* Veterinary Research Laboratory, Stormont, UK, pp. 104–107.

Millar, K.R. and Albyt, A.T. (1984) A comparison of vitamin B_{12} levels in the livers of sheep receiving treatments used to correct cobalt deficiency. *New Zealand Veterinary Journal* 32, 105–108.

Millar, K.R. and Lorentz, P.P (1979) Urinary methylmalonic acid as an indicator of the vitamin B_{12} status of grazing sheep. *New Zealand Veterinary Journal* 27, 90–92.

Millian, N.S. and Garrow, T.A. (1998) Human betaine-homocysteine transferase is a zinc metalloenzyme. *Archives of Biochemistry and Biophysics* 356, 93–98.

Minson, D.J. (1990) *Forages in Ruminant Nutrition*. Academic Press, San Diego, California, pp. 208–229.

Mitsioulis, A., Bansemer, P.C. and Koh, T.-S. (1995) Relationship between vitamin B_{12} and cobalt concentrations in bovine liver. *Australian Veterinary Journal* 72, 70.

Nexo, E. and Olesen, H. (1982) Intrinsic factor, transcobalamin and haptocorrin. In: Dolphin, D.D. (ed.) B_{12}–Volume 2. *Biochemistry and Medicine*. John Wiley & Sons, New York, pp. 57–86.

NRC (1998) *Nutrient Requirements of Swine*, 10th edn. National Academy of Sciences, Washington, DC.

NRC (2005) *Mineral Tolerance of Animals*, 2nd edn. National Academy of Sciences, Washington, DC.

Polak, D.M., Elliot, J.M. and Haluska, M. (1979) Vitamin B_{12} binding proteins in bovine serum. *Journal of Dairy Science* 62, 697–701.

Poston, J.M. (1980) Cobalamin-dependent formation of leucine and β-leucine by rat and human tissue. *Journal of Biological Chemistry* 255, 10067–10072.

Poston, J.M. and Stadman, T.C. (1975) Cobamides as cofactors: methyl cobamides and the synthesis of methionine, methane and acetate. In: Babior, B.M. (ed.) *Cobalamin: Biochemistry and Pathophysiology*. John Wiley & Sons, New York, pp. 111–140.

Price, J. (1991) The relative sensitivity of vitamin B_{12}-deficient propionate and 1-carbon metabolism to low cobalt intake in sheep. In: Momcilovic, B. (ed.) *Proceedings of Seventh International Symposium on Trace Elements in Man and Animals, Dubrovnik*. IMI, Zagreb, pp. 27-14–27-15.

Price, J., Ueno, S. and Wood, S.G. (1993) Recent developments in the assay of plasma vitamin B_{12} in cattle. In: Anke, M., Meissner, D. and Mills, C.F. (eds) *Proceedings of the Eighth International Symposium on Trace Elements in Man and Animals*. Verlag Media Touristik, Gersdof, Germany, pp. 317–318.

Quadros, E.V., Matthews, D.M., Hoffbrand, A.V. and Linnell, J.C. (1976) Synthesis of cobalamin coenzymes by human lymphocytes *in vitro* and the effects of folates and metabolic inhibitors. *Blood* 48, 609–619.

Quirk, M.F. and Norton, B.W. (1987) The relationship between the cobalt nutrition of ewes and the vitamin B_{12} status of ewes and their lambs. *Australian Journal of Agricultural Research* 38, 1071–1082.

Quirk, M.F. and Norton, B.W. (1988) Detection of cobalt deficiency in lactating heifers and their calves. *Journal of Agricultural Science, Cambridge* 110, 465–470.

Refsum, H., Johnston, C., Guttormsen, A.B. and Nexo, E. (2006) Holotranscobalamin in human plasma: determination, determinants and reference values in healthy adults. *Clinical Chemistry* 52, 129–137.

Rice, D.A., McLoughlin, M., Blanchflower, W.J., Goodall, E.A. and McMurray, C.H. (1987) Methyl malonic acid as an indicator of vitamin B_{12} deficiency in sheep. *Veterinary Record* 121, 472–473.

Riedel, B., Fiskerstrand, T., Refsum, H. and Ueland, P.M. (1999) Co-ordinate variations in methylmalonyl-CoA mutase and methionine synthase and the cobalamin factors in human glioma cells during nitrous oxide exposure and the subsequent recovery. *Biochemical Journal* 341, 133–138.

Russel, A.J.F., Whitelaw, A., Moberley, P. and Fawcett, A.R. (1975) Investigation into diagnosis and treatment of cobalt deficiency in lambs. *Veterinary Record* 96, 194–198.

Sanschi, D.E., Berthiaume, R., Matte, J.J., Mustafa, A.F. and Girard, C.L (2005) Fate of supplementary B-vitamins in the gastrointestinal tract of dairy cows. *Journal of Dairy Science* 88, 2043–2054.

Sato, K., Muramatsu, M. and Amano, S. (2002) Application of vitamin B_{12}-targeting site on *Lactobacillus helveticus* B-1 to vitamin B_{12} assay by chemiluminescence method. *Analytical Biochemistry* 308, 1–4.

Schulz, W.J. (1987a) Unsaturated vitamin B_{12} binding capacity in human and ruminant blood serum – a comparison of techniques including a new technique by high performance liquid chromatography. *Veterinary Clinical Pathology* 16, 67–72.

Schulz, W.J. (1987b) A comparison of commercial kit methods for assay of vitamin B_{12} in ruminant blood. *Veterinary Clinical Pathology* 16, 102–106.

Schwab, E.C., Schwab, C.G., Shaver, R.D., Girard, C.L., Putnam, D.E. and Whitehouse, N.L. (2006) Dietary forage and nonfiber carbohydrate contents influence B-vitamin intake, duodenal flow and apparent ruminal synthesis in lactating dairy cows. *Journal of Dairy Science* 89, 174–187.

Schwan, O., Jacobsson, S.-O., Frank, A., Rudby-Martin, L. and Petersson, L.R. (1987) Cobalt and copper deficiency in Swedish Landrace Pelt sheep. *Journal of Veterinary Medicine A* 34, 709–718.

Schwarz, F.J., Stangl, G.L. and Kirchgessner, M. (2000) Cobalt requirement of beef cattle feed intake and growth at different levels of cobalt supply. *Journal of Animal Physiology and Nutrition* 83, 121–131.

Scott, J.M. (1992) Folate–vitamin B_{12} interrelationships in the central nervous system. *Proceedings of the Nutrition Society* 51, 219–224.

Seetheram, B., Bose, S. and Li, N. (1999) Cellular import of cobalamin (vitamin B_{12}). *Journal of Nutrition* 129, 1761–1764.

Shallow, M., Ellis, N.J.S. and Judson, G.J. (1989) Sex-related responses to vitamin B_{12} and trace element supplementation in prime lambs. *Australian Veterinary Journal* 66, 250–251.

Sherrell, C.G. (1984) Cobalt deficiency top-dressing recommendations. *Aglink FPD 814*, Ministry of Agriculture and Fisheries, Wellington, New Zealand.

Smith, E.L. (1948) Presence of cobalt in the anti-pernicious anaemia factor. *Nature, UK* 162, 144–145.

Smith, R.M. (1997) Cobalt. In: O'Dell, B.L. and Sunde, R.A. (eds) *Handbook of Nutritionally Essential Mineral Elements*. Marcel Dekker Inc., New York, pp. 357–388.

Smith, R.M. and Marston, H.R. (1970a) Production, absorption, distribution and excretion of vitamin B_{12} in sheep. *British Journal of Nutrition* 24, 857–877.

Smith, R.M. and Marston, H.R. (1970b) Some metabolic aspects of vitamin B_{12} deficiency in sheep. *British Journal of Nutrition* 24, 879–891.

Smith, S.E., Koch, B.A. and Turk, K.L. (1951) The response of cobalt-deficient lambs to liver extract and vitamin B_{12}. *Journal of Nutrition* 144, 455–464.

Somers, M. and Gawthorne, J.M. (1969) The effect of dietary cobalt intake on the plasma vitamin B_{12} concentration of sheep. *Australian Journal of Experimental Biology and Medical Science* 47, 227–233.

Stangl, G.L., Roth-Maier, D.A. and Kirchgessner, M. (2000a) Vitamin B_{12} deficiency and hyperhomocysteinaemia are partly ameliorated by cobalt and nickel supplementation in pigs. *Journal of Nutrition* 130, 3038–3044.

Stangl, G.L., Schwarz, F.J., Jahn, B. and Kirchgessner, M. (2000b) Cobalt-deficiency-induced hyperhomocysteinaemia and oxidative stress in cattle. *British Journal of Nutrition* 83, 3–6.

Stangl, G.L., Schwarz, F.J., Muller, H. and Kirchgessner, M. (2000c) Evaluation of cobalt requirement of beef cattle based on vitamin B_{12}, folate, homocysteine and methylmalonic acid. *British Journal of Nutrition* 84, 645–653.

Stemme, K., Meyer, U., Flachowsky, G. and Scholz, H. (2006) The influence of an increased cobalt supply to dairy cows on the vitamin B status of their calves. *Journal of Animal Physiology and Animal Nutrition* 90, 173–176.

Sutherland, R.J., Cordes, D.O. and Carthew, G.C. (1979) Ovine white liver disease: a hepatic dysfunction associated with vitamin B_{12} deficiency. *New Zealand Veterinary Journal* 27, 227–232.

Suttle, N.F. (1988) The role of comparative pathology in the study of copper and cobalt deficiencies in ruminants. *Journal of Comparative Pathology* 99, 241–258.

Suttle, N.F. (1995) Relationship between vitamin B_{12} and cobalt concentrations in bovine liver. *Australian Veterinary Journal* 72, 278.

Suttle, N.F. (2000) Minerals in livestock production. *Asian-Australasian Journal of Animal Science* 13, 1–9.

Suttle, N.F., Jones, D., Clark, M. and Coop, R. (1986) Studies on the trace element/infection interface in lambs. In: *Proceedings of Sixth International Conference on Production Disease in Farm Animals, Belfast*. Veterinary Research Laboratory, UK, Northern Ireland, pp. 195–198.

Suttle, N.F., Brebner, J., Munro, C.S. and Herbert, E. (1989) Towards an optimum dose of cobalt in anthelmintics in lambs. *Proceedings of the Nutrition Society* 48, 87A.

Suttle, N.F., Bell, J., Thornton, I. and Agyriaki, A. (2003) Predicting the risk of cobalt deprivation in grazing livestock from soil composition data. *Environmental Geochemistry and Health* 25, 33–39.

Sutton, A.L. and Elliot, J.M. (1972) Effect of ratio of roughage to concentrate and level of feed intake on ovine ruminal vitamin B_{12} production. *Journal of Nutrition* 102, 1341–1346.

Tiffany, M.E. and Spears, J.W. (2005) Differential responses to dietary cobalt in finishing steers fed corn- versus barley-based diets. *Journal of Animal Science* 83, 2580–2589.

Tiffany, M.E., Spears, J.W., Xi, L. and Horton, J. (2003) Influence of dietary cobalt source and concentration on performance, vitamin B_{12} status and ruminal and plasma metabolites in growing and finishing steers. *Journal of Animal Science* 81, 3151–3159.

Tiffany, M.E., Fellner, V. and Spears J.W. (2006) Influence of cobalt concentration on vitamin B_{12} production and fermentation of mixed ruminal microorganisms in continuous culture flow-through fermenters. *Journal of Animal Science* 84, 635–640.

Ulvund, M.J. (1995) Kobaltmangel hos sau. [Cobalt deficiency in lambs.] *Nordiske Veterinaria Tidsskrift* 107, 489–401.

Underwood, E.J. and Filmer, J.F. (1935) The determination of the biologically potent element (cobalt) in limonite. *Australian Veterinary Journal* 11, 84–92.

Underwood, E.J. and Suttle, N.F. (1999) *The Mineral Nutrition of Livestock*, 3rd edn. CAB International, Wallingford, UK.

Vellema, P., Rutten, V.P., Hoek, A., Moll, L. and Wentink, G.H. (1996) The effect of cobalt supplementation on the immune response in vitamin B_{12}-deficient Texel lambs. *Veterinary Immunology and Immunopathology* 55, 151–161.

Vogeser, M. and Lorenzl, S. (2007) Comparison of automated assays for the determination of vitamin B_{12} in serum. *Clinical Biochemisty* 40, 1342–1345.

Wahle, K.W.J. and Duncan, W.R.H. and Garton, G.A. (1979) Propionate metabolism in different species of ruminant. *Annales de Recherche Veterinaire* 10, 362–364.

Walker, C.K. and Elliot, J.M. (1972) Lactational trends in vitamin B_{12} status on conventional and restricted-roughage rations. *Journal of Dairy Science* 55, 474–478.

Watanabe, F., Nakano, Y., Tachikake, N., Saido, H., Tamura, Y. and Yamanaka, H. (1991) Vitamin B_{12} deficiency increases the specific activities of rat liver NADH- and NADPH-linked aquocobalamin reductase isoenzymes involved in coenzyme synthesis. *Journal of Nutrition* 121, 1948–1954.

Wiese, S.C., White, C.L., Williams, I.H. and Allen, J.G. (2007) Relationships between plasma methylmalonic acid and ruminal succinate in cobalt-deficient and repleted-sheep. *Australian Journal of Agricultural Research* 58, 367–373.

Wilson, P.R. and Grace, N.D. (2001) A review of tissue reference values used to assess the trace element status of farmed deer (*Cervus elaphus*). *New Zealand Veterinary Journal* 49, 126–132.

Wuerges, J., Garau, G., Geremia, S., Fedosov, S.N., Petersen, T.E. and Randaccio, L. (2006) Structural basis for mammalian vitamin B_{12} transport by transcobalamin. *Proceedings of the National Academy of Sciences USA* 103, 4386–4391.

Zervas, G. (1988) Use of soluble glass boluses containing Cu, Co and Se in the prevention of trace element deficiencies in goats. *Journal of Agricultural Science, Cambridge* 110, 155–158.

Zinn, R.A., Owens, F.N., Sturat, R.L., Dunbar, J.R. and Norman, B.B. (1987) B-vitamin supplementation of diets for feedlot calves. *Journal of Animal Science* 65, 267–277.

11 Copper

Copper was first shown to be essential for growth and haemoglobin formation in laboratory rats by Hart and Elvejhem in 1928 and subsequently for the prevention of a wide range of clinical and pathological disorders in several monogastric species. A number of naturally occurring wasting diseases in grazing animals were soon found to respond to copper therapy, such as 'salt-sick' of cattle in Florida (Neal et al., 1931), 'lecksucht' of sheep and cattle in the Netherlands (Sjollema, 1933) and 'copper pine' in UK calves (Jamieson and Allcroft, 1947). An ataxic disease of newborn lambs ('swayback') in Western Australia was attributed to copper deficiency in the ewe during pregnancy (Bennetts and Chapman, 1937). A similar disease had troubled flocks consuming seaweed along the coast of Iceland long before it was recognized as swayback (Palsson and Grimsson, 1953).

Subsequently, extensive copper-deficient areas, affecting both crops and livestock, were discovered throughout the world. The importance of an interaction between copper and molybdenum was discovered when a severe diarrhoea ('teart') of cattle, associated with *excessive* molybdenum in the herbage in Somerset, UK, was controlled by massive doses of copper (Ferguson et al., 1938, 1943). Concurrent investigations of another 'area' problem, a chronic poisoning of sheep and cattle in eastern Australia, found abnormally *low* molybdenum concentrations in the herbage and a disorder controllable by supplementing with molybdenum (Dick and Bull, 1945). Underlying both problems was a profound effect of molybdenum on copper metabolism that depended on a further interaction with a third element, sulfur, revealed in a classic series of experiments with sheep that used sulfate (SO_4) as the source (Dick, 1956). This three-way interaction between copper, molybdenum and sulfur is involved in most outbreaks of copper-responsive disorder (RD) in grazing livestock, even those in Icelandic sheep that consumed seaweed rich in SO_4 (Suttle, 1991). Since the mid-2000s, studies with laboratory animals have rapidly advanced our understanding of the complex ways in which this copper is safely absorbed, transported and incorporated into functional enzymes and proteins (Prohaska, 2006). Ironically, copper-RD rarely occurs naturally in non-ruminants, and the emphasis throughout this chapter will be placed on the more vulnerable ruminant animal.

Functions of Copper

Copper is present in and essential for the activity of numerous enzymes (Table 11.1), cofactors and reactive proteins. The essentiality of copper for processes such as reproduction and bone development is presumed

Table 11.1. Some copper-dependent enzymes found in mammalian tissues,[a] their functions and possible consequences of a marked reduction in activity (from Bonham et al., 2002; Prohaska, 2006).

Enzyme	Functions	Abbreviation	Pathognomic significance
Caeruloplasmin[b] (ferroxidase)	$Fe^{2+} \to Fe^{3+}$, hence Fe transport; antioxidant	Cp	Anaemia
Cytochrome c oxidase	Terminal electron-transfer respiratory chain	CCO	Anoxia (causing neuronal degeneration, cardiac hypertrophy)
Diamine oxidase	Oxidative deamination of diamines and their derivatives	DAO	Unknown
Dopamine-β-monooxygenase	Catecholamine metabolism	DBM	Behaviour?
Hephaestin	Export of iron from intestine	HEP	Anaemia
Ferroxidase II	Iron oxidation		Anaemia
Lysyl oxidases	Desmosine cross-linkages in connective tissues	LO	Aortic rupture, joint disorders, osteoporosis
Monoamine oxidase	Oxidative deamination of monoamines	MAO	Cell signalling, leukocyte trafficking
Peptidylglycine-α amidating monooxygenase	Elaboration of numerous biogenic molecules (e.g. gastrin)	PAM	Appetite?
Superoxide dismutases	Intracellular and extracellular dismutation of O_2^- to H_2O_2	SOD1 SOD2	Lipid peroxidation, modulation of vascular tone
Thiol oxidase	Disulfide bond formation	TO	Loss of wool and hair strength
Tyrosinase	Tyrosine $\to$ melanin	TY	Depigmentation

[a]These enzymes are found in varying concentrations and excesses in most tissues. Some are supplied with copper by specific 'chaperone' proteins (e.g. SOD1 and SOD2) and their functions may overlap with each other (e.g. caeruloplasmin (Cp) and HEP) or with those of copper-independent enzymes (e.g. MnSOD). Reductions in activity are not therefore synonymous with dysfunction or copper-responsive disorder.
[b]Multiple forms and/or genes have been identified and may indicate a level of functional importance that is underestimated by survival in specific gene-deletion studies.

to depend on such functions, but only two processes have unequivocal dependency on a specific enzyme: pigmentation (tyrosinase) and connective tissue development (lysyl oxidases) (Suttle, 1987a; Prohaska, 2006). Genes controlling many copper-dependent enzymes and proteins have been identified (e.g. the sheep caeruloplasmin gene, sCp) (Lockhart and Mercer, 1999) and gene-deletion studies with laboratory animals have shown some to be necessary for embryonic survival (e.g. dopamine-β-monooxygenase (DBM) and the transmembrane copper-transporting protein Ctr1). The links between functional molecule and physiological process may be clarified by new techniques that follow the sequence in which the animal tries to avert the consequences of copper deprivation by 'switching on' specific 'copper genes'. Details of three enzyme functions involving copper proteins follow.

Cellular respiration

The ubiquitous haem enzyme complex, cytochrome c oxidase (CCO), is responsible for the terminal electron transfer in the respiratory chain and thus for energy generation in all tissues. Low CCO activity may cause several dysfunctions, including impairment of the respiratory burst in neutrophils that is integral to cellular immunity (Jones and Suttle, 1987). In copper-depleted cattle, a breakdown of basement membranes in the acinar cells of the pancreas is one of the earliest pathological lesions, where a reduction of CCO activity may have secondary effects on glycosylation or sulfation (Fell et al., 1985). CCO activity in the liver (Mills et al., 1976), gut mucosa (Fell et al., 1975; Suttle and Angus, 1978) and neutrophils is also reduced, but there is usually a surplus of CCO in all tissues (Smith et al., 1976) and the respiration rate is not necessarily constrained. Gene-deletion studies in laboratory mice show that animals cannot survive without CCO, but they are equally dependent on DBM (Prohaska, 2006).

Protection from oxidants

Copper may protect tissues against 'oxidant stress' from free radicals, including those generated during respiration, and interact with other nutrients with antioxidant properties as it does so. Neutrophils rely upon the 'respiratory burst' to generate free radicals such as superoxide ($O_2^{\cdot-}$), which kill engulfed pathogens and must have fully functional oxidant defences to avoid 'self-harm'. *In vitro* tests show that blastogenesis (Jones and Suttle, 1987) or phagocytic killing (Boyne and Arthur, 1981; Jones and Suttle, 1981; Xin et al., 1991) is compromised in cultured cells from copper-deprived sheep and cattle. The superoxide dismutases (SODs) SOD1 and SOD2 dispose of $O_2^{\cdot-}$ (to H_2O_2) and their activities in neutrophils are reduced by copper deprivation. Caeruloplasmin (Cp), the predominant copper protein in plasma, is expressed in several tissue types (Prohaska, 2006) and also scavenges free radicals (Saenko et al., 1994). Impaired responses of splenic lymphocytes and neutrophils have been reported *in vitro* in marginally depleted male rats showing *no* reduction in liver copper or plasma Cp concentrations (Bonham et al., 2002). Hepatic activity of the iron-containing haem enzyme catalase is reduced in copper deficiency (Taylor et al., 1988). Catalase metabolizes H_2O_2, source of the dangerous OH^- radical and a modulator of cell signalling.

Iron transport

The historic role of Cp in iron transport has been overshadowed by recent research into hephaestin, a copper protein with 50% homology to Cp and similar properties, including ferroxidase activity. When animals are deprived of copper, hephaestin concentrations in enterocytes decrease and iron concentrations increase, probably because of a lack of ferroxidase activity. Deletion of the *Cp* gene produces only a mild anaemia, suggesting that hephaestin and ferroxidase II may be more important contributors of ferroxidase activity (Prohaska, 2006).

Functions of copper-binding proteins

There are more copper-binding proteins than there are cuproenzymes, though not all have proven functions (Bonham et al., 2002; Prohaska, 2006). The metallothioneins (MTs), especially MTII (Lee et al., 1994), bind copper and may have protective roles during exposure to excess copper. Two clotting agents, factors 5 and 8, are cuproproteins with copper-dependent activity (Prohaska, 2006) and their lack in copper deprivation may impair blood clotting by weakening platelet adhesion (Lominadze et al., 1997). Factor 8 is structurally similar to Cp (Saenko et al., 1994). Two copper-transporter proteins, Crt1 and 2, are integral to the transfer of copper across membranes, facilitating uptake by and efflux from both enterocytes and hepatocytes. Prion proteins may facilitate copper storage in the central nervous system (CNS) and they become deranged in bovine spongiform encephalopathy (Hanlon et al., 2002), but a hypothesis that the substitution of manganese for copper on prion proteins was of pathogenic significance has found no support (Legleiter et al., 2008).

Clinical Signs of Copper Deprivation in Ruminants

A wide range of copper-RD have been associated with low copper status (Suttle, 1987a) and come under the clinical term 'hypocuprosis'. Susceptibility depends on the species and the stage of development at the time of copper deprivation.

Ataxia

Following the pioneering field studies of Bennetts and Chapman (1937), swayback was eventually reproduced experimentally by feeding pregnant ewes diets low in copper or high in molybdenum and sulfur (Suttle, 1988a). Three types of ataxia occur naturally in lambs:

- neonatal (paralysis or ataxia at birth, followed by death; primary anoxic lesions in the brain stem and demyelination of the cerebral cortex);
- delayed (uncoordinated hind-limb movements, stiff and staggering gait, swaying hind quarters, often triggered by flock disturbances); and
- atypical (older lambs stand transfixed, head quivering and apparently blind; primary lesion, cerebral oedema).

The particular vulnerability of the lamb to neonatal ataxia is related to a phase of rapid myelination in the fetal CNS in mid-pregnancy. There is a second phase of spinal cord myelination a few weeks after birth and copper deprivation from late pregnancy causes delayed ataxia. Demyelination is caused by myelin aplasia rather than myelin degeneration and is associated with degeneration of the motor neurons of the brain and spinal cord in sheep (Suttle, 1988a) and goats (Wouda et al., 1986). The lesions of each type of swayback are irreversible, and may commence as early as 6 weeks before birth and continue to develop afterwards. They are arguably the most sensitive clinical consequence of copper deprivation in sheep (Suttle et al., 1970) and swayback may be the only abnormality to develop in a flock. In the goat kid, the delayed form is predominant (Hedger et al., 1964) and cerebella hypoplasia is an additional feature (Wouda et al., 1986). Calves undergo a slow regular myelination of the CNS and do not succumb to neonatal or delayed ataxia (Suttle, 1987a), but deer can be affected (Grace and Wilson, 2002). Adult pregnant yak (Bos grunniens) are affected by a copper-RD called 'shakeback disease' because of an unsteady gait (Xiao-Yun et al., 2006a,b). However, no neuropathological investigations have been conducted and there are no reports of abnormalities in their offspring.

Abnormal wool and hair

When copper deprivation occurs in sheep after myelination is complete, it is the fleece that displays abnormality: wool crimp is progressively lost until fibres emerge almost straight, giving rise to 'stringy' and 'steely' wool with diminished tensile strength and elastic properties. The zonal changes are irreversible, but copper supplementation quickly restores normal properties in new wool growth (Underwood, 1977). Abnormalities are most obvious in the wool of Merino sheep, which is normally heavily crimped, and 'steely wool' remained a problem in South Australia long after the initial link with copper was established (Hannam and Reuter, 1977). Similar abnormalities have been induced in experiments by depriving other breeds of copper (Lee, 1956) and occur in Scottish Blackface lambs on improved hill pasture as an early sign of deprivation (Whitelaw et al., 1979). Changes in the growth and physical appearance of the hair, giving the coat a thin, wavy, harsh appearance, have also been also reported in cattle (Fig. 1.6) and are again an early sign of copper deprivation (Suttle and Angus, 1976).

Depigmentation

In breeds of cattle with highly pigmented coats, loss of coat colour (achromotrichia) is usually the earliest and sometimes the only clinical sign of copper deprivation. Greying of black or bleaching of brown hair is sometimes seen, especially round the eyes (Fig. 11.1). In the Aberdeen Angus, a brownish tinge can be given to the coat and the skin becomes mottled (Hansen et al., 2009). White wool develops in normally black-woolled sheep and pigmentation is so sensitive to changes in copper intake that unpigmented bands can be produced by intermittent copper deprivation (Underwood, 1977). Depigmentation can be rapidly induced

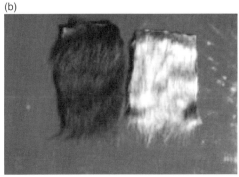

Fig. 11.1. Clinical symptoms of copper deficiency in ruminants include (a) loss of hair or fleece colour, giving a 'spectacled' appearance around the eye; (b) loss of hair or wool strength, causing loss of crimp in wool; and (c) skeletal abnormalities, including osteoporosis and widening of epiphyses.

by dosing sheep with molybdenum and SO_4. Inclusion of a few black-fleeced 'sentinel' sheep in a flock can provide early warning of more widespread dysfunction. Loss of coat colour is not a recognizable feature of yak with 'shakeback' (Xiao-Yun et al., 2006a,b).

Anaemia

Anaemia develops after severe or prolonged copper deprivation. In lambs it is hypochromic and microcytic in type, similar to the anaemia of iron deficiency, while in cows and ewes it may be hypochromic and macrocytic (Suttle, 1987a). Anaemia was induced in Scottish Blackface ewes but not in Cheviot × Border Leicester ewes given similar diets (Suttle et al., 1970), a result that may reflect the greater severity of copper deprivation inducible in the former breed. Haemolysis may contribute to the development of anaemia. Signs of oxidative stress in the form of Heinz bodies in erythrocytes have been reported in the copper-deficient lamb (Suttle et al., 1987a), and low activities of erythrocyte SOD1 have been implicated in the incidence of haemolytic Heinz-body anaemia in hypocupraemic cattle given kale (Barry et al., 1981). The pregnant or lactating yak is unusual in developing hypochromic microcytic anaemia, which is responsive to copper supplementation, as an early consequence of copper deprivation (Xiao-Yun et al., 2006a,b).

Bone disorders

Abnormalities in bone development vary widely both within and between species. The disturbances of endochondral ossification that give rise to uneven bone growth (osteochondrosis) can only affect growing animals and bone morphology will be influenced by the rate of growth, body-weight distribution, movement and even by the rate of hoof growth at the time of copper deprivation. Widening of the epiphyses of the lower-limb bones is a common manifestation in growing cattle (Fig. 11.1c). In the absence of these 'rickets-like' lesions, osteitis fibrosa has been found in the vertebra of copper-deprived calves (Mills et al., 1976). 'Beading' of the ribs is seen in sheep and cattle and may be due to overgrowth of costochondral junctions. A generalized osteoporosis and low incidence of spontaneous bone fractures can occur in grazing cattle and sheep with hypocuprosis (Cunningham, 1950; Whitelaw et al., 1979).

Connective tissue disorders

Natural clinical manifestations of connective tissue dysfunction are rare. Osteochondrosis in young farmed deer has been attributed to copper deprivation (Grace and Wilson, 2002), is accompanied by gross defects in the articular cartilages (Thompson et al., 1994) and may arise through impaired collagen and elastin development. Subperiosteal haemorrhages and imperfect tendon attachments are seen in lambs on molybdenum-rich pastures (Hogan et al., 1971; Pitt et al., 1980). In cattle, lesions in the ligamentum nuchae supporting the neck and scapulae may lead to dislocation of the scapulae, causing the head to droop and creating a 'hump' (B. Ruksan, personal communication, 1989). The abnormal gait associated with severe bovine hypocuprosis – variously described as 'pigeon-toed', 'stiff-legged' or 'bunny–hopping' – is probably due to a combination of bone and connective tissue disorders.

Cardiovascular disorders

Cardiac lesions were a feature of 'falling disease' of cattle in Western Australia and attributed to copper deprivation (Bennetts and Hall, 1939), but the condition has not been reported elsewhere or in other species. The sudden deaths, usually after mild exercise or excitement, were attributed to acute heart failure and the critical lesion was a slow and progressive degeneration of the myocardium with replacement fibrosis and accumulation of iron. Liver and blood copper were extremely low in the affected animals (32 µmol kg^{-1} dry matter (DM) and 1.6 µmol l^{-1} DM, respectively), as was pasture copper (1–3 mg kg^{-1} DM). Cardiac enlargement has found in experimentally deprived calves in one study (Mills et al., 1976) but not in another (Suttle and Angus, 1978). The deprivation was quicker to develop in the latter study due to the use of a small molybdenum supplement.

Scouring or diarrhoea

The copper-responsive diarrhoea widely reported in grazing cattle has been induced experimentally in hypocupraemic Friesian calves (Mills et al., 1976). Histochemical and ultrastructural changes were observed in the small intestine, but no clear relationship between these changes and diarrhoea was apparent (Fell et al., 1975) and such changes were not a feature of other clinically affected calves (Suttle and Angus, 1978). 'Teart scours' occur *before* liver or blood copper reaches subnormal levels and they may be caused by acute, localized, thiomolybdate-induced copper depletion of the intestinal mucosa (Suttle, 1991). Sheep are less inclined to scour on 'teart pasture' than cattle, but diarrhoea can be induced by molybdenum-rich diets (Suttle and Field, 1968; Hogan et al., 1971). Scouring has been observed in goats maintained on Dutch 'teart' pastures.

Susceptibility to infection

Long-term studies of genetic and other causes of mortality on a Scottish hill farm revealed a marked increase in mortality when the pastures were improved by liming and reseeding (Fig. 11.2a). Losses were higher in breeds with low efficiencies of copper utilization and reduced by copper supplementation or genetic selection for high copper status (Fig. 11.2b). Microbial infections were major causes of death, and since there were parallel changes in the incidence of a proven manifestation of copper deprivation – swayback – susceptibility to infection must have been increased by copper deprivation. Subsequent work indicated that perinatal mortality was increased unless copper supplements were given in late pregnancy, but the precise role of infection was not established (Suttle et al., 1987b).

Infertility

Low fertility associated with delayed or depressed oestrus occurs in beef cows grazing copper-deficient pastures, but the relationship with low copper status has been inconsistent (Phillippo et al., 1982). In experiments with penned heifers given a diet of marginal copper content (4 mg) with molybdenum (5 mg) or

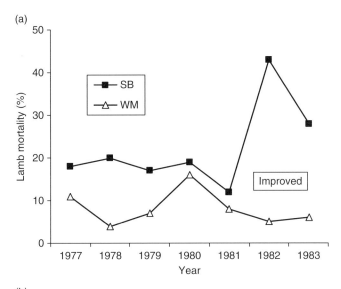

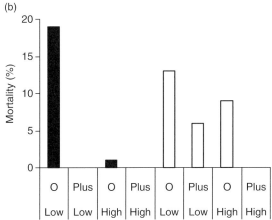

Fig. 11.2. (a) Lamb mortality in two sheep breeds known to absorb copper with low and high efficiency (Scottish Blackface (SB) and Welsh Mountain (WM), respectively) managed as a single flock before and after pasture improvement in 1981–1982. (b) Copper supplementation (O versus plus) reduced losses from both swayback (solid columns) and infectious diseases (open columns) in an inbred cross between the two breeds, selected for low plasma copper (low versus high). Since swayback is a recognized consequence of copper deprivation, susceptibility to infections must be similarly recognized as such (from Woolliams et al., 1986a).

iron (500–800 mg kg^{-1} DM), infertility only occurred in molybdenum-supplemented groups, despite the fact that liver and plasma copper in all treated groups seemed equally low (Humphries et al., 1983; Phillippo et al., 1987b). Both improvements and impairments of fertility have occasionally been reported in normocupraemic cows given parenteral copper (Suttle and Phillippo, 2005). Infertility and abortions have been reported in experimental copper deprivation of ewes (Howell and Hall, 1970).

Growth retardation

Poor growth is a common feature of copper deprivation in grazing sheep (Whitelaw et al., 1979, 1983; Woolliams et al., 1986b), cattle

(Thornton et al., 1972a,b; Whitelaw et al., 1984) and deer (Grace et al., 2005a). Growth retardation in the field is usually associated with mild exposure to molybdenum (Phillippo, 1983) and – like infertility – is mostly a feature of molybdenum-induced copper deprivation under experimental conditions (Humphries et al., 1983; Phillippo et al., 1987a; Gengelbach et al., 1994). Furthermore, growth can be retarded in the absence of any clinical abnormalities and is accompanied by poor feed conversion efficiency (FCE) when molybdenum-induced copper deprivation begins in utero (Legleiter and Spears, 2007). The possibility that both growth retardation and infertility are caused by molybdenum toxicity rather than copper deprivation will be discussed later. Biochemical explanations for growth retardation are unclear, but the low milk yield of beef cows on pastures high in molybdenum is believed to have been responsible for the poor growth of their suckled calves (Wittenberg and Devlin, 1987).

'Copper Value' of Feeds for Ruminants

Faced with such a long list of naturally occurring copper-RDs, the provision of adequate dietary supplies of copper assumes considerable importance.

Absorption

Studies with laboratory animals show that the absorption of copper (A_{Cu}) is facilitated by both specific (Crt1) and non-specific divalent metal transport (DMT) proteins in the intestinal mucosa, with the former being strongly regulated by need (Prohaska, 2006; Hansen et al., 2008a). However, in weaned ruminants A_{Cu} is determined largely by digestive processes in the rumen. These degrade organic and inorganic sulfur sources to sulfide (S^{2-}) (Suttle, 1974a) while failing to digest 30–50% of the organic matter. Much of the copper released during rumen digestion is precipitated as copper sulfide (CuS) and remains unabsorbed (Bird, 1970), while that released during post-ruminal digestion may become partially bound to undigested constituents. Whereas the A_{Cu} coefficient in the young milk-fed lamb is 0.70–0.85, in the weaned lamb it is <0.10 (Suttle, 1974b). The rumen protozoa are particularly important as generators of S^{2-} and their removal by isolation (Ivan, 1988) or the administration of anti-protozoal ionophores (Barrowman and van Ryssen, 1987) increases A_{Cu}. The fractional retention of copper in the liver can reflect the A_{Cu}.

Availability of copper in feeds

The ability of a feed to meet copper requirements or cause chronic copper poisoning (CCP) depends more on A_{Cu} than on the concentration of copper that the feed contains. Feed sources differ widely in availability (Table 11.2) for reasons that are not completely understood. Fresh grass is a poor copper source while brassicas, cereals and certain cereal by-products are good sources of copper for sheep. Little is known about the forms of copper in feeds, but pot ale syrup, a by-product from the distillery industry, contains copper in a poorly absorbable form compared to copper sulfate ($CuSO_4$) (A_{Cu} 0.01 and 0.06, respectively) and distillers' grains (Suttle et al., 1996a). In poultry waste, 64% of the copper can be extracted with EDTA and it has a similar A_{Cu} to that of $CuSO_4$ in sheep (0.08 versus 0.07) (Suttle and Price, 1976). However, the A_{Cu} value of a feed or supplement will depend on interactions with other ration constituents, particularly those high in sulfur, molybdenum and iron (e.g. Suttle and Price, 1976) because A_{Cu} is determined largely by the synchronicity of release of copper and its potential antagonists in the rumen (Suttle, 1991). Thus, A_{Cu} can be increased by feeding sheep once daily rather than every 4 h (Luo et al., 1996). When the diet is enriched with molybdenum as well as sulfur, thiomolybdates (TMs) are formed. These not only complex copper but leave it firmly bound to particulate matter (Allen and Gawthorne, 1987), reducing A_{Cu} still further (Price and Chesters, 1985) until as little as 1% may be absorbed. Little is known about interactions between feeds and

Table 11.2. Estimates of the absorbability of copper (A_{Cu}, %) in natural foodstuffs of low molybdenum content (<2 mg kg^{-1} DM) in Scottish Blackface ewes.

	A_{Cu} (mean ± standard deviation)	Number of estimates
Grazed herbage (July)	2.5 ± 1.09	7
Grazed herbage (September/October)	1.4 ± 0.86	6
Silage	4.9 ± 3.2	7
Hay	7.3 ± 1.8	5
Root brassicas	6.7 ± 0.9	2
Cereals	9.1 ± 0.97	3
Distillers' grains[a]	5.0 ± 0.70	3
Leafy brassicas[b]	12.8 ± 3.20	5

[a]The relatively low value for grain-based by-products is attributable to use of a basal diet supplemented with S (Suttle et al., 1996a).
[b]The high value for a feed high in S may be attributable to the presence of S in non-degradable forms (see Chapter 9).

it is assumed that mixtures of forages or blends of forage with concentrates will have the mean A_{Cu} of the constituent feeds (Suttle, 1994). However, allowing lambs access to fresh straw bedding while they received a 'coarse concentrate' diet reduced hepatic copper retention by 32 and 39% (Crosby et al., 2005; Day et al., 2006).

Outcome of copper–molybdenum–sulfur interactions in forages

The effects of molybdenum and sulfur on A_{Cu} in grasses, hays and silages, as determined by plasma copper repletion rates in hypocupraemic Scottish Blackface sheep, have been described by prediction equations and are illustrated in Fig. 11.3. Profound effects are predicted for both antagonists within commonly encountered ranges, with the first increments in herbage molybdenum and sulfur having the largest inhibitory effects on A_{Cu} (see Chapter 9 for copper and sulfur interactions). Copper–sulfur interactions will rarely escape molybdenum influence and copper–molybdenum interactions will usually be influenced by sulfur. However, the inhibition by sulfur is strongly curvilinear for herbage (Fig. 11.3c) and this waning may explain the lack of effect on liver copper reported when supplements raised dietary sulfur in grazing lambs from 2.7 to 6 g kg^{-1} DM (Grace et al., 1998b). The outcome of antagonism varies between forages, with sulfur per se having an enhanced influence in silages (Fig. 11.3a) and both antagonists having reduced influence in hays (Fig. 11.3b) when compared with fresh grass (Fig. 11.3c). The magnitude of the effect of molybdenum on A_{Cu} from grazed herbage has been confirmed in sheep using stable isotope methodology (Knowles et al., 2000), but not in long-term grazing studies employing a liver copper repletion technique (Langlands et al., 1981); the reasons for this discrepancy have been discussed (Suttle, 1983b). The predictions represented in Fig. 11.3c help to explain two field problems: first, hypocuprosis in lambs grazing improved pastures in which both sulfur and molybdenum concentrations are raised (Fig. 11.2) (Whitelaw et al., 1979; Woolliams et al., 1986b); second, hypocuprosis in cattle transferred from winter feeds to spring pasture, despite a higher copper intake (Jarvis and Austin, 1983). Both copper–sulfur (Suttle, 1974a; Bremner et al., 1987) and copper–molybdenum–sulfur (Suttle, 1977) interactions markedly affect A_{Cu} in concentrate-type diets.

The copper–iron interaction

Antagonisms between copper and iron also influence A_{Cu}. Supplements of iron oxide (Fe_2O_3, providing 800 mg Fe kg^{-1} DM) and simulating the effects of ingesting soil iron have been found to significantly lower A_{Cu} from 0.06 to 0.04 in sheep (Suttle and Peter, 1985); similar

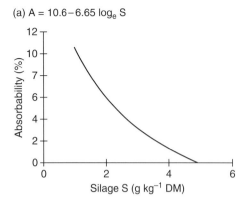

(a) $A = 10.6 - 6.65 \log_e S$

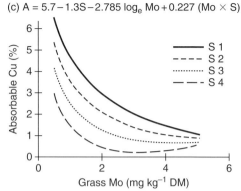

(c) $A = 5.7 - 1.3S - 2.785 \log_e Mo + 0.227 (Mo \times S)$

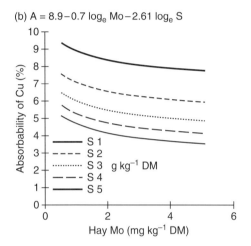

(b) $A = 8.9 - 0.7 \log_e Mo - 2.61 \log_e S$

Fig. 11.3. The adequacy of fresh and conserved grass as sources of copper for ruminants is determined largely by the extent to which sulfur and molybdenum combine to reduce Cu absorption (A_{Cu}), but the interactions vary with forage type: (a) in silages, Mo has a small and little studied effect on A_{Cu} but S reduces A_{Cu} in a logarithmic manner; (b) in hays, the inhibitory effect of Mo is detectable but less than that of S; it is not greatly influenced by interactions with S and A_{Cu} remains relatively high; (c) in fresh grass, A_{Cu} starts low and is further greatly reduced by small increments in Mo and S (data from Suttle, 1983a,b) obtained using the technique outlined by Suttle and Price, 1976).

inhibitions were obtained with iron as iron sulfate ($FeSO_4$) (Fig. 11.4). Both the accelerated depletion (Humphries et al., 1983) and slowed repletion (Rabiansky et al., 1999) of liver copper reserves reported in iron-supplemented cattle probably reflect reductions in A_{Cu}. Copper–iron interactions in both sheep (Suttle et al., 1984) and cattle (Bremner et al., 1987) are in part dependent on sulfur. Soil iron may be capable of reducing A_{Cu} by three mechanisms: the trapping of S^{2-} as iron sulfide (FeS) by soluble iron in the rumen (Suttle et al., 1984); the adsorption of copper by insoluble iron compounds (Suttle and Peter, 1985); and the down-regulation of the non-specific carrier DMT1 (Garrick et al., 2003) by excess soluble iron (Hansen et al., 2008). Bovine copper status has been lowered by as little as 250 mg Fe kg^{-1} DM as saccharated iron carbonate ($FeCO_3$) (Bremner et al., 1987). There is no evidence that the copper–iron interaction can add to the effect of the copper–molybdenum–sulfur interaction in cattle (Humphries et al., 1983; Phillippo et al., 1987a,b; Gengelbach et al., 1994) or sheep (N.F. Suttle, unpublished data) to produce further reductions in A_{Cu}. The copper–iron interaction is not manifested in the pre-ruminant calf (Bremner et al., 1987) or in sheep on concentrate diets that are low in sulfur (Rosa et al., 1986).

A copper–manganese interaction

A study with cattle showed that the addition of manganese (as manganese sulfate, $MnSO_4$, 500 mg kg^{-1} DM) to a corn-based finishing diet for steers exacerbated the hypocupraemia induced by a diet marginal in copper (5.4 mg) and high in sulfur (8 g added as calcium sulfate), but supplemented with molybdenum (2 mg kg^{-1} DM) (Legleiter and Spears, 2007). The antagonism

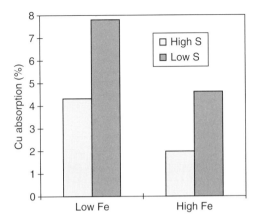

Fig. 11.4. The absorbability of copper (A_{Cu}) in diets for weaned ruminants is significantly affected by interactions between copper and iron, which are partly dependent on sulfur. The effects on A_{Cu} cannot yet be predicted in the manner used for copper–molybdenum–sulfur interactions and are not additive to those of molybdenum–sulfur, but with 800 mg added iron kg^{-1} DM the effect is to roughly halve A_{Cu}.

was stronger in a longer study involving an initial growing phase on corn silage (Hansen et al., 2009) and it was suggested that excess manganese adds to the effect of excess mucosal iron by suppressing DMT – one of the two transporters that carry copper across the gut mucosa – thereby inhibiting A_{Cu} (Hansen and Spears, 2008). Alternatively, it is possible that soluble manganese, like iron, can trap S^{2-} in the rumen and impair A_{Cu} at distal sites. While cattle are not normally exposed to such high levels of manganese as salts, they are exposed to manganese in soils (see Chapter 14). Manganese may be by far the most labile of the soil-borne elements (Brebner et al., 1985).

Other influences upon availability

The inhibitory effect of large zinc supplements (200–400 mg kg^{-1} DM) on hepatic copper retention produced in housed sheep (Bremner et al., 1976) is unlikely to occur across the normal range of forage zinc, but may be elicited during the treatment of facial eczema with large doses of zinc salts (Morris et al., 2006). Small supplements of cadmium can reduce the

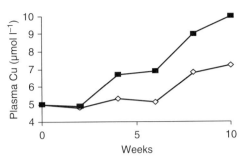

Fig. 11.5. The administration of a bolus (solid symbol) that slowly released anthelmintic hastened the spontaneous recovery in plasma copper in lactating ewes at pasture. Ewes showed evidence of suppressed immunity to gut parasites, which may have indirectly increased copper requirements (Suttle, 1996).

placental transfer of copper in ewes and cows (p. 497), and this interaction may be important where herbage cadmium has been increased by the use of superphosphates high in cadmium. Malabsorptive conditions of humans are associated with poor absorption of copper (Prohaska, 2006) and a similar association in sheep could explain why intestinal parasitism lowers copper status in lactating ewes with suppressed immunity (Fig. 11.5) (Suttle, 1996). Similarly, the low A_{Cu} values in autumn grass (Table 11.2) may partly reflect an autumn rise in larval nematode infection on the grazed pasture. Early reports of inhibitory effects of dietary calcium on A_{Cu} have not been repeated (Underwood and Suttle, 1999).

Sources of Copper and its Antagonists

Copper in pasture

The copper content of pastures and forages varies with the species, strain and maturity of the plant, with certain soil conditions (not pH) and with the fertilizers used (Table 11.3) (McFarlane et al., 1990). Temperate grasses tend to contain less copper than legumes grown in the same conditions (4.7 versus 7.8 mg kg^{-1} DM, respectively), but under tropical conditions the position is reversed

(7.8 versus 3.9 mg kg^{-1} DM, respectively) (Minson, 1990). Copper concentrations can range from 4.5 to 21.1 mg kg^{-1} DM among grass species grown on the same soil (Underwood and Suttle, 1999). Smaller differences between species have been recorded in West Kenya, with *Pennisetum clandestinum* (kikuyu) having the highest and *Chloris gayana* (rhodes grass) the lowest copper concentration (Table 11.4) (Jumba et al., 1995). Copper is unevenly distributed in temperate grasses: leaves contain 35% higher concentrations than stems and values in the whole plant therefore tend to decline during the growing season (Minson, 1990), and the fall can be marked (Jarvis and Austin, 1983). The low copper levels found in hays and silages in Alberta (4.3 ± 2.45 and 5.0 ± 1.98 (standard deviation) mg kg^{-1} DM, respectively) may partly reflect their maturity (Suleiman et al., 1997).

Copper antagonists in pasture

The determinants of A_{Cu}, herbage molybdenum, sulfur and iron, vary in concentration between species (Jumba et al., 1995), managements systems (Whitelaw et al., 1983) and seasons, with molybdenum increasing from 1.5 to 3.3 mg kg^{-1} DM between the first and second cuts of pasture in a recent Norwegian study (Govasmark et al., 2005). Pasture molybdenum increases as soil pH increases, and high concentrations of molybdenum and sulfur can occur on soils high in organic matter (Table 11.3). Temperate legumes tend to be richer in molybdenum than grasses in the same sward and Kincaid et al. (1986) reported a 10-fold difference on a molybdeniferous soil. In a study of fodders and grains grown in British Columbia, the copper to molybdenum ratio ranged from 0.1 to 52.7 (Miltimore and Mason, 1971)! Low molybdenum concentrations in

Table 11.3. The mean (range) of copper, molybdenum, sulfur and iron concentrations in subterranean and/or strawberry clover-tops sampled on two to four occasions from sites of the major soil types in the south east of South Australia (McFarlane et al., 1990).

Soil type	No. of sites	Cu (mg kg^{-1})	Mo (mg kg^{-1})	S (g kg^{-1})	Fe (mg kg^{-1})
Sand/clay	24	6.6 (3.0–14.6)	1.4 (0.1–4.1)	2.5 (1.5–4.0)	160 (45–346)
Red gum	22	9.5^a (4.1–15.9)	1.8 (0.1–5.4)	2.6 (1.5–4.3)	520^a (81–2300)
Groundwater rendzina	20	8.5 (4.2–12.8)	1.6 (0.2–5.4)	2.5 (1.9–5.1)	510^a (200–1000)
Deep sand	16	7.3 (4.0–14.0)	1.1 (0.1–3.8)	2.8 (2.1–4.6)	140 (119–154)
Peat	4	7.2 (4.4–11.4)	8.3^a (4.7–16.2)	3.6^a (2.8–4.5)	110 (48–320)
Calcareous sand	11	5.3 (1.9–9.5)	10.1^a (1.6–21.8)	2.5 (1.7–3.4)	100 (70–130)

Mean values followed by ^{a} differ significantly ($P < 0.05$) from others. Each is comprised of 5–10 samples.

Table 11.4. Species differences in available copper concentrations in four grasses calculated by the prediction equations in Fig. 11.3 from copper, molybdenum and sulfur concentrations in the pastures (from Jumba et al., 1995).

	Herbage concentration			Available Cu (mg kg^{-1} DM)	
Species	Cu (mg kg^{-1} DM)	Mo (mg kg^{-1} DM)	S (g kg^{-1} DM)	Grass equation	Hay equation
Pennisetum purpureum (napier grass)	4.1	0.85	1.1	0.135	0.303
Setaria sphacelata (setaria)	3.9	0.66	1.4	0.141	0.308
Chloris gayana (rhodes grass)	3.5	0.64	2.0	0.107	0.251
Pennisetum clandestinum (kikuyu grass)	5.7	1.50	1.7	0.127	0.383

maize silage grown in the molybdeniferous 'teart' area of Somerset (A. Adamson, personal communication, 1980) raise the possibility of important differences in A_{Cu} between forage crops. Herbage sulfur and protein are correlated and both decrease as the sward matures. 'Improved' pasture species such as kikuyu are relatively high in sulfur, and this counters their relatively high copper content when A_{Cu} is calculated (Table 11.4). Similarly, the low copper content of Alberta hays, mentioned above, is probably offset by increases in A_{Cu} (Table 11.2, Fig. 11.3). Spring pastures are often rich in iron from contaminating soil and a waning iron × copper antagonism has been invoked to explain the recovery from hypocuprosis in cattle grazing in summer (Jarvis and Austin, 1983). Herbage iron is high in overgrazed pastures and in those on soils prone to waterlogging (e.g. red gums and rendzinas, Table 11.3). Values can vary greatly between farms (150–1345 mg Fe kg^{-1} DM) (Nicol et al., 2003) and increase exponentially in autumn (Suttle, 1998b). Contamination of pastures and silages with soil iron is commonplace and should be taken into account when assessing their value as copper sources. Further details of the range of sulfur and iron concentrations found in forages are given in Chapters 9 and 13, respectively.

Composition of concentrate feeds

Species differences among the gramineous grains in copper and antagonist concentrations are relatively small and levels are generally low compared with most other feeds. In Northern Ireland, oats and barley have been found to contain an average of 3.9 and 4.9 mg Cu, 0.25 and 0.30 mg Mo and 0.8–1.5 g S kg^{-1} DM, respectively (Todd, 1972). Similar average levels have been subsequently been found for other grains elsewhere. Leguminous and oilseed meals are rich in copper, with 25–40 mg kg^{-1} DM being found in palm-kernel cake (Chooi et al., 1988), but they are also richer in molybdenum than cereals, containing 1–4 mg kg^{-1} DM. Distillery by-products can be notoriously rich sources of copper, with 86, 138, 44 and 129 mg kg^{-1} DM reported for pot ale syrup, wheat and barley distillery dried grains and ensiled grain residues, respectively, in one study (Suttle et al., 1996a): copper is dissolved from the copper stills during fermentation. Dairy milk and milk products are inherently low in copper (approximately 1 mg kg^{-1} DM), but other feeds of animal origin are richer, with meat meal and fish meal typically containing 5–15 mg kg^{-1} DM. Feeds of animal origin are relatively poor sources of molybdenum. The molybdenum concentration in milk is highly dependent on dietary intake and, in contrast to copper, can be raised several-fold above the normal of about 0.06 mg l^{-1} in ewes or cows. However, milk molybdenum will bypass the rumen and probably have little effect on the copper status of the suckling.

Sources of Available Copper for Non- and Pre-ruminants

Milk and milk replacers

There are wide differences between species in both colostral and main milk copper. Colostrums contain two to three times as much copper as main milk. The cow and goat secrete milk with low copper levels (0.15 mg or 2 μmol l^{-1}) and sows a high level (0.75 mg or 10 μmol l^{-1}). The unique position of the pig reflects the need to sustain large, rapidly growing litters during a brief lactation. Copper secretion in milk is reduced in dams deprived of copper, but cannot be increased above normal levels by dietary copper supplements (Underwood and Suttle, 1999). A_{Cu} in the newborn of all species can proceed by endocytosis and the relatively small amounts of copper ingested are very well absorbed. The apparent absorption of copper (AA_{Cu}) from a milk replacer supplemented with copper at 10 mg kg^{-1} DM was relatively high at 0.39 in calves (Jenkins and Hidiroglou, 1987).

Solid feeds

Variations in A_{Cu} between feeds have been extensively studied, particularly in poultry, although low availability is unlikely to cause natural copper deprivation. Corn gluten meal, cottonseed meal, peanut hulls and soybean meal (SBM) all have low values relative to $CuSO_4$ for poultry (approximately 40%) (Baker and Ammerman, 1995). However, the value obtained for $CuSO_4$ will be greatly influenced by the basal diet. Hepatic retention of copper from supplementary $CuSO_4$ in chicks was found to decrease substantially when soy protein was substituted for casein in the basal diet (Fig. 11.6) (Funk and Baker, 1991). Since low relative absorption of copper (RA_{Cu}) in vegetable proteins is partly attributed to antagonism from phytate, cereals are likely to have low RA_{Cu} values because of their high phytate content – particularly corn, which has little phytase. The addition of phytase to the corn/SBM diets of pigs can raise AA_{Cu} from 0.22 to 0.37 (Adeola et al., 1995), but excess dietary calcium has the opposite effect, enhancing the phytate antagonism (Leach et al., 1990). A feed attribute of low RA_{Cu} will be partly transferred to copper supplements, as is the case with soy protein (Fig. 11.6), but this is only of practical significance in the context of growth promotion (p. 287). Certain animal by-products have low RA_{Cu} (feather meal <1%, meat and bone meal 4–28%), possibly due to heat denaturation during processing, but others (poultry waste, 67%) are good sources (Baker and Ammerman, 1995).

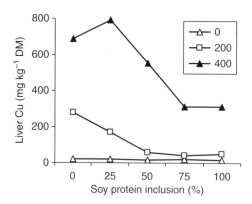

Fig. 11.6. Replacing casein with soy protein in the diet of chicks greatly reduces the availability of supplementary inorganic copper (200 or 400 mg kg^{-1} DM as $CuSO_4$), as measured by hepatic Cu retention (data from Funk and Baker, 1991).

Copper Requirements of Ruminants

Net requirement in sheep and cattle

The components of net requirement in factorial models of minimum need (see Chapter 1) are not easily determined (Underwood and Suttle, 1999). The maintenance component of copper consists mainly of faecal endogenous losses and, since the faeces is a major route for excreting excess absorbed copper, not all losses need to be replaced. The growth component of copper need is relatively small because the major carcass components (muscle, fat and bone) are low in copper and should exclude hepatic copper stores. Muscle is the major contributor to the extra-hepatic pool of copper (Fig. 11.7). On average, muscle contains only 0.8 and 1.2 mg Cu kg^{-1} fresh weight in cattle (Simpson et al., 1981) and sheep (Grace, 1983), respectively, and declines with copper depletion (Simpson et al., 1981). Relatively large amounts of copper are deposited in the fleece (Fig. 11.7), giving a significant requirement for wool growth in sheep (ARC, 1980).

Gross requirements for sheep: influence of diet

The copper requirements of weaned animals vary 10-fold under the powerful influence of interactions with iron, molybdenum and sulfur. These cause A_{Cu} to vary from 0.01 to 0.10, and single models of requirement are wholly unrealistic. Assessments of A_{Cu} in feeds given to inefficient absorbers of copper, such as the Scottish Blackface (Table 11.2 and Fig. 11.3), allow contrasting situations to be budgeted for in, or from, Table 11.5:

- indoor feeding on normal, dry mixed rations or grazing dry-season swards: A_{Cu} 0.06;
- grazing normal green swards or indoor feeding on high-molybdenum forage: A_{Cu} 0.03;
- grazing molybdenum-rich swards (>5 mg Mo kg^{-1} DM): A_{Cu} 0.015; and
- grazing iron-rich (>800 mg Fe kg^{-1} DM) swards: A_{Cu} 0.02.

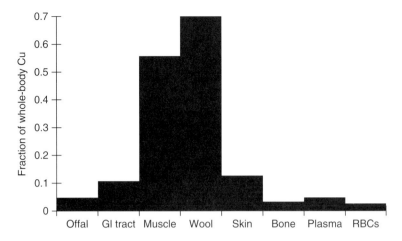

Fig. 11.7. The distribution of copper in the shorn, empty body of a mature sheep. The relatively large amounts in the fleece are also shown (data from Grace, 1983). GI, gastrointestinal; RBC, red blood cell.

Table 11.5. Factorial estimates of the dietary copper requirements of sheep and cattle given three types of diet with contrasting copper absorbabilities[a] (A_{Cu}).

| | Live weight (LW) (kg) | Growth rate or milk yield (kg day^{-1}) | Net requirement[b] (mg day^{-1}) | Food intake[c] (kg DM day^{-1}) | Gross requirement (A_{Cu}) 0.06 | 0.03 | 0.015 |
					(mg kg^{-1} DM)		
Lamb	20	0.1	0.13	0.5	4.3	8.6	17.2
		0.2	0.18	0.7	4.3	8.6	17.2
Ewe (16-week twin fetuses)	75	0	0.63	1.5	7.0	14.0	21.0[d]
Lactating ewe	75	1	0.52	1.5	5.8	11.6	23.2
		3	1.24	2.9	7.1	14.2	28.4
Calf	100	0.5	0.65	3.0	3.6	7.2	14.4
		1.0	0.90	3.6	4.2	8.4	16.8
Cow (40-week fetus)	500	0	4.1	9.1	7.5	15.0	30.0[d]
Lactating	500	20	4.0	15.2	4.4	8.8	17.6
		40	6.0	19.3	5.2	10.4	20.8

[a]Approximate A_{Cu} values for particular feed classes are as follows: 0.06 for roughage and concentrates; 0.03 for normal green swards; 0.015 for molybdenum-rich swards (>2mg Mo kg^{-1} DM); intermediate values apply for brown, dry season swards (0.04) and iron-rich (<800 mg Fe kg^{-1} DM) green swards (0.02).
[b]Components of factorial model: M, 4 µg kg^{-1} LW; G, 0.5 mg kg^{-1} LW gain; L, 0.1 and 0.22 mg l^{-1} for cow and sheep milk, respectively; and F, 4 mg Cu kg^{-1} clean wool growth.
[c]Real food intakes will vary according to diet quality (see table 16.6).
[d]Requirements for late pregnancy allow for build up of a large fetal reserve of copper.

Gross requirements for sheep: influence of genotype

Circumstantial evidence of genetic variation in copper requirements was first indicated by differences in incidence of swayback amongst different breeds under a common management regimen (Fig. 11.2), and later by substantial differences in liver copper between different breeds given the same diet (Woolliams et al., 1982; van der Berg et al., 1983). Breed differences in a given trait are not proof of genetic variation, but evidence was strengthened when crosses between breeds were shown to have a copper status intermediate between those of the parent breeds (Weiner, 1987). Proof came when sire selection for high and low plasma copper within a cross between breeds of relatively high (Welsh Mountain) and low (Scottish Blackface) status produced two lines that matched the copper status of the respective parents (Woolliams et al., 1986b). Comparisons of liver copper between sire families in the Merino (Judson et al., 1994), Suffolk and Texel (Suttle et al., 2002) breeds show that liver copper status can be highly heritable (h^2 >0.6). These genetic differences probably reflect differences in A_{Cu} (Wiener et al., 1978; Woolliams et al., 1983), but differences in excretion may contribute (Woolliams et al., 1982, 1983; Suttle et al., 2002).

The requirements in Table 11.5 are appropriate for Scottish Blackface, but should be reduced by >50% for more efficient absorbers and retainers of copper, such as the Texel.

The weighting of A_{Cu} for effects of the interaction with molybdenum in Fig. 11.3c is similar to that found in studies with another breed under grazing conditions in New Zealand (Knowles et al., 2000), but the baseline A_{Cu} was higher and requirements therefore lower. Allowance for iron–copper antagonism would be prudent.

Gross requirements for cattle

Cattle generally are poor absorbers and A_{Cu} is assumed to be similar to that of Scottish Blackface sheep in the model predictions given in Table 11.5, but outcomes are speculative for mixtures of roughage and concentrate (Suttle, 1994). Small molybdenum and sulfur supplements have similar inhibitory effects on copper absorption in Jersey cattle to those reported for Scottish Blackface sheep (Underwood and Suttle, 1999) and the effects of molybdenum and sulfur on requirement are assumed to be the same as those found in sheep. Increasing dietary copper byt 58% fully countered the depleting effects of a 5mg Mo kg^{-1} DM supplement in steers (Ward and Spears, 1997). Severe hypocupraemia developed in steers on a corn-silage-based diet that was low in molybdenum but contained 1000mg added iron (as FeSO$_4$) kg^{-1} DM, indicating a requirement well above the level of 8mg Cu kg^{-1} DM provided in the basal diet (Mullis et al., 2003). Requirements are therefore increased when dietary iron is high.

The evidence for genetic variations in cattle is growing: Friesian calves have higher liver copper than Galician Blond calves in Spain, while crossbred calves are of intermediate status (Miranda et al., 2006). Requirements may be marginally higher in Jersey than in Friesian cows (Du et al., 1996) and in Simmental than Aberdeen Angus (Mullis et al., 2003; Stahlhut et al., 2005). Within a grazing Aberdeen Angus herd, serum copper concentrations had a heritability of 0.28 (Morris et al., 2006). Lactation in cows does not raise requirements significantly because of the low copper content of milk (approximately 3 μmol l^{-1}).

Deer, goats and other ruminant species

Using the ARC (1980) factorial model with an A_{Cu} coefficient of 0.04 and a growth component of copper of 1.6mg kg^{-1} live-weight gain (LWG), Grace et al. (2008) put the dietary copper requirement for growing hinds at 8mg kg^{-1} DM. Earlier field trials on low-molybdenum farms suggested that this was far more than was needed to avoid growth retardation, but sufficient to avoid hypocupraemia (Fig. 11.8). Although deer seem less susceptible to the copper–molybdenum antagonism than cattle (Mason et al., 1984), the exposure of deer to high-molybdenum pasture increased the severity of hypocupraemia and requirements were not met by copper at 7–8mg kg^{-1} DM. In the absence of data on caprine requirements, they have been assumed to be the same as those of sheep. However, breed and seasonal differences in copper requirements may be found (Osman et al., 2003) and 1-year-old cashmere goats, even in the 'non-growing' season, have shown improved LWG and FCE when copper was added to a diet that contained 7.4mg Cu kg^{-1} DM. This indicates a relatively high need for copper compared with sheep.

Copper Requirements of Non-ruminants

Detailed calculations of needs are superfluous because variation in A_{Cu} between feeds is

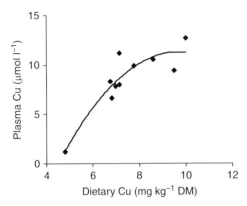

Fig. 11.8. The curvilinear relationship between mean plasma and dietary copper concentrations in deer on ten farms in Canterbury, New Zealand (data from Nicol et al., 2003) indicate a dietary copper requirement to avoid hypocupraemia by the beginning of winter of 7–8 mg kg^{-1} DM, in the absence of antagonism from molybdenum (maximum dietary molybdenum was only 0.7 mg kg^{-1} DM).

small and dietary provision from natural sources generous.

Pigs

A purified diet supplying copper at 6 mg kg^{-1} DM fully meets the requirements of baby pigs (Okonkwo et al., 1979) and ARC (1981) considered 4 mg kg^{-1} DM to be sufficient for growing pigs. All normal diets composed of cereals and the usual protein supplements will meet that need, although it may be higher on pelleted rations if pelleting always reduces AA_{Cu} in the way it did in one study (Larsen et al., 1999). The additional copper requirement of 4–6 mg day^{-1} for lactation (5–7 kg milk day^{-1} containing 0.75 mg Cu kg^{-1}) will be met by the additional food intake (3 kg day^{-1}) over that for late pregnancy if the diet contains only 4 mg Cu kg^{-1} DM and absorptive efficiency is 50%.

Poultry

Definitive data on the minimum copper requirements of chicks for growth or of hens for egg production have not been reported. The NRC (1994) gives the following copper requirements (all mg kg^{-1} DM): starting chicks (0–8 weeks), 5; growing Leghorn-type chickens (8–18 weeks), 4; broilers (all ages), 8; starting turkey poults, 8; growing turkeys, 6; and breeding turkeys, 6. No requirements are given for laying or breeding hens, presumably because their previous low recommendations (3–4 mg kg^{-1} DM) have not been questioned. Natural poultry rations usually contain > 6 mg kg^{-1} DM and should rarely require supplementation.

Horses

The copper requirements of horses are ill-defined, but improvements in angular limb deformities in foals when dietary copper is raised from 8 to 25 mg kg^{-1} DM suggested a relatively high requirement for a non-ruminant species (Barton et al., 1991). Using a factorial model with A_{Cu} at 0.30 and the growth component at 1.0, Grace et al. (1999) placed the need for a 200-kg horse gaining 1 kg day^{-1} at only 4.5 mg Cu kg^{-1} DM. This low requirement is supported by the fact that horses thrive on pastures with 6–9 mg Cu kg^{-1} DM and neither growth nor bone development is enhanced by copper supplementation (Pearce et al., 1999a). Exposure to molybdenum does not impair copper metabolism or raise copper requirements in horses (Strickland et al., 1987; Pearce et al., 1999b).

Metabolism of Copper

Certain aspects of copper metabolism must be described to fully understand the variable effects of copper deprivation and overload on livestock. Adjustment to fluctuations in copper supply is achieved by control of absorption, hepatic storage and biliary copper secretion. The relative importance of each process and outcome in terms of relationships between dietary and liver copper varies widely between species; those faced with endemic risks of copper deprivation (i.e. ruminants) have relatively poor control over absorption and avidly store excess copper, whilst species at no risk (i.e. non-ruminants) have good control over absorption and biliary secretion and

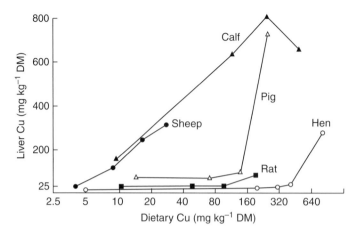

Fig. 11.9. Species differ in the extent to which they store excess dietary copper in their livers. Ruminant species, for which the risk of copper deficiency is ever-present, store copper avidly while non-ruminant species, which are rarely at risk, do not.

thus maintain low liver copper levels over a wide range of dietary copper (Fig. 11.9).

Transport and cellular uptake

The binding of copper to MT in the gut mucosa is an important means of restricting copper uptake in non-ruminants exposed to excess copper (Cousins, 1985) and may contribute to the adaptation of exposed sheep (Woolliams et al., 1983). Normally, copper is passed from Crt1 to vectors for final passage across the basolateral membrane into the portal bloodstream, where most becomes bound to albumin. However, copper transport to the tissues can proceed without albumin (Prohaska, 2006). On reaching the mature liver, copper is taken up by a two-stage process involving binding first to glutathione and then to MT before being partitioned between biliary secretion, synthesis of Cp and storage (Bremner, 1993). In ruminants, Cp normally constitutes a lower proportion of total plasma copper than in non-ruminants (80% versus 95%) and avian species have very little Cp in the bloodstream. The sheep gene (sCp) is not expressed in the fetus (Lockhart and Mercer, 1999). Increases in the hepatic activity of enzymes involved in glutathione synthesis during copper deficiency (Chen et al., 1995) may enhance the hepatic capture of copper, as does exposure to excess selenium (Hartman and van Ryssen, 1997). Extra-hepatic uptake of copper may be achieved by Cp receptors in cell membranes (McArdle, 1992; Saenko et al., 1994), although albumin and amino acids such as histidine facilitate copper uptake in vitro and may deliver copper to transporters such as Ctr1 and Ctr2 (Prohaska, 2006).

Biliary secretion and urinary excretion

Biliary copper secretion is quiescent in the fetus, but is awakened after birth (Prohaslka, 2006) to allow both the entero-hepatic recycling of copper and the excretion of excesses. Losses of biliary copper in sheep can increase more than fivefold as liver copper rises (Fig. 11.10) although breeds may differ in this respect (Suttle et al., 2002). At high copper intakes, cattle limit hepatic storage sooner than sheep by means of biliary secretion (Phillippo and Graca, 1983), although this is not evident in Fig. 11.9. The difference may be explained by a lower threshold for copper storage in lysosomes in the bovine (Lopez-Alonso et al., 2005). Goats retain less copper in their liver than sheep when exposed to high copper intakes (Zervas et al., 1990) and may also have a well-developed capacity

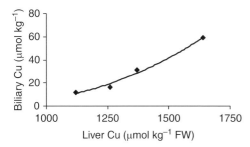

Fig 11.10. The regulation of copper metabolism in grazing Romney ewes. Biliary Cu concentration increased linearly with increases in liver Cu above about 1250 μmol kg^{-1} DM. Any down-regulation of absorption would amplify regulation by diminishing the reabsorption of secreted Cu. Note that no sheep developed clinical signs of toxicity at the highest liver Cu levels (data from Grace et al., 1998a).

for biliary copper secretion. Urinary excretion of copper is normally small, constant and unaffected by copper intake in all species, although it is increased in sheep by exposure to molybdenum (Smith et al., 1968).

Biochemical Consequences of Copper Deprivation in Ruminants

The sequence of changes seen in ruminant species deprived of copper is illustrated in Fig. 11.11 (see also Fig. 3.3).

Depletion

With the liver so well-developed as a storage organ, the first change is a decline in liver copper; the concentration present is a reflection of previous dietary copper supply. The newborn generally begins life with high liver copper (Prohaska, 2006), with values rising to 3–6 mmol kg^{-1} DM in the calf and lamb provided that the copper supply during pregnancy is adequate (Underwood, 1977). In the ovine fetal liver, 15–35% of the copper is associated with MT (Bremner et al., 1977). Restriction of the maternal supply of copper can drastically reduce neonatal reserves, hastening subsequent depletion and revealing effects of maternal breed (compare Weiner

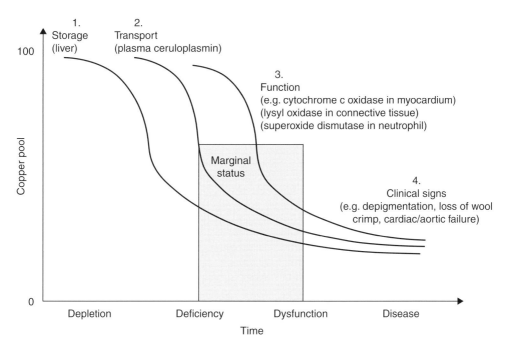

Fig. 11.11. The sequence of biochemical changes leading to the appearance of clinical signs of copper deprivation.

et al., 1984a versus 1984b). Priority is clearly given to the fetus because cows with low liver copper (<0.39 mmol kg^{-1} DM) can carry calves with 5.2 mmol kg^{-1} DM in their livers (Gooneratne and Christensen, 1989). However, with maternal values <0.26 mmol kg^{-1} DM, fetal values declined by up to 50%. The pattern followed by liver copper during depletion is one of exponential decline.

Deficiency

When liver reserves approach exhaustion or cannot be mobilized fast enough to meet the deficit in supply, Cp synthesis decreases and plasma copper falls below normal (Fig. 11.12). Blood copper shows a slower decline because it is some time before erythrocytes become depleted of SOD1, their major copper constituent, and accumulate in sufficient numbers to add to the effect on plasma copper. Extra-hepatic changes vary between tissues, with large reductions found in kidney or brain copper in copper-depleted lambs, adult sheep (Suttle et al., 1970) and growing cattle (Suttle and Angus, 1976; Legleiter et al., 2008), but not in cardiac muscle (Suttle et al., 1970) unless depletion is prolonged (Mills et al., 1976).

Reduction in cuproenzyme activities proceed at various rates at different sites, indicating progress towards a state of local *dysfunction*, but the rate-limiting pathway is often hard to define. For example, a study of field cases of swayback indicated a threshold relationship between brain copper, CCO and pathological change, with brain copper <3 mg kg^{-1} DM critical (Mills and Williams, 1962). However, the relationships were less clear in experimentally induced swayback (Suttle et al., 1970), and early depletion of CCO activity may not be sufficient to constitute a respiratory constraint (Smith et al., 1976). While SOD1 tends to decline in the brain of copper-deprived calves, there is an opposite trend in MnSOD and no change in tissue oxidant capacity (Legleiter et al., 2008). The catecholamine neurotransmitters dopamine and noradrenaline, as well as monoamine oxidase (MAO) and SOD1 activities, are also low in the brainstem of ataxic lambs (O'Dell et al., 1976). Decreases in peptidylglycine-α amidating monooxygenase (PAM) have been reported in the brain of newborn rats from copper-depleted dams. This is of particular interest, given the multiplicity of biogenic molecules dependent on PAM (including the appetite-regulating hormones gastrin and cholecystokinin) and the slow recovery of enzyme activity following copper repletion (Prohaska, 2006). There have been no studies of PAM in the CNS of ruminants.

Dysfunction

Experimentally copper-deprived cattle become severely hypocupraemic before growth is retarded (Suttle and Angus, 1976; Legleiter and Spears, 2007; Hansen et al., 2008), but iron then begins to accumulate in liver (Mills et al., 1976), probably a consequence of low hephaestin activity (Table 11.1). Haemoglobin synthesis may also decline at a late stage, but impairment of iron transport is not necessarily responsible (Suttle, 1987a; Prohaska, 2006). Early depigmentation is caused by low

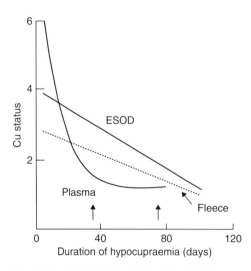

Fig. 11.12. Different time courses for changes in three indices of copper status in sheep given a low-copper diet: erythrocyte superoxide dismutase (ESOD), plasma Cu and fleece Cu. Arrows indicate the onset of clinical signs and units vary (Suttle and McMurray, 1983).

tyrosinase activity (Table 11.1), inhibiting the conversion of tyrosine to melanin in the hair or wool follicle (Holstein et al., 1979). The normal crimp, tensile strength and elasticity of wool are dependent on disulfide groups that provide the cross-linkages in keratin and on the alignment of long-chain fibrillae in the fibre: both are adversely affected in copper deprivation (Marston, 1946) and this may reflect loss of thiol oxidase activity (see Table 11.1). Osteoporosis and a reduction in osteoblastic activity can occur in young housed lambs born of copper-deficient ewes, but without morphological bone changes (Suttle et al., 1972). Leukocyte counts can increase in the copper-deprived heifer, but protection against common pathogens is not compromised (Arthington et al., 1996; Gengelbach et al., 1997); similar results have been obtained in the copper-deficient calf (Stabel et al., 1993). Molybdenum-induced hypocupraemia can lower the antibody response to *Brucella abortus* antigen in cattle (Cerone et al., 1995). However, the precise mechanism(s) whereby resistance to infection is lowered is still unknown and there are many contenders (Bonham et al., 2002).

Disorder

The critical site at which disorder develops probably depends on the metabolic stress placed on the rate-limiting enzyme at critical stages in tissue development, as clearly shown by the development of swayback. With deprivation occurring *in utero* in swayback lambs, low CCO activity in the CNS, particularly in affected neurons of the grey matter of the spinal cord, may cause disorder (Suttle, 1988a). Low CCO activity has been described in the hearts of copper-deprived calves in association with abnormal mitochondrial ultrastructure (Leigh, 1975), but CCO was not necessarily the rate-limiting enzyme. A reduction in the activity of lysyl oxidases could diminish the stability and strength of bone collagen and impair mineralization of the cartilage, with consequent changes in bone morphology and reductions in bone strength in chicks (Opsahl et al., 1982). A similar 'biochemical lesion' in elastin may weaken connective tissue in the copper-deprived calf.

Biochemical Consequences of Molybdenum Exposure in Ruminants

The primary biochemical consequence of low exposure to molybdenum in ruminants is the formation of unabsorbable copper–TM complexes in the gut, which leads to copper deprivation and the sequence of events described above. At high molybdenum to copper intakes (>8 mg Mo kg^{-1} DM in sheep), TMs leave the rumen in absorbable forms (Price et al., 1987). The physiological significance of TMs has provoked controversy in the UK (Suttle, 2008a,b).

Abnormal blood copper distribution

Molybdenum exposure can cause TMs to accumulate in plasma as albumin-bound trichloroacetic acid (TCA)-insoluble (I) copper complexes (Dick et al., 1975; Smith and Wright, 1975a,b), with a much slower clearance rate than normal albumin-bound copper (Mason et al., 1986). The anomalously bound copper can be estimated by difference between the values of plasma treated or not treated with TCA, but with relatively large error (e.g. ± 2 µmol l^{-1}) (Suttle, 2008a). Cattle and deer must be exposed to higher molybdenum to copper ratios than sheep before TCA-I copper is detected (Freudenberger et al., 1987; Wang et al., 1988). It was initially assumed that TCA-I copper is poorly utilized by tissues, but copper–molybdenum–albumin complexes are not excretable, slowly hydrolysed (Mason, 1986) and may serve as a pool of slowly releasable copper in some circumstances (Suttle and Small, 1993). When lambs with high initial liver copper reserves were given a low-copper diet, liver copper depletion was not accelerated by adding tetrathiomolybdate (TTM) to the diet, despite large increases in TCA-I copper. When copper was also added, TTM decreased liver copper accretion but TCA-I copper decreased, presumably because copper reduced the absorption of TTM (Fig. 11.13). A 'flooding'

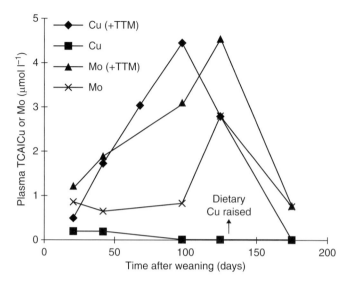

Fig. 11.13. The biochemical consequences of exposing lambs to tetrathiomolybdate (TTM, 5.4 mg Mo kg^{-1} DM) via the diet depend on the copper also provided. Initially, when dietary Cu is low (3.5 mg kg^{-1} DM), excess TTM is absorbed, causing trichloroacetic acid-insoluble Cu (TCAlCu, ♦ and Mo, ▲) to accumulate in plasma. When Cu is raised after 130 days (to 15.5 mg kg^{-1} DM), TCAlCu disappears and plasma Mo is markedly reduced due to insoluble Cu–TTM complex formation in the gut (from Suttle and Field, 1983).

parenteral dose of TTM inactivates SOD1 in sheep (Suttle et al., 1992a) and removes copper from MT in vitro (Allen and Gawthorne, 1988), but pharmacological doses of TTM, safely used to prevent CCP in sheep, do not impair Cp activity (Suttle, 2008a).

Effects on reproduction and growth

Impaired pulsatile release of luteinizing hormone (LH) has been detected in molybdenum-supplemented infertile heifers, but not in those with an equally severe hypocupraemia induced by iron, and previous growth was only retarded in groups given molybdenum (Phillippo et al., 1987b). There are three possible explanations:

- infertility and growth were impaired by molybdenum exposure per se;
- extreme hypocupraemia predisposed to molybdenum toxicity; and
- molybdenum induced a localized copper deprivation, not reflected by liver and plasma copper, that had previously reached a nadir.

The last explanation was (Suttle 1991; 1998), and remains, the most plausible (Suttle, 2008b). Xin et al., (1993) failed to impair LH release or fertility with molybdenum supplementation, despite using rations with lower copper to molybdenum ratios (1.9 versus 0.8). Crucially, this did not induce hypocupraemia or deplete sites of suspected influence on LH release (the hypothalamus and anterior pituitary). Close comparison of the copper-deprived groups (Fig. 11.14) shows that those given molybdenum, with or without iron, had lower liver copper and blood SOD1 concentrations than the unretarded group, depleted by iron alone. Late reductions in SOD1 activity in neutrophils were found in calves from molybdenum-supplemented heifers, but not in equally hypocupraemic offspring from iron-supplemented heifers (Gengelbach et al., 1997). Furthermore, diets low in copper (Mills et al., 1976) and high in manganese (Hansen et al., 2009) can retard cattle growth and produce all other clinical abnormalities associated with moderate exposure to molybdenum, while excess iron can induce copper-RD in the field. Decreases in food

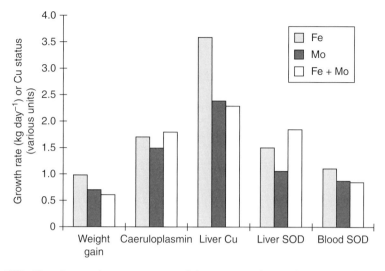

Fig. 11.14. Difficulties of assessing copper status of three groups of severely copper-deficient heifers, given iron, molybdenum or as copper antagonists. Only the groups given Mo grew poorly, although all had equal caeruloplasmin concentrations (Humphries *et al.*, 1983). Poor growth was, however, accompanied by lower liver Cu and blood superoxide dismutase (SOD) activity and suggestions that Cu deprivation of Mo-supplemented groups was not involved are unfounded.

intake and FCE underlie the growth retardation (Mills *et al.*, 1976; Hansen *et al.*, 2009).

Other effects of thiomolybdates

Serial daily injections of TTM have retarded growth and impaired fertility in lambs, but only after months of exposure (Moffor and Rodway, 1991). On the rare occasions when dietary molybdenum concentrations exceed those of copper and allow TM to be slowly and continuously absorbed, TM may be detoxified by albumin binding (Lannon and Mason, 1986). There is the possibility of unabsorbable TCA-I copper complexes forming in the abomasum when parasitic infection causes leakage of plasma into an acid, TM-rich lumen (Ortolani *et al.*, 1993), and exposure to molybdenum may enhance the inflammatory response whereby sheep expel gut parasites (Suttle *et al.*, 1992b; McLure *et al.*, 1999). Entero-hepatic recycling of copper can probably be impaired by TTM since exposure to molybdenum greatly increases faecal endogenous loss (Smith *et al.*, 1968) and increases the rate of liver copper depletion in sheep (Woolliams *et al.*, 1983; Freudenberger *et al.*, 1987), although not in deer (Freudenberger *et al.*, 1987).

Occurrence of Copper-responsive Disorders in Ruminants

Outbreaks of copper-RD are influenced by soil, plant, seasonal and genetic factors.

Soil factors

Copper deprivation has been associated with three different soil anomalies: inherently low copper status, as in the calcareous sands of Western Australia (Beck, 1951) and the heathlands of East Anglia (UK); low plant-available copper status, as in peat soils rich in organic matter (Haynes, 1997); and those enriched with molybdenum, which creates pastures with low animal-available copper. Most copper-RD come into the latter category (Phillippo, 1983). High herbage molybdenum (>20 mg kg^{-1} DM) in the 'teart' pastures of Somerset is attributable to molybdenum enrichment of the underlying Lower Lias clay (Thornton and Alloway, 1974). Some peat soils in New Zealand (Cunningham, 1950), southern Australia (McFarlane et al., 1990), Washington state (Kincaid et al., 1986) and Ireland (Thornton and Alloway, 1974) are high in molybdenum as well as low in available copper. The alluvial plains of Argentina, the marine shales of Derbyshire in the UK (Thornton and Alloway, 1974), Caithness (Scotland) (Suttle, 2008b) and north-east Manitoba (Boila et al., 1984a,b), the grey soils of north-east Saskatchewan (Gooneratne and Christensen, 1989) and the Taupo ash soils of New Zealand (Grace et al., 2005b) all provide sufficient molybdenum to induce hypocupraemia or hypocuprosis. Problem areas can be delineated by soil or stream sediment maps of molybdenum concentration (Thornton and Alloway, 1974; Boila et al., 1984b), but the persistence of hypocupraemia in one such area (Caithness) (Suttle, 2008b) suggests that the antagonism from molybdenum is not readily controlled. In southern Australia, bovine hypocupraemia is associated with iron-rich soils (rendzinas) (McFarlane et al., 1990) and, in parts of England, iron contamination from soil may elicit hypocupraemia in cattle in spring (Jarvis and Austin, 1983). The ingestion of SO_4-rich drinking water from deep aquifers has caused hypocupraemia and growth retardation in heifers and neonatal mortality in their offspring (Smart et al., 1986).

Plant factors

Choice of pasture species and pasture management can affect the uptake of copper and molybdenum from soils (see Chapter 2) and hence the risk of copper-RD. Improvement of Scottish hill pastures by liming and reseeding induced hypocuprosis in lambs (Fig. 11.2) (Whitelaw et al., 1979; 1983) and calves (Whitelaw et al., 1984) by raising herbage molybdenum and sulfur, but may not do so elsewhere (Hannam and Reuter, 1977). Reliance on legumes rather than fertilizer as a source of supplementary nitrogen on 'organic' farms is likely to increase the risk of copper-RD on molybdenum-rich soils because of the propensity of legumes to accumulate molybdenum. Outbreaks of swayback on coastal pastures in Greece have been attributed to the ingestion of SO_4-rich halophytic plants by ewes (Spais, 1959) and a similar pathogenesis explains outbreaks in sheep that graze seaweed in Iceland (Palsson and Grimsson, 1953). Heinz-body anaemia and growth retardation, both responsive to copper supplementation, have been reported in cattle consuming a sulfur-rich brassica (kale) crop (Barry et al., 1981).

Seasonal factors

Peaks in herbage iron in autumn temperate pasture (Jarvis and Austin, 1983) or seasonally waterlogged pastures (McFarlane et al., 1990) are believed to trigger copper-RD, and simultaneous increases in manganese may add to the antagonism. Irrigation of pasture with iron-rich bore-water to overcome seasonal water deficits has been found to induce a copper-responsive diarrhoea in grazing cattle (Campbell et al., 1974). Pasture molybdenum can increase greatly between April and August in Scotland (Suttle and Small, 1993) and there are similar seasonal increases in Norway (Govasmark et al., 2005). However, herbage sulfur declines as a pasture matures (see Chapter 9) and A_{Cu} is probably higher in mature than in fresh green herbage. Induced copper-RD are essentially 'green sward' problems, rarely seen during dry summers or arid regions, and plasma copper is higher in goats in the dry than in the cool season in Oman (Osman et al., 2003). In parts of southern Australia (Hannam and Reuter, 1977) and of the UK (Bain et al., 1986), bovine hypocupraemia is associated with months (and years) of heavy rainfall. Swayback of lambs is more prevalent after mild than after severe winters, possibly because of less supplementary feeding and more soil ingestion reducing A_{Cu} in mild winters (Suttle et al., 1984). However, the occurrence of copper-RD will often remain poorly correlated with current herbage composition and dependent on the sufficiency of copper stores built up during times of generous supply.

Genetic factors

The genetic variation in copper metabolism referred to earlier translates into breed differences between Scottish Blackface and Welsh Mountain sheep in relative susceptibility to copper deprivation (Fig. 11.2). By contrast, the Texel, North Ronaldsay and Suffolk breeds are highly unlikely to suffer from lack of copper. Comparable differences between breeds of cattle are less pronounced and recognition is complicated by breed and seasonal differences in natural coat colour and hence propensity to display the earliest sign of disorder, but the Simmental is gaining a reputation as a relatively vulnerable breed compared with Aberdeen Angus (Gooneratne et al., 1994; Mullis et al., 2003; Stahlhut et al., 2005; Legleiter et al., 2008). Breed differences in plasma copper have been reported in Omani goats, but only in the dry season and in adult females, but in not bucks or kids (Osman et al., 2003). Jersey and Friesian cattle are more susceptible to copper deprivation than some sheep breeds in New Zealand, with a survey showing that where the two species grazed the same property cattle could have 'low' liver copper status, while sheep were of moderate or high status (Korte 1996; cited by Grace et al., 1998a).

Diagnosis of Copper-responsive Disorders in Ruminants

With the singular exception of swayback, clinical signs of copper deprivation cannot be relied upon for diagnosis. Nutritional stresses of various kinds can lead to temporary loss of wool strength or crimp in sheep and they are not specific signs of copper dysfunction. Loss of coat colour has been noted in many vitamin deficiencies and in cobalt-deprived cattle (see p. 225). When cattle shed their winter coat, natural depigmentation occurs and the 'foxy-brown' coat discolouration seen in Friesian cattle, for example, is *not* a sign of copper deprivation (Mee, 1991). Diagnosis rests upon three conditions:

- the presence of clinical symptoms or subclinical loss of production;
- biochemical evidence of subnormal tissue or blood copper; and
- an improvement after treatment with copper when compared with untreated cohorts.

An incomplete profile will leave the diagnosis insecure. The contribution that biochemical analyses can make is summarized in Table 11.6. Note that all diagnostic aids require flexible interpretation and bands to indicate marginal risk of past or future problems.

Dietary copper and its antagonists

The diagnostic value of dietary copper alone can be unreliable because of vast differences in A_{Cu}. The most common approach has been to predict A_{Cu} and hence risk from the dietary copper to molybdenum ratio, but estimates of the 'critical' ratio range from >2:1 for normal growth in beef cattle (Miltimore and Mason, 1971) to approximately 4:1 for swayback avoidance (Alloway, 1973). In an experiment with housed Aberdeen Angus steers, depletion was prevented by raising the copper to molybdenum ratio in a corn-silage-based diet from 1.0 to 1.75, and performance was only marginally compromised. There are several reasons why the copper to molybdenum ratio may be diagnostically imprecise:

- the current ratio may not reflect the previous pasture A_{Cu} and hence liver copper stores;
- the ratio does not allow for the influence of sulfur (see Fig. 11.3);
- the effect of molybdenum will probably vary with season and legume content;
- the inhibitory effect of molybdenum wanes with molybdenum >8 mg kg^{-1} DM (Suttle, 1983b; Grace et al., 2005b);
- other antagonists, notably iron, may also be influential but equally unpredictable; and
- the critical ratio may vary with diet and forage diet type (Fig. 11.3).

Table 11.6. Marginal bands[a] for copper:antagonist concentrations in the diet and copper in the bloodstream and soft tissues of ruminants to aid the diagnosis and prognosis of copper-responsive disorders (Cu-RD) in ruminants on fresh herbage (H) or forage (F) based diets.

Criterion[b]	Diet	Sheep and cattle	Deer[c] and goats	Interpretive limits
Diet Cu:Mo	H	1.0–3.0	0.5–2.0	Diet S >2 g kg^{-1} DM
	F	0.5–2.0	0.3–1.2	Diet Mo <8 mg kg^{-1} DM for sheep, deer and goats, <15 mg kg^{-1} DM for cattle
Diet Fe:Cu	H	50–100	50–100	
Diet Cu (mg kg^{-1} DM)	H	6–8	6–8	Diet Mo <1.5 mg kg^{-1} DM
	F	4–6	4–6	
Liver Cu (µmol kg^{-1} DM)[d,e]	H or F	100–300	180–300	
Plasma Cu (µmol l^{-1})[e]	H or F	3–9	5–8	Diet Mo <8 mg kg^{-1} DM for sheep, deer and goats, <15 mg kg^{-1} DM for cattle: no acute phase response; age >1 week
Blood Cu (µmol l^{-1})	H or F	6–10	6–10	
Hair or wool Cu (mg kg^{-1} DM)	H or F	2–4 (sheep) 4–8 (cattle)	6–9 (deer) 3–5 (goats)	Clean sample, recent growth

[a]Means within marginal band indicate a probability of Cu-RD that increases with proximity to the **bold** limit, while values beyond the band do not. As the *number of* individual values falling within a marginal band increases, so does the likelihood of herd or flock benefit from Cu supplementation.
[b]Alternative criteria such as caeruloplasmin and erythrocyte CuZnSOD activities are not included because they introduce additional problems of interpretation or standardization without adding significantly to the precision of diagnosis.
[c]Some data for cervines from Wilson and Grace (2001).
[d]Divide by 3.0 to obtain values on fresh weight basis.
[e]Multiply by 0.064 to obtain values in mg.

It is noteworthy that molybdenum supplementation (+2 mg kg^{-1} DM) has been associated with growth retardation in growing cattle given corn silage containing 7 mg Cu kg^{-1} DM, whereas it was not recorded in those finished on corn containing less copper (5.4 mg kg^{-1} DM) (Legleiter and Spears, 2007). The effects of different forage to cereal ratios merit study. On the first point, a Norwegian survey showed that the mean copper to molybdenum ratio decreased from 3.8 to 2.0 between pasture cuts, for which the respective ranges were 1.1–8.3 and 0.8–5.2 (Govasmark et al., 2005). In yak, the same copper-RD has been attributed to molybdenum (Xiao-Yun et al., 2006b) and iron exposure (Xiao-Yun et al., 2006a). Limits for dietary iron to copper ratios are also given (Table 11.6), but allowance for sulfur is omitted because sulfur is relatively difficult and costly to assay and there is usually sufficient sulfur in green swards to allow extensive expression of antagonisms from both molybdenum and iron. However, SO$_4$ intakes from drinking water must be allowed for in some areas (Smart et al., 1986). Limitations remain because the effects of iron and molybdenum cannot simply be combined since the two antagonists do not act additively (e.g. in Fig. 11.14).

Liver copper

The early diagnostic threshold (liver copper <25 mg (395 µmol) kg^{-1} DM) is *obsolete*, although still often quoted to exaggerate the importance of surveys, and was reduced long ago to <10 mg kg^{-1} DM for all ruminants (Underwood and Suttle, 1999). The critical value may be influenced by breed and/or yield; Holsteins are suggested to have a lower threshold for liver copper than Jerseys for the maintenance of normal plasma copper (Du et al., 1996). In Friesians, copper-RD has been associated with liver copper <3 mg kg^{-1} DM (Humphries et al., 1983; Phillippo et al., 1987a,b), while subclinical growth retardation occurred in Aberdeen Angus with mean liver copper probably <4 mg kg^{-1} DM (Legleiter and Spears, 2007). Liver copper <100 µmol kg^{-1} DM indicates a high risk and 100–300 µmol kg^{-1} DM a possible risk for copper-RD in bovines, but the lower limit for cervines may be 180 µmol kg^{-1} DM (Grace and Wilson, 2002). Higher thresholds are needed in newborn than in older animals (Table 11.6).

Plasma and caeruloplasmin copper

Plasma hypocupraemia indicates subnormal Cp synthesis, but no copper-RD may be evident with mean or individual plasma copper values of 3–9 µmol l^{-1} in sheep, cattle (Underwood and Suttle 1999) and deer (Nicol et al., 2003). Mean plasma values of 3–4.5 µmol Cu l^{-1} are common in clinically affected sheep (Whitelaw et al., 1979, 1983; Woolliams et al., 1986b), cattle (McFarlane et al., 1991) and growth-retarded deer (Grace et al., 2005b). Infectious diseases and vaccinations increase plasma copper substantially by inducing Cp synthesis, even in initially hypocupraemic animals (Suttle, 1994), thus overestimating true copper status. A simultaneous assay of biochemical markers of acute-phase reactions (low plasma zinc, fibrinogen >1 g l^{-1} or 'positive' haptoglobin) can identify such anomalies. Assays of Cp offer little advantage over plasma copper because the two parameters should be highly correlated (e.g. Legleiter and Spears, 2007; Suttle, 2008a). Furthermore, assay methods for Cp use different diamine substrates and some are prone to interference, giving poor correlations with plasma copper (Laven and Livesey, 2006; Laven et al., 2007; Suttle, 2008a). Analysis of TCA-soluble copper is a reliable surrogate for Cp in most circumstances (Suttle, 2008a) but it can give slightly higher copper values than aqueous dilution prior to detection by atomic absorption (Suttle, 1994). The measurement of TCA-I copper is generally of little diagnostic use (Paynter, 1987): values are usually within background error (e.g. in sheep) (Suttle, 2008b), not raised in cattle by moderate molybdenum exposure (Suttle, 2008a) and sometimes anomalously high (e.g. in goats) (Osman et al., 2003).

In using copper concentrations in blood components to monitor copper status, two things should be borne in mind:

- Normal values for plasma copper and Cp are around 50% lower at birth (Table 11.6) (Suttle, 1994) than in later life, but increase rapidly during the first week in calves (McMurray, 1980), kids Osman et al., 2003) and lambs (Grace et al., 2004) due to an increase in Cp.
- Serum should *not* be used from blood samples of older cattle and sheep because Cp is 'lost' during clotting and values can be significantly lower in serum than in plasma from the same blood sample. Low serum values may reflect loss of copper-containing, Cp-like clotting factors.

In bovine samples, a highly variable and unpredictable proportion (5–30%) of copper and 'Cp activity' is lost (McMurray, 1980; Paynter, 1987; Laven and Livesey, 2006; Laven et al., 2007) and this may have contributed to lack of confidence in the diagnostic merits of serum copper analyses (Vermunt and West, (1994). Loss from ovine samples is smaller and less erratic (2–3 µmol Cu l^{-1}) (Suttle, 1994), but the use of a correction factor to convert serum copper to plasma copper introduces unacceptably large errors at the upper end of the diagnostic range (Laven and Smith, 2008). Serum samples from horses (Paynter, 1987) and deer (Laven and Wilson, 2009) give similar results to those found in plasma samples, but the position of other species is unknown.

Whole-blood and erythrocyte copper

Whole-blood copper is rarely used in diagnosis although it has certain advantages. Approximately one-third of blood copper is contributed by erythrocytes and these have a long half-life. Anomalies caused by the induction of Cp or sudden changes in the copper supply have less impact on blood copper than on plasma copper and none on erythrocyte copper (Arthington et al., 1996), but groups with 'subnormal' blood copper may still not benefit from copper supplementation (Phillippo, 1983). The marginal range given for whole-blood copper (6–10 µmol Cu l^{-1}) was derived from a longitudinal study of plasma, whole blood and liver copper (Koh and Judson 1987). The activity of erythrocyte (E)SOD1 provides additional diagnostic information, with low values confirming a prolonged deficiency (Ward and Spears, 1997). ESOD was more closely correlated with growth rate than Cp in Fig. 11.14 and the mortality described in Fig. 11.2 (see Jones and Suttle, 1987). However, the conventional units of ESOD activity are relative. 'Enzyme equivalence' may vary between laboratories, and erythrocytes must be washed to remove plasma inhibitors and heavily diluted to minimize interference from haemoglobin (Herbert et al., 1991). Tentative diagnostic ranges for one particular kit assay method have been proposed (Herbert et al., 1991), but ESOD1 activity recorded with that kit may be subject to a sulfur × copper interaction at very high sulfur intakes (Grace et al., 1998b).

Neutrophil copper and SOD1

Steers given a diet high in molybdenum (15 mg kg^{-1} DM) but with a copper to molybdenum ratio of 1.0–1.8 have shown reductions in neutrophil copper (from 13.0 to 5.1 pg/10^{7}cells) and neutrophil SOD1 (from 0.65 to 0.31 IU/10^6 cells) after 8 months, by which time mean liver copper was marginal (285 µmol kg^{-1} DM) but plasma copper was normal (Xin et al., 1991). Furthermore, the in vitro killing capacity of the neutrophil was reduced from 27.7% to 17.3%, although the animals were clinically normal. Low neutrophil SOD1 activity has also been found in calves with molybdenum-induced hypocuprosis (5 mg Mo kg^{-1} DM and a dietary copper to molybdenum ratio of 1), but not in equally hypocupraemic, clinically normal calves given supplementary iron (Gengelbach et al., 1997). In contrast to the conventional criteria of copper status, decreases in neutrophil copper or SOD1 may give a warning of impending dysfunction.

Hair and fleece copper

Hair and fleece copper have been widely measured without becoming established as reliable diagnostic aids. Values can reflect suboptimal copper intake in both sheep and cattle, but a slow rate of decline means that low values indicate prolonged deprivation (Fig. 11.12) (Suttle and Angus, 1978). Hair and fleece represent the sample of choice from a grossly deteriorated carcass. Diagnostic interpretation may be influenced by other nutritional deficiencies that influence hair or wool growth and colour (see p. 245).

Other possible criteria

Plasma MAO and liver CCO activities decrease in cattle deprived of copper (Mills et al., 1976), as does liver SOD1, but SOD1 may be no more specific than other tests (Fig. 11.14). Plasma diamine oxidase has shown promise in mice and rats (Prohaska, 2006) but has shown similar changes to Cp and ESOD1 in a field supplementation study with sheep (Kendall et al., 2001) and to copper and Cp in an experiment with copper-deprived cattle (Legleiter and Spears, 2007). Terminal responses in gene expression after prolonged copper deprivation in cattle showed up-regulation of the chaperone protein supplying SOD1 (CCS) in the duodenal mucosa and down-regulation of CCO (measured as its COX1 sub-unit) in the liver, but no change in SOD1 expression in the liver (Fig. 11.15). Early changes in circulating levels of mRNA for CCS merit attention. The rates of decline in tyrosinase and thiol oxidase activity in the bulb of hair or wool fibres should be studied in copper-RD, since in vivo sampling would be a relatively simple and painless. The direct measurement of non-Cp copper (Beattie et al., 2001) and plasma SOD2 and 3 activity may be worth exploring in livestock. A variety of immunological markers are affected by

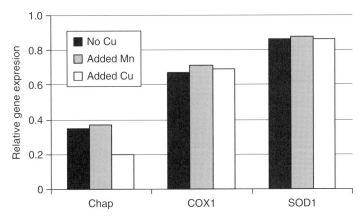

Fig. 11.15. The effects of long-term copper deprivation, with or without excess dietary manganese, on the expression of three 'Cu genes' in cattle showing signs of hypocuprosis. The chaperone gene that facilitates the incorporation of Cu into superoxide dismutase 1 (SOD1) in the duodenal mucosa (*Chap*) is up-regulated by Cu deprivation, but the expression of SOD1 and cytochrome oxidase (COX1) in the liver are not (from Hansen *et al.*, 2008).

subclinical copper deprivation in laboratory animals (Bonham *et al.*, 2002).

Prevention and Control of Copper Deprivation in Ruminants

A variety of methods for providing livestock with sufficient copper have been devised, but they differ in efficacy and suitability.

Pasture fertilizers

The application of copper-containing fertilizers can increase plant yields and raise the copper concentration of herbage and crops, but required levels are often much higher for the grazing animal (see Table 11.5) than for the plant (≤4 mg Cu kg^{-1} DM) (McFarlane, 1989; Haynes, 1997). The amounts of copper required vary with the soil type and climatic conditions. Early Australian experience indicated that a single dressing of 5–7 kg CuSO$_4$ ha^{-1} was effective for 3–4 years, except on calcareous soils (Underwood *et al.*, 1943) and residual benefits have been reported on sandy soils in Australia after 23 years. However, on yellow–brown/yellow–grey earths in New Zealand hill country, the response in herbage copper can last for little more than a year and, in a recent study, 3 kg CuSO$_4$ ha^{-1} raised pasture copper for barely 100 days (Grace *et al.*, 2005a). Fertilizer treatment is unsuitable on soils high in organic matter because copper becomes 'fixed' in unavailable humic acid complexes. It is not advised when induced hypocuprosis occurs on pastures high in copper (McFarlane *et al.*, 1990) and is uneconomical under range conditions.

Dietary supplementation: inorganic sources

Copper deprivation can be economically prevented by the provision of salt licks containing 0.5–1.9% copper on extensive farms and by incorporating CuSO$_4$ into mineral mixtures or concentrates offered to housed cattle. Supplementary copper can also be applied in solution to the forage to maintain a dietary copper to molybdenum ratio >3 or dispensed into the drinking water supply by metering devices. With the latter method, small doses can illicit rapid clinical responses (Farmer *et al.*, 1982), possibly because some copper bypasses the 'hostile' rumen environment. Other copper salts such as the acetate, carbonate (Ledoux et al., 1995b), citrate and chloride (Engle and Spears, 2000) and are no better than $CuSO_4$ in raising liver copper.

Dietary supplementation: 'organic' sources

There is, theoretically, scope for sources that protect copper from rumen antagonisms, but this has only recently been realized. Calves given 5 or 10 mg Cu kg^{-1} DM as a glycinate complex rather than $CuSO_4$ in corn-silage-based diets containing first 2 and later 6 mg supplementary Mo kg^{-1} DM showed 31–44% greater responses in plasma and liver copper (Hansen et al., 2008); glycinate succeeding where other complexes had previously failed (Underwood and Suttle, 1999). However, $CuSO_4$ was equally effective in improving LWG (by 14%) and FCE (by 8%), and the differences in copper status only emerged during the final 28-day phase of maximal molybdenum exposure. Extravagant claims have been made for copper–protein hydrolysate mixtures ('metalosates') and specific amino acid complexes such as Cu_2–lysine but neither source had any consistent advantage over $CuSO_4$ in five studies with cattle and one with sheep (for review, see Underwood and Suttle, 1999). Additional studies with sheep (Luo et al., 1996), steers (Rabiansky et al., 1999; Engle and Spears, 2000; Yost et al., 2002; Dorton et al., 2003; Mullis et al., 2003), kids (Mondal et al., 2007) and broilers (Miles et al., 2003) have failed to demonstrate any consistent or sustained advantage for complexes of copper with amino acids (other than glycine) in liver copper, plasma copper or immune responses, as predicted from early studies showing their instability in vitro (Fig. 2.4, p. 27). Against the tide of opinion in peer-reviewed papers is a study in which copper proteinate stimulated the growth of goat kids more than equivalent supplements given as $CuSO_4$ (Datta et al., 2006), but it is arguable that the growth stimulation produced by both copper sources waned in the case of $CuSO_4$ as subclinical toxicity developed. The commercially driven pursuit of trivial advantages over a cheap and effective source – $CuSO_4$ – should cease and attention be given to predicting when supplementation is needed and how much copper should be given. Dry rations for sheep and cattle rarely need supplementing with copper and the problem is usually over- rather than under-provision of copper.

Oral and parenteral dosage

Discontinuous methods can be effective because all grazing animals have the ability to store copper in their livers during periods of excess intake and to draw on those stores when intakes are inadequate. Drenching with $CuSO_4$ at monthly or longer intervals can be effective (e.g. Palsson and Grimsson, 1959), but not in the face of high molybdenum exposure (>5 mg kg^{-1} DM).

Although delayed swayback is untreatable, it may be preventable (Suttle, 1988a) and oral copper doses of 1 mg kg^{-1} LW are recommended for all suckling lambs after a first clinical case in a flock. In Fig. 11.2b, the reduction in swayback incidence (in 1982) was achieved with parenteral copper given at an average age of 30 ± 10 days (Woolliams et al., 1986a). Copper was given parenterally in late pregnancy the following year and swayback was successfully prevented, but losses from infections were increased unless the lambs were also supplemented with copper. Parenteral administration of copper complexes was introduced 100 years ago and was the method of choice, but problems with tissue reactions for slowly translocated complexes (e.g. methionate) and acute copper toxicity for rapidly translocated complexes (e.g. hydroxyquinoline sulfonate) led to the withdrawal of products (Underwood and Suttle, 1999). The parenteral route is not ideal for the treatment of acute copper-responsive diarrhoea ('scours') where the critical copper deficit is in the gut mucosa, and should not be used in the vain hope of improving immune responses to vaccines (Steffen et al., 1997).

Slow-release oral supplementation

Copper oxide wire particles (CuOWP) of high specific gravity, delivered in capsules by 'balling gun', are retained in the abomasum where copper is released over a period of several weeks (Dewey, 1977; Deland et al., 1979). The optimum dose rate for sheep and cattle is $0.1\,g\,kg^{-1}$ LW (Suttle, 1981) and liver copper stores can remain increased for many months after a single dose (McFarlane et al., 1991). Hypocuprosis has been prevented by dosing lambs (Whitelaw et al., 1983; Woolliams et al., 1986b) and calves (Whitelaw et al., 1984) on improved hill pastures with CuOWP. In farmed deer reared on molybdenum-fertilized pasture in New Zealand, dosing with CuOWP slowed the depletion of liver copper and development of hypocupraemia, but the protection afforded was far less with pasture containing 9.0 rather than $4.4\,mg$ Mo kg^{-1} DM (Wilson and Grace, 2005). Administration is difficult in deer and the practice of routinely dosing CuOWP on normal pastures is not recommended (Nicol et al., 2003).

Although CuO contains copper of intrinsically low availability (25% relative to $CuSO_4$) (Baker, 1999), the large doses that can be safely given as CuOWP make it an effective slow-release source. Efficacy may be reduced where diarrhoea is a problem. In two groups of lambs given the same dose at the same time but grazing different fields, protection from hypocupraemia was afforded for 56 days to a badly scouring group and for over 150 days to another, less-affected group (Suttle and Brebner, 1995). Abomasal parasitism was implicated since less copper is released from CuOWP if abomasal parasitism has raised abomasal pH (Bang et al., 1990). However, CuOWP may have a secondary benefit in the form of anthelmintic activity (Burke et al., 2005). Copper oxide is also effective when given to sheep in coarse powder form (Cavanagh and Judson, 1994) and to cattle in a multi-mineral and vitamin bolus (Parkins et al., 1994). A soluble glass bolus has been devised that releases copper, along with cobalt and selenium, for long periods in cattle (Koh and Judson, 1987; Givens et al., 1988; McFarlane et al., 1991). Slow-release copper sources should only be administered to animals of proven low-copper status. Other methods of copper supplementation should not be used concurrently.

The required frequency of dosing will increase with dietary copper to molybdenum ratios of <3.0. Parasite infections and other causes of diarrhoea should be controlled prior to dosing with CuOWP, which should precede turnout by several weeks.

Genetic selection

The existence of genetic variation in copper metabolism could be exploited to prevent deprivation and lessen the dependence on supplementation where CuRD are endemic. Selection of ram lambs for four generations on the basis of plasma copper has eliminated hypocuprosis in an inbred, vulnerable 'breed' (Woolliams et al., 1986a,b). Alternatively, the use of rams of a 'tolerant' breed, such as the Texel for crossing with the Scottish Blackface, should reduce hypocuprosis problems on hill pastures. Comparisons of liver copper accretion between offspring of Suffolk and Texel sires has revealed a high heritability for liver copper concentration, suggesting scope for selection for high liver copper status based on samples retrieved from the abattoir (Suttle et al., 2002). The heritability of serum copper concentration in an Aberdeen Angus herd (0.28) was sufficiently high to suggest that copper status could be manipulated by breeding (Morris et al., 2006), and the heritability of plasma copper might even higher.

Minimizing antagonisms

Given the prominent influence of antagonisms from iron and molybdenum on the occurrence of copper-RD, prevention can be achieved in part by minimizing exposure to such antagonists. By avoiding the excessive use of lime and improving soil drainage, the antagonism from molybdenum can be reduced. Advantage might also be taken of a lower uptake of molybdenum by corn than grass on 'teart' pastures and the use of corn silage as a winter feed in place of grass silage. By avoiding overgrazing and 'poaching', particularly on 'temporary' pasture, antagonisms from soil iron (and possibly manganese) are minimized. Iron should *never* be used in mineral supplements for grazing or housed, weaned ruminants, even as Fe_2O_3 to give colour.

Occurrence and Diagnosis of Copper Deprivation in Non-ruminants

Copper deprivation rarely causes problems in non-ruminants reared on natural feedstuffs. Where copper intakes are low, enhancement of A_{Cu} probably allows biliary copper to be to be recycled more efficiently and guard against deprivation. Assessment of copper status is complicated by the fact that normal ranges for plasma copper differ greatly between species (Table 11.7).

Horses

The possibility that grazing horses may receive inadequate supplies of copper has received some recent attention because of the vulnerability of ruminants grazing the same pastures, but Thoroughbreds show little or no response in plasma and liver copper to supplementation by oral or parenteral routes in such circumstances (Pearce et al., 1999a; Grace et al., 2002). Equine species can be grouped with other non-ruminants in which uninhibited A_{Cu} lessens the need to store copper (Fig. 11.9). Angular deformities at the pasterns have been claimed to respond to copper supplementation (Barton et al., 1991), but there are other possible causes. Swelling of the distal radial physis in Thoroughbred yearlings did not respond to copper supplements providing 130 mg day^{-1} (Grace et al., 2002). Horses are not adversely affected by molybdeniferous pastures (Pearce et al., 1999b), presumably because they only generate thiomolybdate in their hind guts.

There are no losses of Cp during clotting in equine samples (Paynter, 1987) to complicate the assessment of serum copper, and a normal range of 16.5–20.0 µmol l^{-1} has been suggested (Suttle et al., 1996b). However, a seasonal fall has been seen to take values as low as 9 µmol l^{-1} in grazing fillies in midwinter and may have been attributable to photoperiodism (Pearce et al., 1999a). Like most newborns, the foal begins life with a large hepatic reserve, with as much as 374 ± 130 mg kg^{-1} DM (± standard deviation) being found regardless of whether the mare has been given copper by intramuscular injection during pregnancy (Gee et al., 2000). Most of the foals showed a rapid decline to normal 'adult' values by 160 days of age (21 ± 6 mg kg^{-1} DM), but a small subgroup showed a much slower rate of decline.

Table 11.7. Normal ranges for plasma and liver copper concentrations in some monogastric species.

	Plasma Cu (µmol l^{-1})	Liver Cu (µmol kg^{-1} DM)
Horse	8–12	100–200
Pig	14–16	100–1000
Chicken	3–6	100–500

Pigs

Copper deprivation can be induced by weaning pigs early on to bovine milk or its products. Plasma and liver copper fall simultaneously from the outset, a microcytic, hypochromic anaemia soon develops and the survival time of erythrocytes is shortened (Bush et al., 1956). Iron accumulates in the gut mucosa at a preclinical stage and the piglet is particularly susceptible to the anaemia of iron deprivation, but the rate-limiting factor in copper deprivation may be haem biosynthesis rather than iron transport. The addition of 100 mg Zn kg^{-1} DM to a corn/SBM ration reduced AA_{Cu} from 0.22 to 0.13 in young pigs (Adeola et al., 1995) and a copper-responsive anaemia has been reported in pigs fattened on swill that had been stored in galvanized bins (Pritchard et al., 1985). The swill was high in zinc (2580–5100 mg kg^{-1} DM) and low in copper (3.2–4.7 mg kg^{-1} DM). Other swills being fed in the locality but uncontaminated with zinc contained slightly more copper (3.5–9.0 mg kg^{-1} DM) and caused no problems. It was concluded that swills should be supplemented with copper and not stored in galvanized bins. In severely deficient young pigs, osteoblastic activity became depressed, the epiphyseal plate widened and gross bone disorders developed with fractures and severe deformities, including marked bowing of the forelegs (Baxter et al., 1953). Abnormal elastin development can cause aortic rupture and high mortality (Carnes et al., 1961) for reasons explained below.

Poultry

Experimental copper deprivation in the chick produces a normocytic and normochromic, anaemia and low MAO or CCO activity in the bones may compromise osteoblastic activity, leading to abnormal bone morphology (Rucker et al., 1969,1975). Abnormal elastin development can also cause high mortality due to aortic rupture in chicks under experimental conditions (O'Dell et al., 1961) and turkey poults under commercial conditions (Guenthner et al., 1978). Lysyl oxidase activities (Table 11.1) in plasma, the liver and aorta are subnormal in copper-deprived chicks (and piglets), diminishing the capacity to form desmosine from lysine and the cross-links that give the aorta strength and elasticity (O'Dell, 1976). The particular susceptibility of chicks and piglets to aortic rupture is due to the rapid growth of the aorta in the first few weeks of life. Similar biochemical abnormalities may impair shell formation in the laying hen (Baumgartner et al., 1978) and the air–blood capillary network in the avian lung (Buckingham et al., 1981). Hens deprived of copper lay eggs of abnormal size and shape, either lacking shells or with wrinkled, rough shells due to malformation of the shell membranes (Baumgartner et al., 1978). Egg production and hatchability are markedly reduced and after incubation the embryos from these hens exhibit anaemia, retarded development and a high incidence of haemorrhage (Savage, 1968). Such problems do not occur on natural diets.

Laboratory animals

The lethal effects of aortic rupture have drawn attention away from the role of copper in other parts of the cardiovascular system, including the heart. In copper-deficient mice it is DBM that becomes rate-limiting in the heart (Prohaska, 2006). Other enzymes are also likely to be depleted, including SOD1, but there are compensatory increases in MnSOD activity (Lai et al., 1994). Gene-deletion studies in laboratory animals have shown that while absence of Cp causes iron to accumulate in the liver and elsewhere, the associated anaemia is relatively mild and animals survive (Prohaska, 2006). Increased expression of homologues such as hephaestin (Table 11.1) may provide cover for iron transfer. Immunological markers such as interleukin-2 production and lymphocyte subset phenotyping may be particularly sensitive measures of copper status in non-ruminants (Bonham et al., 2002).

Copper as a Growth Promoter

Pigs

Copper has been widely used at supranutritional levels, usually as $CuSO_4$, to enhance growth and FCE (Braude, 1967). Comparisons of $Cu(OH)_3Cl$ (Cromwell et al., 1998) and citrate (Armstrong et al., 2004) with $CuSO_4$ have shown no effect of copper source on growth promotion. The optimal level lay between 100 and 250 mg Cu kg^{-1} DM, but this figure has varied between studies and may depend on the strength of antagonism from calcium and phytate. Growth promotion may result from improved digestibility or appetite (Zhou et al., 1994a), antibiotic activity, detoxification of goitrogens (see Chapter 12) or a systemic endocrine stimulus. Intravenously administered copper histidinate can stimulate growth, and release of the mRNA for growth hormone is stimulated by both the oral and intravenous routes of copper supplementation, which significantly increase brain copper (Zhou et al., 1994b). 'Supranutritional' copper supplementation is not always a safe procedure. The practice has reduced iron stores in the liver (Bradley et al., 1983) and induced anaemia, copper toxicosis or zinc deficiency. Concurrent supplementation with iron and zinc, each at 150 mg kg^{-1} DM, protected against the adverse effects of copper at 425 mg Cu kg^{-1} DM in one early study (Suttle and Mills, 1966a). Birth and weaning weights can be increased by 9% by similarly supplementing the sow's diet with copper (Cromwell et al., 1993). The environmental pollution caused by the copper-enriched effluents from pig units where copper is used as a growth promoter calls into question the sustainability of this practice (see Chapter 20).

Poultry

The efficacy of $CuSO_4$ as a growth promoter has varied widely. Chick growth was increased by 250 mg Cu kg^{-1} DM in a wheat/fish meal diet, but depressed when a maize/SBM diet was similarly supplemented (Jenkins et al., 1970). By contrast, growth was stimulated by 100–200 mg Cu kg^{-1} DM in a maize/SBM diet, but not in a casein/dextrose diet on which higher availability may have resulted in hepatotoxicity (Fig. 11.6) (Funk and Baker, 1991). With a reduced level of supplementation of a corn/SBM diet (125 mg Cu kg^{-1} DM), chick growth was increased by 3.5–4.9% between 1 and 45 days of age, but an even lower level (63 mg Cu kg^-) given as citrate increased the growth benefit to 9.1% (Ewing et al., 1998). In turkey poults, growth has been retarded by 204 mg Cu kg^{-1} DM in a short-term (10-day) study (Ward et al., 1994) and by 120 mg kg^{-1} DM in a longer-term (24-week) study (Kashani et al., 1986). The brief addition of copper to the drinking water as well as diet is often practised, but is not recommended (Ward et al., 1994).

Copper sources

There are differences in 'copper value' amongst inorganic copper salts: cuprous chloride has a marked (44%) advantage over $CuSO_4$ (Baker and Ammerman, 1995) and cupric citrate is superior to $CuSO_4$ and the oxychloride, both as a growth stimulant and in raising liver copper stores (Ewing et al., 1998). Insoluble copper compounds such as CuO can be poor (Baker, 1999) or variable sources (Koh et al., 1996) for poultry. Neither copper methionine nor copper lysine has shown consistently superior availability over $CuSO_4$ in the studies with pigs or poultry reviewed by Baker and Ammerman (1995), and subsequent studies with 'organic' sources (e.g. Waldroup et al., 2003) do not indicate an advantage for amino acid 'chelates'. Pig growth is stimulated more by copper lysine than $CuSO_4$ added to the diet (Zhou et al., 1994a), but this may reflect a higher palatability rather than availability. Faecal excretion of copper in pigs given copper proteinates or $CuSO_4$ indicates that they provide copper of similar availability (Veum et al., 2004).

Chronic Copper Poisoning in Sheep

Occurrence

In its recent review of copper toxicity in all species, the NRC (2005) concurred that sheep are more susceptible than any other species to CCP, but that susceptibility is not linked to abnormalities of the sCp gene as might be suspected from studies of Wilson's disease in humans (Lockhart and Mercer, 1999). The number of reported CCP cases in the UK rose steadily to about 100 per annum between 1970 and 1990, due partly to the growing popularity of susceptible breeds such as the Texel. The equally susceptible North Ronaldsay (Wiener et al., 1978) can succumb to CCP while grazing, but the disease rarely occurs in pastured sheep unless copper fungicides have been applied in orchards and contaminate the underlying pasture. In parts of Australia and elsewhere, normal copper intakes together with very low levels of molybdenum (0.1–0.2 mg kg^{-1} DM) or the consumption of plants such as *Heliotropium europaeum*, containing hepatotoxic alkaloids (heliotrine and lasiocarpine), cause 'toxaemic jaundice' or 'yellows', both forms of CCP (Howell et al., 1991). Exposure to another hepatotoxin, phomopsin, in cases of lupinosis also increases susceptibility to CCP (White et al., 1994). However, CCP occurs most frequently in housed lambs, milk sheep and pedigree rams receiving large amounts of concentrates and is more likely to occur in sheep housed on slatted floors than in those bedded on straw (Crosby et al., 2004; Day et al., 2006). Complete sheep diets containing copper at >15 mg kg^{-1} air-dry feed can cause CCP (Hartmann, 1975) and it is difficult for feed compounders to keep values consistently below this limit, particularly when copper-rich constituents such as certain distillery by-products (Suttle et al., 1996a) are used. The feeding of palm-kernel cake, containing

25–40 mg Cu kg^{-1} DM, has also caused CCP (Chooi et al., 1988). Cases of CCP have been caused by contamination of pastures with copper-rich pig slurry (Suttle and Price, 1976; Kerr and McGavin, 1991) and feed with copper-supplemented pig or cattle rations, and by the excessive consumption of copper-containing salt licks and the unwise use of copper supplements (oral or parenteral). The addition of monensin (Barrowman and van Ryssen, 1987) and possibly selenium (Hartmans and van Ryssen, 1997) to sheep feeds predisposes to CCP.

Pathogenesis

Whatever the cause, CCP is characterized by two phases: the first (pre-haemolytic) is clinically but not biochemically 'silent':

- Liver copper accretion proceeds over a period of weeks or months until concentrations of 1000–1500 mg kg^{-1} DM or more are reached (Fig. 11.10).
- Glutamate dehydrogenase is released from damaged hepatocytes into the bloodstream but serum copper does not increase.
- γ-Glutamyltransferase activity in plasma indicating chronic damage to the bile ducts increases after prolonged exposure.

High ESOD1 activity may also indicate copper overload (Suttle 1977; Grace et al., 1998a). The second phase is marked by an acute haemolytic crisis and accompanied by a marked rise in serum acid phosphatase and creatine kinase, as well as plasma copper. The detailed pathology of the disease has been reviewed by Howell and Gooneratne (1987).

Subclinical hepatotoxicity

Sheep can show histological (King and Bremner, 1979) or biochemical (Fig. 11.16) (Woolliams et al., 1982) evidence of liver damage at liver copper concentrations as low as 350 mg kg^{-1} DM. Whether such changes compromise liver function at times of heavy metabolic load, such as late pregnancy in the twin-bearing ewe, is unknown, but the precautionary principle should be applied. A marginally toxic range of

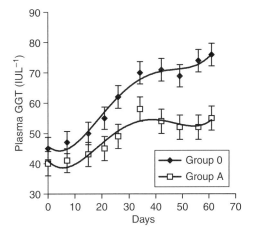

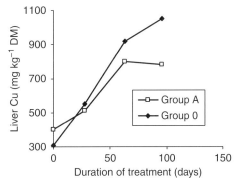

Fig. 11.16. The effects of dietary supplementation with a cocktail of copper antagonists (Group A: molybdenum, 2; sulfur, 3000; zinc, 300 mg kg^{-1} DM) on the accumulation of copper in the liver and a biochemical indicator of liver damage (γ-glutamyl transferase, GGT) in Texel rams given a diet high in Cu and iron (26 and 345 mg kg^{-1} DM, respectively). Group 0 consisted of unsupplemented controls (Suttle and Blake, 2006).

350–1500 mg kg^{-1} DM should replace the commonly used single threshold of 750 mg kg^{-1} DM for liver copper (Table 11.8); marginal bands should also be used to interpret liver enzyme activity and plasma copper.

Diagnosis

The diagnosis of haemolytic CCP rests upon clinical evidence of haemolysis (jaundice and haemoglobinuria) and is confirmed by high liver

Table 11.8. Marginal bands[a] for copper to antagonist concentrations in the diet, copper concentrations in the bloodstream and tissues and serum enzyme activities (indicative of liver or muscle damage) in the diagnosis and prognosis of chronic copper poisoning (CCP) in livestock.

	Sheep	Cattle	Goats	Pigs	Poultry	Interpretive limits
Diet Cu:Mo	10–**20**	30–**60**	30–**60**	NA	NA	
Diet Cu (mg kg^{-1} DM)	12–**36**	50–**300**	30–**100**	250–**750**	200–**400**	
Liver Cu (mmol kg^{-1} DM)[b]	6.4–**16.0**	6.4–**16.0**	6.4–**16.0**	32.0–**48.0**	1.0–**2.0**	
Kidney Cu[c] (mmol kg^{-1} DM)	0.6–**0.8**	0.6–**0.8**	0.6–**0.8**	10.0–**16.0**		
Kidney Fe (mmol kg^{-1} DM)	16–**18**	16–**18**	16–**18**			
Plasma[d] Cu (µmol/l)	20–**25**	20–**25**	20–**25**	40–**60**		No acute-phase response
Serum bilirubin (mmol l^{-1})	4–**20**					
Serum AAT[e]	60–**100**	20–**60**				Also released during muscle damage
Serum GLDH[e]	12–**51**	12–**51**				Consider other possible causes (e.g. liver fluke)
Serum GGT[e]	45–**100**	45–**100**				

AAT, aspartate aminotransferase; Fe, iron; GGT, γ-glutamyltransferase; GLDH, glutamate dehydrogenase.
[a]Values below the lower limit rule out CCP; values above the upper limit are strongly suggestive of CCP; marginal values for several criteria indicate probable CCP.
[b]Multiply by 63.4 to obtain units in mg kg^{-1} DM and by 21.1 for mg kg^{-1} fresh weight.
[c]High kidney but normal liver Cu indicates acute toxicity; high kidney Cu with normal kidney Fe indicates hepatic crisis without haemolytic crisis.
[d]Serum samples should not be used for sheep or cattle because values can be underestimated by up to 6 µmol Cu l^{-1}: the corresponding marginal range for ovine serum Cu is 15.8–19.2 µmol Cu l^{-1}, judging by data of Laven and Smith (2008).
[e]All units of enzyme activity are given as IU $^{-1}$ measured at 30°C.

and kidney copper and high kidney iron (Table 11.8). If liver copper alone is high, the possibility of hepatotoxicity should be considered. Raised 'liver enzyme' activities in surviving, clinically normal cohorts confirm pre-haemolytic liver injury. Colostral intake of γ-glutamyltransferase invalidates its diagnostic use in the first weeks of life. Acute CP, usually caused by maladministration of parenteral or oral copper supplements, is distinguished from CCP by raised kidney copper and iron but relatively low liver copper.

Treatment

Affected sheep can be treated with TTM (Gooneratne et al., 1981), a mediator of the copper–molybdenum–sulfur antagonism (see earlier). The most convenient treatment is a course of three subcutaneous injections of 3.7 mg ammonium salt (ATTM) kg^{-1} LW, given on alternate days, and it is effective in some sheep already in haemolytic crisis (Humphries et al., 1988). Increases in total plasma copper occur after ATTM treatment due to accumulation of TCA-I copper, and analysis of TCA soluble copper is recommended (Suttle, 2008b). The success of treatment can only be assessed by liver enzyme assays in plasma (Suttle et al., 1992a; Suttle, 2008b). There is a temporary, marked fall in blood ESOD after ATTM treatment (Suttle et al., 1992a), but this probably has no important physiological consequences.

Prevention

Antagonists of copper can be used to prevent CCP. Provision of molybdate-containing salt licks has been found to eliminate mortality from toxaemic jaundice (Dick and Bull, 1945) and led to the preventive use of dietary molybdenum plus SO_4 (Suttle, 1977) and parenteral ATTM (Gooneratne et al., 1981). In outbreaks of CCP, unaffected sheep with raised liver enzyme activities should be treated with ATTM: 50 mg given intravenously twice weekly to castrated male sheep has reduced liver copper stores by 52%, but the subcutaneous route is more convenient (Ledoux et al., 1995a). All sheep should be transferred to a diet low in available copper. This can be achieved by replacing the proprietary concentrate with urea-treated whole grain and/or adding gypsum (15 g kg^{-1} DM) and sodium molybdate (19.2 mg kg^{-1} DM) to the forage. The accumulation of copper from concentrates can be greatly reduced by the addition of antagonists such as iron (Rosa et al., 1986) and zinc supplements (Bremner et al., 1976; Akpan et al., 2005) or a 'cocktail' (Fig. 11.16).

It is common for mixtures of antagonists to be added by feed compounders to reduce the risks of CCP, although there is no evidence that additive effects will be obtained. A 'cocktail' of antagonists (molybdenum, zinc and sulfur) reduced but did not prevent liver copper accretion in Texel sheep given a concentrate high in iron (Fig. 11.16). Scope for the increased use of sulfur is limited by the risk of sulfur toxicity (see Chapter 9). With vulnerable breeds or dairy sheep at recurrent risk while housed, the grazing of strips of pasture fertilized with sodium molybdate (0.27 kg ha^{-1}) during summer may lessen the accumulation of copper over the years. In addition, appropriate genetic selection could lower the susceptibility of vulnerable breeds to CCP. Selection of Texel sires for low plasma copper has resulted in lower liver copper in their offspring (J. A. Woolliams, personal communication, 1990) and selection for low liver copper should achieve similar results (Suttle et al., 2002), lessening vulnerability to CCP during intensive fattening. Similar selection within dairy sheep would also lessen their vulnerability to CCP.

Copper Poisoning in Cattle and Goats

Occurrence in cattle

Cattle are at greatest risk of developing CCP prior to weaning when given copper-rich milk substitutes from which copper is well absorbed (Shand and Lewis, 1957). Exposure to 200 mg Cu kg^{-1} DM in a milk substitute from 3 to 45 days of age was harmful to calves, but 50 mg kg^{-1} DM was tolerated for 9 weeks (Jenkins and Hidiroglou, 1989). However, tolerance will depend on maternal copper status and the liver copper concentrations present in the newborn calf. The administration of CuOWP to newborn calves, against proprietary instructions, has contributed to an outbreak of CCP in which the dams also received parenteral copper injections in late pregnancy (Steffen et al., 1997). After weaning, cattle are less vulnerable to CCP and dietary copper levels of up to 900 mg kg^{-1} DM can be fed for several months without causing apparent harm (Felsman et al., 1973). Occasionally CCP is reported in adult cows (Tremblay and Baird, 1991) and its incidence has increased in the UK (Livesey et al., 2002). Between 2000 and 2007 there were 31 ± 9.2 (standard deviation) incidents annually compared to 50 ± 9.3 outbreaks of lead poisoning (VIDA, 2007). The presence of liver fluke infections, which are also on the increase, may be partly responsible, lowering the capacity of the liver to safely store copper (N. F. Suttle, unpublished data). Some instances of CCP have involved massive pollution from mining activities (Gummow et al., 1991). Like sheep, cattle can succumb to acute CP when given excessive amounts by injection (NRC, 2005) and doses should always be carefully scaled according to LW.

Occurrence in goats

Goat kids have tolerated copper doses of 200 mg day^{-1} for 98 days (Cummins et al., 2008) and 60 mg added dietary Cu kg^{-1} DM for 137 days, whereas lambs succumbed to CCP (Zervas et al., 1990). Goats must therefore be classed as a copper-tolerant species alongside cattle. The addition of 20 or 40 mg

Cu kg^{-1} DM to a diet adequate in copper (9.5 kg^{-1} DM) for 84 days *improved* LWG and was more effective when added as proteinate rather than as CuSO$_4$ to a basal diet for goat kids (Datta *et al.*, 2006). This is either a unique example of hyperavailability or lower toxicity for a 'copper chelate'. Given that the diet was low in molybdenum (0.28 mg kg^{-1} DM) and contained 51% bran but no long roughage, it is arguable that the growth stimulation produced by the more available source, CuSO$_4$, waned as subclinical hepatotoxicity developed.

Subclinical hepatotoxicity

The possibility that there are harmful subclinical consequences of excessive copper intakes prior to the development of a haemolytic crisis is gaining support. The performance of finishing steers was lowered by copper supplementation at 20–40 mg kg^{-1} DM and impaired rumen function was implicated (Engle and Spears, 2000). However, no adverse effects were evident in the growing phase and individual liver copper values in the finished groups of 12 steers could well have exceeded 600 mg kg^{-1} DM. Steers dosed with CuOWP digested their feed less well in the second than in the first phase of an experiment, when liver contained 556 and 640 mg Cu kg^{-1} DM, respectively (Arthington, 2005). In the same paper, a dietary copper supplement of 120 mg kg^{-1} DM in a concentrate, offered to heifers with *ad libitum* access to hay, decreased LWG. Plasma copper entered the marginally toxic range (20–25 µmol l^{-1}) in cattle finished on corn-based diets containing monensin, an enhancer of A_{Cu}, and 17 mg Cu kg^{-1} DM (Hansen *et al.*, 2008). The superior milk yield of cows given 75% of the NRC (2001) copper requirement from organic sources rather than 100% from CuSO$_4$ could arguably be attributed to reduced hepatotoxicity. Sustained over-supplementation of dairy cows has been associated with hepatotoxicity in newborn calves that died suddenly with no evidence of haemolysis (Hunter *et al.*, 2010). Cases of sudden death in intensively reared bulls with high kidney copper, moderate liver copper and no evidence of haemolysis raised the possibility of a hepatic crisis brought on by consumption of a copper-rich mineral mixture (C. Low, personal communication, 1996). Copper supplements have also retarded growth in goat kids (Mondal *et al.*, 2007).

Diagnosis

The criteria for a diagnosis of haemolytic CCP are the same for cattle and goats as for sheep (p. 290). The assessment of high liver copper status in the newborn calf requires different diagnostic limits from those used for older cattle because values of 10 mmol Cu kg^{-1} DM have been found in histologically normal livers in the newborn (Hunter *et al.*, 2010). In the absence of corresponding data for caprines, their newborn should be assessed using the limits given for calves. Serum samples should never be used in surveillance work (see earlier) because copper concentrations can be underestimated by as much as 7 µmol l^{-1} (Laven *et al.*, 2007).

Prevention

The prevention of haemolytic CCP and hepatotoxicity requires an end to the common practice of supplementing herds with copper by more than one route, often at unnecessarily high levels, when there is no evidence of low copper status (Steffen *et al.*, 1997; Livesey *et al.*, 2002; Suttle and Phillippo, 2005). Although monensin is now a prohibited additive in Europe, its use elsewhere alongside copper supplements is contraindicated. Hunter *et al.* (2010) concluded that the permitted concentration of copper in milk substitutes in Europe, 30 mg kg^{-1} DM, is far too generous; 10 mg kg^{-1} DM is 10 times more than all pre-ruminants need from a milk-based product!

Chronic Copper Poisoning in Pigs and Poultry

Occurrence

In non-ruminants, CCP follows a similar two-phase pattern to that seen in sheep, except that the haemolytic crisis is less pronounced and less life-threatening while growth retardation is more prominent. The disorder occurs in a minority of pigs given copper supplements as growth stimulants to rations that are not suitably balanced with other minerals, and pigs can be protected by adding zinc and iron to the diet in broadly equivalent concentrations (Suttle and Mills, 1966a,b). Sows have been given similar supplements throughout six gestations without causing copper toxicity, although liver copper levels increased to 1899 mg kg^{-1} DM (Cromwell et al., 1993). The copper tolerance of chicks is influenced by the type of diet, increasing from 200 to 800 mg kg^{-1} DM when dehulled SBM replaces casein as the protein source in diets high in calcium (Funk and Baker, 1991). Copper accretion rises when chicks are given low-calcium diets (Leach et al., 1990). Antagonism from phytate may therefore protect poultry from toxicity, while the addition of phytase to improve the availability of plant phosphorus (see Chapter 6) may lower copper tolerance.

In an experiment in which 120 or 240 mg Cu kg^{-1} DM inhibited growth of turkeys up to 8 weeks of age (Kashani et al., 1986), the corn/SBM diet was fed in mash form, and this may have lessened the protection from phytate (see Chapter 6). Erosion of the gizzard is an additional feature of CCP in poultry and occurs at dietary copper concentrations of around 500 mg kg^{-1} DM (Christmas and Harms, 1979; Jensen and Maurice, 1979). Methionine supplementation lessened the growth retardation caused by excess copper but did not prevent gizzard erosion (Kashani et al., 1986) and it would be interesting to examine the toxicity to the gizzard of a relatively insoluble copper salt such as cupric carbonate. The toxicity of copper to chicks is increased when the diet is marginal in selenium (Jensen, 1975). In the latter study, mortality was associated with exudative diathesis (see Chapter 15) and prevented by raising the dietary selenium from 0.2 to 0.7 mg kg^{-1} DM (with 800 mg Cu kg^{-1} DM).

References

Adeola, O., Lawrence, B.V., Sutton, A.L. and Cline, T.R. (1995) Phytase induced changes in mineral utilization in zinc-supplemented diets for pigs. *Journal of Animal Science* 73, 3384–3391.

Akpan, H.D., Udosen, E.O., Udofia, A.A., Akpan, E.J. and Joshua, A.A. (2005) The effect of phytase and zinc supplementation on palm kernel cake toxicity in sheep. *Pakistan Journal of Nutrition* 4, 148–153.

Allen, J.D. and Gawthorne, J.M. (1987) Involvement of the solid phase rumen digesta in the interaction between copper, molybdenum and sulphur in sheep. *British Journal of Nutrition* 58, 265–276.

Allen, J.D. and Gawthorne, J.M. (1988) Interactions between proteins, thiomolybdates and copper. In: *Proceedings of the 6th Symposium on Trace Elements in Man and Animals*. Plenum, New York, pp. 315–316.

Alloway, B.J. (1973) Copper and molybdenum in swayback pastures. *Journal of Agricultural Science, Cambridge* 80, 521–524.

ARC (1980) *The Nutrient Requirements of Ruminants*. Commonwealth Agricultural Bureaux, Farnham Royal, UK, pp. 184–185.

ARC (1981) *The Nutrient Requirements of Pigs*. Commonwealth Agricultural Bureaux, Farnham Royal, UK, pp. 215–248.

Armstrong, T.A., Cook, D.R., Ward, M.M., Williams, C.M. and Spears, J.W. (2004) Effect of copper source (cupric citrate or cupric sulfate) and concentration on growth performance and fecal copper excretion in weanling pigs. *Journal of Animal Science* 82, 1234–1240.

Arthington, J.D. (2005) Effects of copper oxide bolus administration or high-level copper supplementation on forage utilization and copper status in beef cattle. *Journal of Animal Science* 83, 2894–2900.

Arthington, J.D., Corah, L.R. and Blecha, F. (1996) The effect of molybdenum-induced copper deficiency on acute phase protein concentrations, superoxide dismutase activity, leukocyte numbers and lymphocyte proliferation in beef heifers inoculated with bovine herpes virus-I. *Journal of Animal Science* 74, 211–217.

Bain, M.S., Spence, J.B. and Jones, P.C. (1986) An investigation of bovine serum copper levels in Lincolnshire and South Humberside. *Veterinary Record* 119, 593–595.

Baker, D.H. (1999) Cupric oxide should not be used as a copper supplement for animals or humans. *Journal of Nutrition* 129, 2278–2279.

Baker, D.H. and Ammerman, C.B. (1995) Copper bioavailability. In: Ammerman, C.B., Baker, D.H. and Lewis, A.J. (eds) *Bioavailability of Nutrients for Animals.* Academic Press, New York, pp. 127–156.

Bang, K.S., Familton, A.S. and Sykes, A.R. (1990) Effect of ostertagiasis on copper status in sheep: a study involving use of copper oxide wire particles. *Research in Veterinary Science* 49, 306–314.

Barrowman, P.R. and van Ryssen, J.B.J. (1987) Effect of ionophores on the accumulation of copper in the livers of sheep. *Animal Production* 44, 255–261.

Barry, T.N., Reid, T.C., Millar, K.R. and Sadler, W.A. (1981) Nutritional evaluation of kale (*Brassica oleracea*) diets. 2. Copper deficiency, thyroid function and selenium status in young cattle and sheep fed kale for prolonged periods. *Journal of Agricultural Science, Cambridge* 96, 269–282.

Barton, J., Hurtig, M. and Green, S. (1991) The role of copper in developmental orthopaedic disease in foals. In: *Highlights of Agricultural and Food Research in Ontario*, September 1991. University of Guelph, Guelph, Canada, pp. 14–18.

Baumgartner, S., Brown, D.J., Salevsky, E. Jr. and Leach, R.M. Jr (1978) Copper deficiency in the laying hen. *Journal of Nutrition* 108, 804–811.

Baxter, J.H., Van Wyk, J.J. and Follis, R.H. Jr (1953) A bone disorder associated with copper deficiency. *Bulletin of the Johns Hopkins Hospital* 93, 1–39.

Beattie, J.H., Reid, M.D., Fairweather-Tait, S., Dainty, J.R., Majsak-Newman, G. and Harvey, L. (2001). Selective extraction of blood plasma exchangeable copper for isotope studies of dietary copper absorption. *Analyst* 126, 2225–2229.

Beck, A.B. (1951) *A survey of the copper content of Western Australian pastures.* Leaflet No. 678, Department of Agriculture, Western Australia.

Bennetts, H.W. and Chapman, F.E. (1937) Copper deficiency in sheep in Western Australia: a preliminary account of the aetiology of enzootic ataxia of lambs and on anaemia of ewes. *Australian Veterinary Journal* 13, 138–149.

Bennetts, H.W. and Hall, H.T.B. (1939) 'Falling disease' of cattle in the south-west of Western Australia. *Australian Veterinary Journal* 15, 152–159.

Bird, P.R. (1970) Sulfur metabolism and excretion studies in ruminants. III. The effect of sulfur intake on the availability of copper in sheep. *Proceedings of the Australian Society of Animal Production* 8, 212–218.

Boila, R.J., Devlin, T.J., Drydale. R.A. and Lillie, L.E. (1984a) Geographical variation in the copper and molybdenum contents of forages grown in northwestern Manitoba. *Canadian Journal of Animal Science* 64, 899–918.

Boila, R.J., Devlin, T.J., Drydale, R.A. and Lillie, L.E. (1984b) The severity of hypocupraemia in selected herds of beef cattle in northwestern Manitoba. *Canadian Journal of Animal Science* 64, 919–936.

Bonham, M., O'Connor, J.M., Hannigan, B.M. and Strain, J.J. (2002) The immune system as a physiological indicator of marginal copper status. *British Journal of Nutrition* 87, 393–403.

Boyne, R. and Arthur, J.R. (1981) Effect of selenium and copper deficiency on neutrophil function in cattle. *Journal of Comparative Pathology* 91, 271–276.

Bradley, B.D., Graber, G., Condon, R.J. and Frobish, L.T. (1983) Effects of graded levels of dietary copper on copper and iron concentrations in swine tissues. *Journal of Animal Science* 56, 625–630.

Braude, R. (1967) Copper as a stimulant in pig feeding. *World Review of Animal Production* 3, 69–82.

Brebner, J., Thornton, I., McDonald, P. and Suttle, N.F. (1985) The release of trace elements from soils under conditions of simulated rumenal and abomasal digestion. In: Mills, C.F., Bremner, I. and Chesters, J.K. (eds) *Proceedings of the Fifth International Symposium on Trace Elements in Man and Animals, Aberdeen.* Commonwealth Agricultural Bureau, Farnham Royal, UK, pp. 850–852.

Bremner, I. (1993) Metallothionein in copper deficiency and toxicity. In: Anke, M., Meissner, D. and Mills, C.F. (eds) *Proceedings of the Eighth International Symposium on Trace Elements in Man and Animals.* Verlag Media Touristik, Gersdorf, Germany, pp. 507–515.

Bremner, I., Humphries, W.R., Phillippo, M., Walker, M.J. and Morrice, P.C. (1987) Iron-induced copper-deficiency in calves: dose–response relationships and interactions with molybdenum and sulfur. *Animal Production* 45, 403–414.

Bremner, I., Young, B.W. and Mills, C.F. (1976) Protective effect of zinc supplementation against copper toxicosis in sheep. *British Journal of Nutrition* 36, 551–561.

Bremner, I., Young, B.W. and Mills, C.F. (1977) Copper sheep foetus. *British Journal of Nutrition* 37.

Buckingham, K., Heng-khou, C.S., Dubick, M., Lefevre, M., Cross, C., Julian, L. and Rucker, R. (1981) Copper deficiency and elastin metabolism in avian lung. *Proceedings of the Society for Experimental Biology and Medicine* 166, 310–319.

Burke, J.M., Miller, J.E. and Brauer, D.K. (2005) The effectiveness of copper oxide wire particles as an anthelmintic in pregnant ewes and safety of dosing. *Veterinary Parasitology* 131, 291–297.

Bush, J.A., Jensen, W.N., Athens, J.W., Ashenbrucker, H., Cartwright, G.E. and Wintrobe, M.M. (1956) Studies on copper metabolism. 19. The kinetics of iron metabolism and erythrocyte life-span in copper-deficient swine. *Journal of Experimental Medicine* 103, 701–712.

Campbell, A.G., Coup, M.R., Bishop, W.H. and Wright, D.E. (1974) Effect of elevated iron intake on the copper status of grazing cattle. *New Zealand Journal of Agricultural Research* 17, 393–399.

Carnes, W.H., Shields, G.S., Cartwright, G.E. and Wintrobe, M.M. (1961) Vascular lesions in copper deficient swine. *Federation Proceedings* 20, 118.

Cavanagh, N.A. and Judson, G.J. (1994) Copper oxide powder as a copper supplement for sheep. *Journal of Trace Element and Electrolytes in Health and Disease* 8, 183–188.

Cerone, S., Sansinea, A. and Nestor, A. (1995) Copper deficiency alters the immune response of bovine. *Nutrition Research* 15, 1333–1341.

Chen, Y., Saari, J.T. and Kang, Y.J. (1995) Expression of gamma-glutamyl cysteine synthetase in the liver of copper deficient rats. *Proceedings of the Society for Experimental Biology and Medicine* 210, 102–106.

Chooi, K.F., Hutagalung, R.I. and Wan Mohammed, W.E. (1988) Copper toxicity in sheep fed oil palm products. *Australian Veterinary Journal* 66, 156–157.

Christmas, R.B. and Harms, R.H. (1979) The effect of supplemental copper and methionine on the performance of turkey poults. *Poultry Science* 58, 382–384.

Cousins, R.J. (1985) Absorption, transport and hepatic metabolism of copper and zinc with special reference to metallothionein and caeruloplasmin. *Physiological Reviews* 65, 238–309.

Cromwell, G.L., Monegue, H.J. and Stahly, T.S. (1993) Long-term effects of feeding a high copper diet to sows during gestation and lactation. *Journal of Animal Science* 71, 2996–3002.

Cromwell, G.L., Lindemann, M.D., Monegue, H.J., Hall, D.D. and Orr, D.E. (1998) Tribasic copper chloride and copper sulfate as copper sources for weanling pigs. *Journal of Animal Science* 76, 118–123.

Cummins, K.A., Solaimin, S.G. and Bergen, W.G. (2008) The effect of dietary copper supplementation of fatty acid profile and oxidative stability of adipose tissue depots in Boer × Spanish goats. *Journal of Animal Science* 86, 390–396.

Cunningham, I.J. (1950) Copper and molybdenum in relation to diseases of cattle and sheep in New Zealand. In: McElroy, W.D. and Glass, B. (eds) *Copper Metabolism: A Symposium on Animal, Plant and Soil Relationships, Baltimore, USA*. Johns Hopkins Press, Baltimore, Maryland, pp. 246–273.

Datta, C., Mondal, M.K. and Biswas, P. (2006) Influence of dietary inorganic and organic form of copper salt on performance, plasma lipids and nutrient utilisation of Black Bengal (*Capra indicus*) goat kids. *Animal Feed Science and Technology* 135, 191–209.

Deland, M.P.B., Cunningham, P., Milne, M.L. and Dewey, D.W. (1979) Copper administration to young calves: oral dosing with copper oxide compared with subcutaneous copper glycinate injection. *Australian Veterinary Journal* 55, 493–494.

Dewey, D.W. (1977) An effective method for the administration of trace amounts of copper to ruminants. *Search, Australia* 8, 326–327.

Dick, A.T. (1956) Molybdenum in animal nutrition. *Soil Science* 81, 229–258.

Dick, A.T. and Bull, L.B. (1945) Some preliminary observations on the effect of molybdenum on copper metabolism in herbivorous animals. *Australian Veterinary Journal* 21, 70–72.

Dick, A.T., Dewey, D.W. and Gawthorne, J.M. (1975) Thiomolybdates and the copper–molybdenum–sulfur interaction in ruminant nutrition. *Journal of Agricultural Science, Cambridge* 85, 567–568.

Dorton, K.L., Engle, T.E., Hamar, D.W., Siciliano, P.D. and Yemm, R.S. (2003) Effects of copper source and concentration on copper status and immune function in growing and finishing steers. *Animal Feed Science and Technology* 110, 31–44.

Du, Z., Hemken, R.W. and Harmon, R.J. (1996) Copper metabolism of Holstein and Jersey cows and heifers fed diets high in cupric sulfate or copper proteinate. *Journal of Dairy Science* 79, 1873–1880.

Engle, T.E. and Spears, J.W. (2000) Effects of dietary copper concentration and source on performance and copper status of growing and finishing steers. *Journal of Animal Science* 78, 2446–2451.

Ewing, H.P., Pesti, G.M., Bakalli, R.I. and Menten, J.F.M. (1998) Studies of the feeding of cupric sulfate pentahydrate, cupric citrate and copper oxychloride to broiler chickens. *Poultry Science* 77, 445–448.

Farmer, P.E., Adams, T.E. and Humphries, W.R. (1982) Copper supplementation of drinking water for cattle grazing molybdenum-rich pastures. *Veterinary Record* 111, 193–195.

Fell, B.F., Dinsdale, D. and Mills, C.F. (1975) Changes in enterocyte mitochondria associated with deficiency of copper in cattle. *Research in Veterinary Science* 18, 274–281.

Fell, B.F., Farmer, L.J., Farquharson, C., Bremner, I. and Graca, D.S. (1985) Observations on the pancreas of cattle deficient in copper. *Journal of Comparative Pathology* 95, 573–590.

Felsman, R.J., Wise, M.B., Harvey, R.W. and Barrick, E.R. (1973) Effect of graded levels of copper sulfate and antibiotic on performance and certain blood constituents of calves. *Journal of Animal Science* 36, 157–160.

Ferguson, W.S., Lewis, A.H. and Watson, S.J. (1938) Action of molybdenum in milking cattle. *Nature, UK* 141, 553.

Ferguson, W.S., Lewis, A.H. and Watson, S.J. (1943) The teart pasture of Somerset. The cause and cure of teartness. *Journal of Agricultural Science, Cambridge* 33, 44–51.

Freudenberger, D.O., Familton, A.S. and Sykes, A.R. (1987) Comparative aspects of copper metabolism in silage-fed sheep and deer (*Cervus elaphus*). *Journal of Agricultural Science, Cambridge* 108, 1–7.

Funk, M.A. and Baker, D.H. (1991) Toxicity and tissue accumulation of copper in chicks fed casein and soy-based diets. *Journal of Animal Science* 69, 4505–4511.

Garrick, M.D., Dolan, K.G., Horbinski, C., Ghoio, A.J., Higgins, D., Porcubin, M., Moore, E.G., Hainsworth, L.N., Umbreit, J.N., Conrad, M.E., Feng, L., Lis, A., Roth, J.A., Singleton, S. and Garrick, L.M. (2003) DMT 1: a mammalian transporter for multiple metals. *Biometals* 16, 41–54.

Gee, E.K., Grace, N.D., Firth, E.C. and Fennessy, P.F. (2000) Changes in liver copper concentration of Thoroughbred foals from birth to 160 days of age and the effect of prenatal copper supplementation. *Australian Veterinary Journal* 78, 347–353.

Gengelbach, G.P., Ward, J.D. and Spears, J.W. (1994) Effects of dietary copper, iron and molybdenum on growth and copper status of beef cows and calves. *Journal of Animal Science* 72, 2722–2727.

Gengelbach, G.P., Ward, J.D. and Spears, J.W. (1997) Effect of copper deficiency and copper deficiency coupled with high dietary iron or molybdenum on phagocytic function and response of calves to a respiratory disease challenge. *Journal of Animal Science* 75, 1112–1118.

Givens, D.I., Zervas, G., Simpson, V.R. and Telfer, S.B. (1988) Use of soluble rumen boluses to provide a supplement of copper for suckled calves. *Journal of Agricultural Science, Cambridge* 110, 199–204.

Gooneratne, S.R. and Christensen, D.A. (1989) A survey of maternal copper status and foetal tissue copper concentrations in Saskatchewan bovine. *Canadian Journal of Animal Science* 69, 141–150.

Gooneratne, S.R., Howell, J. and Gawthorne, J.M. (1981) Intravenous administration of thiomolybdate for the prevention of chronic copper poisoning in sheep. *British Journal of Nutrition* 46, 457–467.

Gooneratne, S.R., Symonds, H.W., Bailey, J.V. and Christensen, D.A. (1994) Effects of dietary copper, molybdenum and sulfur on biliary copper and zinc secretion in Simmental and Angus cattle. *Canadian Journal of Animal Science* 74, 315–325.

Govasmark, E., Steen, A., Bakken, A.K., Strom, T., Hansen, S. and Bernhoft, A. (2005) Copper, molybdenum and cobalt in herbage and ruminants from organic farms in Norway. *Acta Agriculturae Scandinavica* 55, 21–30.

Grace, N.D. (1983) Amounts and distribution of mineral elements associated with the fleece-free empty body weight gains of the grazing sheep. *New Zealand Journal of Agricultural Research* 26, 59–70.

Grace, N.D. and Wilson, P.R. (2002) Trace element metabolism, dietary requirements, diagnosis and prevention of deficiencies in deer. *New Zealand Veterinary Journal* 50, 252–259.

Grace, N.D., Knowles, S.O., Rounce, J.R., West, D.M. and Lee, J. (1998a) Effect of increasing pasture copper concentrations on the copper status of grazing Romney sheep. *New Zealand Journal of Agricultural Research* 41, 377–386.

Grace, N.D., Rounce, J.R., Knowles, S.O. and Lee, J. (1998b) Changing dietary sulfur intakes and the copper status of grazing lambs. *New Zealand Journal of Agricultural Research* 40, 329–334.

Grace, N.D., Pearce, S.G., Firth, E.C. and Fennessy, P.F. (1999) Content and distribution of macro- and trace-elements in the body of young pasture-fed horses pasture-fed horses. *Australian Veterinary Journal* 77, 172–176.

Grace, N.D., Gee, E.K., Firth, E.C. and Shaw S.L. (2002) Digestible energy intake, dry matter digestibility and mineral status of grazing New Zealand thoroughbred yearlings. *New Zealand Veterinary Journal* 50, 63–69.

Grace, N.D., Wilson, P.R. and Quinn, A.K. (2005a) The effects of copper-amended fertiliser and copper oxide wire particles on the copper status of farmed red deer (*Cervus elaphus*) and their progeny. *New Zealand Veterinary Journal* 53, 31–38.

Grace, N.D., Wilson, P.R. and Quinn, A.K. (2005b) Impact of molybdenum on the copper status of red deer (*Cervus elaphus*). *New Zealand Veterinary Journal* 53, 31–38.

Grace, N.D., Castillo-Alcala, F. and Wilson, P.R. (2008) Amounts and distribution of mineral elements associated with live-weight gains of grazing red deer (*Cervus elaphus*). *New Zealand Journal of Agricultural Research* 51, 439–449.

Guenthner, E., Carlson, C.W. and Emerick, R.J. (1978) Copper salts for growth stimulation and alleviation of aortic rupture losses in turkeys. *Poultry Science* 57, 1313–1324.

Gummow, B., Botha, C.J., Basson, C.J. and Bastianello, S.S. (1991) Copper toxicity in ruminants: air pollution as a possible cause. *Onderstepoort Journal of Veterinary Research* 58, 33–39.

Hanlon, J., Minks, E., Hughes, C., Weavers, E. and Rogers, M. (2002) Metallothionein in bovine spongiform encephalopathy. *Journal of Comparative Pathology* 127, 280–289.

Hannam, R.J. and Reuter, D.J. (1977) The occurrence of steely wool in South Australia, 1972–75. *Agricultural Record* 4, 26–29.

Hansen, S.L. and Spears, J.W. (2008) Impact of copper deficiency in cattle on proteins involved in iron metabolism. *FASEB Journal* 22, 443–445.

Hansen, S.L., Schlegel, P., Legleiter, L.R., Lloyd, K.E. and Spears, J.W. (2008) Bioavailability of copper from copper glycinate in steers fed high dietary sulfur and molybdenum. *Journal of Animal Science* 86, 173–179.

Hansen, S.L., Ashwell, M.S., Legleiter, L.R., Fry, R.S., Lloyd, K.E. and Spears, J.W. (2009) The addition of high manganese to a copper-deficient diet further depresses copper status and growth of cattle. *British Journal of Nutrition* 101, 1068–1078.

Hartman, F. and van Ryssen, J.B.J. (1997) Metabolism of selenium and copper in sheep with and without sodium bicarbonate supplementation. *Journal of Agricultural Science, Cambridge* 128, 357–364.

Hartmann, J. (1975) The frequency of occurrence of copper poisoning and the role of sheep concentrates in it merits enquiry. *Tijdschrift voor Diergeneeskunde* 100, 379–382.

Haynes, R.J. (1997) Micronutrient status of a group of soils in Canterbury, New Zealand, as measured by extraction with EDTA, DTPA and HCl, and its relationship with plant response to applied Cu and Zn. *Journal of Agricultural Science, Cambridge* 129, 325–333.

Hedger, R.S., Howard, D.A. and Burdin, M.L. (1964) The occurrence in goats and sheep in Kenya of a disease closely similar to swayback. *Veterinary Record* 76, 493–497.

Herbert, E., Small, J.N.W., Jones, D.G. and Suttle, N.F. (1991) Evaluation of superoxide dismutase assays for the routine diagnostic assessment of copper status in blood samples. In: Momcilovic, B. (ed.) *Proceedings of the Seventh International Symposium on Trace Elements in Man and Animals, Dubrovnik*. IMI, Zagreb, pp. 5-15–5-16.

Hogan, K.G., Money, D.F.L., White, D.A. and Walker, R. (1971) Weight responses of young lambs to copper and connective tissue lesions associated with grazing pasture of high molybdenum content. *New Zealand Journal of Agricultural Research* 14, 687–701.

Holstein, T.J., Fung, R.Q., Quevedo, W.C. and Bienieki, T.C. (1979) Effect of altered copper metabolism induced by mottled alleles and diet on mouse tyrosinase. *Proceedings of the Society for Experimental Biology and Medicine* 162, 264–268.

Howell, J.M. and Gooneratne, S.R. (1987) The pathology of copper toxicity. In: Howell, J.M. and Gawthorne, J.M. (eds) *Copper in Animals and Man. Volume II*. CRC Press Ltd, Boca Raton, Florida, pp. 53–78.

Howell, J.M. and Hall, G.A. (1970) Infertility associated with experimental copper deficiency in sheep, guinea-pigs and rats. In: Mills, C.F. (ed.) *Trace Element Metabolism in Animals*. E. & S. Livingstone, Edinburgh, pp. 106–109.

Howell, J.M., Deol, H.S., Dorling, P.R. and Thomas, J.B (1991) Experimental heliotrope and copper intoxication in sheep: morphological changes. *Journal of Comparative Pathology* 105, 49–74.

Humphries, W.R., Phillippo, M., Young, B.W. and Bremner, I. (1983) The influence of dietary iron and molybdenum on copper metabolism in calves. *British Journal of Nutrition* 49, 77–86.

Humphries, W.R., Morrice, P.C. and Bremner, I. (1988) A convenient method for the treatment of chronic copper poisoning in sheep using subcutaneous ammonium tetrathiomolybdate. *Veterinary Record* 123, 51–53.

Hunter, A.G., Martineau, H.M., Thomson, J.R., Leitch, S. and Suttle, N. (2010) Copper-induced hepatopathy associated with mortality in artificially-reared Jersey calves *Veterinary Record* (in press).

Ivan, M.M. (1988) The effect of faunation on rumen solubility and liver content of copper in sheep fed low or high copper diets. *Journal of Animal Science* 66, 1498–1501.

Jamieson, S. and Allcroft, R. (1950) Copper pine of calves. *British Journal of Nutrition* 4, 16–31.
Jarvis, S.C. and Austin, A.R. (1983) Soil and plant factors limiting the availability of copper to a beef suckler herd. *Journal of Agricultural Science, Cambridge* 101, 39–46.
Jenkins, K.J. and Hidiriglou, M. (1989) Tolerance of the calf for excess copper in milk replacer. *Journal of Dairy Science* 72, 150–156.
Jenkins, N.K., Morris, T.R. and Valamotis, D. (1970) The effect of diet and copper supplementation on chick growth. *British Poultry Science* 11, 241–248.
Jensen, L.S. (1975) Precipitation of a selenium deficiency by high dietary levels of copper and zinc (in fowls). *Proceedings of the Society for Experimental Biology and Medicine* 149, 113–116.
Jensen, L.S. and Maurice, V. (1979) Influence of sulfur amino acids on copper toxicity in chicks. *Journal of Nutrition* 109, 91–97.
Jones, D.G. and Suttle, N.F. (1981) Some effects of copper deficiency on leukocyte function in sheep and cattle. *Research in Veterinary Science* 31, 151–156.
Jones, D.G. and Suttle, N.F. (1987) Copper and disease resistance. In: *Trace Substances in Environmental Health – XXI*. University of Missouri, Columbia, Missouri, pp. 514–525.
Judson, J.G., Walkley, J.R., James, P.J., Kleeman, D.O. and Ponzoni, R.W. (1994) Genetic variation in trace element status of Merino sheep. *Proceedings of the Australian Society of Animal Production* 20, 438.
Jumba, I.O., Suttle, N.F., Hunter, E.A. and Wandiga, S.O. (1995) Effects of soil origin and mineral composition and herbage species on the mineral composition of forages in the Mount Elgon region of Kenya. 2. Trace elements. *Tropical Grasslands* 29, 47–52.
Kashani, A.B., Samie, H., Emerick, R.J. and Carlson, C.W. (1986) Effect of copper with three levels of sulfur-containing amino acids in diets for turkeys. *Poultry Science* 65, 1754–1759.
Kerr, L.A. and McGavin, H.D. (1991) Chronic copper poisoning in sheep grazing pastures fertilized with swine manure. *Journal of the American Veterinary Medical Association* 198, 99–101.
Kendall, N.R., Jackson. D.W., Mackenzie. A.M., Illingworth. D.V., Gill. I.M. and Telfer. S.B. (2001) The effect of a zinc, cobalt and selenium soluble glass bolus on the trace element status of extensively grazed sheep over winter. *Animal Science* 73, 163–169.
Kincaid, R.L., Gay, C.C. and Krieger, R.I. (1986) Relationship of serum and plasma copper and caeruloplasmin concentrations of cattle and the effects of whole sample storage. *American Journal of Veterinary Research* 47, 1157–1159.
King, T.P. and Bremner, I. (1979) Autophagy and apoptosis in liver during the pre-haemolytic phase of chronic copper poisoning in sheep. *Journal of Comparative Pathology* 89, 515–530.
Knowles, S.O., Grace, N.D., Rounce, J.R., Litherland, D.M. and Lee, J. (2000) Dietary Mo as an antagonist to Cu absorption: stable isotope (^{65}Cu) measurements in grazing sheep. In: Roussel, A.M., Anderson, R.A. and Favrier, A. E. (eds) *Proceedings of the 10th International Symposium on Trace Elements in Man and Animals, Etian, France*. Springer, pp. 717–722.
Koh, T.-S. and Judson, G.J. (1987) Copper and selenium deficiency in cattle: an evaluation of methods of oral therapy and an observation of a copper–selenium interaction. *Veterinary Record Communications* 11, 133–148.
Koh, T.-S., Peng, R.K. and Klasing, K.C. (1996) Dietary copper level affects copper metabolism during lipopolysaccharide-induced immunological stress in chicks. *Poultry Science* 75, 867–872.
Lai, C.-C., Huang, W.-H., Askari, A., Wang, Y., Sarvazyan, N., Klevay, L.M. and Chin, T.H. (1994) Differential regulation of superoxide dismutase in copper deficient rat organs. *Free Radical Biology and Medicine* 16, 613–620.
Langlands, J.P., Bowles, J.E., Donald, G.E., Smith, A.J. and Paull, D.R. (1981) Copper status of sheep grazing pastures fertilized with sulfur and molybdenum. *Australian Journal of Agricultural Research* 32, 479–486.
Lannon, B. and Mason, J. (1986) The inhibition of bovine ceruloplasmin oxidase activity *in vivo* and *in vitro*: a reversible interaction. *Journal of Inorganic Biochemistry* 26, 107–115.
Laven, R.A. and Livesey, C.T. (2006) An evaluation of the effect of clotting and processing on the recovery of copper from bovine blood. *Veterinary Journal* 171, 295–300.
Laven, R.A. and Smith, S.L. (2008) Copper deficiency in sheep: an assessment of relationship between concentration of copper in serum and plasma. *New Zealand Veterinary Journal* 56, 334–338.
Laven, R.A. and Wilson, P.R. (2009) Comparison of concentrations of copper in plasma and serum from farmed red deer (*Cervus elaphus*). *New Zealand Veterinary Journal* 57, 166–169.

Laven, R.A., Lawrence, K.E. and Livesey, C.T. (2007) The assessment of blood copper status in cattle: a comparison of measurements of caeruloplasmin and elemental copper in serum and plasma. *New Zealand Veterinary Journal* 55, 171–176.

Leach, R.M. Jr, Rosenblum, C.I., Amman, M.J. and Burdette, J. (1990) Broiler chicks fed low-calcium diets. 2. Increased sensitivity to copper toxicity. *Poultry Science* 69, 1905–1910.

Ledoux, D.R., Henry, P.R. and Ammerman, C.B. (1995a) Response to high dietary copper and duration of feeding time on tissue copper concentration in sheep. *Nutrition Research* 16, 69–78.

Ledoux, D.R., Pott, E.B., Henry, P.R., Ammerman, C.B., Merritt, A.M. and Madison, J.B. (1995b) Estimation of the relative bioavailability of inorganic copper sources for sheep. *Animal Feed Science and Technology* 15, 1803–1813.

Lee, H.J. (1956) The influence of copper deficiency on the fleece of British breeds of sheep. *Journal of Agricultural Science, Cambridge* 47, 218–244.

Lee, J., Treloar, B.P. and Harris, P.M. (1994) Metallothionein and trace element metabolism in sheep tissues in response to high and sustained zinc dosages. 1. Characterisation and turnover of metallothionein isoforms. *Australian Journal of Agricultural Research* 45, 303–320.

Legleiter, L.R. and Spears, J.W. (2007) Plasma diamine oxidase: a biomarker of copper deficiency in the bovine. *Journal of Animal Science* 85, 2198–2204.

Legleiter, L.R., Spears, J.W. and Liu, H.C. (2008) Copper deficiency in the young bovine results in dramatic decreases in brain copper concentration but does not alter brain prion protein biology. *Journal of Animal Science* 86, 3069–3078.

Leigh, L.C. (1975) Changes in the ultrastructure of cardiac muscle in steers deprived of copper. *Research in Veterinary Science* 18, 282–287.

Livesey, C.T., Bidewell, C.A., Crawshaw, T.R. and David, G.P. (2002) Investigation of copper poisoning in adult cows by the Veterinary Laboratories Agency. *Cattle Practice* 10, 289–294.

Lockhart, P.J. and Mercer, J.F.B. (1999) Cloning and expression analysis of the sheep ceruloplasmin gene. *Gene* 236, 251–257.

Lominadze, D., Saari, J.T., Miller, F.N., Catalfamo, J.L. and Schuschke, D.A. (1997) Von Willebrand factor restores platelet thrombogenesis in copper-deficient rats. *Journal of Nutrition* 127, 1320–1327.

Lopez-Alonso, M., Prieto, F., Miranda, M., Castillo, C., Hernandez, J.R. and Benedito, J.L. (2005) Intracellular distribution of copper and zinc in the liver of copper-exposed cattle from northwest Spain. *The Veterinary Journal* 170, 332–338.

Luo, X.G., Henry, P.R., Ammerman, C.B. and Madison, J.B. (1996) Relative bioavailability of copper in a copper–lysine complex or copper-sulfate for ruminants as affected by feeding regimen. *Animal Feed Science and Technology* 57, 281–289.

Marston, H.R. (1946) Nutrition and wool production. *Proceedings of a Symposium on Fibrous Proteins*. Society of Dyers and Colourists, Leeds, UK.

Mason, J. (1986) Thiomolybdates: mediators of molybdenum toxicity and enzyme inhibitors. *Toxicology* 42, 99–109.

Mason, J., Williams, S., Harrington, R. and Sheahan, B. (1984) Some preliminary studies on the metabolism of ^{99}Mo-labelled compounds in deer. *Irish Veterinary Journal* 38, 171–175.

Mason, J., Woods, M. and Poole, D.B.R. (1986) Accumulation of copper and albumin *in vivo* after intravenous trithiomolybdate administration. *British Veterinary Journal* 41, 108–113.

McArdle, H. (1992) The transport of iron and copper across the cell membrane: different mechanisms for different metals? *Proceedings of the Nutrition Society* 51, 199–209.

McFarlane, J.D. (1989) The effect of copper supply on vegetative and seed yield of pasture legumes and the field calibration of a test for detecting copper deficiency I. Subterranean clover. *Australian Journal of Agricultural Research* 40, 817–132.

McFarlane, J.D., Judson, J.D. and Gouzos, J. (1990) Copper deficiency in ruminants in the south east of Australia. *Australian Journal of Experimental Agriculture* 30, 187–193.

McFarlane, J.D., Judson, J.D., Turnbull, R.K. and Kempe, B.R. (1991) An evaluation of copper-containing soluble glass pellets, copper oxide particles and injectable copper as supplements for cattle and sheep. *Australian Journal of Experimental Agriculture* 31, 165–174.

McClure, S.J., McClure, T.J. and Emery, D.L. (1999) Effects of molybdenum intake on primary infection and subsequent challenge by the nematode parasite *Trichostrongylus colubriformis* in weaned Merino lambs. *Research in Veterinary Science* 67, 17–22.

McMurray, C.H. (1980) Copper deficiency in ruminants. In: *Biological Roles of Copper.* Ciba Foundation 79 (New series). Elsevier, New York, pp. 183–207.
Mee, J.F. (1991) Coat colour and copper deficiency in cattle. *Veterinary Record* 129, 536.
Miles, R.D., Henry, P.R., Sampath, V.C., Shivazad, M. and Comer, C.W. (2003) Relative bioavailability of novel chelates of manganese and copper for chicks. *Journal of Applied Poultry Research* 12, 417–423.
Mills, C.F. and Williams, R.B. (1962) Copper concentration and cytochrome-oxidase and ribonuclease activities in the brains of copper-deficient lambs. *Biochemical Journal* 85, 629–632.
Mills, C.F., Dalgarno, A.C. and Wenham, G. (1976) Biochemical and pathological changes in tissues of Friesian cattle during the experimental induction of copper deficiency. *British Journal of Nutrition* 35, 309–311.
Miltimore, J.E. and Mason, J.L. (1971) Copper to molybdenum ratio and molybdenum and copper concentrations in ruminant feeds. *Canadian Journal of Animal Science* 51, 193–200.
Minson, D.J. (1990) *Forages in Ruminant Nutrition.* Academic Press, San Diego, California, pp. 208–229.
Miranda, M., Cruz, J.M., Lopez-Alonso, M. and Benedito, J.L. (2006) Variations in liver and blood copper concentrations in young beef cattle raised in north-west Spain: associations with breed, sex, age and season. *Animal Science* 82, 253–258.
Moffor, F.M. and Rodway, R.G. (1991) The effect of tetrathiomolybdate on growth rate and onset of puberty in ewes lambs. *British Veterinary Journal* 147, 421–431.
Mondal, M.K., Biswas, P., Roy, B. and Mazumdar, D. (2007) Effect of copper sources and levels on serum lipid profiles in Black Bengal (*Capra hircus*) kids. *Small Ruminant Research* 67, 28–35.
Morris, C.A., Amyes, N.C. and Hickey, S.M. (2006) Genetic variation in serum copper concentration in Angus cattle. *Animal Science* 82, 799–803.
Mullis, L.A., Spears, J.W. and McGraw, R.L. (2003) Effects of breed (Angus v Simmental) and copper and zinc source on mineral status of steers fed high dietary iron. *Journal of Animal Science* 81, 318–322.
Neal, W.M., Becker, R.B. and Shealy, A.L. (1931) A natural copper deficiency in cattle rations. *Science* 74, 418–419.
Nicol, A.M., Keeley, M.J., Giuld, C.D.H., Isherwood, P. and Sykes, A.R. (2003) Liveweight gain and copper status of young deer treated or untreated with copper oxide wire particles on ten deer farms in Canterbury. *New Zealand Veterinary Journal* 51, 14–20.
NRC (1994) *Nutrient Requirements of Poultry*, 9th edn. National Academy of Sciences, Washington, DC.
NRC (2001) *Nutrient Requirements of Dairy Cows*, 5th edn. National Academy of Sciences, Washington, DC.
NRC (2005) *Mineral Tolerances of Animals*, 2nd edn. National Academy of Sciences, Washington, DC.
O'Dell, B.L. (1976) Biochemistry and physiology of copper in vertebrates. In: Prasad, A.S. (ed.) *Trace Elements in Human Health and Disease.* Academic Press, New York, pp. 391–413.
O'Dell, B.L., Hardwick, B.C., Reynolds, G. and Savage, J.E. (1961) Connective tissue defect in the chick resulting from copper deficiency. *Proceedings of the Society for Experimental Biology and Medicine* 108, 402–405.
O'Dell, B.L., Smith, R.M. and King, R.A. (1976) Effect of copper status on brain neurotransmitter metabolism in the lamb. *Journal of Neurochemistry* 26, 451–455.
Okonkwo, A.C., Ku, P.K., Miller, E.R., Keahey, K.K. and Ullrey, D.E. (1979) Copper requirement of baby pigs fed purified diets. *Journal of Nutrition* 109, 939–948.
Opsahl, W., Zeronian, H., Ellison, M., Lewis, D., Rucker, R.B. and Riggins, R.S. (1982) Role of copper in collagen cross-linking and its influence on selected mechanical properties of chick bone and tendon. *Journal of Nutrition* 112, 708–716.
Ortolani, E., Knox, D.P., Jackson, F., Coop, R.L. and Suttle, N.F. (1993) Abomasal parasitism lowers liver copper status and influences the Cu × Mo × S antagonism in lambs. In: Anke, M., Meissner, D. and Mills, C.F. (eds) *Proceedings of the Eighth International Symposium on Trace Elements in Man and Animals.* Verlag Media Touristik, Gersdorf, Germany, pp. 331–332.
Osman, N.I.E.D., Johnson, E.H., Al-Busaidi, R.M. and Suttle, N.F. (2003) The effects of breed, neonatal age and pregnancy on the plasma copper status of goats in Oman. *Veterinary Research Communications* 27, 219–229.
Palsson, P.A. and Grimsson, H. (1953) Demyelination in lambs from ewes which feed on seaweeds. *Proceedings of the Society for Biology in Medicine* 83, 518–522.
Parkins, J.J., Hemingway, R.G., Lawson, D.C. and Ritchie, N.S. (1994) The effectiveness of copper oxide powder as a component of a sustained-release multi-trace element and vitamin rumen bolus system for cattle. *British Veterinary Journal* 150, 547–553.
Paynter, D.I. (1987) The diagnosis of copper insufficiency. In: Howell, J.McC. and Gawthorne, J.M. (eds) *Copper in Animals and Man. Volume II.* CRC Press Ltd, Boca Raton, Florida, pp. 101–119.

Pearce, S.G., Firth, E.C., Grace, N.D. and Fennessy, P.F. (1999a) The effect of copper supplementation on the copper status of pasture-fed young thoroughbreds. *Equine Veterinary Journal* 30, 204–210.

Pearce, S.G., Firth, E.C., Grace, N.D. and Fennessy, P.F. (1999b) The effect of high pasture molybdenum concentrations on the copper status of grazing horses in New Zealand. *New Zealand Journal of Agricultural Research* 42, 93–99.

Phillippo, M. (1983) The role of dose response trials in predicting trace element deficiency disorders. *Occasional Publication of the British Society of Animal Production* 7, pp. 51–59.

Phillippo, M. and Graca, D.S. (1983) Biliary copper secretion in cattle. *Proceedings of the Nutrition Society* 42, 46A.

Phillippo, M., Humphries, W.R., Lawrence, C.B. and Price, J. (1982) Investigation of the effect of copper therapy on fertility in beef suckler herds. *Journal of Agricultural Science, Cambridge* 99, 359–364.

Phillippo, M., Humphries, W.R. and Garthwaite, P.H. (1987a) The effect of dietary molybdenum and iron on copper status and growth in cattle *Journal of Agricultural Science, Cambridge* 109, 315–320.

Phillippo, M., Humphries, W.R., Atkinson, T., Henderson, G.D. and Garthwaite, P.H. (1987b) The effect of dietary molybdenum and iron on copper status, puberty, fertility and oestrus cycles in cattle. *Journal of Agricultural Science, Cambridge* 109, 321–336.

Pitt, M., Fraser, J. and Thurley, D.C. (1980) Molybdenum toxicity in the sheep: epiphysiolysis, exostoses and biochemical changes. *Journal of Comparative Pathology* 90, 567–576.

Price, J. and Chesters, J.K. (1985) A new bioassay for assessment of copper bioavailability and its application in a study of the effect of molybdenum on the distribution of available copper in ruminant digesta. *British Journal of Nutrition* 53, 323–336.

Price, J., Will, M.A., Paschaleris, G. and Chesters, J.K. (1987) Identification of thiomolybdates in digesta and plasma from sheep after administration of ^{99}Mo-labelled compounds into the rumen. *British Journal of Nutrition* 58, 127–138.

Pritchard, G.C., Lewis, G., Wells, G.A.H. and Stopforth, A. (1985) Zinc toxicity, copper deficiency and anaemia in swill-fed pigs. *Veterinary Record* 117, 545–548.

Prohaska, J.R. (2006) Copper. In: Filer, L.J. and Ziegler, E.E. (eds) *Present Knowledge in Nutrition*, 7th edn. International Life Science Institute–Nutrition Foundation, Washington, DC.

Rabiansky, P.A., McDowell, L.R., Velasquez-Pereira, J., Wilkinson, N.S., Percival, S.S., Martin, F.G., Bates, D.B., Johnson, A.B., Batra, T.R. and Salgado-Madriz, E. (1999) Evaluating copper lysine and copper sulfate sources for heifers. *Journal of Dairy Science* 82, 2642–2650.

Rosa, I.V., Ammerman, C.B. and Henry, P.R. (1986) Interrelationships of dietary copper, zinc and iron on performance and tissue mineral concentration in sheep. *Nutrition Reports International* 34, 893–902.

Rucker, R.B., Parker, H.E. and Rogler, J.C. (1969) Effect of copper deficiency on chick bone collagen and selected bone enzymes. *Journal of Nutrition* 98, 57–63.

Rucker, R.B., Riggins, R.S., Laughlin, R., Chan, M.M., Chen, M. and Tom, K. (1975) Effect of nutritional copper deficiency on the biomechanical properties of bone and arterial elastin metabolism in the chick. *Journal of Nutrition* 105, 1062–1070.

Saenko, E.L., Yaroplov, A.I. and Harris, E.D. (1994) Biological functions of caeruloplasmin expressed through copper-binding sites. *Journal of Trace Elements in Experimental Medicine* 7, 69–88.

Savage, J.E. (1968) Trace minerals and avian reproduction. *Federation Proceedings* 27, 927–931.

Shand, A. and Lewis, G. (1957) Chronic copper poisoning in young calves. *Veterinary Record* 69, 618–620.

Simpson, A.M., Mills, C.F. and McDonald, I. (1981) Tissue copper retention or loss in young growing cattle. In: Howell, J.McC., Gawthorne, J.M. and White, C.L. (eds) *Proceedings of the Fourth International Symposium on Trace Element Metabolism in Man and Animals*. Australian Academy of Sciences, Canberra, pp. 133–136.

Sjollema, B. (1933) Kupfermangel als Ursache von Krankheiten bei Pflanzen und Tieren. *Biochemische Zeitschrift* 267, 151–156.

Smart, M.E., Cohen, R., Christensen, D.A. and Williams, C.M. (1986) The effects of sulfate removal from the drinking water on the plasma and liver copper and zinc concentrations of beef cows and their calves. *Canadian Journal of Animal Science* 66, 669–680.

Smith, B.S.W. and Wright, H. (1975a) Effect of dietary molybdenum on copper metabolism: evidence for the involvement of molybdenum in abnormal binding of copper to plasma proteins. *Clinica Chimica Acta* 62, 55–62.

Smith, B.S.W. and Wright, H. (1975b) Copper:molybdenum interactions: effects of dietary molybdenum on the binding of copper to plasma proteins in sheep. *Journal of Comparative Pathology* 85, 299–305.

Smith, B.S.W., Field, A.C. and Suttle, N.F. (1968) Effect of intake of copper, molybdenum and sulfate on copper metabolism in the sheep. III. Studies with radioactive copper in male castrated sheep. *Journal of Comparative Pathology* 78, 449–461.

Smith, R.M., Osborne-White, W.S. and O'Dell, B.L. (1976) Cytochromes in brain mitochondria from lambs with enzootic ataxia. *Journal of Neurochemistry* 26, 1145–1148.

Spais, A.G. (1959) *Askeri veteriner dergisi* 135, 161.

Stabel, J.R., Spears, J.W. and Brown, T.T. (1993) Effect of copper deficiency on tissue, blood characteristics and immune function of calves challenged with infectious bovine rhinotracheitis virus and pasteurella haemolytica. *Journal of Animal Science* 71, 1247–1255.

Stahlhut, H.S., Whishant, C.S. and Spears, J.W. (2005) Effect of chromium supplementation and copper status on performance and reproduction of beef cows. *Animal Feed Science and Technology* 128, 266–275.

Steffen, D.J., Carlson, M.P. and Casper, H.H. (1997) Copper toxicosis in suckling beef cattle with improper administration of copper oxide boluses. *Journal of Veterinary Diagnostic Investigation* 9, 443–446.

Strickland, K., Smith, F., Woods, M. and Mason, J. (1987) Dietary molybdenum as a putative copper antagonist in the horse. *Equine Veterinary Journal* 19, 50–54.

Suleiman, A., Okine, E. and Goonewardne, L.A. (1997) Relevance of National Research Council feed composition tables in Alberta. *Canadian Journal of Animal Science* 77, 197–203.

Suttle, N.F. (1974a) Effects of organic and inorganic sulfur on the availability of dietary copper to sheep. *British Journal of Nutrition* 32, 559–568.

Suttle, N.F. (1974b) Effects of age and weaning on the apparent availability of dietary copper to young lambs. *Journal of Agricultural Science, Cambridge* 84, 255–261.

Suttle, N.F. (1977) Reducing the potential toxicity of concentrates to sheep by the use of molybdenum and sulfur supplements. *Animal Feed Science and Technology* 2, 235–246.

Suttle, N.F. (1981) Effectiveness of orally administered cupric oxide needles in alleviating hypocupraemia in sheep and cattle. *Veterinary Record* 108, 417–420.

Suttle, N.F. (1983a) Assessing the mineral and trace element status of feeds. In: Robards, G.E. and Packham, R.G. (eds) *Proceedings of the Second Symposium of the International Network of Feed Information Centres*. Commonwealth Agricultural Bureaux, Farnham Royal, UK, pp. 211–237.

Suttle, N.F. (1983b) Effects of molybdenum concentration in fresh herbage, hay and semi-purified diets on the copper metabolism of sheep. *Journal of Agricultural Science, Cambridge* 100, 651–656.

Suttle, N.F. (1987a) The nutritional requirement for copper in animals and man. In: Howell, J. McC. and Gawthorne, J.M. (eds) *Copper in Animals and Man. Volume I*. CRC Press, Boca Raton, Florida, pp. 21–44.

Suttle, N.F. (1987b) Safety and effectiveness of cupric oxide particles for increasing liver copper stores in sheep. *Research in Veterinary Science* 42, 219–223.

Suttle, N.F. (1988a) Relationships between the trace element status of soils, pasture and animals in relation to the growth rate of lambs. In: Thornton, I. (ed.) *Geochemistry and Health*. Science Reviews Limited, Northwood, UK, pp. 69–79.

Suttle, N.F. (1988b) The role of comparative pathology in the study of copper and cobalt deficiencies in ruminants. *Journal of Comparative Pathology* 99, 241–258.

Suttle, N.F. (1991) The interactions between copper, molybdenum and sulfur in ruminant nutrition. *Annual Review of Nutrition* 11, 121–140.

Suttle, N.F. (1994) Meeting the copper requirements of ruminants. In: Garnsworthy, P.C. and Cole, D.J.A. (eds) *Recent Advances in Animal Nutrition*. Nottingham University Press, Nottingham, UK, pp. 173–188.

Suttle, N.F. (1996) Non-dietary influences on mineral requirements of sheep. In: Masters, D.G. and White, C.L. (eds) *Detection and Treatment of Mineral Nutrition Problems in Sheep*. ACIAR Monograph, Canberra, pp. 31–44.

Suttle, N.F. (2008a) Relationships between the concentrations of tri-chloroactic acid-soluble copper and caeruloplasmin in the serum of cattle from areas with different soil concentrations of molybdenum. *Veterinary Record* 162, 237–240.

Suttle, N.F. (2008b) Lack of effect of parenteral thiomolybdate on ovine caeruloplasmin activity: diagnostic implications. *Veterinary Record* 162, 593–594.

Suttle, N.F. and Angus, K.W. (1976) Experimental copper deficiency in the calf. *Journal of Comparative Pathology* 86, 595–608.

Suttle, N.F. and Angus, K.W. (1978) Effects of experimental copper deficiency on the skeleton of the calf. *Journal of Comparative Pathology* 82, 137–145.

Suttle, N.F. and Blake, J.S. (2006) Effects of adding a 'cocktail' of copper antagonists, molybdenum, sulfur and zinc, on liver copper accumulation in Texel rams given a concentrate diet. *Proceedings of the British Society of Animal Science* 72, 63.

Suttle, N.F. and Brebner, J. (1995) A putative role for larval nematode infection in diarrhoeas which did not respond to anthelmintic drenches. *Veterinary Record* 137, 311–316.

Suttle, N.F. and Field, A.C. (1968) The effect of intake of copper, molybdenum and sulfate on copper metabolism in sheep 1. Clinical condition and blood copper distribution in the pregnant ewe. *Journal of Comparative Pathology* 78, 351–363.

Suttle, N.F. and Field, A.C. (1983) Effects of dietary supplements of thiomolybdates on copper and molybdenum metabolism in sheep. *Journal of Comparative Pathology* 93, 379–389.

Suttle, N.F. and McMurray, C.H. (1983) Use of erythrocyte copper:zinc superoxide dismutase activity and hair or fleece copper concentrations in the diagnosis of hypocuprosis in ruminants. *Research in Veterinary Science* 35, 47–52.

Suttle, N.F. and Mills, C.F. (1966a) Studies of the toxicity of copper to pigs. 1. Effects of oral supplements of zinc and iron salts on the development of copper toxicosis. *British Journal of Nutrition* 20, 135–148.

Suttle, N.F. and Mills, C.F. (1966b) Studies of the toxicity of copper to pigs. 2. Effect of protein source and other dietary components on the response to high and moderate intakes of copper. *British Journal of Nutrition* 20, 149–161.

Suttle, N.F. and Peter, D.W. (1985) Rumen sulfide metabolism as a major determinant of the availability of copper to ruminants. In: Mills, C.F., Bremner, I. and Chesters, J.K. (eds) *Proceedings of the Fifth International Symposium on Trace Elements in Man and Animals, Aberdeen.* Commonwealth Agricultural Bureau, Farnham Royal, UK, pp. 367–370.

Suttle, N.F. and Phillippo, M. (2005) Effect of trace element supplementation on the fertility of dairy herds. *Veterinary Record* 156, 155–156.

Suttle, N.F. and Price, J. (1976) The potential toxicity of copper-rich animal excreta to sheep. *Animal Production* 23, 233–241.

Suttle, N.F. and Small, J.N.W. (1993) Evidence of delayed availability of copper in supplementation trials with lambs on molybdenum-rich pasture. In: Anke, M., Meissner, D. and Mills, C.F. (eds) *Proceedings of the Eighth International Symposium on Trace Elements in Man and Animals.* Verlag Media Touristik, Gersdorf, Germany, pp. 651–655.

Suttle, N.F., Field, A.C. and Barlow, R.M. (1970) Experimental copper deficiency in sheep. *Journal of Comparative Pathology* 80, 151–162.

Suttle, N.F., Angus, K.W., Nisbet, D.I. and Field, A.C. (1972) Osteoporosis in copper-depleted lambs. *Journal of Comparative Pathology* 82, 93–97.

Suttle, N.F., Abrahams, P. and Thornton, I. (1984) The role of a soil × dietary sulfur interaction in the impairment of copper absorption by soil ingestion in sheep. *Journal of Agricultural Science, Cambridge* 103, 81–86.

Suttle, N.F., Jones, D.G., Woolliams, C. and Woolliams, J.A. (1987a) Heinz body anaemia in lambs with deficiencies of copper or selenium. *British Journal of Nutrition* 58, 539–548.

Suttle, N.F., Jones, D.G., Woolliams, J.A. and Woolliams, C. (1987b) Copper supplementation during pregnancy can reduce perinatal mortality and improve early growth in lambs. *Proceedings of the Nutrition Society* 46, 68A.

Suttle, N.F., Brebner, J., Small, J.N. and McLean, K. (1992a) Inhibition of ovine erythrocyte superoxide dismutase activity (ESOD; EC1.14.1.1) *in vivo* by parenteral ammonium tetrathiomolybdate. *Proceedings of the Nutrition Society* 51, 145A.

Suttle, N.F., Knox, D.P., Angus, K.W., Jackson, F. and Coop, R.L. (1992b) Effects of dietary molybdenum on nematode and host during *Haemonchus contortus* infection in lambs. *Research in Veterinary Science* 52, 230–235.

Suttle, N.F., Brebner, J. and Pass, R. (1996a) A comparison of the availability of copper in four whisky distillery by-products with that in copper sulfate for lambs. *Animal Science* 62, 689–690.

Suttle, N.F., Small, J.N.W., Collins, E.A., Mason, D.K. and Watkins, K.L. (1996b) Serum and hepatic copper concentrations used to define normal, marginal and deficient copper status in horses. *Equine Veterinary Journal* 28, 497–499.

Suttle, N.F., Lewis, R.M. and Small, J.N. (2002) Effects of breed and family on rate of copper accretion in the liver of purebred Charollais, Suffolk and Texel lambs. *Animal Science* 75, 295–302.

Taylor, C.G., Bettger, W.J. and Bray, T.M. (1988) Effect of dietary zinc or copper deficiency on the primary free radical defence system in rats. *Journal of Nutrition* 118, 613–621.

Thompson, K.G., Audige, L., Arthur, D.G., Juhan, A.F., Orr, M.B., McSporran, K.D. and Wilson, P.R. (1994) Osteochondrosis associated with copper deficiency in young farmed red deer and wapiti × red deer hybrids. *New Zealand Veterinary Journal* 42, 137–143.

Thornton, I. and Alloway, B.J. (1974) Geochemical aspects of the soil–plant–animal relationship in the development of trace element deficiency and excess. *Proceedings of the Nutrition Society* 33, 257–266.

Thornton, I., Kershaw, G.F. and Davies. M.K. (1972a) An investigation into copper deficiency in cattle in the southern Pennines. 1. Identification of suspect areas using geochemical reconnaissance followed by blood copper surveys. *Journal of Agricultural Science, Cambridge* 78, 157–163.

Thornton, I., Kershaw, G.F. and Davies, M.K. (1972b) An investigation into copper deficiency in cattle in the southern Pennines. 2. Response to copper supplementation. *Journal of Agricultural Science, Cambridge* 78, 165–171.

Todd, J.R. (1972) Copper, molybdenum and sulfur contents of oats and barley in relation to chronic copper poisoning in housed sheep. *Journal of Agricultural Science, Cambridge* 79, 191–195.

Tremblay, R.R.M. and Baird, J.D. (1991) Chronic copper poisoning in two Holstein cows. *Cornell Veterinarian* 81, 205–213.

Underwood, E.J. (1977) *Trace Elements in Human and Animal Nutrition*, 4th edn. Academic Press, London.

Underwood, E.J. and Suttle, F. (1999) Copper. In: *The Mineral Nutrition of Livestock*, 3rd edn. CAB International, Wallingford, UK, pp. 283–342.

Underwood, E.J., Robinson, T.J. and Curnow, D.H. (1943) The influence of topdressing with copper sulfate on the copper content and the yield of mixed pasture at Gingin. *Journal of the Department of Agriculture of Western Australia* 20, 80–87.

Van der Berg, R., Levels, F.H.R. and van der Schee, W. (1983) Breed differences in sheep with respect to the accumulation of copper in the liver. *Veterinary Quarterly* 5, 26–31.

Veum, T.L., Carlson, M.S., Bollinger, D.W. and Ellersieck, M.R. (2004) Copper proteinate in weanling pig diets for enhancing growth performance and reducing faecal copper excretion compared with copper sulfate. *Journal of Animal Science* 82, 1062–1070.

Vermunt, J.J. and West, D.M (1994) Predicting copper status in beef cattle using serum copper concentrations. *New Zealand Veterinary Journal* 42, 194–195.

VIDA (2007) *Veterinary Investigation Surveillance Report, 2007*. Veterinary Laboratories Agency, Weybridge, UK.

Waldroup, P.W., Fritts, C.A. and Yan, F. (2003) Utilization of Bio-Mos© mannan oligosaccharide and Bioplex© copper in broiler chicks. *International Journal of Poultry Science* 2, 44–52.

Wang, Z.Y., Poole, D.B.R. and Mason, J. (1988) The effects of supplementation of the diet of young steers with Mo and S on the intracellular distribution of copper in liver and on copper fractions in blood. *British Veterinary Journal* 114, 543–551.

Ward, J.D. and Spears, J.W. (1997) Long-term effects of consumption of low-copper diets with or without supplemental molybdenum on copper status, performance and carcass characteristics of cattle. *Journal of Animal Science* 75, 3057–3065.

Ward, T.L., Watkins, K.L. and Southern, L.L. (1994) Interactive effects of dietary copper and water copper level on growth, water uptake and plasma and liver copper concentrations of poults. *Poultry Science* 73, 1306–1311.

White, C.L., Masters, D.G., Paynter, D.I., Howell, J. McC., Roe, S.P., Barnes, M.J. and Allen, J.G. (1994) The effects of supplementary copper and a mineral mix on the development of lupinosis in sheep. *Australian Journal of Agricultural Research* 45, 279–291.

Whitehead, D.C. (1966) *Nutrient Minerals in Grassland Herbage*. Commonwealth Bureau of Pastures and Field Crops, Publication No. I (mimeographed).

Whitelaw, A., Armstrong, R.H., Evans, C.C. and Fawcett, A.R. (1979) A study of the effects of copper deficiency in Scottish Blackface lambs on improved hill pasture. *Veterinary Record* 104, 455–460.

Whitelaw, A., Russel, A.J.F., Armstrong, R.H., Evans, C.C. and Fawcett, A.R. (1983) Studies in the prophylaxis of induced copper deficiency in sheep grazing reseeded hill pastures. *Animal Production* 37, 441–448.

Whitelaw, A., Fawcett, A.R. and McDonald, A.J. (1984) Cupric oxide needles for the prevention of bovine hypocuprosis. *Veterinary Record* 115, 357.

Wiener, G. (1987) The genetics of copper metabolism in man and animals. In: Howell, J. McC. and Gawthorne, J.M. (eds) *Copper in Man and Animals. Volume I*. CRC Press, Boca Raton, Florida, pp. 45–61.

Wiener, G., Suttle, N.F., Field, A.C., Herbert, J.G. and Woolliams, J.A. (1978) Breed differences in copper metabolism in sheep. *Journal of Agricultural Science, Cambridge* 91, 433–441.

Wiener, G., Wilmut, I., Woolliams, C. and Woolliams, J.A. (1984a) The role of the breed of dam and the breed of lamb in determining the copper status of the lamb. 1. Under a dietary regime low in copper. *Animal Production* 39, 207–217.

Wiener, G., Wilmut, I., Woolliams, C. and Woolliams, J.A. (1984b) The role of the breed of dam and the breed of lamb in determining the copper status of the lamb. 2. Under a dietary regime moderately high in copper. *Animal Production* 39, 219–227.

Wilson, P.R. and Grace, N.D. (2001) A review of tissue reference values used to assess the trace element status of farmed red deer (*Cervus elaphus*). *New Zealand Veterinary Journal* 49, 126–132.

Wilson, P.R. and Grace, N.D. (2005) Effects of molybdenum on tissue copper concentrations and response to copper wire particle supplementation in deer. *Proceedings of the Deer Branch of the New Zealand Veterinary Association*, pp. 34–36.

Wittenberg, K.M. and Devlin, T.J. (1987) Effects of dietary molybdenum on productivity and metabolic parameters of lactating beef cows and their offspring. *Canadian Journal of Animal Science* 67, 1055–1066.

Woolliams, J.A., Suttle, N.F., Wiener, G., Field, A.C. and Woolliams, C. (1982) The effect of breed of sire on the accumulation of copper in lambs, with particular reference to copper toxicity. *Animal Production* 35, 299–307.

Woolliams, J.A., Suttle, N.F., Wiener, G., Field, A.C. and Woolliams, C. (1983) The long-term accumulation and depletion of copper in the liver of different breeds of sheep fed diets of different copper content. *Journal of Agricultural Science, Cambridge* 100, 441–449.

Woolliams, C., Suttle, N.F., Woolliams, J.A., Jones, D.G. and Wiener, G. (1986a) Studies on lambs from lines genetically selected for low or high plasma copper status. 1. Differences in mortality. *Animal Production* 43, 293–301.

Woolliams, J.A., Woolliams, C., Suttle, N.F., Jones, D.G. and Wiener, G. (1986b) Studies on lambs from lines genetically selected for low or high plasma copper status. 2. Incidence of hypocuprosis on improved hill pasture. *Animal Production* 43, 303–317.

Wouda, W., Borst, G.H.A. and Gruys, E. (1986) Delayed swayback in goat kids, a study of 23 cases. *The Veterinary Quarterly* 8, 45–56.

Xiao-Yun, S., Guo-Zhen, D. and Hong, L. (2006a) Studies of a naturally occurring molybdenum-induced copper deficiency in the yak. *The Veterinary Journal* 171, 352–357.

Xiao-Yun, S., Guo-Zhen, D., Ya-Ming, C. and Bao-Li, F. (2006b) Copper deficiency in yaks on pasture in western China. *Canadian Veterinary Journal* 47, 902–905.

Xin, Z., Waterman, D.F., Hemken, R.W. and Harmon, R.J. (1991) Effects of copper status on neutrophil function, superoxide dismutase and copper distribution in steers. *Journal of Dairy Science* 74, 3078–3085.

Xin, Z., Silvia, W.J., Waterman, D.F., Hemken, R.W. and Tucker, W.B. (1993) Effect of copper status on luteinizing hormone secretion in dairy steers. *Journal of Dairy Science* 76, 437–444.

Yost, G.P., Arthington, J.D., McDowell, L.R., Martin, F.G., Wilkinson, N.S. and Swenson, C.K. (2002) Effect of copper source and level on the rate and extent of copper repletion in Holstein heifers. *Journal of Dairy Science* 85, 3297–3303.

Zervas, G., Nikolau, E. and Mantzios, A. (1990) Comparative study of chronic copper poisoning in lambs and young goats. *Animal Production* 50, 497–506.

Zhou, W., Kornegay, E.T., van Laar, H., Swinkels, J.W.G.M., Wong, E.A. and Lindemann, M.D. (1994a) The role of feed consumption and feed efficiency in copper-stimulated growth. *Journal of Animal Science* 72, 2385–2394.

Zhou, W., Kornegay, E.T., Lindemann, M.D., Swinkels, J.W.G.M., Welten, M.K. and Wong, E.A. (1994b) Stimulation of growth by intravenous injection of copper in weanling pigs. *Journal of Animal Science* 72, 2395–2403.

12 Iodine

Iodine was considered to be unique among the mineral elements in that deprivation could cause a clinical abnormality – enlargement of the thyroid gland in the neck or 'goitre' – that was easily recognized and specific for that deficiency. Attempts to treat goitre can be traced back to 3000BC, when a Chinese emperor successfully used seaweed as a remedy. The relationship between iodine nutrition and goitre only emerged in the l9th century with the isolation of iodine from seaweed and the discovery that iodine is highly concentrated in the normal thyroid gland but dilute in goitrous thyroids. In 1927, a putative hormone, containing 65% iodine by weight, was isolated from thyroid tissue and called 'thyroxine' (see Harington, 1953). Treatment and prevention of goitre with thyroid extracts or iodized salt became routine. Extensive 'goitrous' areas were eventually discovered on every continent (Fig. 12.1), often associated with an environmental deficiency of iodine, and these can still cause problems in livestock (Singh et al., 2005). However, the link between geochemical iodine status and incidence of goitre is not straightforward (Stewart and Pharaoh, 1996), being complicated, for one thing, by the presence of organic, goitrogenic factors in foods (reviewed by Stoewsand, 1995). Exposure to certain types of goitrogen can induce goitre that cannot be cured or prevented by iodine supplementation. Subsequent discoveries of the adverse effects of iodine deficiency on brain development (Hetzel, 1991), of implied effects on embryonic viability (Sargison et al., 1998) and of selenium deprivation on iodine metabolism (see Chapter 15) (Beckett and Arthur, 1994) meant that the terms 'iodine deficiency' and 'goitre' were not synonymous, and 'iodine deficiency disorder' (IDD) replaced 'goitre' (Hetzel and Welby, 1997) for describing the varied consequences of iodine deprivation.

A link between thyroid activity and basal metabolic rate led early workers to conclude that requirements for iodine across species as diverse as the rat (Levine et al., 1933) and cow (Mitchell and McClure 1937) are remarkably similar when expressed in terms of calorie intake or heat production (Table 12.1). These estimates agreed with those expressed conventionally as dietary iodine concentrations and based on early field observations of goitre incidence in poultry, pigs and sheep (0.1–0.2 mg kg^{-1} dry matter (DM)) (Orr and Leitch, 1929), but were substantially lower for cows. Environmental stressors that affect basal metabolic rate, such as extreme cold, place stress on the thyroid gland and thyroid activity controls productivity in all classes of livestock, particularly small ruminants (Todini, 2007).

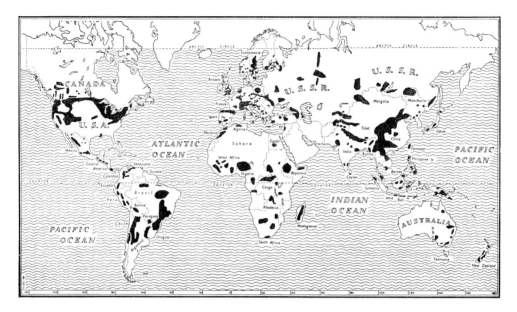

Fig. 12.1. Areas of the world where goitre was endemic (map by courtesy of the Iodine Education Bureau, London).

Table 12.1. Minimum iodine requirements of farm animals (from Mitchell and McClure, 1937).

Animal	Weight (kg)	Heat production (kcal)	I (µg day^{-1})	Requirement (mg kg^{-1} DM)
Poultry	2.3	225	5–9	0.05–0.09
Sheep	50	2,500	50–100	0.05–0.10
Pigs	68	4,000	80–160	0.03–0.06
Milking cow (16 l/day)	454	20,000	400–800	0.30–0.60

Function of Iodine

Iodine has only one known but vital function as a constituent of thyroid hormones. Structural studies of the thyroxine molecule have revealed that its four atoms of iodine are attached in two similar positions on linked 'outer' and 'inner' tyrosine rings (OR and IR). Thyroxine therefore became known as 3,3',5,5'-tetraiodothyronine (T4), a relatively inactive prohormone requiring removal of one atom by two OR-deiodinase enzymes (5'ORD types I and II) to generate the active hormone, 3,3',5-triiodothyronine (T3). Both T4 and T3 can be inactivated by an IR deiodinase (type III), and physiological activity of thyroid hormones is regulated peripherally by the three deiodinases rather than by the thyroid gland (Fig. 12.2) (for reviews, see Kulper et al., 2005; Todini, 2007). The rate of gene transcription is controlled by T3 (Bassett et al., 2003), which thus influences protein synthesis in all cells and has major effects on fetal development (Erenberg et al., 1974; Hopkins, 1975). Thyroid hormones have a thermoregulatory role, increasing cellular respiration and energy production, and have widespread effects on intermediary metabolism, growth, muscle function, immune defence and circulation. These are particularly important in facilitating the change from the fetal to the free-living stage (Symonds and Clarke, 1996). The seasonality of reproduction in ewes is related to seasonal changes in thyroid activity (Follett and Potts, 1990) and the required cooperation from the male is probably facilitated by a thyroid response to change in day length (Todini, 2007). As a determinant of metabolic rate, T3 interacts with hormones such as insulin, growth

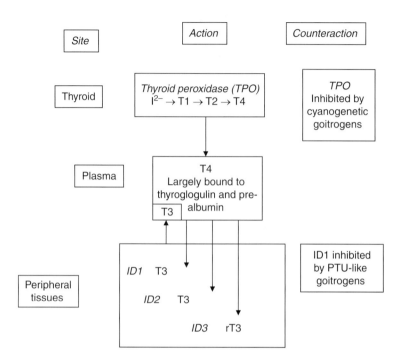

Fig. 12.2. The principal steps in the regulation of thyroid hormone metabolism: incorporation of iodide (I^{2-}) into mono-, di- and tetraiodothyronines (T1, T2 and T4) in the thyroid, catalysed by thyroperoxidase (TPO); transportation of bound T4 to peripheral tissues and its activation to T3 by two outer-ring deiodinases (ID1 and 2); release of T3 into the bloodstream. T3 is also formed by similar processes in the thyroid gland, but extra-thyroidal synthesis is more important. All enzymes are selenium dependent and up-regulated during primary iodine (or T3) deprivation. Secondary T3 deprivation can arise through inhibition of TPO by thiocyanates and inhibition of ID1 by goitrins formed from ingested precursors, causing up-regulation of the inhibited enzymes. The relative importance of ID1 and 2 varies between organs and species. Low circulating concentrations of free T3 stimulate the release of thyrotrophic hormones from the brain. Excess T3 induces a third selenium deiodinase that removes iodine from the inner tyrosine ring of T4 (ID3) to form an inactive analogue of reverse T3 (rT3).

hormone and corticosterone (Ingar, 1985), and regulatory proteins of exocrine origin. For example, leptin production by adipose tissue is induced by thyrotrophic hormones and controls appetite (Menedez et al., 2003); thyroidectomy raises plasma leptin in the ovine fetus (O'Connor et al., 2007) and alters the leptin response to underfeeding in the horse (Buff et al., 2007). In the long term, life is unsustainable without a thyroid gland.

Metabolism of Iodine

Like all the anionic elements, iodine is absorbed very efficiently from the gastrointestinal tract and this enables any iodine secreted prior to absorptive sites to be extensively recycled. Whereas phosphorus is recycled via saliva in ruminants, iodine is recycled via secretion into the abomasum (Miller et al., 1974). Unheated soybeans contain a heat-labile factor that can induce goitre by impairing the intestinal recycling of iodine (Hemken, 1960). Absorbed iodine is transported in the bloodstream loosely bound to plasma proteins. A small proportion of extra-thyroidal iodine circulates in free ionic form, like chloride, and accumulates in soft tissues such as muscle and liver when excess iodine is consumed (Downer et al., 1981). Recycling of thyroidal iodine occurs via the iodide pool. Excess dietary iodine is excreted predomi-

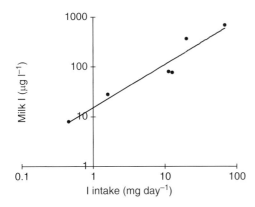

Fig 12.3. Iodine concentrations in cow's milk are directly proportional to their iodine intake (from Underwood and Suttle, 1999).

nantly via urine as iodide, but in lactating animals significant amounts can be secreted in milk (Miller et al., 1965; Hillman and Curtis, 1980). The slope of the relationship between intake and milk secretion of iodine (Fig. 12.3) indicates recovery of about 30% of ingested iodine in milk with yields of 30 l day^{-1} and recovery may be even higher in sheep since milk iodine can be much higher in sheep than in cows grazing the same pasture (50 versus 20 μg l^{-1}) (Grace et al., 2001).

T4 Synthesis and Storage

Approximately 80% of iodine in the mammalian body is found in the thyroid gland because 90% of the iodine entering that organ goes no further (Hetzel and Welby, 1997). Captured iodine is used to iodinate tyrosine to form mono- (T1) and diiodotyrosine (T2), and two molecules of T2 are then used to form T4, with the help at each stage of thyroperoxidase (Fig. 12.2) (Howie et al., 1995). The efficiency with which iodine is captured varies according to need and is regulated by the secretion of two hormones with thyrotropin-releasing and T4-stimulating properties (TRH and TSH). Together, these hormones determine the levels of T4 under feedback control from the circulating levels of free (f) T4 and T3, the tiny fractions (<1%) of predominantly globulin- or albumin-bound, plasma T4 and T3 pools (Beckett and Arthur, 1994; Todini, 2007).

Environmental temperature affects iodine uptake by the thyroid gland. In lactating goats, the retention of ^{125}I was about 10-fold higher (at 10% of the dose) 50 days after an oral dose when animals were kept at 5°C rather than 33°C (Lengemann, 1979). At the lower temperature, less ^{125}I was excreted in milk (2.6% versus 16.8% of dose) but more was lost in urine (71.2% versus 52.5%).

Activation of T4

T4 is taken up by intracellular receptors on inner mitochondrial membranes and in the cell nucleus, where it is activated by two selenium-dependent deiodinases: ID1, which can be inhibited by propylthiouracil (PTU), and ID2, which is relatively insensitive to PTU (Fig. 12.2) (Hetzel and Welby, 1997). There are major differences between ruminants and non-ruminants, with ID1 being particularly important in brown adipose tissue (BAT) in the newborn ruminant and serving as a source of T3 for other tissues (Nicol et al., 1994). In animals of normal iodine status, 80% or more of T3 can be formed extra-thyroidally, principally in the liver and kidneys (Ingar, 1985) but also in the skin (Hetzel and Welby, 1997; Villar et al., 2000), testes (Brzezinska-Slebodzinska et al., 2002) and mammary gland. In early lactation in cows, there is a fall in ID2 activity in the liver and a rise in mammary tissue, accompanied by decreases in plasma T4 and T3 concentrations (Pezzi et al., 2003). In newborn mammals, serum T4 and T3 concentrations rise immediately after birth (Rowntree et al., 2005) following the ingestion of colostrum rich in T4 and T3 (Pezzi et al., 2003). This probably reflects the importance of T4 and T3 to survival in the neonate. In mare's milk, T3 rises to a peak value of 0.71 nmol l^{-1} 4 days after parturition and there are similar increases in ID1 and 2 activities (Slebodzinski et al., 1998). In addition to possible roles in lactogenesis, ingested T3 may promote development of the gut mucosa. In the chick embryo, there are large increases in T3 and ID activity after 19 days development, when the lungs begin to function and glucocorticoid regulation changes (Darras et al., 1996). Selenium deprivation can induce

IDD and disrupt the activity of ID1 in BAT but the brain and pituitary are protected, being able to increase ID2 activity during iodine deprivation regardless of selenium status (Mitchell et al., 1997). When weanling rats were exposed to low temperatures (4°C for 18 h), ID2 activity in BAT increased. However, the increase was slowed and eventually greatly diminished by selenium deprivation (Arthur et al., 1992).

Sources of Iodine

Forages

Plant iodine concentrations are highly variable due to species and strain differences, climatic and seasonal conditions and the capacity of the soil to provide iodine; interactions between these factors are also important. Soils high in iodine such as boulder clays and alluvial soils generally produce plants richer in the element than iodine-low soils, such as those derived from granites (Groppel and Anke, 1986). However, correlations between soil iodine and plant iodine are often poor because plant species and strains differ widely in their ability to absorb and retain soil iodine (Table 12.2). Botanical species vary in iodine uptake and there are interactions between species and seasonal effects. Rapid seasonal declines in plant iodine have been noted for five different species in Germany, but the effect was proportionally lower in leguminous than in gramineous species (Groppel and Anke, 1986). Elsewhere, pasture iodine has ranged from 0.16 to 0.18 mg kg^{-1} DM and from 0.06 to 0.14 mg kg^{-1} DM in white clover and grasses, respectively, in the Netherlands (Hartmans, 1974), and means of 0.2–0.3 mg kg^{-1} DM have been reported for mixed swards in Wales (Alderman and Jones, 1967) and Northern Ireland (McCoy et al. 1997). Proximity to the sea has a major influence on plant iodine, as illustrated in Fig. 12.4. Dilution of the iodine deposited from marine sources over winter may contribute to the marked seasonal decline in plant iodine between spring and summer pastures (Table 12.3). Plant iodine can be much higher in the leaf than in the stem of swedes (0.28 versus <0.05 mg kg^{-1} DM, respectively) (Knowles and Grace, 2007) and white clover (Crush and Carradus, 1995), but not in kale (Knowles and Grace, 2007). The only artificial fertilizer naturally high in iodine is Chilean nitrate of soda and applications of this nitrogenous fertilizer can double or treble plant iodine. Heavy applications of seaweed can increase pasture iodine 10- to 100-fold (Gurevich, 1964), but nitrogenous fertilizers generally reduce herbage iodine by increasing plant growth (Alderman and Jones, 1967).

Soil iodine

The addition of soil to a pelleted lucerne diet for 63 days has been found to significantly increase serum iodine in sheep, clearly demonstrating that soil iodine can be absorbed (Grace, 2006). Extremely high values recorded for pas-

Table 12.2. Iodine content of New Zealand pasture species grown on soils of different type and iodine concentration (from Johnson and Butler (1957).

	Herbage I (µg kg^{-1} DM)		
	Silt loam	Sandy loam	Hill soil
Total soil I (µg kg^{-1} DM)	8530	3000	1760
Perennial ryegrass	1500	1600	1350
Short-rotation ryegrass	150	168	230
Cocksfoot	175	258	225
Paspalum dilatatum	1280	1280	1700
White clover	500	725	800

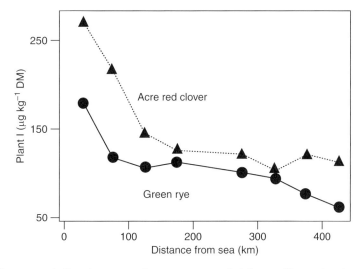

Fig. 12.4. Iodine concentrations in crops and grasses are greatly influenced by marine deposition and therefore decrease with distance from the sea (data from Groppel and Anke, 1986).

ture iodine in the spring (15–60 mg kg^{-1} DM) have been attributed to soil contamination and linked to year-to-year variations in the incidence of iodine-responsive mortality in lambs (Clark et al., 1998). Ingested soil can be an important source of iodine for grazing or foraging livestock.

Other foodstuffs

Animal feedstuffs vary widely as sources of iodine and the differences first reported in 1952 were mostly confirmed 30 years later (Table 12.4). Cereals and oilseed meals are poor sources of iodine, containing on average 0.05 mg kg^{-1} DM, but are influenced by soil

Table 12.3. Effects of plant species and sampling time on plant iodine concentrations (μg kg^{-1} DM) in the German Democratic Republic (Groppel and Anke, 1986).

	24 April	16 June
Meadow fescue	184±24	20±4
Green rye	305±108	43±20
Green wheat	215±53	18±4
Acre red clover	294±34	103±37
Lucerne	358±81	149±37

iodine status. Fish meals are exceedingly rich in iodine; animal protein sources are generally intermediate, but can be increased by iodine supplementation of the source animals. Seaweed may contain as much 4–6 g I kg^{-1} DM, and its inclusion in the rations of cows and hens greatly increases the milk and egg iodine. Sheep that graze seaweed along the coastlines of northern Europe usually have a surfeit of dietary iodine.

Water

The iodine content of drinking water largely reflects the iodine content of surrounding rocks and soils and correlates with iodine concentrations in local plants of an area (Fuge, 2005). In some parts of the world, iodine concentrations in the drinking water have been inversely correlated with the incidence of endemic goitre in man but in other parts there can be no such correlation (Stewart and Pharaoh, 1996) and similar inconsistencies no doubt exist for farm animals. Sea water is rich in iodine, containing > 50 μg I l^{-1}, and deposition of marine iodine influences drinking water iodine. In one study in East Germany, values fell from 7.6 ± 4.4 (mean ± standard deviation)

Table 12.4. Confirmation from the German Democratic Republic (GDR) (Groppel and Anke, 1986) of the wide differences in iodine concentrations (µg kg^{-1} DM) between animal foodstuffs reported by the Chilean Iodine Education Bureau (CIEB) in 1952.

Feedstuff	CIEB range	Feedstuff	GDR Mean	n	Standard deviation
Hays and straws	100–200	Hay	136	13	61
		Cereal straw	368	5	283
Cereal grains	40–90	Maize	44	9	18
		Barley	95	69	69
Oilseed meals	100–200	Rapeseed (ext.)	67	7	30
		Soybean (ext.)	97	8	33
Meat meals	100–200	Meat meals	–	–	–
Milk products	200–400	Dried skimmed milk	376	8	196
Fish meals	800–8000	Fish meal	6688	16	2879

to 1.1 ± 0.9 µg l^{-1} as distance from the coast increased from <50 to >400 km (Groppel and Anke, 1986). Exceptionally, drinking water can contain more iodine than sea water and this may be sufficient to suppress thyroid activity if other supplementary sources of iodine are consumed (Pedersen et al., 2007) However, drinking water does not normally contain sufficient iodine to contribute significantly to iodine intake.

Milk and milk products

Iodine concentrations of milk are extremely variable in all species due to the ease with which iodine crosses the mammary barrier (Fig. 12.3). Winter milk in the UK contains over twice as much iodine as summer milk (200 versus 90 µg l^{-1}) (MAFF, 2000), largely due to the over-supplementation of concentrates given to the housed dairy cow (Phillips et al., 1988). Milk iodine is also influenced by stage of lactation: colostrum is two to three times richer in iodine than main milk and concentrations decline towards the end of lactation. Milk iodine is increased by exogenous iodine from iodophors used to maintain dairy hygiene, but their decreased use is believed to have contributed largely to a reduction in milk iodine in New Zealand from 435 to 120 µg l^{-1} between 1977–1978 and 1987–1988 (Sutcliffe, 1990). Iodine differs from most minerals in milk in that it is associated with milk fat, and milk products such as dried skimmed milk, buttermilk and whey usually contain less iodine than the milk from which they are derived.

Sources of Goitrogens

Goitrogens occur more widely than was once thought and are normally divided into two categories according to their mode of action and counteraction, although there may be some overlap (Freyberger and Ahr, 2005).

Thyroperoxidase inhibitors

Plants can contain one or more of about 100 different glucosinolates, along with β-glucosidases such as myrosinase, that release HCN after structural damage to plant tissue, during harvesting or processing (Mawson et al., 1993). HCN is converted to thiocyanate (SCN) or isothiocyanate and the process can continue during digestion, through the activity of myrosinases produced by microbes that inhabit the gut. Thiocyanates inhibit thyroperoxidase (Fig. 12.2). White clover is an example of a common cyanogenic (CG) pasture component, yielding HCN concentrations that vary widely between strains (0.1–1.1 mg kg^{-1} DM) and seasons, decreasing by 75% between spring and early summer (Crush and Carradus, 1995). Brassica species of both leafy and root types are renowned for their CG properties. Kale (*B. oleracea* var. *acephala*) contains sinigrin, 10.4 mg kg^{-1} DM, as its predominant glucosinolate (Kushadet *et al.*, 1999) and swedes (*B. napus*) are also rich in glucosinolates. Both kale and swedes are important 'winter gap' feeds for outwintered ruminants. The use of cassava as a livestock feed is limited by its content of the glucosinolates linamarin and lotaustralin (Khajarern *et al.*, 1977), but HCN concentrations in both leaf and root can be reduced by appropriate processing and drying. Mustard cake is a goitrogen-rich byproduct of the extraction of oils from mustard seed (*B. juncea*) (Pattanaik *et al.*, 2001). The effects of CG goitrogens are attenuated by iodine supplementation.

Deiodinase inhibitors

The synthetic goitrogen PTU can induce IDD by inhibiting ORD1 in mammals, and inhibition can therefore take place outside the thyroid (Fig. 12.2). Plants can contain similar substances, such as the glucosinolate goitrin (vinyl thio oxazolidinethione, VTO) or its precursors (e.g. progoitrin in oilseed rape (*Brassica campestris* and *B. napus*)). Rapeseed meal (RSM) can contain 44 mmol progoitrin kg^{-1} DM as its major goitrogenic component (Taraz *et al.*, 2006) but meals prepared from 'low' glucosinolate varieties may contain only 2–5 mmol total glucosinolates kg^{-1} DM (Hill, 1991). Iodine supplementation does not counter the effects of ORD1 inhibitors (Emanuellson, 1994), and short-term exposure to them may cause goitre which is not an IDD. However, thiocyanates are formed during the hydrolysis of compounds such as progoitrin and the deiodinase inhibitor thus yields thyroperoxidase inhibitors whose effects can be countered by supplementary iodine.

Goitrogenicity

CG and PTU-like goitrogens have additive effects and the goitrogenicity of a feed or ration depends on the concentrations of each type present, their fate during digestion and the ratio of CG to iodine. CG and non-CG strains of white clover (*Trifolium repens*), all grown in the same area in New Zealand, contained 0.2 and 0.04 mg I kg^{-1} DM, respectively, indicating that goitrogenicity may be exaggerated by the concentration of CG alone (Johns, 1956). However, in pot trials in which both HCN and iodine concentrations varied widely between strains, no correlation between the two parameters was found (Crush and Carradus, 1995). The concentration of goitrogen precursors is not reduced by minimizing the available sulfate concentration in the soil, but may be reduced when nitrogen (Alderman and Stranks, 1967; McDonald *et al.*, 1981) or phosphate fertilizers (Wheeler and Vickery, 1989) are applied.

Genetic selection

In the early years of plant breeding, selection for disease and pest resistance also increased goitrogenicity, and the most popular strains of white clover in New Zealand were rich in HCN. Popularity and cyanogenicity are not always bedfellows: ladino types of *T. repens* are low in HCN (0.17 mg kg^{-1} DM) but popular in the USA (Crush and Carradus, 1995). The pendulum swung and strains were selected for 'low' goitrogen content, particularly oilseed rape, widely cultivated for edible oil and generating large quantities of RSM, a valuable protein supplement for livestock (Bell, 1984). Canadian plant breeders led the way in producing low-glucosinolate cultivars, yielding a product called Canola meal containing <30 mmol glucosinolate kg^{-1} DM; such products will have relatively small effects on iodine status (Emanuellson, 1994). Winter varieties grown in middle and western Europe are a different proposition and products derived from them require treatment to remove glucosinolates. Selection for high plant iodine status holds little promise (Crush and Carridus, 1995).

Metabolism of Goitrogens

Glucosinolates are less goitrogenic in ruminants than non-ruminants because they are extensively destroyed in the gastrointestinal tract

(Hill, 1991; Emanuelson, 1994). However, goitrin and SCN excretion in cow's milk rises when RSM is fed, a clear indication of incomplete destruction. Excretion of SCN is accompanied by a reduction in milk iodine and relatively small increases in milk iodine when iodine supplements are fed. This apparently paradoxical response may arise from inhibition by SCN of iodine uptake by the mammary gland and conservation of iodine in the lactating animal (Iwarson, 1973), but only at the expense of any suckled offspring. Increases in TSH secretion in cows on high-glucosinolate diets (Laarveld et al., 1981) indicate that, qualitatively, a net impairment of T3 function can occur, but the magnitude of the effect may be less in lactating than non-lactating animals, especially in sheep which secrete far more iodine in milk than cows. The browse legume Leucaena leucocephala contains mimosine, which is degraded to a goitrogen in the rumen and thus causes thyroid enlargement in steers in some locations, although not on the island of Hawaii, where local cattle possess a rumen microorganism that detoxifies the goitrogen. Transfer of a rumen inoculum from Hawaii to other locations prevented the induction of IDD in stock consuming L. leucocephala (Jones and Megarrity, 1983).

Iodine Requirements

Early estimates of daily iodine requirements, converted into dietary concentrations assuming a normal daily DM consumption, lay between 0.05 and 0.10 mg kg^{-1} DM for poultry, pigs and sheep and 0.03–0.06 mg kg^{-1} DM for milking cows (Table 12.1). However, the minimum 'iodine requirement' depends on the criterion of adequacy: requirements for growth are not necessarily those for reproduction and lactation, or for maintenance of thyroid structure and circulating levels of T4. Furthermore, critical levels for all species are likely to vary with environmental temperature, rate and stage of production and the efficiency of energy utilization. The ARC (1980) adopted a physiological approach by estimating requirements for ruminants that sustained TSR with an assumed capture efficiency of 33% for iodine reaching the thyroid. This approach overestimates need because it ignores the recycling of iodine and it has been suggested that functional assessment of hypothyroidism (i.e. TRH or TSH induction) should be used to define iodine requirements (Underwood and Suttle, 1999).

Non-ruminants

Highly variable iodine requirements have been reported for growth in early feeding trials with chicks and laying birds (0.03–1.0 mg kg^{-1} DM), but requirements for normal thyroid structure were more consistent (approximately 0.3 mg kg^{-1} DM) (Underwood and Suttle, 1999) and similar to requirements for both growth and normal thyroid development in pheasants and quail (Scott et al., 1960). Exposure of broilers to RSM high or low in glucosinolates can double thyroid size without affecting growth but iodine supplementation may not reduce thyroid size (Schone et al., 1992). Soaking the diets in myrosinase, a process that would generate CG, produced fivefold increases in thyroid size that were reduced by iodine supplementation. Heating myrosinase-treated RSM diets also increased goitrogenicity. NRC (1994) recommended concentrations of 0.3–0.4 mg I kg^{-1} DM for growing and laying birds.

Growing pigs on cereal and soybean meal rations require between 0.09 and 0.15 mg I kg^{-1} DM to avoid thyroid hyperplasia (Cromwell et al., 1975; Schöne et al., 1988). Growth of 6–7-week-old pigs was only slightly and inconsistently compromised when the diet contained 0.03–0.04 mg I kg^{-1} DM after prior depletion of iodine reserves, but thyroid weight was invariably and markedly reduced by adding 0.1 mg I kg^{-1} DM (Sihombing et al., 1974). NRC (1998) recommended a level of 0.14 mg I kg^{-1} DM for all pig rations. For pigs and poultry, requirements based on minimum thyroid weight exaggerate iodine requirements and iodine needs may vary with processes such as soaking and pelleting.

Ruminants

Daily TSR-based iodine requirements predict a substantial effect of season (ARC, 1980). When converted to dietary iodine, the mean requirements are $0.11\,mg\,kg^{-1}$ DM in summer and $0.54\,mg\,kg^{-1}$ DM in winter for sheep, with a similar winter value for cattle ($0.52\,mg\,kg^{-1}$ DM). Alternative estimates of requirement, derived from the relationship between dietary iodine and incidence of disorder, suggest a lesser effect of seasonally low temperatures. Mee et al. (1995) reported a significant reduction in thyroid weight in peri-natal calves (from 22.9 to 15.8 g) when the pregnant cow's iodine intake from a grass silage diet, containing 0.15–$0.21\,mg\,I\,kg^{-1}$ DM, was increased by about $25\,mg^{-1}\,head^{-1}\,day^{-1}$. McCoy et al. (1997) reported a 120% increase in thyroid weight in calves born to housed cows given experimental diets containing 0.06 rather than $0.27\,mg\,I\,kg^{-1}$ DM during late pregnancy. However, the survival of the calves was not affected by iodine supplementation, confirming the results of the earlier field trials in Northern Ireland with winter-calving herds (Mee et al., 1995). This suggests that iodine supply in the field trials was marginal and unlikely to be five times higher than the summer requirement, as suggested by the ARC (1980). Herbage iodine levels of 0.18–$0.27\,mg\,kg^{-1}$ DM can sustain normal growth in cattle (Wichtel et al., 1996), milk yield in cows (Grace and Waghorn, 2005) and wool growth in sheep (Barry et al., 1983) in the summer in New Zealand.

Influence of Goitrogens on Iodine Requirements

Organic goitrogens

The influence of organic goitrogens on iodine requirements is poorly defined in ruminants. The ARC (1980) suggested an increase in iodine provision to $2\,mg\,kg^{-1}$ DM in the presence of substantial quantities of goitrogens. Thyroid weight increased linearly from 67.4 to $120.8\,mg\,kg^{-1}$ live weight (LW) as the glucosinolate concentration in the lambs' diet (from RSM) increased up to $17.5\,mmol\,kg^{-1}$ DM but neither food intake or growth rate was affected (Hill et al., 1990). The principal concentrate component of the diet contained 0.8 mg added $I\,kg^{-1}$ DM but the goitre may not have been wholly iodine-responsive. Administration of $6\,mmol\,KCNS\,day^{-1}$ by vaginal pessary for the last two-thirds of pregnancy in ewes lowered their plasma T3 and T4 concentrations by 11% ($P<0.05$) without affecting values in the newborn lamb or subsequent lamb growth (Donald et al., 1993). Until further information becomes available, the ARC (1980) recommendation of $2\,mg\,I\,kg^{-1}$ DM for goitrogen-rich diets should be maintained as a safe allowance for ruminants.

Feeding RSM ($80\,g\,kg^{-1}$ DM) of a high glucosinolate type to weaned pigs increased requirements from around 0.1 to $0.5\,mg\,I\,kg^{-1}$ DM (Schöne et al., 1988). The addition of $50\,mmol\,KCNS\,kg^{-1}$ DM to the diet of young pigs induced goitre and a precipitous fall in serum protein-bound iodine (PBI) (from 45 to $5\,\mu g\,l^{-1}$ in 21 days), but the challenge was severe and values recovered almost as quickly when the goitrogen was removed from the low-iodine diet ($0.04\,mg\,kg^{-1}$ DM) (Sihombling et al., 1974). In broilers, early growth was inhibited by $>250\,g$ RSM (providing $>4.2\,mmol$ glucosinolates) kg^{-1} DM but by the finishing stage the tolerable level of RSM had doubled (dietary I unspecified but no supplement was used) (Taraz et al., 2006). The thyrotoxicity of thiouracil-type goitrogens cannot be remedied by increasing the iodine supply.

Inorganic goitrogens

Human goitre has been associated with high arsenic and fluoride intakes, the consumption of hard waters high in calcium and magnesium and drinking water high in nitrates. However, the evidence for an antagonism between fluorine and iodine in livestock is unconvincing (see Chapter 18). Antithyroid activity has been demonstrated for several cations and anions when ingested in excess, including sodium, chloride and especially perchlorate. Inorganic goitrogens may increase urinary iodine excretion or impair iodine uptake by the thyroid and – like organic goitrogens – they are of particular significance when dietary iodine is marginal. Evidence for the effects of high or low intakes of cobalt on goitre incidence in livestock is equivocal (Underwood and Suttle, 1999).

Biochemical, Histological and Immunological Signs of Iodine Deprivation

A simplified sequence of biochemical events during simple or CG-induced iodine deprivation is given in Fig. 12.5.

Depletion

In the initial phase, circulating levels of iodine in the bloodstream decrease, and both milk and urinary iodine decrease. In the thyroid, colloid is depleted and thyroid iodine declines below the normal level for mammals of 2–5 g kg^{-1} DM. Although the thyroid is only small (approximately 4 g dry weight in the adult cow) the amounts of iodine stored can be substantial (8–16 g) relative to minimum daily need (0.3 mg day^{-1} in heifers) (McCoy et al., 1997) and the depletion phase can therefore be prolonged. Enhanced activity in the thyroid gland during iodine deprivation causes oxidative stress and both cattle and sheep respond by increasing cytosolic glutathione peroxidase activity in the gland (Fig. 12.6) (Zagrodski et al., 1998); the subsequent course of events may depend on selenium status. The average hen's egg contains some 4–10 μg of iodine, most of which is in the yolk. This amount can be increased as much as 100-fold by feeding excess iodine, but is reduced during iodine depletion.

Deficiency

During the deficiency phase, there is hyperplasia of the cuboidal epithelium, iodine concentration in the enlarging thyroid is reduced further and the forms of iodine present change disproportionally in attempts to more efficiently use the diminishing supply of iodine. The ratios of T3 to T4 and T1 to T2 both increase, and the extent of these changes is evident from a comparison of thyroid glands of goats from an endemic goitre area with those from goats from an area with abundant environmental iodine (Fig. 12.7) (Karmarkar et al., 1974). In the bloodstream, serum T4 decreases, causing the serum T4 to

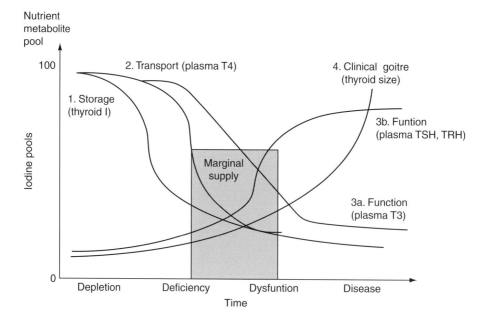

Fig. 12.5. A general scheme for the sequence of biochemical changes in livestock deprived of iodine; where thiouracil-type goitrogens are involved, impaired conversion of T4 to T3 in and beyond the thyroid changes the picture.

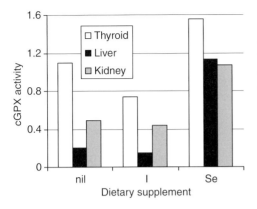

Fig 12.6. The effects of prolonged iodine and selenium deficiencies, singly or combined, on cytosolic glutathione peroxidase (cGPX) in three tissues of lambs, expressed as a fraction of activities in a fully supplemented group. In the thyroid, large increases in activity in the group depleted of iodine alone ('Se') and minimal decreases with selenium depletion ('I') indicate adaptations to protect the thyroid from peroxidation. Sensitivity and efficacy of the former adaptation is indicated by the fact that iodine depletion was not sufficient to cause thyroid enlargement or alter plasma thyroid hormone concentrations (data from Voudouri et al., 2003).

T3 ratio to decrease. Induction of a hypothyroid state in pigs by the administration of methimazole causes increases in ID2 in the thyriod and in skeletal muscle (Wassen et al., 2004).

Dysfunction

The approach of dysfunction is indicated by hypertrophy of the cuboidal epithelium lining the thyroid follicles and increases in the thyrotrophic hormones TRH and TSH in the bloodstream (e.g. in lambs, Clark, 1998; in the horse, Johnson et al., 2003). Cellular and humoral immune responses in goitrous goats have been improved by treatment with colloidal iodine or T4, both given by injection (Singh et al., 2005), but these may have been secondary to improved appetite.

Biochemical Changes Induced by Impaired Deiodination

Inhibition of peripheral T4/T3 conversion by exposure to thiouracil goitrogens or selenium deprivation soon causes T4:T3 ratios in serum to increase (Arthur et al., 1988), but the exposure required can be considerable. In cashmere goats given oral doses of PTU (1.1–17.5 mg kg^{-1} LW on a logarithmic scale) the smallest dose markedly inhibited ID1 activity in the thyroid, but >4.4 mg was required to raise T4:T3 and induce thyroid hyperplasia (Fig. 12.8). A compensatory two- to threefold increase in ID2 activity in the skin indicated one means by which animals can increase the peripheral

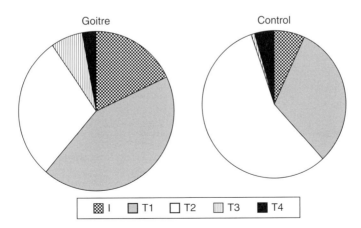

Fig. 12.7. The distribution (%) of labelled iodine (I) and the thyronines mono-, di-, tri- and tetraiodothyronine (T1–T4) in the thyroid glands of 10 goats from goitre areas and 10 from control areas, after intravenous administration of ^{151}I (from Karmarkar et al., 1974).

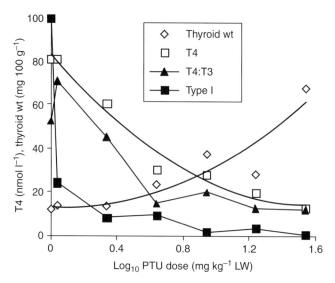

Fig. 12.8. Responses of cashmere goats to oral doses of propylthiouracil (PTU: a goitrin that inhibits deiodination) for 2 months. Hepatic type I deiodinase activity (% of untreated group) was markedly inhibited and the plasma tetraiodothyronine (T4) to triiodothyronine (T3) ratio was increased by the smallest dose. Much larger doses are needed to cause thyroid enlargement. Both plasma T4 and the T4:T3 ratio was decline as the thyroid enlarges (from Villar *et al.*, 1998).

production of T3 and the limitations of measuring deiodinase activity at single site (Villar *et al.*, 2000). The cashmere goat is less susceptible to thiouracil goitrogens than cattle (Villar *et al.*, 1998), but the ranking of sheep is unknown. When both CG and PTU goitrogens are ingested, biochemical changes may be different and the relative changes in ID1 and 2 may indicate the predominant goitrogen in the diet.

In pigs given RSM (i.e. both goitrogen types), extra-thyroidal responses also accompanied those occurring within the gland (Spiegel *et al.*, 1993). The presence of CG in the RSM was confirmed by a rise in serum SCN and a sixfold rise in thyroid weight, but liver size doubled. Serum T4 decreased and T3 concentrations were maintained; T4:T3 ratios therefore decreased and hepatomegaly was believed to have increased T3 synthesis from T4 and preserved the growth rate. In broilers however, feeding more than 190 g RSM (high in glucosinolates) kg^{-1} DM can cause hepatomegaly and lower serum T3 while leaving T4 unaffected (Taraz *et al.*, 2006). Such results begin to explain the diagnostic limitations of measurements of thyroid size and serum T4 discussed later.

Clinical Manifestations of Thyroid Dysfunction

Goitre

The degree of thyroid enlargement increases with the degree and duration of iodine deprivation and indicates an attempt to compensate for insufficient production of thyroid hormones. This and other manifestations occur whether the impaired hormonogenesis arises from absolute or conditioned (i.e. goitrogen-induced) iodine deprivation. Goitre is manifested predominantly in the newborn animal, which is usually delivered by a clinically normal dam. In the adult, enlargement of the thyroid can be present in otherwise healthy animals (Villar *et al.*, 1998).

Impaired fertility

Infertility or sterility and poor conception rates, due to irregular or suppressed oestrus, are allegedly features of thyroid dysfunction in cows and have been attributed to the enhanced losses of iodine at peak lactation, when mating usually occurs (Hemken, 1960). Male fertility is also affected, with a decline in libido and a deterioration in semen quality being characteristic of iodine deprivation in bucks (Patanaik et al., 2001), rams, bulls and stallions. In the Brown Leghorn male deprived of iodine, the testes remain small and without spermatozoa. Thyroidectomy reduces egg production in hens and the seasonal cycle of egg production in poultry is related to seasonal variations in thyroid activity.

Impaired embryo and fetal development

In a study of pregnant sheep given a diet very low in iodine (5–10 µg kg^{-1} DM), fetal brain maturation was impaired and lambs were stillborn and without wool (Potter et al., 1981, 1982). The brain abnormalities were reversed by giving iodine during the third trimester. Fetal development may be arrested at any stage by thyroid dysfunction, leading to early death and resorption (e.g. in ewes, Sargison et al., 1998), abortion or subnormal birth weights, but such abnormalities are not unequivocal evidence of iodine deprivation (e.g. in calves, Smyth et al., 1992). Neonatal mortality increases in lambs from ewes fed on kale or swedes (Andrews and Sinclair, 1962; Knowles and Grace, 2007) and 'memory effects' of exposure to white clover in early pregnancy have been implicated as a cause of raised lamb mortality in New Zealand (Sargison et al., 1998). The need for thyroid hormones to produce lung surfactants may be an important determinant of neonatal survival (Erenberg et al., 1974; Symonds and Clarke, 1996). Poultry can withstand considerable iodine deprivation without marked loss of production. In one study, it took 2 years of feeding an iodine-deficient diet to retard embryo development, prolong hatching time and decrease hatchability in hens (Rogler, 1958).

Post-natal mortality and growth retardation

Experimental iodine deprivation in utero lowers the basal metabolic rate and resistance to hypothermia: sucking behaviour is impaired in the newborn lamb (Potter et al., 1982). In mildly iodine-deficient areas where the occurrence of goitre is low and variable, productivity can still be seriously impaired. In a flock of Polwarth ewes in Tasmania showing only minimal thyroid enlargement, a high lamb mortality of 36% was reported (King, 1976) and the findings were confirmed elsewhere (Knights et al., 1979; Andrewartha et al., 1980; McGowan, 1983). Given the importance of thermogenesis from BAT to survival in animals born into a cold environment, iodine deprivation may increase susceptibility to cold stress (Arthur et al., 1992). There is, however, little convincing evidence as yet that marginal iodine deprivation impairs survival in the calf (McCoy et al., 1995, 1997; Mee et al., 1995). The addition of goitrogens to a pig diet low in iodine has been found to severely restrict growth and stunt skeletal development (Sihombing et al., 1974), while the inclusion of RSM in a natural diet for weanling pigs reduced the growth rate by 50% (Lüdke and Schöne, 1988). Iodine supplementation can increase growth rate and food intake in castrated and intact male goats given a diet high in goitrogens (Pattanaik et al., 2001).

Disorders of the integument

Changes in the skin and its outgrowths are common features of thyroid dysfunction. Pigs (and calves) born to iodine-deprived mothers have thick pulpy skins due to subcutaneous oedema, and this early sign of deficiency, recognized in various species in goitrous regions, has been reproduced experimentally in piglets (Sihombling et al., 1974). Myxoedema also develops when weanling pigs are given goitrogen-rich diets (Lüdke and Schöne, 1988). Milder deficiency is reflected in minor changes, such as rough, dry skin and a harsh coat in pigs and scanty wool and hairiness of the fleece in sheep. Thyroid insufficiency in the young lamb permanently impairs the quality of the adult fleece, since the normal development of the wool-producing, secondary follicles is highly dependent on thyroid activity (Ferguson et al., 1956). Administration of methylthiouracil to cashmere goats reduces the proportion of active secondary hair follicles and delays the onset of moult (Rhind and McMillen, 1996). Thyroidectomy of the ram reduces wool growth, but replacement therapy only needed to raise T4 levels in the bloodstream to 30% of normal to restore yield (Maddocks et al., 1985). Supplementation of the ewe with iodine can increase wool production by 6% (Statham and Koen, 1981). Thyroid insufficiency can be very conspicuous in some types of bird, particularly the male in which the comb decreases in size, moulting is inhibited and the characteristic plumage is lost. Broilers exposed to CG show incomplete feathering (Schone et al., 1992).

Low milk yield

Reductions in milk yield are a conspicuous feature of iodine deprivation in the dairy cow and can probably occur in all mammals. The hypothyroid state induced by the prolonged feeding of goitrogen-rich foods is accompanied by loss of appetite leading to impaired growth and/or depressed milk yield (Hill, 1991). The earlier in post-natal life that maternal exposure to goitrogens begins, the greater the impact on milk yield and hence lamb growth. However, the overall relationships between serum T4 or T3 and milk yield in cows are *negative* (for supporting references, see Pezzi et al., 2003), possibly due to the confounding effect of stage of lactation, redistribution of thyronines in the body and their secretion into milk.

Occurrence of Iodine Deficiency Disorders

Geochemical factors

The occurrence of IDD was traditionally indicated by maps showing the incidence of goitre, one of the most widespread of all mineral deficiency diseases (Fig. 12.1), but important local variations in goitre incidence can be found within small countries (Statham and Bray, 1975; Wilson, 1975), which cannot be shown on such large-scale maps. A deficiency of plant-available soil iodine is the primary reason for the occurrence IDD in most areas. In regions that have been subjected to comparatively recent (Pleistocene) glaciation, the low iodine status of soils can be attributed to the removal of iodine-rich surface soils and insufficient time for subsequent replenishment with airborne iodine (Hetzel and Welby, 1997). In other 'goitre regions', the existence of low-iodine soils can generally be attributed to loss by leaching, scant marine deposition, low rainfall or combinations of these factors (Fuge, 2005). Hercus et al. (1925) estimated that 22–50 mg of iodine per acre (1 acre = 0.405 hectares) fell annually in rain on the Atlantic coastal plain, compared with only 0.7 mg per acre in the inland Great Lakes region of North America where goitre was endemic. Goitre is a problem in the highland interior of Papua New Guinea (Walton and Humphrey 1979), but a coastal location is less beneficial if it does not face the prevailing wind. Soils low in organic matter (e.g. sandy soils) tend to be low in iodine because they trap little of any deposited iodine.

Soil–plant interactions

In areas of marginal soil iodine status, the incidence and severity of IDD is influenced by climatic and seasonal conditions. These affect the grazing animal directly and indirectly through changes in pasture productivity and degree of contamination with marine or soil iodine. Goitre incidence in lambs is higher in 'good' than in 'poor' seasons in Tasmania where luxuriant herbage growth in 'good', high-rainfall seasons may reduce soil iodine ingestion (Statham and Bray, 1975). However, responses to iodine supplementation in outwintered ewes have been seen to be greater in a 'wet' than in a 'dry' year (Sargison et al., 1998). Increases in the iodine status of outwintered ewes between mating and lambing (Sargison et al., 1998) may reflect autumn increases in herbage iodine and mask a threat to embryonic survival caused by iodine deprivation in early pregnancy. The four- to fivefold reduction in pasture iodine that can accompany the advancing maturity of a sward (Table 12.2) means that the risk of IDD tends to increase as the grazing season progresses in herds that calve year-round.

Secondary iodine disorders

The occurrence of goitre can depend on dietary selenium supply (see Chapter 15) and the consumption of goitrogen-containing plants. Where the animal's diet is composed largely of highly goitrogenic feeds, the incidence of goitre can be high. Figure 12.9 shows the dramatic effect of feeding kale or swedes to ewes in late pregnancy on the incidence of neonatal goitre. When goitrogen exposure occurs in early pregnancy, the conception rate may be decreased without evidence of neonatal goitre (Sargison et al., 1998) and the use of kale to supplement autumn calving cows may adversely affect the conception rate. The feeding of mustard cake has induced IDD in goats (Pattanaik et al., 2001). Barbari goats (18.8 kg initial LW) reared on a concentrate containing mustard-cake goitrogens gained twice as much weight (5.4 versus 2.6 kg) when supplemented with just 75 µg I day^{-1} for 180 days as those given no supplement. However, growth in lambs fattened on rape is rarely improved by iodine supplementation, despite evidence of thyroid enlargement (Fitzgerald, 1983). It is possible that interception of depositing iodine by broad-leaved species raises concentrations of iodine in the leaf in autumn. Inclusion of as little as 80 g kg^{-1} DM of a high-glucosinolate RSM in pig rations will induce hypothyroidism and suppress growth, effects that supplementary iodine can only partially alleviate (Fig. 12.10). The feeding of cassava can induce IDD, but diets containing 500 mg HCN kg^{-1} DM have been fed to sows throughout gestation and lactation without adverse effects on sow or litter; bitter varieties pose the greatest risk of inducing iodine deficiency.

Genetic factors

Although lambs from hill breeds of sheep have better thermoregulatory powers than lowland breeds, the difference is probably attributable to differences in birth coat characteristics rather than differences in iodine metabolism (Todini, 2007). Associations between plasma T4 and wool growth rate, body size or reaction to food restriction are not matched by reports of breed differences in susceptibility to IDD (Todini 2007).

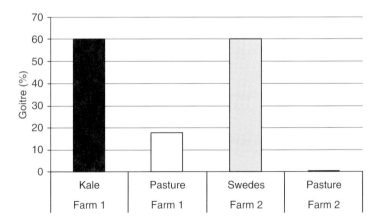

Fig. 12.9. The incidence of goitre in newborn lambs can greatly increase when ewes are fed on kale or swedes, rather than pasture, for the latter half of pregnancy. Mean iodine (mg kg^{-1} DM) was <0.05 in kale leaf and 0.15 in pasture on farm 1, and <0.05 in swede root and 0.11 in pasture on farm 2. Goitre incidence on pasture did not relate to pasture iodine (data from Grace et al., 2001).

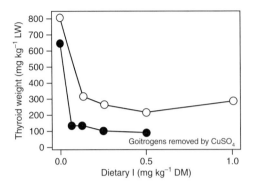

Fig. 12.10. Thyroid enlargement in pigs given diets containing some thiouracil-type goitrogens (as rapeseed meal) cannot be fully controlled by iodine supplementation, but their removal by treatment with copper sulfate allows full recovery (data from Schöne et al., 1988).

Diagnosis of Iodine Deficiency Disorders

Diagnosis of IDD has been based on morphological, histological or biochemical indices of iodine status, but all of those in common use are of questionable precision (Clark et al., 1988).

Thyroid morphology, histology and composition

Measurements involving the thyroid are widely used, but they can only confirm one particular IDD, albeit the most conspicuous: goitre. Absence of neonatal goitre does not rule out IDD (Sargison et al., 1998). Severe goitre can be diagnosed from a visibly or palpably enlarged thyroid, but milder forms require determination of thyroid weight. Early studies indicated that the thyroid status of newborn lambs was normal if the fresh gland weighed <1.3g, doubtful between 1.3 and 2.8g and 'hyperplastic' if >2.8g (Sinclair and Andrews, 1961); corresponding values for calves were <10, 10–13 and >13g (Wilson, 1975). However, considerable variation in size can occur in histologically normal calf thyroids (McCoy et al., 1997). Hyperplasia may be a more reliable index than thyroid weight (Smyth et al., 1992), but can also be found in healthy calves (Mee et al., 1995; McCoy et al., 1997); the terms 'colloid goitre' and 'hyperplastic goitre' are therefore diagnostically misleading. Thyroid weight can vary greatly in newborn lambs in an outbreak of goitre and caution is needed when making a flock diagnosis on the basis of individual thyroid weights. In a New Zealand field study, an individual thyroid weight to live weight ratio >0.8 (g kg^{-1}) indicated a 90% probability that the lamb came from an iodine-responsive flock, with a marginal band of 0.4–0.8 (Knowles and Grace, 2007). A mean thyroid weight to live weight ratio >0.4 (g kg^{-1}) was previously associated with iodine-responsive mortality (Clark et al., 1988). Conventional interpretation of thyroid iodine concentrations has also been called into question. Early in this century, hyperplasia was associated with thyroid iodine <1g kg^{-1} DM (Marine and Williams, 1908) and a critical concentration of 1.2g kg^{-1} DM was subsequently reported for several farm species (Andrews et al., 1948; Sinclair and Andrews, 1958). However, lower values have since been found in histologically normal calf thyroids (Wilson, 1975; McCoy et al., 1995, 1997).

Iodine in the bloodstream

Serum iodine or PBI and butanol-extractable iodine normally consist largely of T4 in bound forms and reflect the stage of depletion reached rather than the onset of dysfunction. PBI values of 30–40μg l^{-1} have been taken as 'normal' for adult sheep and cattle, but a value of 39μg l^{-1} in peri-parturient ewes was reported in an outbreak of neonatal goitre (Statham and Koen, 1981). Lower normal PBI values have been recorded in the domestic fowl and horse (Irvine, 1967). Misleadingly high PBI values are found after the administration of diiodosalicylic acid, which becomes strongly bound to serum proteins while remaining poorly available for T4 synthesis (Miller and Ammerman, 1995). The measurement of total serum iodine has come into vogue as a simple measure of current intake but does not indicate thyroid hormone function (Mee et al., 1995), particularly when CG are being consumed. Since exposure to such goitrogens reduces uptake of iodine by the thyroid and secretion in milk, serum iodine should be high for a given iodine intake. In the study of Knowles and

Grace (2007) (Fig. 12.9), serum iodine tended to be higher in ewes at lambing, 10 days after kale feeding, than in pastured ewes (50 ± 12 versus 39 ± 5 µg l^{-1}) despite a far higher incidence of goitre. A high ratio of serum iodine to T4 may indicate exposure to CG goitrogens.

T4 assays

Serum T4 can indicate past iodine nutrition and stage of depletion, but is a poor indicator of current thyroid dysfunction or disorder. In lactating ewes, a mean plasma T4 of <50 nmol l^{-1} was said to indicate thyroid insufficiency from field studies in Australia (Wallace et al., 1978); that threshold was lowered slightly for more general use in New Zealand, but flocks with a 'normal' mean serum T4 (<47 nmol l^{-1}) at mating were subsequently found to be iodine responsive (Clark et al., 1998). Interpretation is influenced by a number of factors that affect serum T4: the season (higher values in winter than summer due to photoperiodism) (Todini 2007); pregnancy (values fall in late pregnancy); early lactation (at a time of negative energy balance, serum T4 falls to 20–40 µmol l^{-1} in healthy dairy cows, and similar changes occur in goats) (Pezzi et al., 2003; Todini, 2007); birth (serum T4 in normal newborn animals is relatively high and may remain so for several weeks) (Clark et al., 1998); infection (serum T4 is decreased by intestinal parasitism) (Prichard et al., 1974); food intake (there are generally positive associations between food intake and plasma T4, but appetite can be either cause or effect, being strongly driven in species such as deer) (Todini, 2007).

In newborn lambs, T4 values decline to adult levels after about 8 weeks and a ratio of <1 in lamb to ewe T4 concentrations in serum at birth may indicate deficiency in the offspring (Andrewartha et al., 1980). Similarly, T4 values in newborn calves from heifers given adequate dietary iodine are three times higher than those of their dams, but the critical ratio may be higher than in sheep (<2.5) (McCoy et al., 1997). Serum T4 values in the newborn calf and piglet can fall by 50% within a week (Mostyn et al., 2006). Where the cause of hypothyroidism is impaired conversion of T4 to T3, as in selenium deprivation or ID1 inhibition by goitrogens, serum T4 may be particularly misleading. However, low plasma T4 concentrations have been found to accompany clinically significant goitre in pigs induced by RSM (e.g. Schöne et al., 1988, 1992). Assays of fT4 are rarely utilized.

T3 assays

T3 assays are preferred by some as a measure of dysfunction, particularly in selenium-induced IDD. Wichtel et al. (1996) found significantly lower T3 concentrations in a dairy herd that responded to selenium than in an unresponsive herd (1.81 versus 2.06 nmol l^{-1}). T4 concentrations were high in untreated heifers (66 and 99 nmol l^{-1}), but were reduced by giving selenium. Poor correlations between mean serum T3, measured at scanning and late pregnancy, with iodine-responsive lamb mortality have been found in the New Zealand trials (Clark et al., 1988). However, T3 is affected by many things other than iodine status, including: feed and water restriction (↓) (Blum and Kunz, 1981); cold (↑) (sheep, Nazifi et al., 2003) and heat stress (↓) (beef cattle, Christopherson et al., 1979; sheep, Nazifi et al., 2003; poultry, (Sahin and Kocuk, 2007); parasitic infection (↓) (Hennessy and Pritchard, 1981); protein (↓) (Ash et al., 1985) and iron deficiencies (↓) (Beard et al., 1989); pregnancy (↓) (Todini et al., 2007); birth (the early reduction in serum T3 in calves is proportionately greater than that seen in T4; see Fig. 12.11); and ageing (serum T3 falls from 2.8 ± 0.2 to 1.7 ± 0.1 µg l^{-1} (mean ± standard error) between 1 and 5 years of age in goats (Todini, 2007). It is also genotype-dependent in piglets (Mostyn et al., 2006).

Iodine in urine and milk

The rates of iodine excretion in milk and urine reflect iodine intakes, and milk iodine surveys indicate the 'goitre status' of an area. Alderman and Stranks (1967) suggested that dietary iodine supply for dairy cows may be suboptimal with milk iodine <25 µg l^{-1}, but

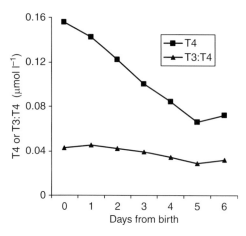

Fig. 12.11. There are characteristic reductions is serum tetra- and triiodothyronine (T4 and T3) in many species in the first week after birth. Here the effect is shown for calves and is greater for T3 than T4, resulting in a decreased in T3:T4 ratio (data from Rowntree et al., 2005).

milk yield was not increased by supplementing a New Zealand herd with milk iodine $<20\,\mu g\ l^{-1}$ (Grace and Waghorn, 2005). Higher thresholds (possibly $>50\,\mu g\ l^{-1}$) (McCoy et al., 1997) would certainly be needed for colostrum iodine in studies of peri-natal calf mortality. Average ewe milk iodine can predict the field incidence of goitre in the lambs, determined by palpable thyroid enlargement, and concentrations $<80\,\mu g\ l^{-1}$ point to a possible dysfunctional state (Table 12.5). Where thyroid dysfunction is induced by CG, urine iodine will probably be deceptively high and milk iodine deceptively low.

Conclusion

The biochemical confirmation of IDD in livestock, like the definition of iodine requirements, needs a new approach. Neither mean serum T4 or T3 nor T4:T3 ratios have predicted IDD in newborn lambs in New Zealand dose–response trials (Clark et al., 1998). Simple microplate ELISA assays for TSH are now widely used to predict goitre prevalence in infants (Sullivan et al., 1997) and should be adopted for livestock. Adaptive increases in ORD activity clearly indicate imminent thyroid dysfunction and have potential diagnostic value in livestock. Interpretation of conventional indices of iodine status can be improved by the adoption of marginal ranges, assessments of the variability in a sampled population and use of more than one index of status. Response to iodine supplementation will depend on the proportion of individuals that lie within or beyond the limits of marginality shown in Table 12.6.

Prevention and Control

Disorders caused by simple or CG-induced iodine deprivation can be prevented and cured by iodine supplementation. Those caused by thiouracil-type goitrogens or selenium deprivation require other measures.

Continuous methods

With stall- or hand-fed animals, supplementation is accomplished by the use of iodized salt licks, the feeding of kelp (seaweed) or incorporating iodine into mineral mixtures or concentrates. With the exception of diiodosalicylic acid, the commonly used iodine salts – such as potassium iodide (KI) or calcium iodate $(Ca(IO_3)_2)$ – have a uniformly high availability to both ruminants and non-ruminants (Miller and Ammerman, 1995). Diiodosalicylic acid has only 15% of the value of KI for cattle, but may be more available to non-ruminants. Iodides are subject to considerable losses of iodine by volatilization and by leaching, particularly in hot, humid climates, unless suitably stabilized (Kelly, 1953). Cuprous iodide is stable and readily absorbed despite a low solubility and is favoured by some workers. Orthoperiodates (e.g. $Ca_5(IO_6)_2$) are also stable and not toxic at

Table 12.5. Relation of ewe milk iodine to thyroid enlargement in lambs (from Mason, 1976).

Percentage of lambs with enlarged thyroids	Ewe milk I ($\mu g\ l^{-1}$)
23–69	23–64
12–22	71–98
0–4	98–111

Table 12.6. Indices of marginal iodine status[a] in livestock.

	Condition	Pigs	Poultry	Sheep	Cattle
Diet (mg/kg DM)	Summer	**0.04**–0.08	**0.06**–0.08	**0.10**–0.15	**0.075**–0.10
	Winter	NA (+)	NA (+)	**0.20**–0.30	**0.15**–0.20
	Goitrogens[b]	**0.14**–0.28	**0.18**–0.24	**0.75**–1.00	**0.75**–1.00
Thyroid – wet weight (g kg^{-1} LW)	Newborn	NA	NA	0.4–**0.9**	0.5–**1.0**
	Weanling	**0.2**–0.4	**0.06**–0.10	0.4–**0.8**	
Iodine (g kg^{-1} DM)	Adult				0.1–**0.2**
Serum PBI (µg l^{-1})	Adult	20–33	**30**–40	**30**–40	**30**–40
Serum T4 (µmol l^{-1})	D:NB	**40**–60	NA	**1.0**–1.5	**2.0**–2.5
	Adult	**20**–30	5–9	**20**–30	**1.8**–2.0
Serum T3 (µmol l^{-1})	D:NB	0.6–0.8	NA	2–**3.4**	<**4**
	Adult	NA	1.5–2.0	1.0–**1.7**	<**2**
Milk (µg l^{-1})		NA	NA	**70**–100	**15**–30

D, dam; NB, newborn; NA, not available (but requirements may be raised (+) in the free-ranging, cold-stressed animal); PBI, protein-bound iodine; T3, triiodothyronine; T4, tetraiodothyronine.
[a]Mean values for sample falling within the ranges given indicate a possibility of response to iodine supplementation (or goitrogen removal) if sustained. If values for ≥10% of individuals sampled lie beyond the limit in **bold**, supplementation with iodine will probably be beneficial.
[b]Three- to fivefold increases in iodine requirement can be caused by the ingestion of diets rich in cyanogenetic goitrogens.

the levels required in salt licks or mineral mixtures (0.1 g I kg^{-1} DM). Consumption of all free-access sources can be extremely variable (see Chapter 2). Herbage iodine can be increased by the application of fertilizers containing iodine but uptake of iodine is poor, especially where liming is practised, unless iodate is used instead of iodide (Whitehead, 1975).

Discontinuous methods

Dosing or drenching with iodine salts can control goitre but is time-consuming and costly, unless animals are being handled for other reasons. Two oral doses of 280 mg KI or 360 mg potassium iodate (KIO$_3$), after months 3 or 4 of pregnancy, have prevented neonatal mortality and associated goitre in lambs from flocks outwintered on goitrogenic kale. A single 1 ml intramuscular injection of an iodized poppy-seed oil containing 40% (w/v) iodine, administered at mating (Clark et al., 1998) or 7–9 weeks before lambing, can also be effective (Sinclair and Andrews, 1958, 1961; Statham and Koen, 1981). Iodine injections can reduce goitre incidence and improve survival to weaning (McGowan, 1983; Clark et al., 1998), improve wool yield in the year after treatment and control goitre at a further lambing (Statham and Koen, 1981). The recommended dose of iodized poppy oil for cows is only 4 ml (i.e. not proportional to body weight vis-à-vis sheep).

Slow-release methods

Slow-release intraruminal devices have been developed, one of which lasts for several years by employing a simple diffusion principle and an iodide-filled, polyethylene sack that lodges in the rumen (Mason and Laby, 1978). Rigid capsules can achieve a similar result (Siebert and Hunter, 1982) and efficacy has been demonstrated (Ellis and Coverdale, 1982; Rogers et al., 1998). Slow-release preparations for parenteral use in sheep and cattle are now available and raise iodine status for up to a year (Grace and Waghorn, 2005).

Countering thiouracil-type goitrogens

Soaking RSM in copper sulfate ($CuSO_4$) solution (25 g l^{-1}) and drying (Ludke and Schöne, 1988) or supplementation of diets with 250 mg Cu kg^{-1} DM can eliminate both VTO and the related indolyl compound (ITO), restoring normal growth and thyroid size in weaned pigs (Schöne et al., 1988) and broilers (Schone et al., 1992), whereas supplementary iodine alone can be ineffective. The RSM goitrogens are predominantly of the thiouracil type (11.9 g VTO and 3.8 g ITO kg^{-1} DM).

Iodine Toxicity

Iodine is a cumulative, chronic poison and the reports of tolerable doses for all sources naturally vary. Exposure to excess iodine paradoxically results in hypothyroidism because of feedback inhibition of T3 synthesis. However, there is a wide margin between the required (<0.5 mg kg^{-1} DM) and the tolerated concentration (<50 mg kg^{-1} DM) for species other than the horse.

Tolerance

Beef cattle given diets containing 100–200 mg I kg^{-1} DM for 3–4 months have been found to grow poorly and develop respiratory problems. The minimum safe concentration ranges were considered to be 25–50 mg kg^{-1} DM (Newton et al., 1974). The only natural dietary circumstances that present a risk of iodine toxicity are where large proportions of seaweed, cultured marine algae or their derivatives are fed since these can contain 1–11 g I kg^{-1} DM (Indergaard and Minsaas, 1991). As with many minerals given in excess, the smaller livestock species appear to be more tolerant than large animals, but this is partly because large-animal experiments usually last longer than those with pigs and poultry. Nevertheless, iodine tolerances have been placed at 300–400 mg kg^{-1} DM for pigs and poultry and 50 mg kg^{-1} DM for sheep and cattle (NRC, 2005). The horse appears to be exceptionally vulnerable to iodine toxicity, with goitre being reported in mares and their foals given large amounts of seaweed (McDowell, 1992). The tolerance of equines has been placed as low as 5 mg kg^{-1} DM. Periods of high soil iodine ingestion may suppress serum T4 in ewes (Clark et al., 1998). High iodine intakes that are tolerated by the dairy cow may increase milk iodine sufficiently to increase the risk of thyrotoxicosis in humans (Phillips et al., 1988).

Pharmacological side effects

Toxicoses have been caused by misuse of ethylenediamine dihydriodide (EDDI) in the oral treatment of foot rot and soft tissue lumpy jaw, particularly in the USA. Tolerances of bolus doses of EDDI equivalent to 100 mg I kg^{-1} DM feed have been reported after 13 months' exposure. The onset of lactation reduces serum iodine levels and the risk of toxicity because extra food is provided and much of the excess iodine can be secreted in milk (Fish and Swanson, 1983). The signs of toxicity are associated with iodine intakes (as EDDI) of 74–402 mg day^{-1} and are similar to those for other sources of iodine. They include nasal and lachrymal discharge, conjunctivitis, coughing, broncho-pneumonia, hair loss and dermatitis (Hillman and Curtis, 1980). The metabolic rate is elevated, as indicated by a raised body temperature and heart rate. Blood glucose and urea concentrations are raised and immunosuppression has been reported (Haggard et al., 1980). When EDDI was incorporated into a free-access salt mixture (1.25 g I kg^{-1}) for a herd of beef cows at pasture, consumption of the mixture ranged from 20 to 70 g animal-unit^{-1} day^{-1} and provided sufficient iodine (24–88 mg day^{-1}) to reduce the incidence of foot rot without harming calf performance. Mean serum iodine did not exceed 0.46 mg l^{-1} (Maas et al., 1984). Total serum iodine increases logarithmically as the dose increases, reaching about 4 mg l^{-1} when 0.77 mg EDDI kg^{-1} LW is given daily for 28 days (Maas et al., 1989) and providing a rough index of exposure. Milk iodine concentrations can exceed 2 mg l^{-1} and also confirm iodosis, but serum T4 can remain normal (Hillman and Curtis, 1980). Neonatal losses do not present with typical morphological or histological evidence of goitre, although serum T4 is decreased (Fish and Swanson, 1983).

References

Alderman, G. and Jones, D.I.H. (1967) The iodine content of pastures. *Journal of the Science of Food and Agriculture* 18, 197–199.

Alderman, G. and Stranks, M.H. (1967) The iodine content of bulk herd milk in summer in relation to estimated dietary iodine intake of cows. *Journal of the Science of Food and Agriculture* 18, 151–153.

Andrewartha, K.A., Caple, I.W., Davies, W.D. and McDonald, J.W. (1980) Observations on serum thyroxine concentrations in lambs and ewes to assess iodine nutrition. *Australian Veterinary Journal* 56, 18–21.

Andrews, E.D. and Sinclair, D.P. (1962) Goitre and neonatal mortality in lambs. *Proceedings of the New Zealand Society of Animal Production* 22, 123–132.

Andrews, F.N., Shrewsbury, C.L., Harper, C., Vestal, C.M. and Doyle, L.P. (1948) Iodine deficiency in newborn sheep and swine. *Journal of Animal Science* 7, 298–310.

ARC (1980) *The Nutrient Requirements of Ruminants*. Commonwealth Agricultural Bureaux, Farnham Royal, UK, pp. 184–185.

Arthur, J.R. (1997) Selenium biochemistry and function. In: Fischer, P.W., L'Abbé, M.R., Cockell, K.A. and Gibson, R.S. (eds) *Proceedings of the Ninth International Symposium on Trace Elements in Man and Animals (TEMA 9)*. NRC Research Press, Ottawa, pp. 1–5.

Arthur, J.R., Morrice, P.C. and Beckett, G.J. (1988) Thyroid hormone concentrations in selenium-deficient and selenium-sufficient cattle. *Research in Veterinary Science* 45, 122–123.

Arthur, J., Nicol, F., Guo, Y. and Trayhurn, P. (1992) Progressive effects of selenium deficiency on the acute, cold-induced stimulation of type II deiodinase activity in rat brown adipose tissue. *Proceedings of the Nutrition Society* 51, 63A.

Ash, C.P.J., Crompton, D.W.T. and Lunn, P. G. (1985) Endocrine responses of protein-malnourished rats infected with *Nippostrongylus brasiliensis* (Nematoda). *Parasitology* 91, 359–368.

Barry, T.N., Duncan, S.J., Sadler, W.A., Millar, K.R. and Sheppard, A.D. (1983) Iodine metabolism and thyroid hormone relationships in growing sheep fed kale (*Brassica oleracea*) and ryegrass (*Lolium perenne*)–white clover (*Trifolium repens*) fresh forage diets. *British Journal of Nutrition* 49, 241–254.

Bassett, J.H.D., Harvey, C.B. and Williams, G.R. (2003) Mechanisms of thyroid hormone receptor-specific nuclear and extra-nuclear actions. *Molecular and Cellular Endocrinology* 213, 1–11.

Beard, J., Tobin, B. and Green, W. (1989) Evidence for thyroid hormone deficiency in iron-deficient anemic rats. *Journal of Nutrition* 119, 772–778.

Beckett, G.J. and Arthur, J. (1994) The iodothyronine deiodinases and 5′deiodination. *Baillieres Clinical Endocrinology and Metabolism* 8, 285–304.

Bell, J.M. (1984) Nutrients and toxicants in rapeseed meal: a review. *Journal of Animal Science* 58, 996–1010.

Blum, J.W. and Kunz, P. (1981) The effects of fasting on thyroid hormone levels and kinetics of reverse tri-iodothyronine in cattle. *Acta Endocrinologica* 98, 234–239.

Brzezinska-Slebodzinska, E., Slebodzinski, A.B. and Kowalska, K. (2002) Evidence for the presence of 5′-deiodinase in mammalian seminal plasma and for the increase in enzyme activity in the prepubertal testis. *International Journal of Andrology* 23, 218–224.

Buff, P.R., Messer, N.T., Cogswell, A.M., Wilson, D.A., Johnson, P.J., Keisler, D.H. and Ganjam, V.K. (2007) Induction of pulsatile secretion of leptin in horses following thyroidectomy. *Journal of Endocrinology* 193, 353–359.

Chilean Iodine Educational Bureau (1952) *Iodine Content of Foods*. CIEB, London, 183 pp.

Christopherson, R.J., Gonyon, H.W. and Thompson, J.R. (1979) Effects of temperature and feed intake on plasma concentrations of thyroid hormones in beef cattle. *Canadian Journal of Animal Science* 59, 655–661.

Clark, R.G. (1998) Thyroid histological changes in lambs from iodine supplemented and untreated ewes. *New Zealand Veterinary Journal* 46, 216–222.

Clark, R.G., Sargison, N.D., West, D.M. and Littlejohn, R.P. (1998) Recent information on iodine deficiency in New Zealand sheep flocks. *New Zealand Veterinary Journal* 46, 216–222.

Cromwell, G.L., Sihombing, D.T.H. and Hays, V.W. (1975) Effects of iodine level on performance and thyroid traits of growing pigs. *Journal of Animal Science* 41, 813–818.

Crush, J.R. and Caradus, J.R. (1995) Cyanogenesis potential and iodine concentration in white clover (*Trifolium repens* L.) cultivars. *New Zealand Journal of Agricultural Research* 38, 30–316.

Darras, V.M., Kotanen, S.P., Geris, K.L., Berghman, L.R. and Kuhn, E.R. (1996) Plasma thyroid hormone levels and iodothyronine deiodinase activity following acute glucocorticoid challenge in embryonic development compared with post-hatch chickens. *General and Comparative Endocrinology* 104, 203–212.

Donald, G.E., Langlands, J.P., Bowles, J.E. and Smith, A.J. (1993) Subclinical selenium insufficiency. 4. Effects of selenium, iodine and thiocyanate supplementation of grazing ewes on their selenium and iodine status and on the status and growth of their lambs. *Australian Journal of Experimental Agriculture* 33, 411–416.

Downer, J.V., Hemken, R.W., Fox, J.D. and Bull, L.J. (1981) Effect of dietary iodine on tissue iodine content of the bovine. *Journal of Animal Science* 52, 413–417.

Ellis, K.J. and Coverdale, O.R. (1982) The effects on new-born lambs of administering iodine to pregnant ewes. *Proceedings of the Australian Society of Animal Production* 14, 660.

Emanuelson, M. (1994) Problems associated with feeding rapeseed meal to dairy cows. In: Garnsworthy, P.C. and Cole, D.J.A. (eds) *Recent Advances in Animal Nutrition – 1994*. Nottingham University Press, Nottingham, UK, pp. 189–214.

Erenberg, A., Omori, K., Menkes, J.N., Oh, W. and Fisher, D.A. (1974) Growth and development of the thyroidectomised ovine foetus. *Pediatrics Research* 8, 783–789.

Ferguson, K.A., Schinckel, P.G., Carter, H.B. and Clarke, W.H. (1956) The influence of the thyroid on wool follicle development in the lamb. *Australian Journal of Biological Sciences* 9, 575–585.

Fish, R.E. and Swanson, E.W. (1983) Effects of excessive iodide administered in the dry period on thyroid function and health of dairy cows and their calves in the periparturient period. *Journal of Animal Science* 56, 162–172.

Fitzgerald, S. (1983) The use of forage crops for store lamb fattening. In: Haresign, W. (ed.) *Sheep Production*. Butterworths, London, pp. 239–286.

Follett, B.K. and Potts, C. (1990) Hypothyroidism affects reproductive refractoriness and the seasonal oestrus period in Welsh Mountain ewes. *Journal of Endocrinology* 127, 103–109.

Freyberger, A. and Ahr, H.-J. (2005) Studies on the goitrogenic mechanism of action of N,N,N',N' tetra-methylthiourea. *Toxicology* 217, 169–175.

Fuge, R. (2005) Soils and iodine deficiency. In: Selinus, O. and Alloway, B.J. (eds) *Essentials of Medical Geology*. British Geological Survey, Edinburgh, pp. 417–434.

Grace, N.D. (2006) Effect of ingestion of soil on the iodine, copper, cobalt (vitamin B_{12}) and selenium status of grazing sheep. *New Zealand Veterinary Journal* 54, 44–46.

Grace, N.D. and Waghorn, G.C. (2005) Impact of iodine supplementation of dairy cows on milk production and iodine concentrations in milk. *New Zealand Veterinary Journal* 53, 10–13.

Grace, N.D., Knowles, S.O. and Sinclair, G.R. (2001) Effect of pre-mating iodine supplementation of ewes fed pasture or a brassica crop pre-lambing on the incidence of goitre in newborn lambs. *Proceedings of the New Zealand Society of Animal Production* 61, 164–167.

Groppel, B. and Anke, M. (1986) Iodine content of feedstuffs, plants and drinking water in the GDR. In: Anke, M., Boumann, W., Braunich, H., Bruckner, B. and Groppel, B. (eds) *Spurenelement Symposium Proceedings*, Vol. 5. *Iodine*. Friedrich Schiller University, Jena, Germany, pp. 19–28.

Gurevich, G.P. (1964) Soil fertilization with coastal iodine sources as a prophylactic measure against endemic goiter. *Federation Proceedings* 23, T511–T514.

Haggard, D.L., Stowe, H.D., Conner, G.H. and Johnson, D.W. (1980) Immunological effects of experimental iodine toxicosis in young cattle. *American Journal of Veterinary Research* 41, 539–543.

Harington, C.R. (1953) *The Thyroid Gland, Its Chemistry and Physiology*. Oxford University Press, London.

Hartmans, J. (1974) Factors affecting the herbage iodine content. *Netherlands Journal of Agricultural Science* 22, 195–206.

Hemken, R.W. (1960) Iodine. *Journal of Dairy Science* 53, 1138–1143.

Hennessey, D.R. and Pritchard, R.K. (1981) Functioning of the thyroid gland in sheep infected with *Tricostrongylus colubroformis*. *Research in Veterinary Science* 30, 87–92.

Hercus, C.E., Benson, W.N. and Carter, C.L. (1925) Endemic goitre in New Zealand, and its relation to the soil-iodine. *Journal of Hygiene* 24, 321–402.

Hetzel, B.S. (1991) The international public health significance of iodine deficiency. In: Momcilovic, B. (ed.) *Proceedings of the Seventh International Symposium on Trace Elements in Man and Animals, Dubrovnik*. IMI, Zagreb, pp. 7-1–7-3.

Hetzel, B.S. and Welby, M.C. (1997) Iodine. In: O'Dell, B.L. and Sunde, R.A. (eds) *Handbook of Nutritionally Essential Mineral Elements*. Marcel Dekker, New York, pp. 557–582.

Hill, R. (1991) Rapeseed meal in the diet of ruminants. *Nutrition Abstracts and Reviews, Series B* 61, 139–155.

Hill, R., Vincent, I.C. and Thompson, J. (1990) The voluntary food intake and weight gain of lambs given concentrate foods containing rapeseed meals with a range of glucosinolate contents. *Animal Production,* 50, 587.

Hillman, D. and Curtis, A.R. (1980) Chronic iodine toxicity in dairy cattle: blood chemistry, leucocytes and milk iodine. *Journal of Dairy Science* 63, 55–63.

Hopkins, P.S. (1975) The development of the foetal ruminant. In: McDonald, I.W. and Warner, A.C.I. (eds) *Metabolism and Digestion in the Ruminant. Proceedings of the IVth International Symposium on Ruminant Physiology,* 7th edn. University of New England Publishing Unit, Armidale, Australia, pp. 1–14.

Howie, F., Walker, S.W., Akesson, B., Arthur, J.R. and Beckett, G.J. (1995) Thyroid extracellular glutathione peroxidase: a potential regulator of thyroid hormone synthesis. *Biochemical Journal* 308, 713–717.

Indergaard, M. and Minsaas, J. (1991) Animal and human nutrition. In: Gury, M.D. and Blunden, G. (eds) *Seaweed Resources in Europe: Uses and Potential.* John Wiley & Sons, New York.

Ingar, S.H. (1985) The thyroid gland. In: Wilson, J.D. and Foster, D.W. (eds) *Williams Textbook of Endocrinology.* Saunders, Philadelphia, Pennsylvania, pp. 682–815.

Irvine, C.H.G. (1967) Protein-bound iodine in the horse. *American Journal of Veterinary Research* 28, 1687–1692.

Iwarsson, K. (1973) Rapeseed meal as a protein supplement for dairy cows. 1. The influence of certain blood and milk parameters. *Acta Veterinaria Scandinavica* 14, 570–594.

Johns, A.T. (1956) The influence of high-production pasture on animal health. In: *Proceedings of the 7th International Grassland Congress, Palmerston North, New Zealand,* pp. 251–261.

Johnson, J.M. and Butler, G.W. (1957) Iodine content of pasture plants. 1. Method of determination and preliminary investigation of species. *Physiologia Plantarum* 10, 100–111.

Johnson, P.J., Messer, N.T., Ganjam, V.K., Thompson, D.L., Erefsal, K.R., Loch, W.E. and Ellersieck, M.R. (2003) Effects of propylthiouracil and bromocryptine on serum concentrations of thyrotropin and thyroid hormones in normal female horses. *Equine Veterinary Journal* 35, 296–301.

Jones, R.J. and Meggarity, R.G. (1983) Comparative responses of goats fed on *Leucena leucephola* in Australia and Hawaii. *Australian Journal of Agricultural Research* 34, 781–790.

Karmarkar, M.G., Deo, M.G., Kochupillai, N. and Ramalingaswami, V. (1974) Pathophysiology of Himalayan endemic goiter. *American Journal of Clinical Nutrition* 27, 96–103.

Kelly, F.C. (1953) Studies on the stability of iodine compounds in iodized salt. *Bulletin of the World Health Organization* 9, 217–230.

Khajarern, S., Khajarern, J.M., Kitpanit, N. and Muller, Z.O. (1977) Cassava in the nutrition of swine. In: Nestel, B. and Graham, M. (eds) *Cassava as Animal Feed. Proceedings of a Workshop Held at the University of Guelph.* International Development Research Centre, Ottawa, pp. 56–64.

King, C.F. (1976) Ovine congenital goitre associated with minimal thyroid enlargement. *Australian Journal of Experimental Agriculture and Animal Husbandry* 16, 651–655.

Knights, G.I., O'Rourke, P.K. and Hopkins, P.S. (1979) Effects of iodine supplementation of pregnant and lactating ewes on the growth and maturation of their offspring. *Australian Journal of Experimental Agriculture and Animal Husbandry* 19, 19–22.

Knowles, S.O. and Grace, N.D. (2007) A practical approach to managing the risks of iodine deficiency in flocks using thyroid-weight:birthweight ratios of lambs. *New Zealand Veterinary Journal* 55, 314–318.

Kulper, G.G., Kester, M.H., Peeters, R.P. and Visser, T.J. (2005) Biochemical mechanisms of thyroid hormone deiodination. *Thyroid* 15, 787–798.

Kushadet, M.M., Brown, A., Kurilich, A.C., Juvik, J.A., Klein, B.P., Wallig, M.A. and Jeffery, E.H. (1999) Variation in glucosinolates in vegetable crops of *Brassica oleracea. Journal of Agricultural and Food Chemistry* 47, 1541–1548.

Laarveld, B., Brockman, R.P. and Christensen, D.A. (1981) The effects of Tower and Midas rapeseed meals on milk production and concentrations of goitrogens and iodine in milk. *Canadian Journal of Animal Science* 61, 131–139.

Lengemann, F.W. (1979) Effect of low and high ambient temperatures on metabolism of radioiodine by the lactating goat. *Journal of Dairy Science* 62, 412–415.

Levine, H., Remington, R.E. and von Kolnitz, H. (1933) Studies on the relation of diet to goiter. 2. The iodine requirement of the rat. *Journal of Nutrition* 6, 347–354.

Lüdke, H. and Schöne, F. (1988) Copper and iodine in pigs diets with high glucosinolate rapeseed meal. I. Performance and thyroid hormone status. *Animal Feed Science and Technology* 22, 33–43.

Maas, J., Davis, L.E., Hempsteed, C., Berg, J.N. and Hoffman, K.A. (1984) Efficacy of ethylenediamine dihydriodide in the prevention of naturally occurring foot rot in cattle. *American Journal of Veterinary Research* 45, 2347–2350.

Maas, J., Berg, J.N. and Petersen, R.G. (1989) Serum distribution of iodine after oral administration of ethylenediamine dihydriodide in cattle. *American Journal of Veterinary Research* 50, 1758–1759.

Maddocks, S., Chandrasekhar, Y. and Setchell, B.P. (1985) Effect on wool growth of thyroxine replacement in thyroidectomised lambs. *Australian Journal of Biological Science* 38, 405–410.

MAFF (2000) Iodine in milk. *Food surveillance information sheet No. 198*, Joint Food Safety and Standards Group.

Marine, D. and Williams, W.W. (1908) The relation of iodine to the structure of the thyroid gland. *Archives of Internal Medicine* 1, 349–384.

Mason, R.W. (1976) Milk iodine content as an estimate of the dietary iodine status of sheep. *British Veterinary Journal* 132, 374–379.

Mason, R.W. and Laby, R. (1978) Prevention of ovine congenital goitre using iodine-releasing intramuscular devices: preliminary results. *Australian Journal of Experimental Agriculture and Animal Husbandry* 18, 653–657.

Mawson, R., Heaney, R.K., Piskula, M. and Kozlowska, H. (1993) Rapeseed meal-glucosinolates and their nutritional effects. Part I Rapeseed production and chemistry ogf glucosinolates. *Die Nahrung* 37, 131–140.

McCoy, M.A., Smyth, J.A., Ellis, W.A. and Kennedy, D.G. (1995) Parenteral iodine and selenium supplementation in stillbirth perinatal weak calf syndrome. *Veterinary Record* 136, 124–126.

McCoy, M.A., Smyth, J.A., Ellis, W.A., Arthur, J.R. and Kennedy, D.G. (1997) Experimental reproduction of iodine deficiency in cattle. *Veterinary Record* 141, 544–547.

McDonald, R.C., Manley, T.R., Barry, T.N., Forss, D.A. and Sinclair, A.G. (1981) Nutritional evaluation of kale (*Brassica oleracea*) diets. 3. Changes in plant composition induced by fertility practices, with special reference to SMCO and glucosinolate concentrations. *Journal of Agricultural Science, Cambridge* 19, 27–33.

McGowan, A.C. (1983) The use of 'Lipiodol' for sub-clinical iodine deficiency in livestock. *Proceedings of the New Zealand Society of Animal Production* 43, 135–136.

Mee, J.F., Rogers, P.A.M. and O'Farrell, K.J. (1995) Effect of feeding a mineral-vitamin supplement before calving on the calving performance of a trace element deficient dairy herd. *Veterinary Record* 137, 508–512.

Menendez, C., Baldelli, R., Camina, J.P., Escudero, B., Peino, R., Dieguez, C. and Casanueva, F.F. (2003) TSH stimulates leptin secretion by a direct effect on adipocytes. *Journal of Endocrinology* 176, 7–12.

Miller, E.R. and Ammerman, C.B. (1995) Iodine bioavailability. In: Ammerman, C.B., Baker, D.H. and Lewis, A.J. (eds) *Bioavailability of Nutrients for Animals*. Academic Press, New York, pp. 157–168.

Miller, J.K., Swanson, E.W. and Hansen, S.M. (1965) Effects of feeding potassium iodide, 3,5 diiodosalicylic acid or L-thyroxine on iodine metabolism of lactating dairy cows. *Journal of Dairy Science* 48, 888–894.

Miller, J.K., Swanson, E.W., Spalding, G.E., Lyke, W.A. and Hall, R.F. (1974) The role of the abomasum in recycling of iodine in the bovine. In: Hoekstra, W.G., Suttie, J.W., Ganther, H.E. and Mertz, W. (eds) *Proceedings of the Second International Symposium on Trace Elements in Man and Animals*. University Park Press, Baltimore, Maryland, pp. 638–640.

Mitchell, H.H. and McClure, F.J. (1937) *Mineral Nutrition of Farm Animals*. Bulletin No. 99, National Research Council, Washington, DC.

Mitchell, J.H., Nicol, F., Beckett, G.J. and Arthur, J.A. (1997) Selenium and iodine deficiencies: effects on brain and brown adipose tissue selenoenzyme activity and expression. *Journal of Endocrinology* 155, 255–263.

Mostyn, A., Sebert, S., Litten, J.C., Perkins, K.S., Laws, J., Symonds, M.E. and Clarke, L. (2006) Influence of porcine genotype on the abundance of thyroid hormones and leptin in sow milk and its impact on growth, metabolism and expression of key adipose tissue genes in offspring. *Journal of Endocrinology* 190, 631–639.

Newton, G.L., Barrick, E.R., Harvey, R.W. and Wise, M.B. (1974) Iodine toxicity, physiological effects of elevated dietary iodine on calves. *Journal of Animal Science* 38, 449–455.

Nazifi, S., Saeb, M., Rowghani, E. and Kaveh, K. (2003) The influences of thermal stress on serum biochemical parameters of Iranian fat-tailed sheep and their correlation with triiodothyronine (T_3), thyroxine (T_4) and cortisol concentrations. *Comparative Clinical Pathology* 12, 135–139.

Nicol, F., Lefrane, H., Arthur, J.R. and Trayhurn, P. (1994) Characterisation and postnatal development of 5′-deiodinase activity in goat perirenal fat. *American Journal of Physiology* 267, R144–R149.

NRC (1994) *Nutrient Requirements of Poultry*, 9th edn. National Academy of Sciences, Washington, DC.

NRC (1998) *Nutrient Requirements of Swine*, 10th edn. National Academy of Sciences, Washington, DC.

NRC (2005) *Mineral Tolerance of Animals*, 2nd edn. National Academy of Sciences, Washington, DC.

O'Connor, D.M., Blache, D., Hoggard, N., Brookes, E., Wooding, F.B.P., Fowden, A.L. and Forhead, A.J. (2007) Developmental control of plasma leptin meesenger ribonucleic acid in the ovine foetus during late gestation: role of glucocorticoids and thyroid hormones. *Endocrinology* 176, 7–12.

Orr, J.B. and Leitch, I. (1929) *Iodine in Nutrition. A Review of Existing Information*. Special Report Series, Medical Research Council, UK, No. 123, 108 pp.

Pattanaik, A.K., Khan, S.A., Varshney, V.P. and Bedi, S.P.S. (2001) Effect of iodine level in mustard (*Brassica juncea*) cake-based concentrate supplement on nutrient utilisation and serum thyroid hormones of goats. *Small Ruminant Research* 41, 51–59.

Pedersen, I.B., Laurberg, P., Knudsen, N., Jorgensen, T., Perrild, H., Ovesen, L. and Rasmussen, L.B. (2007) An increased incidence of overt hypothyroidism after iodine fortification of salt in Denmark: a prospective population study. *Journal of Clinical Endocrinology and Metabolism* 92, 3122–3127.

Pezzi, C., Accorsi, P.A., Vigo, D., Govoni, N. and Gaiani, R. (2003) 5′-Deiodinase activity and circulating thyronines in lactating cows. *Journal of Dairy Science* 86, 152–158.

Phillips, D.I.W., Nelson, M., Barker, D.J.P., Morris, J.A. and Wood, T.J. (1988) Iodine in milk and the incidence of thyrotoxicosis in England. *Clinical Endocrinology* 28, 61–66.

Potter, B.J., McIntosh, G.H. and Hetzel, B.S. (1981) The effect of iodine deficiency on foetal brain development in the sheep. In: Hetzel, B.S. and Smith, R.M. (eds) *Fetal Brain Disorders – Recent Approaches to the Problem of Mental Deficiency*. Elsevier/North Holland Biomedical Press, Amsterdam, pp. 119–147.

Potter, B.J., Mano, M.T., Belling, G.B., McIntosh, G.H., Hua, C., Cragy, B.G., Marshall, J., Wellby, M.L. and Hetzel, B.S. (1982) Retarded fetal brain development resulting from severe iodine deficiency in sheep. *Applied Neurobiology* 8, 303–313.

Prichard, R.K., Hennessy, D.R. and Griffiths, D.A. (1974) Endocrine responses of sheep to infection with *Trichostrongylus colubriformis*. *Research in Veterinary Science* 17, 182–187.

Rhind, S.M. and McMillen, S.R. (1996) Effects of methylthiouracil treatment on the growth and moult of cashmere fibre in goats. *Animal Science* 62, 513–520.

Rogers, P.A.M., Lynch, P.J., Porter, W.L. and Bell, G.D (1998) A slow-release iodine, selenium and cobalt cattle bolus. *Cattle Practice* 6, 129–132.

Rogler, J.C. (1958) Effects of iodine on hatchability and embryonic development. *Dissertation Abstracts* 18, 1925–1926.

Rowntree, J.E., Hill, G.M., Hawkins, D.R., Link, J.E., Rincker, M.J., Bednar, G.W. and Kreft, R.A. Jr (2004) Effect of selenium on selenoprotein activity and thyroid hormone metabolism in beef and dairy cows and calves. *Journal of Animal Science* 82, 2995–3005.

Sahin, K. and Kocuk, O. (2007) Selenium supplementation in heat-stressed poultry. *CAB Reviews: Perspectives in Agriculture, Veterinary Science and Natural Resources* 2, 1–8.

Sargison, N.D., West, D.M. and Clark, R.G. (1997) An investigation of the possible effects of subclinical iodine deficiency on ewe fertility and perinatal lamb mortality. *New Zealand Veterinary Journal* 45, 208–211.

Sargison, N.D., West, D.M. and Clark, R.G. (1998) Effects of iodine deficiency on ewe fertility and perinatal lamb mortality. *New Zealand Veterinary Journal* 46, 72–75.

Schöne, F., Lüdke, H., Hennig, A. and Jahreis, G. (1988) Copper and iodine in pig diets with high glucosidolate rapeseed meal. II. Influence of iodine supplements for rations with rapeseed meal untreated or treated with copper ions on performance and thyroid hormone status of growing pigs. *Animal Feed Science and Technology* 22, 45–59.

Schöne, F., Jahreis, G., Richter, G. and Lange, R. (1992) Evaluation of rapeseed meals: effects of iodine supply and glucosinolate degradation by myrosinase or copper. *Journal of the Science of Food and Agriculture* 61, 245–252.

Scott, M.L., van Tienhoven, A., Holm, E.R. and Reynolds, R.E. (1960) Studies on the sodium, chlorine and iodine requirements of pheasants and quail. *Journal of Nutrition* 71, 282–288.

Siebert, B.D. and Hunter, R.A. (1982) In: Hacker, C.B. (ed.) *Nutritional Limits to Animal Production from Pasture*. Commonwealth Agricultural Bureaux, Farnham Royal, UK, pp.409–425.

Sihombing, D.T.H., Cromwell, G.L. and Hays, V.W. (1974) Effects of protein source, goitrogens and iodine level on performance and thyroid status of pigs. *Journal of Animal Science* 39, 1106–1112.

Sinclair, D.P. and Andrews, E.D. (1958) Prevention of goitre in new-born lambs from kale-fed ewes. *New Zealand Veterinary Journal* 6, 87–95.

Sinclair, D.P. and Andrews, E.D. (1961) Deaths due to goitre in new-born lambs prevented bv iodized poppy-seed oil. *New Zealand Veterinary Journal* 9, 96–100.

Singh, J.L., Sharma, M.C., Kumar, M., Gupta, G.C. and Kumar, S. (2005) Immune status of goats in endemic goitre and its therapeutic treatment. *Small Ruminant Research* 63, 249–255.

Slebodzinski, A.B., Brzezinska-Slebodzinska, E., Nowak, J. and Kowalska, K. (1998) Triiodothyronine (T3), insulin and characteristics of 5'-monodeiodinase (5'-MD) in mare's milk from parturition to 21 days post-partum. *Reproduction Nutrition and Development* 38, 235–244.

Smyth, J.A., McNamee, P.T., Kennedy, D.G., McCullough, S.J., Logan, E.F. and Ellis, W.A. (1992) Stillbirth/ perinatal weak calf syndrome: preliminary pathological, microbiological and biochemical findings. *Veterinary Record* 130, 237–240.

Spiegel, C., Bestetti, G.E., Rossi, G.L. and Blum, J.W. (1993) Normal circulating tri-iodothyronine concentrations are maintained despite severe hypothyroidism in growing pigs fed rapeseed presscake meal. *Journal of Nutrition* 123, 1554–1561.

Statham, M. and Bray, A.C. (1975) Congenital goitre in sheep in southern Tasmania. *Australian Journal of Agricultural Research* 26, 751–768.

Statham, M. and Koen, T.B. (1981) Control of goitre in lambs by injection of ewes with iodized poppy seed oil. *Australian Journal of Experimental Agriculture and Animal Husbandry* 22, 29–34.

Stewart, A.G. and Pharaoh, P.O.D. (1996) Clinical and epidemiological correlates of iodine deficiency disorders. In: Appleton, J.D., Fuge, R. and McCall, G.J.H. (eds) *Environmental Geochemistry and Health*. Geological Society Special Publication No.113, London, pp. 223–230.

Stoewsand, G.S. (1995) Bioactive organosulfur phytochemicals in Brassica oleracea – a review. *Food Chemistry and Toxicology* 33, 537–543.

Sullivan, K.M., May, W., Nordenberg, D., Houston, R. and Maberly, G.F. (1997) Use of thyroid stimulating hormone testing in newborns to identify iodine deficiency. *Journal of Nutrition* 127, 55–58.

Sutcliffe, E. (1990) Iodine in New Zealand milk. *Food Technology in New Zealand* 25, 32–38.

Symonds, M.E. and Clarke, L. (1996) Influence of thyroid hormones and temperature on adipose tissue development and lung maturation. *Proceedings of the Nutrition Society* 55, 561–569.

Taraz, Z., Jalali, S.M.A. and Rafeie, F. (2006) Effects of replacement of soybean meal with rapeseed meal on organ weight, some blood biochemical parameters and performance of broiler chicks. *International Journal of Poultry Science,* 5, 1110–1115.

Todini, L. (2007) Thyroid hormones in small ruminants: effects of endogenous, environmental and nutritional factors. *Animal* 1, 997–1008.

Todini, L., Malfatti, A., Valbonesi, A., Trabalza-Marinucci, M. and Debenedetti, A. (2007) Plasma total T3 and T4 concentrations in goats at different physiological stages as affected by energy intake. *Small Ruminant Research* 68, 285–290.

Underwood, E.J. and Suttle, N.F. (1999) *The Mineral Nutrition of Livestock*, 3rd edn. CAB International, Wallingford, UK.

Villar, D., Rhind, S.M., Dicks, P., McMillen, S.R., Nicol, F. and Arthur, J.R. (1998) Effect of propylthiouracil-induced hypothyroidism on thyroid hormone profiles and tissue deiodinase activity in cashmere goats. *Small Ruminant Research* 29, 317–324.

Villar, D., Nicol, F., Arthur, J.R., Dicks, P., Cannavan, A., Kennedy, D.G. and Rhind, S.M. (2000) Type II and type III monodeiodinases in the skin of untreated and propylthiouracil-treated cashmere goats. *Research in Veterinary Science* 68, 119–123.

Voudouri, A., Chadio, S.E., Menegatos, J.G., Zervas, G.P., Nicol, F. and Arthur, J.R. (2003) Selenoenzyme activities in selenium- and iodine-deficient sheep. *Biological Trace Element Research* 94, 213–224.

Wallace, A.L.C., Gleeson, A.R., Hopkins, P.S., Mason. R.W. and White, R.R. (1978) Plasma thyroxine concentrations in grazing sheep in several areas of Australia. *Australian Journal of Biological Science* 31, 39–41.

Walton, E.A. and Humphrey, J.D. (1979) Endemic goitre of sheep in the highlands of Papua New Guinea. *Australian Veterinary Journal* 55, 43–44.

Wassen, F.W., Klootwijk, W., Kaptein, E., Duncker, D.J., Visser, T.J. and Kulper, G.G. (2004) Characteristics and thyroid state-dependent regulation of iodothyronine deiodinases in pigs. *Endocrinology* 145, 4251–4263.

Wheeler, J.L. and Vickery, P.J. (1989) Variation in HCN potential among cultivars of white clover (*Trifolium repens* L). *Grass and Forage Science,* 44, 107–109.

Whitehead, D.C. (1975) Uptake by perennial ryegrass of iodide, elemental iodine and iodate added to soil as influenced by various amendments. *Journal of the Science of Food and Agriculture* 26, 361–367.

Wichtel, J.J., Craigie, A.L., Freeman, D.A., Varela-Alvarez, H. and Williamson, N.B. (1996) Effect of selenium and iodine supplementation on growth rate and on thyroid and somatotropic function in dairy calves at pasture. *Journal of Dairy Science* 79, 1865–1872.

Wilson, J.G. (1975) Hypothyroidism in ruminants with special reference to foetal goitre. *Veterinary Record* 97, 161–164.

Zagrodzki, P., Nicol, F., McCoy, M.A., Smyth, J.A., Kennedy, D.G., Beckett, G.J. and Arthur, J.R. (1998) Iodine deficiency in cattle: compensatory changes in thyroidal selenoenzymes. *Research in Veterinary Science* 64, 209–211.

13 Iron

Iron is by far the most abundant trace element in the body and its value as a dietary constituent has been appreciated for over 2000 years. A connection between dietary iron supply and disorders of the blood was proposed in the 16th century, but its physiological basis was not discerned until horse haemoglobin (Hb) was shown to contain 0.335% iron (Zinoffsky, 1886) and the finding was confirmed in other species. Stockman (1893) ended a long controversy by showing that Hb concentrations in anaemic women could be rapidly increased by supplementation with salts as diverse as ferrous citrate and ferrous sulfide. Approximately 60% of iron in the body is present as Hb in the bloodstream, and rapid expansion of red cell mass greatly increases demands for iron and risk of anaemia in the first few weeks of life in many species of domestic animal, particularly the pig. For many years nutritional interest in iron was focused on its role in Hb formation, but a broader concept of the physiological significance of iron developed when its presence in the ubiquitous enzyme group of cytochrome oxidases (COX) was established. Iron supplies are rarely inadequate for older farm livestock, in which the major problem is presented by excess dietary iron. By contrast, iron-deficiency anaemia is the sixth most prevalent health risk to humans according to World Health Organization statistics (Stoltzfus, 2001), and most recent research on iron nutrition has been conducted with this in mind.

Functions of Iron

Hb is a complex of the protoporphyrin haem and globin. The haem molecule contains four atoms of iron, one in the centre of each of its four linked porphyrin rings. Hb is packaged in erythrocytes and allows the tissues to 'breathe' by carrying oxygen from the lungs to the tissues as oxyhaemoglobin via arterial blood, and returning carbon dioxide as carboxyhaemoglobin via the venous circulation. Myoglobin is a simpler, less abundant iron porphyrin found in muscle, where its higher affinity for oxygen completes the transfer of oxygen from Hb into the cell. The ability of iron to change between the divalent and trivalent state allows other haemoproteins, cytochromes a, b and c, to participate in the electron-transfer chain – on which cell respiration depends – as COX. The discovery of iron-containing flavoprotein enzymes has extended the core relationship of iron to other basic metabolic pathways. By activating or assisting enzymes such as succinate dehydrogenase, iron is involved at every stage of the tricarboxylic acid (Krebs) cycle, and iron-containing catalase and peroxidases remove potentially dangerous products of metabolism. Connective tissue development is influenced by iron-activated hydroxylases (O'Dell, 1981).

Dietary Sources of Iron

Most feeds for farm animals contain high though variable concentrations of iron, depending on the plant species, growing conditions and the degree of contamination from soil and other exogenous sources.

Grains and seeds

Most cereal grains contain 30–60 mg Fe kg^{-1} dry matter (DM) when sampled in the field and species differences appear to be small, although 10 and 20 mg kg^{-1} DM have been recorded for Egyptian-grown maize and barley, respectively (Abou-Hussein et al., 1970). Purchased cereals contain more iron and, in the UK, average values of 100, 100, 120 and 140 mg kg^{-1} DM have been reported for barley, maize, oats and wheat, respectively (MAFF, 1990). The higher values recorded for cereal by-products – 480 mg kg^{-1} DM in maize gluten feed, 220 mg kg^{-1} DM in wheat feed and 2600 mg kg^{-1} DM in rice-bran meal – reflect both the uneven distribution of iron in the grain and contamination during processing. Leguminous seeds and oilseed meals commonly contain 100–200 mg Fe kg^{-1} DM. The iron inherent in feeds is present as organic complexes, with some bound to phytate and some stored in phytosiderophores as ferritin. Wheat contains iron mostly as monoferric phytate (Morris and Ellis, 1976) and soybeans can contain substantial amounts as ferritin (Lonnerdahl, 2009).

Animal sources

Cow's milk in full lactation averages about 0.5 mg Fe kg^{-1} fresh weight (equivalent to only 4 mg kg^{-1} DM) and similar low iron concentrations are found in milks from other species, although sow's milk contains three times more, as does bovine colostrum (Tables 2.3 and 2.4). Most of the iron in secreted milks is present as the glycoprotein, lactoferrin. Feeds of animal origin, other than milk and milk products, are rich sources of iron and Liebscher (1958; cited by Kolb, 1963) gave the following representative figures: blood meal, 3108; fish meal, 381; meat meal, 439; dried skimmed milk, 52 mg kg^{-1} DM. When non-dairy products were included in cereal rations as protein supplements overall iron intakes were substantially increased, but their use is now restricted. Dried skimmed milk, whey and buttermilk powders used to vary greatly in iron content due to contamination during processing, but are now usually low in iron compared with most farm feeds.

Forages

Pasture iron shows marked seasonal fluctuations with peaks in spring and autumn, and iron values can range from 70–111 to 2300–3850 mg kg^{-1} DM in New Zealand (Campbell et al., 1974) and southern Australia. High values are attributable to soil contamination, which is most likely to occur on soils prone to waterlogging. Lower values of between 90 and 110 mg kg^{-1} DM have been reported for Egyptian green-cut clover and clover hay (Abou-Hussein et al., 1970) and values <30 mg kg^{-1} DM have been reported for some grasses on poor sandy soils in Australia (E.J. Underwood, unpublished data).

Absorption of Iron

The evaluation of feeds as iron sources using animals requires a preliminary understanding of the ways in which retention of iron is regulated. Neither urinary nor faecal excretion of iron plays any significant part in iron homeostasis (Kreutzer and Kirchgessner, 1991). In whole-body ^{59}Fe retention studies, little isotope is lost from the body once unabsorbed iron from a bolus dose has passed through the

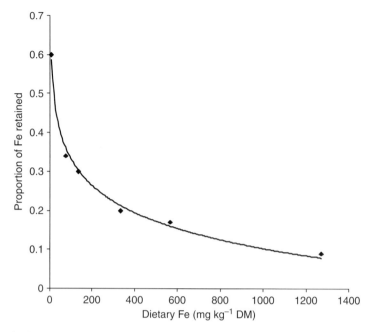

Fig.13.1. Iron is absorbed according to need and retention of ^{59}Fe from a labelled wheat meal fell exponentially as previous dietary iron concentration increased in rats (data from Fairweather-Tait and Wright, 1984).

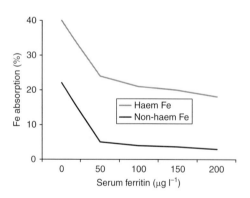

Fig. 13.2. The effect of body iron stores, as indicated by serum ferritin concentration, on the absorption of haem and non-haem iron in humans. Whole-body retention of Fe was measured after dual labelling with ^{55}Fe and ^{59}Fe in human volunteers. Subjects with the lowest iron stores absorbed Fe with the highest efficiency regardless of Fe source, but haem Fe was preferentially absorbed regardless of body stores (data from Roughead and Hunt, 2000).

digestive tract (Roughead and Hunt, 2000). Negligible endogenous losses of iron allow retention to be equated to absorption (Figs 13.1 and 13.2).

Absorption

Iron is absorbed primarily from the duodenum by a two-stage process involving mucosal uptake and serosal transfer. At the mucosal stage, haem and non-haem iron are processed and regulated differently. Haem is probably taken up by a haem carrier protein, HCP1, and its iron released by haem oxygenase into the intracellular pool (West and Oates, 2008). Non-haem iron in the Fe^{3+} state is reduced in the apical membrane and taken up by the divalent metal transporter (DMT1) (Garrick et al., 2003). Serosal transfer is accomplished by hephaestin (a copper-dependent homologue of caeruloplasmin) and a transport protein, ferroportin, in the basolateral membrane.

Iron is absorbed according to need and absorption (A_{Fe}) is apparently affected by factors such as age and the iron status of the body as well as by dietary source, but these factors are sometimes confounded. Thus A_{Fe} is maximal in young, iron-deprived animals such as the anaemic baby pig (Matrone et al., 1960; Ullrey et al., 1960) and during pregnancy. Expression of DMT1 in enterocytes is increased if animals are fed diets low in iron (Moos et al., 2002) and A_{Fe} is facilitated by the migration from the crypts of mucosal cells with increased iron-binding properties.

The sensitivity of mucosal control is demonstrated by data for ^{59}Fe retention following a standard test meal (Fig. 13.2): retention fell from 60% to 9% of the dose as the iron concentration of the previous diet increased from 8 to 1270 mg kg^{-1} DM. Feeding the lowest iron level for 24 h was sufficient to fully enhance absorption. The influence of body iron stores has been confirmed in human volunteers by a negative exponential relationship between ^{59}Fe retention and serum ferritin concentrations (Fig. 13.2). The same study confirmed the long-established superiority of haem iron over non-haem iron as an iron source, while showing that absorption from both sources was influenced by body iron status (Fig. 13.2). However, increases in dietary iron only decreased the efficiency of absorption of non-haem iron. Non-haem iron is trapped in the mucosa as ferritin, making little or no further progress at high iron intakes, and the shedding of iron-loaded mucosal cells causes faecal ferritin excretion to increase.

Nutritive Value of Iron in Feeds

The absorbability of iron in a feed ($^{max}A_{Fe}$) (i.e. its potential iron value) will be greater than the proportion absorbed when, as is often the case, iron is fed in excess of need. The major factors affecting $^{max}A_{Fe}$ are the chemical forms of iron present in the feed and the proportions of certain minerals and organic constituents that reduce or enhance A_{Fe}. Iron is well absorbed from human milk and this has been attributed partly to the high absorbability of lactoferrin and partly to the presence of two enhancers of iron absorption, citrate and lactose. However, infants absorb iron poorly from milk replacers based on bovine milk (Lonnerdal, 1988). The absorption of non-haem iron is greatly enhanced by ascorbic acid (e.g. Biesegel et al., 2007), while calcium (e.g. Minihane and Fairweather-Tait, 1998), phytate and polyphenols (e.g. Hurrell et al., 2003) reduce A_{Fe}. In view of the stimulatory effect of vitamin C on A_{Fe}, the relative biological value (RBV) of iron ascorbate has been tested in man and was found to exceed that of ferrous sulfate (FeSO$_4$): some laboratories now standardize RBV against the organic source. In vitro tests of iron value have the advantage of being immune from animal influences and now provide a closer simulation of digestion in vivo (Hemalatha et al., 2007), but they must be calibrated against results from animal tests and this has yet to be done in farm animals.

Milk iron

Matrone et al. (1960) estimated that 53–59% of inorganic iron added to cow's milk was utilized for Hb formation in anaemic baby pigs fed less iron than they required and, in similar circumstances, 72–82% of ingested iron was apparently absorbed (Ullrey et al., 1960). The similar iron requirements of baby pigs reared on condensed milk and solid diets (Hitchcock et al., 1974) suggests that A_{Fe} in the bovine milk replacer was similar to that in the corn/soybean meal (SBM) diet used. Of 100 mg Fe kg^{-1} DM added to a milk replacer (as FeSO$_4$), only 35% was apparently absorbed (Jenkins and Hidiroglou, 1987). Iron is probably well absorbed from the milk of sheep and goats, given that anaemia is rarely seen in suckled offspring. High vitamin C concentrations in the milks of grazing animals may enhance A_{Fe}, but pasteurization and dehydration of bovine milk may lower the A_{Fe} in milk replacers by destroying vitamin C. Young farm animals gradually eat more solid feed and drink less milk with the approach of weaning and nothing is known of the effects of this transition on A_{Fe}. However, de-phytinized cereal porridges made with bovine milk have been found to provide far less A_{Fe} to human volunteers than those made with water (Hurrell et al., 2003).

Could the 'culprit' be lactoferrin? Dosing calves with iron-saturated, bovine lactoferrin failed to prevent a post-natal fall in blood Hb, whereas the same amount of iron as $FeSO_4$ was effective (Kume and Tanabe, 1994).

Non-haem iron for chicks

The combined effects of three antagonists – calcium, phytate and polyphenols – may account for the low RBV (compared to $FeSO_4$) recorded in oilseed meals and derivatives given to chicks. Values of 0.39 and 0.56 have been found for soybean protein concentrate (SBC) (Biehl et al., 1997) and cottonseed meal (Boling et al., 1998), respectively. However, RBV for the SBC, added to a casein/dextrose diet, was not improved by the addition of phytase (Biehl et al., 1997). In cereals, A_{Fe} may vary between plant species. Wheat iron was surprisingly of equal value to iron chloride ($FeCl_3$) for chicks and this was attributed to the high A_{Fe} of monoferric phytate (Morris and Ellis, 1976). Furthermore, rats were able to retain 60% of the iron from their wheat meal in Fig. 13.1. However, an RBV of only 0.20 was accorded to ground maize (Chausow and Czarnecki-Maulden, 1988) and this is consistent with the low RBV generally found for maize-based diets in humans. Although iron is absorbed better from genetically modified (GM), low-phytate than high-phytate maize (Mendoza et al., 1998), there is no difference in A_{Fe} from inorganic iron supplements added to GM or wild-type maize in humans (Mendoza et al., 2001).

Non-haem iron for pigs

Little is known about the value of iron in pig feeds, but the addition of phytase has improved A_{Fe} in a corn/SBM ration for growing pigs (Stahl et al., 1999). The release of phosphorus and iron from SBM during in vitro incubation with phytase show different dose-dependencies (Stahl et al., 1999), and the release of zinc and iron from feeds by heat treatment is dissimilar (Hemalatha et al., 2007), suggesting a complex relationship between phytate and A_{Fe}. The effect of de-phytinization with phytase may depend on the source of phytate and the particular phytase used (e.g. acid or alkaline) and may vary between animal species.

Non-haem iron in ruminants

While the pre-formed, tetra-ferric phytate is poorly available to milk-fed calves (Bremner et al., 1976), weaned ruminants should be protected from any phytate × iron antagonism by the action of rumen microbial phytases (see Chapter 6) and those at pasture by a plentiful supply of vitamin C.

Haem iron

Haem iron is better absorbed than the iron in inorganic salts or non-haem complexes by humans (Fig. 13.2), but the degree of superiority (three- to fivefold) increases as body iron stores increase and may be overestimated if recent iron intake exceeds requirement, because more severely non-haem iron absorption is inhibited by high dietary or body iron (Roughead and Hunt, 2000). Low RBV (0.22–0.35 relative to $FeSO_4$) has been reported for three blood meals fed to chicks, but higher values (0.40 and 0.70) were reported for Hb per se (Henry and Miller, 1995), suggesting that heat processing of blood by-products may lower the RBV of haem iron. The availability of haem iron to humans is reduced by calcium supplements (Roughead et al., 2005), but haem iron now makes little contribution to livestock rations.

Iron Requirements

Requirements for dietary iron vary according to live weight (LW), dietary composition and the criterion of adequacy used, being higher for maximum Hb than growth because mild anaemia does not retard growth. Since the growth requirement for iron is predominantly for Hb and increase in red cell mass constitutes a progressively smaller component of

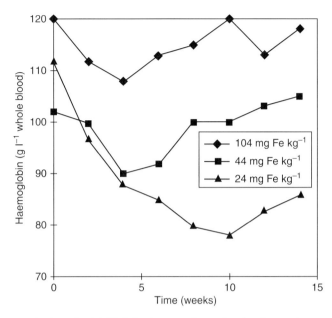

Fig. 13.3. Young animals given fixed, initially inadequate dietary iron levels can show spontaneous recoveries from any anaemia (data for calves from Bremner et al., 1976). This is because Fe requirements for each unit of weight gain are constant, while appetite increases as animals grow. The Fe concentration needed in the diet therefore decreases with age.

weight gain, needs in terms of dietary iron concentration will decline as individuals of all species grow to mature body weight. Spontaneous recoveries in blood Hb occur on diets initially low enough in iron to induce anaemia in calves (Fig. 13.3) and lambs (Bassett et al., 1995).

Pigs

Baby pigs became anaemic when reared on bovine milk containing 1.4 mg Fe kg^{-1} DM (Matrone et al., 1960) or a 'synthetic milk' containing 25 mg kg^{-1} DM (Ullrey et al., 1960) and needed between 40 and 100 mg iron to be added (from FeSO$_4$) for full growth and Hb repletion. Similarly, 50 mg supplemental iron kg^{-1} DM was needed for maximum live-weight gain (LWG), but 100 mg kg^{-1} DM increased Hb and serum iron in baby pigs fed on condensed cow's milk containing 2 mg Fe kg^{-1} DM. Requirements on milk and dry solid diets were similar, but were reduced if the solid diet was soaked (Hitchcock et al., 1974). A corn/SBM diet containing 52 mg kg^{-1} DM failed to replete anaemic baby pigs, but an additional 50 mg iron (as FeSO$_4$) allowed a rapid rise to a normal blood Hb (12 g l^{-1}) (Stahl et al. 1999). A linear relationship between LWG and supplementary iron (as glycinate) has been reported over the range 0–120 mg Fe kg^{-1} DM, but the response was maximal at the penultimate of three levels (90 mg kg^{-1} DM) in a 40-day trial starting at 7.4 kg LW (Feng et al., 2007). NRC (1998) requirements decreased from 100 mg Fe kg^{-1} DM for baby pigs to 40 mg Fe kg^{-1} for older, fattening pigs. This would be expected from a constant iron requirement for live-weight gain (LWG$_{Fe}$) and food intakes that increased from 1 to 2.5 kg day^{-1} between 20 and 120 kg LW. Needs for lactation are higher than in other mammals and the NRC (1998) recommends 80 mg Fe kg^{-1} DM for gilts and sows. Where copper sulfate (CuSO$_4$) is added to pig rations at the rate of 150–250 mg Cu kg^{-1} DM as a growth stimulant, iron (and zinc) requirements are increased (Suttle and Mills, 1966; Bradley et al., 1983) but it is hard to say by how much. Although iron supplements can detoxify gossypol and prevent poisoning on diets rich in cottonseed meal (Buitrago et al., 1970), iron requirements do not increase because gossypol does not impair iron metabolism.

Poultry

Early estimates of iron requirements for growth varied from 40 (Hill and Matrone, 1961) to 75–95 mg kg^{-1} DM (Fig. 13.4) on fibre and phytate-free diets and 75–80 mg kg^{-1} DM in one containing SBC (Davis et al., 1962) for the first 21–28 days of life. In the same period, a supplement of 38.5 mg iron (as FeSO$_4$) to a 'low-iron' diet containing 46.5 mg Fe kg^{-1} DM (mostly from SBC) was optimal for blood Hb (Aoyagi and Baker, 1995). Although growth rate increases substantially between 21 and 42 days, so too does food consumption, and the uniformly high provision of 80 mg Fe kg^{-1} DM for broilers up to 8 weeks of age recommended by the NRC (1994) is increasingly generous. The NRC's scaled iron reduction for turkey poults from 80 to 50 mg kg^{-1} DM between 4 and 16 weeks of age is nearer the mark. An average hen's egg contains about 1 mg iron and the daily demand for dietary iron therefore rises with the onset of lay, particularly with corn-based diets rich in calcium in which A_{Ca} may be as low as 0.1. The required iron concentrations for energy-rich and energy-lean diets given by the NRC (1994), 56 and 38 mg kg^{-1} DM, respectively, do not appear overgenerous. Commercial laying rations should cover these requirements, especially under free-range conditions or where ground limestone or oyster shell is on offer as these frequently contain 0.2–0.5% iron. However, dietary iron requirements for egg production remain ill-defined.

Sheep and cattle

Normal growth and blood Hb values were maintained in milk-fed calves given supplemental iron at 30 mg day^{-1} for 40 weeks from birth (Matrone et al., 1957). Iron at 40 mg kg^{-1} DM was sufficient to prevent all but a mild anaemia in veal calves, provided that the supplemental iron was in a soluble form (Bremner and Dalgarno, 1973). In experiments with weaned growing-finishing lambs, 10 mg kg^{-1} DM was inadequate and the minimum requirement lay between 25 and 40 mg kg^{-1} DM (Lawlor et al., 1965). Definitive studies with adult animals have not been reported but their requirements will be low, particularly during lactation because relatively little iron is secreted in milk. The additional need of a dairy cow yielding 40 kg milk day^{-1} is met by just 16 mg Fe kg^{-1} DM when an additional 10 kg DM is fed to sustain peak production.

Iron Transport and Storage

Assessing the effects of a given dietary iron supply on iron status requires a knowledge of the normal regulation of iron transport and storage.

Transport

Absorbed iron from both haem and non-haem sources enters a common mucosal pool prior

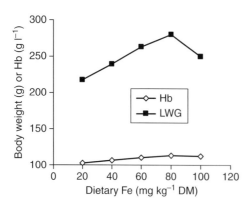

Fig. 13.4. Responses in body weight (LWG) and blood haemoglobin (Hb) to increasing dietary iron concentration in chicks given a semi-purified diet for 21 days. Body weight is the more sensitive index of requirement, which is 80 mg Fe kg^{-1} DM (data for group given 5 mg Cu kg^{-1} DM from McNaughton and Day, 1979).

to delivery at the serosal surface, where it is oxidized to Fe^{3+} and binds to apotransferrin, a non-haem glycoprotein that binds two atoms of iron per mole. The principal mucosal ferroxidases are the molybdeno-enzyme xanthine dehydrogenase and the cuproporotein hephaestin. Caeruloplasmin may also function as a ferroxidase and facilitate iron transfer, although its contribution is probably smaller than was once thought. Circulating transferrin provides the total iron-binding capacity (TIBC) of plasma and the degree of unsaturation (UIBC) reflects the proportion of iron-free apotransferrin present. Receptors for transferrin in cell membranes carry iron to the cell interior by endocytosis, where it is reduced and incorporated into functional forms. However, up-regulation of DMT1 in the erythron during iron deprivation and expression of isoforms in the human placenta and embryo (Chong et al., 2005) suggest that there are other routes of entry for iron into cells. Transferrin is also involved in the recycling of iron from aged erythrocytes via the reticuloendothelial system (Liebold and Guo, 1992). There, regulatory iron-responsive binding proteins, sensitive to free iron concentrations, receive iron from transferrin and partition it between functional and storage pathways. An important new regulatory protein, hepcidin, has recently come to the fore. Hepcidin is synthesized mainly in the liver. It is released into plasma and controls both iron absorption and systemic recycling (Ganz 2004; Loreal et al., 2005). However, it is responsive not only to iron status, but also to anoxia and inflammation.

Storage

Ferritin and haemosiderin are the main iron-storage compounds of the body. Ferritin is a non-haem protein (globulin) that can take up to 2000 iron atoms into its molecular core, and is found particularly in the gut and liver. Haemosiderin is the predominant storage form when a high iron status is attained. It contains 35% iron, primarily in a colloidal form as ferric hydroxide, and probably results from the aggregation of ferritin molecules and the subsequent denaturation of their protein constituent (Kent and Bahu, 1979). A high positive correlation between serum ferritin concentrations and body iron stores exists in humans, with $1 \mu g \, l^{-1}$ equivalent to approximately 8–10 mg stored iron, so that ferritin estimation is a useful indicator of iron stores (Baynes, 1997). Ferritin is not the inert molecule it was once thought to be, and there are 'gated' pores or channels in membranes down which ferritin can flow (Liu and Thiel, 2005).

Biochemical and Physiological Manifestations of Iron Deprivation

The general sequence of changes that occur when animals fail to consume the required amounts of iron is illustrated in Fig. 13.5.

Depletion

The first change is a depletion of iron stored as ferritin or haemosiderin in the liver, kidneys, spleen and mucosa. These changes are accompanied by decreases in serum ferritin. The duration of the depletion phase is determined by the size of the initial iron reserve – which can vary from 20 to 540 mg iron in the newborn calf (Charpentier et al., 1966, cited by ARC, 1980) – and the shortfall in daily dietary iron supply. In a depletion experiment with calves given a milk diet containing only 5 mg Fe kg^{-1} DM, the blood iron pool actually *increased* by 750 mg and the muscle iron pool by 250 mg in 3 months, with the liver and diet contributing iron in roughly equal measure. In the depletion phase, a marked redistribution of body iron takes place.

Deficiency

The second phase is marked by a fall in serum iron and raised TIBC (and UIBC) in the serum. Initially anaemic piglets given only 25 mg Fe kg^{-1} DM showed a further, steady decline in blood Hb and mean cell volume (MCV) and a smaller delayed fall in mean corpuscular Hb

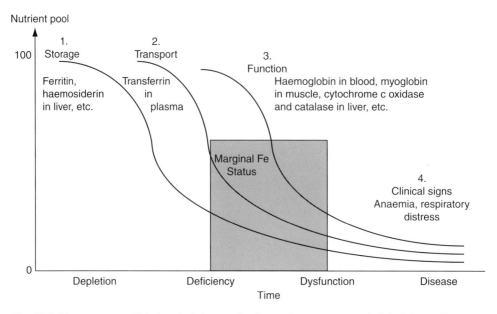

Fig. 13.5. The sequence of biochemical changes leading to the appearance of clinical signs of iron deprivation.

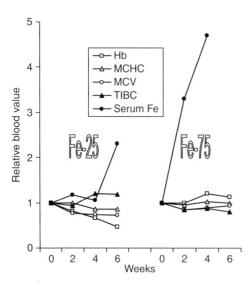

Fig. 13.6. Development and reversal of iron deprivation in anaemic piglets given a synthetic milk diet containing 25 or 75 mg Fe kg^{-1} DM. Changes are shown relative to initial values of haemoglobin (Hb), 73 g l^{-1}; mean corpuscular Hb concentration (MCHC), 30.2%; mean cell volume (MCV), 50.6 u3; serum Fe, 220 μg l^{-1}; total iron-binding capacity (TIBC), 6.56 mg l^{-1} (data from Ullrey et al., 1960).

concentration (MCHC), while the addition of 50 mg Fe kg^{-1} DM reversed those effects (Fig. 13.6). Hypoferraemia is the most sensitive index of deficiency, but this only becomes clear in the contrast between the two diets, being masked by a late rise in serum iron in the low-iron group. Increases in serum transferrin receptors indicate a concerted drive to maintain iron status (Baynes, 1997). In iron-deprived pigs and calves, a reduction in blood catalase activity precedes a fall in Hb (Grassmann and Kirchgessner, 1973), but a surplus of activity of this and other enzymes, such as COX, may explain why enzyme activities decrease before growth is retarded (ARC, 1980). Synthesis of haemoglobin and myoglobin fails to maintain their concentrations in blood and muscle, respectively, but growth may still not be retarded in calves (ARC, 1980), lambs (Bassett et al., 1995) or pigs (ARC, 1981). Mild anaemia (70 g Hb l^{-1}) did not impair oxygen transport or consumption and did not increase the heart rate in penned calves during unaccustomed exercise (Bremner et al., 1976). At the molecular level, the expression of DMT1 and hepcidin increase.

Dysfunction

Dysfunction is indicated when the growth rate decreases. This may not occur until the activity of iron-dependent enzymes is so diminished that anaerobic glycolysis and lactate–glucose recycling increase, leading to inefficient glucose utilization (Hostettler-Allen et al., 1993). The order and extent of biochemical changes varies between species. In iron-deprived pigs, myoglobin concentrations and COX activities are much reduced (Gubler et al., 1957), but in calves myoglobin has priority over Hb for scarce supplies of iron (ARC, 1980). In iron-deprived chicks, Hb values are reduced earlier than muscle myoglobin or liver COX and SDH activities (Davis et al., 1962). Metallothionein concentrations in the blood cells increase, particularly in the reticulocytes, during iron deprivation, reflecting an increase in erythropoietic activity (Robertson et al., 1989). Iron-deprived, mildly anaemic animals may absorb increased amounts of potentially toxic metals (e.g. lead and cadmium (see Chapter 18)) and essential elements (e.g. copper and manganese (see Chapters 11 and 14)) as secondary consequences of the increase in mucosal DMT1.

Clinical Manifestations of Iron Deprivation

Anaemia

Prolonged iron deprivation causes loss of appetite, poor growth, lethargy, blanching of the visible mucous membranes, an increased respiration rate and, when severe, high mortality. These signs are preceded and largely caused by the development of anaemia, shown by a marked fall in packed cell volume (PCV or haematocrit). In the study illustrated in Fig. 13.6, PCV fell from an initial 24.3% to 13.3%, only one-third of the normal value. The anaemia is hypochromic and microcytic because newly released erythrocytes are small and have less than the normal concentration of Hb (Fig. 13.6), compounding the effects of a decrease in the number of erythrocytes in circulation. Unseen lies a normoblastic, hyperplastic bone marrow containing little haemosiderin. The early onset of anaemia provides a point of contrast with the late development of anaemia in cobalt and copper deprivation (Chapters 10 and 11). Piglets are particularly vulnerable and piglet anaemia was long ago recognized as an uncomplicated iron-responsive condition (McGowan and Crichton, 1924; Hart et al., 1930). Sometimes referred to as 'thumps' because of the laboured spasmodic breathing, the condition usually develops within 2–4 weeks of birth, by which time Hb may have fallen to as low as $30–40\,g\,l^{-1}$ and PCV is greatly reduced. Mortality is high, but survivors begin a slow spontaneous recovery at about 6–7 weeks of age. Anaemic piglets are listless and flabby; their appetite is poor and their growth subnormal. Their skins become wrinkly and their hair rough, and the incidence of diarrhoea is high. Severe, debilitating anaemia can be produced experimentally in calves within 8–10 weeks by rearing them exclusively on cow's milk (Blaxter et al., 1957).

Cardiac hypertrophy

Post-mortem examination of anaemic piglets shows the heart to be dilated, with an excess of pericardial fluid, and the lungs to be oedematous (Seamer, 1956). Cardiac hypertrophy is found in all livestock species after iron deprivation, including the chick (Aoyagi and Baker, 1995). Hypertrophy probably results from a combination of increased heart rate and defective cellular metabolism, but can arise from other causes such as copper deprivation.

Hypothyroidism

Evidence of hypothyroidism has been reported in severely iron-deprived rats (PCV 16% as against a normal 41%) and thermoregulation was impaired when they were cold-stressed (Beard et al., 1989). The simultaneous treatment of infants with iron and iodine is more effective in treating goitre than treatment with

Cognitive dysfunction

Iron-deficiency anaemia in children has been associated with loss of cognitive function, but a thorough review of studies with children and animals concluded that the primary defect was in motor activity (McCann and Ames, 2007). However, studies in mice indicated that a 'marginal' iron intake lowered brain iron and diminished startle responses and maze performance. Furthermore, the defects were not completely reversed by subsequent iron supplementation (Kwik-Uribe et al., 2000). The anaemia seen in many farm animals shortly after birth may occasionally have residual effects on behaviour, and spontaneous recovery in iron status may not always prevent long-term cognitive dysfunction.

Susceptibility to infection

Decreased resistance to infection in iron deprivation has been described and the complex underlying mechanisms extensively and critically reviewed (Weinberg, 1984; Herschko, 1993). Both innate and cellular immunity have been compromised by iron deprivation, but the degree of anaemia established in experiments exceeds that likely to be found on farms and mild iron deficiency can occasionally improve resistance to infection (Hershko, 1993). Iron deprivation alters the balance between pro- and anti-inflammatory cytokines, significantly reducing interleukin-10 and interferon-δ (Kuvibidia and Warrier, 2004), and its effect on the course of a specific infection may depend partly on the degree to which a particular pathogen stimulates an inflammatory response. Iron-deprived piglets exhibit greater susceptibility to the endotoxin of Escherichia coli (Osborne and Davis, 1968) than healthy animals.

Occurrence of Iron Deprivation

Iron deprivation is of limited practical significance in farm livestock other than piglets. Primary iron deprivation has never been demonstrated unequivocally in grazing animals, a reflection of the normally high inherent and contaminant iron contents of pastures. Confinement increases the possibility of iron deprivation in young sucking animals or animals reared largely on a diet of milk or milk products, when extraneous sources of iron are minimal. High egg production in poultry is sometimes accompanied by anaemia, but it is not iron responsive and is therefore not a consequence of iron deprivation.

Piglet anaemia

Severe anaemia occurs wherever pigs are housed in concrete pens and denied access to soil or pasture unless preventive measures are taken, but does not occur in pigs reared outdoors. Newborn piglets absorb colostrum proteins intact into the bloodstream and plasma volume increases markedly during the first 12 h of nursing without significant change in red blood cell volume causing a physiological anaemia (Underwood and Suttle, 1999). This is compounded by an iron deficit brought about by a combination of factors: unusually low body iron stores at birth, compared with newborn of other species; absence of the polycythaemia of birth common to other animal species; large litter size; very rapid early growth rate compared with that of the lamb or calf. Piglet live weight can increase fourfold by 3 weeks and 10-fold by 8 weeks of age, requiring the retention of 7–11 mg Fe day^{-1}, whereas only 1 mg day^{-1} is obtained from milk alone (Venn et al., 1947). Older pigs do not develop iron-responsive anaemia unless copper is used to stimulate growth.

Lamb and calf anaemia

Anaemia in lambs and calves is far less prevalent and less severe than in piglets, but a transient, mild anaemia has been reported in young, naturally reared calves (30% <90g Hb l⁻¹) (Hibbs et al., 1963; Atyabi et al., 2006) and in lambs reared indoors (Ullrey et al., 1965; Green et al., 1993, 1997) and responds to iron supplementation in both species (Carleson et al., 1961; Ullrey et al., 1965; Kume and Tanabe, 1994). Twin calves are more likely to develop anaemia than single calves because they compete for a limited maternal supply of iron (Kume and Tanabe, 1994) and the same may be true of twin lambs. Weight gain was not improved by iron supplementation in one study with lambs, even when the anaemia in untreated lambs (minimum Hb value $81\,g\,l^{-1}$) lasted for 4 weeks (Bassett et al., 1995), but weaning weight was improved by 1 kg in another larger study with 525 lambs (Green et al., 1997) and also increased in the calves studied by Carleson et al. (1961). The variable growth responses of anaemic populations probably reflect the small proportion of clinically affected individuals (e.g. 4–14%) (Green et al., 1993). The use of vegetable protein concentrates in milk replacers and urea to replace vegetable protein in the rations of weaned ruminants may lower iron status.

Veal production

The induction of anaemia has been regarded as a necessary feature of veal production in which energy-rich milk substitutes low in iron are fed to closely confined calves to produce 'white' meat. The tolerance of anaemia in such animals is demonstrated by the fact that Hb values can fall to 50% of normal while rapid growth rates are sustained. However, the practice raises serious welfare questions as well as problems in sustaining a marginal iron supply for animals whose need for iron, in terms of dietary concentration, is steadily decreasing. An alternative practice involving lesser iron deprivation, more freedom of movement and the production of 'pink meat' has gained support (Moran, 1990).

Infections as a cause of anaemia

Microbial pathogens also need iron and they have iron-responsive binding proteins in their outer membranes to ensure that they get it (Hershko, 1993). Host defence mechanisms include the production of antibodies to the iron-responsive binding proteins of pathogens such as *Pasteurella haemolytica* (Confer et al., 1995). Iron is redistributed in an attempt to deplete the pathogen of iron, with serum ferritin increasing and plasma iron decreasing as part of the acute-phase response (Weinberg, 1984; Herschko, 1993). This causes a secondary anaemia by depriving the erythropoietic tissues of iron. The UIBC of lactoferrin in milk increases during infections such as mastitis, and bactericidal properties associated with iron deprivation have been confirmed in mice (Terraguchi et al., 1995) and weanling pigs (Wang et al., 2006). Parenteral administration of complexed iron has enhanced survival in chicks infected with fowl typhoid (*Salmonella gallinarum*) (Smith and Hill, 1978). Parasitic infestation involving severe blood loss can also produce a secondary iron-deficiency anaemia. Such effects have been reported for various parasites, including *Bunostomum* and *Trichostrongylidae* (Kolb, 1963). Infection by a blood-sucking parasite of the ovine abomasum (*Haemonchus contortus*) has been found to cause losses of up to 24 ml erythrocytes per day in lambs (Abbot et al., 1988).

Diagnosis of Iron-responsive Disorders

Whole blood criteria

The risk for disorder is initially assessed by measuring blood Hb and/or the highly correlated PCV, and a positive diagnosis is strengthened by the presence of low MCHC (calculated by dividing Hb by PCV) and low MCV (calculated by dividing PCV by erythrocyte count). Normal Hb values are commonly stated as 110–120 g l⁻¹ in pigs, poultry and cattle and 100–110 g l⁻¹ in sheep and goats, but these values vary with age. In healthy pigs, Hb values decline from 125 to 85 g l⁻¹ between birth and 6–8 weeks of age, despite iron supplementation, but gradually increase to 135 g l⁻¹ after 5–6 months (Miller, 1961). In calves, the postnatal fall can last for 2 months (Atyabi et al.,

2006). Sheep show a lesser fall and incomplete recovery of birth values (Ullrey et al., 1965). Sex differences are insignificant except in poultry, where slightly higher Hb values occur in cocks than in hens, particularly those in lay. Some reduction in Hb (and PCV) can be tolerated (e.g. in piglets) (Schrama et al., 1997), but values <70 g l^{-1} are a cause for concern and values 50–60% below normal induce clinically obvious pallor (e.g. in lambs) (Green et al., 1993).

Serum and liver iron

Low serum iron and increased TIBC and UIBC in serum can provide additional confirmation of iron deficiency, but individual variability in normal values can be large, as indicated by the means and extreme values reported in Table 13.1. There are significant differences between the cow and her newborn calf, the latter having a higher TIBC that can remain elevated for up to 2 months. By contrast, serum iron is similar to the maternal value at birth but declines thereafter (Atyabi et al., 2006). However, these changes may be partly developmental rather than indicative of iron deprivation. Iron deprivation would also be indicated by low liver iron: a marginal band of 75–100 mg kg^{-1} DM separates the deprived from the normal piglet (Ullrey et al., 1960) and similar bands may be applicable in other species. Serum ferritin is of limited diagnostic value in all species because values become minimal before anaemia develops. Serum iron values in poultry vary with age and egg laying, averaging 2.25 mg l^{-1} (range 1.73–2.90) in immature pullets and rising to 6.7 to 9.4 mg l^{-1} at the onset of lay (Ramsay and Campbell 1954); diagnostic limits must change accordingly (Table 13.2).

Differential diagnosis

Diagnostic guidelines are given in Table 13.2, but it is important to distinguish anaemia caused by dietary iron deprivation and that associated with infection or malnutrition. Anaemia of infection would be indicated by high rather than low serum ferritin, normal MCHC and normal MCV: there may be evidence of a febrile response, including raised blood levels of other acute-phase proteins and, in parasitic infections, faecal excretion of oocysts or eggs. Green et al. (1993) reported that abnormally high plasma copper coupled with coccidial oocysts in faeces was suggestive of infection in lamb anaemia.

Treatment of Iron Deprivation

Injection of iron by the intramuscular route is widely practised (Patterson and Allen, 1972) using an iron-dextran or a dextrin–ferric oxide compound of high stability to minimize muscle damage. In piglets, a dose of 2 ml iron-dextran (200 mg iron) at 3 days old restores Hb values to those found at birth and maintains them throughout the nursing period. A smaller dose (100 mg) may delay the recovery

Table 13.1. Normal serum iron, total iron-binding capacity (TIBC) and percentage unsaturation (UIBC) in livestock (from Underwood and Morgan, 1963; Kaneko, 1993; Puls, 1994).

	Serum Fe (µmol l^{-1})	TIBC (µmol l^{-1})	UIBC (%)
Cattle	17.4 ± 5.2	40.8 ± 10.0	74
Sheep	34.5 ± 1.25	59.8 ± 3.2	49
Goats	17–36[a]	NA	NA
Pigs	21.7 ± 5.9	56.8 ± 6.8	53
Poultry	25–50[a]	NA	NA
Horses	13–40[a]	59.1 ± 5.7	66

NA, not available.
[a]Data from Puls (1994).

Table 13.2. Marginal bands for assessing the risk of iron deprivation (D) or excess (E)[a] from population means for biochemical criteria in domestic livestock.

Livestock	D/E	Diet Fe (mg kg^{-1} DM)	Liver Fe (mmol kg^{-1} DM)[b]	Serum Fe (µmol l^{-1})[b]	Serum ferritin (µg l^{-1})
Cattle	D	40–60	1.79–2.68	8.9–17.9	10–30
	E	1000–**4000**	>17.9	10.7–**32.2**	>**80**
Pigs	D^N	60–80	0.54–0.89	2.7–10.7	5–10
	D^A	30–50	1.79–2.68	14.3–17.9	35–55
	E	1000–**2500**	>17.9	>**26.7**	–
Poultry	D	35–45	1.79–2.68	10–15	–
	E	200–**2000**	17.9–**107.2**	>53.6	–
Sheep	D	30–50	1.25–1.79	15–20	–
	E	600–**1200**	>17.9	>**39**	–

A, adult; N, neonate. Marginal values are probably lower for neonates than for adults of other species. The presence of individual values close to the **bold** limit in a sampled group with a normal mean value may justify intervention.
[a]Tolerable E values will be far higher if contaminant soil iron contributes largely to the dietary iron level.
[b]Multiply by 55.9 to obtain liver values in mg kg^{-1} DM and serum values in µg l^{-1}.

in Hb without significantly reducing the growth rate, appetite or antibody responses to vaccination (Schrama et al., 1997), and an additional injection at 21 days has been found to confer no benefits when the pre-starter ration contains 80 mg Fe kg^{-1} DM (Bruininx et al., 2000). Anaemia in the neonatal lamb (Ullrey et al., 1965) and calf (Carleston et al., 1961) is controlled in a similar manner to piglet anaemia, with intramuscular doses of 200 mg iron being used for lambs (Green et al., 1993, 1997) and 500 mg for calves (Hibbs et al., 1963). Daily oral supplementation with 20–40 mg iron as FeSO$_4$ is also effective in newborn calves (Kume and Tanabe, 1994).

Prevention of Iron Deprivation

Oral route

The earliest and simplest preventive procedure for piglet anaemia is to provide a small amount of soil or pasture sods that the piglets can consume regularly. Supplementary iron can be given orally, but must be given within 2–4 days of age and usually again at 10–14 days. Inorganic iron given to the sow during pregnancy and lactation cannot be relied upon to significantly increase iron stores of the piglet at birth or the iron content of the sow's milk (ARC, 1981), but supplementation of the sow's diet with 2 g Fe kg^{-1} DM as FeSO$_4$ can control piglet anaemia when the offspring have access to the mother's faeces (Gleed and Sansom, 1982). The anaemia of weaned pigs given high dietary levels of copper (250 mg kg^{-1} DM) as a growth stimulant can be prevented by adding FeSO$_4$ to such rations (Gipp et al., 1974; Bradley et al., 1983), with 150 mg Fe kg^{-1} DM (with zinc) sufficient to offset the effect of 450 mg Cu kg^{-1} DM in one study (Suttle and Mills, 1966).

Inorganic sources

Inorganic iron salts vary little in availability provided they are soluble, including ferric salts that were once believed to be of intrinsically poor availability. Ferric ammonium citrate and ferric succinate have marginally higher RBV than FeSO$_4$ (119 and 107 respectively versus 100) and are infinitely superior to FeCO$_3$ (RBV 2) for chicks (Fritz et al., 1970), but FeCO$_3$ has occasionally been accorded a higher value (Henry and Miller, 1995). Ferric oxide (Fe$_2$O$_3$), used as a colouring agent in mineral mixes, is among the poorest of inorganic iron sources although it is capable of impairing A_{Cu} (Suttle and Peter, 1985) and should not be used. Many calcium, phosphorus and magnesium sources are heavily contaminated with iron, sometimes indicated by a rust-coloured appearance, but the RBV is exceedingly low, while di- and monocalcium phosphates contain iron with an RBV of 0.62 (Deming and Czarnecki-Maulden, 1989). Soluble inorganic sources of iron have the disadvantage that they promote lipid oxidation and can introduce unwanted sensory changes to a feed (Hurrell, 2002). Milk replacers can be supplemented with micro-encapsulated $FeSO_4$ to minimize adverse interactions with lipids (Lysionek et al., 2001), but cost may prohibit use for livestock. Soluble inorganic iron need not and should not be added to compound livestock feeds beyond the 'starter' stage in any livestock rearing programme, particularly in hot and/or humid environments.

Organic sources

The provision of iron in chelated forms such as deferoxamine and deferiprone plays an important role in the treatment and prevention of iron-responsive disorders in humans (Herschko et al., 2005) and NaFeEDTA is far superior to $FeSO_4$ (Mendoza et al., 2001). However, there is no convincing evidence in peer-reviewed papers that either specific amino acid or mixed peptide iron chelates have a higher RBV than $FeSO_4$ for livestock. Early, unconfirmed reports that giving such chelates to sows during gestation and lactation promote growth, Hb formation and liver iron stores of the piglets by raising sows' milk iron may be partially attributed to consumption by the piglet of iron-rich sows' faeces and may occur whatever iron source is used (Underwood and Suttle, 1999). A novel iron complex has been claimed to have a higher value than $FeSO_4$ for growing pigs (Yu et al., 2000), but design of the relevant experiment has made comparison difficult. In chicks, iron methionine has been accorded a low RBV of 0.88 (Cao et al.,1996). In the context of human nutrition, there is no convincing evidence that iron glycinate possesses a high RBV (Fox et al. 1998; Hurrell, 2002). Ferrous ascorbate has a high RBV, but the fumarate is of equal value to $FeSO_4$, though less likely to taint food (Hurrell, 2002). The most effective way of enhancing A_{Fe} is to attack organic antagonists. Antagonism from phytate can probably be reduced by supplements of citric acid.

Toxicity of Iron

Mechanisms

Free iron is cytotoxic because of its high reduction–oxidation potential and ability to generate free oxygen radicals via Haber–Weiss and Fenton reactions (Jenkins and Kramer, 1988). For this reason it is transported and stored tightly bound to protein. When tissue stores become excessive during chronic iron overload, sufficient reactive iron may be present to cause peroxidative damage at sites such as the liver (Kent and Bahu, 1979). The underlying pathogenic mechanism is peroxidative damage to lipid membranes (Gordeuk et al., 1987) and susceptibility to iron overload increases with age (Wu et al., 1990).

Interaction with vitamin E and polyunsaturated fatty acids

The risk of iron-induced peroxidation will depend on the antioxidant status of the animal and particularly its vitamin E status (Ibrahim et al., 1997). Vitamin E deficiency increases susceptibility to iron toxicity following the intramuscular injection of iron in piglets (Patterson and Allen, 1972) and conversely iron overload will reduce liver concentrations of vitamin E (Omara and Blakley, 1993). Iron-induced lipid peroxidation is also increased by the ingestion of readily peroxidized lipid sources such as polyunsaturated fatty acids (PUFA) (Muntane et al., 1995; Ibrahim et al., 1997). Pasture iron concentrations can rise to high levels in spring and autumn and the ingestion of iron-rich soil with PUFA-rich grass in spring may present a hazard for grazing animals.

Adventitious sources of iron

High liver iron may arise from exposure to plant toxins (e.g. lupinosis) (Gardiner, 1961), from the accelerated breakdown of erythrocytes after the haemolytic crisis in copper poisoning and presumably in conditions such as vitamin E deficiency or brassica poisoning, in which erythrocyte lifespan is shortened. Copper deprivation also results in hepatic iron accumulation (see Chapter 11).

Tolerance of dietary iron

There is a high tolerance towards dietary iron in all species. Protection is afforded by the down-regulation of hepcidin, which blocks iron transport out of the mucosal cell and indirectly down-regulates iron uptake by DMT1. Tolerance of excess iron will therefore probably be reduced if the diet is low in copper or manganese because DMT1 is up-regulated in such circumstances (see Chapters 11 and 14). Calves reared on milk substitute have tolerated Fe (as $FeSO_4$) at 2 g kg^{-1} DM for 6 weeks, but spleen and liver iron were considerably raised (to 1.30 and 1.55 g kg^{-1} DM, respectively) and a further increase of iron to 5 g kg^{-1} DM depressed appetite and growth (Jenkins and Hidiroglou, 1987). Tolerance of iron has been exploited by giving pigs up to 3.2 g Fe kg^{-1} DM for 4 months to combat gossypol poisoning; they did not develop iron toxicity (Buitrago et al., 1970). Similarly massive daily doses (up to 0.1 g kg^{-1} LW) have been used in sheep to combat sporidesmin intoxication (Munday and Manns, 1989); the mechanism of protection is believed to involve inhibition of copper absorption. Tolerable iron concentrations for livestock, based on short-term studies, have been set as high as 1000 mg kg^{-1} DM in cattle and poultry and 3000 mg kg^{-1} DM in pigs (NRC, 1980) but 800 mg kg^{-1} DM can depress appetite in the newly hatched chick (Cao et al., 1996) and 1 g kg^{-1} DM (as citrate) has been seen to depress LWG in lambs (Rosa et al., 1986). As with all cumulative poisonings, the total available dietary iron intake rather than concentration will determine the risk of chronic disorder. If iron is present in soluble forms, a marginal band of 0.75–1.25 g kg^{-1} DM separates acceptable from potentially harmful concentrations for weaned ruminants and poultry.

For pigs the equivalent band is 2500–3500 mg kg^{-1} DM. Where exogenous iron sources of low RBV such as oxides, hydroxides, carbonate or soil provide most of the iron, tolerable levels will be much higher.

Lower iron concentrations may cause liver injury in the longer term, particularly if vitamin E status is low or if there are endogenous sources of excess iron (NRC, 2005).

Adverse interactions

The grazing animal cannot avoid ingesting excess iron in the soil that contaminates relatively bare pasture in both wet and dry conditions. The diarrhoea that developed on pastures irrigated with iron-rich bore water in New Zealand was associated with copper depletion (Campbell et al., 1974). Excess iron down-regulates DMT1 expression in the duodenal mucosa and this is likely to cause secondary inhibition of other elements that are partially (e.g. copper) or wholly (e.g. manganese) dependent on the same carrier protein for absorption (Hansen and Spears, 2008). Although ingested soil iron is in a predominantly inert and insoluble form, sufficient iron can be retained to have a hepatotoxic effect in weaned lambs when concentrations in the liver reach 1 g kg^{-1} DM (N.F. Suttle, unpublished data). In calves on milk replacer, peroxidative changes occur in heart and muscle rather than in liver lipids after iron exposure (Jenkins and Kramer, 1988). High iron intakes have not always caused reductions in liver copper in calves (McGuire et al., 1985) or sheep and have even increased the toxicity of copper to sheep (Rosa et al., 1986), possibly because the diets were low in sulfur (see Chapter 11).

Diagnosis

The clinical signs for chronic iron toxicity will differ from those associated with acute iron toxicity, namely anorexia, vascular congestion and irritation in the gastrointestinal tract. No abnormalities were recorded at post-mortem examination by Jenkins and Hidiroglou (1987), who concluded that the primary cause of poor performance in calves given milk replacers high in iron was poor appetite and feed conversion efficiency. In a study with calves, blood indices such as serum iron, TIBC and UIBC did not reflect excessive iron intakes (1 g kg^{-1} DM) well (McGuire et al., 1985), and reliance must be placed on high liver iron in association with hepatopathy.

References

Abbott, E.M., Parkins, J.J. and Holmes, P.J. (1988) Influence of dietary protein on the pathophysiology of haemonchosis in lambs given continuous infections. *Research in Veterinary Science* 45, 41–49.

Abou-Hussein, E.R.M., Raafat, M.A., Abou-Raya, A.K. and Shalaby, A.S. (1970) Manganese, iron and cobalt content in common Egyptian feedstuffs. *United Arab Republic Journal of Animal Production* 10, 245–254.

ARC (1980) *The Nutrient Requirements of Ruminants*. Commonwealth Agricultural Bureaux, Farnham Royal, UK, pp. 184–185.

ARC (1981) *The Nutrient Requirements of Pigs*. Commonwealth Agricultural Bureaux, Farnham Royal, UK, pp. 215–248.

Atyabi, N., Gharagozloo, F. and Nassiri, S. (2006) The necessity of iron supplementation for normal development of commercially reared suckling calves. *Comparative Clinical Pathology* 15, 165–168.

Aoyagi, S. and Baker, D.H. (1995) Iron requirements of chicks fed a semi-purified diet based on casein and soy protein concentrate. *Poultry Science,* 74, 412–415.

Bassett, J.M., Burrett, R.A., Hanson, C., Parsons, R. and Wolfensohn, S.E. (1995) Anaemia in housed lambs. *Veterinary Record* 136, 137–140.

Baynes, R.D. (1997) The soluble form of the transferrin receptor and iron status. In: Fisher, P.W.F., L'Abbé, M.R., Cockell, K.A. and Gibson, R.S. (eds) *Trace Elements in Man and Animals – 9 – Proceedings of the Ninth International Symposium*. NRC Research Press, Ottawa, pp. 471–475.

Beard, J., Tobin, B. and Green, W. (1989) Evidence for thyroid hormone deficiency in iron-deficient, anaemic rats. *Journal of Nutrition* 119, 772–778.

Biehl, R.R., Emmert, J.L. and Baker, D.H. (1997) Iron bioavailability in soybean meal as affected by supplemental phytase and 1 alpha-hydoxycholecalciferol. *Poultry Science* 76, 1424–1427.

Biesegel, J.M., Hunt, J.R., Glahn, R.P., Welch, R.M., Menkir, A. and Maziya-Dixon, B.B. (2007) Iron bioavailability from maize and beans: a comparison of human measurements with Caco-2 cell and algorithm predictions. *American Journal of Clinical Nutrition* 86, 388–396.

Blaxter, K.L., Sharman, K.L. and MacDonald, A.M. (1957) Iron deficiency anaemia in calves. *British Journal of Nutrition* 11, 234–246.

Boling, S.D., Edwards, H.M. 3rd, Emmert, J.L., Biehl, R.R. and Bafer, D.H. (1998) Bioavailability of iron in cottonseed meal, ferric sulfate and two ferrous sulfate by-products of the galvanising industry. *Poultry Science* 77, 1388–1392.

Bradley, B.D., Graber, G., Condon, R.J. and Frobish, L.T. (1983) Effects of graded levels of dietary copper on copper and iron concentrations in swine tissues. *Journal of Animal Science* 56, 625–630.

Bremner, I. and Dalgarno, A.C. (1973) Iron metabolism in the veal calf. The availability of different iron compounds. *British Journal of Nutrition* 29, 229–243.

Bremner, I., Brockway, J.M., Donnelly, H.T. and Webster, A.J.F. (1976) Anaemia and veal calf production. *Veterinary Record* 99, 203–205.

Buitrago, J.A., Clawson, A.J. and Smith, F.H. (1970) Effects of dietary iron on gossypol accumulation in and elimination from porcine liver. *Journal of Animal Science* 31, 554–558.

Bruininx, E.M.A.M., Swinkels, J.W.G.M., Parmentier, H.K., Jetten, C.W.J., Gentry, J.L. and Schrama, J.W. (2000) Effects of an additional iron injection on growth and humoral immunity of weanling pigs. *Livestock Production Science* 67, 31–39.

Campbell, A.G., Coup, M.R., Bishop, W.H. and Wright, D.E. (1974) Effect of elevated iron intake on the copper status of grazing cattle. *New Zealand Journal of Agricultural Research* 17, 393–399.

Cao, J., Lo, X.J., Henry, P.R., Ammerman, C.B., Littell, R.C. and Miles, D. (1996) Effect of dietary iron concentration, age and length of iron feeding on feed intake and tissue iron concentrations of broiler chicks for use as a bioassay of supplemental iron sources. *Poultry Science* 75, 495–504.

Carleson, R.H., Swenson, M.J., Ward, G.M. and Booth, N.H. (1961) Effects of intramuscular iron dextran in newborn lambs and calves. *Journal of the American Veterinary Medical Association* 139, 457–461.

Chausow, D.G. and Czarnecki-Maulden, G.L. (1988) The relative bioavailability of iron from feedstuffs of plant and animal origin to the chick. *Nutrition Research* 8, 175–179.

Chong, W.S., Kwan, P.C., Chan, P.Y., Chiu, T.K., Cheung, T.K. and Lau, T.K. (2005) Expression of divalent metal transporter1 (DMT1) isoforms in first trimester human placenta and embryonic tissues. *Human Reproduction* 20, 3532–3538.

Confer, A.W., McGraw, R.D., Dierham, J.A., Morton, R.J. and Panciera, R.J. (1995) Serum antibody responses of cattle to iron-regulated outer membrane proteins of *Pasteurella haemolytica*. *Veterinary Immunology and Immunopathology* 47, 101–110.

Davis, P.N., Norris, L.C. and Kratzer, F.H. (1962) Iron deficiency studies in chicks using treated isolated soybean protein diets. *Journal of Nutrition* 78, 445–453.

Deming, J.G. and Czarnecki-Maulden, G.L. (1989) Iron bioavailability in calcium and phosphorus sources. *Journal of Animal Science* 67 (Suppl. 1), 253 (abstract).

Fairweather-Tait, S.J. and Wright, A.J.A. (1984) The influence of previous iron intake on the estimation of bioavailability of iron from a test meal given to rats. *British Journal of Nutrition* 51, 185–191.

Feng, J., Ma, W.Q., Xu, Z.R., Wang, Y.Z. and Liu, J.X. (2007) Effects of iron glycine chelate on growth, haematological and immunological characteristics in weanling pigs. *Animal Feed Science and Technology* 134, 261–272.

Fox, T.E., Eagles, J. and Fairweather-Tait, S.J. (1998) Bioavailability of iron glycine as a fortificant in foods. *American Journal of Clinical Nutrition* 67, 664–668.

Fritz, J.C., Pla, G.W., Roberts, T., Boehne, J.W. and Hove, E.L. (1970) Biological availability in animals of iron from common dietary sources. *Journal of Agricultural and Food Chemistry* 18, 647–652.

Ganz, T. (2004) Hepcidin in iron metabolism. *Current Opinions in Haematology* 11, 251–254.

Gardiner, M.R. (1961) Lupinosis – an iron storage disease of sheep. *Australian Veterinary Journal* 37, 135–140.

Garrick, M.D., Dolan, K.G., Horbinski, C., Ghoio, A.J., Higgins, D., Porcubin, M., Moore, E.G., Hainsworth, L.N., Umbreit, J.N., Conrad, M.E., Feng, L., Lis, A., Roth, J.A., Singleton, S. and Garrick, L.M. (2003) DMT 1: a mammalian transporter for multiple metals. *Biometals* 16, 41–54.

Gipp, W.F., Pond, W.G., Kallfelz, F.A., Tasker, J.B., Campen, D.R., van Krook, L. and Visek, W.J. (1974) Effect of dietary copper, iron and ascorbic acid levels on hematology, blood and tissue copper, iron and zinc concentrations and copper and iron metabolism in young pigs. *Journal of Nutrition* 104, 532–541.

Gleed, P.T. and Sansom, B.F. (1982) Ingestion of iron in sows faeces by piglets reared in farrowing crates with slatted floors. *British Journal of Nutrition* 47, 113–117.

Gordeuk, V.R., Bacon, B.R. and Brittenham, G.M. (1987) Iron overload: cause and consequences. *Annual Reviews of Nutrition* 7, 485–508.

Grassmann, E. and Kirchgessner, M. (1973) Katalase Aktivitat des Blutes von Saugferkeln und Mastkalbern bei mangelnder Eisenversorgung. *Zentralblatt fur Veterinarmedizin A* 20, 481–486.

Green, L.E., Berriatua, E. and Morgan, K.L. (1993) Anaemia in housed lambs. *Research in Veterinary Science* 54, 306–311.

Green, L.E., Graham, M. and Morgan, K.L. (1997) Preliminary study of the effect of iron dextran on a non-regenerative anaemia of housed lambs. *Veterinary Record* 140, 219–222.

Gubler, C.J., Cartwright, G.E. and Wintrobe, M.M. (1957) Studies on copper metabolism. 20. Enzyme activities and iron metabolism in copper and iron deficiencies. *Journal of Biological Chemistry* 224, 533–546.

Hansen, S.L. and Spears, J.W. (2008) Impact of copper deficiency in cattle on proteins involved in iron metabolism. *FASEB Journal* 22, 443–445.

Hart, E.B., Elvehjem, C.A. and Steenbock, H. (1930) A study of the anemia of young pigs and its prevention. *Journal of Nutrition* 2, 277–294.

Hemalatha, S., Platel, K. and Srinivasin, K. (2007) Influence of heat processing on the bioaccessibility of zinc and iron from cereals and pulses consumed in India. *Journal of Trace Elements in Medicine and Biology* 21, 1–7.

Henry, P.R. and Miller, E.R. (1995) Iron bioavailability. In: Ammerman, C.B., Baker, D.H. and Lewis, A.J. (eds) *Bioavailability of Nutrients for Animals*. Academic Press, New York, pp. 169–200.

Hershko, C. (1993) Iron, infection and immunity. *Proceedings of the Nutrition Society* 52, 165–174.

Hershko, C., Link, C., Konijn, A.M. and Cabantchik, Z.I. (2005) Objectives and mechanism of iron chelation therapy. *Annals of the New York Academy of Sciences* 1054, 124–135.

Hibbs, J.R., Conrad, H.R., Vandersall, J.H. and Gale, C. (1963) Occurrence of iron deficiency anaemia in dairy calves at birth and its alleviation by iron dextran injection. *Journal of Dairy Science* 46, 1118–1124.

Hill, C.H. and Matrone, G. (1961) Studies on copper and iron deficiencies in growing chickens. *Journal of Nutrition* 73, 425–431.

Hitchcock, J.P., Ku, P.K. and Miller, E.R. (1974) Factors influencing iron utilization by the baby pig. In: Hoekstra, W.G., Suttie, J.W., Ganther, H.E. and Mertz, W. (eds) *Trace Element Metabolism in Animals – 2*. University Park Press, Baltimore, Maryland, pp. 598–600.

Hostettler-Allen, R., Tappy, L. and Blum, J.W. (1993) Enhanced insulin-dependent glucose utilization in iron-deficient veal calves. *Journal of Nutrition* 123, 1656–1667.

Hurrell, R.F. (2002) Fortification: overcoming technical and practical barriers. *Journal of Nutrition* 132, 806S–812S.

Hurrell, R.F., Reddy, M.B., Juillerat, M.-A. and Cook, J.D. (2003) Degradation of phytic acid in cereal porridges by human subjects. *American Journal of Clinical Nutrition* 77, 1213–1219.

Ibrahim, W., Lee, V.-S., Ye, C.-C., Szabo, J., Bruckner, G. and Chow, C.K. (1997) Oxidative stress and antioxidant status in mouse liver: effects of dietary lipid, vitamin E and iron. *Journal of Nutrition* 127, 1401–1406.

Jenkins, K. and Hidiroglou, M. (1987) Effect of excess iron in milk replacer on calf performance. *Journal of Dairy Science* 70, 2349–2354.

Jenkins, K. and Kramer, J.K.G. (1988) Effect of excess dietary iron on lipid composition of calf liver, heart and skeletal muscle. *Journal of Dairy Science* 71, 435–441.

Kaneko, J.J. (1993) *Clinical Biochemistry of Domestic Animal*, 4th edn. Academic Press, New York.

Kent, G. and Bahu, R.M. (1979) Iron overload. In: MacSween, R.N.M., Anthony, P.P. and Schewr, P.J. (eds) *Pathology of the Liver*. Churchill Livingstone, Edinburgh, pp. 148–163.

Kolb, E. (1963) The metabolism of iron in farm animals under normal and pathologic conditions. *Advances in Veterinary Science and Comparative Medicine* 8, 49–114.

Kreutzer, M. and Kirchgessner, M. (1991) Endogenous iron excretion: a quantitative means to control iron metabolism. *Biological Trace Element Research* 29, 77–92.

Kume, S.-E. and Tanabe, S. (1994) Effect of twining and supplemental iron-saturated lactoferrin on iron status of newborn calves. *Journal of Dairy Science* 77, 3118–3123.

Kuvibidi, S. and Warrier, R.P. (2004) Differential effects of iron deficiency and underfeeding on serum levels of interleukin-10, interleukin-12 p40 and interferon-gamma in mice. *Cytokine* 26, 73–81.

Kwik-Uribe, C.L., Golub, M.S. and Keen, C.L. (2000) Chronic marginal iron intakes during early development in mice alter brain iron concentrations and behaviour despite postnatal iron supplementation. *Journal of Nutrition* 130, 2040–2048.

Lawlor, M.J., Smith, W.H. and Beeson, W.M. (1965) Iron requirement of the growing lamb. *Journal of Animal Science* 24, 742–747.

Liebold, E.A. and Guo, B. (1992) Iron-dependent regulation of ferritin and transferrin receptor expression by the iron-responsive element binding protein. *Annual Review of Nutrition* 12, 345–368.

Liu, X. and Thiel, E.C. (2005) Ferritin as an iron concentrator and chelator target. *Annals of the New York Academy of Sciences* 1054, 136–140.

Lonnerdal, B. (1988) Trace elements in infancy: a supply/demand perspective. In: Hurley, L.S., Keen, C.L., Lonnerdal, B. and Rucker, R.B. (eds) *Proceedings of Sixth International Symposium on Trace Elements in Man and Animals*. Plenum Press, New York, pp. 189–195.

Lonnerdal, B. (2009) Soybean ferritin: implications for iron status of vegetarians. *American Journal of Clinical Nutrition* 89, 1680S–1685S.

Loreal, O., Haziza-Pigeon, C., Troadec, M-B., Detvaud, L., Turlin, B., Courselaud, B., Ilyin, G. and Brissot, P. (2005) Hepcidin in iron metabolism. *Current Protein and Peptide Science* 6, 279–291.

Lysionek, A.E., Zubilliga, M.B., Saigueiro, M.J., Caro, R.A., Weill, R. and Boccio, J.R. (2001) Bioavailability study of dried microencapsulated ferrous sulfate – SFE 171 – by means of a prophylactic-preventive method. *Journal of Trace Elements in Medicine and Biology* 15, 255–259.

MAFF (1990) *UK Tables of the Nutritive Value and Chemical Composition of Foodstuffs*. In: Given, D.I. (ed.) Rowett Research Services, Aberdeen, UK.

Matrone, G., Conley, C., Wise, G.H. and Waugh, R.K. (1957) A study of iron and copper requirements of dairy calves. *Journal of Dairy Science* 40, 1437–1447.

Matrone, G., Thomason, E.L. Jr and Bunn, C.R. (1960) Requirement and utilization of iron by the baby pig. *Journal of Nutrition* 72, 459–465.

McCann, J.C. and Ames, B.N. (2007) An overview of evidence for a causal relation between iron deficiency during development and deficits in cognitive behaviour. *American Journal of Clinical Nutrition* 85, 931–945.

McGowan, J.P. and Crichton, A. (1924) Iron deficiency in pigs. *Biochemical Journal* 18, 265–272.

McGuire, S.O., Miller, W.J., Gentry, R.P., Neathery, M.W., Ho, S.Y. and Blackmon, D.M. (1985) Influence of high dietary iron as ferrous carbonate and ferrous sulfate on iron metabolism in young calves. *Journal of Dairy Science* 68, 2621–2628.

McNaughton, J.L. and Day, E.J. (1979) Effect of dietary Fe:Cu ratios on haematological and growth responses of broiler chickens. *Journal of Nutrition* 109, 559–564.

Mendoza, C., Viteri, F.E., Lonnerdahl, B., Raboy, V., Young, K.A. and Brown, K.H. (2001) Absorption of iron from unmodified maize and genetically altered low-phytate maize fortified with ferrous sulfate or sodium iron EDTA. *American Journal of Clinical Nutrition* 73, 80–85.

Mendoza, C., Viteri, F.E., Lonnerdahl, B., Young, K.A., Raboy, V. and Brown, K.H. (1998) Effect of genetically modified low-phytic acid maize on absorption of iron from tortillas. *American Journal of Clinical Nutrition* 68, 1123–1127.

Miller, E.R. (1961) Sheep haematology from birth to maturity. *Journal of Animal Science* 20, 890–897.

Minihane, A.M. and Fairweather-Tait, S.J. (1998) Effect of calcium supplementation on daily nonheme-iron absorption and long-term iron status. *American Journal of Clinical Nutrition* 68, 96–102.

Moos, T., Trinder, D. and Morgan, E.H. (2002) The effect of iron status on DMT 1 expression in duodenal enterocytes from β_2-microglogulin knock-out mice. *American Journal of Physiology. Gastrointestinal and Liver Physiology* 46, G687–G694.

Moran, J. (1990) Growing calves for pink veal. A guide to rearing, feeding and managing calves in Victoria. *Department of Agriculture Technical Report* 176, Melbourne.

Morris, E.R. and Ellis, R. (1976) Isolation of monoferric phytate from wheat bran and its biological value as an iron source to the rat. *Journal of Nutrition* 106, 753–760.

Munday, R. and Manns, E. (1989) Protection by iron salts against sporidesmin intoxication in sheep. *New Zealand Veterinary Journal* 37, 65–68.

Muntare, J., Mitjavila, M.T., Rodriguez, M.C., Puig-Parellada, P., Fernandez, Y. and Mitjavila, S. (1995) Dietary lipid and iron status modulate lipid peroxidation in rats with induced adjuvant arthritis. *Journal of Nutrition* 125, 1930–1937.

NRC (1994) *Nutrient Requirements of Poultry*, 9th edn. National Academy of Sciences, Washington, DC.

NRC 1998) *Nutrient Requirements of Swine*, 10th edn. National Academy of Sciences, Washington, DC.

NRC (2001) *Nutrient Requirements of Dairy Cows*, 7th edn. National Academy of Sciences, Washington, DC.

NRC (2005) *Mineral Tolerance of Animals*, 2nd edn. National Academy of Sciences, Washington, DC.

Omara, F.O. and Blakley, B.R. (1993) Vitamin E is protective against iron toxicity and iron-induced hepatic vitamin E depletion in mice. *Journal of Nutrition* 123, 1649–1655.

Osborne, J.C. and Davis, J.W. (1968) Increased susceptibility to bacterial endotoxin of pigs with iron-deficiency anemia. *Journal of the American Veterinary Medical Association* 152, 1630–1632.

Patterson, D.S.P. and Allen, W.M. (1972) Biochemical aspects of some pig muscle disorders. *British Veterinary Journal* 128, 101–111.

Ramsay, W.N.M. and Campbell, E.A. (1954) Iron metabolism in the laying hen. *Biochemical Journal* 58, 313–317.

Robertson, A., Morrison, J.N., Wood, A.M. and Bremner, I. (1989) Effects of iron deficiency on metallothionein I concentrations in blood and tissues of rats. *Journal of Nutrition* 119, 439–445.

Rosa, I.V., Ammerman, C.B. and Henry, P.R. (1986) Interrelationships between dietary copper, zinc and iron on performance and tissue mineral concentration in sheep. *Nutrition Reports International* 34, 893–902.

Roughead, Z.K. and Hunt, J.R. (2000) Adaptation in iron absorption: iron supplementation reduces non-haem Fe but not haem-Fe absorption. *American Journal of Clinical Nutrition* 72, 982–989.

Roughead, Z.K., Zito, C.A. and Hunt, J.R. (2005) Inhibitory effects of dietary calcium on the initial uptake and subsequent retention of haem and non-haem Fe in humans: comparisons using an intestinal lavage method. *American Journal of Clinical Nutrition* 82, 589–597.

Schrama, J.W., Schouten, J.M., Swinkels, J.W.G.M., Gentry, J.L., Reiling, G. de V. and Parmentier, M.K. (1997) Effect of haemoglobin status on humoral immune response of weanling pigs of different coping styles. *Journal of Animal Science* 75, 2588–2596.

Seamer, J. (1956) Piglet anaemia. A review of the literature. *Veterinary Reviews and Annotations* 2, 79–93.
Smith, I.M. and Hill, R. (1978) The effect on experimental mouse typhoid of chelated iron preparations in the diet. In: Kirchgessner, M. (ed.) *Proceedings of Third International Symposium on Trace Elements in Man and Animals*. Arbitskreis Tierernährungsforchung, Weihenstephan, Germany, pp. 383–386.
Stahl, C.H., Han, Y.M., Roneker, R.R., House, W.A. and Lei, X.G. (1999) Phytase improves iron bioavailability for haemoglobin synthesis in young pigs. *Journal of Animal Science* 77, 2135–2142.
Stockman, R. (1893) The treatment of chlorosis by iron and some other drugs. *British Medical Journal* 1, 881–885, 942–944.
Stoltzfus, R.J. (2001) Defining iron-deficiency anemia in public health terms: time for reflection. *Journal of Nutrition* 131, 565S–567S.
Suttle, N.F. and Mills, C.F. (1966) Studies of the toxicity of copper to pigs. 1. Effects of oral supplements of zinc and iron salts on the development of copper toxicosis. *British Journal of Nutrition* 20, 135–148.
Suttle, N.F. and Peter, D.W. (1985) Rumen sulphide metabolism as a major determinant of the availability of copper to ruminants. In: Mills, C.F., Bremner, I. and Chesters J.K. (eds) *Proceedings of the Fifth International Symposium on Trace Elements in Man and Animals, Aberdeen*. Commonwealth Agricultural Bureau, Farnham Royal, UK, pp. 367–370.
Terraguchi, S., Shin, K., Ozawa, K., Nakamura, S., Fukuwatori, Y., Tsyuki, S., Namahira, H. and Shimanura, S. (1995) Bacteriostatic effect of orally administered bovine lacto-ferrin on proliferation of *Clostridium* species in the gut of mice fed on bovine milk. *Applied and Environmental Microbiology* 61, 501–506.
Ullrey, D.E., Miller, E.R., Thompson, O.A., Ackermann, I.M., Schmidt, D.A., Hoefer, J.A. and Luecke, R.W. (1960) The requirement of the baby pig for orally administered iron. *Journal of Nutrition* 70, 187–192.
Ullrey, D.E., Miller, E.R., Long, C.H. and Vincent, B.H. (1965) Sheep haematology from birth to maturity. 1. Erythrocyte population, size and haemoglobin concentration. *Journal of Animal Science* 24, 141–145.
Underwood, E.J. and Morgan, E.H. (1963) Iron in ruminant nutrition. 1. Liver storage iron, plasma iron and total iron-binding capacity levels in normal adult sheep and cattle. *Australian Journal of Experimental Biology and Medical Science* 41, 247–253.
Underwood, E.J. and Suttle, N.F. (1999) Iron. In: *The Mineral Nutrition of Livestock*, 3rd edn. CAB International, Wallingford, UK, pp 375–396.
Venn, J.A.J., McCance, R.A. and Widdowson, E.M. (1947) Iron metabolism in piglet anemia. *Journal of Comparative Pathology and Therapeutics* 57, 314–325.
Wang, Y.-Z., Shan, T.-Z., Xu, Z.-R., Feng, J. and Wang, Z.Q. (2006) Effects of lactoferrin (LF) on performance, intestinal microflora and morphology of weanling pigs. *Animal Feed Science and Technology* 135, 263–272.
Weinberg, E.D. (1984) Iron withholding: a defence against infection and neoplasia. *Physiological Reviews* 64, 65–102.
West, A.R. and Oates, P.S. (2008) Mechanisms of heme iron absorption: current questions and controversies. *World Journal of Gastroenterology* 14, 4101–4110.
Wu, W.-H., Meydani, M., Meydani, S.N., Burklund, P.M., Blumbery, J.B. and Munro, H.N. (1990) Effect of dietary iron overload on lipid peroxidation, prostaglandin synthesis and lymphocyte proliferation in young and old rats. *Journal of Nutrition* 120, 280–289.
Yu, B., Huang, W.-J. and Chiou, P.W.-S. (2000) Bioavailability of iron from amino acid complex in weanling pigs. *Animal Feed Science and Technology* 86, 39–52.
Zimmerman, M., Adou, P., Torresani, T. and Hurrell, R. (2000) Iron supplementation in goitrous children improves their response to oral iodized oil. *European Journal of Endocrinology* 142, 217–223.
Zinoffsky, O. (1886) Ueber die Grosse des Hamoglobinmoleculs. *Hoppe-Seylers Zeitschriftur Physiologische Chemie* 10, 16.

14 Manganese

Manganese is the fifth most abundant metal on Earth and was shown to be essential for growth and fertility in laboratory animals in the 1930s. The need for manganese in livestock nutrition became evident when two diseases of poultry, perosis ('slipped tendon') and nutritional chondrodystrophy, were prevented by feeding manganese supplements (Schaible *et al.*, 1938), and breeder hens deprived of manganese produced chick embryos with malformed skeletons (Lyons and Insko, 1937). Bone development was also severely affected in pigs deprived of manganese (Johnson, 1943), but some years elapsed before adverse effects were demonstrated in cattle (Bentley and Phillips, 1951) and sheep (Lassiter and Morton, 1968). By this time, manganese had been found to be widely distributed at very low concentrations in mammalian tissues and necessary for the normal development of bone and reproductive processes in both sexes. Subsequent studies with laboratory animals showed that severe manganese deprivation could impair immunity (Hurley and Keen, 1987) and central nervous system function (Hurley, 1981). Interest in manganese in ruminant nutrition has been rekindled by confirmation that manganese deprivation of the heifer in pregnancy causes neonatal chondrodystrophy (Hansen *et al.*, 2006b), by suggestions – albeit ill-founded – that NRC (2001) requirements for cattle have been grossly underestimated (Weiss and Socha, 2005) and by the discovery that excess manganese can exacerbate copper deprivation in cattle (Hansen *et al.*, 2006a). However, the emphasis in this chapter is placed on the manganese nutrition of poultry since this is the species in which manganese deprivation is most likely to be encountered on farms.

Functions of Manganese

The functions of manganese can be linked to metalloenzymes that are activated by the element. Arginase appears to be an exception.

Pyruvate carboxylase

A specific biochemical role for manganese in intermediary energy metabolism was confirmed when pyruvate carboxylase was discovered to be a manganese metalloprotein (Scrutton *et al.*, 1966, 1972). Manganese is probably required for normal lipid and carbohydrate metabolism through the activity of pyruvate carboxylase; the fat accumulation seen in manganese deprivation is also a feature of biotin deficiency and biotin activates the same enzyme. Defects in lipid and carbohydrate metabolism have been reported in manganese-deprived rats and guinea pigs and a diet low in manganese can reduce fat deposition in pigs (Plumlee et al., 1956) and newborn goat kids (Anke et al., 1973b).

Superoxide dismutase

A second function was identified when a superoxide dismutase (MnSOD) was isolated from chicken liver mitochondria (Gregory and Fridovich, 1974) and subsequently found to contain 2 mg Mn mol^{-1}. Located primarily in the mitochondria, MnSOD complements cytosolic CuZnSOD in protecting cells from damage by reactive oxygen species, notably the superoxide radical O_2^- (see Fig. 15.1). The human gene for MnSOD, *SOD2*, has been identified. Its inducibility by cytokines such as tumour necrosis factor α (Wong and Goeddel, 1988) suggests that it is a 'stress-responsive' gene (St. Clair, 2004) needed for added protection against oxidative stress associated with inflammatory responses to some infections. Manganese deprivation lowers MnSOD activity in the heart (Davis et al., 1992) and increases the peroxidative damage caused by high dietary levels of polyunsaturated fatty acids (PUFA) (Malecki and Greger, 1996). However, compensatory increases in CuZnSOD suggest overlapping roles and possible interactions between dietary copper and manganese. Mitochondria are responsible for 60% of cellular O_2 consumption and may be particularly vulnerable to free-radical damage at times of high metabolic activity (Leach and Harris, 1997). Tissue activities of MnSOD are low in lambs at birth (Paynter and Caple, 1984), probably reflecting the low oxidative stress associated with the protected lifestyle of the fetus. The subsequent need to respire is associated with a marked rise in lung MnSOD by 4 weeks of age. Most of the manganese in the ovine heart is present as MnSOD (Paynter and Caple, 1984), which is by far the most dominant dismutase in this organ and elsewhere.

Glycosyltransferase

Manganese is needed for the synthesis of the mucopolysaccharides in cartilage through its activation of glycosyltransferase. Impaired glycosyltransferase activity reduces the synthesis of glycosaminoglycan and oligosaccharide side chains in animals deprived of manganese (Leach and Harris, 1997). Manganese-deprived chicks have less proteoglycan in the cartilage of the tibial growth plate than manganese-replete chicks and the carbohydrate composition of monomers is changed (Liu et al., 1994). In laying hens, subnormal egg production and poor shell formation may result from impaired mucopolysaccharide synthesis (Hill and Mathers, 1968) and a reduction in the hexosamine content of the shell matrix (Longstaff and Hill, 1972). Manganese is also involved in the formation of prothrombin, a glycoprotein, again through its activation of glycosyltransferases. The clotting response from vitamin K is reduced in manganese-deprived chicks (Doisey, 1973).

Manganese and reproduction

Studies of manganese distribution among tissues of the reproductive tract of normal and anoestrous ewes has revealed a possible role for manganese in the functioning of the corpus luteum (Hidiroglou, 1975). The hypothesis that lack of manganese inhibits the synthesis of cholesterol and certain sex hormones, thus causing infertility (Doisey, 1973), is not supported by recent evidence that plasma cholesterol levels and conception rate are unaffected in heifers sufficiently manganese-deprived to give birth to deformed calves (Hansen et al., 2006a,b). Manganese deprivation has been found to restrict testicular growth relative to body growth in ram lambs despite the fact that manganese concentrations in the testes were not depleted. The effect may have been caused indirectly by hormonal influences upon the testes (Masters et al., 1988).

Dietary Sources of Manganese

Seeds and grains

Seeds and seed products vary widely in manganese, due partly to inherent species differences.

Table 14.1. Mean manganese concentrations (mg kg^{-1} DM) in cereal grains and their mill products from three countries. Values in parentheses are standard deviations and indicate wide variation.

Source	Maize	Barley	Wheat	Oats	Sorghum	Wheat bran	Wheat pollard
North America	5	14	31	36	–	108	101
Australia	8	15	37	43	16	133	100
UK	5.9	18.5	35.6	45.2	8.3	88.7	–
	(3.0)	(3.6)	(15.2)	(15.4)	(0.5)	(33.3)	

Barley commonly contains 15–28 mg Mn kg^{-1} dry matter (DM), maize grain far less and wheat or oats far more (Table 14.1). Species differences amongst lupin seeds are striking: the white lupin (*Lupinus albus*) can contain 817–3397 mg Mn kg^{-1} DM, some 10–15 times more than other common lupin species growing on the same sites (Gladstones and Drover, 1962; White *et al.*, 1981), while *Lupinus angustifolius* may contain as little as 6 mg Mn kg^{-1} DM (White *et al.*, 1981). Field beans (*Vicia faba*) are low in manganese (7.8 ± 1.5 mg kg^{-1} DM) (MAFF, 1990). Manganese is highly concentrated in the outer layers of grains. This means that the inclusion of the mill products bran and pollard (middlings) (Table 14.1) in compound feeds markedly increases dietary manganese.

Protein supplements and sources

Protein concentrates of animal origin, such as blood meal, meat meal, feather meal and fish meal, contain less manganese (0.2–20 mg kg^{-1} DM) than protein supplements of plant origin, such as linseed, rapeseed and soybean meal (SBM) (35–55 mg kg^{-1} DM) (MAFF, 1990). Manganese is one of the least abundant elements in milk, rarely exceeding 0.1 mg l^{-1} in any species, except in colostrum. Bovine milk contains 20–40 μg l^{-1} and concentrations cannot be raised by increasing manganese intake (Underwood and Suttle, 1999). Milk by-products are therefore low in manganese (<1 mg kg^{-1} DM). The manganese in bovine milk is mostly associated with the casein fraction, associated with low-molecular-weight proteins, whereas in human milk it occurs mostly in the whey fraction bound to lactoferrin (Lonnerdahl *et al.*, 1985).

Forages

Pastures vary markedly in manganese about a high mean value of 86 mg kg^{-1} DM, and only 3% of the grass samples reported on by Minson (1990) had <20 mg Mn kg^{-1} DM. Mixed New Zealand pastures usually contain 140–200 mg Mn kg^{-1} DM, but >400 mg Mn kg^{-1} DM has been found in some areas (Grace, 1973). Differences due to species and state of maturity are generally small (Minson, 1990), but may be masked by high values arising from soil contamination. Soils usually contain 300–1100 mg Mn kg^{-1} DM (e.g. Suttle *et al.*, 2003), much more than the pasture they support. Contamination can also occur during sample processing if mills with steel blades are used. Increases in soil pH markedly decrease manganese uptake by plants, whether they are cereals (Fig. 14.1), grasses (Fig. 2.2) or legumes (Mitchell, 1957). Low manganese levels have been reported for maize silages from Canada (approximately 22 mg kg^{-1} DM) (Miltimore *et al.*, 1970; Buchanan-Smith *et al.*, 1974), the UK (14.6 ± 7.2 mg kg^{-1} DM) (MAFF, 1990) and USA (17.1 ± 20.3 mg kg^{-1} DM) (Berger, 1995). Lucerne hays from Canada, the USA and the UK contain, on average, 43.7, 22.7 mg and Mn kg^{-1} DM, respectively, but values vary widely (see Table 2.1).

Nutritive Value of Manganese in Feeds

Control of absorption

The absorbability of manganese as a feed attribute ($^{max}A_{Mn}$) (see p. 27) can be masked by the control of manganese absorption (A_{Mn}) exerted by the animal, by a marked interaction with dietary

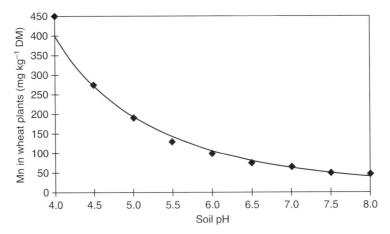

Fig. 14.1. Manganese concentrations in crops and pastures fall markedly when acid soils are limed (data for wheat plants in coordinated Food and Agriculture Organization trials) (Silanpaa, 1982).

iron and by age effects. In rats on a diet almost devoid of manganese and low in iron, the A_{Mn} coefficient was 0.57, decreasing to 0.20 with 48 or 188 mg manganese added as manganese carbonate ($MnCO_3$) (Fig. 14.2). However, with the diet high in iron (276 mg kg^{-1} DM), A_{Mn} was greatly reduced and the regulation of A_{Mn} was far more effective, declining to 0.064 in a diet high in both iron and manganese. Iron supplementation reduced mucosal cell manganese by >50% (Davis et al., 1992) and probably down-regulates the shared divalent metal transporter (DMT1) (see Chapter 1). In humans, A_{Mn} is inversely related to serum ferritin, a measure of iron stores (see Chapter 13), and high iron status may reduce the transfer of iron and manganese across the gut mucosa (Finley, 1999), again by down-regulating DMT1. Similar regulation and interactions probably occur in farm animals (Hansen and Spears, 2008) and under such circumstances the assessment of $^{max}A_{Mn}$ is difficult. Although chicks have been found to retain <5% of an oral dose of ^{54}Mn when the diet contained only 4 mg Mn kg^{-1} DM (Mathers and Hill, 1967), retention still fell when stable manganese was added to the dose. A_{Mn} also fell from 4.5% to 2.8% as the chicks grew from 2 to 5 weeks of age. Since a purified diet of similar manganese concentration has been seen to give a low incidence of perosis indicating a state of deprivation (Hill, 1967), there would appear to be limited scope for absorbing manganese according to need in some poultry rations.

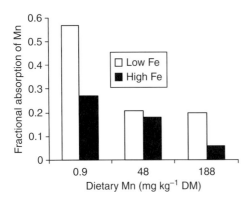

Fig. 14.2. The fractional absorption of manganese in rats decreases non-linearly as dietary Mn concentration rises, particularly if the diet is high (276 mg kg^{-1} DM) rather than low (19 mg kg^{-1} DM) in iron (data from Davis et al., 1992).

Milk and milk products

Manganese in milk and milk products is well absorbed by the suckling or artificially reared animal. Calves reared on whole milk (Carter et al., 1974) or milk substitute (Kirchgessner and Neese, 1976) containing 0.75 mg Mn kg^{-1} DM have been found to retain 60% and

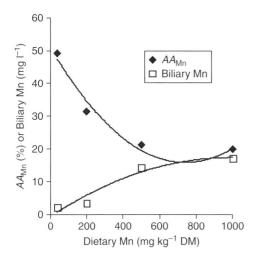

Fig. 14.3. The relationship between apparent absorption of manganese (AA_{Mn}), biliary Mn and dietary Mn concentration in calves fed milk replacers (from Jenkins and Hidiroglou, 1991). Homeostasis is achieved by reductions in absorptive efficiency and increases in biliary Mn secretion. The former is the more important at low Mn intakes, but makes little contribution with intakes >500 mg kg⁻¹ DM.

40%, respectively, of ingested manganese, but a supplement of 15 mg Mn kg⁻¹ DM reduced that figure to 16.3% (Carter et al., 1974). In calves reared on milk replacers containing 90–100 mg Fe kg⁻¹ DM, apparent absorption of manganese (AA_{Mn}) at 7 weeks of age decreased from 50% to 20% as the added manganese increased from 40 to 1000 mg kg⁻¹ DM (Fig. 14.3). However, retention (often closely related to AA) was only 4.9% over a period of 9 weeks from a milk replacer containing 14.2 mg Mn kg⁻¹ DM (Suttle, 1979). Artificially reared piglets absorbed 35–68% of manganese from milk formula feeds low in manganese (Atkinson et al., 1993; Rutherfurd et al., 2006). Milk and milk products thus have an intrinsically high A_{Mn}.

Grains

Manganese is poorly absorbed from grains and this has been illustrated by studies on human volunteers (Johnson et al., 1991). Intrinsically and extrinsically labelled foods were consumed with a basal meal low in manganese. Absorptive efficiencies (percentage of dose) were 1.7, 2.2, 3.8, 5.2 and 8.9 for sunflower seeds, wheat, spinach, lettuce and manganese chloride ($MnCl_2$), respectively. Low A_{Mn} in grains was attributed to the formation of complexes with phytate and/or fibre. Furthermore, the method of labelling did not affect the differences between sources, indicating that the determinants of A_{Mn} inherent in grains could equally affect the A_{Mn} of inorganic manganese supplements. However, the higher absorption of inorganic than feed manganese persists in diets rich in phytate and/or fibre. When manganese sulfate ($MnSO_4$) was added to a maize/SBM diet for chicks, A_{Mn} increased from only 2.8% to 10.3% (Wedekind et al., 1991). This range of values agrees well with that reported for humans (Johnson et al., 1991) and may be applicable to all non-ruminant livestock. A low A_{Mn} of 0.5% has been reported for the 45 mg Mn kg⁻¹ DM in a corn/SBM diet given to young pigs, but inhibition of A_{Mn} by both dietary iron (347 mg kg⁻¹ DM) and the prior injection of iron-dextran were implicated (Finley et al., 1997). In a trial in young pigs (initial live weight 9.4 kg), AA_{Mn} was a high 50% from a corn/SBM diet containing 45 mg Mn kg⁻¹ DM, but was reduced to 30% by 100 mg supplementary zinc (Adeola et al., 1995). The removal of a trace element supplement that added 127 mg Fe kg⁻¹ DM and 127 mg Zn kg⁻¹ DM to the fattening ration of pigs increased biliary and bone manganese (Shelton et al., 2004). Supplementation with phytase has had variable effects, reducing AA_{Mn} in the latter study unless zinc was also added, but improving it in chicks (Biehl et al., 1995). In rats, A_{Mn} was reduced by supplementing a low-manganese diet with a saturated fat, stearic acid, but not with unsaturated fats (Finley and Davis, 2001).

Effects of calcium and phosphorus

The addition of calcium and phosphorus to the diet have long been held to lower A_{Mn}, but such effects were obtained with large excesses of both calcium and phosphorus (e.g. Schaible et al.,

1938). More recently, the addition of 8g kg^{-1} DM *excess* inorganic phosphorus has been found to reduce A_{Mn} by about 50% in chicks at dietary manganese levels ranging from the sub-optimal (<12mg kg^{-1} DM) (Baker and Odoho, 1994) to excessive (1000mg kg^{-1} DM) (Wedekind et al., 1991) in either a semi-purified (Baker and Odoho, 1994) or a maize/SBM (Wedekind et al., 1991) diet. Although excess calcium was added with phosphorus in both studies to avoid calcium to phosphorus imbalances (a common but questionable experimental strategy), calcium alone had no effect on A_{Mn} (Wedekind et al., 1991). Provided such gross over-feeding of inorganic phosphorus is avoided, the major impairment of manganese utilization in practice may come from interactions between calcium and organic feed constituents. It would be surprising if the adverse effects of phytate (Biehl et al., 1995) were not dependent upon dietary calcium in practical rations for chicks (see Chapters 4 and 6); the A_{Mn} in such feeds may not exceed 0.05 unless added or inherent phytase is present. The fortification of a synthetic milk with calcium glycerophosphate does not lower A_{Mn} in piglets (Atkinson et al., 1993).

Forages

There have been no reports relating to A_{Mn} in forages for ruminants, either as relative or absolute measurements. Since the main antagonists of manganese absorption – phytate and fibre – are broken down in the rumen, A_{Mn} is probably higher than that achievable on practical diets by non-ruminants. A_{Mn} may reach 10–20% when manganese intakes are low, although only about 10% of manganese in the rumen of sheep may be in a soluble form (Bremner, 1970). The soil manganese that often contaminates pasture may be extensively released in the rumen; large amounts of soluble manganese have been recovered from the effluents of continuous cultures of rumen microbes, supplemented with two soils of relatively low manganese content (a chalk and a Weald loam) (Brebner et al., 1985). In its factorial model of the manganese requirements of dairy cows, the NRC (2001) assumed an A_{Mn} coefficient of 0.065–0.075 for feed manganese.

Manganese Requirements

Estimates of manganese requirements vary with species, stage of development, the chemical form of manganese ingested (inherent or added) and the chemical composition of the whole diet, particularly of poultry rations. Criteria of adequacy are also important, in particular whether the 'requirement' is for optimal growth, leg condition or reproduction. However, the attainment of a maximal tissue manganese concentration is a poor index of need. Factorial models have rarely been used to derive requirements because of a dearth of reliable numerical values for model components.

Poultry

Growth

Requirements are much higher in poultry than pigs because of a lower absorptive efficiency. Furthermore, there is a widely promoted view (e.g. Bao et al., 2007) that manganese (and other mineral) requirements are far higher than those listed by the NRC (1994) – which are based largely on data from the 1950s – because growth rates have doubled during the intervening years. However, food intakes have also greatly increased and manganese requirements, in dietary concentration terms, are more likely to reflect subsequent increases in feed conversion efficiency (FCE). The NRC (1994) recommends 60mg Mn kg^{-1} DM for all stages of broiler growth. This differs from earlier guidelines, which recommended 60mg Mn kg^{-1} as a maximum for starter chicks. Consequently, the first increment employed in many experiments was 60mg kg^{-1} DM, added to diets already containing 20–30mg kg^{-1} DM! The growth requirement therefore remains ill-defined, but in four such studies (i.e. with 20–30mg Mn kg^{-1} DM) (Sands and Smith, 1999; Li et al., 2004; Yan and Waldroup, 2006; Lu et al., 2007) broiler growth was *not* improved by manganese supplementation in either the starter (1–21 days) or growth (22–49 days) stages.

Bao et al. (2007) used smaller manganese increments to determine manganese requirements, but increased copper, iron and zinc simultaneously and used organic mineral sources. Assuming that interactions between minerals were minimal at the low levels used and that organic sources do not have superior availability (see below), the manganese requirement for Cobb broilers in the growing stage was 34–55 mg kg^{-1} DM. An earlier study showed that broiler growth on maize/SBM diets containing 37.5 mg Mn kg^{-1} DM was not improved by adding manganese (Wedekind et al., 1991). Although the addition of saturated fat to the diet may increase the need for manganese, the above evidence suggests that modern broilers require far less manganese than recommended by the NRC (1994).

One reason for this may be that the gradual reduction in calcium and phosphorus requirements over the years has raised A_{Mn}. If the manganese requirement is determined largely by the growth rate, requirements should rise between the starter and growth stages as growth rates double. This does not happen (Fig. 14.4; Lu et al., 2007), possibly and partly because small reductions in dietary calcium and phosphorus for the grower chick raise A_{Mn}.

Leg abnormalities

The increased growth rates of modern broilers have led to considerable increases in the incidence of leg abnormalities such as tibial dyschondroplasia (Whitehead et al., 2003; Dibner et al., 2007). Lack of manganese is a causal factor and recent broiler studies have indicated that the manganese requirement to minimize the incidence of swelling of the tibiotarsal joint is higher than that for growth (Fig. 14.4). However, the abnormality was not eliminated by manganese supplementation and the requirement varied from 21–81 (Li et al., 2004) to 119–219 mg Mn kg^{-1} DM (Lu et al., 2007) in two studies with the same broiler strain (Arbor Acres). Furthermore, the incidence of leg abnormalities, which were not sufficiently severe to adversely affect performance, may have been exacerbated by the use of diets high in PUFA. Choline deficiency was the first (Jukes, 1940) of many factors other than manganese, not all nutritional, that have been shown to influence the incidence of tibial dyschondroplasia and related disorders (Whitehead et al., 2003).

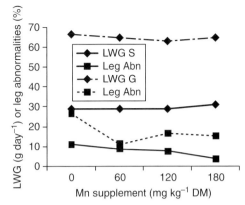

Fig. 14.4. Supplementation of basal diets for starter (S) and grower (G) chicks, containing manganese at 23 and 21 mg kg^{-1} DM, respectively. Manganese had no effect on growth (live-weight gain; LWG), but reduced the incidence of leg abnormalities (Abn; swollen tibiotarsal joint). The faster growth of G than S chicks did not increase the manganese requirement, which is higher for normal leg development than for growth (from Li et al., 2004).

Needs of other classes

The NRC (1994) gives manganese requirements of 17–25 kg^{-1} DM for white-egg-laying strains (depending on food intake) and 33 mg kg^{-1} DM for breeding hens. Requirements for brown-egg-laying strains are some 10% lower at all stages because of their higher food intake. The requirements given for turkeys are increased from previous guidelines, along with those of broilers, to a uniform 60 mg Mn kg^{-1} DM.

Pigs

The manganese requirement for satisfactory reproduction is substantially higher than that for body growth. Pigs have grown normally on semi-purified diets containing as little as 0.5–1.5 mg MN kg^{-1} DM from normal weaning to market weights (Johnson, 1943; Plumlee et al., 1956). A maize diet containing 12 mg Mn kg^{-1} DM has been found to be adequate for growth and skeletal development, with only a slight improvement in reproductive performance with additional manganese (40 mg kg^{-1}

DM) (Grummer et al., 1950). Similar results from other groups suggest that the maximum requirement for growing pigs on commercial rations is probably around 4 mg kg^{-1} DM, but higher (10 mg kg^{-1} DM) for sows on natural diets (NRC, 1998).

Cattle

The manganese requirements of the pre-ruminant are obviously met by normal milk and are therefore <1 mg kg^{-1} DM. In older cattle, requirements are substantially lower for growth than for optimal reproductive performance. Early studies showed that 10 mg Mn kg^{-1} DM was adequate for growth but marginal for maximum fertility (Bentley and Phillips, 1951). Recent studies have confirmed that 8 mg Mn kg^{-1} DM is adequate for the growth of finishing steers (Legleiter et al., 2005) and 16 mg kg^{-1} DM for normal growth and fertility in growing heifers (Hansen et al., 2006a), but not for development of the unborn calf (Hansen et al., 2006b). This last finding appears to contradict earlier work in which identical-twin heifers given practical rations containing 16 and 21 mg Mn kg^{-1} DM for 2.5–3.5 years remained healthy, fertile and showed no response to manganese supplementation (Hartmans, 1974). The combined requirements for growth and pregnancy in the heifer clearly represent a period of maximum demand: the outcome of such studies may depend on the timing of conception in relation to manganese depletion and the growth achieved by the pregnant heifer. In other studies, 20 mg Mn kg^{-1} DM has been found to be adequate for both the growth and reproduction of heifers (Rojas et al., 1965; Howes and Dyer, 1971). The NRC (2001) used a factorial model to estimate that dairy cows require 17–18 mg Mn kg^{-1} DM. This range may be marginal for the heifer that conceives well before it is fully grown, and 20–25 mg kg^{-1} DM is recommended for such stock. Recommendations of 28–49 mg kg^{-1} DM based on a different model with an unrealistically elevated maintenance requirement (Weiss and Socha, 2005) are too high. Few pastures or forages will fail to meet NRC (2001) requirements.

Sheep and goats

A definitive study of the requirements of growing sheep showed that 13 mg Mn kg^{-1} DM is adequate for live-weight gain and wool growth, but slightly more (16 mg kg^{-1} DM) is needed for testicular growth (Masters et al., 1988). These values are similar to those recommended for cattle. This is surprising, given that there is as much manganese in the fleece of a mature sheep as in the rest of its body (Fig. 14.5) (Grace, 1983).

Data on goats are meagre and their requirements are ill-defined. Female goats fed on diets containing 20 mg Mn kg^{-1} DM in the first year and 6 mg Mn kg^{-1} DM in the second have been found to grow as well as those receiving an additional 100 mg Mn kg^{-1} DM, but reproductive performance was greatly impaired at the lower level (Anke et al., 1973a). This confirms the lower requirement for growth than for fertility, but leaves the minimum requirement for either purpose undefined.

Metabolism

Absorption, transport and cellular uptake

There are parallels between the metabolism of manganese and iron. Manganese in human milk becomes largely bound to lactoferrin receptors at the brush border, as does iron (Davidson and Lonnerdahl, 1989). Mechanisms of manganese absorption from bovine milk and solid diets are also probably shared with iron, involving uptake by the non-specific apical transporter DMT1 and binding to ferritin (Arrenondo and Nunez, 2005). Absorbed manganese and iron are both transported in plasma by transferrin (Davidsson et al., 1989) and are taken up by transferrin receptors in the liver. Parenterally administered ^{54}Mn has a higher turnover than absorbed dietary ^{54}Mn and is therefore an unphysiological marker of manganese metabolism (Davis et al., 1992). Expression of DMT1 at sites such as the kidney and lung provides another locus for interactions between manganese and iron (Garrick et al., 2003).

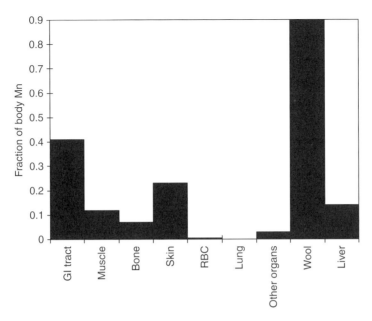

Fig. 14.5. The distribution of manganese in the shorn, empty body of mature sheep and the relatively large amount removed in the fleece (data from Grace, 1983). GI, gastrointestinal; RBC, red blood cells.

Excretion

Surplus manganese can be excreted partially via the bile, since little is reabsorbed at high manganese intakes. Biliary manganese increases linearly in chicks when a deficient diet is supplemented with 7 or 14 mg Mn kg^{-1} DM (Halpin and Baker, 1986), but curvilinearly in pre-ruminant calves when the manganese level in their milk replacer is increased from 40 to 1000 mg kg^{-1} DM (Fig. 14.3). Faecal endogenous loss (FE_{Mn}) is greatly increased by increases in dietary manganese in rats on both low- and high-iron diets (Davis et al., 1992). When A_{Mn} and FE_{Mn} are so clearly regulated, the derivation of appropriate values for minimum FE_{Mn} is difficult. Extrapolation of a linear relationship between the intake and faecal excretion of manganese in dry and lactating dairy cows gives a high value of 151 mg day^{-1} for FE_{Mn} (and 0.26 for AA_{Mn}), equivalent to >0.2 mg kg^{-1} live weight (Weiss and Socha, 2005). However, a linear extrapolation from what is probably a non-linear relationship (Fig. 14.3) to low manganese intakes is a potential source of error, and the estimate did indeed carry an enormous error (151 ± 41 mg).

Tissue distribution and storage

Manganese is one of the least abundant trace elements in all livestock tissues, with concentrations in the whole body ranging from 0.6 to 3.9 mg kg^{-1} fresh weight (FW) in the carcasses of calves and sheep (Suttle, 1979; Grace, 1983). In adult sheep, manganese concentrations are highest in the liver (4.2 mg kg^{-1} FW), followed by the pancreas (1.7 mg kg^{-1} FW) and kidneys (1.2 mg kg^{-1} FW); elsewhere, values were <0.3 mg kg^{-1} FW (Grace, 1983). In quantitative terms, the gastrointestinal tract and skin contain most of the manganese in the sheep carcass (Fig. 14.5). Masters et al. (1988) found no evidence of manganese storage in the liver or bone of lambs given diets ranging from 13 to 45 mg Mn kg^{-1} DM. In a grazing study, however, increasing dietary levels of manganese from 123 to 473 mg kg^{-1} DM (i.e. up to 30 times the requirement) increased tissue manganese by 25% in most organs and by 260% in the digestive tract (Grace and Lee, 1990).

In chicks, liver, kidney and skeletal manganese levels become less and less responsive to dietary manganese as the dietary concentration

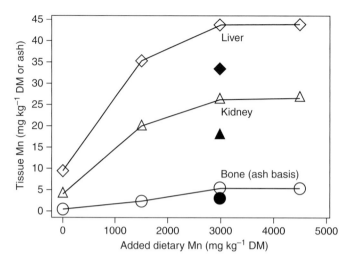

Fig. 14.6. Poultry can store excess manganese in their tissues, but high dietary concentrations are required to cause appreciable increases, particularly in bone (data from two experiments (open and closed symbols) of Halpin and Baker, 1986).

exceeds 1000 mg Mn kg^{-1} DM (Fig. 14.6) (Halpin and Baker, 1986). Curvilinear responses indicate that a comparison of linear slopes to gauge A_{Mn} by the retention method would give unreliable values. Because of the large contribution of the skeleton to body mass, a small rise in bone manganese could constitute a significant 'passive' reserve (Leach and Harris, 1997) and it would be surprising if any skeletal reserve in the pullet did not become available with the onset of lay and the accompanying bone resorption. There is little evidence that the fetus builds up a hepatic manganese reserve (Graham et al., 1994), but very high liver manganese (411–943 mg kg^{-1} DM) has been found in newborn calves following supplementation of the dam (Howes and Dyer, 1971).

Biochemical Changes in Manganese Deprivation

The sequence of biochemical changes that precedes the development of clinical signs of manganese deprivation in livestock is illustrated in Fig. 14.7. The changes differ from those for other elements in that the depletion and deficiency phases are not easily distinguished by changes in tissue manganese at the most accessible sites (blood, liver and bone) (see Chapter 3).

Subnormal manganese concentrations in the bones and reduced activities of alkaline phosphatase in the blood and soft tissues of deficient chicks and ducks have been found (Van Reen and Pearson, 1955). However, bone alkaline phosphatase activity is not always subnormal, even when the characteristic bone changes of manganese deprivation are visible. Reductions in MnSOD activity have been related to structural changes in mitochondria and cell membranes in the livers of laboratory animals (Underwood and Suttle, 1999) and hearts of broiler chicks (Luo et al., 1993). Masters et al. (1988) noted significant reductions in MnSOD in the heart and lung of their most depleted group. They speculated that in circumstances of oxidant stress (e.g. from PUFA) or depletion of other dietary antioxidants (e.g. copper, selenium and vitamin E), manganese depletion may exacerbate oxidative stress.

Clinical Manifestations of Manganese Deprivation

The signs of manganese deprivation vary with species, the degree and duration of deprivation and the age and productive function of the animal at the onset of deprivation.

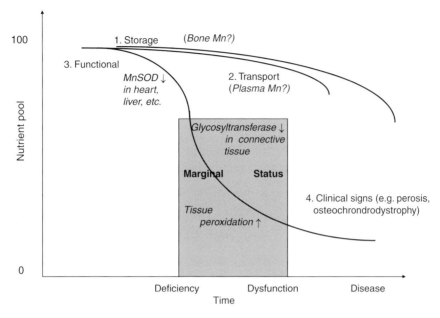

Fig. 14.7. The sequence of pathophysiological changes in livestock given an inadequate manganese supply differs from those shown by most other minerals (see Fig. 3.2): there are few stores to deplete and little decrease in the assumed transport pool (plasma) before functional pools of manganoenzymes are depleted and dysfunction occurs, with the epiphyseal plate as the predilection site.

Skeletal abnormalities

Manganese deprivation in chicks (Gallup and Norris, 1939a), poults and ducklings is manifested as the disease 'perosis' or 'slipped tendon' and in chick embryos by chondrodystrophy. Perosis is characterized by enlargement and malformation of the tibiometatarsal joint, twisting and bending of the tibia, thickening and shortening of the long bones and slipping of the gastrocnemius tendon from its condyles. With increasing severity of the condition, chicks are reluctant to move, squat on their hocks and soon die. Manganese deprivation in the breeder hen is manifested by shortened, thickened wings and legs in the embryo and a 'parrot beak', which results from a shortening of the lower mandible and globular contour of the head; mortality is high and the underlying defect is decreased endochondral growth or chondrodystrophy (Liu et al., 1994). In manganese-deprived pigs, skeletal abnormalities are characterized by crooked and shortened legs with enlarged hock joints, which together cause lameness (Johnson, 1943; Plumlee et al., 1956). In sheep, lack of manganese causes difficulty in standing and joint pain, with poor locomotion and balance (Lassiter and Morton, 1968). In goats there are tarsal joint excrescences, leg deformities and ataxia (Anke et al., 1973a,b). When heifers are deprived of manganese, their calves are of low birth weight and show disproportionate dwarfism, superior brachygnathism ('undershot' lower mandible), swollen joints and an unsteady gait (Hansen et al., 2006b).

Ataxia

Ataxia in the offspring of manganese-deprived animals was first observed in the chick (Caskey and Norris, 1940), but has also been observed in kids born of deficient goats (Anke et al., 1973a). The abnormality has been intensively studied in laboratory animals. Manganese deprivation in rats during late pregnancy produces an irreversible congenital defect in the fetus and neonate, characterized by ataxia and loss of equilibrium due to impaired vestibular function. A structural defect in the inner ear is caused by impaired synthesis of mucopolysaccharide and hence impaired bone development of the skull, particularly the otoliths of embryos (Underwood and Suttle, 1999). Neonatal ataxia is thus a secondary consequence of skeletal defects.

Reproductive disorders

Defective ovulation and testicular degeneration were observed in the earliest experiments with severely manganese-deprived rats and mice, and in similar subsequent studies in poultry reproductive function was impaired in both males and females (Gallup and Norris, 1939b). Depressed or delayed oestrus and poor conception rates have been associated with experimental manganese deprivation in cows, goats and ewes (Underwood and Suttle, 1999), but were not seen in manganese-deprived heifers in a recent study (Hansen et al., 2006a). In rams, manganese deprivation caused a reduction in testicular growth relative to body growth (Masters et al., 1998).

Occurrence of Manganese Deprivation

Outbreaks of manganese deprivation are largely confined to intensively reared avian species. Naturally occurring congenital chondrodystrophies in calves have been associated with low tissue manganese (e.g. Hidiroglou et al., 1990; Valero et al., 1990; Staley et al., 1994), although unequivocal manganese responsiveness has yet to be demonstrated. In ruminants, manganese deprivation is most likely to occur in livestock grazing pastures on alkaline soils (Suttle, 2000) or reared on maize silage with maize or barley supplements. Dams bearing their first progeny while still growing are probably the most vulnerable, as they are under experimental conditions (Hansen et al., 2006b). However, manganese-responsive disorders are rarely reported in the field and grazing sheep and cattle can develop and reproduce normally even in areas where plant-growth responses can be obtained from the application of manganese-containing fertilizers. Ewes in south Australia (Egan, 1972) and cows in south-west England (Wilson, 1966) on low-manganese pastures have occasionally shown improved conception rates when given supplementary manganese. High dietary iron is probably a risk factor for manganese deprivation in all species.

Diagnosis of Manganese Deprivation

There are no agreed biochemical criteria for supporting a diagnosis of manganese deprivation.

Blood and plasma

Reported blood and plasma manganese values are extremely variable, reflecting individual variability, analytical inadequacies (Hidiroglou, 1979) and probably physiological influences. Blood manganese is higher in the newborn calf than in the dam (23–28 versus 15–17 µg l^{-1}) (Weiss and Socha, 2005), but is not significantly lowered in the manganese-deprived heifer (Hansen et al., 2006a). Plasma manganese does not reflect manganese intake in steers on basal diets containing 8–128 mg Mn kg^{-1} DM (range of means 10–13.5 µg l^{-1}) (Legleiter et al., 2005); similar results have been found in heifers given 15.8–65.8 mg Mn kg^{-1} DM, except that the range of plasma manganese was higher (16–20 µg l^{-1}) and rose to that level in unsupplemented and supplemented groups after the synchronization of oestrus. In ram lambs given diets containing 13 or 45 mg kg^{-1} DM, plasma manganese has been seen to rise significantly from 1.85 to 2.74 µg l^{-1} (Masters et al., 1988); values <2 µg l^{-1} may be indicative of deprivation in young lambs, but the different order of these values compared to those in cattle should be noted.

Tissues

Liver manganese is frequently measured because the liver is by far the richest of the

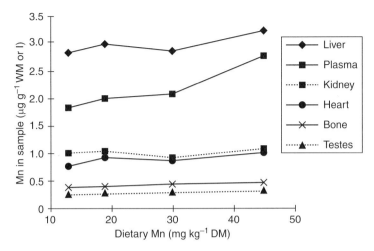

Fig. 14.8. In lambs, manganese concentrations in the heart are relatively more responsive to Mn deprivation than those in other soft tissues (data from Masters *et al.*, 1988).

tissues in manganese (Fig 14.8). However, in a study by Masters *et al.* (1988) in lambs, decreases in dietary manganese resulted in no significant changes in liver manganese about a mean of approximately 3 mg kg^{-1} FW, whereas the heart (and lung) both showed significant depletions from far lower maximal concentrations (Fig. 14.8). Although liver manganese was found to be linearly related to dietary manganese in heifers (Legleiter *et al.*, 2005), the range in dietary manganese was much wider (8–248 mg kg^{-1} DM). Within that range, the first increment to 28 mg kg^{-1} DM did not increase liver manganese. The diagnostic merits of heart manganese should be examined more closely, with levels of <0.2 mg kg^{-1} FW provisionally indicative of deprivation. Such low concentrations would place great demands on analytical precision.

Integument

Attempts to measure hair manganese as an index of status have given conflicting results and widely varying individual values, but wool and feather manganese may be more useful. The wool of lambs fed on a low-manganese diet for 22 weeks had on average only 6.1 mg Mn kg^{-1}, compared to 18.7 mg kg^{-1} DM in control lambs (Lassiter and Morton, 1968).

Similarly, the skin and feathers of pullets fed on a low-manganese diet for several months averaged 1.2 mg Mn kg^{-1} DM, compared to 11.4 mg kg^{-1} DM in comparable birds on a manganese-rich diet (Mathers and Hill, 1968). The use of standard washing, solvent extraction and drying procedures should lessen variability and improve the diagnostic value of assaying manganese in appendages of the skin. Toe, horn and hoof samples appear to not to have been used to monitor manganese status and may yield useful information. Guidelines for the assessment of manganese status in farm livestock are summarized in Table 14.2.

'Functional manganese'

Responses to manganese deprivation at the molecular level may ultimately be required to give unequivocal biochemical confirmation of manganese dysfunction. In iron-induced manganese deficiency in rats, MnSOD decreases more in the heart than in the liver (Davis *et al.*, 1992), and the use of mRNA for heart MnSOD in broiler chicks to assess manganese status has been advocated (Li *et al.*, 2004; Lu *et al.*, 2007). However, diagnostic limits cannot yet be suggested and reliance must still be placed on conventional

Table 14.2. Marginal bands[a] for mean manganese concentrations in diet, blood and tissues for assessing risk of manganese disorders in farm livestock.

	Diet (mg kg^{-1} DM)	Blood (µg l^{-1})	Serum (µg l^{-1})	Liver (mg kg^{-1} DM)	Bone (mg kg^{-1} DM)	Heart (mg kg^{-1} DM)	Pancreas (mg kg^{-1} DM)
Cattle	8–20	<20	<20	<9.0	1.0–1.4^R	–	3.0–5.0
Sheep	8–20	12–20	1.8–2.0	8–9	0.3–0.4^{MF}	0.6–0.7	4–8
Goat	10–20	–	–	3–6	5.0–6.0^R	1.5–2.0	–
Pigs	6–20	10–12	3–4	6–9	2.0–2.3^{RCT}	–	2.0–3.0
Poultry	**30–40**	**30–50**	5–10	6–12	4.0–6.0^{TF}	–	10–15

C, cortical bone used for assay; F, fat-free; M, metatarsal; R, rib; T, tibia.
[a]Mean values below bands or individual values close to the lower (**bold**) limit indicate the probability of a positive response to manganese supplements; mean values within bands indicate the possibility of such responses. Multiply by 0.0182 to obtain values in mmol and µmol, respectively.

measures of manganese status, as summarized in Table 14.2, despite their severe limitations.

Control of Manganese Deprivation

Manganese deprivation in poultry can be prevented by the incorporation of manganese salts or oxides into mineral supplements, and even the needs of the developing chick embryo can thus be met. The importance of manganese 'availability' in inorganic supplements was indicated by early work showing that the manganese in two ores was relatively unavailable to poultry (Underwood and Suttle, 1999). The commonly used sources are MnSO$_4$, manganese oxide (MnO) and MnCO$_3$; of these, MnSO$_4$ has the highest 'availability' and values of the other two sources relative to MnSO$_4$ have been assessed at 30% and 55%, respectively, for poultry and 35% and 30%, respectively, for sheep (Henry, 1995). Higher levels of MnCO$_3$ and MnO should be used to correct dietary deficits, but the margin may be small for some sources of MnO (Henry et al., 1989). Insoluble sources may be better suited to hot and humid environments, causing less caking and peroxidation in the feed than more soluble salts. Most practical rations for pigs supply adequate manganese without supplementation. In areas where 'simple' or 'conditioned' manganese-responsive disorders are suspected in cattle, dietary supplements of MnSO$_4$ at the rate of 4 g for cows, 2 g for heifers and 1 g day^{-1} for calves are sufficient for treatment and prevention. Treatment of pastures with 15 kg MnSO$_4$ ha^{-1} has also been effective in the Netherlands (Hartmans, 1974), but is rarely practised. Supplementation of cattle rations based on maize and maize silage with 20 mg Mn kg^{-1} DM is recommended.

Organic Manganese Supplements

The years from 2000 to 2010 have seen a flurry of publications on the merits of organic sources of manganese. However, the results are hard to interpret because of the wide variety of complexes tested, the unusual circumstances of some tests and the variety of responses used in the tests.

Locus of effect

One thing that has been established is that any effect of a manganese amino acid source is unlikely to be attributable to resistance to antagonisms occurring in or beyond the stomach. Li et al. (2004) studied 12 different manganese complexes both in vitro and in vivo, including five methionine complexes, two 'proteinates' and five manganese amino acids. The in vitro studies used the method of Brown and Zeringue (1994) (see Fig. 2.4) and confirmed their general conclusion: none of the complexes was stable at pH 2 or 5. Any effects attributable to a manganese amino acid source must therefore occur after diet preparation and before such peptide complexes reach the acid stomach environment.

Purity of source

The same study (Li et al., 2004) showed that manganese–peptide complexes can provide far more than just manganese and amino acids. Manganese accounted for 6.5–17.4% of a given complex and the remainder was not necessarily amino acids, which varied from 1.5–51%: other constituents included iron (up to 2.5%), copper (1.1%) and zinc (5.3%). It cannot be assumed that any response to such manganese complexes is attributable to its manganese content, particularly when supranutritional levels of manganese are provided (e.g. 3 g kg^{-1} DM) (Smith et al., 1995).

Test conditions

Using methionine-supplemented controls to allow for the effect of accompanying amino acids, Henry et al. (1989) reported increases of 8% and 33% for manganese–methionine over manganese when assessed by increases in tibia and kidney manganese, respectively, with dietary manganese reaching 2.1 g kg^{-1} DM. But why should the storage site make such a difference? The kidneys are often a repository for unwanted or unusable trace elements and high kidney manganese from a supplement given at 'pharmacological' rather than nutritional levels is neither proof nor a quantitative test of superior availability.

Experimental protocols generally address problems that currently trouble the livestock industry and conditions are set to maximize the chances of getting a positive response to manganese. In poultry, reductions in the high fat content of the broiler carcass and its tendency to peroxidation (and hence taint formation) have been sought. In Li et al.'s (2004) study, the diet contained 7.1–7.4% PUFA to maximize peroxidation. Whatever the outcome on extreme PUFA-rich diets, results may not be applicable to normal rations. Conditions of heat stress are sometimes employed (e.g. Smith et al., 1995; Sands and Smith, 1999). A manganese–proteinate complex had a relative availability of 142% for broilers when compared with MnSO$_4$ under heat-stress conditions, but only 120% under thermoneutral conditions; bone manganese was the response criterion (Smith et al., 1995). However, the daily ration is also 'heat-stressed' and effects of the source may reflect interactions between the feed and the environment; diet palatability may have been affected at the massive manganese concentrations used.

Advantages associated with the lower solubility of a chelate have been suggested under humid conditions (Miles et al., 2003), but they would also be provided by a relatively insoluble inorganic source such as MnO or MnCO$_3$. A 15.81% advantage for a manganese chelate with a methionine analogue (2-hydroxy-4-(methylthio)-butanoic acid; HMB) over MnSO$_4$ was claimed in a broiler study from slopes ratio comparisons of responses in tibia manganese over a supplementary manganese range of 100–800 mg kg^{-1} DM (Yan and Waldroup, 2006), but there were no differences between the two manganese sources until the highest level was reached. Studies with dairy cows suggest that a small proportion (11%) of ingested HMB can escape rumen degradation and be absorbed (Lapierre et al., 2007) and the chelated manganese may well enter the body's methionine pool, as does selenium given as selenomethionine (see Chapter 15). If this were the case, increases in tissue manganese given as a manganese–HMB complex would not reflect the availability of manganese for enzyme synthesis.

Indices of response

In poultry or pig rations, organic manganese sources have generally failed to give better growth rates or FCE than inorganic sources that provide the same concentration of manganese (Sands and Smith, 1999; Sawyer et al., 2006; Yan and Waldroup, 2006; Bao et al., 2007; Lu et al., 2007), but the net can be – and usually is – cast much wider. Thus, Sands and Smith (2002) proceeded to look at biochemical markers of lipid metabolism in heat-stressed broilers, but only investigated responses to a manganese–proteinate that had failed to alleviate the effects of heat stress on performance. While serum insulin and non-esterified fatty acids (NEFA) were reduced by raising the supplementary level from 60 to 240 mg Mn kg^{-1} DM, there was no advantage over adding *no* manganese and *no* interaction with heat stress.

Li *et al.* (2004) compared responses to inorganic and 15 organic manganese sources in bone manganese, heart manganese and heart MnSOD and its mRNA. A significant advantage in 'availability' was established for just one of the 15 complexes with only one of the indices (MnSOD mRNA). Differences in relative availability were widened when a high-calcium basal diet (18.5 g kg^{-1} DM) (Lu *et al.*, 2007) reflected chelation strength. However, the commonly used manganese–methionine was a weak chelator and did not possess enhanced availability in a diet of normal calcium concentration for poultry. Tests at such high dietary calcium concentrations are only of relevance to the manganese nutrition of laying hens. The advantage for manganese–methionine described earlier (Henry *et al.*, 1989) depends on the site of manganese accretion used to evaluate availability, and with tibia manganese it was commercially insignificant. In a comparison of the ability of different sources to reduce leg abnormalities in starter and grower broilers, MnSO$_4$ was just as effective as three organic manganese sources in reducing swelling of the tibiotarsal joint by 50%. A residual 14% of cases were unresponsive to manganese from any source (Li *et al.*, 2004).

Sawyer *et al.* (2006) measured discoloration in pork chops on retail display. If they waited long enough (6–7 days, but not 4 days) manganese supplementation at 350 mg kg^{-1} DM reduced colour loss, but the improvement was greater with MnSO$_4$ than with a manganese–amino acid complex.

Conclusion

Recent studies confirm the conclusion drawn in the previous review (Underwood and Suttle, 1999):

- The addition of organic manganese sources to the diet of poultry is unlikely to improve health or production when compared with an inorganic salt or oxide.
- If benefits cannot be obtained in a species with low natural capacity for A_{Mn} then they are unlikely to be obtained in either pigs or ruminants.
- Precious research and development resources should no longer be wasted in chasing trivial advantages for manganese chelates.

Toxicity

Ruminants

Animals dependent on pastures and forages are occasionally exposed to high manganese intakes that are potentially toxic. When sheep grazing New Zealand pastures containing 140–200 mg Mn kg^{-1} DM were given pellets providing 250 or 500 mg Mn day^{-1}, as MnSO$_4$, the growth rate was significantly depressed and some reduction in heart and plasma iron was evident, despite the fact that the pastures contained 1100–2200 mg Fe kg^{-1} DM (Grace, 1973). In a later study (Grace and Lee, 1990), a similar manganese supplement lowered liver iron and increased pancreatic zinc levels without retarding growth. A mutual metabolic antagonism between manganese and iron at the absorptive site has been known about for many years (Matrone *et al.*, 1959). The anorexic effect of high manganese intakes is less when the element is given continuously by diet (Black *et al.*, 1985) rather than by bolus administration, which briefly raises ruminal manganese concentrations to very high levels. Furthermore, MnSO$_4$ may be more reactive than the forms in or on manganese-rich grass, where most manganese comes from soil contamination. Sheep can tolerate 3000 mg Mn kg^{-1} DM for 21 days when given as MnO (Black *et al.*, 1985). However, the lability of soil manganese is surprisingly high (Brebner *et al.*, 1985) and there is new evidence that 500 mg Mn kg^{-1} DM, albeit as MnSO$_4$, can exacerbate copper deprivation (Hansen *et al.*, 2006a,b) and old evidence that soil ingestion exacerbates copper deprivation (Suttle *et al.*, 1984). The possibility that the soil effect is dependent on a manganese–copper interaction as well as iron–copper antagonism requires investigation.

In a study by Jenkins and Hirdiroglou (1991), neither appetite nor growth rate was decreased by 500 mg Mn kg^{-1} DM in a calf milk replacer. However, anaemia was induced, despite the presence of supplementary iron at 100 mg kg^{-1} DM. Supplementary manganese at 1000 mg kg^{-1} DM was lethal: no specific abnormalities were evident at post-mortem examination, but liver manganese was increased from 7.2 to 26.7 mg kg^{-1} DM (Jenkins and Hidiroglou, 1991).

The tolerable manganese levels of 2600 mg kg^{-1} DM stated for weaned calves (NRC, 2005) and 3000 and 4500 mg kg^{-1} DM for weaned lambs (Wong-Valle et al., 1989) are too high and should be halved.

Tolerance will depend on the dietary level of iron and possibly copper and molybdenum (Hansen et al., 2006a).

Pigs and poultry

The toxic level of manganese for pigs has been variously estimated at 500, 1250 and 4000 mg kg^{-1} DM. This range probably reflects the variable influence of iron, fibre and phytate on manganese absorption (Underwood and Suttle, 1999). High manganese intakes impair A_{Fe} and haemoglobin synthesis (ARC, 1981) and the adverse interaction probably arises from downregulation of DMT1 in the gastric mucosa. The tolerable limit of 1 g Mn kg^{-1} DM set by the ARC (1981) is a sensible compromise.

Poultry are tolerant towards manganese, no doubt reflecting a relatively poor A_{Mn}, and the safe limit has been placed at 2 g Mn kg^{-1} DM (NRC, 2005). With a corn/SBM diet, however, 1.5 g Mn kg^{-1} DM has been found to decrease body weight (but not food intake) in male broiler chicks by 21 days of age, whether the source was inorganic or organic (Miles et al., 2003). An iron supplement of 50 mg kg^{-1} DM was adequate for normal rations, but may have been wanting at the high manganese intakes employed. Dietary calcium supplements were lower than those used in earlier work, although adequate for broilers (see Chapter 4). The tolerable levels of manganese will vary under the influence of interactions with dietary iron, calcium, phytate and fibre.

References

Adeola, O., Lawrence, B.V., Sutton, A.C. and Cline, T.R. (1995) Phytase-induced changes in mineral utilisation in zinc-supplemented diets for pigs. *Journal of Animal Science* 73, 3384–3391.

Anke, M., Groppel, B., Reisseg, W., Ludke, H., Grun, M. and Dittrich, G. (1973a) Manganmangel beim Wiederkauer 3. Manganmangelbedingte Fortpflanzungs-, Skelett- und Nervenstorungen bei weiblichen Wiederkauern und ihren Nachkommen. *Archiv für Tierernahrung* 23, 197–211.

Anke, M., Hennig, A., Groppel, B., Dittrich, G. and Grun, M. (1973b) Manganmangel beim Wiederkauer. 4. Der Einfluss des Manganmangels auf den Gehalt neugeborener Lammer an Fett, Protein, Mangan, Asche, Kalzium, Phosphor, Zink und Kupfer. *Archiv für Tierernahrung* 23, 213–223.

ARC (1981) *The Nutrient Requirements of Pigs*. Commonwealth Agricultural Bureaux, Farnham Royal, UK, pp. 215–248.

Arredondo, M. and Munez, M.T. (2005) Iron and copper metabolism. *Molecular Aspects of Medicine* 26, 313–327.

Atkinson, S.A., Shah, J.K., Webber, C.E., Gibson, I.L. and Gibson, R.S. (1993) A multi-element isotopic tracer assessment of true fractional absorption of minerals from formula with additives of calcium, phosphorus, zinc, copper and iron in young piglets. *Journal of Nutrition* 123, 1586–1593.

Baker, D.H. and Odohu, G.W. (1994) Manganese utilisation in the chick: effects of excess phosphorus on chicks fed manganese deficient diets. *Poultry Science* 73, 1162–1165.

Bao, Y.M., Choct, M., Iji, P.A. and Bruerton, K. (2007) Effect of organically complexed copper, iron, manganese and zinc on broiler performance, mineral excretion and accumulation in tissues. *Journal of Applied Poultry Research* 16, 448–455.

Bentley, O.G. and Phillips, P.H. (1951) The effect of low manganese rations upon dairy cattle. *Journal of Dairy Science* 34, 396–403.

Berger, L.L. (1995) Why do we need a new NRC data base? *Animal Feed Science and Technology* 53, 99–107.

Biehl, R.R., Baker, D.H. and DeLuca, H.F. (1995) 1 α-Hydroxylated cholecalciferol compounds act additively with microbial phytase to improve phosphorus, zinc and manganese utilization in chicks fed soy-based diets. *Journal of Nutrition* 125, 2407–2416.

Black, J.R., Ammerman, C.B. and Henry, P.R. (1985) Effect of quantity and route of manganese monoxide on feed intake and serum manganese of ruminants. *Journal of Dairy Science* 68, 433–436.

Brebner, J., Thornton, I., McDonald, P. and Suttle, N.F. (1985) The release of trace elements from soils under conditions of simulated rumenal and abomasal digestion. In: Mills, C.F., Bremner, I. and Chesters J.K. (eds) *Proceedings of the Fifth International Symposium on Trace Elements in Man and Animals.* Commonwealth Agricultural Bureau, Farnham Royal, UK, pp. 850–852.

Bremner, I. (1970) Zinc, copper and manganese in the alimentary tract of sheep. *British Journal of Nutrition* 24, 769–783.

Brown, T.F. and Zeringue, L.K. (1994) Laboratory evaluations of solubility and structural integrity of complexed and chelated trace mineral supplements. *Journal of Dairy Science* 77, 181–189.

Buchanan-Smith, J.G., Evans, E. and Poluch, S.O. (1974) Mineral analyses of corn silage produced in Ontario. *Canadian Journal of Animal Science* 54, 253–256.

Carter, J.C., Miller, W.J., Neathery, M.W., Gentry, R.P., Stake, P.E. and Blackmon, D.M. (1974) Manganese metabolism with oral and intravenous ^{54}Mn in young calves as influenced by supplemental manganese. *Journal of Animal Science* 380, 1284–1290.

Caskey, C.D. and Norris, L.C. (1940) Micromelia in adult fowl caused by manganese deficiency during embryonic development. *Proceedings of the Society for Experimental Biology and Medicine* 44, 332–335.

Davidson, L. and Lonnerdahl, B. (1989) Fe-saturation and proteolysis of human lactoferrin: effect on brush-border receptor-mediated uptake of Fe and Mn. *American Journal of Physiology–Gastrointestinal and Liver Physiology,* 257, G930–G934.

Davidson, L., Lonnerdahl, B., Sandstrom, B., Kunz, C. and Keen, C.L. (1989) Identification of transferrin as the major plasma carrier protein for manganese introduced orally or intravenously or after *in vitro* addition in the rat. *Journal of Nutrition* 119, 1461–1464.

Davis, C.D., Wolf, T.L. and Greger, J.L. (1992) Varying levels of dietary manganese and iron affect absorption and gut endogenous losses of manganese by rats. *Journal of Nutrition* 122, 1300–1308.

Dibner, J., Richards, J.D., Kitchell, M.I. and Quiroz, M.A. (2007) Metabolic challenges and early bone development. *Journal of Applied Poultry Research* 16, 126–137.

Doisey, E.A. Jr (1973) Micronutrient controls on biosynthesis of clotting proteins and cholesterol. In: Hemphill, D. D. (ed.) *Trace Substances in Environmental Health – 6.* University of Missouri, Columbia, Missouri, p. 193.

Egan, A.R. (1972) Reproductive responses to supplemental zinc and manganese in grazing Dorset Horn ewes. *Australian Journal of Experimental Agriculture and Animal Husbandry* 12, 131–135.

Finley, J.W. (1999) Manganese absorption and retention by young women is associated with serum ferritin concentration. *American Journal of Clinical Nutrition* 70, 37–43.

Finley, J.W. and Davis, C.D. (2001) Manganese absorption and retention in rats is affected by the type of dietary fat. *Biological Trace Element Research* 82, 143–158.

Finley, J.W., Caton, J.S., Zhou, Z. and Davison, K.L. (1997) A surgical model for determination of true absorption and biliary excretion of manganese in conscious swine fed commercial diets *Journal of Nutrition,* 127, 2334–2341.

Gallup, W.D. and Norris, L.C. (1939a) The amount of manganese required to prevent perosis in the chick. *Poultry Science* 18, 76–82.

Gallup, W.D. and Norris, L.C. (1939b) The effect of a deficiency of manganese in the diet of the hen. *Poultry Science* 18, 83–88.

Garrick, M.D., Dolan, K.G., Horbinski, C., Ghoio, A.J., Higgins, D., Porcubin, M., Moore, E.G., Hainsworth, L.N., Umbreit, J.N., Conrad, M.E., Feng, L., Lis, A., Roth, J.A., Singleton, S. and Garrick, L.M. (2003) DMT 1: a mammalian transporter for multiple metals. *Biometals* 16, 41–54.

Gladstones, J.S. and Drover, D.P. (1962) The mineral composition of lupins. 1. A survey of the copper, molybdenum and manganese contents of lupins in the south west of Western Australia. *Australian Journal of Experimental Agriculture and Animal Husbandry* 2, 46–53.

Grace, N.D. (1973) Effect of high dietary Mn levels on the growth rate and the level of mineral elements in the plasma and soft tissues of sheep. *New Zealand Journal of Agricultural Research* 16, 177–180.

Grace, N.D. (1983) Amounts and distribution of mineral elements associated with fleece-free empty body weight gains in grazing sheep. *New Zealand Journal of Agricultural Research* 26, 59–70.

Grace, N.D. and Lee, J. (1990) Effect of Co, Cu, Fe, Mn, Mo, Se and Zn supplementation on the elemental content of soft tissues and bone in sheep grazing ryegrass/white clover pasture. *New Zealand Journal of Agricultural Research* 33, 635–647.

Graham, T.W., Thurmond, M.C., Mohr, F.C., Holmberg, C.A., Anderson, M.L. and Keen, C.L. (1994) Relationship between maternal and foetal liver copper, iron, manganese and zinc concentrations and foetal development in California Holstein dairy cows. *Journal of Veterinary Diagnostic Investigation* 6, 77–87.

Gregory, E.M. and Fridovich, I. (1974) Superoxide dismutases: properties, distribution, and functions. In: Hoekstra, W.G., Suttie, J.W., Ganther, H.E. and Mertz, W. (eds) *Trace Element Metabolism in Animals – 2*. University Park Press, Baltimore, Maryland, pp. 486–488.

Grummer, R.H., Bentley, O.G., Phillips, P.H. and Bohstedt, G. (1950) The role of manganese in growth, reproduction, and lactation in swine. *Journal of Animal Science* 9, 170–175.

Halpin, K.M. and Baker, D.H. (1986) Long-term effects of corn soyabean meal, wheat bran and fish meal on manganese utilisation in the chick. *Poultry Science* 65, 1371–1374.

Hansen, S.L. and Spears, J.W. (2008) Impact of copper deficiency in cattle on proteins involved in iron metabolism. *FASEB Journal* 22, 443–445.

Hansen, S.L., Spears, J.W., Lloyd, K.E. and Whisnant, C.S. (2006a) Growth, reproductive performance and manganese status of heifers fed varying concentrations of manganese. *Journal of Animal Science* 84, 3375–3380.

Hansen, S.L., Spears, J.W., Lloyd, K.E. and Whisnant, C.S. (2006b) Feeding a low manganese diet to heifers during gestation impairs foetal growth and development. *Journal of Dairy Science* 89, 4304–4311.

Hartmans, J. (1974) Tracing and treating mineral disorders in cattle under field conditions. In: Hoekstra, W.G., Suttie, J.W., Ganther, H.E. and Mertz, W. (eds) *Trace Element Metabolism in Animals – 2*. University Park Press, Baltimore, Maryland, pp. 261–273.

Henry, P.R. (1995) Manganese bioavailability. In: Ammerman, C.B., Baker, D.H. and Lewis, A.J. (eds) *Bioavailability of Nutrients for Animals*. Academic Press, New York, pp. 239–256.

Henry, P.R., Ammerman, C.B. and Miles, R.D. (1989) Relative bioavailability of manganese in a manganese–methionine complex for broiler chicks. *Poultry Science* 68, 107–112.

Hidiroglou, M. (1975) Mn uptake by the ovaries and reproductive tract of cycling and anestrous ewes. *Canadian Journal of Physiology and Pharmacology* 53, 969–972.

Hidiroglou, M. (1979) Manganese in ruminant nutrition. *Canadian Journal of Animal Science* 59, 217.

Hidiroglou, M., Ivan, M., Bryoa, M.K., Ribble, C.S., Janzen, E.D., Proulx, J.G. and Elliot, J.I. (1990) Assessment of the role of manganese in congenital joint laxity and dwarfism in calves. *Annales de Recherches Vétérinaires* 21, 281–284.

Hill, R. (1967) Vitamin D and manganese in the nutrition of the chick. *British Journal of Nutrition* 21, 507–512.

Hill, R. and Mathers, J.W. (1968) Manganese in the nutrition and metabolism of the pullet. 1. Shell thickness and manganese content of eggs from birds given a diet of low or high manganese content. *British Journal of Nutrition* 22, 635–643.

Howes, A.D. and Dyer, I.A. (1971) Diet and supplemental mineral effects on manganese metabolism in newborn calves. *Journal of Animal Science* 32, 141–145.

Hurley, L.S. (1981) Teratogenic effects of manganese, zinc and copper in nutrition. *Physiological Reviews* 61, 249–295.

Hurley, L.S. and Keen, C.L. (1987) Manganese. In: Mertz, W. (ed.) *Trace Elements in Human and Animal Nutrition*, Vol II. Academic Press, New York, p. 185.

Jenkins, K.J. and Hidiroglou, M. (1991) Tolerance of the pre-ruminant calf for excess manganese or zinc in milk replacer. *Journal of Dairy Science* 74, 1047–1053.

Johnson, P.E., Lykken, G.I. and Kortnta, E.D. (1991) Absorption and biological half-life in humans of intrinsic and extrinsic ^{54}Mn tracers from foods of plant origin. *Journal of Nutrition* 121, 711–717.

Johnson, S.R. (1943) Studies with swine on rations extremely low in manganese. *Journal of Animal Science* 2, 14–22.

Jukes, T.H. (1940) Effect of choline and other supplements on perosis. *Journal of Nutrition* 20, 445–458.

Kirchgessner, M. and Neese, K.R. (1976) Copper, manganese and zinc content of the whole body and in carcase cuts of veal calves at different weights. *Zeitschrift für Lebensmittel-Untersuchung und-Forschung* 161, 1–12.

Lapierre, H., Vazquez-Anon, M., Parker, D., Dubreuil, P. and Lobley G.E. (2007) Absorption of 2-hydroxy-4-methylthiobutanoate in dairy cows. *Journal of Dairy Science* 90, 2937–2940.

Lassiter, J.W. and Morton, J.D. (1968) Effects of a low manganese diet on certain ovine characteristics. *Journal of Animal Science* 27, 776–779.

Leach, R.M. Jr and Harris, E.D. (1997) Manganese. In: O'Dell, B.L. and Sunde, R.A. (eds) *Handbook of Nutritionally Essential Mineral Elements*. Marcel Dekker, New York, pp. 335–356.

Legleiter, L.R., Spears, J.W. and Lloyd, K.E. (2005) Influence of dietary manganese on performance, lipid metabolism and carcass composition of growing and finishing steers. *Journal of Animal Science* 83, 2434–2439.

Li, S., Luo, X., Crenshaw, T.D., Kuang, X., Shao, G. and Yu, S. (2004) Use of chemical characteristics to predict the relative bioavailability of supplemental organic manganese sources for broilers. *Journal of Animal Science* 82, 2352–2363.

Liu, A.C.-H., Heinrichs, B.S. and Leach, R.M. Jr (1994) Influence of manganese deficiency on the characteristics of proteoglycans of avian epiphyseal growth plate cartilage. *Poultry Science* 73, 663–669.

Longstaff, M. and Hill, R. (1972) The hexosamine and uronic acid contents of the matrix of shells of eggs from pullets fed on diets of different manganese content. *British Poultry Science* 13, 377–385.

Lonnerdal, B., Keen, C.L. and Hurley, L.S. (1985) Manganese-binding proteins in human and cow's milk. *American Journal of Clinical Nutrition* 41, 550–559.

Lu, L., Luo, X.G., Ji, C., Liu, B. and Yu, S.X. (2007) Effect of manganese supplementation on carcass traits, meat quality and lipid oxidation in broilers. *Journal of Animal Science* 85, 812–822.

Luo, X.G., Su, Q., Huang, J.C. and Liu, J.X. (1993) Effects on manganese (Mn) deficiency on tissue Mn-containing superoxide dismutase (MnSOD) activity and its mitochondrial ultrastructures in broiler chicks fed a practical diet. *Veterinaria et Zootechnica Sinica* 23, 97–101.

Lyons, M. and Insko, W.M. Jr (1937) *Chondrodystrophy in the Chick Embryo Produced by Manganese Deficiency in the Diet of the Hen*. Bulletin No. 371, Kentucky Agricultural Experiment Station.

MAFF (1990) *UK Tables of the Nutritive Value and Chemical Composition of Foodstuffs*. In: Givens, D.I. (ed.) Rowett Research Services, Aberdeen, UK.

Malecki, E.A. and Greger, J.L. (1996) Manganese protects against heart mitochondrial lipid peroxidation in rats fed high levels of polyunsaturated fatty acids. *Journal of Nutrition* 126, 27–33.

Masters, D.G., Paynter, D.I., Briegel, J., Baker, S.K. and Purser, D.B. (1988) Influence of manganese intake on body, wool and testicular growth of young rams and on the concentration of manganese and the activity of manganese enzymes in tissues. *Australian Journal of Agricultural Research* 39, 517–524.

Mathers, J.W. and Hill, R. (1967) Factors affecting the retention of an oral dose of radioactive manganese by the chick. *British Journal of Nutrition* 21, 513–517.

Mathers, J.W. and Hill, R. (1968) Manganese in the nutrition and metabolism of the pullet. 2. The manganese contents of the tissues of pullets given diets of high or low manganese content. *British Journal of Nutrition* 22, 635–643.

Matrone, G., Hartman, R.H. and Clawson, A.J. (1959) Studies of a manganese–iron antagonism in the nutrition of rabbits and baby pigs. *Journal of Nutrition* 67, 309–317.

Miles, R.D., Henry, P.R., Sampath, V.C., Shivazad, M. and Comer, C.W. (2003) Relative bioavailability of novel chelates of manganese and copper for chicks. *Journal of Applied Poultry Research* 12, 417–423.

Miltimore, J.E., Mason, J.L. and Ashby, D.L. (1970) Copper, zinc, manganese and iron variation in five feeds for ruminants. *Canadian Journal of Animal Science* 50, 293–300.

Minson, D. J. (1990) *Forages in Ruminant Nutrition*. Academic Press, San Diego, California, pp. 208–229.

Mitchell, R.L. (1957) The trace element content of plants. *Research, UK,* 10 357–362.

NRC (1985) *Nutrient Requirements of Sheep*, 6th edn. National Academy of Sciences, Washington, DC.

NRC (1994) *Nutrient Requirements of Poultry*, 9th edn. National Academy of Sciences, Washington, DC.

NRC (1998) *Nutrient Requirements of Swine*, 10th edn. National Academy of Sciences, Washington, DC.

NRC (2000) *Nutrient Requirements of Beef Cattle*, 7th edn. National Academy of Sciences, Washington, DC.

NRC (2001) *Nutrient Requirements of Dairy Cows*, 7th edn. National Academy of Sciences, Washington, DC.

NRC (2005) *Mineral Tolerance of Animals*, 2nd edn. National Academy of Sciences, Washington, DC.

NRC (2007) *Nutrient Requirements of Horses*, 6th edn. National Academy of Sciences, Washington, DC.

Paynter, D. and Caple, I.W. (1984) Age-related changes in activities of the superoxide dismutase enzymes in tissues of the sheep and the effect of dietary copper and manganese on these changes. *Journal of Nutrition* 114, 1909–1916.

Plumlee, M.P., Thrasher, D.M., Beeson, W.M., Andrews, F.N. and Parker, H.E. (1956) The effects of a manganese deficiency upon the growth, development, and reproduction of swine. *Journal of Animal Science* 15, 352–367.

Rojas, M.A., Dyer, I.A. and Cassatt, W.A. (1965) Manganese deficiency in the bovine. *Journal of Animal Science* 24, 664–667.

Rutherfurd, S.M., Darragh, A.J., Hendriks, W.H., Prosser, C.G. and Lowry, D. (2006) Mineral retention in three-week-old piglets fed goat and cow milk infant formulas. *Journal of Dairy Science* 89, 4520–4526.

Sands, J.S. and Smith, M.O. (1999) Broilers in heat stress conditions: effects of dietary manganese proteinate or chromium picolinate supplementation. *Journal of Applied Poultry Research* 8, 280–287.

Sands, J.S. and Smith, M.O. (2002) Effects of dietary manganese proteinate or chromium picolinate supplementation on plasma insulin, glucagons, glucose and serum lipids in broiler chickens reared under thermoneutral or heat stress conditions. *International Journal of Poultry Science* 1, 145–149.

Sawyer, J.T., Tittor, A.W., Apple, J.K., Morgan, J.B., Maxwell, C.V., Rakes, L.K. and Fakler, T.M. (2006) Effects of supplemental manganese on performance of growing–finishing pigs and pork quality during retail display. *Journal of Animal Science* 85, 1046–1053.

Schaible, P.J., Bandemer, S.L. and Davidson, J.A. (1938) *The Manganese Content of Feedstuffs and its Relation to Poultry Nutrition.* Technical Bulletin No. 159, Michigan Agricultural Experiment Station.

Scrutton, M.C., Utter, M.F. and Mildvan, A.S. (1966) Pyruvate carboxylase. 6. The presence of tightly bound manganese. *Journal of Biological Chemistry* 241, 3480–3487.

Scrutton, M.C., Griminger, P. and Wallace, J.C. (1972) Pyruvate carboxylase: bound metal content of the vertebrate liver enzyme as a function of diet and species. *Journal of Biological Chemistry* 247, 3305–3313.

Shelton, J.L., Southern, L.L., LeMieux, F.M., Bidner, T.D. and Page, T.G. (2004) Effects of microbial phytase, low calcium and phosphorus, and removing the dietary trace mineral premix on carcass traits, pork quality, plasma metabolites, and tissue mineral content in growing-finishing pigs. *Journal of Animal Science* 82, 2630–2639.

Silanpaa, M. (1982) *Micronutrients and the Nutrient Status of Soils: A Global Study.* Food and Agriculture Organization, Rome, p. 67.

Smith, M.O., Sherman, I.L., Miller, L.C. and Robbins, K.R. (1995) Relative biological availability of manganese from manganese proteinate, manganese sulphate and manganese oxide in broilers reared at elevated environmental temperatures. *Poultry Science* 74, 702–707.

St Clair, D. (2004) Manganese superoxide dismutase: genetic variation and regulation. *Journal of Nutrition* 134, 3190S–3191S.

Staley, G.P., Van der Lugt, J.J., Axsel, G. and Loock, A.H. (1994) Congenital skeletal malformations in Holstein calves associated with putative manganese deficiency. *Journal of the South African Veterinary Association* 65, 73–78.

Suttle, N.F. (1979) Copper, iron, manganese and zinc concentrations in the carcases of lambs and calves and the relationship to trace element requirements for growth. *British Journal of Nutrition* 42, 89–96.

Suttle, N.F. (2000) Minerals in livestock production. *Asian-Australasian Journal of Animal Science* 13, 1–9.

Suttle, N.F., Abrahams, P. and Thornton, I. (1984) The role of a soil × dietary sulfur interaction in the impairment of copper absorption by soil ingestion in sheep. *Journal of Agricultural Science, Cambridge* 103, 81–86.

Suttle, N.F., Bell, J., Thornton, I. and Agyriaki, A. (2003) Predicting the risk of cobalt deprivation in grazing livestock from soil composition data. *Environmental Geochemistry and Health* 25, 33–39.

Underwood, E.J. and Suttle, F. (1999) *The Mineral Nutrition of Livestock*, 3rd edn. CAB International, Wallingford, UK.

Valero, G., Alley, M.R., Badcoe, L.M., Manktellow, B.W., Merral, M. and Lowes, G.S. (1990) Chondrodystrophy in calves associated with manganese deficiency. *New Zealand Veterinary Journal* 38, 161–167.

Van Reen, R. and Pearson, P.B. (1955) Manganese deficiency in the duck. *Journal of Nutrition* 55, 225–234.

Wedekind, K.J., Titgemeyer, E.C., Twardock, R. and Baker, D.H. (1991) Phosphorus but not calcium affects manganese absorption and turnover in chicks. *Journal of Nutrition* 121, 1776–1786.

Weiss, W.P. and Socha, M.T. (2005) Dietary manganese for dry and lactating cows. *Journal of Dairy Science* 88, 2517–2523.

White, C.L., Robson, A.D. and Fisher, H.M. (1981) Variation in the nitrogen, sulfur, selenium, cobalt, manganese, copper and zinc contents of grain from wheat and two lupin species grown in a Mediterranean climate. *Australian Journal of Agricultural Research* 32, 47–59.

Whitehead, C.C., Fleming, R.H., Julian, R. and Sorensen, P. (2003) Skeletal problems associated with selection for increased production. In: Muir, W. and Aggrey, S. (eds) *Poultry Genetics, Breeding and Biotechnology.* CAB International, Wallingford, UK, pp. 29–52.

Wilgus, H.S. Jr, Norris, L.C. and Heuser, G.F. (1937) The role of manganese and certain other trace elements in the prevention of perosis. *Journal of Nutrition* 14, 155–167.

Wilson, J.G. (1966) Bovine functional infertility in Devon and Cornwall: response to manganese supplementation. *Veterinary Record* 79, 562–566.

Wong, G.H.W. and Goeddel, D.V. (1988) Induction of manganous superoxide dismutase by tumour necrosis factor: possible protective mechanism. *Science* 242, 941–944.

Wong-Valle, J., Henry, P.R., Ammerman, C.B. and Rao, P.V. (1989) Estimation of the relative bioavailability of manganese sources for sheep. *Journal of Animal Science* 67, 2409–2414.

Yan, F. and Waldroup, P.W. (2006) Evaluation of Mintrex manganese as a source of manganese for young broilers. *International Journal of Poultry Science* 5, 708–713.

15 Selenium

Introduction

For many years biological interest in selenium was confined to its toxic effects on animals. Two naturally occurring diseases of livestock, 'blind staggers' and 'alkali disease', occurring on the Great Plains of North America, were attributed to acute and chronic selenium poisoning, respectively (see Moxon, 1937). These discoveries stimulated investigations of selenium in soils, plants and animal tissues to determine minimum toxic intakes and effective control measures. Eventually, selenium was shown to be an essential nutrient, with minute amounts in inorganic form (selenite) preventing the development of liver necrosis in rats (Schwarz and Foltz, 1957) and exudative diathesis (ED) in chicks on torula yeast diets (Patterson et al., 1957). Within 2 years, a muscular degeneration that occurred naturally in lambs and calves in parts of Oregon and New Zealand was found to respond to selenium therapy. Subsequently, selenium deprivation was found to impair the growth, health and fertility of livestock in areas of many countries and an interrelationship with vitamin E became a fruitful field of research (Thompson and Scott, 1969). A specific biochemical role for selenium emerged with the discovery that glutathione peroxidase (GPX) is a selenoprotein (SeP) (Rotruck et al., 1973) and correlations between GPX activity in tissues and selenium intake were demonstrated.

The essentiality of selenium has since taken on a far more complex perspective with the identification of numerous biologically active SePs (Behne and Kyriakopoulos, 2001), each with its own apparent selenium requirement (Lei et al., 1998; Sunde et al., 2005) and some of which facilitate important interactions with another trace element, iodine (see Chapter 12) (Beckett and Arthur, 2005). For selenium more than any other element, the dietary form determines metabolic fate. However, two myths have been nurtured: that inorganic selenium is poorly utilized by livestock and that a certain form of organic selenium, selenomethionine (SeMet), is simultaneously hyperavailable and relatively non-toxic. An alternative hypothesis is supported here: that the predominant form of selenium in natural feeds, SeMet, can have a lower and more variable availability for important functions than inorganic selenium. The subject of selenium in nutrition and health has occupied whole conferences and books (Hatfield et al., 2006; Surai, 2006) and inspired one government to enforce the selenization of its general fertilizers (Merja, 2005). All of these have addressed human rather than animal concerns, many of which remain unresolved.

©N. Suttle 2010. *Mineral Nutrition of Livestock*, 4th Edition (N. Suttle)

Functions of Selenium

As many as 25 mammalian *SeP* genes have now been identified (Beckett and Arthur, 2005). Some of the SePs that they encode are listed in Table 15.1 and many are expressed in the brain (Schweizer *et al.*, 2004). Most SePs protect the tissue in which they occur from reactive oxygen species (ROS) – by-products of essential oxidative processes – and are believed to play important roles in cell signalling and transcription. If peroxidation gets out of control, it can initiate chain reactions of ROS generation and cause tissue damage. The task of terminating such reactions and protecting against peroxidation is shared by other tissue enzymes (e.g. the superoxide dismutases CuZnSOD and MnSOD, catalase, glutathione-S-transferase) and by non-enzyme scavengers such as vitamin E (MacPherson, 1994). The biochemical and clinical abnormalities caused by a lack of selenium therefore show variable responses to vitamin E. Vitamin E functions as a lipid-soluble antioxidant in cell membranes while selenium, as GPX, is water soluble and acts primarily as an intracellular antioxidant. The complex interplay between antioxidants is illustrated in Fig. 15.1.

Glutathione peroxidases

Several different, immunologically distinct peroxidases utilize glutathione (GSH) as a reducing substrate and their multiplicity and ubiquity reflects the importance of controlling peroxidation. The first peroxidase to be identified and studied in detail was cytosolic GPX1. This is the most abundant GPX in the body, accounting for most of the selenium in erythrocytes and liver. With 4 g atoms Se mol^{-1} in selenocysteine (SeCys) residues (Flohe *et al.*, 1973), the tetrameric enzyme catalyses the reduction of hydrogen peroxide (H_2O_2) or hydroperoxides

Table 15.1. Selenoproteins that have been purified and/or cloned, their location and possible functions (after Beckett and Arthur, 2005).

Nomenclature	Selenoprotein	Principal location	Function
GPX1	Cytosolic glutathione peroxidases (GPX)	Tissue cytosol, red blood cells	Storage, antioxidant
GPX2	Phospholipid hyperoxide GPX	Intracellular membranes, particularly testes	Intracellular antioxidant
GPX3	Plasma GPX	Plasma, kidney, lung	Extracellular antioxidant
GPX4	Gastrointestinal GPX	Intestinal mucosa	Mucosal antioxidant
GPX5	Epididymal GPX	Epididymis	Weak antioxidant
SPS-2	Selenophosphate synthetase 2	Ubiquitous	SeCys biosynthesis
ID1 or ORD1	Iodothyronine 5′-deiodinase type I	Liver, kidney, muscle[a]	Conversion of T4 to T3
ID2 or ORD2	Iodothyronine 5′-deiodinase type II	BAT	Conversion of T4 to T3
ID3 or ORD3	Iodothyronine 5′-deiodinase type III	Placenta	Conversion of T4 to rT3
TR1 and 2	Thioredoxin reductase 1 and 2	Kidney, brain	Redox cycling
SePN	Selenoprotein N	Muscle	Cell proliferation
SePP	Selenoprotein P	Plasma	Transport, metal detoxifier
SePR	Selenoprotein R	Liver, kidney	Methionine sulfoxide reductase
SePW	Selenoprotein W	Muscle	Antioxidant, calcium-binding
MCSeP	Mitochondrial capsular selenoprotein	Sperm mitochondrial capsule	Store for GPX4

BAT, brown adipose tissue; GPX, glutathione peroxidase; rT3, reverse T3; SeCys, selenocysteine; T3, triiodothyronine; T4, thyroxine.
[a]Presence of ID1 in thyroid varies between species, none being found in farm livestock.

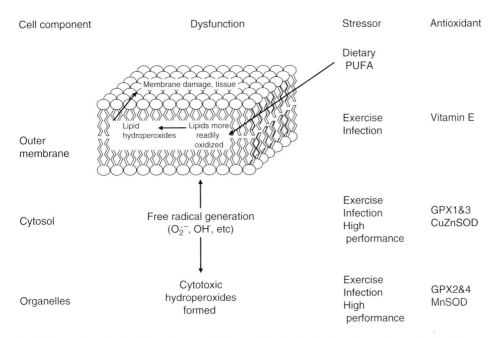

Fig. 15.1. A schematic of cell membrane oxidative metabolism/antioxidant interactions, showing the interplay between glutathione peroxidases (GPX1, 2 and 4) and other micronutrients in protecting membranes and organelles from peroxidation. CuZnSOD, copper/zinc-dependent superoxide dismutase; MnSOD, manganese-dependent superoxide dismutase; $O_2^{.-}$, superoxide anion; $OH^{.}$, hydroxyl radical; PUFA, polyunsaturated fatty acids.

formed from fatty acids and other substances according to the following general reactions, which can create oxidized GSH (GSSG) from either GSH or reduced GSH (GSSH):

$$H_2O_2 + 2GSSH \rightarrow GSSG + 2H_2O$$

$$ROOH + 2GSH \rightarrow ROH + GSSG + H_2O$$

However, blood and tissue GPX1 activities fall to almost undetectable levels without obvious pathological or clinical changes and GPX1 became regarded as a non-essential storage SeP (Sunde, 1994). Animals can survive without the gene for GPX1, although changes are triggered in other SePs that may be functionally more important (Behne and Kyriakopoulos, 2001), such as:

- GPX2, located in the intestinal mucosa where it may afford protection from without (from dietary hydroperoxides) and within, when metabolic activity is intense;
- GPX3, a tetrameric plasma or extracellular peroxidase, synthesized principally in the kidneys and lungs where may it protect delicate structures such as the proximal renal tubule and alveolar capillaries; and
- GPX4, a phospholipid hydroperoxidase monomer associated with intracellular membranes that is 'spared' during selenium deprivation, suggesting a rate-limiting importance (Weitzel et al., 1990).

GPX4 may be responsible for the substitutive relationship between selenium and vitamin E (Arthur and Beckett, 1994a). Testes are rich in GPX4 activity (Lei et al., 1998), which has a dual role as an antioxidant monomer and structural polymer of GPX4 (Behne and Kyriakopoulos, 2001). Substrate (GSH) concentrations in plasma are too low for GPX3 to have a major extracellular protective role, but the mRNA for GPX3 is expressed in mammary tissue (Bruzelius et al., 2007) and the peroxidase provides some of the of selenium in milk.

There is cooperation between the different GPX, non-selenium-containing GSH transferases and vitamin E in assembling robust defences against tissue peroxidation. This may be particularly important in thyrocytes, in which H_2O_2 is produced for thyroid hormone synthesis (Korhle et al., 2005).

Iodothyronine 5'-deiodinase

The membrane-bound SeP iodothyronine 5'-deiodinase type I (ID1) transforms thyroxine (T4) to triiodothyronine (T3), the physiologically active form of the hormone (Arthur and Beckett, 1994b), and selenium deprivation can increase plasma T4 to T3 ratios in cattle (Arthur et al., 1988; Awadeh et al., 1998; Contreras et al., 2005) and sheep (Donald et al., 1994b). A second deiodinase (type II or ID2) can also form T3 from T4, but ID2 is under feedback control from T4 and therefore liable to be doubly inhibited in selenium deprivation. Species differ in the way in which they generate T3, with ID1 being the predominant deiodinase in ruminants and ID2 in non-ruminants (Nicol et al., 1994). While ID1 can be located primarily in the thyroid, liver and kidneys, none is present in the thyroid in farm livestock, which generate most T3 outside the thyroid (for sheep, see Voudouri et al., 2003). Thermogenesis in the newborn is heavily dependent on T3 generation in brown adipose tissue (BAT) and ID2 is abundant in BAT but also in found in brain; ID2 is also expressed in bone cells. A third selenium-containing deiodinase (ID3) has been found in the placenta (Salvatore et al., 1995). Selenium deprivation may thus indirectly influence the basic metabolic rate and a wide range of physiological processes, including parturition and survival during cold stress (Fig. 15.2), with adverse effects on production. Depletion of ID2 in the pituitary of the selenium-deprived rat may reduce growth by inhibiting the release of growth hormones, but concurrent iodine deprivation attenuates that growth inhibition by increasing ID2 activity (Moreno-Reyes et al., 2006).

Thioredoxin reductases

Two ubiquitous SeP thioredoxin reductases (TR1 and TR2) have now been identified, with a third isoform occurring predominantly in testis. TRs are flavoprotein oxidoreductases with

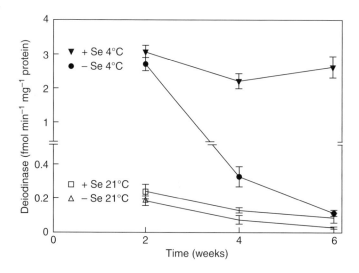

Fig. 15.2. The interaction between selenium metabolism and cold stress in rats. Exposure to low temperatures causes an early increase in deiodinase activity in brown adipose tissue to combat increased oxidative stress in the tissue, but the adaptation cannot be sustained if the diet is low in Se (Arthur et al., 1992).

diverse roles, including reduction–oxidation regulation and hence the regulation of transcription (Korhle *et al.*, 2005), recycling of vitamins C and E and absorption of minerals such as calcium (Moreno-Reyes *et al.* 2006). By maintaining muscle protein in an active, reduced form, TRs may protect muscle from dystrophy (Surai, 2006). Bone turnover decreases in the selenium-deprived rat and TRs may protect osteoblasts and osteoclasts from oxidant stress caused by ROS produced during the bone remodelling process (Moreno-Reyes *et al.*, 2006).

Selenophosphate synthetase 2

Another enzyme, selenophosphate synthetase (SPS)-2, has the pivotal role of catalysing the incorporation of selenide into an SeP, called SePP, the principal constituent of plasma selenium. The SePP molecule has 10 SeCys residues and serves as a selenium transporter while possessing antioxidant properties. Deletion of the *SePP* gene has lethal consequences in mice and local expression in tissues such as the testes and brain is vital (Renko *et al.*, 2008). SePP has the potential to complex heavy metals and may incidentally provide protection from the toxicity of elements such as cadmium and mercury (Hill, 1974).

Other forms and functions

Enough similar monomeric, small-molecular-weight SePs have now been identified to take up nearly half the alphabet. SePW was the first to be isolated from the heart and muscles, where concentrations decline during selenium depletion. SePW is conserved in brain and has a calcium-binding domain, providing a possible link to the calcification that occurs in the dystrophic muscles of animals deprived of selenium. SePN may play a role in cell proliferation and early development (Surai, 2006).

Specific Clinical Manifestations of Selenium Deprivation

The clinical consequences of selenium deprivation vary widely within and between species and some may reflect cellular vulnerability to oxidant stress at critical stages of development. Detailed descriptions of specific clinical disorders are given in the previous edition (Underwood and Suttle, 1999). They have not changed much, are readily diagnosed and treated, and are simply summarized in Table 15.2.

Non-specific Selenium-responsive Disorders

The term 'non-specific' here refers to disorders that can clearly respond to selenium treatment (selenium-responsive disorders, SeRD), but that resemble other conditions with different aetiologies. They can occur with or without specific clinical disorders.

Table 15.2. A summary of specific clinical disorders that respond to selenium supplementation.

Disorder	Description and consequence	Predilection site	Affected species	Responsiveness to vitamin E
White muscle disease	Dystrophy of striated muscle: myopathy	Skeletal muscle, heart, gizzard	A, B, Ca, Ce, E, O, P	Rare in ruminants, ++ in A, P
Exudative diathesis	Increased permeability of capillaries: oedema, swelling and bruising	Thorax, neck, wing	A > P	++
Hepatosis	Necrosis	Liver	P	+
Mulberry heart disease	Microangiopathy	Heart > brain	P	++ on PUFA-rich diets

A, avian; B, bovine; Ca, caprine; Ce, cervine; E, equine; O, ovine; P, porcine; PUFA, polyunsaturated fatty acids.

Reproductive disorders

Selenium deprivation reduces egg production in hens and conception rate, litter size and piglet survival in sows (Underwood and Suttle, 1999). In boars, sperm development is particularly sensitive to selenium supply since a level that is adequate for growth (0.06 mg Se kg^{-1} dry matter (DM)) has been associated with low sperm motility, a high proportion of 'tail abnormalities' and low fertilization rates (Marin-Guzman et al., 1997). Similar observations have been made in cockerels (Surai et al., 1998). In ewes, high embryonic mortality around implantation has been attributed to a lack of selenium (Hartley, 1963; Wilkins and Kilgour, 1982). In a New Zealand study, some 20–50% of ewes were infertile unless additional selenium was given before mating, but neither vitamin E nor a synthetic antioxidant afforded protection (Table 15.3). Fertility seems less vulnerable to selenium deprivation in grazing cattle (Wichtel et al., 1998), but has been improved by selenium supplementation in housed heifers (MacPherson et al., 1987). In cattle, delayed expulsion of the afterbirth (retained placenta) can be a selenium-responsive condition (Underwood and Suttle, 1999). However, herd incidence can vary tremendously from year to year with no clear indication of cause, although dietary vitamin E supply is one factor (Bourne et al., 2006). Selenium supplementation has reduced the incidence of endometritis and cystic ovaries (Harrison et al., 1984) and indirect improvements in fertility may result from decreases in these and other reproductive disorders, including retained placenta. Male fertility may be adversely affected and reduced viability of semen has been reported in selenium-deprived bulls (Slaweta et al., 1988).

Neurodegenerative disorders

Neurodegenerative disorders have rarely been reported in farm livestock, but transgenic mice lacking the ability to express certain SePPs develop neurological abnormalities, and it has been suggested that the priority normally given to selenium supply to the brain preserves its integrity (Schweizer et al., 2004). Results of cognitive behaviour tests in humans correlate with blood selenium status and similar effects in livestock may have passed unnoticed.

Peri-natal and post-natal mortality

Selenium supplementation of the pregnant ewe has improved lamb viability (Table 15.3) (Kott et al., 1983). The probability of survival increased from 0.61 to 0.91 in the first 5 days of life in one study, but administration of an iodine antagonist (thiocyanate) during pregnancy prevented the response (Donald et al., 1994b). Iodine deprivation up-regulates ID2 in tissues (Moreno-Reyes et al., 2006). Selenium deprivation may have restricted the thermogenic response to cold stress by lowering ID2 activity in BAT (see Chapter 12). The survival of older lambs can also be compromised by selenium deprivation. Injecting 5-month-old lambs with selenium has been found to reduce mortality from 27% to 8% (Hartley, 1967) and treatment of Merino lambs regularly from marking time reduced mortality from 17.5% to 0% (McDonald, 1975). In Japanese quail, the viability of newly hatched chicks was impaired by selenium deprivation, and hatchability is regarded as the most sensitive criterion of deprivation in poultry (Underwood and Suttle, 1999).

Table 15.3. The effects of pre-mating treatment with selenium, vitamin E or synthetic antioxidant on ewe fertility (200 ewes/group) (from Hartley, 1963).

	Untreated	Selenium	Vitamin E	Antioxidant
Barren ewes (%)	45	8	50	43
Lambs born/100 ewes lambing	105	120	109	112
Lamb mortality (%)	26	15	16	14
Lambs marked/100 ewes lambing	43	93	46	54

Lowered disease resistance

Claims that susceptibility to infection is an early consequence of selenium deprivation far outnumber proven cases of cause and effect. Selenium supplementation has shortened episodes of bovine mastitis without reducing their frequency, but injections of barium selenate decreased the incidence of mastitis in dairy goats (Sanchez et al., 2007). Adding selenium yeast to the diet has decreased episodes of diarrhoea in calves (Guyot et al., 2007), but an unusual beef breed, the double-muscled Belgian Blue, was used in this study and a probiotic response to yeast cannot be ruled out. Inverse relationships have been reported between blood GPX1 and *Staphylococcus aureus* infections of the udder across a high range of mean blood selenium (90–360 μg l^{-1}) (Jukola et al., 1996), but the relationship may not have been causal. Prolonged selenium depletion does not impair resistance to viral infection in calves (Reffett et al., 1988) or nematode infection in lambs (Jelinek et al., 1988; McDonald et al., 1989). Marginal selenium depletion has lowered the resistance of chicks to the protozoan parasite *Eimeria tenella* (Colnago et al., 1994). Deaths and lesions in chickens following *Escherichia coli* infection have been reduced from 86% to 21% by adding 0.4 mg inorganic selenium to a diet marginal in selenium (0.05 mg kg^{-1} DM) just 1 day before challenge, but was ineffective if the birds were chilled (Larsen et al., 1997).

Blood disorders

Heinz-body anaemias have been reported in selenium-deprived steers (Morris et al., 1984) and lambs (Suttle et al., 1987) and attributed to peroxidative damage due to low activities of GPX1 in erythrocytes. Post-parturient haemoglobinuria can develop in dairy cows turned out to graze spring pasture and there are circumstances in which low erythrocyte GPX1 activities are believed to be a contributory cause but the disease is not a specific consequence of selenium deprivation (Ellison et al., 1986).

Subclinical Selenium Deprivation in Sheep and Cattle

The term 'subclinical' is used to describe a group of abnormalities that have no distinctive pathological features and are mild in effect. Such SeRD were first identified in parts of Australia and New Zealand as selenium-responsive 'ill-thrift' in lambs at pasture and in beef and dairy cattle of all ages (Hartley, 1967; McDonald, 1975). The underlying dysfunctions are unknown but, in economic terms, 'ill-thrift' is probably the most important manifestation of selenium deprivation.

Wool production

Wool production is the most sensitive index of selenium deprivation (Gabbedy, 1971) and is often improved by selenium supplementation (e.g. Slen et al., 1961; McDonald, 1975). In a 4-year study in New South Wales, fleece yields from ewes were increased by 3.8–7.5% each year with selenium supplementation (Langlands et al., 1991a) and their lambs produced 9.5% more wool without themselves being supplemented (Langlands et al., 1990). The increase in wool yield was greater in ewes that reared a lamb to weaning than in those that did not, reflecting the priority given to the demands of pregnancy. Large responses (22%) in greasy wool yield were reported at two successive shearings in Western Australia following selenium supplementation (Whelan et al., 1994b). There is a correlation between wool and body growth (see Chapter 9), but responses to selenium may be more precisely measured in wool yield.

Growth retardation

Selenium supplementation has improved lamb growth in many countries (Underwood and Suttle, 1999). The response is progressive and, by weaning, a 10.8% improvement was recorded in New South Wales flocks (Langlands et al., 1990). Responses were greater in Merino than in cross-bred lambs and at the higher of two widely different stocking rates, occurring without improvements in ewe fecundity (Langlands et al., 1991b). The degree of growth retardation in lamb flocks in New Zealand has been found to increase exponentially as their mid-trial mean blood selenium status declined (Grace and Knowles, 2002). Vitamin E does not stimulate growth in these situations and the underlying defect may be lack of deiodinase activity rather than poor antioxidant defence. By contrast, cattle can grow without restriction on pastures that provide insufficient selenium for sheep (Langlands et al., 1989). Growth responses to selenium in cattle have been recorded in herds showing concomitant hypocupraemia and maximal responses were only obtained if copper was given with selenium (Gleed et al., 1983; Koh and Judson, 1987). Growth is also retarded by selenium deprivation in broiler chickens (Jianhua et al., 2000).

Loss of milk yield

Milk production from grazing dairy cows can be impaired by selenium deprivation, but only when blood and pasture selenium concentrations are exceedingly low (Tasker et al., 1987). The main effect is a reduction in milk fat yield (Fraser et al., 1987; Knowles et al., 1999). This raises the possibility of an effect on lipid metabolism via pancreatic dysfunction, but this has yet to be studied in selenium-deprived ruminants. Pancreatic acinar cells are the first to show abnormalities in cattle deprived of copper (see Chapter 11). The effects on milk yield in sheep and goats have not been specifically studied, but growth retardation in the suckling (e.g. Langlands et al., 1990) could be a secondary effect of reduced milk yield in the nursing ewe.

Cumulative effects

The combined effects of improvements at different stages of production can be substantial. In early New Zealand trials, highly significant weight gains in surviving lambs were observed. When combined with the increased number of lambs born and surviving, an improvement of 1513 kg in the total weaned lamb live weight (LW) was obtained by supplementing a group of 201 ewes with selenium (Hartley, 1967). In eastern Australia, the increase in total annual fleece production of the Se-treated groups (40 lambs) over untreated groups (33 surviving lambs) was 47 kg, or 39% (McDonald, 1975).

Dietary Sources of Selenium

The provision of adequate dietary supplies of selenium is clearly important, but is complicated by the fact that selenium concentrations in feeds and forages vary exceptionally widely, depending on the plant species, season of sampling and selenium status of the soil. The dominant form of selenium in ordinary feeds and forages is protein-bound SeMet, which accounts for 55–65% of the selenium in cereals and more in protein concentrates. There are smaller proportions of SeCys (5–15%) and other seleno-compounds (Whanger, 2002).

Forages

Forage selenium can range widely both within and between types on a given farm (Table 15.4), making it difficult to assess total selenium intake in ruminants. In the UK, the average selenium concentration in fresh grass is 0.05 ± 0.02 (standard deviation, $n = 71$) mg kg^{-1} DM (MAFF, 1990), but in Australia and New Zealand, where selenium-responsive disorders have occurred more frequently, forage selenium may be 50% lower (Langlands et al., 1989; Whelan et al., 1994b; Grace and Knowles, 2002). Legumes tend to contain less selenium than grasses, but the difference diminishes as soil selenium status declines (Minson, 1990) and lucerne is an exception to the rule. The importance of species is strikingly illustrated by 'accumulator' plants that survive in seleniferous areas where other species cannot, with some accumulating selenium in macro- rather than micro-quantities in so doing (p. 409).

Table 15.4. Variation in selenium concentrations in forages sampled throughout the year on a Louisiana farm (Kappel et al., 1984).

Feed (no. of samples)	Mean Se (µg kg^{-1} DM)	Standard deviation	Range
Bahia grass (15)	61	20	27–98
Bermuda grass (5)	173	158	71–448
Ryegrass/oats (9)	72	19	45–105
Corn silage (10)	59	25	27–100
Sorghum silage (6)	57	12	39–74
Alfalfa hay (9)[a]	295	197	51–954

[a]Some of the alfalfa hays were bought in.

Cereals and legume seeds

Cereal grains and other seeds vary widely in selenium content between and within countries. The average values for wheat grown in the UK, the USA and Canada have been calculated at 0.03, 0.37–0.46 and 0.76 mg kg^{-1} DM, respectively (Adams et al., 2002). Differences between grains are relatively small, but they have been ranked in the order wheat > rice > maize > barley > oats (Lyons et al., 2005). Linseed meal is generally rich in selenium compared with extracted soybean meal (SBM) and rapeseed meal (0.82 ± 0.05, 0.30 ± 0.19 and 0.14 ± 0.09 mg kg^{-1} DM, respectively) (MAFF, 1990), whereas lupinseed meals can be extremely low in selenium (<0.02) (Moir and Masters, 1979). Cereal by-products such as maize gluten feed, rice-bran meal and distillers' dark grains are good sources of selenium (0.24 ± 0.09, 0.16 ± 0.06 and 0.18 ± 0.18 mg kg^{-1} DM, respectively) (MAFF, 1990).

Animal protein supplements

The only protein concentrates now in common use that add substantial amounts of selenium to compounded rations are of marine origin, with salmon and herring meals being rich sources (1.9 mg Se kg^{-1} DM) (Miltimore et al., 1975) and tuna-fish meal containing as much as 5.1–6.2 mg Se kg^{-1} DM (Scott and Thompson, 1971). Animal protein concentrates of non-marine origin contain less selenium and can be highly variable (0.11–1.14 mg Se kg^{-1} DM) (Moir and Masters, 1979), but in the UK values for two products, blood and feather meals, have been relatively consistent (0.6±0.08 and 0.76 ± 0.06 mg Se kg^{-1} DM, respectively) (MAFF, 1990).

Colostrum and milk

As with other minerals (see Chapter 2), selenium concentrations are much higher in colostrums than in main milks, with respective mean values for unsupplemented groups of 75 and 40 µg l^{-1} for sows (Mahan and Peters, 2004), 70 and 40 µg l^{-1} for cows (Awadeh et al., 1998) and 81 and 23 µg l^{-1} for ewes (Meneses et al., 1994). Colostral and milk selenium can be doubled by parenteral selenium supplementation of the ewe at lambing, but the effect can be short-lived (14 days) (Meneses et al., 1994; Rock et al., 2001). The similarity in milk selenium concentrations in the different species masks the fact that, when expressed on a DM basis, they are higher in cows (approximately 0.32 mg kg^{-1} DM) than in the sow and ewe (approximately 0.2 mg kg^{-1} DM).

Selenium Absorption

For most elements, the nutritional value of a given source is heavily dependent on the efficiency with which it is absorbed. For selenium, however, this is only half the story. Inorganic selenium as selenite can be passively absorbed, while selenate shares an active absorptive pathway with molybdate and sulfate (SO$_4$) and may be vulnerable to antagonism from these anions (Cardin and Mason, 1975). By contrast, SeMet and SeCys follow the routes of methionine and cysteine across the small intestinal mucosa and are absorbed as intact molecules (Fig. 15.3) (Vendeland et al., 1992). The value of a selenium source will therefore be influenced by dietary protein supply and scope for microbial protein synthesis.

Non-ruminants

Early studies with rats showed that a high proportion (0.85) of the selenium ingested as selenite was absorbed. Selenite became widely used as the highly absorbable reference source in evaluating organic selenium sources, although selenate generally has a higher availability – the margin being 1.3-fold in poultry (Henry and Ammerman, 1995). There are no published results for true absorption (A_{Se}) or apparent absorption (AA_{Se}) of selenium in feeds for poultry, but an early balance study in young pigs given a corn/SBM diet found AA_{Se} of 0.72 and 0.64, respectively, when selenite or seleniferous corn provided a 0.2 mg Se kg^{-1} DM supplement (Groce et al., 1973). The difference in favour of inorganic selenium was significant. A stable isotope study in human volunteers reported an A_{Se} of 0.95 and 0.50 from selenate and selenite, respectively, from an infant-formula diet, but retention was similar (Van Dael et al., 2001).

Ruminants

Selenium is absorbed by pre-ruminants with a similar high efficiency to that seen in non-ruminants. In calves given milk replacer, AA_{Se} decreased exponentially from 0.90 as dietary selenium was increased but remained high (0.6) with selenium intake 100 times that required (Jensen and Hidiroglou, 1986) suggesting little, if any, homeostatic control over selenium absorption. With the development of a functioning rumen, AA_{Se} declines to 0.30–0.59 (Table 15.5), but low (even negative) values are obtained if feeds low in selenium are fed to sheep of high selenium status (Langlands et al., 1986). Ingested selenium leaves the rumen mostly in the bacterial fraction (Koenig et al., 1997) and studies with mice suggest that selenium in microbial protein is less available to mice than inorganic selenium (Serra et al., 1997). Some feed selenium amino acids may escape undegraded from the rumen. In sheep given a standard pelleted diet, ^{75}Se-labelled selenite and ^{75}SeMet are equally well absorbed (Ehlig et al., 1967), but less SeMet may be retained when given intravenously (Fig. 15.4). When a roughage or concentrate diet was fed (Koenig et al., 1997), absorption of ^{75}Se was apparently lower from SeMet with the roughage diet. In lambs given a concentrate diet relatively low in selenium and sulfur (0.07 and 1.6 g kg^{-1} DM, respectively), ^{75}SeMet had a high AA_{Se} of 0.8 (White et al., 1998). Additions of SO_4 lowered AA_{Se} in a selenite-supplemented corn silage diet from 0.39 to 0.29 in lactating cows (Ivancic and Weiss, 2001) and reduced the uptake of inorganic selenium by rumen microbes in sheep (van Ryssen et al., 1998). Low A_{Se} may have been responsible for the relatively low selenium status of steers given sulfur-rich molasses and selenium yeast supplements (Arthington, 2008).

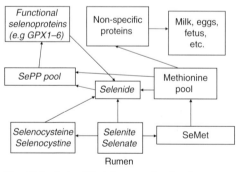

Fig. 15.3. A simplified diagram showing the contrasting movements of selenomethionine (SeMet), selenocysteine and inorganic selenium after the ingestion of food by ruminants (after Waschulewski and Sunde, 1988; Burk et al., 2001). In the rumen, inorganic Se can be incorporated into selenoamino acids. Transsulfuration or transamination is required to allow SeMet to enter the Se-transport (selenoprotein P, SePP) pool and then be incorporated into functional proteins such as GPX, glutathione peroxidases.

Table 15.5. The apparent absorption of selenium (AA_{Se}) by pre-ruminant calves and sheep. The data suggest that AA_{Se} decreases with the development of a rumen, increasing dietary selenium and roughage feeding.

Diet	Se source	Animal	Diet Se (mg kg^{-1} DM)	AA_{Se}	Reference
Milk replacer	Selenite	Calf	10.0	0.61	Jenkins and Hidiroglou (1986)
	Inherent	Calf	0.2	0.90	Jenkins and Hidiroglou (1986)
Roughage	Inherent	Sheep	0.37	0.42	Koenig et al. (1997)
	Inherent	Sheep	0.16	0.50	Langlands et al. (1986) expt. 3[a]
	Inherent	Sheep	0.15	0.41	Krishnamurti et al. (1989)
	Inherent	Sheep	0.10	0.45	Krishnamurti et al. (1997)
	Inherent	Sheep	0.01	0.59	Krishnamurti et al. (1997)
Concentrate	Inherent	Sheep	0.27	0.53	Koenig et al. (1997)
	Mixed	Sheep	0.24	0.67	Langlands et al. (1986) expt. 1
	Inherent	Sheep	0.06	0.63	Alfaro et al. (1987)

[a]Large effects of previous selenium intake and selenium status on AA_{Se} here suggest possible errors in making comparisons between studies.

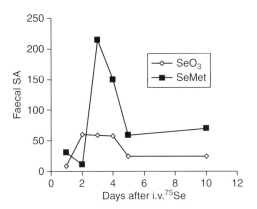

Fig. 15.4. Confirmation of the distinctive metabolism of selenite (SeO$_3$) and selenomethionine (SeMet) is provided by data for faecal specific radioactivity (SA) after intravenous (i.v.) doses of each source in sheep continuously fed a pelleted diet. Note the poorer retention of SeMet (data from Langlands et al., 1986).

Metabolism

Discordance between A_{Se} and available selenium is attributable to a major difference in post-absorptive metabolism between SeMet and other sources of selenium (Fig. 15.3) (Burk et al., 2001). This has 'knock-on' effects on selenium retention, excretion and both placental and mammary transfer.

Separate pathways

SeMet enters the methionine pool and a variable proportion goes where methionine rather than selenium is required, but partial conversion to SeCys via a lyase and adenosylmethionine is possible (NRC, 2005). SeCys is rapidly incorporated into SePP in the liver and exported into the plasma (Davidson and Kennedy, 1993), from where it is taken up and incorporated into one of many functional SePs in the tissues. Selenite and selenate are reduced to selenide and incorporated into SePP. Oral and parenteral doses of ^{75}SeMet are similarly metabolized; after passage through the liver, clearance from the bloodstream is very slow (half-life in plasma of 12 days). Most of the label is retained in muscle (White et al., 1988) and the ^{75}Se retained in the liver and kidneys is associated with protein (Ehlig et al., 1967). By contrast, clearance of SeCys or inorganic ^{75}Se is rapid. Incorporation of SeCys into erythrocyte GPX1 occurs at erythropoiesis and there is a lag before the newly 'packaged' GPX1 is released into the bloodstream. Labelled ^{75}SeMet, on the other hand, can be incorporated into erythrocytes as methionine in haemoglobin (Beilstein and Whanger, 1986). Some transfer of selenium from SeMet to SeCys occurs during 'transsulfuration' (see Chapter 9) or transamination (Fig. 15.3). Unless and until that happens, SeMet (but not SeCys) is influenced by the supply and demand for methionine. If the diet is deficient in methionine, supplementation with SeMet can increase tissue selenium while GPX activity declines (Waschulewski and Sunde, 1988). When methionine requirements are high, as in early lactation and in suckled offspring, one would expect both the milk and newly deposited muscle to be enriched with SeMet at the expense of GPX in both the dam and offspring. In ruminants, selenium metabolism will be indirectly affected by the degradable sulfur and nitrogen supply and other factors that affect rumen microbial protein synthesis (see Chapter 9). Reviews of the regulation of SeP can be found elsewhere (Hatfield et al., 2006).

Retention

Selenium accumulates in tissues at rates that vary with dietary selenium concentration, source and tissue site. This is shown in Fig. 15.5, which uses data from a survey designed to show the effect of natural variations in selenium concentrations in locally grown crops across the USA on the selenium status of the fattened pig (Mahan et al., 2005). At most experimental stations, selenite was added to provide $0.3\,mg\,Se\,kg^{-1}$ DM. After calculating the contribution of organic selenium from natural feeds by difference, highly significant regressions were found with body selenium accretion, with hair (particularly rich in sulfur amino acids) selenium increasing faster than liver or loin selenium with increases in dietary organic selenium. The intercepts shown for the three slopes represent selenium derived from the inorganic selenium supplement (0.3). From a similar contribution, organic selenium gave 3.3 times more selenium in the loin and 1.7 times more selenium in the hair, but less in the liver (0.67) than inorganic selenium.

The liver is probably exceptional because it processes all forms of selenium from the portal blood flow. In weaned lambs given 2.9 mg Se as seleniferous wheat, in addition to 0.2 mg kg^{-1} DM from other components, selenium concentrations in muscle, heart and wool increased linearly with time to triple after 56 days; levels in the liver and kidneys increased faster to reach a plateau after 14–28 days (Taylor, 2005). Large and complex effects of high intakes of inherent versus inorganic selenium (65 µg kg^{-1} LW) on tissue selenium accretion have been demonstrated in cattle (Lawler

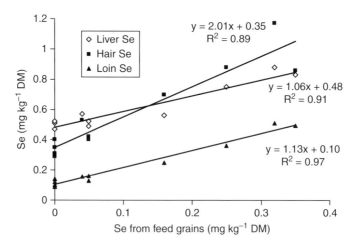

Fig. 15.5. The relationship between the natural (organic) selenium concentration in pig rations and the selenium concentration found in liver, loin muscle and hair at commercial slaughter in collaborative US trials (Mahan et al., 2005) selecting those in which a standard supplement of 0.3 mg inorganic Se kg^{-1} DM was added. Assuming that the intercepts represent the contribution to Se accretion coming from the inorganic source, organic Se causes relatively large increases in hair and muscle Se, but not in liver Se.

et al., 2004). The kidney was the richest tissue in selenium, containing three- to sixfold higher concentrations than the poorest (muscle), but was less affected by selenium source than other tissues.

Responses in tissue (or blood) selenium do not always correspond to changes in functional SeP. Young pigs given supplements of selenite have been found to have higher tissue selenium but lower serum GPX3 activity than those given selenium yeast (Mahan et al., 1999). Giving cows selenium yeast rather than selenite lowers the slope of the relationship between blood GPX and blood selenium, but not that between liver and blood selenium (Knowles et al., 1999). Selenium that accumulates during times of plenty is mobilized during selenium depletion at widely different rates depending on the speciation of tissue selenium, particularly the proportions of SeMet to SeCys, and metabolic activity of the tissue (Juniper et al., 2008). Selenium that has accumulated as SeMet in tissue proteins cannot be mobilized without catabolism of the protein. Thus, selenium in muscle cannot be regarded as a 'store' in the same sense that copper in the liver is a store.

Excretion

Selenium can be lost from the body by exhalation of methylselenol (CH_3SeH), urinary excretion or faecal endogenous excretion (FE_{Se}). Biliary secretion can amount to 28% of selenium intake in sheep: although most is reabsorbed, the remainder contributes significantly to FE_{Se} at low organic selenium intakes in sheep (Langlands et al., 1986) and cattle (Koenig et al., 1991). Injected inorganic selenium accumulates principally in the liver and is extensively secreted via the bile in sheep (Archer and Judson, 1994), but estimates of total FE_{Se} have been twice as high when the body selenium pool is labelled with $^{75}SeMet$ than with $^{75}Se_2O_3$ (Langlands et al., 1986), at 17% as opposed to 9% of selenium intake. In selenium-loaded pigs, biliary selenium secretion increases as dietary selenium increases, irrespective of dietary source (selenite or selenium yeast) (Kim and Mahan, 2001). Losses of selenium in urine vary widely from <10% (Krishnamurti et al., 1989) to 40–50%, irrespective of source (organic, Langlands et al., 1986; inorganic, Ivancic and Weiss, 2001), and were three times higher in sheep of high than of low selenium status when given the same diet (Langlands et al., 1986). Supplementation with SeMet or selenite (0.26 mg kg^{-1} DM) produces similar increases in urine selenium concentration in lactating cows (Juniper et al., 2006). When a selenite supplement for pigs was raised from 0.1 to 0.5 mg Se kg^{-1} DM, 80% of the extra selenium was excreted via the urine (Groce et al., 1971).

Maternal transfer

Supplementation with selenium during pregnancy increases selenium status in the newborn lamb (Langlands et al., 1990; Rock et al., 2001), calf (Pehrson et al., 1999; Gunter et al., 2003) and piglet (Mahan and Peters, 2004; Yoon and McMillan, 2006), but the parenteral route is more effective than the oral route. The concentrations of selenium in colostrum and main milk can both be raised by inorganic selenium supplements in sheep (Meneses et al., 1994; Rock et al., 2001), goats (Zachara et al., 1993) and cattle (Conrad and Moxon, 1979; Rowntree et al., 2004), but SeMet is by far the more effective. Adding selenium yeast (Se-Y) to the diet raises milk selenium in the dairy cow in a dose-dependent manner with a slope eight times that obtained with selenite or a 'chelated' selenium source (Givens et al., 2004), but the effect of source was only marginal in ewes given 0.3 mg Se kg^{-1} DM. The crucial but as yet unanswered question is this: is SeMet continuing to behave distinctly and is the increase in milk selenium matched by an increase in selenium value or functional availability?

Selenium Availability

Functional tests

Selenium sources appear to vary widely in 'availability', but early studies in chicks showed that ranking depended on the index of response (Fig. 15.6) (Cantor et al., 1975a,b). There is a striking difference in relative 'availability' between selenoamino acids: SeCys has a similar value to selenite in all species, regardless of test method, whereas SeMet is highly variable and is sometimes *less* available than selenite (Henry and Ammerman, 1995). The true value of a selenium source can become clearer if results are separated according to method into 'functional' and 'non-functional' availability tests. An inferiority of SeMet for poultry becomes apparent when selenium sources are evaluated by methods that measure incorporation into a functional SeP (e.g. GPX) or resistance to certain diseases (ED) (Table 15.6). Since selenium sources do not differ greatly in absorbability or retention, they must differ in availability for SeP synthesis. From the earlier description of the effect of selenium form on metabolism, it is clear that SeMet can accumulate in tissues in forms that have a relatively low functional availability (Waschulewski and Sunde, 1988). As far as the selenium nutrition of livestock is concerned, the low functional availability of SeMet indicates that it may be part of the problem in terms of SeRD, rather than a panacea.

Functional tests present their own problems of interpretation because requirements for different functions vary. In comparisons of selenium in buckwheat bran, SeMet and selenite in selenium-deprived rats, selenite was generally the most effective in repleting the rats, particularly when liver thioredoxin

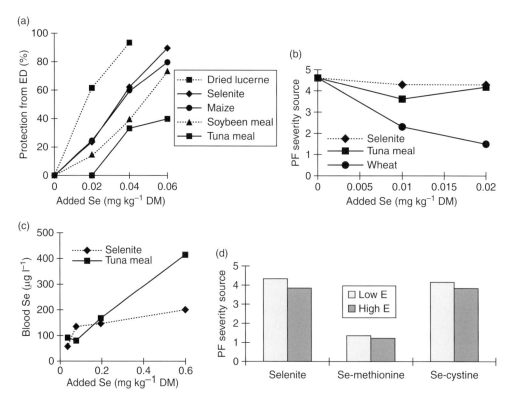

Fig. 15.6. The nutritive value of different selenium sources for chicks is determined by the test system as well as the source (data from Cantor *et al.*, 1975a,b). (a) Protection against exudative diathesis (ED) is high from lucerne meal and low from a fish meal with selenite intermediate. (b) Protection against pancreatic fibrosis (PF) is equally poor with selenite and a fish meal, but good with a cereal source. (c) The ability to raise blood selenium is higher for a fish meal than for selenite. (d) Protection against PF is good with selenomethionine, the major source of selenium in cereals, and may explain the good performance of wheat in (b).

Table 15.6. The impact of test method on the 'availability' of selenoamino acids relative to selenite in poultry. 'Functional' tests are those involving either glutathione peroxidase (GPX) assays or resistance to exudative diathesis; non-specific tests involve selenium assays in blood, tissues, whole body or products (test results are quoted on pp. 315–316 of Henry and Ammerman, 1995).

	Test method	Number	Mean (SE)	Coefficient of variation (%)	Range
Selenocyst(e)ine	Functional	4	1.05 (0.15)	15	0.74–1.47
	Non-specific	11	1.06 (0.04)	4	0.89–1.23
Selenomethionine	Functional	13	0.79 (0.08)[a]	10	0.37–1.69
	Non-specific	13	1.03 (0.05)[a]	5	0.72–1.33

SE, standard error.
[a]$P = 0.022$ for statistical significance of difference between the functional and non-specific test method.

reductase was the yardstick (Reeves et al., 2005). The dose–response profile of each function must be known if valid assessments of availability are to be made. For example, plateaus for GPX3 in plasma and GPX1 and 4 in the thyroid and pituitary were reached with 0.2 mg Se kg^{-1} DM, but >0.3 mg Se kg^{-1} DM was required to achieve plateaus in GPX1 and 4 in the liver, heart and lungs of young pigs (Lei et al., 1998). The majority of studies involve only two levels of selenium supplementation – the highest being the level permitted by the US Food and Drug Administration (FDA) (e.g. 0.3 mg Se kg^{-1} DM) – and functional availability cannot be accurately compared in such trials (e.g. Mahan and Peters, 2004). Furthermore, studies with rats indicate that there is down-regulation of plasma GPX3 and both erythrocyte and liver GPX1 in pregnancy and lactation, with plateau dietary selenium falling from 0.075–0.1 to 0.05–0.075 mg Se kg^{-1} DM (Sunde et al., 2005), thus lowering the limit to dietary selenium below which valid comparisons of availability are attainable. The relative availability of selenium sources cannot be measured unless either none has induced plateau GPX activity at the chosen site and level of selenium supplementation or response curves are compared.

Non-specific tests

Tests that involve protection from selenium deprivation (e.g. ED incidence) of necessity involve the use of special diets low in selenium. It is much easier to conduct tests with selenium added in different forms to diets already adequate in selenium (e.g. Henry et al., 1988), but they may indicate relative availability for excretion or milk secretion. Since SeMet can partially bypass selenium-specific regulatory mechanisms for dealing with excesses, its selenium may continue accumulating in the body while selenium from other sources is being increasingly excreted when provided at levels far above requirements. Enhanced transfer of selenium from SeMet to the newborn or mother's milk reflects the rate of transfer of methionine across the placental and mammary barriers, rather than the intrinsic availability of the selenium. However, greater placental transfer of SeMet probably leaves the mother with less selenium compared with inorganic sources, and benefits in maternal blood GPX1 from giving SeMet or selenium yeast rather than selenite have rarely been found. Although there may sometimes be a benefit in GPX status in newborns from dams given SeMet (e.g. Rock et al., 2001; Gunter et al., 2003), the functional availability of extra SeMet in milk to the suckling may be relatively low.

Figure 15.7 shows that large increases in milk selenium produced by supplementing the sow with selenium yeast rather than selenite are not accompanied by greater activities of GPX3 in sow and piglet serum at weaning. Given the 20% higher intakes of selenium by piglets nursed by sows given selenium yeast, a reduction in functional availability of SeMet-enriched milk is

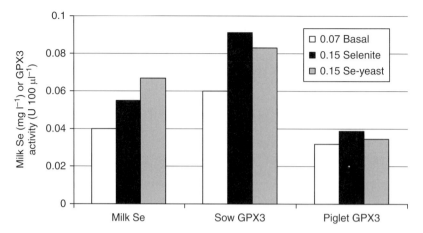

Fig. 15.7. Supplementing sows with selenium yeast (Se-yeast) gives greater increases in milk Se but smaller increases in serum glutathione peroxidase (GPX)3 activity in sows and their offspring at weaning when compared with selenite (providing 0.15 mg added Se kg^{-1} DM) (data from Mahan and Peters, 2004).

indicated that is in line with the low functional availability of SeMet for solid diets (Table 15.6) and with estimates of A_{Se} for cow's milk compared with selenite given to rats (relative value 0.82) (Mutanen et al., 1986). The 'disadvantage' for SeMet is of the same order as the 'advantage' over selenite in terms of relative transfer efficiencies of selenium into milk methionine (+10– +20%) (Givens et al., 2004). The high efficiency of conversion of milk protein to muscle protein in the neonate probably leads to high levels of sequestration of SeMet into muscle at the expense of SeP synthesis.

Tissue selenium accretion in the newborn and milk selenium secretion overestimate physiological availability to the supplemented animal.

Problems with tracers

Tracers for dietary selenium are likely to give false estimates of fluxes from the gut to the body in sheep. Alfaro et al. (1987) found an AA_{Se} of 0.65 for ^{75}Se given orally as selenious acid, whereas stable selenium balance data showed an AA_{Se} of 0.30 for the basal diet of hay plus concentrates. Krishmamurti et al. (1989) found a TA_{Se} of 0.2 for a hay by isotope dilution after parenteral administration of labelled inorganic ^{75}Se, but stable balance data showed an AA_{Se} of 0.41. In cattle given a grass hay, exceedingly low TA_{Se} coefficients of 0.10–0.16 have been reported (Koenig et al., 1991), but the stable isotope study required preliminary tissue enrichment with 4 mg ^{76}Se as selenite, orally, for 5 days. The partition of absorbed inorganic selenium between body compartments will differ from that of organic, dietary selenium, invalidating the principle on which isotope dilution techniques rest (Chapter 2, p. 28).

Selenium Requirements

The selenium intake required to avoid clinical and subclinical SeRD cannot be fixed because it will increase as the oxidant stress presented by a given combination of diet and environment increases, a stress that is itself an outcome of a struggle between multifactorial pro- and antioxidant forces. Thus, there is probably a 'hierarchy' of requirements for a particular rate of production (Table 15.7), but the precise extent to which stressors increase the need for selenium is ill-defined. If the objectives are changed to the requirement for optimal activity of a particular SeP then a different hierarchy of requirements can be produced that is far higher than the requirement for growth in a stress-free laboratory environment (Sunde et al., 2005).

Table 15.7. A suggested hierarchy of selenium requirements and critical blood glutathione peroxidase activities in farm livestock under field circumstances that provide different degrees of oxidant stress.

Vitamin E	PUFA	Temperature stress	Se requirement (mg kg^{-1} DM)	Critical GPX1 (U g^{-1} Hb)	Field circumstances
High	Low	Low	0.02–0.03	20–30	Mid-season grass pasture, vitamin E-supplemented pigs/poultry
	Medium	Low	0.03–0.04	30–40	Spring grass, legume sward
Low	Low	Low	0.04–0.06	40–60	Brown sward, poor hay, grain feeding
	High	Low	0.1–0.2	80–100	PUFA-supplemented diets for production of 'healthy' foods
High	Medium	High	0.04–0.06	40–60	Outdoor spring calving, lambing or farrowing; pigs/poultry in hot weather
Medium	Medium	Medium	0.06–0.08	60–80	High-yielding cow with mastitis

GPX, glutathione peroxidase; PUFA, polyunsaturated fatty acids.

Sheep

From an analysis of the relationship between field disorders and forage selenium, the ARC (1980) concluded that the level of selenium to avoid white muscle disease (WMD) was 0.03–0.05 mg kg^{-1} DM. Shifting the target to optimal growth and narrowing the context to Michigan and hay/grain diets, a similar conclusion was reached by Ullrey (1974). From an analyses of responses in fleece yield, the most sensitive index of need, less selenium was found to be required by sheep grazed on Australian pastures (<0.03 mg kg^{-1} DM) (Langlands, 1991a; Whelan et al., 1994b), unless there is an additional need for fetal growth in the ewe lamb (Wilkins et al., 1982). Indoor experiments using semi-purified diets have given slightly higher requirements for growth of 0.04–0.06 mg Se kg^{-1} DM (Oh et al., 1976), but the digestibility of the basal diet was probably high.

Factorial estimates indicate that selenium requirements increase as digestibility increases (Underwood and Suttle, 1999), falling between 0.03 and 0.06 mg Se kg^{-1} DM, the grazed and indoor extremes referred to above.

Cattle

Field experience with grazing stock in Australia and New Zealand suggests that 0.02–0.03 mg Se kg^{-1} DM can be marginal for growth (Langlands et al., 1981; Wichtel et al., 1994, 1996). A similar pasture selenium concentration has been associated with a reduced milk-fat yield (Knowles et al., 1999), but a whole herd of 80 was screened for blood selenium concentration and the 35 cows of lowest selenium status were used in the trial. In five trials with steers newly introduced to feedlot conditions given alfalfa hay, corn silage and a supplement providing approximately 0.05 mg kg^{-1} DM, neither performance nor health was improved by parenteral injection of selenium (25–50 mg), with or without vitamin E (340 IU) (Droke and Loerch, 1989). Factorial estimates (Table 15.8) appear satisfactory in such circumstances where vitamin E is in adequate supply, but, for higher-yielding cows fed indoors on diets that require massive vitamin E supplements (e.g. in the USA) (Weiss, 1998), selenium responses may be obtained with dietary selenium of approximately 0.05 mg kg^{-1} DM. The requirements for such stock could be twice those given in Table 15.7.

Goats

Rations of hay and concentrate, containing 0.027 and 0.04 mg kg^{-1} DM, were insufficient to maintain normal blood GPX activity and resistance to mastitis in five flocks of dairy goats (Sanchez et al., 2007). In assessing selenium status in goats from GPX activity in blood it is

Table 15.8. Factorially derived minimum estimates of selenium requirements for cattle settled at pasture and of moderate growth rate or yield. High-yielding cows, housed and fed energy-dense diets, may need up to twice as much selenium (see text for explanation).

	Live weight (kg)	LWG (kg day^{-1})	DMI (kg day^{-1})		Net requirement (mg kg^{-1} DM)		
			$Q^* = 0.5$	$Q^* = 0.7$	µg day^{-1}	$Q^* = 0.5$	$Q^* = 0.7$
Calves	100	0.5	2.8	1.7	50	0.018	0.029
		1.0	–	2.4	75	–	0.031
	200	0.5	4.3	2.6	75	0.017	0.029
		1.0	6.4	3.6	100	0.016	0.027
	300	0.5	5.6	3.4	100	0.018	0.029
		1.0	8.3	4.7	125	0.015	0.027
	Milk yield (kg day^{-1})						
Beef and dairy cows	500	0	–	–	–	–	–
		10	11.4	7.3	255	0.022	0.035
		20	17.4	11.2	385	0.022	0.034
		30	–	15.1	515	–	0.034

Q^*, a measure of forage quality or digestibility with 0.5 = poorly and 0.7 = highly digestible diets.
All thresholds are higher if the source of exposure is largely organic in nature.

noteworthy that the selenium equivalence (activity µg^{-1} Se) falls halfway between that found in sheep and cattle (Suttle, 1989).

Pigs

Piglets from sows on diets deficient in selenium but adequate in vitamin E (100 IU) have maintained satisfactory growth rates, feed intakes and feed conversion efficiency with diets containing 0.06 mg Se kg^{-1} DM (Glienke and Ewan, 1977). A similar growth requirement has been indicated more recently by Mahan et al. (1999), but the same concentration, 0.06 mg Se kg^{-1} DM, has been associated with high perinatal mortality (Van Vleet et al., 1973). Comparisons of responses of serum GPX3 activity to graded selenium supplements at the growing and finishing stages have suggested that 30% less selenium is required for maximum activity at the later stage (Mahan et al., 1999). With purified diets for gilts, 0.03–0.05 mg Se (with 22 IU vitamin E) kg^{-1} DM is adequate for the first pregnancy (Piatkowski et al., 1979), but animals can become progressively depleted. Mahan and Peters (2004) showed a gradual reduction in GPX1 and GPX3 in sows by their fourth parity on a diet containing 0.06–0.08 mg Se kg^{-1} DM. Selenium supplementation improved litter size but not total weaned litter weight, suggesting that the selenium supply was marginal. In boars, 0.06 mg Se kg^{-1} DM is insufficient for optimum male fertility (Marin-Guzman et al., 1997).

Requirements should be met for all stages of production by diets contains 0.1 mg supplementary inorganic Se kg^{-1} DM and adequate vitamin E.

Selenium requirements are raised if the diet is high in PUFA (Nolan et al., 1995), but the extent of the increase is unknown.

Poultry

The influence of vitamin E on selenium requirements has been clearly demonstrated. The growth requirement of chicks decreased from 0.05 when the purified diet contained no vitamin E to 0.02 and <0.01 mg Se with vitamin E added at 10 and 100 mg kg^{-1} DM, respectively (Thompson and Scott, 1969). More than 0.06 mg Se kg^{-1} DM is required in SeMet-rich natural feeds to give complete protection from ED (Fig. 15.6a). By contrast, the requirement of vitamin E-*replete* chicks for complete protection from pancreatic fibrosis is less when selenium is given as SeMet rather than selenite (Thompson and Scott, 1970). As much as 0.28 mg Se kg^{-1} DM is required to prevent gizzard and heart myopathies in turkey poults on diets marginal in vitamin E *and* sulfur amino acids (Scott *et al.*, 1967). A similar 'requirement' has been estimated for maximal liver GPX1 activity in turkey poults for the first 27 days of life, but growth was not impaired with as little as 0.007 mg Se kg^{-1} DM in a diet adequate in vitamin E and sulfur amino acids (Hadley and Sunde, 1997). In laying hens, 0.05 mg Se kg^{-1} DM maintained production with no added vitamin E or antioxidants in the diet, but hatchability was suboptimal (Latshaw *et al.*, 1977). Furthermore, hens are marginally deprived by a diet containing 0.085 Se mg kg^{-1} DM, and subsequently select a selenium-supplemented diet in preference to that marginal regimen, when given a choice, until they are repleted (Zuberbuehler *et al.*, 2006).

As with pigs, requirements should be met for all stages of production by diets containing 0.1 mg supplementary inorganic Se kg^{-1} DM and adequate vitamin E.

Horses

The selenium requirements of horses probably vary with the degree of oxidant stress as they do in other species, but horses differ from ruminants in that they do not saturate PUFA prior to absorption and may be more prone to vitamin E deficiency and thus have a higher need for selenium. Cases of nutritional muscular dystrophy have been observed in foals in areas where selenium-responsive diseases occur in grazing sheep and cattle (Andrews *et al.*, 1968; Caple *et al.*, 1978) and associated with the consumption of feeds usually containing <0.04 mg Se kg^{-1} DM (Underwood and Suttle, 1999). Mares grazing pastures that vary in selenium from 0.01 to 0.07 mg Se kg^{-1} DM have been seen to maintain plasma selenium between 0.6 and 0.8 µmol l^{-1} for 180 days, but values declined to 0.2 µmol l^{-1} thereafter (Wichtel *et al.*, 1998). However, no health benefits have been reported in cohorts given supplementary Se. No reductions in plasma GPX3 were observed in mature geldings when dietary selenium was reduced from 0.26 to 0.06 mg kg^{-1} DM (Shellow *et al.*, 1985).

A safe allowance of 0.10 mg Se kg^{-1} DM is suggested for growing and mature horses, but this may increase several-fold for the exercising racehorse.

Biochemical Consequences of Selenium Deprivation

The sequence of biochemical changes in selenium-deprived livestock differs in many respects from the general model described earlier (see Chapter 3) and is illustrated in Fig. 15.8.

Depletion

Insufficient selenium intakes lead to reductions in selenium concentrations and GPX1 activity in erythrocytes and tissues, with the decline in erythrocytes delayed and slowed compared to that in tissues. Since most of the selenium in liver and blood is in storage form (GPX1), changes in liver and blood selenium status, whether of selenium or GPX, reflect depletion more than deficiency. Selenium concentrations in milk and eggs are sensitive to changes in selenium intake, and reductions in milk and egg selenium concentrations provide opportunities for conserving selenium when intakes fall.

Deficiency

Selenium is present in plasma at less than half the concentration found in erythrocytes, but responds more rapidly to deprivation in both its GPX3 and SePP components. Reductions in serum selenium occur early in the course of deprivation and, in the study shown in Fig. 15.9, the reduction was complete *before* disease (ED) was apparent in chicks. Pancreatic GPX activity declines prior to the earliest signs of acinar cell atrophy (Bunk and Combs, 1980), but the relatively low enzyme activity in that organ may

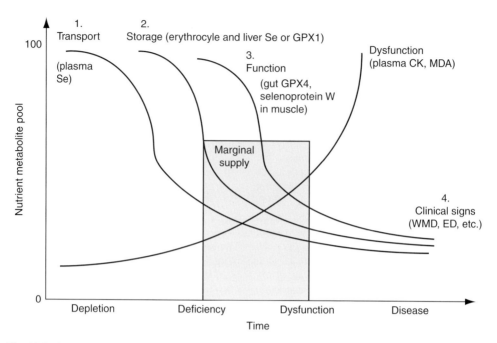

Fig. 15.8. A general scheme for the sequence of biochemical changes in livestock deprived of selenium. Rapidity of transition from one phase to another is influenced by sources of oxidant stress, the status of other antioxidants (notably vitamin E) and, in some species, by the Se source. Glutathione peroxidase (GPX)1 and GPX4 are two of several GPXs. Creatine kinase (CK) and malondialdehyde (MDA) are abnormal constituents of plasma. White muscle disease (WMD) and exudative diathesis (ED) are two of many possible clinical endpoints in different species. See also Chapter 1 and related text.

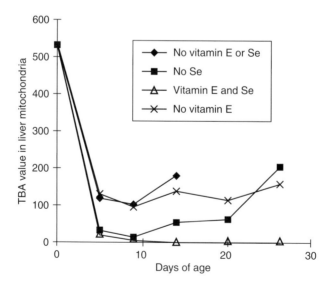

Fig. 15.9. Tissue peroxidation is influenced by interactions of selenium with other antioxidants, such as vitamin E, and with sources of oxidant stress: thus, thiobarbiturate (TBA)-like products of peroxidation are high in liver mitochondria of chicks deprived of vitamin E and Se at hatching (a stressor) and show only partial disappearance before disease (exudative diathesis) occurs after 13 days; combined supplementation with vitamin E and Se is needed to prevent peroxidation (data from Noguchi *et al.*, 1973b).

dispose it to dysfunction (Whitacre et al., 1987). The up-regulation of enzymes that can compensate for loss of GPX activity, such as the rise in glutathione-S-transferase activity seen in the liver and heart of selenium-deprived pigs (Lei et al., 1998), may be a sign of imminent dysfunction. Dysfunction may only arise when GPX2 and GPX4 are depleted to such an extent that such adaptations are overwhelmed.

Dysfunctional deiodination

Plasma T3 concentrations fall in peri-parturient cows of marginal selenium status showing no clinical response to selenium supplementation (Awadeh et al., 1998; Contreras et al., 2005) and both T3 and T4 decline in pregnant ewes on a low-selenium diet (<0.02 mg Se kg^{-1} DM) (Rock et al., 2001). In laying hens deprived of selenium, the reduction in plasma T3 is accompanied by a fall in T4 (Zuberbuehler et al., 2006). Similar changes occurred in chicks deprived of selenium and were accompanied by a reduction in hepatic ID1 activity. (Jianhua et al., 2000). Impaired deiodinase activity is clearly an early biochemical sign of deprivation in many species and may be the rate-limiting function in growth retardation.

In rats, selenium deprivation induces thyrotrophin (TSH) and ID1 expression in an attempt to preserve thyroidal SeP (Bernamo et al., 1995) and these changes may precede changes in the T3:T4 ratio. Response to selenium deprivation may depend on the concurrent supply of iodine. In a study by Voudouri et al. (2003) combined deficiencies of selenium and iodine were required to cause changes in lambs in the form of *raised* hepatic ID1 and thyroid weight; cerebellar ID3 activity, T3, T4 and TSH remained unaffected by either single or combined deficiencies. Although mild, the degree of iodine depletion was sufficient to cause large increases in erythrocyte and thyroidal cytosolic GPX.

Dysfunction in antioxidant defence

Although ROS can be individually identified (Arthur et al., 1988), it is easier to measure the end products of lipid peroxidation such as malondialdehyde (MDA) and other thiobarbiturate active substances (McMurray et al., 1983). An indicator of liver peroxidation, F_2-isoprostanes, increased in the selenium-deprived pig, although growth was not impaired (Lei et al., 1998). Exhalation of ethane and pentane may also indicate peroxidation. Hatching is a time of maximal free-radical generation (Surai, 2006) and Fig. 15.9 shows the effect of selenium and vitamin E on the subsequent spontaneous reduction in peroxidation in liver mitochondria. Initially, only groups given vitamin E showed full recovery from peroxidation, but, as time passed and chicks became more deprived of selenium, both antioxidant supplements were eventually needed to eliminate peroxidation and the associated disorder (ED). Selenium requirements to prevent hepatic microsomal peroxidation *in vitro* and to prevent disease (pancreatic fibrosis) in chicks are similar (Combs and Scott, 1974). There is a high risk of lipid peroxidation occurring in semen because spermatozoa are rich in PUFA and very active, but MDA data indicate that vitamin E is far more protective than selenium (Surai et al., 1998). Lambs given a semi-purified diet low in selenium (0.03 mg kg^{-1} DM) for 10 months have been found to develop subclinical myopathy, despite a marginal blood selenium status (112 µg l^{-1}) (Voudouri et al., 2003); the authors suggested that oxidative stress was responsible and this may have been exacerbated by the high-PUFA diet employed (3.6% corn oil). Reduced concentrations of the calcium-binding SePW in muscle may cause mitochondrial calcium overload and calcification (Tripp et al., 1993).

Immunological Consequences of Selenium Deprivation

Certain components of the immune response can be compromised early in selenium deprivation (Stabel and Spears, 1993). Both vitamin E and selenium can improve antibody responses in non-ruminants and ruminants but their specificity is uncertain (sham injections have been rarely used in controls) and additivity is variable (Stabel and Spears, 1993; MacPherson, 1994). Raising dietary selenium concentration from 0.06 to 0.1 mg kg^{-1} DM increased interleukin-1 concentrations in the plasma of weaned lambs (Qin et al., 2007). In chicks, a selenium supplement can raise antibody titres to injected sheep red blood cells regardless of whether the chicks were chilled (Larsen et al., 1997), whereas improvements in mortality were temperature-dependent. Cold stress reduces vitamin E but not selenium concentrations in the livers of broilers (Ozkan et al., 2007), and it is possible that where selenium has improved resistance to infection, it might sometimes have been covering for an increased need for vitamin E. Increases in immunoglobulin G in blood and colostrum have been reported in calves from selenium-supplemented dams, but without associated improvements in survival (Swecker et al., 1995; Awadeh et al., 1998). Inverse relationships between somatic cell counts (SCC) and GPX1 in blood have been reported (Erskine et al., 1989b) and selenium supplementation lowers SCC after experimental infection of the udder (Erskine et al., 1989a). However, treatment with selenium did not lower SCC in a grazing herd of exceedingly low selenium status that was selenium-responsive in terms of milk-fat yield (Knowles et al., 1999). In selenium-deprived cows, bactericidal activity towards S. aureus and GPX activity of neutrophil leukocytes isolated from the mammary gland were both reduced, but no correlation between SCC and blood selenium was found (Grace et al., 1997). Cellular immunity may appear to be compromised by selenium deprivation in in vitro culture, particularly if mitogenic stressors are applied (see MacPherson, 1994). However, the changes are non-specific and influenced by the nature of the challenge, and may overestimate dysfunction in vivo (Chesters and Arthur, 1988). Although selenium supplementation has improved certain responses of polymorphonuclear cells from sows to stimulation in vitro, vitamin E had more widespread effects (Wuryastuti et al., 1993). Only one case of selenium-responsive disease incidence has so far been reported: mastitis in dairy goats (Sanchez et al., 2007).

Diagnosis of Selenium-responsive Disorders

The diagnosis of SeRD presents considerable difficulties because few, if any, of the subclinical, clinical and pathological consequences are specific and pathogenesis is multifactorial. Biochemical confirmation presents two further complications: selenium is present in samples in functional and non-functional forms and it is no longer safe to assume that blood and tissue selenium are invariably well correlated with GPX activity – in plasma they can show opposing trends (Fig. 15.10) (Arthington, 2008). 'Critical' levels can be dependent on the response criteria and the productive state of the animals under investigation. For selenium, more than for any other element, the best diagnosis is usually afforded by a positive response to selenium supplementation. However, the numbers and/or precision of response measurement required to obtain significant responses are often high (see Chapter 19).

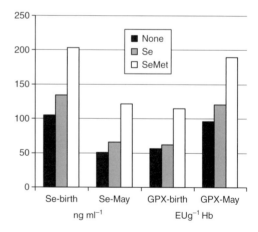

Fig. 15.10. Effect of measure of selenium status on changes occurring after birth in suckler calves. Blood Se shows a declining status on all treatments, while erythrocyte glutathione peroxidase (GPX)1 shows an improving status in all groups (data from Gunter et al., 2003). The authors' inference that only selenomethionine (SeMet) prevented a decline in status applies only to data for blood Se, not for GPX1.

Indices of Functional Selenium Status

GPX in blood and serum

GPX1 activity can provide a more reliable indicator of low selenium status than blood selenium, particularly in the neonate. The advantage is illustrated by results from a study with beef suckler calves (Fig. 15.10): blood GPX1 activity increased rapidly in all groups after birth, indicating an adequate selenium supply from pasture that contained 0.1 mg Se kg^{-1} DM. By contrast, blood selenium decreased, irrespective of treatment, falsely indicating marked depletion. Although plasma GPX3 is probably superior to plasma (or serum) selenium as an indicator of low status, relationships between selenium intake and enzyme activity differ widely both within and between species and must be characterized before they can be reliably interpreted. Figure 15.11 shows a threefold variation in the specific activity (SA) of GPX3 in pigs in experiments from the same laboratory and involving similar diets (relatively low in selenium). In a comparison of responses of different species to selenium supplements, initial SAs were 2.90, 1.46 and 6.97 U μg^{-1} Se, respectively, in cattle, sheep and horses; after 42 days' supplementation with selenite or selenate, the corresponding values were 1.26, 1.29 and 2.98 U μg^{-1} Se (Podoll et al., 1992). In dairy goats, mean GPX activities of 43–45 IU g^{-1} Hb have been associated with susceptibility to mastitis (Sanchez et al., 2007).

Differences in SA probably arise through physiological and dietary influences on the SePP and SeMet concentrations in plasma and physiological limits to the GPX3 response. In chicks, plasma GPX3 responds to dietary selenium up to 0.4 mg kg^{-1} DM (Cantor and Tarino, 1982); in mature geldings, however, it did not respond to selenium supplementation when the grass-hay plus concentrate diet contained 0.06 mg Se kg^{-1} DM (Shellow et al., 1985). If the down-regulation of GPX activity during pregnancy and lactation seen in rats (Sunde et al., 2005) also occurs in livestock, lower limits would be required for reproducing and lactating females. However, in the sows reported on in Fig. 15.11, GPX3 activity (not shown) as well as SA *increased* during pregnancy and lactation, and therefore showed no evidence of down-regulation. Up-regulation of GPX activities in the face of simple or goitrogen-induced iodine deprivation is a more likely complication (see Chapter 12). Results for lactating cows suggests that a plateau for blood GPX1 is reached with <0.15 mg Se kg^{-1} DM (Juniper et al., 2006). Results for GPX1 assays can vary between laboratories and should be standardized (Underwood and Suttle, 1999): assay temperatures may range from 25°C to 37°C and pH from 6.8 to 7.4, increasing recorded activity more than twofold, yet both are rarely specified in papers.

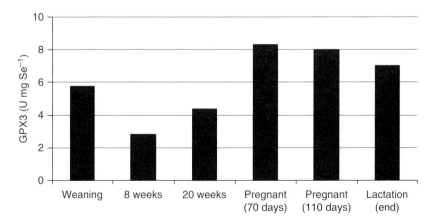

Fig. 15.11. Specific activity of serum glutathione peroxidase (GPX)3 in pigs at different stages of development on diets with 0.06–0.08 mg Se kg^{-1} (data from Kim and Mahan, 2001; Mahan and Peters, 2004).

Indices of dysfunction

Indices of peroxidation have been used experimentally but not diagnostically to assess the incidence of selenium-responsive disease. Reductions in erythrocyte MDA in the peri-parturient cow given a vitamin E supplement indicate both the potential and the non-specificity of such tests; however, a selenium supplement was without effect (Brzezinska-Slebodzinska et al., 1994). Where animals respond to selenium because of a secondary impairment of iodine metabolism, low concentrations of T3 (<2 nmol l^{-1}) may be present prior to treatment and rise afterwards, but the T3 to T4 ratio may not be helpful (Wichtel et al., 1996) and analytical costs can be prohibitive. The increases in T3 associated with cold stress and changes associated with parturition may complicate interpretation. In a study of cold-stressed broilers, neither doubling a dietary selenite supplement of 0.15 mg kg^{-1} DM nor adding SeMet affected the twofold increase in T3, but fresh liver selenium in the basal group was high at 0.45 µg kg^{-1} (Ozkan et al., 2007). Indices of peroxidation and deiodination have yet to be calibrated against other indices of selenium status.

Indices of disorder

Creatine kinase activities can be increased >100 times in WMD due to leakage of intracellular enzymes through damaged muscle membranes. In subclinical cases, however, creatine kinase can decline markedly after the attainment of peak values (N.F. Suttle and D.W. Peter, unpublished data) and cannot differentiate between selenium and vitamin E as the primary or sole deprivation (Table 15.9). The use of specific lactic dehydrogenase isoenzymes can differentiate between damage to cardiac and skeletal muscle, but not the specific disease (e.g. cardiac WMD or mulberry heart disease (MHD)). Raised plasma activities of enzymes found only in liver (e.g. glutamate dehydrogenase, ornithine carbamoyl transferase) can be valuable in the diagnosis of hepatosis dietetica (HD), but activities are raised in other hepatopathies (e.g. copper toxicity). Abnormalities in activity of the ubiquitous aspartate aminotransferase (AAT) cannot localize the site of damage and values may be only moderately increased in animals with marked clinical symptoms of WMD (Oksanen, 1967). Furthermore, selenium responsiveness is not necessarily accompanied by raised AAT (Stevens et al., 1985).

Limitations of Non-functional Indices of Selenium Status

Blood and plasma selenium in sheep

A 'bent-stick' analysis of relationships with wool yield in individual grazing, pregnant Merino ewes has indicated diagnostic limits of <0.89 µmol (70 µg) and <0.51 µmol l^{-1} for blood and plasma selenium, respectively; wool yield became

Table 15.9. Mean (± standard error) blood selenium and erythrocyte glutathione peroxidase (GPX)1, plasma aspartate aminotransferase (AAT), lactate dehydrogenase (LDH) and creatine kinase (CK) activities in normal and white muscle disease (WMD) lambs on low-selenium hay plus oat diets with and without selenium (from Whanger et al., 1977). Only partial control of the disease is obtained by selenium administration.

Blood components or incidence	With Se		Without Se	
	Normal	WMD	Normal	WMD
WMD incidence (%)	–	44	–	56
CK (U µl^{-1})	0.1 ± 0.02^a	3.4 ± 1.1^b	0.1 ± 0.01^a	5.0 ± 0.9^b
LDH (U µl^{-1})	0.61 ± 0.07^a	3.89 ± 1.10^b	0.56 ± 0.06^a	4.6 ± 1.04^b
AAT (U ml^{-1})	53 ± 6^a	748 ± 260^b	62 ± 5^a	635 ± 92^b
GPX1 (U mg^{-1} Hb)	59 ± 7^a	66 ± 5^a	2.2 ± 0.5^b	3.6 ± 1.0^b
Blood selenium (mg l^{-1})	0.14 ± 0.03	0.15 ± 0.04	<0.02	<0.02

Hb, haemoglobin.
Means within a column not sharing a common superscript letter are significantly different ($P<0.01$).

Table 15.10. Marginal limits[a] or bands[b] for indices of selenium deprivation in domestic livestock given diets adequate in vitamin E.

	Blood (µmol l^{-1})	Serum (µmol l^{-1})	Liver (µmol kg^{-1} FW)	Muscle (µmol kg^{-1} FW)	Diet (mg kg^{-1} DM)
Sheep, adult	**<0.50**	**<0.25**	**0.25**–0.45	**0.30**–0.40	**0.03**–0.05
Lamb or lambing	**<0.9**	**<0.5**	0.75–1.35	–	–
Cattle and deer, adult	**0.15**–0.25	**0.10**–0.12	**0.20**–0.30	**0.25**–0.30	**0.02**–0.04
Cattle, neonate	**0.30**–0.50	–	**0.80**–0.90	–	–
Pigs, sow	**1.00**–1.26	**0.90**–1.30	–**0.8**–1.3	**0.63**–0.95	**0.05**–0.07
Pigs, piglet	–	**0.50**–0.71	–	–	–
Poultry, chick	**0.20**–0.25	**0.90**–1.30	**1.0**–2.0	**0.80**–1.00	**0.06**–0.10
Horses, mature[c]	<2.3	<1.6	–	–	<0.085

Conversion factors: SI to elemental, ×79.0 (for µg); elemental to SI, ×12.665 (for µmol).
[a]Proximity of the mean for a population sample to the lower (**bold**) limit indicates increasing probability of benefits in health and production from selenium supplementation.
[b]The further an individual value falls below the 'bent stick' threshold or the higher the proportion of such values in a population sample, the greater the likelihood of selenium responsiveness.
[c]Derived from Calamari *et al.* (2009).

increasingly restricted below these limits (Langlands *et al.*, 1991a) (Table 15.10). Similar thresholds have been derived in weaned lambs (Whelan *et al.*, 1994a). For barren ewes, the corresponding diagnostic thresholds are much lower (<0.51 and <0.25 µmol Se l^{-1} for blood and plasma) (Langlands *et al.*, 1991a). In New Zealand, the likelihood and size of a positive growth response being obtained to selenium supplementation in lambs has been found to increase exponentially as mean blood selenium falls below 0.20 µmol l^{-1}, a much lower threshold than that indicated with wool yield as the arbiter. Furthermore, small but significant improvements in ewe fertility have been obtained by pooling data from five large hill flocks in which mean initial blood and plasma selenium status exceeded commonly used limits of normality and values rose spontaneously in untreated ewes (Munoz *et al.*, 2009).

Ideally, a diagnosis of 'selenium responsiveness' should depend on the proportion of the flock (or herd) in which selenium status falls below the critical threshold for the most sensitive of recordable traits.

Blood and plasma selenium in cattle and deer

In New South Wales, growth rarely improved following selenium treatment and the few beef herds that responded (three out of 21) were not distinguishable on the basis of blood selenium, which was usually <0.25 µmol l^{-1} (Langlands *et al.*, 1989). In New Zealand, plasma selenium values as low as 0.12 µmol l^{-1} have been tolerated in heifers (Wichtel *et al.*, 1996) and dairy herds have shown no depression of milk-fat yield until mean blood selenium fell below 0.15 µmol l^{-1} (Fraser *et al.*, 1987; Tasker *et al.*, 1987). However, collated data from all trials indicate improved milk yield at higher blood selenium levels (Wichtel, 1998). Mean plasma selenium was 0.15 µmol l^{-1} in calves from a herd with a low incidence of WMD (10–14%) (Hidiroglou *et al.*, 1985). Fertility in the grazing dairy cow is not unduly sensitive to selenium status (Wichtel *et al.*, 1994).

Thus, growing and lactating cattle at pasture in the Antipodes tolerate a lower blood and plasma selenium status than sheep. Critical values may be lower in cattle than sheep because there is less 'drain' of SeMet and SeCys into appendages (i.e. no fleece). Deer are also a relatively tolerant species (Mackintosh *et al.*, 1989). The marginal bands for blood and plasma selenium given in Table 15.10, based on the above trials, probably increase as demand for selenium increases. Higher thresholds may be required by cows in the USA than by their counterparts in New Zealand because they generally yield twice as much milk fat each day, but there are no equivalent dose–response data whereby higher limits may be estimated.

Blood and plasma selenium in pre- or non-ruminants

While serum selenium is a popular measure of status in pigs because of its sensitivity to changes in selenium supply (Lei et al., 1998), that sensitivity can present problems. The main constituent, SePP, is a 'negative' acute-phase marker and concentrations of both SePP and total serum selenium decline during sepsis (Hollenbach et al., 2008). The precise level of serum selenium attained at a given dietary selenium concentration may be inversely related to growth rate in young pigs, while GPX3 remains unaffected. Serum selenium levels in racehorses increase by 10% after a training jog (Gallagher and Stowe, 1980), indicating a possible response to exercise. In growing boars that developed selenium-responsive infertility, mean serum selenium was found to have declined to 0.42 µmol (33 µg) l^{-1} (Marin-Guzman et al., 1997). In piglets, the ratio of GPX3 activity to selenium in serum can double between birth and weaning (see data from Yoon and McMillan, 2006). Furthermore, the relationship between blood selenium and GPX1 activity in the adult is affected by dietary selenium source, judging from a recent study in horses (Calamari et al., 2009). Blood selenium overestimates functional status when selenium yeast, as opposed to selenite, is fed. Blood and plasma selenium may be unreliable measures of selenium status in all neonates.

Milk selenium

A curvilinear relationship between concentrations of selenium in milk and blood from dairy cows has been described (Grace et al., 2001) and diagnostic limits for milk selenium have been extrapolated from the established relationship for blood selenium. However, the relationship changes when cows are given SeMet because it is preferentially transferred to milk (Knowles et al., 1999). Interpretation of blood selenium is also affected, with GPX1 activity µg^{-1} Se (i.e. SA) in blood falling by 15% in cows given SeMet rather than selenite. With the increased use of SeMet in compound feeds in some countries, bulk milk selenium can no longer be relied upon to indicate functional selenium status in lactating cows and would be equally unreliable as an index of status in the ewe, doe or sow given SeMet.

Tissue selenium

Although tissue selenium reflects dietary selenium intake from a particular source, diagnostic limits have not been widely studied in the context of SeRD in the field and probably: (i) increase with the proportion of dietary selenium provided by SeMet; and (ii) decrease if the diet is lacking in dietary protein. The excessively low tissue selenium values typical of WMD in pigs are compared with those of normal animals in Table 15.11; similar changes occur in affected foals (Caple et al., 1978). Subnormal tissue selenium concentrations in pigs affected with MHD or HD have also been reported (Table 15.11) (Van Vleet et al., 1970; Simesen et al., 1979). In a Canadian study cardiac failure, myocardial necrosis and calcification in aborted calves were associated with lower liver selenium concentrations than in those of unaffected fetuses (mean 4.4–5.5 versus 7.5 µmol kg^{-1} WM) (Orr and Blakley, 1997), but values were still high compared with those for neonates given in Table 15.10. Low liver selenium values of 0.07–0.4 µmol kg fresh weight (FW) have been reported in deer with WMD (Wilson and Grace, 2001).

Occurrence of Selenium-responsive Disorders

Selenium status of soils and feeds

Soil geochemistry and mapping have defined broad areas of low soil selenium status and high risk for SeRD in man and/or livestock (Combs, 2001; Fordyce 2005; Broadley

Table 15.11. Selenium concentrations (µg kg^{-1} FW; mean ± standard deviation) in skeletal muscles, livers and hearts of pigs with three different selenium-responsive diseases compared with tissues from normal pigs (Pedersen and Simesen, 1977).

Se-responsive disease	Liver	Heart	Muscle
Nutritional muscular dystrophy	66 ± 17[b]	38 ± 18[b]	32 ± 16[ab]
Hepatosis dietetica	68 ± 52	51 ± 41	–
Mulberry heart disease	141 ± 67	98 ± 44	–
Normal pigs	300 ± 100	164 ± 60	
	670 ± 53[ab]	210 ± 20[ab]	104 ± 12[ab]

[a]Data for normal pigs will vary, reflecting selenium intakes; those for liver and heart from Lindberg (1968) are relatively high.
[b]Data converted from a dry weight basis, using a conversion factor of 0.2 for muscle and heart, 0.3 for liver.

et al., 2006). These include soils of granitic or volcanic origin, such as the pumice soils of New Zealand and the tablelands of northern New South Wales; of mountainous countries of northern Europe, such as Finland and Sweden (Selinus, 1988); of Denmark, Siberia and much of China, stretching in a zone from the north-east to south-central regions (Fordyce, 2005). The effects of soil origin can be confounded by those of climate and altitude. The selenium status of forages is inversely related to altitude, probably because of the leaching effect of rainfall on the soil and the diluting effect of lusher growth. There is a negative relationship between rainfall and selenium status in cattle (Langlands et al., 1981) and sheep (Langlands et al., 1991a). Selenium levels in crops and forages are also affected by soil acidity, with higher selenium uptake on alkaline than on acid soils (Ullrey, 1974). The prevalence of low blood selenium levels in sheep (Anderson et al., 1976) and cattle (Arthur et al., 1979) in Scotland is attributable to a combination of granitic soils, high altitude, high rainfall and slightly acidic soils. There are fears that the selenium status of soils, crops, pastures and people are in decline because of reduced inputs of anthropomorphic selenium from atmospheric deposition and crude fertilizers (Broadley et al., 2006), but the routine selenium supplementation of livestock feeds has probably arrested and even reversed that decline in developed countries such as the UK. Furthermore, the local incidence of SeRD often correlates poorly with soil or dietary selenium status because of the multifactorial nature of the underlying dysfunctions. This is illustrated in Fig. 15.12, Table 15.7 and the following text.

Vitamin E

Selenium-responsive disorders such as WMD used to be commonplace in pigs in the mid west of the USA, with mortalities of 15–20% being recorded in Michigan. However, the high selenium concentration needed to prevent disorders (0.10–0.15 mg kg^{-1} DM) (Ullrey, 1974) indicate that vitamin E deficiency had increased the risk of dysfunction. Supplies of vitamin E to ruminants grazing green pastures are plentiful and may explain the tolerance of selenium deprivation in New South Wales (Langlands et al., 1989), but vitamin E supplies fall during the dry season. Conservation of forage as hay can also lead to loss of vitamin E, particularly if material is badly weathered (Miller et al., 1995). The treatment of grains and straws with alkali and also the storage of moist grain in silos with propionic acid as a preservative can destroy vitamin E, predisposing livestock fed on such feeds to SeRD (Allen et al., 1975).

Vitamin E is poorly transported via the placenta and offspring rely on colostrum for

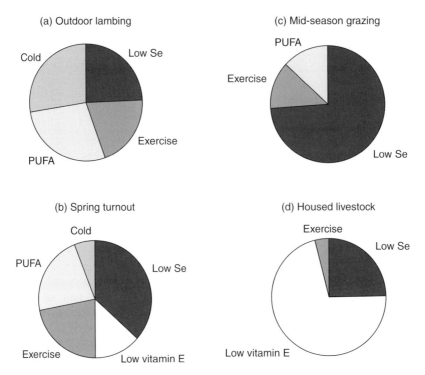

Fig. 15.12. A hypothetical illustration of the different relative contributions of selenium and vitamin E deprivation, oxidant stress (from cold exposure, muscular exercise or dietary polyunsaturated fatty acids (PUFA)) to the risk of selenium-responsive disorders developing in sheep or cattle when housed or at pasture: (a) lambs born outdoors, green sward, cold spring; (b) growing cattle turned out on to lush spring pasture; (c) continuously grazed sheep or cattle, mild temperatures; (d) housed sheep or cattle.

their early supply. The mother, therefore, secretes large quantities of vitamin E in colostrum and maternal plasma tocopherol concentrations can halve around parturition (Goff and Stabel, 1990), possibly increasing the mother's need for selenium. Low vitamin E intake in early life may also increase the risk of selenium deprivation in newborn, leaving them slow to suckle. In the USA, incidence of mastitis in cows can be reduced by massive supplements of vitamin E (2000–4000 IU day^{-1}) (Weiss et al., 1997). Selenium responses in SCC and udder infections may reflect a primary lack of vitamin E. Inclusion rates for expensive vitamin E in mineral/vitamin supplements may often be too low to eliminate responses to selenium (Swain et al., 2000) (Fig. 15.12), but the efficiency of substitution between selenium and vitamin E is poor and supranutritional supplements of one should not be used to overcome a deficiency of the other.

Dietary oxidants in ruminants

After absorption from the diet, PUFA form unstable lipid hydroperoxides, a source of ROS, in the tissues and high intakes of PUFA induce myopathy (Rice and Kennedy, 1988). Most cases of WMD in calves and acute paralytic myoglobinuria in older cattle occur shortly after turn out on to spring pastures, which are rich in PUFA (Fig. 15.13) (McMurray et al., 1983). The myopathy occurs in the face of a rapid rise in plasma tocopherol and is selenium-responsive (McMurray and McEldowney, 1977; Arthur, 1988), but supplementation of rations with PUFA can reduce plasma tocopherol concentrations (Wang et al., 1996). High PUFA intakes probably increase the risk of SeRD in livestock and are thought to have contributed to the development of WMD in goat kids (Rammell et al., 1989). Although hydrogenation of PUFA occurs in the rumen, increases in the unsaturated

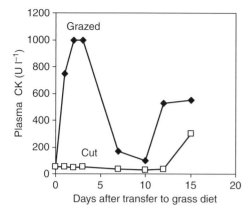

Fig. 15.13. When two groups of calves of very low selenium status were transferred from a semi-purified diet to grass, either grazed or cut and fed indoors, only individuals from the grazed group showed evidence of myopathy in the form of increased plasma creatine kinase (CK) activities. When the indoor group was turned out after 12 days, a muted increase in CK was seen. Thus, a combination of low selenium status, transition to grass and exercise seemed necessary to precipitate myopathy (data for the worst pair of grazed calves from Arthur, 1988).

linolenic acid in plasma at turnout show that, in the short term, hydrogenation is far from complete (McMurray et al., 1983). Plasma tocopherol in samples submitted to veterinary diagnostic laboratories in the UK is low in spring and may reflect initially poor saturation in the rumen of high PUFA intakes from spring grass (Small et al., 1996). The higher risk of WMD on legume than on grass diets may be related to the higher rate of passage of legumes from the rumen, which allows more PUFA to escape hydrogenation. Adding PUFA-rich supplements, such as marine algae, to the diets of livestock to increase the unsaturated fat content of human foods, using forms that resist rumen degradation in the ruminant, may increase the risk of SeRD.

Dietary oxidants in pigs and poultry

The unhatched chick meets its need for energy largely by the β-oxidation of fatty acids and yolk lipids are used for the synthesis of membrane phospholipids by the developing embryo. It has been hypothesized that oxidant stress is high during embryo development and that the addition of PUFA to the hen's diet may increase peroxidation, leading to selenium-responsive reductions in hatchability (Surai, 2006). The addition of PUFA (50 g fish oil kg^{-1} DM) to a cereal-based, broiler-breeder ration, containing 0.08 mg Se kg^{-1} DM, reduced hatchability from 75% to 57% and chick birth weight by 1.1 g, and increased 0- to 4-day mortality. However, the addition of 0.5 mg selenium as selenium yeast did not improve hatchability (Pappas et al., 2006) and responsiveness to selenium forms with higher functional availability should be studied.

Endogenous oxidants

The generation of ROS increases with metabolic rate and the need for protection from GPX (and vitamin E) increases with metabolic activity (Fig. 15.13; Arthur, 1998). In calves on a low-selenium diet, the change to a PUFA-rich grass diet alone did not induce myopathy, but grazing, with the inevitable increase in exercise, did. The mingling of piglet litters can precipitate WMD and the physical exertion of establishing a social order has been implicated (Ullrey, 1974). The importance of endogenous ROS is indicated by the uneven distribution of myopathic lesions amongst fibre types, with type I (slow-twitch, oxidative) fibres being more vulnerable than type II (fast-twitch, glycolytic). Lack of previous exercise may be as important to the incidence of myopathy at turnout as sudden exercise because there is less stimulus for the development of resistant type II muscle fibres in the confined animal (McMurray et al., 1983). Sudden exercise and associated stressors can cause capture myopathy in deer (Montane et al., 2002). The generation of free radicals in neutrophils undergoing a respiratory burst after phagocytosis may explain both the impaired phagocytic killing in vitro by cells from selenium-depleted calves (Boyne and Arthur, 1979) and the occasional reports of impaired immunocompetence in vivo. Selenium in SeMet may afford relatively poor protection against such 'explosive' intracellular events. Prolonged heat stress increases tissue peroxidation: the exposure of Japanese quail to 34°C

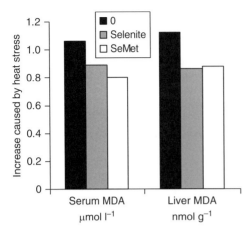

Fig. 15.14. Prolonged heat stress in Japanese quail caused two indices of peroxidation – malondialdehyde (MDA) concentrations in serum and liver – to increase when no selenium supplement was added to a basal diet marginal in selenium and vitamin E (0). The addition of selenium (0.3 mg kg^{-1} DM) as selenite or selenomethionine (SeMet) afforded only partial protection (from Sahin et al., 2008).

for 8 h day^{-1} for 90 days greatly increased MDA not only in serum and liver, but also in the egg yolk (Sahin et al., 2008). Supplementation of the corn/SBM diet, marginal in selenium (0.05 mg kg^{-1} DM) and vitamin E (25 IU added kg^{-1}) and also heat-exposed, with either selenite or SeMet (0.15 or 0.3 mg kg^{-1} DM) significantly reduced MDA (Fig. 15.14) and improved egg production, but protection was far from complete. Cold stress does not increase tissue peroxidation (Ozkan et al., 2007).

Toxins

Fungal toxins such as aflatoxin and ochratoxin are fairly common contaminants of livestock feeds. They cause harm by increasing lipid peroxidation in the tissues (Surai, 2006). A selenium plus vitamin E-responsive myopathy occurs in sheep grazing lupin stubbles contaminated with the fungus *Diaporthe toxica* in Western Australia, but only certain selenium forms (SeMet) or methods of administration (subcutaneous vitamin E) are protective (Smith and Allen, 1997). Combined supplementation with selenium and vitamin E decreases the toxicity of aflatoxin and ochratoxin to broiler chicks (Anil Kumar et al., 2005). Specific protective effects of moderate selenium supplementation against mycotoxicoses should not be hard to demonstrate and are conspicuous by their absence.

Iodine status and cold stress

If the biochemical dysfunction is one of poor deiodination rather than peroxidation, a new set of interactants including goitrogens is introduced and these may influence the occurrence of 'ill-thrift'. If oxidant stress is minimal, impairment of selenium-dependent thyroid hormone function may become health-limiting through a reduction in ID1 or ID2 activity in BAT. This is most likely to occur when animals give birth outdoors in cold, wet and windy conditions. Extremes of temperature and humidity can increase WMD incidence in piglets (Ullrey, 1974). The hypothesis that wool growth was being suppressed by selenium-induced iodine depletion was tested and rejected when fortnightly injections of T4 or T3 did not improve wool production (Donald et al., 1994a), but the irregular pattern of administration was unphysiological and T3 concentrations in the blood were decreased by T3 supplementation. Cold stress has increased blood T3 from 0.97 to 1.89 ng ml^{-1} in broilers on a diet with 0.15 mg added selenium (as selenite), but neither doubling the supplement nor changing it to an organic source (Sel-Plex) affected the response (Ozkan et al., 2007). When selenium deprivation is induced in rats, however, susceptibility to cold stress and the T3 response is clearly affected by a selenium–iodine interaction (Fig. 15.2) (Arthur et al., 1992).

High dietary sulfur

High sulfur concentrations have been implicated in the aetiology of WMD (Underwood and Suttle, 1999). A hypothesis that the high sulfur concentrations in molasses might lower selenium status and liver selenium has recently been tested and supported, insofar as selenium status was lower in forage-fed steers supplemented with corn rather than sugarcane molasses (Arthington, 2008).

Prevention and Control of Selenium Deprivation

Selenium deprivation can be prevented in a number of ways, with the method of choice depending on the conditions of husbandry.

Soil and foliar treatments

Soils or foliage can be treated with selenite solutions, but fertilizer selenium is poorly absorbed by most plants, especially from acid soils, and the residual effects can be small after a first cut (Underwood and Suttle, 1999). The use of a less soluble form of selenium (barium selenate, $BaSeO_4$) at 10g Se ha^{-1} in prill (granular) form may be effective for 3 years, but the required improvements in selenium status of the grazing animal can take 6 weeks to materialize (Whelan et al., 1994b). Mixtures with the more soluble sodium selenate (Na_2SeO_4) solve this problem and a range of products are now available (White and Broadley, 2009). A single application can raise blood serum and milk selenium in cattle for at least 12 months (Grace and West, 2006). The selenium content of grains can be increased by soil or foliar treatment, with the latter resulting in seven- and ninefold increases in rice selenium when applied at 29g Se ha^{-1} as selenite and selenate, respectively (Chen et al., 2002; cited by White and Broadley, 2009).

Dietary supplementation

It is possible to raise selenium intakes by importing grains and forages from high-selenium areas, but such a system requires controlled analyses of the feeds being blended. Selenium-containing mineral supplements have been routinely added to compound feeds to provide 0.15mg Se kg^{-1} DM, following a long-established practice from New Zealand (Andrews et al., 1968). Supplementation to 0.3mg kg^{-1} DM is now permitted in the USA and is highly effective in raising the selenium status of gilts and their offspring (Mahan and Kim, 1996). Trace mineralized salt, fortified with Na_2SeO_3 at the rate of 26–30mg Se kg^{-1} and offered freely to ewes and lambs, decreases the incidence of WMD (or plasma creatine kinase) without raising tissue selenium above levels found naturally (Underwood and Suttle, 1999), but may not protect all individuals. A compressed, selenized salt block, containing 11.8mg Se kg^{-1}, has been found to significantly raise the mean blood and plasma selenium in ewes and lambs, but left 7–33% of ewes unprotected (Langlands et al., 1990). The problem of individual variations in block consumption cannot be overcome by adding more selenium (Money et al., 1986), although this may raise average block consumption by the flock (Langlands et al., 1990). Selenite supplements have never failed to treat SeRD, including poultry diseases such as ED (Nogouchi et al, 1973; Cantor et al., 1975b) and WMD (Cantor and Tarino, 1982; Cantor et al., 1982), although this sometimes gets forgotten amidst all the hype surrounding 'organic' selenium supplements. Given the uncertainty and small scale of many production responses to selenium, selenized mineral supplements given in loose or block form can be a cost-effective option.

Parenteral supplementation

Subcutaneous injections or oral drenching, usually with Na_2SeO_3 in doses providing 0.1mg Se kg^{-1} LW at 1–3 monthly intervals, have been the most common means of preventing SeRD in grazing livestock (e.g. Meneses et al., 1994). Doses can be administered when animals are yarded for other management procedures such as anthelmintic dosing or vaccination and selenium can be combined with other drugs in a single product (e.g. Field et al., 1988). The frequency of dosing may be higher for the anthelmintic than is necessary for the accompanying selenium, but when given in the prescribed amounts does not result in excessive or dangerous selenium concentrations in edible tissues (e.g. Cooper et al., 1989). It is, however, vital that selenium be given at responsive times (i.e. at mating, late pregnancy and prior to weaning) and the optimum oral dose may be closer to 0.2 than 0.1 mg kg^{-1} LW (Langlands et al., 1990). Pig diseases such as MHD and HD have been prevented by the injection of 0.06 mg Se kg^{-1} LW, together with vitamin E, into the sow and baby pig (Van Vleet et al., 1973). The parenteral administration of insoluble BaSeO$_4$ at the high but safe dose of 1 mg Se kg^{-1} LW is effective in lambs (Zachara et al., 1993), ewes (Munoz et al., 2009), pregnant cows (Contreras et al., 2005), horses (Wichtel et al., 1998) and dairy goats (Sanchez et al., 2007). In meat animals, however, the recommended site of injection and withdrawal period must be observed with care to avoid problems of high residues (Archer and Judson, 1994). A new micro-encapsulated product providing selenium at a similar dose rate together with vitamin B$_{12}$ (see Grace et al., 2006) can maintain the selenium status of ewes and their lambs on selenium-deficient pastures for >300 days following a pre-mating injection.

Slow-release oral methods

In early Australian studies, heavy ruminal pellets consisting of 95% finely divided iron and 5% elemental selenium released sufficient selenium to maintain adequate blood values for several months and prevent WMD in sheep (Underwood and Suttle, 1999). The pellets could not be given to young lambs and sometimes provided limited and variable individual protection to older sheep (e.g. Whelan et al., 1994b), but formulations have improved (Donald et al., 1994c). In beef cattle, losses of 7–56% of administered pellets were observed in three out of 21 herds in one study (Langlands et al., 1989). Regurgitation occurred within minutes of dosing and was often repeated upon re-dosing, but was neither observed nor suspected in dairy heifers given two 30 g pellets containing 10% elemental selenium and growth was improved in a single-herd study (Wichtel et al., 1994). Efficacy was demonstrated for 2–3 years in beef cows in a Canadian study (Hidiroglou et al., 1985). The regurgitation of heavy pellets is not a problem in sheep (Langlands et al., 1990). Controlled-release capsules can deliver selenium and anthelmintic simultaneously, protecting lambs for at least 180 days (Grace et al., 1994). Sustained protection can also be provided by giving soluble glass boluses to cattle (Koh and Judson, 1987), goats (Zervas, 1988) and sheep (Milar and Meads 1988; Kendall et al., 2001). Slow-release devices, whether orally or parenterally administered, are so effective that they should not be given with other sources of supplementary selenium and dosing may not need to be repeated every year.

Organic Selenium Supplements

Three forms of organic selenium are in commercial use as supplements for livestock: selenium yeast (Se Y,) SeMet and selenium-metalosates (mixtures of inorganic selenium and amino acids). The first two can be considered together because selenium enrichment in yeast is largely as SeMet. Since functional tests of availability (e.g. Table 15.6) show a tendency for SeMet to be less available than inorganic selenium, there is unlikely to be any advantage to the supplemented animal by providing these more expensive organic sources. Although tissue selenium concentrations are usually higher after supplementation with SeMet sources than with selenite or selenate, no health or production advantages have been reported. It is noteworthy that selenium-responsive disorders arise in the field on feeds and forages containing SeMet as the major selenium source.

Organic selenium for non-ruminants

Neither selenium yeast nor inorganic selenium has improved growth or meat quality in pigs when added to a diet marginal in selenium (0.06 mg kg^{-1} DM) (Mahan et al., 1999). The enhanced transfer of SeMet across placental, mammary and oviduct barriers might be of benefit to offspring if there are corresponding increases in their functional SeP. Although selenium yeast has raised selenium concentrations in sow's serum and milk, and piglet tissues more than with the same addition as selenite (Mahan, 2000; Mahan and Peters, 2004), GPX3 activity in both dam and offspring is lower and similar results have been obtained with blood GPX1 in the piglet at birth (Yoon and McMillan, 2006). In heat-stressed broilers, SeMet and selenite are similar in the partial protection that they afford against peroxidation (Fig. 15.14), but the deliberate addition of oxidized fat to the diet does not harm egg production or increase the need for selenium (Leeson et al., 2008). Only one unequivocal short-term benefit has been reported, when SeMet restored the appetite and liver GPX activities of selenium-deprived chicks faster than selenite (Bunk and Combs, 1980). In a recent study comparing selenite, Se Y and B-Traxim selenium in broiler breeder and laying hens, hatchability was *lower* with selenium-Y than with selenite, providing an additional 0.1 mg Se kg^{-1} DM; selenite also gave *higher* plasma GPX3 in hens than either organic selenium source (Leeson et al., 2008). In adult horses, supplementation with selenium yeast rather than selenite increased the proportion on SeMet in the blood but did not increase GPX1 (Calamari et al., 2009).

Organic selenium for ruminants

Differences between organic and inorganic selenium supplements should be less in ruminants than non-ruminants because inorganic selenium is partially incorporated into selenoamino acids during anaerobic fermentation in the rumen, just as it is during yeast culture. In comparisons of SeMet and selenite, neither source improved beef cow nor calf performance on pastures providing 0.1 mg Se kg^{-1} DM, when compared with an unsupplemented group (Gunter et al., 2003). Similar increases in blood GPX1 have been produced by SeMet and selenite in the dairy cow (Knowles et al., 1999; Ortman and Pehrson, 1999b; Weiss, 2005). The two sources have been equally effective in increasing milk-fat yield from a selenium-deprived New Zealand dairy herd (Knowles et al., 1999), and similar increases in plasma selenium and GPX3 have been produced in beef steers on molasses-supplemented diets (Arthington, 2008). Although organic selenium sources have increased blood GPX1 activity more than selenite providing the same amount of selenium in weaned lambs, plasma interleukin was increased equally by both sources (Qin et al., 2007). Providing selenium as SeMet greatly increases milk selenium concentrations in lactating cows (Conrad and Moxon 1979; Fisher et al., 1995; Knowles et al., 1999; Ortman and Pehrson, 1999b), but selenium-metalosate does not (Givens et al., 2004). Possible benefits from selenium yeast supplementation to the consumer of milk, eggs and other edible products are discussed in Chapter 20.

Selenium Toxicity

Selenium is the most toxic of the essential trace elements. Problems can arise naturally or from the careless administration of selenium supplements and accidental acute and chronic problems have become the more common, even in seleniferous areas (O'Toole and Raisbeck, 1995). The study of Mahan et al. (2005) revealed that it was common practice for experimental stations (i.e. relatively well-informed establishments) to add selenium to pig rations at the maximum permitted by the FDA, even in seleniferous areas.

Natural Occurrence of Selenium Poisoning

Selenium poisoning occurs in the livestock in seleniferous areas of many countries, including the Great Plains of the USA, Canada and Russia and also parts of China and Israel due to the weathering of selenium-rich rocks (Fordyce, 2005). For example, geological outcrops of shales, rich in selenium, cause selenosis in Ireland (McLoughlin et al., 1989) and the risk of selenium poisoning can be broadly predicted by geochemical mapping (Combs, 2001). The vegetation growing on selenium-rich soils also plays a part and plant species can be divided into three types: containing high ('macro', >1 g kg^{-1} DM), intermediate (0.5–1.0 g kg^{-1} DM) or relatively low ($\leq$ 0.02 g kg^{-1} DM) selenium concentrations. These are represented, respectively, by primary accumulator or converter plants such as

Astragalus, Haplopappus and Stanleya genera; secondary absorbers such as Atriplex and some Brassica species; and normal forage species (e.g. Pascopyrum, Poasecunda, Trifolium) and cereals (Rosenfeld and Beath, 1964; Fordyce, 2005). Accumulator plants play a dual role by absorbing selenium from forms that are relatively unavailable to other species and returning it to the soil in organic forms that are more available to other plant species. The predominant forms of selenium in accumulator plants are the soluble organic compounds methylselenocysteine and selenocystathionine (Fordyce, 2005), while SeMet predominates in normal range species (Whanger, 2002). The accumulation of selenium is favoured by low rainfall, which has been implicated in outbreaks of selenosis in India (Underwood and Suttle, 1999). The direct ingestion of seleniferous soil may contribute to the selenium burden, but has not been specifically investigated.

Clinical Manifestations of Chronic Selenosis

Grazing livestock

The name 'alkali disease' has been given to a disease among livestock grazing seleniferous pasture characterized by dullness and lack of vitality, emaciation, roughness of coat, loss of hair, soreness and sloughing of the hooves; also by stiffness and lameness due to joint erosion between the long bones. Animals affected by a related disorder called 'blind staggers' collapse suddenly and die (Rosenfeld and Beath, 1964). 'Blind staggers' is associated more with the ingestion of selenium accumulators and 'alkali disease' with the consumption of grain and grass in seleniferous areas. Atrophy of the heart ('dish-rag' heart), cirrhosis of the liver and anaemia were reported in early studies of 'blind staggers' but the symptoms have never been reproduced experimentally, leading O'Toole and Raisbeck (1995) to suggest that other factors such as polioencephalomalacia, alkaloid poisoning and starvation – rather than selenosis – are responsible. The lesions of experimental chronic selenosis are confined to the integument and the lameness and pain from the condition of the hooves can be so severe that affected animals are unwilling to move, resulting in death from thirst and starvation. A histological study (O'Toole and Raisbeck, 1995) showed that the primary lesions were epidermal, with tubules in the stratum medium of the hoof becoming replaced by islands of parakeratotic cellular debris and the germinal epithelium becoming disorganized. These features distinguish selenosis from laminitis, a common disorder of cattle and horses in which the lesions are found in the dermis. Tail hair loss is associated with dyskeratosis and atrophic hair follicles. Organic selenium readily passes the placental and mammary barriers, so that calves and foals in seleniferous areas may be born with deformed hooves or develop them during the suckling period.

Pigs and poultry

Growing pigs exposed to high intakes of selenium can develop similar hoof lesions to those seen in ruminants. If selenium intake is sufficiently high to severely reduce appetite then central nervous system lesions predominate, taking the form of bilateral malacia of grey matter in the spinal cord (Goehring et al., 1984b). Alopecia and hoof lesions can develop within 12 weeks of exposure to 10 mg kg^{-1} DM inorganic selenium in a corn/SBM diet (Kim and Mahan, 2001). In all species, the incidence and severity of hoof lesions may be influenced by the quality and quantity of the diet and the related rate of hoof growth. Impaired embryonic development was featured in selenosis in pigs given a diet containing 10 mg Se kg^{-1} DM as selenite. The conception rate was reduced and an increased proportion of piglets were dead, small or weak at birth (Wahlstrom and Olson, 1959).

Growing chicks eat less food and grow slowly when given seleniferous diets. Egg production and hatchability are reduced in laying hens, with hatchability being particularly affected (Ort and Latshaw, 1978). A proportion of the fertile eggs from hens on high-selenium diets produce grossly deformed embryos, characterized by missing eyes and beaks and distorted wings and feet.

Biochemical Manifestations of Chronic Selenosis

Biochemical events follow the general scheme outlined in Fig. 3.2 (p. 41).

Accretion phase

During initial exposure to excess inorganic selenium there is a steady rise in tissue selenium concentrations over weeks or months – depending on the rate and continuity of intake – until saturation values are reached and excretion in the urine and faeces begins to keep pace with absorption. Saturation may be reached with 20–30 mg Se kg^{-1} DM in the liver, kidneys, hair/fleece and hooves of clinically affected animals but healthy animals can attain higher values during chronic exposure (Davis et al., 2006). A diagnosis of selenosis cannot rest on tissue selenium concentrations alone. At toxic selenium intakes the relationship between blood selenium and GPX1 activity breaks down. Blood selenium continues to increase, particularly with organic selenium sources (Kim and Mahan, 2001), while GPX1 and hair selenium values plateau (Fig. 15.15) (Goehring et al., 1984a; Ullrey, 1987; Kim and Mahan, 2001). Ewe lambs exposed to selenate or seleniferous wheat (3–15 mg kg^{-1} DM) exhibit proliferation of the jejunal mucosa and increases in liver weight and protein content (Neville et al., 2008). Where exposure has been discontinuous, separating hair or hoof into proximal, medial and distal portions may give a better measure of exposure than composite samples (J. Small and N.F. Suttle, unpublished data). The selenium concentrations in milk and eggs are particularly sensitive to high selenium intake. Values ranging between 0.16 and 1.27 mg l^{-1} have been reported for cow's milk from seleniferous rural areas in the USA (Rosenfeld and Beath, 1964). Increases of five- to ninefold (up to nearly 2 mg Se kg^{-1} DM) have been obtained in egg selenium when 8 mg Se as selenite kg^{-1} DM is added to hens' diets for 42–62 weeks (Arnold et al., 1973). Increases from 3.6 to 8.4 mg Se in the yolk and from 11.3 to 41.3 mg Se kg^{-1} DM in the white have been reported in eggs from hens that had their dietary selenium raised from 2.5 to 10.0 mg kg^{-1} (Moxon and Poley, 1938).

Dysfunction phase

The mechanisms that become disordered in selenium toxicosis remain unknown and there are no specific biochemical changes that indicate the onset of dysfunction. Increased thiol oxidation, reduction–oxidation cycling (Stewart et al., 1999) and lipid peroxidation (Kaur et al., 2003) may occur but substitution of selenium for sulfur, particularly in the disulfide bonds that give strength to keratin, has also been suggested (NRC, 2005). Indices of tissue damage such as plasma AAT rise little and late (e.g. Kim and Mahan, 2001). In beef cows exposed to 12 mg Se kg^{-1} DM from 80–10 days of gestation, diminished mitogenic responses of lymphocytes in vitro have been reported in the absence of clinical abnormalities (Yaeger et al., 1998). The biochemical changes accompanying selenosis are translated into the diagnostic guidelines in Table 15.12.

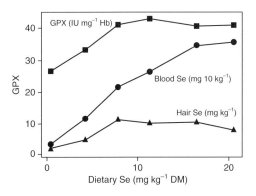

Fig. 15.15. Selenium concentrations in the hair and blood and glutathione peroxidase (GPX) activity in the blood of pigs exposed to excess Se via the diet (data from Goehring et al., 1984a).

Selenium Tolerance

The tolerance of livestock towards high selenium intakes varies with the form in which the selenium is ingested, the duration and continuity of

Table 15.12. Marginal bands[a] for biochemical indices of selenium status for use in assessing risk of chronic selenosis.

Species	Diet (mg kg^{-1} DM)	Blood (mg l^{-1})	Serum/plasma (mg l^{-1})	Liver (mg kg^{-1} FW)	Hair/fleece (mg kg^{-1} DM)	Hoof (mg kg^{-1} DM)
Cattle	5–**10**	1.5–**2.0**	2.5–**3.5**	4.0–**9.0**	5–**10**	2–**5**
Sheep	5–**20**	1.0–**2.0**	2.0–**3.0**	15–**25**	4–**8**	4–**8**
Pigs	3–**5**	2–**3**	0.7–**1.0**	1.5–**3.0**	4–**6**	3–**4**
Poultry	3–**5**	–	0.15–**0.30**	2.0–**6.0**	–	–

[a]Values above the band indicate a definite risk of selenosis; values within bands indicate high previous selenium intakes that, if continued, will probably lead to selenosis the closer they are to the upper, **bold** limit.

the exposure, the criterion of tolerance, animal genotype and interactions between these factors. Tolerances are therefore better expressed as ranges than thresholds. Genetic differences in susceptibility to selenosis have been reported with 'red'-haired pigs, which are more susceptible than black- or white-haired breeds (Wahlstrom et al., 1984), possibly because their food consumption is higher.

Source

Elemental selenium is well tolerated because of its insolubility and poor absorption and selenides are less toxic than soluble selenites or selenates, regardless of species. However, tolerances towards inorganic sources are of somewhat academic interest. Selenium occurs in feeds and forages largely in organic forms, such as SeMet, which is metabolized quite differently to inorganic sources and is generally less toxic, despite producing higher tissue selenium concentrations than equivalent concentrations given as selenite in poultry (Scott and Thompson, 1971). High tolerances have also been reported in pigs given seleniferous grain (Goehring et al., 1984a,b) or selenium yeast (Kim and Mahan, 2001). With a corn/SBM diet, 8 mg Se kg^{-1} DM as selenite was sufficient to impair appetite and growth during 5 weeks' exposure (Goehring et al., 1984a), but was harmless when given as seleniferous wheat and oats (Goehring et al., 1984b). In older pigs, hoof lesions developed with 15 mg Se as selenium-Y compared to 5–10 mg kg^{-1} DM as selenite (Kim and Mahan, 2001). Higher tissue and erythrocyte selenium concentrations have been found in the pigs given Se Y, indicating that selenium from its SeMet accumulates partly as undegraded methionine in a safe but poorly available form. The difference in tissue selenium accretion between inorganic and organic sources varies widely between organs, and is most marked in muscle and least marked in the kidneys in young growing pigs (Goehring et al., 1984a, b; Kim and Mahan, 2001), steers (Lawler et al., 2004) and mature sheep (Davis et al., 2006). This suggests that selenium accretion reflects different rates of methionine accretion in tissues.

Species differences

In non- or pre-ruminants exposed to inorganic selenium, tolerances are similar to those reported above for pigs. The supplementation of chicks via drinking water, equivalent to 7 mg Se kg^{-1} DM, has been found to impair growth and appetite within 7 days (Cantor et al., 1984); in hens, 5 mg Se kg^{-1} DM is borderline for hatchability (Ort and Latshaw, 1978). In calves given selenite in a milk replacer for 45 days, growth and FCE are impaired with 10 mg but not 5 mg added Se kg^{-1} DM (Jenkins and Hidiriglou, 1986). Ruminants are more tolerant because of their lower efficiency of selenium absorption: wethers have tolerated 10 mg inorganic Se kg^{-1} DM for a year (Cristaldi et al., 2005), ewe lambs 15 mg Se kg^{-1} DM for 84 days (Neville et al., 2008) and ewes 20 mg Se kg^{-1} DM for 72 weeks (Davis et al., 2006), showing neither biochemical or histological evidence of tissue damage. In steers given the equivalent of 5, 10 and 25 mg kg^{-1} DM supplementary selenium for 120 days as selenite, only one out of four developed hoof lesions at the highest intake (O'Toole and Raisbeck, 1995). Pregnant beef cows tolerated 12 mg Se kg^{-1} DM for 3 months (Yaeger et al., 1998).

The precise tolerance of grazing cattle and horses on seleniferous range is difficult to establish because selenium intakes from forages, varying widely in selenium concentration, palatability and accessibility, are ill-defined. Where consumption is restricted to non-accumulator plants, containing about 5 mg Se kg^{-1} DM, clinical selenosis may appear after prolonged exposure. In the O'Toole and Raisbeck (1995) study, SeMet exposure at 20 mg kg^{-1} DM induced hoof lesions in four out of five steers, indicating higher toxicity than for inorganic selenium. If the steers' diet allowed little growth, a high proportion of the SeMet, given in bolus doses, might have been deposited in the hoof. High dietary protein contents reduce the toxicity of seleniferous diets, probably in part through the diluting effects of microbial sulfur amino acid synthesis and higher protein deposition in tissues or secretion as milk. Tolerances within and between species may vary inversely with the rate at which selenium can be 'deposited' in methionine-rich products such as muscle, milk and eggs. Hens tolerate 10 mg Se kg^{-1} DM from seleniferous grain, but embryonic development is adversely affected (Poley and Moxon, 1938).

Lactation

The specific tolerances of lactating animals have not been studied, but they may be higher than those of non-lactating animals because milk selenium secretion serves as a 'sink' for the disposal of large amounts of selenium ingested in organic forms. In a study of cows yielding 31 kg milk day^{-1}, 12% of the selenium ingested as selenium yeast in a total mixed ration, containing 6.1 mg added Se kg^{-1} DM, was secreted and maximal blood selenium was only 0.72 µg l^{-1} compared to 1.51 µg l^{-1} in beef cattle on a similar diet (Juniper et al., 2008).

Prevention and Control of Chronic Selenosis

Selenium poisoning is incurable and early preventive methods involved the administration of substances that were themselves potentially toxic (e.g. bromobenzene, arsenic) (Underwood and Suttle, 1999). Preventive measures now rely on treatment of the soil or diet, but each has practical limitations. It may be simpler to utilize the more seleniferous areas for grain production and export the grain to low-selenium areas and countries (see Chapter 20).

Soil treatments

Raising the soil inorganic sulfur to selenium ratio can depress selenium uptake by plants in some circumstances. Continued heavy dressings with superphosphate (containing calcium sulfate (CaSO$_4$)) to increase pasture yields have been implicated as a cause of selenium deficiency in grazing sheep and cattle in parts of New Zealand (Andrews et al., 1968) and Australia (McDonald, 1975). Applications of 1000 kg gypsum (CaSO$_4$) ha^{-1} has greatly reduced the selenium concentrations in sugarcane in India (Dhillon and Dhillon, 1991). The addition of sulfur or gypsum to soils in a seleniferous area in North America has not reduced selenium uptake by cereals, probably because these soils are often already high in SO$_4$.

Dietary treatments

The protective effects of several heavy metals against inorganic selenium sources have been demonstrated in chicks (Hill, 1974) and in the case of copper, this may be attributable to the formation of unavailable copper selenides in the gut. However, protection against toxicity from organic selenium has not been studied and no information is available on protective effects in ruminants. Daily doses of a proprietary, high-SO$_4$ mixture (30 g) have been found to protect buffaloes at different stages of selenium toxicity (Degnala disease) (Arora et al., 1975). Since SeMet is of relatively low toxicity, protection may be afforded by supplements such as SO_4 that increase methionine synthesis in the rumen and 'trap' inorganic selenium in a less toxic form. Provision of a molasses–urea block may also lower the toxicity of ingested selenium by increasing microbial protein (and hence methionine) synthesis. Any protein supplement that increases growth rate or milk yield should 'dilute' the effect of high selenium uptake.

Acute Selenium Toxicity

Acute selenium poisoning is characterized by salivation, respiratory distress, oedema and pulmonary congestion, reflecting circulatory failure and degenerative changes in the heart, liver and kidneys (Rosenfeld and Beath, 1964). Toxicity has occurred at injected doses of 0.4–0.6 mg Se kg^{-1} LW in sheep where there was no history of selenium deficiency and in cattle where other stressors (weaning and vaccination) were present (Underwood and Suttle, 1999). Selenite and selenate are more acutely toxic than elemental selenium or selenides and careful scaling of parenteral therapeutic doses to body weight is recommended. Since accumulator plants can contain >4 g Se kg^{-1} DM in toxic organic forms, 'blind staggers' may be attributable to acute selenosis. In a study with groups of five pigs, dietary exposure to the same concentration of selenium as selenate, SeMet or *Astragalus bisulcatus* has caused poliomyelomalacia, but lesions were least likely and slowest to develop in the SeMet group (two cases after 9 and 24 days) and fastest in the group given the seleniferous plant (four cases in <5 days) (Panter *et al.*, 1996).

References

Adams, M.I., Lombi, E., Zhao, F.-J. and McGrath, S.P. (2002) Evidence for low selenium concentrations in bread-making wheat. *Journal of the Science of Food and Agriculture* 82, 1160–1165.
Alfaro, E., Neathery, M.W., Miller, W.J., Gentry, R.P., Crowe, C.T., Fielding, A.S., Etheridge, R.E., Pugh, D.G. and Blackmon, D.M. (1987) Effects of ranging amounts of dietary calcium on selenium metabolism in dairy calves. *Journal of Dairy Science* 70, 831–830.
Allen, W.M., Bradley, R., Berrett, S., Parr, W.H., Swannack, K., Barton, C.R.Q. and MacPhee, A. (1975) Degenerative myopathy with myoglobinuria in yearling cattle. *British Veterinary Journal* 131, 292–306.
Anderson, P.H., Barrett, S. and Patterson, D.S.P. (1976) Some observations on 'paralytic myoglobinuria' of cattle in Britain. *Veterinary Record* 99, 316–318.
Andrews, E.D., Hartley, W.J. and Grant, A.B. (1968) Selenium-responsive diseases of animals in New Zealand. *New Zealand Veterinary Journal* 16, 3–17.
Anil Kumar, P., Sathyanarayana, M.L., Vijayasarathi, S.K., Sreenivasa Gowda, R.N. and Rao, S. (2005) Effect of vitamin E and selenium on serum biochemical parameters in broiler chicken fed with aflatoxin and ochratoxin. *Indian Veterinary Journal* 82, 522–525.
ARC (1980) *The Nutrient Requirements of Ruminants*. Commonwealth Agricultural Bureaux, Farnham Royal, UK, pp. 184–185.
Archer, J.A. and Judson, G.J. (1994) Selenium concentrations in tissues of sheep given a subcutaneous injection of barium selenate or sodium selenate. *Australian Journal of Experimental Agriculture* 34, 581–588.
Arnold, R.I., Olson, O.E. and Carlson, C.W. (1973) Dietary selenium and arsenic additions and their effects on tissue and egg selenium. *Poultry Science* 52, 847–854.
Arora, S.P., Parvinder, K., Khirwar, S.S., Chopra, R.C. and Ludri, R.S. (1975) Selenium levels in fodders and its relationship with Degnala disease. *Indian Journal of Animal Science* 28, 249.
Arthington, J.D. (2008) Effects of supplement type on measures of growth and selenium status in yearling steers. *Journal of Animal Science* 86, 1472–1477.
Arthur, J.R. (1988) Effects of selenium and vitamin E status on plasma creatine kinase activity in calves. *Journal of Nutrition* 118, 747–755.
Arthur, J.R. (1998) Free radicals and diseases of animal muscle. In: Reznick, A.Z. (ed.) *Oxidative Stress in Skeletal Muscle*. Birkhauser Verlag, Basel, pp. 321–330.

Arthur, J.R. and Beckett, G.J. (1994) Roles of selenium in type I iodothyronine 5′-deiodinase and in thyroid hormone and iodine metabolism. In: Burk, R.F. (ed.) *Selenium in Biology and Human Health.* Springer-Verlag Inc., New York, pp. 93–115.

Arthur, J.R., Price, J. and Mills, C.F. (1979) Observations on the selenium status of cattle in the north-east of Scotland. *Veterinary Record* 104, 340–341.

Arthur, J.R., Morrice, P.C. and Beckett, G.J. (1988) Thyroid hormone concentrations in selenium deficient and selenium-sufficient cattle. *Research in Veterinary Science* 45, 122–123.

Arthur, J., Nicol, F., Guo, Y. and Trayhurn, P. (1992) Progressive effects of selenium deficiency on the acute, cold-induced stimulation of type II deiodinase activity in rat brown adipose tissue. *Proceedings of the Nutrition Society* 51, 63A.

Awadeh, F.T., Kincaid, R.L. and Johnson, K.A. (1998) Effect of level and source of dietary selenium on concentrations of thyroid hormones and immunoglobulins in beef cows and calves. *Journal of Animal Science* 76, 1204–1215.

Beckett, G.J. and Arthur, J.R. (2005) Selenium and endocrine systems. *Journal of Endocrinology* 184, 455–465.

Behne, D. and Kyriakopoulos, A. (2001) Mammalian selenoproteins. *Annual Review of Nutrition* 21, 453–473.

Beilstein, M.A. and Whanger, P.D. (1986) Chemical forms of selenium in rat tissues after administration of selenite or selenomethionine. *Journal of Nutrition* 116, 1711–1719.

Bernamo, G., Nicol, F., Dyer, J.A., Sunde, R.A., Beckett, G.J., Arthur, J.R. and Hesketh, J.E. (1995) Tissue-specific regulation of selenoenzyme gene expression during selenium deficiency in rats. *Biochemical Journal* 311, 425–430.

Bourne, N., Laven, R., Wathes, D.C., Martinez, T. and McGowan, M. (2006) A meta-analysis of the effects of vitamin E supplementation on the incidence of retained foetal membranes in dairy cows. *Theriogenology* 67, 494–501.

Boyne, R. and Arthur, J.R. (1979) Alterations of neutrophil function in selenium deficient cattle. *Journal of Comparative Pathology* 89, 151–158.

Broadley, M.R., White, P.J., Bryson R.J., Meacham, M.C., Bowen, H.C. and Johnson, S.E. (2006) Biofortification of UK food crops with selenium. *Proceedings of the Nutrition Society,* 65, 169–181.

Bruzelius, K., Hoac, T., Sundler, R., Onning, G. and Akesson, B. (2007) Occurrence of selenoprotein enzyme activities and mRNA in bovine mammary tissue. *Journal of Animal Science* 90, 918–927.

Brzezinska-Slebodzinska, E., Miller, J.K., Quigley, J.D. and Moore, J.R. (1994) Antioxidant status of dairy cows supplemented pre-partum with vitamin E and selenium. *Journal of Dairy Science,* 77, 3087–3095.

Bunk, M.J. and Combs, G.F. (1980) Effect of selenium on appetite in the selenium-deficient chick. *Journal of Nutrition* 110, 743–749.

Burk, R.F., Hill, K.E. and Motley, A.K. (2001) Plasma selenium in specific and non-specific forms. *Biofactors* 14, 107–114.

Calamari, L., Ferrari, A. and Bertin, G. (2009) Effect of selenium source and dose on selenium status of horses. *Journal of Animal Science* 87, 167–178.

Cantor, A.H. and Tarino, J. (1982) Comparative effects of inorganic and organic dietary sources of selenium on selenium levels and selenium-dependent glutathione peroxidase activity in blood of young turkeys. *Journal of Nutrition* 112, 2187–2196.

Cantor, A.H., Langevin, M.L., Noguchi, T. and Scott, M.L. (1975a) Efficacy of selenium in selenium compounds and feedstuffs for prevention of pancreatic fibrosis in chicks. *Journal of Nutrition* 105, 106–111.

Cantor, A.H., Scott, M.L. and Noguchi, T. (1975b) Biological availability of selenium in feedstuffs and selenium compounds for prevention of exudative diathesis in chicks. *Journal of Nutrition* 105, 96–105.

Cantor, A.H., Moorhead, P.D. and Musser, M.A. (1982) Comparative effects of selenite and selenomethionine upon nutritional muscular dystrophy, selenium-dependent glutathione peroxidase and tissue selenium concentrations of turkey poults. *Poultry Science* 61, 478–484.

Cantor, A.H., Nash, D. and Johnson, T.H. (1984) Toxicity of selenium in drinking water of poultry. *Nutrition Reports International* 29, 683–688.

Caple, I.W., Edwards, S.J.A., Forsyth, W.M., Whitely, P., Selth, R.H. and Fulton, L.J. (1978) Blood glutathione peroxidase activity in horses in relation to muscular dystrophy and selenium nutrition. *Australian Veterinary Journal* 54, 57–60.

Chesters, J.K. and Arthur, J.R. (1988) Early biochemical defects caused by dietary trace element deficiencies. *Nutrition Research Reviews* 1, 39–56.

Colnago, G.L., Jensen, L.S. and Long, P.L. (1994) Effect of selenium and vitamin E on the development of immunity to coccidiosis in chickens. *Poultry Science* 63, 1136–1143.

Combs, G.F. Jr (2001) Selenium in global food systems. *British Journal of Nutrition* 85, 517–547.

Combs, G.F. and Scott, M.L. (1974) Dietary requirements for vitamin E and selenium measured at the cellular level in the chick. *Journal of Nutrition* 104, 1292–1296.

Conrad, H.R. and Moxon, A.L. (1979) Transfer of dietary selenium to milk. *Journal of Animal Science* 62, 404–411.

Contreras, P.A., Wittwer, F., Matamoros, R., Mayorga, I.M. and van Schaik, G. (2005) Effect of grazing pasture with a low selenium content on the concentration of triiodothyronine and thyroxine in serum ans glutathione peroxidase activity in erythrocytes in cows in Chile. *New Zealand Veterinary Journal* 53, 77–80.

Cooper, B.S., West, D.M. and Pauli, J.V. (1989) Effects of repeated oral doses of selenium in sheep. *New Zealand Veterinary Journal* 37, 37.

Cristaldi, L.A., McDowell, L.R., Buergelt, C.D., Davis, P.A., Wilkinson, N.S. and Martin, F.G. (2005) Tolerance of inorganic selenium by wether sheep. *Small Ruminant Research* 56, 205–213.

Davidson, W.B. and Kennedy, D.G. (1993) Synthesis of [^{75}Se] selenoproteins is greater in selenium-deficient sheep. *Journal of Nutrition* 123, 689–694.

Davis, P.A., McDowell, L.R., Wilkinson, N.S., Buergelt, C.D., Van Alstyne, R. Weldon, R.N. and Marshall, T.T. (2006) Tolerance of inorganic selenium by range-type ewes during gestation and lactation. *Journal of Animal Science* 84, 660–668.

Dhillon, K.S. and Dhillon, S.K. (1991) Accumulation of selenium in sugar cane (*Sacharum officinarum Linu*) in seleniferous areas of Punjab, India. *Environmental Geochemistry and Health* 13, 165–170.

Donald, G.E., Langlands, J.P., Bowles, J.E. and Smith, A.J. (1994a) Subclinical selenium deficiency. 5. Selenium status and the growth and wool production of sheep supplemented with thyroid hormones. *Australian Journal of Experimental Agriculture* 34, 13–18.

Donald, G.E., Langlands, J.P., Bowles, J.E. and Smith, A.J. (1994b) Subclinical selenium deficiency. 6. Thermoregulatory ability of perinatal lambs born to ewes supplemented with selenium and iodine. *Australian Journal of Experimental Agriculture* 34, 19–24.

Donald, G.E., Langlands, J.P., Bowles, J.E., Smith, A.J. and Burke, G.L. (1994c) Selenium supplements for grazing sheep. 3. Development of an intraruminal pellet with an extended life. *Animal Feed Science and Technology* 40, 295–308.

Droke, E.A. and Loerch, S.C. (1989) Effects of parenteral selenium and vitamin E on performance, health and humoral immune response of steers new to the feedlot environment. *Journal of Animal Science* 67, 1350–1359.

Ehlig, C.F., Hogue, D.E., Allaway, W.H. and Ham, D.J. (1967) Fate of selenium from selenite or selenomethionine, with or without vitamin E, in lambs. *Journal of Nutrition* 92, 121–126.

Ellison, R.S., Young, B.J. and Read, D.H. (1986) Bovine post-parturient haemoglobinuria: two distinct entities in New Zealand. *New Zealand Veterinary Journal* 34, 7–10.

Erskine, R.J., Eberhart, R.J., Grasso, P.J. and Scholz, R.W. (1989a) Induction of *Escherichia coli* mastitis in cows fed selenium-deficient or selenium-supplemented diets. *American Journal of Veterinary Research* 50, 2093–2100.

Erskine, R.J., Eberhart, R.J. and Hutchinson, L.J. (1989b) Blood selenium concentrations and glutathione peroxidase activities in dairy herds with high and low somatic cell counts. *Journal of American Veterinary Medical Association* 190, 1417–1421.

Field, A.C., Suttle, N.F., Brebner, J. and Gunn, G. (1988) An assessment of the efficacy and safety of selenium and cobalt included in an anthelmintic for sheep. *Veterinary Record* 123, 97–100.

Fisher, D.D., Saxton, S.W., Elliot, R.D. and Beatty, J.M. (1995) Effects of selenium source on selenium status of lactating cows. *Veterinary Clinical Nutrition* 2, 68–74.

Flohe, L., Gunzler, W.A. and Schock, H.H. (1973) Glutathione peroxidase: a selenoenzyme. *FEBS Letters* 32, 132–134.

Fordyce, F. (2005) Selenium deficiency and toxicity in the environment. In: Selinus, O. (ed.) *Essentials of Medical Geology: Impacts of the Natural Environment on Public Health*. British Geological Survey, pp. 373–415.

Fraser, A.J., Ryan, T.J., Sproule, R., Clark, R.G., Anderson, D. and Pederson, E.O. (1987) The effect of selenium on milk production in dairy cattle. *Proceedings of the New Zealand Society of Animal Production* 47, 61–64.

Gabbedy, B.J. (1971) Effect of selenium on wool production, body weight and mortality of young sheep in Western Australia. *Australian Veterinary Journal* 47, 318–322.

Gallagher, K. and Stowe, H.D. (1980) Influence of exercise on serum selenium and peroxide reduction system of standard thoroughbreds. *American Journal of Veterinary Research* 41, 1333–1335.

Givens, D.I., Allison, R., Cottrill, B. and Blake, J.S. (2004) Enhancing the selenium content of bovine milk through alteration of the form and concentration of selenium in the diet of the dairy cow. *Journal of Science in Food and Agriculture* 84, 811–817.

Gleed, P.T., Allen, W.M., Mallenson, C.B., Rowlands, G.J., Sanson, B.F., Vagg, M.J. and Caswell, R.D. (1983) Effects of selenium and copper supplementation in the growth of beef steers. *Veterinary Record* 113, 388–392.

Glienke, L.R. and Ewan, R.C. (1977) Selenium deficiency in the young pig. *Journal of Animal Science* 45, 1334–1340.

Goehring, T.B., Palmer, I.S., Olson, O.E., Libal, G.W. and Wahlstrom, R.C. (1984a) Effects of seleniferous grains and inorganic selenium on tissue and blood composition and growth performance of rats and swine. *Journal of Animal Science* 59, 725–732.

Goehring, T.B., Palmer, I.S., Olson, O.E., Libal, G.W. and Wahlstrom, R.C. (1984b) Toxic effects of selenium on growing swine fed corn–soya bean meal diets. *Journal of Animal Science* 59, 733–737.

Goff, J.P. and Stabel, J.R. (1990) Decreased plasma retinol, α-tocopherol and zinc concentrations during the periparturient period: effect of milk fever. *Journal of Dairy Science* 73, 3195–3199.

Grace, G.D. and Knowles, S.O. (2002) A reference curve using blood selenium concentration to diagnose selenium deficiency and predict growth responses in lambs. *New Zealand Veterinary Journal* 50, 163–165.

Grace, N.D. and West, D.M. (2006) Effect of Se-amended fertilisers on the Se status of grazing dairy cattle. *Proceedings of New Zealand Society of Animal Production* 66, 182–186.

Grace, N.D., Venning, M. and Vincent, G. (1994) An evaluation of a controlled release system for selenium in lambs. *New Zealand Veterinary Journal* 42, 63–65.

Grace, D.D., Knowles, S.O. and Lee, J. (1997) Relationships between blood selenium concentrations and milk somatic cell counts in dairy cows. *New Zealand Veterinary Journal* 45, 171–172.

Grace, N.D., Ankenbauer-Perkins, K., Alexander, A.M. and Marchant, R.M. (2001) Relationship between selenium concentration or glutathione peroxidase activity and milk selenium concentrations in dairy cows. *New Zealand Veterinary Journal* 49, 24–28.

Grace, N.D., Knowles, S.O. and West, D.M. (2006) Dose response effects of long-acting injectable vitamin B_{12} plus selenium (Se) on the vitamin B_{12} and Se status of ewes and their lambs. *New Zealand Veterinary Journal* 54, 67–72.

Gries, C.L. and Scott, M.L. (1972) Pathology of selenium deficiency in the chick. *Journal of Nutrition* 102, 1287–1296.

Groce, A.W., Miller, E.R., Hitchcock, J.P., Ullrey, D.E. and Magee, W.T. (1973) Selenium balance in the pig as affected by selenium source and vitamin E. *Journal of Animal Science* 37, 942–947.

Gunter, S.A., Beck, P.A. and Phillips, J.M. (2003) Effects of supplementary selenium source on the performance and blood measurements in beef cows and their calves. *Journal of Animal Science* 81, 856–864.

Guyot, H., Spring, P., Andrieu, S. and Rollin, F. (2007) Comparative responses to sodium selenite and organic selenium supplements in Belgian Blue cows and calves. *Livestock Science* 111, 259–263.

Hadley, K.B. and Sunde, R.A. (1997) Determination of dietary selenium requirement in female turkey poults using glutathione peroxidase. In: *Proceedings of the Ninth International Symposium on Trace Elements in Man and Animals (TEMA 9)*. NRC Research Press, Ottawa, pp. 59–60.

Harrison, J.H., Hancock, D.D. and Conrad, H.R. (1984) Vitamin E and selenium for reproduction in the dairy cow. *Journal of Dairy Science* 67, 123–132.

Hartley, W.J. (1963) Selenium and ewe fertility. *Proceedings of the New Zealand Society of Animal Production* 23, 20–27.

Hartley, W.J. (1967) Levels of selenium in animal tissues and methods of selenium administration. In: Muth, O.H. (ed.) *Selenium in Biomedicine*. Avi Publishing, Westport, Connecticut, pp. 79–96.

Hatfield, D.L., Berry, M.J. and Gladyshev, V.N. (2006) *Selenium: Its Molecular Biology and Role in Human Health*. Springer, New York.

Henry, P.R. and Ammerman, C.B. (1995) Selenium bioavailability. In: Ammerman, C.B., Baker, D.H. and Lewis, A.J. (eds) *Bioavailability of Nutrients for Animals*. Academic Press, New York, pp. 303–331.

Henry, P.R., Echevarria, M.G., Ammerman, C.B. and Rao, P.V. (1988) Estimation of the relative bioavailability of inorganic selenium sources for ruminants using tissue uptake of selenium. *Journal of Animal Science* 66, 2306–2312.

Hidiroglou, M., Proulx, J. and Jolette, J. (1985) Intraruminal selenium for control of nutritional muscular dystrophy in the dairy cow. *Journal of Dairy Science* 68, 57–66.

Hill, C.H. (1974) Reversal of selenium toxicity in chicks by mercury, copper, and cadmium. *Journal of Nutrition* 104, 593–598.

Hollenbach, B., Morgenthaler, N.G., Struck J., Alonso, C., Bergmann, A., Köhrle, J. and Schomburg, L. (2008) New assay for the measurement of selenoprotein P as a sepsis biomarker from serum. *Journal of Trace Elements in Medecine and Biology* 22, 24–32.

Ivancic, J. and Weiss, W.P. (2001) Effect of dietary sulfur and selenium concentrations on selenium balance in lactating Holstein cows. *Journal of Dairy Science* 84, 225–232.

Jelinek, P.D., Ellis, T., Wroth, R.H., Sutherland, S.S., Masters, H.G. and Pettersen, D.S. (1988) The effect of selenium supplementation on immunity, and the establishment of an experimental *Haemonchus contortus* infection, in weaner merino sheep fed a low selenium diet. *Australian Veterinary Journal* 65, 214–217.

Jenkins, K.J. and Hidiriglou, M. (1986) Tolerance of the pre-ruminant calf for selenium in milk replacer. *Journal of Dairy Science* 69, 1865–1870.

Jensen, L.S. (1968) Selenium deficiency and impaired reproduction in Japanese quail. *Proceedings of the Society for Experimental Biology and Medicine* 128, 970–972.

Jianhua, H., Ohtsuka, A. and Hayashi, K. (2000) Selenium influnces growth via thyroid hormone status in broiler chickens. *British Journal of Nutrition* 84, 727–732.

Jukola, E., Hakkarainan, J., Saloniemi, H. and Sankari, S. (1996) Blood selenium, vitamin E, vitamin A and β carotene and udder health, fertility treatments and fertility. *Journal of Dairy Science* 79, 838–845.

Juniper, D.T., Phipps, R.H., Jones, A.K. and Bertin, G. (2006) Selenium supplementation of lactating dairy cows: effect on selenium concentration in blood, milk, urine and feces. *Journal of Dairy Science* 89, 3544–3551.

Juniper, D.T., Phipps, R.H., Ramos-Morales, E. and Bertin, G. (2008) Selenium persistency and speciation in the tissues of lambs following the withdrawal of a high-dose selenium-enriched yeast. *Animal* 2, 375–380.

Kappel, L.C., Ingraham, R.H., Morgan, E.B., Dixon, J.M., Zeringue, L., Wilson, D. and Babcock, D.K. (1984) Selenium concentrations in feeds and effects of treating pregnant Holstein cows with selenium and vitamin E on blood selenium values and reproductive performance. *American Journal of Veterinary Research* 45, 691–694.

Kaur, R.S., Sharma, S. and Rampal, S. (2003) Effect of subchronic selenium toxicosis on lipid peroxidation, glutathione redox cycle and antioxidant enzymes in calves. *Veterinary and Human Toxicology* 45, 190–192.

Kendall, N.R., Jackson, D.W., Mackenzie, A.M., Illingworth, D.V., Gill, I.M., and Telfer, S.B.(2001) The effect of a zinc, cobalt and selenium soluble glass bolus on the trace element status of extensively grazed sheep over winter. *Animal Science* 73,163–169

Kim, Y.Y. and Mahan, D.C. (2001) Comparative effects of high dietary levels of organic and inorganic selenium on selenium toxicity of growing-finishing pigs. *Journal of Animal Science* 79, 942–948.

Knowles, S.O., Grace, N.D., Wurms, K. and Lee, J. (1999) Significance of amount and form of dietary selenium on blood milk and casein selenium concentrations in grazing cows. *Journal of Dairy Science* 82, 429–437.

Koenig, K.M., Buckley, W.T. and Shelford, J.A. (1991) Measurement of endogenous faecal excretion and true absorption of selenium in dairy cows. *Canadian Journal of Animal Science* 71, 167–174.

Koenig, K.M., Rode, L.M., Cohen, R.D.H. and Buckley, W.T. (1997) Effects of diet and chemical form of selenium on selenium metabolism in sheep. *Journal of Animal Science* 75, 817–127.

Koh, T.-S. and Judson, G.J. (1987) Copper and selenium deficiency in cattle: an evaluation of methods of oral therapy and an observation of a copper–selenium interaction. *Veterinary Research Communications* 11, 133–148.

Korhle, J., Jakob, F., Contempre, B. and Dumont, J.E. (2005) Selenium, the thyroid and the endocrine system. *Endocrine Reviews* 26, 944–984.

Kott, R.W., Ruttie, J.L. and Southward, G.M. (1983) Effects of vitamin E and selenium injections on reproduction and pre-weaning lamb survival in ewes consuming diets marginally deficient in selenium. *Journal of Animal Science* 57, 553–558.

Krishnamurti, C.R., Ramberg, C.F. and Shariff, M.A. (1989) Kinetic modelling of selenium metabolism in non-pregnant ewes. *Journal of Nutrition* 119, 1146–1155.

Krishnamurti, C.R., Ramberg, C.F., Shariff, M.A. and Boston, R.C. (1997) A compartmental model depicting short-term kinetic changes in selenium metabolism in ewes fed hay containing normal or inadequate levels of selenium. *Journal of Nutrition* 127, 95–102.

Langlands, J.P., Wilkins, J.F., Bowles, J.E., Smith, A.J. and Webb, R.F. (1981) Selenium concentration in the blood of ruminants grazing in northern New South Wales. 1. Analysis of samples collected in the National Brucellosis Eradication Scheme. *Australian Journal of Agricultural Research* 32, 511–521.

Langlands, J.P., Donald, G.E., Bowles, J.E. and Smith, A.J. (1986) Selenium excretion in sheep. *Australian Journal of Agricultural Research* 37, 201–209.

Langlands, J.P., Donald, G.E., Bowles, J.E. and Smith, A.J. (1989) Selenium concentration in the blood of ruminants grazing in northern New South Wales 3. Relationship between blood concentration and the response in liveweight of grazing cattle given a selenium supplement. *Australian Journal of Agricultural Research* 40, 1075–1083.

Langlands, J.P., Donald, G.E., Bowles, J.E. and Smith, A.J. (1990) Selenium supplements for grazing sheep. 1. A comparison between soluble salts and other forms of supplement. *Animal Feed Science and Technology* 28, 1–13.

Langlands, J.P., Donald, G.E., Bowles, J.E. and Smith, A.J. (1991a) Subclinical selenium insufficiency. 1. Selenium status and the response in liveweight and wool production of grazing ewes supplemented with selenium. *Australian Journal of Experimental Agriculture* 31, 25–31.

Langlands, J.P., Donald, G.E., Bowles, J.E. and Smith, A.J. (1991b) Subclinical selenium insufficiency. 2. The response in reproductive performance of grazing ewes supplemented with selenium. *Australian Journal of Experimental Agriculture* 31, 33–35.

Langlands, J.P., Donald, G.E., Bowles, J.E. and Smith, A.J. (1991c) Subclinical selenium insufficiency. 3. The selenium status and productivity of lambs born to ewes supplemented with selenium. *Australian Journal of Experimental Agriculture* 31, 37–43.

Larsen, C.T., Pierson, F.W. and Gross, W.B. (1997) Effect of dietary selenium on the response of stressed and unstressed chickens to *Escherichia coli* challenge and antigen. *Biological Trace Element Research* 58, 169–176.

Latshaw, J.D., Ort, J.F. and Diesem, C.D. (1977) The selenium requirements of the hen and effects of a deficiency. *Poultry Science* 56, 1876–1881.

Lawler, T.L., Taylor, P., Filey, J.W. and Canton, J.S. (2004) Effects of supranutritional and organically bound selenium on performance, carcase characteristics and selenium distribution in finishing beef steers. *Journal of Animal Science* 82, 1488–1493.

Leeson, S., Namkung, H., Caston, L., Durosoy, S. and Schlegel, P. (2008) Comparison of selenium levels and sources and dietary fat quality in diets for broiler breeders and layer hens. *Poultry Science* 87, 2605–2612.

Lei, X.G., Dann, H.M., Ross, D.A., Cheng, W.-S., Combs, G.F. and Roneker, K.R. (1998) Dietary selenium supplementation is required to support full expression of three selenium-dependent glutathione peroxidases in various tissues of weanling pigs. *Journal of Nutrition* 128, 130–135.

Lyons, G., Oritz-Monasterio, I., Stangoulis, J. and Graham, R. (2005) Selenium concentrationsin wheat grain: is there sufficient genotypic variation to use in breeding? *Plant and Soil* 269, 369–380.

Mackintosh, C.G., Gill, J. and Turner, K. (1989) Selenium supplementation of young deer (*Cervus elaphus*). *New Zealand Veterinary Journal* 137, 143–145.

MacPherson, A. (1994) Selenium, vitamin E and biological oxidation. In: Garnsworthy, P.C. and Cole, D.J.A. (eds) *Recent Advances in Nutrition*. Nottingham University Press, Nottingham, UK, pp. 3–30.

MacPherson, A., Kelly, E.F., Chalmers, J.S. and Roberts, D.J. (1987) The effect of selenium deficiency on fertility in heifers. In: Hemphill, D.D. (ed.) *Proceedings of the 21st Annual Conference on Trace Substances in Environmental Health*. University of Missouri, Columbia, Missouri, pp. 551–555.

MAFF (1990) *UK Tables of the Nutritive Value and Chemical Composition of Foodstuffs*. In: Givens, D.I. (ed.) Rowett Research Services, Aberdeen, UK.

Mahan, D.C. (2000) Effects of organic and inorganic selenium sources on sow colostrum and milk selenium content. *Journal of Animal Science* 78, 100–105.

Mahan, D.C. and Kim, Y.Y. (1996) Effect of inorganic and organic selenium at two dietary levels on reproductive performance and tissue selenium concentrations in first parity gilts and their progeny. *Journal of Animal Science* 74, 2711–2718.

Mahan, D.C. and Peters, J.C. (2004) Long-term effects of dietary organic and inorganic selenium sources and levels on reproducing sows and their progeny. *Journal of Animal Science* 82, 1343–1358.

Mahan, D.C., Cline, T.R. and Richert, B. (1999) Effects of dietary levels of selenium-enriched yeast and sodium selenite as selenium sources fed to growing-finishing pigs on performance, tissue selenium, serum glutathione peroxidase activity, carcass characteristics and loin quality. *Journal of Animal Science* 7, 2172–2179.

Mahan, D.C., Brendemuhl, J.H., Carter, S.D., Chiba, L.I., Crenshaw, T.D., Cromwell, G.L., Dove, C.R., Harper, A.F., Hill, G.M., Hollis, G.R., Kim, S.W., Lindemann, M.D., Maxwell, C.V., Miller, P.S., Nelssen, J.L., Richert, B.T., Southern, L.L., Stahly, T.S., Stein, H.H., van Heugten, E., and Yen, J.T. (2005) Comparison of dietary selenium fed to grower-finisher pigs from various regions of the United Staes on resulting tissue Se and loin mineral concentrations. *Journal of Animal Science* 83, 852–857.

Marin-Guzman, J., Mahan, D.C., Chung, Y.K., Pate, J.L. and Pope, W.F. (1997) Effects of dietary selenium and vitamin E on boar performance and tissue responses, semen quality and subsequent fertilization rates in mature gilts. *Journal of Animal Science* 75, 2994–3003.

McDonald, J.W. (1975) Seleinium-responsive unthriftiness of young Merino sheep in Central Victoria. *Australian Veterinary Journal* 51, 433–435.

McDonald, J.W., Overend, D.J. and Paynter, D.I. (1989) Influence of selenium status in Merino weaners on resistance to trichostrongylid infection. *Research in Veterinary Science* 47, 319–322.

McLaughlin, J.G., Cullen, J. and Forristal, T. (1989) Blood selenium concentrations in cattle on seleniferous pastures. *Veterinary Record* 124, 426–427.

McMurray, C.H. and McEldowney, P.K. (1977) A possible prophylaxis and model for nutritional degenerative myopathy in young cattle. *British Veterinary Journal* 133, 535–542.

McMurray, C.H., Rice, D.A. and Kennedy, S. (1983) Nutritional myopathy in cattle: from a clinical problem to experimental models for studying selenium, vitamin E and polyunsaturated fatty acid interactions. In: Suttle, N.F., Gunn, R.G., Allen, W.M., Linklater, K.A. and Wiener, G. (eds) *Trace Elements in Animal Production and Veterinary Practice*. Occasional Publication No. 7, British Society of Animal Production, Edinburgh, pp. 61–76.

Meneses, A., Batra, T.R. and Hidiroglou, M. (1994) Vitamin E and selenium in milk of ewes. *Canadian Journal of Animal Science* 71, 567–569.

Millar, K.R. and Meads, W.J. (1988) Selenium levels in the blood, liver, kidney and muscle of sheep after the administration of iron/selenium pellets or soluble-glass boluses. *New Zealand Veterinary Journal* 36, 8–10.

Miller, G.Y., Bartlett, P.C., Erskine, R.J. and Smith, K.L. (1995) Factors affecting serum selenium and vitamin E concentrations in dairy cows. *Journal of American Veterinary Medicine Association* 206, 1369–1373.

Miltimore, J.E., van Ryswyk, A.L., Pringle, W.L., Chapman, F.M. and Kalnin, C.M. (1975) Selenium concentrations in British Columbia forages, grains and processed feeds. *Canadian Journal of Animal Science* 55, 101–111.

Minson, D.J. (1990) *Forages in Ruminant Nutrition*. Academic Press, San Diego, California, pp. 208–229.

Moir, D.C. and Masters, H.G. (1979) Hepatosis dietetica, nutritional myopathy, mulberry heart disease and associated hepatic selenium levels in pigs. *Australian Veterinary Journal* 55, 360–364.

Money, D.F.L., Meads, W.J. and Morrison, L. (1986) Selenised compressed salt blocks for selenium deficient sheep. *New Zealand Veterinary Journal* 34, 81–84.

Montane J., Marco, I., Manteca, X., Lopez, J. and Laving, S. (2002) Delayed acute capture myopathy in three roe deer. *Journal of Veterinary Medecine. Series A* 49, 93–98.

Moreno-Reyes, R., Egrise, D., Boelaert, M., Goldman, S. and Meuris, S. (2006) Iodine deficiency mitigates growth retardation and osteopenia in selenium-deficient rats. *Journal of Nutrition* 136, 595–600.

Morris, J.G., Chapman, H.C., Walker, D.F., Armstrong, J.B., Alexander, J.D., Miranda, R., Sanchez, A., Sanchez, B., Blair-West, J.R. and Denton, D.A. (1984) Selenium deficiency in cattle associated with Heinz body anaemia. *Science* 223, 291–293.

Moxon, A.L. (1937) *Alkali Disease or Selenium Poisoning*. Bulletin No. 311, South Dakota Agricultural Experiment Station.

Moxon, A.L. and Poley, W.E. (1938) The relation of selenium content of grains in the ration to the selenium content of poultry carcass and eggs. *Poultry Science* 17, 77–80.

Munoz, C., Carson, A.F., McCoy, M.A., Dawson, L.E.R., Irwin, D., Gordon, A.G. and Kilpatrick, D.J. (2009) Effects of supplementation with barium selenate on the fertility, prolificacy and lambing performance of hill sheep. *Veterinary Record* 164, 265–271.

Mutanen, M.P., Aspila, P. and Mykkanen, H.M. (1986) Bioavailability to rats of selenium in milk of cows fed sodium selenite or selenited barley. *Annals of Nutrition and Metabolism* 30, 183–188.

Neville, T.L., Ward, M.A., Reed, J.J., Soto-Navarro, S.A., Julius, S.L., Borowiicz, P.P., Taylor, J.P., Redmer, D.A., Reynolds, L.P. and Caton, J.S. (2008) Effects of level and source of dietary selenium on maternal and fetal body weight, visceral organ mass, cellularity estimates, and jejunal vascularity in pregnant ewe lambs. *Journal of Animal Science* 86, 890–901.

Nicol, F., Lefrane, H., Arthur, J.R. and Trayhurn, P. (1994) Characterisation and post-natal development of 5'-deiodinase activity in goat perinatal fat. *American Journal of Physiology* 267, R144–R149.

Noguchi, T., Cantor, A.H. and Scott, M.L. (1973) Mode of action of selenium and vitamin E in prevention of exudative diathesis in chicks. *Journal of Nutrition* 103, 1502–1511.

Nolan, M.R., Kennedy, S., Blanchflower, W.J. and Kennedy, D.G. (1995) Lipid peroxidation, prostacyclin and thromboxane A2 in pigs depleted of vitamin E and selenium and supplemented with linseed oil. *British Journal of Nutrition* 74, 369–380.

NRC (1994) *Nutrient Requirements of Poultry*, 9th edn. National Academy of Sciences, Washington, DC.

NRC (1998) *Nutrient Requirements of Swine*, 10th edn. National Academy of Sciences, Washington, DC.

NRC (2001) *Nutrient Requirements of Dairy Cows*, 7th edn. National Academy of Sciences, Washington, DC.

NRC (2005) *Mineral Tolerances of Animals*, 2nd edn. National Academy of Sciences, Washington, DC.

Oh, S.H., Pope, A.L. and Hoekstra, W.G. (1976) Dietary selenium requirement of sheep fed a practical-type diet as assessed by tissue glutathione peroxidase and other criteria. *Journal of Animal Science* 42, 984–992.

Oksanen, H.E. (1967) Selenium deficiency: clinical aspects and physiological responses in farm animals. In: Muth, O.H. (ed.) *Selenium in Biomedicine*. Avi Publishing, Westport, Connecticut, pp. 215–229.

Orr, J.P. and Blakley, B.R. (1997) Investigation of the selenium status of aborted calves with cardiac failure and myocardial necrosis. *Journal of Veterinary Diagnostic Investigation* 9, 172–179.

Ort, J.F. and Latshaw, J.D. (1978) The toxic level of sodium selenite in the diet of laying chickens. *Journal of Nutrition* 114–1120.

Ortman, K. and Pehrson, B. (1999) Effect of selenate as a feed additive to dairy cows in comparison to selenite and selenium yeast. *Journal of Animal Science* 77, 3365–3370.

O'Toole, D. and Raisbeck, M.F. (1995) Pathology of experimentally induced chronic selenosis (alkali disease) in yearling cattle. *Journal of Veterinary Diagnostic Investigation* 7, 364–373.

Ozkan, S., Malayoğlu, H.B., Yalçın, S., Karadas, F., Koçtürk, S., Cabuk, M., Oktay, G., Ozdemir, S., Ozdemir, E. and Ergül, M. (2007) Dietary vitamin E (α-tocopherol acetate) and selenium supplementation from different sources: performance, ascites-related variables and antioxidant status in broilers reared at low and optimum temperatures. *British Poultry Science* 48, 580–593.

Panter, K.E., Hartley, W.J., James, L.F., Mayland, H.F., Stegelmeier, B.L. and Kechele, P.O. (1996) Comparative toxicity of selenium from seleno-DL-methionine, sodium selenate and *Astragalus bisulcatus* in pigs. *Fundamental and Applied Toxicology* 32, 217–223.

Pappas, A.C., Acamovic, T., Sparks, A.C., Surai, P.F. and McDevitt, R.M. (2006) Effects of supplementing broiler breeder diets with organoselenium compounds and polyunsaturated fatty acids on hatchability. *Poultry Science* 85, 1584–1593.

Patterson, E.L., Milstrey, R. and Stokstad, E.L.R. (1957) Effect of selenium in preventing exudative diathesis in chicks. *Proceedings of the Society for Experimental Biology and Medicine* 95, 617–620.

Pedersen, K.B. and Simesen, M.G. (1977) Om tilskud of selen og vitamin E-selen mangelsyndromet hos svin. *Nordisk Veterinaermedicin* 29, 161–165.

Pehrson, B., Ortman, K., Madjid, N. and Trafikowska, U. (1999) The influence of dietary selenium as selenium yeast or sodium selenite on the concentration of selenium in the milk of suckler cows and on the selenium status of their calves. *Journal of Animal Science*, 77, 3371–3376.

Piatkowski, T.L., Mahan, D.C., Cantor, A.H., Moxon, A.L., Cline, J.H. and Grifo, A.P. Jr (1979) Selenium and vitamin E in semi-purified diets for gravid and nongravid gilts. *Journal of Animal Science* 48, 1357–1365.

Podoll, K.L., Bernard, J.B., Ullrey, D.B., DeBar, S.R., Ku, P.K. and Magee, W.T. (1992) Dietary selenate versus selenite for cattle, sheep and horses. *Journal of Animal Science* 70, 1965–1970.

Poley, W.E. and Moxon, A.L. (1938) Tolerance levels of seleniferous grains in laying rations. *Poultry Science* 17, 72–76.

Qin, S., Gao, J. and Huang, K. (2007) Effects of different selenium sources on tissue selenium concentrations, blood GSH-Px activities and plasma interleukin levels in finishing lambs. *Biological Trace Element Research* 116, 91–102.

Rammell, C.G., Thompson, K.G., Bentley, G.R. and Gibbons, M.W. (1989) Selenium, vitamin E and polyunsaturated fatty acid concentrations in goat kids with and without nutritional myodegeneration. *New Zealand Veterinary Journal* 37, 4–6.

Reeves, P.G., Leary, P.D., Gregoire, B.R., Finley, J.W., Lindlauf, J.E. and Johnson, L.K. (2005) Selenium bioavailability from buckwheat bran in rats fed a modified purified AIN-93G torula yeast diet. *Journal of Nutrition* 135, 2627–2633.

Reffett, J.K., Spears, J.W. and Brown, T.T. (1988) Effect of dietary selenium and vitamin E on the primary and secondary immune response in lambs challenged with parainfluenza virus. *Journal of Animal Science* 66, 1520–1528.

Renko, R., Werner, M., Renner-Muller, I., Cooper, T.G., Yeung, C.H., Hollenbacy, B., Scharpf, M., Korhle, J., Schomburg, L. and Schweizer, U. (2008) Hepatic selenoprotein P (SePP) expression restores selenium transport and prevents infertility and motor-incoordination in Sepp-knockout mice. *Biochemical Journal* 409, 741–749.

Rice, D.A. and Kennedy, S. (1988) Assessment of vitamin E, selenium and polyunsaturated fatty acid interactions in the aetiology of disease in the bovine. *Proceedings of the Nutrition Society* 47, 177–184.

Rock, M.J., Kincaid, R.L and Carstens, C. (2001) Effects of prenatal source and level of dietary selenium on passive immunity and thermometabolism of newborn lambs. *Small Ruminant Research* 40, 129–138.

Rosenfeld, I. and Beath, O.A. (1964) *Selenium: Geobotany, Biochemistry, Toxicity and Nutrition.* Academic Press, New York, p. 411.

Rotruck, J.T., Pope, A.L., Ganther, H.E., Swanson, A.B., Hafeman. D.G. and Hoekstra, W.G. (1973) Selenium: biochemical role as a component of glutathione peroxidase. *Science, USA* 179, 588–590.

Rowntree, J.E., Hill, G.M., Hawkins, D.R., Link, J.E., Rincker, M.J., Bednar, G.W. and Kreft, R.A. Jr (2004) Effect of selenium on selenoprotein activity and thyroid hormone metabolism in beef and dairy cows and calves. *Journal of Animal Science* 82, 2995–3005.

Sahin, N., Onderci, M., Sahin, K. and Kucuk, O. (2008) Supplementation with organic or inorganic selenium in heat-stressed quail. *Biological Trace Element Research* 122, 229–237.

Salvatore, D., Low, S.C., Berry, M., Maia, A.C., Harney, J.W., Croteau, W., St. Germain, D.L. and Larsen, P.J. (1995) Type 3 iodothyronine deiodinase cloning *in vitro:* expression and functional analysis of the selenoprotein. *Journal of Clinical Investigation* 96, 2421–2430.

Sanchez, J., Montes, P., Jimenez, A. and Andres, S. (2007) Prevention of clinical mastitis with barium selenate in dairy goats from a selenium-deficient area. *Journal of Dairy Science* 90, 2350–2354.

Schwarz, K. and Foltz, C.M. (1957) Selenium as an integral part of factor 3 against dietary necrotic liver degeneration. *Journal of the American Chemical Society* 79, 3292–3293.

Schweizer, U., Schimburg, L. and Savaskan, N.E. (2004) The neurobiology of selenium: lessons from transgenic mice. *Journal of Nutrition* 134, 707–710.

Scott, M.L. and Thompson, J.N. (1971) Selenium content of feedstuffs and effects of dietary selenium levels upon tissue selenium in chick and poults. *Poultry Science* 50, 1742–1748.

Scott, M.L., Olson, G., Krook, L. and Brown, W.R. (1967) Selenium-responsive myopathies of myocardium and of smooth muscle in the young poult. *Journal of Nutrition* 91, 573–583.

Selinus, O. (1988) Biogeochemical mapping of Sweden for geomedical and environmental research. In: Thornton, I. (ed.) *Proceedings of the Second International Symposium on Geochemistry and Health.* Science Reviews, Norwood, UK, pp. 13–20.

Serra, A.B., Serra, S.D., Shinchi, K. and Fujihara, T. (1997) Bioavailability of rumen bacterial selenium in mice using tissue uptake technique. *Biological Trace Element Research* 58, 255–261.

Shellow, J.S., Jackson, S.G., Baker, J.P. and Cantor, A.H. (1985) The influence of dietary selenium levels on blood levels of selenium and glutathione peroxidase activity in the horse. *Journal of Animal Science* 61, 590–594.

Simesen, M.G., Nielsen, H.E., Danielsen, V., Gissel-Nielsen, G., Hjarde, W., Leth, T. and Basse, A. (1979) Selenium and vitamin E deficiency in pigs 2. Influence on plasma selenium, vitamin E, ASAT and ALAT and on tissue selenium. *Acta Veterinaria Scandinavica* 20, 289–305.

Slaweta, R., Wasowiez, W. and Laskowska, T. (1988) Selenium content, glutathione peroxidase activity and lipid peroxide level in fresh bull semen and its relationship to motility of spermatozoa after freezing and thawing. *Journal of Veterinary Medicine, Animal Physiology, Pathology and Clinical Veterinary Medicine* 35, 455–460.

Slen, S.B., Demuriren, A.S. and Smith, A.D. (1961) A note on the effects of selenium on wool growth and body gains in sheep. *Canadian Journal of Animal Science* 41, 263–265.

Small, J.N.W., Burke, L., Suttle, N.F., Bain, M.S., Edwards, J.G. and Lewis, C.J. (1996) Seasonal fluctuations in subnormal tocopherol concentrations detected in bovine and ovine samples submitted by Veterinary Investigation Cenres (VIC) throughout Great Britain during 1995. In: *Proceedings of the XIX World Buiatrics Congress, Edinburgh, Volume 2.* British Cattle Veterinary Association, pp. 413–416.

Smith, G.M. and Allen, J.G. (1997) Effectiveness of α-tocopherol and selenium supplements in preventing lupinosis-associated myopathy in sheep. *Australian Veterinary Journal* 75, 341–437.

Stabel, J.R. and Spears, J.W. (1993) Role of selenium in immune responsiveness and disease resistance. In: Kurfield, D.M. (ed.) *Human Nutrition – A Comprehensive Treatise*, Vol. 8. *Nutrition and Immunology.* Plenum Press, New York, pp. 333–355.

Stevens, J.B., Olsen, W.C., Kraemer, R. and Archaublau, J. (1985) Serum selenium concentrations and glutathione peroxidase activities in cattle grazing forages of various selenium concentrations. *American Journal of Veterinary Research* 46, 1556–1560.

Stewart, M.S., Spallholz, J.E., Neldner, K.H. and Pence, B.C. (1999) Selenium compounds have disparate abilities to impose oxidative stress and induce apoptosis. *Free Radical Biology and Medicine* 26, 42–48.

Sunde, R.A., Evenson, J.K., Thompson, K.M. and Sachdev, S.W. (2005) Dietary selenium requirements based on glutathione peroxidase-1 activity and mRNA levels and other selenium-dependent parameters are not increased by pregnancy in rats. *Journal of Nutrition* 135, 2144–2150.

Surai, P.F. (2006) *Selenium in Animal and Human Health*. Nottingham University Press, Nottingham, UK.

Surai, P.F., Kostjuk, I., Wishart, A.M., McPherson, A., Speak, B.K., Noble, R.C., Ionov, I. and Kutz, E. (1998) Effect of vitamin E and selenium supplementation of cockerel diets on glutathione peroxidase activity, and lipid peroxidation susceptibility in sperm, testes and liver. *Biological Trace Element Research* 64, 119–132.

Suttle, N.F. (1989) Predicting risks of mineral disorders in goats. *Goat Veterinary Society Journal* 10, 19–27.

Suttle, N.F., Jones, D.G., Woolliams, C. and Woolliams, J.A. (1987) Heinz body anaemia in copper and selenium deficient lambs grazing improved hill pastures. *British Journal of Nutrition* 58, 539–548.

Swain, B.K., Johri, T.S. and Majumdar, S. (2000) Effect of supplementation of vitamin E, selenium and their different combinations on the performance and immune response of broilers. *British Poultry Science* 41, 287–292.

Swecker, W.S., Thatcher, C.D., Eversole, D.E., Blodgett, D.J. and Schurig, G.G. (1995) Effect of selenium supplementation on colostral IgG concentration in cows grazing selenium deficient pastures and on post-suckle serum IgG concentration in their calves. *American Journal of Veterinary Research* 56, 450–453.

Tasker, J.B., Bewick, T.D., Clark, R.G. and Fraser, A.J. (1987) Selenium response in dairy cattle. *New Zealand Veterinary Journal* 35, 139–140.

Taylor, J.B. (2005) Time-dependent influence of supranutritional organically bound selenium on selenium accumulation in growing wether lambs. *Journal of Animal Science* 83, 1186–1193.

Thompson, J.N. and Scott, M.L. (1969) Role of selenium in the nutrition of the chick. *Journal of Nutrition* 97, 335–342.

Thompson. J.N. and Scott, M.L. (1970) Impaired lipid and vitamin E absorption related to atrophy of the pancreas in selenium deficient chicks. *Journal of Nutrition* 100, 797–809.

Tripp, M.J., Whanger, P.D. and Schmitz, J.A. (1993) Calcium uptake and ATPase activity of sarcoplasmic reticulum vesicles isolated from control and selenium deficient lambs. *Journal of Trace Elements and Electrolytes in Health and Disease* 7, 75–82.

Ullrey, D.J. (1974) The selenium deficiency pattern in animal agriculture. In: Hoekstra, W.G. and Ganther, H.E. (eds) *Proceedings of the Second International Symposium on Trace Elements in Man and Animals, Wisconsin*. University Park Press, Baltimore, Maryland, pp. 275–294.

Ullrey, D.J. (1987) Biochemical and physiological indicators of selenium status in animals. *Journal of Animal Science* 65, 1712–1726.

Underwood, E.J. and Suttle, N.F. (1999) *The Mineral Nutrition of Livestock*, 3rd edn. CAB International, Wallingford, UK.

Van Dael, P., Davidsson, L., Monoz-Box, R., Fay, L.B. and Barclay, D. (2001) Selenium absorption and retention from a selenite- or selenate-fortified milk-based formula in men measured by a stable-isotope technique. *British Journal of Nutrition* 85, 157–163.

Van Ryssen, J.B.J., van Malsen, P.M. and Hartman, F. (1998) Contribution of dietary suphur to the interaction between selenium and copper in sheep. *Journal of Agricultural Science, Cambridge* 130, 107–114.

Van Vleet, J.F., Carlton, W. and Olander, H.J. (1970) Hepatosis dietetica and mulberry heart disease associated with selenium deficiency in Indiana swine. *Journal of the American Veterinary Medicine Association* 157, 1208–1219.

Van Vleet, J.F., Meyer, K.B. and Olander, H.J. (1973) Control of selenium-vitamin E deficiency in growing swine by parenteral administration of selenium–vitamin E preparations to baby pigs or to pregnant sows and their baby pigs. *Journal of the American Veterinary Medical Association* 163, 452–456.

Vendeland, S.C., Butler, J.A. and Whanger, P.D. (1992) Intestinal absorption of selenite, selenate and selenomethionine in the rat. *Journal of Nutritional Biochemistry* 3, 359–365.

Vouduri, A., Chadio, S.E., Mengatos, J.G., Zervas, G.P., Nicol, F. and Arthur, J.R. (2003) Selenoenzyme activities in selenium- and iodine-deficient sheep. *Biological Trace Element Research* 94, 213–224.

Wahlstrom, R.C. and Olson, O.E. (1959) The effect of selenium on reproduction in swine. *Journal of Animal Science* 18, 141–145.

Wahlstrom, R.C., Goehring, T.B., Johnson, D.D., Libal, G.W., Olson, O.E., Palmer, I.S. and Thaler, R.C. (1984) The relationship of hair colour to selenium content of hair and selenosis in swine. *Nutrition Reports International* 29, 143–148.

Wang, Y.H., Leibholz, J., Bryden, W.L. and Fraser, D.R. (1996) Lipid peroxidation status as an index to evaluate the influence of dietary fats on vitamin E requirements of young pigs. *British Journal of Nutrition* 75, 81–95.

Waschulewski, I.H. and Sunde, R.A. (1988) Effect of dietary methionine on utilisation of dietary selenium from dietary selenomethionine for glutathione peroxidase in the rat. *Journal of Nutrition* 118, 367–374.

Weiss, W.P. (1998) Requirements of fat-soluble vitamins for dairy cows: a review. *Journal of Dairy Science* 81, 2493–2501.

Weiss, W.P. and Hogan, J.S. (2005) Effect of selenium source on selenium status, neutrophil function and response to inflammatory endotoxin challenge of dairy cows. *Journal of Dairy Science* 88, 4366–4374.

Weiss, W.P., Hogan, J.S., Todhunter, D.A. and Smith, K.L. (1997) Effect of vitamin E supplementation in diets with a low concentration of selenium on mammary gland health of dairy cows. *Journal of Dairy Science* 80, 1728–1737.

Weitzel, F., Ursini, F. and Wendel, A. (1990) Phospholipid hydroperoxide glutathione peroxidase in various mouse organs during selenium deficiency and repletion. *Biochemica et Biophysica Acta* 1036, 88–94.

Whanger, P.D. (2002) Selenocompounds in plants and animals and their biological significance. *Journal of the American College of Nutrition* 21, 223–232.

Whanger, P.D., Weswig, P.H., Schmitz, J.A. and Oldfield, J.E. (1977) Effects of selenium and vitamin E on blood selenium levels, tissue glutathione peroxidase activities and white muscle disease in sheep fed purified or hay diets. *Journal of Nutrition* 107, 1298–1307.

Whelan, B.R., Barrow, N.J. and Peter, D.W. (1994a) Selenium fertilizers for pastures grazed by sheep. I. Selenium concentrations in whole blood and plasma. *Australian Journal of Agricultural Research* 45, 863–875.

Whelan, B.R., Barrow, N.J. and Peter, D.W. (1994b) Selenium fertilizers for pastures grazed by sheep. II. Wool and liveweight responses to selenium. *Australian Journal of Agricultural Research* 45, 875–886.

White, P.J. and Broadley, M.R. (2009) Biofortification of crops with seven mineral elements often lacking in human diets – iron, zinc, copper, calcium, magnesium, selenium and iodine. *New Phytologist* 182, 49–84.

White, C.L., Cadwalader, T.K., Hoekstra, W. and Pope, A.L. (1998) The metabolism of ^{75}Se-selenomethionine in sheep given supplementary copper and molybdenum. *Journal of Animal Science* 67, 2400–2408.

Whiteacre, M.E., Combs, G.F., Combs, S.B. and Parker, R.S. (1987) Influence of dietary vitamin E on nutritional pancreatic atrophy in selenium-deficient chicks. *Journal of Nutrition* 117, 460–467.

Wichtel, J.J. (1998) A review of selenium deficiency in grazing ruminants 2: towards a more rational approach to diagnosis and prevention. *New Zealand Veterinary Journal* 46, 54–58.

Wichtel, J.J., Craigie, A.L., Varela-Alvarez, H. and Williamson, N.B. (1994) The effect of intra-ruminal selenium pellets on growth rate, lactation and reproductive efficiency in dairy cattle. *New Zealand Veterinary Journal* 42, 205–210.

Wichtel, J.J., Craigie, A.L., Freeman, D.A., Varela-Alvarez, H. and Williamson, N.B. (1996) Effect of selenium and iodine supplementation on growth rate and on thyroid and somatotropic function in dairy calves at pasture. *Journal of Dairy Science* 79, 1865–1872.

Wichtel, J.J., Grace, N.D. and Firth, E.C. (1998) The effect of injectable barium selenate on the selenium status of horses on pasture. *New Zealand Veterinary Journal* 42, 205–210.

Wilkins, J.F. and Kilgour, R.J. (1982) Production responses to selenium in northern New South Wales. 1. Infertility in ewes and associated production. *Australian Journal of Experimental Agriculture and Animal Husbandry* 22, 18–23.

Wilkins, J.F., Kilgour, R.J., Gleeson, A.C., Cox, R.J., Geddes, S.J. and Simpson, I.H. (1982) Production responses to selenium in northern New South Wales. 2. Liveweight gain, wool production and reproductive performance in young Merino ewes given selenium and copper supplements. *Australian Journal of Experimental Agriculture and Animal Husbandry* 22, 24–28.

Wilson, P.R. and Grace, N.D. (2001) A review of tissue reference values used to assess the trace element status of farmed deer (*Cervus elaphus*). *New Zealand Veterinary Journal* 49, 126–132.

Wuryastuti, H., Stowe, H.D., Bull, R.W. and Miller, E.R. (1993) Effects of vitamin E and selenium on immune responses of peripheral blood, colostrum and milk leukocytes of sows. *Journal of Animal Science* 71, 2464–2472.

Yaeger, M.J., Neiger, R.D., Holler, L., Fraser, T.L., Hurley, D.J. and Palmer, I.S. (1998) The effect of selenium toxicosis on pregnant beef cattle. *Journal of Veterinary Diagnostic Investigation* 10, 268–273.

Yoon, I. and McMillan, E. (2006) Comparative effects of inorganic and organic selenium on selenium transfer from sows to nursing pigs. *Journal of Animal Science* 84, 1729–1733.

Zachara, B.A., Frafikowska, V., Lejman, H., Kimber, C. and Kaptur, M. (1993) Selenium and glutathione peroxidase in blood of lambs born to ewes injected with barium selenate. *Small Ruminant Research* 11, 135–141.

Zervas, G.P. (1988) Use of soluble glass boluses containing Cu, Co and Se in the prevention of trace-element deficiencies in goats. *Journal of Agricultural Science, Cambridge* 110, 155–158.

Zuberbuehler, C.A., Messelkommer, R.E., Arnold, M.M., Forrer, R.S. and Wenk, C. (2006) Effects of selenium depletion and selenium repletion by choice feeding on selenium status of young and old laying hens. *Physiology and Behavior* 87, 430–440.

16 Zinc

Introduction

The first unequivocal evidence that zinc is necessary for growth and health was obtained in laboratory animals (Todd *et al.*, 1934) at a time when associations were being noted between zinc and carbonic anhydrase, an enzyme eventually shown to contain zinc (Keilin and Mann, 1940). In domesticated species, experimental zinc deprivation was produced first in chicks (O'Dell and Savage, 1957) and pigs (Stevenson and Earle, 1956) and later in lambs (Ott *et al.*, 1964) and calves (Mills *et al.*, 1967). In all species, the disorder was characterized by loss of appetite, growth depression, abnormalities of the skin or its appendages and reproductive failure. In the field, zinc supplementation cured and prevented parakeratosis, a thickening and hardening of the skin found in pigs given grain-based diets, rich in calcium (Tucker and Salmon, 1955). Zinc-responsive disorders were also induced by excess dietary calcium in poultry and were eventually attributed to a potent and complex interaction with the major source of phosphorus in feed grains, phytate (*Py*) (Ziegler *et al.*, 1961). Zinc-responsive disorders were much harder to find in ruminants, which have the capacity to degrade *Py* in the rumen.

Continuing biochemical studies revealed >300 zinc-dependent enzymes of diverse structure and function (Vallee and Falchuk, 1993) and a far greater number of functional zinc proteins (Coleman, 1992), making it difficult to identify the rate-limiting factor in zinc-responsive disorders. The quest to understand the pathogenesis of zinc deprivation continues and has led to the measurement of free intracellular zinc at femtomolar concentrations (Outten and O'Halloran, 2001), while therapeutic properties have been found for zinc at near-molar concentrations (Hill *et al.*, 2001). Fear of zinc deprivation is being exploited commercially by unfounded claims for the enhanced bioavailability of a variety of 'organic' zinc supplements.

Functions of Zinc

The functions of zinc are numerous and would only partially be covered by a list of zinc metal-loenzymes, such as that given in the previous edition (Underwood and Suttle, 1999), even if the list were complete. Zinc is required for the structural and functional integrity of over 2000 transcription factors and almost every signalling and metabolic pathway is dependent on one or more zinc-requiring proteins (Beattie and Kwun, 2004; Cousins *et al.*, 2006). Tetrahedral coordination of zinc to cysteine and histidine residues creates 'zinc-finger' domains in DNA-binding proteins (Berg, 1990) and intracellular oxidant stress may enhance the lability of zinc within the zinc fingers (Webster et al., 2001). The most important functions in livestock are those that become limiting to health and production when they are deprived of zinc. There are four outstanding, interconnected candidates.

Gene expression

Fetal growth is especially affected by lack of zinc (Hurley, 1981) and this may reflect the roles that zinc plays in DNA synthesis and nucleic acid and protein metabolism. Zinc regulates genes involved in signal transduction, responses to stress, reduction–oxidation reactions, growth and energy utilization, including a T-cell cytokine receptor and uroguanylin (Cousins et al., 2003). Teratogenic defects may mostly derive from the primary effects of zinc deprivation on gene expression – effects that are most marked when cells are rapidly dividing, growing or synthesizing (Chesters, 1992).

Appetite control

Since the role of zinc in the regulation of appetite was reviewed by O'Dell and Reeves (1989), advances have continued to come from studies with rats. The pattern as well as the rate of food intake can be affected and acute zinc deprivation on diets almost devoid of zinc causes rats to nibble at food rather than 'meal eat'. Deprivation on diets containing a little added zinc (3–6 mg Zn kg^{-1} dry matter (DM)) avoids that complication, but the description of such diets as 'marginal' in zinc (e.g. Kwun et al., 2007) is inappropriate because the diets unequivocally provide insufficient zinc for normal growth and development. Zinc deprivation increases the expression of the gene for the appetite-regulating hormone cholecystokinin (Cousins et al., 2003), but the role of zinc is probably multifactorial. There is also increased expression of leptin, a cytokine hormone whose secretion from adipocytes acts as a satiety signal (Kwun et al., 2007), and reduced expression of pyruvate kinase (Beattie et al., 2008), which is highly regulated by insulin. Changes in the concentrations of neurotransmitters in the brain have been reported in zinc deprivation, but may be secondary consequences of appetite reduction (Kwun et al., 2007). In what might more accurately be called acute zinc deprivation, the reduction in appetite is selective, with carbohydrate being avoided and protein and fat preferred (Kennedy et al., 1998).

Fat absorption

When low plasma and liver concentrations of vitamin E as well as the fat-soluble vitamin A were reported in zinc-deprived animals, attention became focused on the possible role of zinc in fat absorption. The pancreas secretes zinc-dependent phospholipase A_2 (Kim et al., 1998). This hydrolyses phosphatidylcholine, facilitating its absorption and the formation of chylomicrons, which are crucial for the absorption of fat micelles (Noh and Koo, 2001). Duodenal infusion of hydrolysed phosphatidylcholine greatly increases the absorption of fats and the fat-soluble vitamins A and E in well-grown rats that are 'marginally deprived' of zinc. The partial alleviation of the effects of heat stress in quail by supranutritional levels of zinc (Sahin and Kucuk, 2003) may be mediated, in part, by secondary improvements in vitamin status.

Antioxidant defence

In addition to the contribution of a superoxide dismutase, CuZnSOD, in protecting cells from the superoxide radical, in vitro studies suggest that zinc may afford protection from iron-induced lipid peroxidation by blocking iron-binding sites at cell surfaces, acting synergistically with vitamin E (Zago and Oteiza, 2001). Zinc deprivation increases the susceptibility of endothelial cells to oxidant stress (Beattie and Kwun, 2004). In humans, zinc supplementation decreases the expression of anti-inflammatory cytokines (Prasad, 2007), but it is unclear whether this is beneficial to health. Zinc is also a potent inducer of metallothionein (MT), a zinc- and thiol-rich metalloprotein that scavenges free radicals and could act as a reduction–oxidation sensor or active signalling switch in cells (Beattie and Trayhurn, 2002).

Clinical Consequences of Zinc Deprivation

In all species zinc deprivation is characterized by inappetence, retardation or cessation of growth and lesions of the integument and its outgrowths – hair, hoof, horn, wool or feathers.

Anorexia

Loss of appetite is an early sign of zinc deprivation and there are changes in the pattern of food intake from 'meal-eating' to 'nibbling' in ruminants as well as in rats (for lambs, see Droke et al., 1993a). This extreme sensitivity of appetite to nutrient supply is unique to zinc and may reflect the pivotal role of zinc in cell replication, possibly indicating fine control of either substrate supply or demand. If impairment of appetite is bypassed by force-feeding a low-zinc diet, the demise of the animal is hastened (Flanagan, 1984). Pair-feeding studies show that many of the adverse effects of severe zinc deprivation are secondary to a loss of appetite (e.g. Miller et al., 1967), including those on male fertility (Neathery et al., 1973b; Martin and White, 1992). Paired pattern-feeding is therefore an essential feature of in vivo studies of the pathogenesis of zinc deprivation.

Abnormalities of skin and appendages

Thickening, hardening and fissuring of the skin (parakeratosis) is a late sign of zinc deprivation in all species. Predilection sites vary between species as follows: feet and feathers in the chick (Sunde, 1972, 1978); the extremities in young pigs (Tucker and Salmon, 1955); the muzzle, neck, ears, scrotum and back of the hind limbs in calves (Miller et al., 1965); the hind limbs and teats in the dairy cow (Schwarz and Kirchgessner, 1975); and around the eyes, above the hoof and on the scrotum in lambs (Ott et al., 1964). The lesions can also affect stratified epithelia lining the tongue and oesophagus and are similar to those induced by vitamin A deficiency (Smith et al., 1976). The rate of healing of artificially inflicted skin wounds is retarded in the zinc-deprived animal (Miller et al., 1979) and wounds caused by ectoparasites or other skin infections will probably exacerbate the effects of parakeratosis. In horned lambs, the normal ring structure disappears from new horn growth and the horns are ultimately shed, leaving soft spongy outgrowths that continually haemorrhage (Mills et al., 1967); changes in the structure of the hooves can also occur. Wool fibres lose their crimp, become thin and loose and the whole fleece may be shed. Posthitis and vulvitis can occur in zinc-deprived lambs, with enlargement of the sebaceous glands (Demertzis, 1972). Excessive salivation is an early sign peculiar to ruminants (Mills et al., 1967; Apgar et al., 1993) and may reflect a combination of copious saliva production and difficulty in swallowing.

Skeletal disorders

The size and strength of the femur is reduced in baby pigs deprived of zinc, but comparisons with pair-fed controls indicate that the changes are consequences of a reduced feed intake (Table 16.1) (Miller et al., 1968a). Thickening and shortening of the long bones has also been reported in chicks (O'Dell and Savage, 1957; O'Dell et al., 1958). Lack of zinc during embryonic growth grossly disturbs skeletal development and severe abnormalities of the head, limbs and vertebrae have been found in chick embryos from the eggs of severely deprived hens (Kienholz et al., 1961). Bowing of the hind limbs, stiffness of the joints and swelling of the hocks occur in calves deprived of zinc (Miller and Miller, 1962).

Reproductive disorders

Lack of dietary zinc reduced litter size in early studies with pigs (Hoekstra et al., 1967) and impaired hatchability in marginally deprived hens

Table 16.1. Incorporation of pair-fed controls in studies of zinc deprivation shows that some abnormalities (e.g. poor growth) are due largely to poor appetite, while others (e.g. parakeratosis and raised white cell count) are due solely to lack of zinc (data for the baby pig from Miller et al., 1968a).

| Dietary Zn (mg kg^{-1} DM) | 12 | | 100 | | 100 |
Food intake	Ad lib		Restricted		Ad lib
Live-weight gain (g day^{-1})	25	<	88	<	288
Food consumption (g day^{-1})	170	=	170	<	430
Gain/food	0.14	<	0.57	<	0.67
Parakeratosis (%)	100	>	0	=	0
Serum Zn (mg l^{-1})	0.22	<	0.57	<	0.67
Alkaline phosphatase (sigma units)	0.4	<	5.5	<	7.0
Leukocyte count (10^3/mm^3)	17.0	>	12.9	=	11.8
Liver alcohol dehydrogenase (ΔOD min^{-1} mg^{-1} protein)	0.060	=	0.061	=	0.063

OD, optical density.

(Kienholz et al., 1961). Hypogonadism occurs in zinc-deprived bull calves (Pitts et al., 1966), kids (Miller et al., 1964) and ram lambs (Underwood and Somers, 1969). In a study of lambs, spermatogenesis practically ceased within 20 weeks on a on a diet containing only 2.4 mg Zn kg^{-1} DM, but recovered completely during a repletion period (Martin and White, 1992); it was proposed that anorexia reduced the secretion of gonadotrophin-releasing hormone from the hypothalamus. Male goats that were sexually mature when severely deprived of zinc have shown reductions in testicular size and loss of libido (Neathery et al., 1973b). However, when chronic deficiencies of zinc and/or vitamin A have been compared in the goat, spermatogenesis was improved by supplements of vitamin A but not zinc, despite the induction of zinc-responsive skin lesions (Chhabra and Arora, 1993). Mild zinc deprivation in pregnant ewes does not cause congenital malformations, although the numbers born and their birth weights may be slightly reduced (Egan, 1972; Masters and Fels, 1980; Mahmoud et al., 1983). Feeding a diet very low in zinc during pregnancy reduces survival of the newborn lamb and pregnancy toxaemia may occur as a secondary consequence of anorexia in the ewe (Apgar et al., 1993).

Natural Sources of Zinc

The provision of adequate dietary supplies of zinc can be a matter of life or death. Information on the zinc content of feeds is summarized in Table 16.2.

Forages

A high proportion of all pasture zinc values recorded worldwide lie between 25 and 50 mg kg^{-1} DM (range 7 to >100 mg kg^{-1} DM) (Minson, 1990). The main influences are soil zinc status and sward maturity (Minson, 1990). Towards the low extreme, herbage yields may be increased by zinc fertilizers (Masters and Somers, 1980). Pasture zinc can fall by almost 50% in successive cuts, irrespective of the amount of zinc fertilizer used (Underwood and Suttle, 1999). Hays, therefore, tend to be low in zinc (13–25 mg kg^{-1} DM for lucerne) and silages slightly richer (12–45 mg kg^{-1} DM for maize) (Table 16.2). Improved mixed pastures in New Zealand have been found to contain more zinc in the North than the South Island (38 versus 22 mg kg^{-1} DM) (Grace, 1972), with the range embracing values found in Scotland in uncontaminated pastures (Mills and Dalgarno, 1972). Differences between grass species contribute little to variations in forage zinc, although rhodes grass is often low in zinc (e.g. Jumba et al., 1995). Legume species can vary widely at both high and low soil zinc status (Minson, 1990).

Table 16.2. Mean zinc concentrations (± standard deviation) found in common livestock forages and feeds in the UK (from MAFF, 1990) and elsewhere[a] (more data are given in Chapter 2).

Roughages	Zn	Concentrates	Zn	By-products	Zn
Straw[a]	14	Barley	33 (8.5)	Wheat feed	104 (22.7)
Grass		Maize	19 (2.6)	Rice bran	77 (17.0)
Fresh[a]	36	Oats	26 (3.7)	Brewers' grains	73 (12.5)
Dried	33 (7.1)	Wheat	26 (7.0)	Distillers' grains	55 (4.7)
Grass		Maize gluten	80 (10.3)	Rapeseed meal	82 (9.1)
Hay	21 (2.7)	Cassava meal	12 (0.6)	Meat and bone meal	130 (43)
Dried	28 (6.6)	Linseed meal	66 (0.8)	Feather meal	152 (3)
Kale	27	Soybean meal	49 (9.9)		
Maize silage	45 (13.9)	Sorghum	14 (1.0)		
		Field beans	38 (2.7)		

[a]See text for sources and countries.

Concentrates

The zinc content of cereal grains varies little among plant species, but is greatly influenced by soil zinc status. Wheat, oats, barley and millet contain 26–35 mg Zn kg^{-1} DM on average, with slightly lower values common in maize (corn). Cereals grown on low-zinc soils can contain <20 mg Zn kg^{-1} DM, but applications of zinc fertilizers can triple values (Cakmak, 2008). Cereal zinc has been found to be concentrated in the outer layers of the grain and in wheat from a low-zinc area, containing 16 mg Zn kg^{-1} DM, the bran contained 49 mg Zn kg^{-1} DM, pollard 41 mg Zn kg^{-1} DM and white flour 5 mg Zn kg^{-1} DM (Underwood and Suttle, 1999). Cereal straws usually contain only a third of the concentration found in the grain and frequently <12 mg Zn kg^{-1} DM (White, 1993). Vegetable and animal protein sources are invariably higher in zinc than cereals, with feather meal outstandingly so (Table 16.2).

Milk and milk products

Normal cow's milk contains 3–5 mg Zn l^{-1}, most of which is associated with casein. Both dried skimmed milk and buttermilk are therefore good sources of zinc, commonly with between 30 and 40 mg Zn kg^{-1} DM. Colostrum is even richer in zinc (14 mg l^{-1}) and the large amounts of zinc secreted after birth by cows may contribute to the temporary fall in plasma zinc around parturition (Goff and Stabel, 1990). Supplementation of the mother will not increase milk zinc unless the diet lacks zinc (see Miller, 1970, for data on the cow).

Metabolism of Zinc

Certain features of the metabolism of zinc must be understood if the nutritional value of feeds as zinc sources is to be accurately measured.

Excretion and retention

Excretion of zinc occurs predominantly via pancreatic secretions and the faeces, with little zinc voided in urine (e.g. Schryver et al., 1980). However, evidence for regulation of faecal endogenous loss (FE_{Zn}) is limited. In mature sheep given hay, direct estimates of FE_{Zn} are approximately 100 µg Zn kg^{-1} live weight (LW) and are unaffected by zinc intake above the marginal (Suttle et al., 1982). In rats, FE_{Zn} is reduced by zinc intakes below the marginal, although making a smaller contribution to homeostasis than the increase in efficiency of zinc absorption (A_{Zn}) (Underwood and Suttle, 1999). With little regulation of excretion, zinc retention is closely related to zinc absorption.

Absorption

Zinc is absorbed according to need by an active saturable process at normal dietary zinc concentrations. The process principally occurs in the duodenum (Davies, 1980). Sheep and cattle can absorb with a maximal efficiency ($^{max}A_{Zn}$) of 0.75 (Underwood and Suttle, 1999). In non-ruminants, dietary antagonists may set a ceiling on $^{max}A_{Zn}$; one of the most potent of these antagonists is Py, which forms unabsorbable complexes with zinc. Figure 16.1 suggests that broiler chicks can retain (and therefore absorb) at least 64% of the zinc from a diet based on genetically modified barley almost devoid of Py, whereas the maximum retention from a normal, wild-type barley, providing 3.5g Py kg^{-1} DM, is barely 45% of zinc intake (Linares et al., 2007). However, the effect of Py on zinc retention almost disappears if dietary zinc is increased by a mere 20mg kg^{-1} DM as chicks proceed to absorb zinc according to need. Similarly, apparent absorption of zinc (AA_{Zn}) decreased from 0.47 to 0.22 of zinc intake as zinc concentration in a milk replacer for calves was increased from 40 to 1000mg kg^{-1} DM (Jenkins and Hidirolglou, 1991). The relatively high AA_{Zn} recorded at high zinc levels here and on a corn/alfalfa diet (Neathery et al., 1980) may be attributable to a second absorptive process, passive A_{Zn}, and/or mucosal binding, resulting from induction of the metal-binding protein MT, which limits A_{Zn} (Cousins, 1996). Enhancement of A_{Zn} during zinc depletion and inhibition during zinc overload both occur within a week of changes in the zinc supply (Miller et al., 1967; Stake et al., 1975).

Relative Biological Value of Zinc in Feeds

The scant information about the 'zinc value' of feeds comes mostly from assessments of relative availability or relative biological value (RBV) (Baker and Ammerman, 1995). Relative measures of zinc availability might appear to be adjusted automatically for effects of basal diet but they are not (Fig. 16.2), and the RBV of zinc in soybean meal (SBM) has been widely underestimated by comparisons using basal diets low in Py (Edwards and Baker, 2000).

Effect of phytate

Characterizing the effect of Py on the RBV of zinc presents additional difficulties because the antagonism increases with dietary calcium. In one study, a small increase from 6.0 to 7.4g Ca kg^{-1} DM reduced the RBV of zinc for chicks by 3.8-fold (Wedekind et al., 1994a). The A_{Zn} reported in cereal meals given to rats (0.5) (House and Welch, 1989) probably

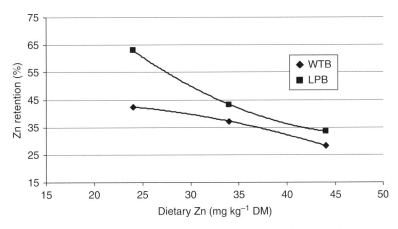

Fig. 16.1. Zinc retention in broiler chicks is determined by need, up to a limit set principally by dietary phytate. Data are from a comparison of genetically modified barley (low-phytate barley, LPB) and high-phytate wild-type barley (WPB) diets (Linares et al., 2007).

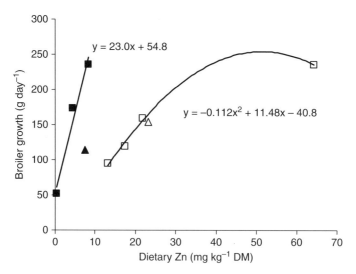

Fig. 16.2. Problems in assessing the relative biological value (RBV) of zinc from the growth responses of broilers to inorganic zinc added to an egg-albumen-based diet (EA, solid symbols) or phytate (*Py*)-rich soy diet (open symbols). First, conventional 'slopes ratio' comparisons would underestimate the value of inorganic zinc added to the *Py*-rich diet if a linear slope was fitted to a curvilinear response. Second, the RBV for zinc provided by a soybean meal supplement (triangles) is only lower than that of inorganic zinc if added to the EA diet (▲, from Edwards and Baker, 2000).

overestimates the zinc value of cereals included in compound feeds for pigs and poultry because of the inevitably low dietary calcium level in a meal consisting solely of cereals.

Effect of response index

A study with rats illustrated in Fig. 16.3 shows that RBV is also influenced by the index of response. An important result not shown is that when 10 instead of 5 mg Zn kg^{-1} DM was added to the diets, body weight after 24 days was doubled and all groups ate more food. Thus, the experimental model was a valid test of zinc responsiveness. Using the ratio of mean responses method (Baker and Ammerman, 1995) gives two different assessments for the RBV of zinc added to the albumin compared with the casein diet: one for body weight (100%) and one for other indices (132–145%). For whey, RBV was 100% according to body weight and tissue zinc, but 142% with serum zinc as the arbiter! Two explanations have been offered for these effects. First, the predominant casein in bovine milk, α_2, may have formed strong ligands with zinc that impaired absorption. Second, the high cysteine content of albumin may have favoured A_{Zn}. However, the similarity of growth on casein- and albumin-based diets suggests that enhancement of A_{Zn} was not accompanied by enhanced value to the animal. One possibility is that some zinc was absorbed from albumin diets as zinc–cysteine complexes that were not degraded but accumulated in the tissues and remained functionally unavailable for other, growth-enhancing purposes. As with selenium (see Chapter 15), the RBV of zinc should reflect the functional value of the absorbed element.

Effect of dietary zinc concentration

Non-linear relationships with dietary zinc (e.g. Fig. 16.1 and Fig. 16.2) mean that the maximum dietary zinc concentration selected for an experiment will sometimes influence estimates of RBV by the linear slopes ratio method (Franz *et al.*, 1980). Thus, RBV can be influenced by the chosen basal diet, index of response and range of dietary zinc concentrations, as well as attributes of the feed under test. The task of measuring A_{Zn} has been greatly

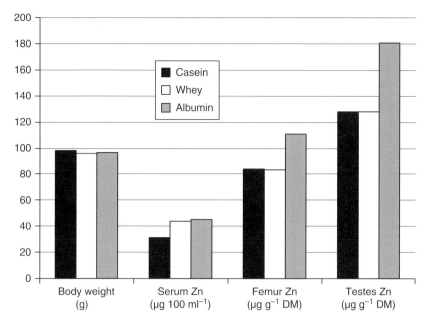

Fig. 16.3. The effects of three different protein sources on four measures of response in rats to a zinc sulfate supplement (5 mg Zn kg^{-1} DM). The rats are on a purified diet (given a value of 100) containing virtually no inherent zinc. Albumin favours zinc accretion in the femur and testes, but this does not benefit growth. Whey shares with albumin the ability to enhance serum zinc, again without growth benefit (data from Roth and Kirchgessner, 1985).

oversimplified and many previous estimates of RBV bear little or no relationship to absolute functional availability (see Chapter 2). Py-phosphorus must be reported if results are to have commercial significance. It is astonishing to often find only non-Py-phosphorus reported for test diets (e.g. Batal et al., 2001).

Feed Phytate: a Major Determinant of Zinc Value for Non-ruminants

The sad fact is that little is known about the absolute effects of feed Py on A_{Zn}, except that they are large. Comparison of the slopes for the different types of diet in Fig. 16.2 indicates that substituting soy-protein concentrate for egg albumin *halved* A_{Zn} for broiler chicks. Comparison of the maximal values for low and high cereal-Py diets in Fig. 16.1 suggests that cereal Py halved A_{Zn} for chicks. Had corn been simultaneously substituted for the purified energy source in the soy diet in the Fig. 16.2 study, the effect of diet on the RBV of zinc in added SBM would probably have been greater. More estimates of maximal efficiencies of zinc retention (e.g. Fig. 16.1) or AA_{Zn} for natural, Py-rich rations are urgently needed in pigs as well as poultry.

A role for modelling

There is a limit to the number of basic feeds and diets that can be tested and the problem lends itself to solution by mathematical modelling. Increasingly complex models for predicting the effects of calcium and Py on A_{Zn} have been proposed (Davies et al., 1985; Wing et al., 1997). The influence of calcium and Py on requirement can be crudely allowed for by using molar ratios of (calcium × Py) or Py to zinc (Fig. 16.4). In the study featured in Fig. 16.1, the optimum Py to zinc molar ratio for broiler growth on wild-type barley lay between 10 and 15. Complete models or formulae would have to allow for the beneficial influence of dietary phytases and the influence of dietary protein per se (Roth and Kirchgessner, 1985).

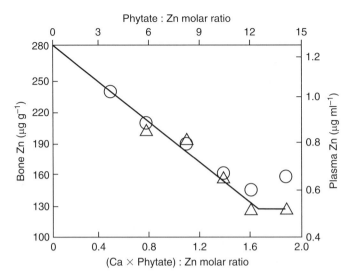

Fig. 16.4. The zinc requirements of pigs are better expressed in relationship to dietary phytate and calcium than as zinc concentrations. However, optimum levels depend on the chosen criterion of adequacy: plasma Zn (circle) or bone Zn (triangle) (Wedekind et al., 1994b).

Zinc Value in Feeds for Horses

Hind-gut fermenters such as the horse are vulnerable to the antagonism between zinc and Py. Schryver et al. (1980) reported an AA_{Zn} of only 0.07 for gelding ponies given a pelleted diet containing 50% corn and 35 mg Zn kg^{-1} DM, but an AA_{Zn} of 0.27 from a zinc oxide (ZnO) supplement (125 mg kg^{-1} DM). With cereals replaced by alfalfa, the corresponding AA_{Zn} for orally administered ^{65}Zn was 0.16 for both treatments. It was concluded that zinc was poorly utilized from Py-rich diets. There is no reason to think that horses cannot absorb zinc from forage with the same high efficiency as sheep from grass or roughages; an A_{Zn} of 0.75 was obtained for a low-zinc forage given to sheep (12 mg kg^{-1} DM) (Suttle et al., 1982).

Enhancing Zinc Absorbability for Non-ruminants

The A_{Zn} in Py-rich feeds can be improved by the factors that improve phosphorus absorption (see Chapter 6):

- minimizing dietary calcium and phosphorus supplements;
- decreasing seed Py by plant breeding (Linares et al., 2007);
- increasing the exposure of Py to plant phytases (e.g. by cereal choice and treatments such as soaking);
- adding microbial phytases (Lei et al., 1993; Adeola et al., 1995; Ao et al., 2007);
- adding hydroxylated vitamin D_3 (OHD$_3$) to enhance the efficacy of endogenous phytase (Roberson and Edward, 1994); and
- manipulating animal genotypes to introduce a capacity to degrade Py (see Chapter 20).

Combinations of treatments such as phytase and OHD$_3$ can have additive effects (Fig. 16.5). In one study, the addition of 1200 IU phytase or OHD$_3$ released 38% of the 13 mg Zn kg^{-1} DM present in a chick diet, most of which came from soy-protein concentrate (dextrose-replaced corn), while the combined supplements appeared to free all zinc. It is possible that phytase releases dietary zinc in the stomach while OHD$_3$ protects secreted zinc in the intestine. Brans can be rich in phytase and their inclusion in the diet is likely to improve A_{Zn} (Roberson et al., 2005). However, improvements in A_{Zn} will not be fully expressed if the diet provides zinc in excess of requirements.

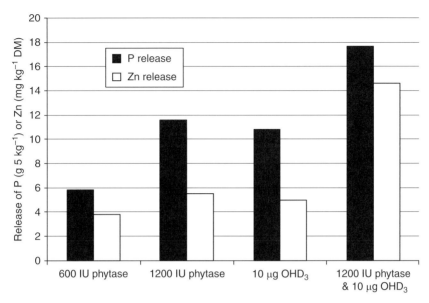

Fig. 16.5. Phytase and hydroxylated vitamin D_3 (OHD$_3$) analogue supplements to phytate-rich diets release zinc and phosphorus in a proportionate (ratio of P to Zn) and additive manner to the broiler chick. Data are from experiments 3 (P) and 5 (Zn) of Biehl *et al.* (1995) in which corn/soy and soy concentrate, respectively, provided the phytate.

Availability of Zinc in Milk and Milk Products

The study illustrated in Fig. 16.3 was prompted by evidence that zinc is less well absorbed from infant milk 'replacers' based on bovine milk than from human milk itself, but it has possible implications for the zinc nutrition of young livestock. Could the formation of ligands between α_2 casein and zinc lower A_{Zn} in the suckled calf or lamb on milk replacer? Is whey a superior source of zinc for pigs? The AA_{Zn} of 0.47 recorded for a milk replacer given to calves (Jenkins and Hidiroglou, 1991) probably underestimates AA_{Zn} because it contained 40 mg Zn kg^{-1} DM, considerably more zinc than the probable requirement. The incorporation of soy protein in milk replacers for the pre-ruminant lowers A_{Zn} substantially, judging from the relatively low plasma and liver zinc concentrations sustained (Xu *et al.*, 1997).

Deriving Dietary Zinc Requirements

Requirements have been estimated in one of three ways: associations between field disorder and dietary zinc concentration, feeding trials and factorial modelling. There is a dearth of information on the maintenance components of net requirement and on the A_{Zn} coefficient to convert net to gross requirement in factorial models. A solitary direct estimate of 27 µg kg^{-1} LW has been reported for FE_{Zn} in cattle (Hansard *et al.*, 1968), but losses of that order would be consistent with the total faecal zinc excretion observed in young calves (Miller *et al.*, 1968b) or dairy cows (Schwarz and Kirchgessner, 1975) given purified diets very low in zinc if 0.25 of their dietary zinc intakes had remained unabsorbed. In non-ruminants, the reabsorption of secreted zinc may be restricted on diets with a high *Py* to zinc ratio, thus increasing FE_{Zn} (Suttle, 1983; Oberleas, 1996). Whatever the species, factorial models indicate that requirements will decrease with the approach to mature body weight, but increase with growth rate and feed conversion efficiency (FCE) in the growing animal. In non-ruminants, the dominant effects are those of dietary calcium and *Py* concentrations via their influences on A_{Zn}. There is no evidence yet of antagonistic factors in forages or concentrates that limit A_{Zn} for ruminant livestock (P_y being rapidly hydrolysed in the rumen; see Chapter 6), but that situation may change if ruminants are fed like non-ruminants on high levels of ground corn. Whatever the combination of animal and diet, the estimated 'requirement' depends on the chosen index of adequacy.

Zinc Requirements for Pigs

Feeding trials

Early experiments placed the zinc requirement for growth at around 46 mg Zn kg^{-1} DM in weanling pigs given a corn/SBM diet and slightly less to eliminate parakeratosis (Smith et al., 1960; Miller et al., 1970). These results were confirmed when unsupplemented corn/SBM rations for starter pigs with 37–42 mg Zn kg^{-1} DM caused hypozincaemia and parakeratosis (Wedekind et al., 1994b). However, diets with 27 mg inherent Zn kg^{-1} DM were optimal for growth at the finishing stage (Hill et al., 1986; Wedekind et al.,1994b).

Factorial estimates

Results of a hypothetical, restricted factorial model (Table 16.3) help to explain why requirement falls with age. Requirement as a dietary concentration is dominated by the net need for growth, which varies little between species at around 24 mg kg^{-1} LW gain. Its contribution to total requirement is predicted to fall by 60% between 10 and 120 kg LW. The requirements given in Table 16.3 are roughly half those recommended by NRC (1998) after the starter (20 kg LW) starter stage. Assuming that the average corn/SBM starter and finisher rations contain 5 and 4 g P_y kg^{-1} DM, these requirements translate to required P_y to zinc molar ratios of <15. Pigs given diets containing protein from animal sources, such as meat meal, fish meal or whey, will have lower zinc needs unless dietary copper levels are very high. Since excess calcium increases the zinc requirement (Underwood and Suttle, 1999), the finding that a corn/SBM ration containing 30–34 mg Zn kg^{-1} DM and 16 g Ca kg^{-1} DM was marginal for sows (Hoekstra et al., 1967) overestimates the zinc requirements for reproduction.

Zinc Requirements for Poultry

Chick growth

Early trials placed the requirement at 35 mg Zn kg^{-1} DM for cereal/SBM diets relatively high in calcium (approximately 10 g Ca kg^{-1} DM). However, the requirement has increased over the years (Underwood and Suttle, 1999) and the latest studies indicate a requirement of

Table 16.3. Hypothetical factorial requirements (R) of growing pigs for zinc, assuming a net growth (G) requirement of 24 mg Zn kg^{-1} live-weight gain and the apparent absorption (AA) values given in parentheses.

Live weight (kg)	Live-weight gain (kg day^{-1})	Food intake (kg day^{-1})	Zn requirement	
			Net (mg day^{-1})	Gross[a] (mg kg^{-1} DM)
10	0.3	0.6	8.0	53 (0.25)
20	0.4	1.1	9.6	35 (0.25)
40	0.6	1.9	14.4	30 (0.25)
80	0.8	2.7	19.2	24 (0.30)
120	1.0	3.1	24.0	22 (0.35)

[a]In the absence of data for faecal endogenous loss of zinc in pigs, a simplified model where R = G / AA rather than a complete model (see Chapter 1) is used to generate gross needs. AA is assumed to be low initially, but to rise slightly towards the fattening stage as dietary soybean meal and calcium concentrations decrease and antagonism from phytate decreases.

Table 16.4. Zinc requirements (mg kg^{-1} DM) for broilers, poults and laying birds on corn/soybean[a] rations.

	Age	Broiler growth	Age	Poult growth[b]	Laying hen
Starter	(0–1 weeks)	60–65	0–4	65–70	–
	(1–3 weeks)	45–50	8–12	50–60	50–60
Grower	(4–8 weeks)	30–35	20–24	35–40	–

[a]Requirements are likely to be up to 20% lower for wheat-based or phytase-supplemented diets, provided that the diet does not contain excess calcium.
[b]Based on a comparison of broiler and poult requirements by Dewar and Downie, 1984.

40–50 mg kg^{-1} DM for optimum growth up to 21 days of age (Mohanna and Nys, 1999; Huang et al., 2007; Linares et al., 2007). Using serum 5′-nucleotidase activity or pancreas zinc transporter-2 mRNA as indices, Huang et al. (2007) concluded that 84 mg Zn kg^{-1} DM was optimal, but the justification for these as 'preferred indices' of need was unconvincing. With growth rates of 20 g day^{-1} and FCE of 60% now commonplace, it is tempting to conclude that zinc requirements have continued to increase. However, most of the data come from feeding trials in newly hatched chicks and may have limited applicability. The chick has a poor capacity to digest Py in its first week of life and this is accompanied by low zinc retention (p. 62; Olukosi et al., 2007). In its first few days after hatching, the 1970s chick required >50 mg Zn kg^{-1} DM to prevent feather fraying, although the incidence varied between strains (Underwood and Suttle, 1999). Lower requirements of 22.4 mg kg^{-1} DM have been reported for chicks on diets of unspecified but presumably low Py content, containing soy-protein concentrate but no cereal (e.g. Batal et al., 2001); they have little practical value other than to highlight the importance of Py. Requirements for growth on conventional corn/SBM rations are summarized in Table 16.4. These decline with age, in contrast to the latest NRC (1994) recommendation of 40 mg kg^{-1} DM for all growth stages.

Future trials should explore nutritional strategies for lessening exposure to Py during the first week of life, such as using wheat to replace corn and feeding mash. In addition, trials during the grower stage (>8 weeks old) are needed and would probably show requirements declining to 30 mg Zn kg^{-1} DM or less. As dietary calcium levels are reduced to optimize the utilization of Py-phosphorus (see Chapter 6), zinc requirements for all stages of growth should decline.

Poult growth

The zinc requirements of turkey poults are higher than those of chickens. In one study, the increase in growth requirement was only 20% (Dewar and Downie, 1984), far less than the 75% indicated by NRC (1994); the former has been used to generate the requirements in Table 16.4. The trend will be for zinc requirements to decrease, as in growing chicks, as antagonism from Py is reduced.

Egg production

Although requirements decrease as body weight increases, high recommendations for laying (50 mg kg^{-1} DM) and breeding hens (65 mg kg^{-1} DM) (NRC, 1994) are probably justified because of a low A_{Zn} on diets high in calcium.

Zinc Requirements for Sheep and Cattle

Field experience

Early studies of associations between zinc-responsive disorders in cattle and dietary zinc suggested high but variable estimates of need. Foot lesions have developed in housed young bulls on 30–50 mg Zn kg^{-1} DM and skin lesions in cattle on forages with 18–83 mg Zn kg^{-1} DM

(ARC, 1980). In the latter cases, the accuracy of some feed zinc analyses may be questionable. Although growth retardation in beef cattle has been associated with range pastures containing 12–25 mg Zn kg^{-1} DM (Mayland et al., 1980), beef and dairy cattle can generally graze pastures with <20 mg Zn kg^{-1} DM and remain healthy (Price and Humphries, 1980; N.D. Grace, personal communication, 1998).

Feeding trials with cattle

Studies with weaned calves on practical diets have shown early growth responses to zinc supplementation on a corn silage diet containing 24 mg Zn kg^{-1} DM (Spears, 1989), a hay/ground-corn diet containing 17 mg Zn kg^{-1} DM (Engle et al., 1997) and a corn silage diet containing 33 mg Zn kg^{-1} DM (Spears and Kegley, 2002). In the last study, growth improvement was marginal and independent of source: plasma zinc remained normal in all groups, but food intake tended to be higher in the zinc-supplemented groups, raising the possibility that the palatability of the diet was improved. However, it is possible that the escape of undegraded phytate from the rumen in animals on corn-based diets raises the zinc requirement above the level that satisfies grazing animals.

Data in dairy cows remain scarce. Lactating cows given semi-purified diets containing 17 or 22 mg Zn kg^{-1} DM remained healthy (Neathery et al., 1973a; Schwarz and Kirchgessner, 1975), whereas those on a diet with 6 mg Zn kg^{-1} DM developed lesions on feet and teats (Schwarz and Kirchgessner, 1975); zinc requirement probably fell between those limits. Suggestions that a level of 149 mg Zn kg^{-1} DM was inadequate for veal calves given a soybean protein in a milk replacer (Xu et al., 1997) are misplaced since growth, plasma zinc and liver zinc were all normal in the experiment.

Feeding trials with sheep

Early researchers used semi-purified diets to induce zinc deficiency and responses to zinc supplementation gave relatively low estimates of requirement that varied according to the criterion of adequacy. Requirements for optimal wool growth and plasma zinc (14 mg kg^{-1} DM) were approximately twice those for LW gain (Fig. 16.6), although the latter varied from <9 to 33 mg kg^{-1} DM between studies (Underwood and Suttle, 1999). Requirements for male fertility were thought to be particularly high since testicular growth and spermatogenesis in ram lambs are markedly subnormal with dietary zinc at 17 mg kg^{-1} DM but normal at 32 mg kg^{-1} DM (Ott et al., 1965; Underwood and Somers, 1969). However, such estimates of requirement are dependent on the spacing of treatments and, with small increments, optimal male fertility was obtained with 14 mg Zn kg^{-1} DM, the same as that for optimal wool growth (Martin and White, 1992). Sheep commonly graze pastures or consume feeds indoors with <20 mg Zn kg^{-1} DM and remain perfectly healthy (Grace, 1972; Pond, 1983; Masters and Fels, 1985; White et al., 1991). Sheep break down whole grain better than cattle and antagonism from undegraded Py may consequently be unimportant in sheep.

Factorial models for sheep and cattle

The factorial estimates of requirement, the derivation of which was described in detail in the last edition, are reproduced in Tables 16.5 and 16.6. No new data have appeared to

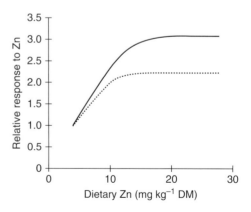

Fig. 16.6. Responses in wool growth (—) and live-weight gain (····) when zinc is added to a semi-purified diet low in zinc show that the optimum zinc level is much higher for wool growth (White, 1993).

Table 16.5. Factorial estimates of minimum zinc requirements for sheep at different DMI.

	Live weight (kg)	Production rate (kg day^{-1})	Requirement (mg day^{-1}) Net[a]	Requirement (mg day^{-1}) Gross[b]	DMI (kg day^{-1})	Zn requirement (mg kg^{-1} DM)[c]
Growth	5	0.15	4.6	5.4	0.2	27.0
	10	0.15	5.6	8.0	0.4	20.0
	20	0.15	6.6	9.4	0.64–0.57	14.7–16.5
	40	0.075	6.8	9.7	1.0–0.6	9.7–16.2
		0.150	8.6	12.3	1.4–0.8	8.8–15.4
		0.300	12.2	17.4	1.1	15.8
Adult	50	–	6.0	8.6	0.8–0.5	10.8–17.2
Pregnancy (late)	75	2 fetuses	11.9	17.0	1.7–1.4	10.0–12.1
Lactation	75	1	15.7	22.4	1.9–1.5	11.8–14.9
		2	22.9	32.7	2.8–2.2	11.7–14.9
		3	30.1	43.0	3.7–2.4	11.6–17.9

[a]Net requirement consists of: maintenance, 0.1 mg Zn kg^{-1} live weight; growth, 24 mg Zn kg^{-1} live weight gain; pregnancy, 0.28 and 1.5 mg Zn day^{-1} for mid and late stages; wool growth, 115 mg Zn kg^{-1} fresh weight; milk, 7.2 mg Zn kg^{-1}.
[b]The absorption coefficient is 0.70 for all but the youngest (milk-only diet) lambs.
[c]In ranges, higher values correspond to a low DMI on highly digestible diets.

Table 16.6. Factorial estimates of minimum zinc requirements for cattle on fresh or conserved grass or lucerne diets[a].

	Live weight (kg)	Production (kg day^{-1})	Requirement (mg day^{-1}) Net[b]	Requirement (mg day^{-1}) Gross[c]	DMI (kg day^{-1})	Zn requirement (mg kg^{-1} DM)[d]
Growth	40	0.5	16.0	18.8	1.0	18.7
	100	0.5	22.0	31.4	4.0–2.2	7.9–14.3
		1.0	34.0	48.6	4.5–2.8	10.8–17.4
	200	0.5	32.0	45.7	6.0–3.5	7.6–13.1
		1.0	44.0	62.9	6.5–4.0	9.7–15.7
	300	0.5	43.0	61.4	8.0–4.5	7.7–13.6
		1.0	54.0	77.1	8.4–5.4	10.1–14.3
Adult	500	0	50.0	71.4	7.6–4.6	9.4–15.5
Pregnancy	500	90–100 days	51.1	73.0	9.0–6.6	8.1–11.1
		180–270 days	56.3	80.4	11.2–6.9	7.2–11.7
Lactation	500	10	90.0	128.6	11.7–8.3	11.0–15.5
		20	129.5	185.0	17.8–12.7	10.4–14.6
		30	170.0	242.9	19.2–15.5	12.7–15.7

[a]On corn or corn silage-based diets, A_{Zn} may be as low as 0.35 and requirements twice as high as those in the table.
[b]Net requirement consists of: maintenance, 0.1 mg Zn kg^{-1} live weight; growth, 24 mg Zn kg live-weight gain; pregnancy, 1.1 and 6.3 mg Zn day^{-1} at mid and late stages; milk, 4 mg Zn kg^{-1}.
[c]The absorption coefficient (A_{Zn}) is 0.70 for all but the youngest (milk-fed) calves, for which the assumed value was 0.85.
[d]In ranges, higher values correspond to a low DMI on highly digestible diets.

support changes to model components and the estimates are in line with most feeding trial and field observations. For a given class of animal, minimum requirements increase when the required level of performance rises and highly digestible diets are fed. Even so, 18 mg Zn kg^{-1} DM is theoretically sufficient for all but the milk-fed animal.

Zinc Requirements for Goats

Clinical zinc deprivation has been produced in a quarter of goats given a diet consisting partly of wheat straw and containing 15 mg Zn kg^{-1} DM for 171–200 days (Chhabra and Arora, 1993). It is therefore likely that the requirements for the non-lactating goat are of the same order as those for sheep and cattle (i.e. approximately 20 mg kg^{-1} DM). Growth rate and FCE have been improved in cashmere goats during the cashmere growing period when zinc was added to a ration containing 22 mg Zn kg^{-1} DM – less so with 15 than with 30 mg Zn added (as zinc sulfate, ZnSO4) (Jia et al., 2008), suggesting that requirement lies between 37 and 52 mg Zn kg^{-1} DM.

Zinc Requirements for Horses

The zinc requirements of stabled horses are much higher than those of pastured horses because cereal Py lowers A_{Zn}, but the adverse effect may be partially offset if phytase-rich wheat bran is also included. Horses, like ruminants, thrive on pastures containing 15–20 mg Zn kg^{-1} DM, but stabled horses given grains should be allowed 30–40 mg Zn kg^{-1} DM.

Zinc Homeostasis

The biochemical consequences of zinc undernutrition can only be appreciated if the normal regulation of zinc status is understood.

Control of plasma zinc

Absorbed zinc is transported in the portal bloodstream loosely bound to plasma albumin, which accounts for about two-thirds of plasma zinc. Zinc is also present in plasma as an α-2 macroglobulin and as traces of MT. Induction of hepatic MT synthesis by zinc arriving in the liver plays a key role in removing zinc from the portal plasma and partitioning it between various pathways (Bremner, 1993; Lee et al., 1994).

Glucocorticoids and cytokines reduce plasma zinc and increase hepatic zinc by inducing MT synthesis (Cousins, 1996). Plasma zinc is higher in unfed than in fed pigs, with the difference widening as zinc intakes fall (Wedekind et al., 1994b). Plasma zinc may thus be affected by many factors other than dietary zinc. It is reduced by the following conditions:

- microbial infection (Corrigall et al., 1976; Orr et al., 1990);
- 24-h fasting (in Japanese quail (Harland et al., 1974), but raising concerns that food deprivation prior to sampling might lower plasma zinc in other species);
- parturition in cows (Goff and Stabel, 1990), particularly after difficult calvings (mean plasma zinc 0.38 ± 0.14 compared to 0.68 ± 0.23 mg l^{-1} at 18–24 h after normal parturition) (Dufty et al., 1977); and
- hyperthermic stress in cattle (Wegner et al., 1973) and Japanese quail (Sahin and Kucuk, 2003).

In the latter study, serum zinc fell from 31.3 to 24.9 µmol l^{-1}. Note the high normal values in avian species on a diet containing adequate zinc (36 mg kg^{-1} DM) and that hypozincoemia was not fully restored by trebling zinc intake.

Some of these hypozincaemic effects of stressors may be secondary effects of reduced food intake.

Cellular uptake of zinc

Mechanisms of zinc uptake by cells and tissues beyond the liver may involve uptake of albumin-bound zinc by endocytosis (Rowe and Bobilya, 2000) and processing by a family of zinc transporters (which had grown to ten members when counted by Cousins et al. (2006)) and Zip proteins. Some of these are stimulated by high and others by low zinc intakes. A rare glimpse of the complexity of the intracellular regulation of a trace element is given for zinc in Fig. 16.7. Zinc distribution throughout the body is illustrated in Fig. 16.8: muscle is rich in zinc and normally provides the largest body pool, followed by bone, but both are exceeded by wool in sheep. Most of the zinc in the bloodstream (80%) is present in the erythrocytes, which contain about 1 mg Zn/10^6 cells: of this, over 85% is present as

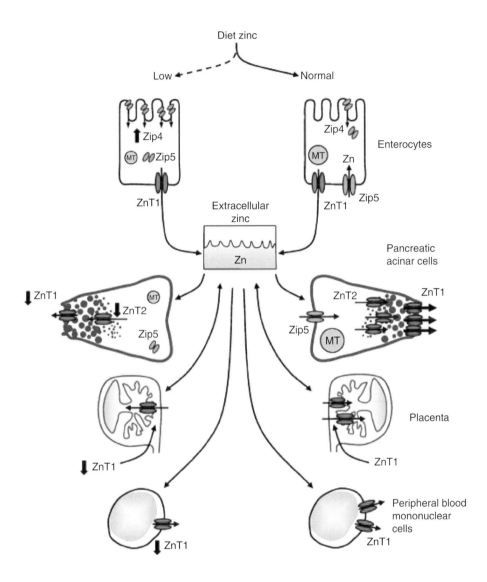

Fig. 16.7. Changes in murine zinc transporter (T), Zip protein and metallothionein (MT) expression in response to dietary Zn content. Enterocytes increase Zip4 expression when dietary Zn is low, with more Zip4 localized to the apical membrane to enhance Zn acquisition from the diet. With an adequate Zn supply, enterocytes express more MT and Zip5 is localized to the basolateral membrane. Low dietary Zn decreases ZnT1 and ZnT2 expression in pancreatic acinar cells, causes internalization of Zip5 and a reduction in MT, and decreases ZnT1 expression in placenta (visceral yolk sac) and peripheral blood mononuclear cells. These events increase intestinal absorption and reduce Zn loss from pancreatic and intestinal secretions while conserving Zn in cells with a high turnover, such as those of the immune system (Lichten and Cousins, 2000).

carbonic anhydrase and about 5% as CuZnSOD. Reticulocytes are particularly rich in zinc and MT (approximately $6\,mg/10^6$ cells) (Cousins, 1996). Mutant mice lacking the genes for MTI and II have increased susceptibility to both deficiency and toxicity of zinc (Kelly et al., 1996). Although MT has a high affinity for zinc, other elements such as cadmium and copper also induce MT synthesis and have higher affinities for the metalloprotein. MT therefore lies at the heart of the

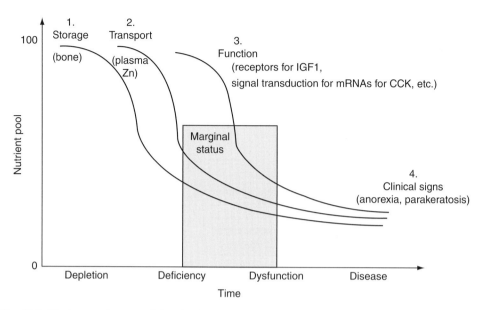

Fig. 16.8. The temporal stages of zinc deprivation leading to the appearance of clinical signs. See also Chapter 3. IGF1, insulin-like growth factor 1, CCK, cholecystokinase.

many reported interactions between zinc, cadmium and copper (Fig. 2.7, p. 33).

Storage

The capacity to store zinc is poorly developed and short-term studies have shown little change in tissue zinc when the dietary zinc supply is varied over a moderate range (for pigs, see Miller et al., 1968a). The largest tissue zinc responses to excessive supply are seen in bone (for sheep, see Stake et al., 1975; for poultry, see Wedekind et al., 1992). There is no accumulation of a fetal hepatic reserve of zinc in late pregnancy, at least in sheep (Williams et al., 1978).

Biochemical Manifestations of Zinc Deprivation

The sequence of biochemical changes that occurs when livestock receive diets that fail to meet minimum needs is illustrated in Fig. 16.8. This has been modified from the general model (see Chapter 3).

Depletion

Severe experimental zinc deprivation is associated with biochemical changes in the blood and tissues that occur so soon (e.g. Mills et al., 1967) that the existence and reliability of a depletion phase in which stores are mobilized is questionable. However, significant amounts of zinc may be redistributed from the large pools in muscle and bone during chronic depletion, thus delaying the onset of deficiency. Such factors may explain the tolerance of newborn lambs to maternal zinc depletion (Masters and Moir, 1983). Feeding chicks diets with 300 mg Zn kg^{-1} DM rather than 10.6 mg Zn kg^{-1} DM for 8 days delays the onset of growth retardation from day 5 to day 8 after a switch to a diet devoid of zinc (Emmert and Baker, 1995).

Deficiency

Animals respond to an inadequate zinc supply by up-regulating several intestinal zinc transporters, detectable by mRNA abundance assays (Fig. 16.7; Cousins et al., 2006). Despite these adaptations, an early decline in plasma or serum zinc has been observed in lambs, calves and baby pigs deprived of zinc. In lambs, the growth rate is maintained until plasma zinc falls below 0.4mg (6.2µmol) l^{-1} (Mills et al., 1967; White et al., 1994), which therefore represents a threshold for growth-related dysfunction (White, 1993). The first abnormality to be observed in an enzyme was subnormal carbonic anhydrase activity in the blood of zinc-deprived calves (Miller and Miller, 1962). Serum (or plasma) alkaline phosphatase (AP) activities decrease in the zinc-deprived baby pig, lamb (Saraswat and Arora, 1972), ewe (Apgar, 1979) and cow (Kirchgessner et al., 1975) and AP activities also decrease in the bones of turkey poults (Starcher and Kratzer, 1963) and cows (Kirchgessner et al., 1975). Some zinc enzymes are present in tissues in excess of requirements and reduced activity would not necessarily cause dysfunction. Mild maternal zinc depletion lowers MTI in the plasma, erythrocytes and soft tissues of offspring (Morrison and Bremner, 1987).

Dysfunction

The assay of apoenzymes of acetyl cholinesterase and mannosidase may bring a new dimension to the assessment of zinc status. While plasma activities of the holoenzyme are maintained in pregnant sheep during prolonged, severe zinc deprivation, apoenzyme levels increase and can be detected by the in vitro reconstitution of enzyme activity (Fig. 16.9) (Apgar et al., 1993). Such changes presumably indicate an attempt by the ewe to minimize the risk of enzyme dysfunction. Monitoring changes in the differential display of mRNA (e.g. Huang et al., 2007) may also help to identify the rate-limiting, zinc-dependent process. Increases in mRNA for cholecystokinase may indicate attempts to limit the need for zinc by constraining appetite (Cousins, 1997). Activities of CuZnSOD are reduced in the testes, intestine and liver of the zinc-deprived goat (Chhabra and Arora, 1993). As a state of dysfunction develops, zinc concentrations in most soft tissues decline slightly, but the fall can be greatest in pancreas and liver in the pig (Swinkels et al., 1996) and chick (Huang et al., 2007).

Disorder

The low plasma vitamin A values reported in early work on zinc-deprived pigs, lambs and goats (Underwood and Suttle, 1999) probably reflect a general reduction in the efficiency of fat-soluble vitamin absorption, rather than reductions food intake or liver ADH activity, and may cause night blindness (Chhabra and Arora, 1993). Poor capacity to absorb fats and essential fatty acids may contribute to clinical abnormalities, particularly those affecting the skin. Teratogenic effects of maternal zinc deprivation in the rat have been related to impaired activity of fetal thymidine kinase and impaired DNA synthesis in the central nervous system (Hurley, 1981). However, thymidine kinase is not a zinc-containing enzyme and other enzymes (e.g. DNA polymerases) are also reduced in activity. The fundamental dysfunction may be in gene expression, through impairment in the G1 phase of the cell cycle in zinc-deprived animals (Chesters and Boyne, 1997).

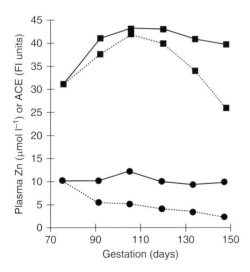

Fig. 16.9. The activity of angiotensin-converting enzyme (ACE) (■) declines at a later stage during zinc deprivation (broken line) than plasma Zn (●) concentration in pregnant ewes. Zn-supplemented ewes are represented by solid line (data from Apgar et al., 1993).

Disease resistance

Severe zinc deprivation impairs or changes certain components of the immune system in laboratory animals (Underwood and Suttle, 1999) and humans (Prasad, 2007). However, in marginally depleted, mature mice, immunopathology was *secondary* to loss of appetite (Beach *et al.*, 1981). In addition, severely or marginally zinc-deprived lambs have shown loss of appetite, poor growth and skin lesions *before* increased susceptibility to infection (Droke and Spears, 1993). Furthermore, susceptibility to pneumonic *Pasteurella haemolytica* infection is not increased by zinc deprivation (Droke *et al.*, 1993b). Supplementation with zinc does not improve the humoral immune responses of marginally zinc-deficient pigs to injected sheep red blood cells (Cheng *et al.*, 1998). In one study, responses of heifers to a subcutaneous injection of phytohaemagglutinin were impaired by zinc deprivation *before* any loss of appetite or fall in plasma zinc (Engle *et al.*, 1997), but in finishing steers the opposite was found, with zinc proteinate improving growth without affecting *in vitro* or *in vivo* immune responses (Spears and Kegley, 2002). Similarly, in chicks responding in terms of growth to zinc supplementation, antibody responses to sheep red blood cells were not zinc responsive (Mohanna and Nys, 1999). Therefore, there is no evidence that the immune system is more sensitive to zinc deprivation than other zinc-dependent functions.

Accumulation of heavy metals

Studies with rats marginally deprived of zinc have revealed a markedly increased uptake of mercury (Kul'kova *et al.*, 1993). Liver and kidney mercury levels increased two- to threefold when dietary zinc was reduced from 40 to 6 mg kg^{-1} DM with 10 mg Hg kg^{-1} DM present. Marginal zinc deprivation may therefore be important in areas of heavy metal pollution.

Diagnosis of Zinc-responsive Disorders

Severe disorders can readily be diagnosed from the combined evidence of clinical and pathological disorders and the biochemical defects just described, but the diagnosis of early stages or milder forms presents difficulties. Determination of dietary zinc can be helpful in ruminants, but not in pigs or poultry. The diagnostic value of the biochemical abnormalities just described increases the later they occur following zinc depletion (Fig. 16.8).

Plasma and serum zinc

Given the hypozincaemic effects of varied stressors, the widely quoted normal range (0.8–1.2 mg or 12.3–18.5 μmol Zn l^{-1}) for sample means is too high, frequently giving 'subnormal' values in healthy stock (in pigs, Carlson *et al.*, 2004; in sheep, Grace, 1972). The overdiagnosis of zinc deprivation can be prevented by using a lower limit of normality of 0.6 mg (10 μmol) Zn l^{-1} and a marginal band for mean serum values of 0.4–0.6 mg Zn l^{-1} in pigs and sheep (Table 16.7), and giving little weight to isolated low values in a sampled population. Since serum iron declines and copper rises under the influence of most stressors, abnormal zinc to copper ratios with normal zinc to iron ratios should distinguish false 'lows'.

Serum metalloproteins and enzymes

Erythrocyte (and plasma) MT concentrations are reduced by zinc deprivation but unaffected by the stressors that affect plasma zinc (Bremner, 1991), and have been investigated as an alternative measure of zinc status. Attention has now switched to mononuclear cell MT mRNA levels, which also correlate with zinc status (Sandoval *et al.*, 1997b; Cao and Cousins, 2000). However, it is unlikely that such sophisticated methods will be used to diagnose the relatively rare and easily avoided zinc disorders in livestock. Serum AP activity follows a similar time course to serum zinc during zinc deprivation in sheep (Apgar, 1979) and pigs (Swinkels *et al.*, 1996). However, serum AP levels are affected by the release of AP from the gut and bone in various disease states, making AP activity no more reliable than plasma zinc. Assay of apoenzymes,

Table 16.7. Marginal bands[a] for assessing the risk of zinc deprivation in livestock from the most diagnostically useful biochemical indices of zinc status.

	Diet (mg kg^{-1} DM)	Serum (mg l^{-1})	'Coat' (mg kg^{-1} DM)	Pancreas (mg kg^{-1} WM)	Bone (mg kg^{-1} DMff)
Cattle and goat	10–20	0.4–0.6	75–100	20–25	50–70 (r)
Pig	30–50[b]	0.4–0.7	140–150	30–35	80–90 (t)
Poultry	50–70[b]	0.8–1.4	200–275	20–30	100–160 (t) (45–60 in toe)
Sheep	10–20	0.4–0.6	80–100	<18	Unknown

ff, fat-free; r, rib; t, tibia.
[a]Mean values for a population sample below the given ranges for more than one criterion indicate probable benefits from zinc supplementation in sufficient individuals to merit interventions. Values within bands indicate the possibility of future benefits if zinc status does not improve.
[b]For cereal–vegetable protein diets, need decreases with age and the dietary phytate and phytate to zinc molar ratio should be <15.
For units in μmol, multiply values by 0.0153.

angiotensin-converting enzyme and mannosidase may bring specificity to the assessment of zinc status.

Other indices

Bone, hair, wool and feathers are particularly rich in zinc (100–200 mg kg^{-1} DM) and concentrations in the appendages of clinically affected animals are reduced (Table 16.8). However, individual variability is high and values are affected by age, breed, sampling site and season (Miller et al., 1965; Grace and Sumner, 1986; White et al., 1991). The zinc concentrations in the male sex organs and secretions are normally high and reflect zinc status. Values of 105 ± 4.4 and 74 ± 5.0 mg Zn kg^{-1} DM have been reported for the testes of normal and infertile zinc-deprived rams, respectively (Underwood and Somers, 1969). Similar reductions have been reported in clinically affected goats (Miller et al., 1964), but not in young calves (Miller et al., 1967). At post-mortem, assays of bone zinc can also add to the diagnostic template summarized in Table 16.8, but liver zinc can increase greatly during infection giving false 'normal' or 'high' values. Although less rich in zinc than bone (normal values 70–80 mg kg^{-1} DM), the toe is readily sampled in poultry and zinc concentrations reflect zinc intake (Linares et al., 2007). An immediate improvement in appetite following zinc supplementation provides good diagnostic evidence of zinc-responsive disorder, but a recent report that FCE rather than appetite can be the first abnormality to affect zinc-deprived cattle (Engle et al., 1997) complicates matters.

Occurrence of Zinc Deprivation

Large areas of zinc-deficient soils exist in many countries in which yields of pastures and crops are improved by applications of zinc fertilizers (Alloway, 2004; Cakmak, 2008). Forage and grain zinc are concurrently increased, but even without these interventions few reports of clinical signs of zinc deprivation in grazing animals have appeared and it is generally assumed that the forages and grains grown on zinc-deficient soils contain enough zinc for the needs of ruminants. This assumption is only occasionally unsafe and the predominant risks of disorder occur amongst non-ruminants.

Non-ruminants

Commercial maize-based rations containing plant protein sources, such as SBM, sesame, lupin and cottonseed meals, may not provide sufficient zinc because of the 'chelating' effects of Py, particularly for young, rapidly growing

Table 16.8. Effects of zinc deprivation (D) on tissue zinc concentrations in the growing animal vary from organ to organ and from species to species when compared with zinc-replete controls (C).

		Plasma or serum (µmol l⁻¹)	Packed erythrocytes (µmol l⁻¹)	Muscle (mmol kg⁻¹ DM)	Liver (mmol kg⁻¹ DM)	Pancreas (mmol kg⁻¹ DM)	Appendages (mmol kg⁻¹ DM)	Reference[c]
Calf	D	9.3	28.5	1.46	1.57	–	1.39	1
	C	13.0[b]	31.4	1.41	1.72	–	1.80[b]	1
Chick	D	26.0	153.0	0.74[a]	0.85	1.07	2.17	2
	C	32.1	154.5	0.74[a]	0.80	1.39[b]	1.87	2
		(Whole blood)						
Kid	D		23.3	2.35	2.13	–	1.27	1
	C		28.5[b]	2.31	2.35	–	1.53[b]	1
Pig	D	3.4	–	–	1.64	1.56	–	3
	C	13.2	–		3.01[b]	3.00[b]	–	3
Lamb	D	3.3	–	2.00	1.11	0.85	1.51	4
				2.60	1.62	1.35	1.83	4

Multiply by 65.4 to convert from µmol l⁻¹ or mmol kg⁻¹ DM to µg l⁻¹ or mg kg⁻¹ DM.
[a]Thigh muscle (in poultry, breast muscle had 60% less zinc).
[b]Significant effects of zinc deprivation: controls not pair-fed.
[c]1, Miller et al. (1967); 2, Savage et al. (1964); 3, Miller et al. (1968a); 4, Ott et al. (1964).

animals on diets containing more calcium than they require (e.g. 1% Ca for broiler chicks). Zinc nutrition can be pushed from the marginal to inadequate by excesses of other minerals. Mortality and lesions similar to those of parakeratosis and rectifiable by zinc supplements have been observed in growing pigs given 250 mg Cu kg⁻¹ DM as a growth stimulant (O'Hara et al., 1960; Suttle and Mills, 1966). The need to routinely supplement plant-based diets with zinc should be questioned because the zinc, P_y and phytase concentrations in seed and grains vary appreciably with their source. Natural corn/SBM mixtures can provide sufficient zinc for growing pigs (e.g. Wedekind et al., 1994b), but the true risk of deficiency in diets balanced with respect to other minerals is unknown and almost certainly exaggerated. The risk of zinc deprivation in pigs and poultry is minimal when dietary calcium and copper are not abnormally high and is decreased by agents of de-phytinization. The introduction of low-P_y varieties of soy and maize to improve phosphorus utilization (see Chapter 6) and phytase additives may eliminate the risk of zinc deprivation altogether. The risk of deprivation is lower when cereals with higher phytase contents than maize (e.g. wheat) are used.

Ruminants

Early reports of typical abnormalities responsive to zinc therapy in cattle in 'British Guiana' (Legg and Sears, 1960) and Greece (Spais and Papasteriadis, 1974) on pastures containing 18–83 mg Zn kg⁻¹ DM proved to be false indicators of the general risk of zinc deprivation. Heavy ewe and lamb mortality have been reported in a flock maintained on hay from irrigated pastures of Chloris gayana (Rhodes grass) that were low in zinc (20 mg kg⁻¹ DM) and protein (95 g kg⁻¹ DM) (Mahmoud et al., 1983); the problem was controlled by injecting zinc, but again may be atypical. Even mild zinc deprivation is rarely reported in grazing sheep and cattle. Fertility has been marginally improved by supplemental zinc in some years in Dorset Horn ewes on south Australian pastures containing about 20 mg Zn kg⁻¹ DM (Egan, 1972). In western Australia, where pastures frequently contain <20 mg Zn kg⁻¹ DM in the autumn and winter (Masters and Somers, 1980), one study found that the administration of intra-ruminal zinc pellets (see Masters and Moir, 1980) before mating and during pregnancy increased the number of lambs born and reared and produced a small increase in lamb birth and weaning weights (Masters and Fels, 1980). However, these responses were not repeatable (Masters and Fels, 1985). Early pregnancy may be a period of maximum susceptibility to zinc deprivation. Forages in Idaho have been found to contain insufficient zinc (often <17 mg kg^{-1} DM) for suckler cows and their calves (Mayland et al., 1980), but reports of natural zinc deprivation and hypozincaemia remain rare despite the widespread distribution of pastures containing no more zinc (e.g. Price and Humphries, 1980). It is possible that livestock grazing green swards have plentiful reserves of vitamins A and E that offset chronic zinc-induced reductions in vitamin absorption. Exposure to excess copper and cadmium may increase risk of disorder by increasing the mucosal binding of zinc by MT (Bremner, 1993). Inhibition of A_{Zn} from low-zinc diets by excess selenium suggests that risk may also be raised in seleniferous regions (House and Welch, 1989).

Prevention and Control of Zinc Disorders

Housed livestock

With pigs, poultry and housed cattle, zinc deprivation can be prevented readily and cheaply by adding zinc to mineral supplements or whole mixed rations. Supplementation with 50 mg Zn kg^{-1} DM is more than sufficient except for pigs receiving 250 mg Cu kg^{-1} DM as a growth stimulant, in which case an additional 150 mg Zn kg^{-1} DM may be necessary (Suttle and Mills, 1966). An alternative approach for non-ruminants, as discussed earlier, is to raise zinc availability in the diet by adding phytase (e.g. Adeola et al., 1995) and thus removing the primary antagonist, Py. However, the use of pharmacological concentrations of zinc (3 g kg^{-1} DM) in rations for nursery pigs may inhibit phytase activity (Augsberger et al., 2004).

Grazing livestock

Several methods are available for the prevention and control of zinc deprivation in grazing sheep and cattle. Treatment of soil with zinc-containing fertilizers is feasible and, in early studies in Western Australia, applications of 5–7 kg ZnSO$_4$ ha^{-1} repeated every 2–3 years were sufficient, but amounts vary with the environment. Under extensive range conditions where fertilizer applications are uneconomic, the provision of salt licks containing 1–2% zinc should provide sufficient cover. Zinc supplements must be consumed regularly because zinc is not stored well. A heavy intra-ruminal zinc pellet can release sufficient zinc to overcome seasonal deficiencies in sheep (Masters and Moir, 1980) and has been found to maintain normal plasma and wool zinc concentrations and plasma AP activities as effectively as a dietary zinc supplement in young wethers consuming a diet low in zinc (3.8 mg kg^{-1} DM) (Masters and Fels, 1980). A soluble glass rumen bolus has raised plasma zinc in 8-month-old outwintered lambs 100 days after dosing without improving growth (Kendall et al., 2001).

Inorganic zinc sources

The common forms of zinc used to supplement animal rations are the oxide (ZnO) and feed-grade sulfate (ZnSO$_4$.7H$_2$O). Comparisons of RBV in the chick have shown that feed-grade sources of ZnO have 44 or 61% of the value possessed by analytical-grade ZnSO$_4$, depending on index (tibia zinc or growth) (Wedekind and Baker, 1990; Wedekind et al., 1992); similar results have been obtained with sheep (Sandoval et al., 1997a,b) and horses (Wichert et al., 2002). The assessment of RBV can be affected by the choice of basal diet (Fig. 16.2) as well as by the response index (Fig. 16.3), but the RBV of three feed-grade ZnO sources in a soy-based diet varied from 30% to 90%, regardless of response index, compared with analytical-grade ZnSO$_4$; two feed-grade sources of ZnSO$_4$ had RBVs of 87–88% (Edwards and Baker, 1999). In similar trials with sheep, three different ZnO sources had RBVs of 106%, 87% and 79% (Sandoval et al., 1997a).

Such differences in RBV between ZnO and ZnSO$_4$ sources are important because they influence the margin of superiority, if any, attributed to an organic source with which they are compared and also estimates of requirement. Tetrabasic zinc chloride has a similar RBV to $ZnSO_4$ for chicks (Batal et al., 2001).

Organic zinc sources for pigs and poultry

Previous reviews have provided no evidence for a consistent benefit from providing zinc as organic complexes rather than simple inorganic salts (Fig. 16.10) (Baker and Ammerman, 1995; Spears, 1996; Underwood and Suttle, 1999), and little has appeared since. Zinc lysine and $ZnSO_4$ are of equal value to pigs (Cheng et al., 1998); zinc methionine and $ZnSO_4$ are equally effective in improving growth and zinc status in 1-day-old broiler chicks (Mohanna and Nys, 1999); and $ZnSO_4$ and zinc picolinic acid are equally and marginally effective in reducing the effects of heat stress and attendant reductions in vitamin status in laying quail (Sahin et al., 2005). Although improvements in skin thickness, collagen content and resistance to tear have been reported in broilers given supplementary organic zinc in a diet that contained sufficient inorganic zinc for optimal growth (Rossi et al., 2007), responses to further additions of inorganic zinc were not tested. Occasional benefits in individual studies could well reflect the low quality of the feed-grade inorganic zinc source used for comparison rather than any intrinsic value of complexation. The fact that phytase supplements can improve the efficacy of a zinc proteinate in broiler chicks suggests that chelation does not eliminate the antagonism from Py (Ao et al., 2007). An improvement in 17-day-old broiler performance has been recorded when part of an excessive $ZnSO_4$ supplement (40 of 140 mg Zn kg^{-1} DM) was replaced with a zinc–amino acid chelate (Hudson et al., 2004), but the change may have simply improved the palatability of the diet. Additions of 100 mg Zn kg^{-1} DM as $ZnSO_4$ or a novel chelate to a sow diet already adequate in zinc (120 mg kg^{-1} DM) have not improved sow health, piglet performance and tissue zinc at weaning (Payne et al., 2006). Differences in intestinal villus heights point to pharmacological effects and there was a tendency for food intake to increase by 20% in supplemented sows. It is noteworthy that providing zinc as the sulfate, citrate or picolinic acid in the study reported in Fig. 16.3 had no effect on zinc status with any of the three protein diets.

Organic zinc sources for ruminants

Given the lack of evidence for benefits from feeding organic sources in non-ruminants, it would be surprising if there were any benefits in ruminants, which face lesser dietary constraints on A_{Zn}. Indeed, earlier reviews showed no consistent advantage (Baker and Ammerman, 1995; Spears, 1996; Underwood and Suttle, 1999). There was a tendency for two zinc proteinates to support higher growth rates than ZnO in finishing steers on ground-corn diets in the continuation of a study cited earlier (Spears and Kegley, 2002), but the advantage was small (<7%) and larger differences in RBV have been found between inorganic sources: there were unexplained effects of zinc source on some carcass characteristics. Feeding the zinc proteinate increases soluble zinc concentrations in rumen liquor, and greater uptake of zinc from zinc proteinate than $ZnSO_4$ by ovine rumen and omasal epithelia in vitro has also been reported, but no zinc was found in the serosal side buffer

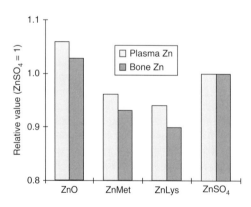

Fig. 16.10. Providing pigs with zinc complexed with lysine (ZnLys) or methionline (ZnMet) resulted in lower Zn concentrations in plasma and bone than when inorganic sources (the oxide, ZnO, or sulfate, $ZnSO_4$) were added to a marginally inadequate maize/soybean meal diet (data from Wedekind et al., 1994b). Thus, claims for enhanced bioavailability of 'chelated' sources were not supported.

(i.e. none was absorbed) (Wright et al., 2008). Feeding zinc as a liquid zinc–methionine complex to steers had only transient beneficial effects on performance (Wagner et al., 2008).

Several studies in cattle have looked at the effects of wholly or partially replacing inorganic minerals with organic mineral, without unsupplemented control groups. The basal diets usually contained more copper, manganese and zinc than the NRC (2001) generously recommend and the supplements alone were sufficient to meet those total requirements. However, no useful nutritional information can be gained from experiments that lack unsupplemented control groups. A rare 2-year study in beef suckler herds with unsupplemented controls and pasture low in one element (zinc, 16 mg kg^{-1} DM), has reported on the effects of multi-mineral supplementation and 50% substitution of organic for inorganic sources – but the complexity of the results underlines the importance of controls (Ahola et al., 2004). Responses varied between years and indices of performance, with a tendency ($P<0.08$) for minerals given as proteinates to improve pregnancy rate in the first year. In both years, however, the weight of calf weaned per cow treated was significantly greater in *unsupplemented* than in supplemented cows, regardless of mineral source. The lack of unsupplemented controls in many commercially sponsored experiments is probably intentional, avoiding possible negative answers to two questions that the sponsor would rather not pose: 'Is mineral supplementation necessary?' and 'Does an organic source offer any benefit?' The supplementation of beef cattle on hay low in copper and marginal in zinc with organic copper and zinc in another 2-year study did not improve pregnancy rates or calf performance (Muehlenbein et al., 2001).

Pharmacological Responses to Zinc

Growth of early-weaned pigs

Numerous reports exist of large zinc supplements, providing 2–4 g Zn kg^{-1} DM, improving the growth and health of early-weaned ('nursery') piglets (Cromwell, 2001). These 'pharmacological' supplements are believed to stabilize bacterial and digestive activity in the gut, reducing the incidence of diarrhoea and improving growth (Zang et al., 1995), but growth responses can be obtained in the absence of diarrhoea. Responses can be additional to those obtained from a concurrent antibacterial agent, but not to those obtained with pharmacological copper supplements (Smith et al., 1997). Large zinc supplements change the activity of digestive enzymes in the pancreas, intestinal mucosa and lumen, reduce villus height and increase mucin production in the caecum (Hedemann et al., 2006) and MT mRNA in the intestinal mucosa, but not in lymphocytes (Carlson et al., 2004). Lymphocyte responses to *in vitro* mitogen stimulation are enhanced (Davis et al., 2004), but activity of phytase supplements is reduced in pigs and chicks by 0.8–1.0 g Zn kg^{-1} DM (Augspurger et al., 2004). A collaborative study involving nine research stations across the USA has sought to define the optimum level of supplementation (Fig. 16.11) (Hill et al., 2001). Growth responses tended to be larger and the optimum zinc concentration higher in piglets weaned at 15 rather than 20 days of age. The inclusion of massive zinc supplements is now routine in starter rations for piglets in the USA but not in Europe, where supplementation is limited to 0.5 g kg^{-1} DM on environmental protection grounds.

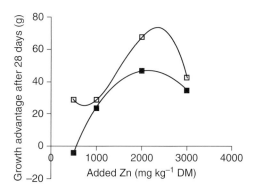

Fig. 16.11. The effects of pharmacological additions of inorganic zinc on the growth of nursery pigs weaned at 15 (thick) or 21 (thin line) days of age (from Hill et al., 2001).

Organic versus inorganic zinc

Manufacturers of organic zinc complexes have widely supported studies intended to demonstrate that the hyperavailability of organic sources allows growth responses to be obtained in nursery pigs with less zinc and thus cause less harm to the environment by lowering faecal zinc excretion. A generic experimental design is used in which hyperavailability is assumed and organic sources are used to provide less zinc than an inorganic source (e.g. Case and Carlson, 2002; Carlson et al., 2004). Thus, direct comparisons cannot be made, but the quadratic relationship between dietary zinc and growth (Fig. 16.11) guarantees that lower concentrations will give relatively large growth responses, probably not significantly different from the higher level of inorganic zinc. Thus, an illusion of equal efficacy at much lower organic zinc concentrations is created! It is disappointing to find such poorly designed trials accepted in peer-reviewed papers. Comparing sources at the same low concentration, zinc proteinate (Carlson et al., 2004), zinc methionine analogue and ZnO (Case and Carlson, 2002) did not improve nursery pig growth at 0.5g Zn kg^{-1} DM, although a zinc–polysaccharide complex did (Case and Carlson, 2002). Faecal zinc is determined by dietary zinc concentration rather than source in pigs (Case and Carlson, 2002) and poultry (Bao et al., 2007) supporting the view that most organic sources are *not* hyperavailable. Health problems in nursery pigs are better addressed by weaning later.

Facial eczema

High zinc intakes prevent facial eczema in sheep and cattle in New Zealand, a disease characterized by liver damage, photosensitization, weight loss and death. The disease occurs in sheep and cattle grazing pastures infected with the fungus *Pithomyces chartarum*, which produces the hepatotoxic mycotoxin sporidesmin. The degree of protection afforded by oral zinc against the liver damage and photosensitivity increases with daily intakes of up to 2g in sheep and 23mg Zn kg^{-1} LW in cattle. Such intakes are some 50 times the normal zinc requirements in the absence of the mycotoxin and they induce signs of zinc toxicity (Towers et al., 1975; Smith et al., 1977), which initially limited their usefulness. However, an intraruminal bolus has been developed that affords safe and prolonged protection by releasing about 0.8g of zinc from ZnO per day for about 6 weeks (Munday et al., 1997) and other essential trace elements can be provided in the same bolus (Se and Co: Grace et al., 1997).

Zinc Toxicity

Livestock exhibit considerable tolerance to high intakes of zinc, the extent of tolerance depending partly on the species but mainly on the nature of the diet, especially its calcium, copper, iron and cadmium content. Weanling pigs can clearly be fed for several weeks on diets containing 2–3g Zn kg^{-1} DM, but growth and appetite are depressed at 4–8g kg^{-1} DM and mortality is high. However, raising the dietary calcium from 7 to 11g kg^{-1} DM ameliorates the toxic effects of 4g Zn kg^{-1} DM (Underwood and Suttle, 1999). Broilers and layer hens tolerate 1–2g Zn kg^{-1} DM and show only slight growth and appetite depression at 4g Zn kg^{-1} DM (Oh et al., 1979). In the laying hen, very high zinc intakes (20g Zn kg^{-1} DM) have been used for short periods (14 days) to reduce food intake and induce moult without adverse long-term consequences (Stevenson and Jackson, 1984). The growth rate of pre-ruminant calves is reduced when their milk replacer contains >500mg kg^{-1} DM (added as ZnO) (Jenkins and Hidiriglou, 1991), but high concentrations of zinc are produced in the digesta on such highly digestible diets. Weaned ruminants are slightly less tolerant of high zinc intakes than non-ruminants, probably due to the lack of protection from *Py* and vulnerability of the rumen microflora. Diets containing 1g Zn, as ZnO, have been found to reduce weight gain and decrease FCE in lambs; 1.5g depressed feed consumption and 1.7g kg^{-1} DM induced a depraved appetite or pica, characterized by excessive salt consumption and wood chewing (Ott et al., 1966a). The higher zinc levels reduced volatile fatty acid production in the rumen and the acetic to propionic ratio (Ott et al., 1966b).

References

Adeola, O., Lawrence, B.V., Sutton, A.C. and Cline, T.R. (1995) Phytase-induced changes in mineral utilisation in zinc-supplemented diets for pigs. *Journal of Animal Science* 73, 3384–3391.

Ahola, J.K., Baker, D.S., Burns, P.D., Mortimer, R.G., Enns, R.M., Whittier, J.C., Geary, T.W. and Engle, T.E. (2004) Effect of copper, zinc and manganese supplementation and source on reproduction, mineral status and performance over a two-year period. *Journal of Animal Science* 82, 2375–2383.

Alloway, B.J. (2004) *Zinc in Soils and Crop Nutrition*. IZA Publications, International Zinc Association, Brussels, pp. 1–116.

Ao, T., Pearce, L.L., Pescatore, A.J., Cantor, A.H., Dawson, K.A., Ford, M.J. and Shafer, B.L. (2007) Effects of organic and phytase supplementation in a maize–soybean meal diet on the performance and tissue zinc content of broiler chicks. *British Poultry Science* 48, 690–695.

Apgar, J. (1979) Alkaline phosphatase activity and zinc level in plasma as indicators of zinc status in pregnant and lactating sheep. *Nutrition Reports International* 19, 371–376.

Apgar, J., Everett, G.A. and Fitzgerald, J.A. (1993) Dietary zinc deprivation effects parturition and outcome of pregnancy in the ewe. *Nutrition Research* 13, 319–330.

ARC (1980) *The Nutrient Requirements of Ruminants*. Commonwealth Agricultural Bureaux, Farnham Royal, UK, pp. 184–185.

Augspurger, N.R., Spencer, J.D., Webel, D.M. and Baker, D.H. (2004) Pharmacological zinc levels reduce the phosphorus-releasing efficacy of phytase in young pigs and chickens. *Journal of Animal Science* 82, 1732–1739.

Baker, D.H. and Ammerman, C.B. (1995) Zinc bioavailability. In: Ammerman, C.B., Baker, D.H. and Lewis, A.J. (eds) *Bioavailability of Nutrients for Animals*. Academic Press, New York, pp. 367–398.

Bao, Y.M., Choct, M., Iji, P.A. and Bruerton, K. (2007) Effect of organically complexed copper, iron, manganese and zinc on broiler performance, mineral excretion and accumulation in tissues. *Journal of Applied Poultry Research* 16, 448–455.

Batal, A.B., Parr, T.M. and Baker, D.H. (2001) Zinc bioavailability in tetrabasic zinc chloride and the zinc requirement of young chicks fed a soya concentrate diet. *Poultry Science* 80, 87–90.

Beach, R.S., Gershwin, M.E. and Hurley, L.S. (1981) Nutritional factors and autoimmunity. I. Immunopathology of zinc deprivation in New Zealand mice. *Journal of Immunology* 126, 1999–2006.

Beattie, J.H. and Kwun, I.-S. (2004) Is zinc deficiency a risk factor in atherosclerosis? *British Journal of Nutrition* 91, 177–181.

Beattie, J.H. and Trayhurn, P. (2002) Metallothioneins and oxidative stress. *Nutritional Sciences* 5, 228–233.

Beattie, J.H., Gordon, M.-J., Rucjlidge, G.J., Reid, M.D., Duncan, G.J., Horgan, G.W., Cho, Y.-E. and Kwun, I.-S. (2008) Aorta protein networks in marginal and acute zinc deficiency. *Proteonomics* 8, 2126–2135.

Berg, J.M. (1990) Zinc fingers and other metal-binding domains: elements for interactions between molecules. *Journal of Biological Chemistry* 265, 6513–6516.

Biehl, R.R., Baker, D.H. and DeLuca, H.F. (1995) 1α-Hydroxylated cholecalciferol compounds act additively with microbial phytase to improve phosphorus, zinc and manganese utilization in chicks fed soy-based diets. *Journal of Nutrition* 125, 2407–2416.

Bremner, I. (1991) A molecular approach to the study of copper and zinc metabolism. In: Momcilovic, B. (ed.) *Proceedings of the Seventh International Symposium on Trace Elements in Man and Animals*. IMI, Zagreb, pp. 1-1–1-3.

Bremner, I. (1993) Metallothionein in copper deficiency and copper toxicity. In: Anke, M., Meissner, D. and Mills, C.F. (eds) *Proceedings of the Eighth International Symposium on Trace Elements in Man and Animals*. Verlag Media Touristik, Gersdorf, Germany, pp. 507–515.

Cakmak, I. (2008) Enrichment of cereal grains with zinc: agronomic or genetic biofortification. *Plant and Soil* 302, 1–17.

Cao, J. and Cousins, R.J. (2000) Metallothionein mRNA in monocytes and peripheral blood mononuclear cells and in cells from dried blood spots increases after zinc supplementation of men. *Journal of Nutrition* 130, 2180–2187.

Carlson, M.S., Boren, C.A., Wu, C., Huntingdon, C.E., Bollinger, D.W. and Veum, T.L. (2004) Evaluation of various inclusion rates of organic zinc either as polysaccharide or proteinate complex on the growth performance, plasma and excretion of pigs. *Journal of Animal Science*, 82, 1359–1366.

Case, C.L. and Carlson, M.S. (2002) Effect of feeding organic and inorganic sources of additional zinc on growth performance and zinc balance in pigs. *Journal of Animal Science* 80, 1917–1924.

Cheng, J., Kornegay, E.T. and Schell, T. (1998) Influence of lysine on the utilisation of zinc from zinc sulfate and a zinc–lysine complex by young pigs. *Journal of Animal Science* 76, 1064–1074.

Chesters, J.K. (1992) Trace element–gene interactions. *Nutrition Reviews* 50, 217–223.

Chesters, J.K. and Boyne, R. (1997) Interactions of mimosine and zinc deficiency on the transit of BHK cells through the cell cycle. In: Fischer, P.W.F., L'Abbe, M.R., Cockell, K.A. and Gibson, R.S. (eds) *Proceedings of the Ninth International Symposium on Trace Elements in Man and Animals (TEMA 9)*. NRC Research Press, Ottawa, pp. 61–62.

Chhabra, A. and Arora, S.P. (1993) Effect of vitamin A and zinc supplement on alcohol dehydrogenase and superoxide dismutase activities of goat tissues. *Indian Journal of Animal Sciences* 63, 334–338.

Coleman, J.E. (2002) Zinc proteins: enzymes, storage proteins, transcription factors and replication proteins. *Annual Reviews of Biochemistry*, 61, 897–946.

Corrigall, W., Dalgarno, A.C., Ewen, L.A. and Williams, R.B. (1976) Modulation of plasma copper and zinc concentrations by disease status in ruminants. *Veterinary Record* 99, 396–397.

Cousins, R.J. (1996) Zinc. In: Filer, L.J. and Ziegler, E.E. (eds) *Present Knowledge in Nutrition*, 7th edn. International Life Science Institute–Nutrition Foundation, Washington, DC.

Cousins, R.J. (1997) Differential mRNA display, competitive polymerase chain reaction and transgenic approaches to investigate zinc-responsive genes in animals and man. In: Fischer, P.W.F., L'Abbe, M.R., Cockell, K.A. and Gibson, R.S. (eds) *Proceedings of the Ninth International Symposium on Trace Elements in Man and Animals (TEMA 9)*. NRC Research Press, Ottawa, pp. 849–852.

Cousins, R.J., Blanchard, R.K., Moore, J.B., Cui, L., Green, C.L., Liuzzi, J.P., Cao, J. and Bobo, J.A. (2003) Regulation of zinc metabolism and genomic outcomes. *Journal of Nutrition* 133, 1521S–1526S.

Cousins, R.J., Liuzzi, J.P. and Lichten, L.A. (2006) Mammalian zinc transport, trafficking and signals. *Journal of Biological Chemistry* 281, 24085–24089.

Cromwell, G.L. (2001) Antimicrobial and promicrobial agents. In: Lewis, J. and Southern, L.L. (eds). *Swine Nutrition*, 2nd edn. CRC Press, Boca Raton, Florida, p. 401.

Davies, N.T. (1980) Studies on the absorption of zinc by rat intestine. *British Journal of Nutrition* 43, 189–203.

Davies, N.T., Carswell, A.J.P. and Mills, C.F. (1985) The effect of variation in dietary calcium intake on the phytate–zinc interaction in rats. In: Mills, C.F., Bremner, I. and Chesters, J.K. (eds) *Proceedings of the Fifth International Symposium on Trace Elements in Man and Animals*. Commonwealth Agricultural Bureaux, Farnham Royal, UK, pp. 456–457.

Davis, M.E., Brown, D.C., Maxwell, C.V., Johnson, Z.B., Kegley, E.B. and Dvorak, R.A. (2004) Effect of phosphorylated mannans and pharmacological additions of zinc oxide on growth and immunocompetence of weanling pigs. *Journal of Animal Science* 82, 581–587.

Demertzis, P.N. and Mills, C.F. (1973) Oral zinc therapy in the control of infectious pododermatitis in young bulls. *Veterinary Record* 93, 219–222.

Dewar, W.A. and Downie, J.N. (1984) The zinc requirements of broiler chicks and turkey poults fed on purified diets. *British Journal of Nutrition* 51, 467–477.

Droke, E.A. and Spears, J.W. (1993) In vitro and in vivo immunological measurements in growing lambs fed diets deficient, marginal or adequate in zinc. *Journal of Nutrition* 123, 71–90.

Droke, E.A., Spears, J.W., Armstrong, J.D., Kegley, E.B. and Simpson, R. (1993a) Dietary zinc affects serum concentrations of insulin and insulin-like growth factor I in lambs. *Journal of Nutrition* 123, 13–19.

Droke, E.A., Spears, J.W., Brown, T.T. and Quereshi, M.A. (1993b) Influence of dietary zinc and dexamethasone on immune responses and resistance to *Pasteurella haemolytica* challenge in growing lambs. *Nutrition Research* 13, 1213–1216.

Dufty, J.H., Bingley, J.B. and Cove, L.Y. (1977) The plasma zinc concentration of nonpregnant, pregnant and parturient Hereford cattle. *Australian Veterinary Journal* 53, 519–522.

Edwards, H.M. III and Baker, D.H. (1999) Zinc bioavailability in soybean meal. *Journal of Animal Science* 78, 1017–1021.

Edwards, H.M. III and Baker, D.H. (2000) Bioavailability of zinc in several sources of zinc oxide, zinc sulfate and zinc metal. *Journal of Animal Science* 77, 2730–2735.

Egan, A.R. (1972) Reproductive responses to supplemental zinc and manganese in grazing Dorset Horn ewes. *Australian Journal of Experimental Agriculture and Animal Husbandry* 12, 131–135.

Emmert, J.L. and Baker, D.H. (1995) Zinc stores in chickens delay the onset of deficiency symptoms. *Poultry Science* 74, 1011–1021.

Engle, T.E., Nockels, C.F., Kimberling, C.V., Weaber, D.L. and Johnson, A.B. (1997) Zinc repletion with organic and inorganic forms of zinc and protein turnover in marginally zinc-deficient calves. *Journal of Animal Science* 75, 3074–3081.

Flanagan, P.R. (1984) A model to produce pure zinc deficiency in rats and its use to demonstrate that dietary phytate increases the excretion of endogenous zinc. *Journal of Nutrition* 114, 493–502.

Franz, K.B., Kennedy, B.M. and Fellers, D.A. (1980) Relative bioavailability of zinc from selected cereals and legumes using rat growth. *Journal of Nutrition* 110, 2272–2283.

Goff, J.P. and Stabel, J.R. (1990) Decreased plasma retinol α-tocopherol and zinc concentration during the periparturient period: effect of milk fever. *Journal of Dairy Science* 73, 3195–3199.

Grace, N.D. (1972) Observations on plasma zinc levels in sheep grazing New Zealand pastures. *New Zealand Journal of Agricultural Research* 15, 284–288.

Grace, N.D. and Sumner, R.M.W. (1986) Effect of pasture allowance, season and breed on the mineral content and rate of mineral uptake by wool. *New Zealand Journal of Agricultural Research* 29, 223–230.

Grace, N.D., Munday, R., Thompson, A.M., Towers, N.R., O'Donnell, K., McDonald, R.M., Stirnemann, M. and Ford, A.J. (1997) Evaluation of intraruminal devices for combined facial eczema control and trace element supplementation in sheep. *New Zealand Veterinary Journal* 45, 236–238.

Hansard, S.L., Mohammed, A.S. and Turner, J.W. (1968) Gestation age effects upon maternal–foetal zinc utilization in the bovine. *Journal of Animal Science* 27, 1097–1102.

Harland, B.F., Fox, M.R.S. and Fry, B.E. Jr (1974) Changes in plasma zinc related to fasting and dietary protein intake of Japanese quail. *Proceedings of the Society for Experimental Biology and Medicine* 145, 316–322.

Hedemann, M.S., Jensen, B.B. and Poulsen, H.D. (2006) Influence of dietary zinc and copper on digestive enzyme activity and intestinal morphology in weaned pigs. *Journal of Animal Science* 84, 3310–3320.

Hill, D.A., Peo, E.R., Lewis, A.J. and Crenshaw, J.D. (1986) Zinc–amino acid complexes for swine. *Journal of Animal Science* 63, 121–130.

Hill, G.M., Mahan, D.C., Carter, S.D., Cromwell, G.L. and Ewan, R.C. (2001) Effect of pharmacological concentrations of zinc oxide with or without the inclusion of an antibacterial agent on nursery pig performance. *Journal of Animal Science* 79, 934–941.

Hoekstra, W.G., Faltin, E.C., Lin, C.W., Roberts, H.F. and Grummer, R.H. (1967) Zinc deficiency in reproducing gilts fed a diet high in calcium and its effect on tissue zinc and blood serum alkaline phosphatase. *Journal of Animal Science* 26, 1348–1357.

House, W.A. and Welch, R.M. (1989) Bioavailability of and interactions between zinc and selenium in rats fed wheat grain intrinsically labelled with ^{65}Zn and ^{75}Se. *Journal of Nutrition* 119, 916–921.

Huang, Y.L., Lu, L., Luo, X.J. and Liu, B. (2007) An optimal dietary zinc level of broiler chicks fed a corn–soybean meal diet. *Poultry Science* 86, 2582–2589.

Hudson, B.P., Dozier, W.L. III and Wilson, J.L. (2004) Broiler live performance response to dietary zinc source and the influence of zinc supplementation in broiler breeder diets. *Animal Feed Science and Technology* 118, 329–335.

Hurley, L.S. (1981) Teratogenic effects of manganese, zinc and copper in nutrition. *Physiological Reviews* 61, 249–295.

Jenkins, K.J. and Hidiriglou, M. (1991) Tolerance of the pre-ruminant calf to excess manganese and zinc in a milk replacer. *Journal of Dairy Science* 74, 1047–1053.

Jia, W., Jia, Z., Zhang, W., Wang, R., Zwhang, S. and Zhu, X. (2008) Effects of dietary zinc on performance, nutrient digestibility and plasma zinc status in Cashmere goats. *Small Ruminant Research* 80, 68–72.

Jumba, I.O., Suttle, N.F., Hunter, E.A. and Wandiga, S.O. (1995) Effects of soil origin and mineral composition and herbage species on the mineral composition of forages in the Mount Elgon region of Kenya. 2. Trace elements. *Tropical Grasslands* 29, 47–52.

Keilin, D. and Mann, T. (1940) Carbonic anhydrase. Purification and nature of the enzyme. *Biological Journal* 34, 1163–1176.

Kelly, E.J., Quaife, C.J., Froelick, G.J. and Palmiter, R.D. (1996) Metallothionein I and II protect against zinc deficiency and toxicity in mice. *Journal of Nutrition* 126, 1782–1790.

Kendall, N.R., Jackson, D.W., Mackenzie, A.M., Illingworth, D.V., Gill, I.M. and Telfer, S.R. (2001) The effect of a zinc, cobalt and selenium soluble glass bolus on the trace element status of extensively grazed sheep. *Animal Science* 73, 163–170.

Kennedy, K.J., Rains, T.M. and Shay, N.F. (1998) Zinc deficiency changes preferred macronutrient intake in subpopulations of Sprague-Dawley outbred rats and reduces hepatic pyruvate kinase gene expression. *Journal of Nutrition* 128, 43–49.

Kienholz, E.W., Turk, D.E., Sunde, M.L. and Hoekstra. W.G. (1961) Effects of zinc deficiency in the diets of hens. *Journal of Nutrition* 75, 211–221.

Kim, E.-S., Noh, S.K. and Koo, S.I. (1998) Marginal zinc efficiency lowers the lymphatic absorption of α-tocopherol in rats. *Journal of Nutrition* 128, 265–270.

Kirchgessner, M., Schwarz, W.A. and Roth, H.P. (1975) Zur Aktivitat der alkalischen Phosphatase in Serum und Knochen von zinkdepletierten und -repletierten Kuhen. *Zeitschrift fur Tierphysiologie, Tierernahrung und Futtermittelkunde* 35, 19l–200.

Kul'kova, J., Bremner, I., McGaw, B.A., Reid, M. and Beather, J.H. (1993) Mercury–zinc interactions in marginal zinc deficiency. In: Anke, M., Meissner, D. and Mills, C.F. (eds) *Proceedings of the Eighth International Symposium on Trace Elements in Man and Animals.* Verslag Media Touristik, Gersdorf, Germany, pp. 635–637.

Kwun, I.-S., Cho, Y.-E., Lomeda, R.-A.R., Kwon, S.-T., Kim, Y. and Beattie, J.H. (2007) Marginal zinc deficiency in rats decreases leptin expression independently of food intake and corticotrophin-releasing hormone in relation to food intake. *British Journal of Nutrition* 98, 485–489.

Lee, J., Treloar, B.P. and Harris, P.M. (1994) Metallothionein and trace element metabolism in sheep tissues in response to high and sustained zinc dosages. 1. Characterisation and turnover of metallothionein isoforms. *Australian Journal of Agricultural Research* 45, 303–320.

Legg, S.P. and Sears, L. (1960) Zinc sulfate treatment of parakeratosis in cattle. *Nature, UK* 186, 1061–1062.

Lei, X.G., Ku, P.K., Miller, E.R., Ullrey, D.E. and Yokoyama, M.T. (1993) Supplemental microbial phytase improved bioavailability of dietary zinc to weanling pigs. *Journal of Nutrition* 123, 1117–1123.

Linares, L.B., Broomhead, J.N., Guaiume, E.A., Ledoux, D.R., Veum, T.L. and Raboy, V. (2007) Effects of a low phytate barley (*Hordeum vulgare* L.) on zinc utilisation in young broiler chicks. *Poultry Science* 86, 299–308.

MAFF (1990) *UK Tables of the Nutritive Value and Chemical Composition of Foodstuffs.* In: Givens, D.I. (ed.) Rowett Research Services, Aberdeen, UK.

Mahmoud, O.M., El Samani, F., Bakheit, R.O. and Hassan, M.A. (1983) Zinc deficiency in Sudanese desert sheep. *Journal of Comparative Pathology* 93, 591–595.

Martin, G.B. and White, C.L. (1992) Effects of dietary zinc deficiency on gonadotrophin secretion and testicular growth in young male sheep. *Journal of Reproduction and Fertility* 96, 497–507.

Masters, D.G. and Fels, H.E. (1980) Effect of zinc supplementation on the reproductive performance of grazing Merino ewes. *Biological Trace Element Research* 2, 281–290.

Masters, D.G. and Fels, H.E. (1985) Zinc supplements and reproduction in grazing ewes. *Biological Trace Element Research* 7, 89–93.

Masters, D.G. and Moir, R.J. (1980) Provision of zinc to sheep by means of an intraruminal pellet. *Australian Journal of Experimental Agriculture and Animal Husbandry* 20, 547–552.

Masters, D.G. and Moir, R.J. (1983) Effect of zinc deficiency on the pregnant ewe and developing foetus. *British Journal of Nutrition* 49, 365–372.

Masters, D.G. and Somers, M. (1980) Zinc status of grazing sheep: seasonal changes in zinc concentrations in plasma, wool and pasture. *Australian Journal of Experimental Agriculture and Animal Husbandry* 26, 20–24.

Mayland, H.F., Rosenau, R.C. and Florence, A.R. (1980) Grazing cow–calf responses to zinc supplementation. *Journal of Animal Science* 51, 966–974.

Miller, E.R., Luecke, R.W., Ullrey, D.E., Baltzer, B.V., Bradley, B.L. and Hoefer. J.A. (1968) Biochemical, skeletal and allometric changes due to zinc deficiency in the baby pig. *Journal of Nutrition* 95, 278–286.

Miller, E.R., Liptrap, D.O. and Ullrey, D.E. (1970) Sex influence on zinc requirement of swine. In: Mills, C.F. (ed.) *Trace Element Metabolism in Animals – 1*. Livingstone, Edinburgh, pp. 377–379.

Miller, J.K. and Miller, W.J. (1962) Experimental zinc deficiency and recovery of calves. *Journal of Nutrition* 76, 467–474.

Miller, W.J. (1970) Zinc nutrition of cattle: a review. *Journal of Nutrition,* 53, 1123–1135.

Miller, W.J., Pitts, W.J., Clifton, C.M. and Schmittle, S.C. (1964) Experimentally produced zinc deficiency in the goat. *Journal of Dairy Science* 47, 556–559.

Miller, W.J., Morton, J.D., Pitts, W.J. and Clifton, C.M. (1965) The effect of zinc deficiency and restricted feeding on wound healing in the bovine. *Proceedings of the Society for Experimental Biology and Medicine* 118, 427–431.

Miller, W.J., Powell, G.W., Pitts, W.J. and Perkins, H.F. (1965) Factors affecting zinc content of bovine hair. *Journal of Dairy Science* 48, 1091–1095.

Miller, W.J., Blackmon, D.M., Gentry, R.P., Powell, G.W. and Perkins, H.E. (1966) Influence of zinc deficiency on zinc and dry matter content of ruminant tissues and on excretion of zinc. *Journal of Dairy Science* 49, 1446–1453.

Miller, W.J., Blackmon, D.M., Gentry, R.P., Pitts, W.J. and Powell, G.W. (1967) Absorption, excretion and retention of orally administered zinc-65 in various tissues of zinc deficient and normal goats and calves. *Journal of Nutrition* 92, 71–78.

Miller, W.J., Martin, Y.G., Gentry, R.P. and Blackmon, D.M. (1968) ^{65}Zn and stable zinc absorption, excretion and tissue concentrations as affected by type of diet and level of zinc in normal calves. *Journal of Nutrition* 94, 391–401.

Mills, C.F. and Dalgarno, A.C. (1972) Copper and zinc status of ewes and lambs receiving increased dietary concentrations of cadmium. *Nature, UK* 239, 171–173.

Mills, C.F., Dalgarno, A.C., Williams, R.B. and Quarterman, J. (1967) Zinc deficiency and the zinc requirements of calves and lambs. *British Journal of Nutrition* 21, 751–768.

Minson, D.J. (1990) *Forages in Ruminant Nutrition*. Academic Press, San Diego, California, pp. 208–229.

Mohanna, C. and Nys, Y. (1999) Effect of zinc content and sources on the growth, body zinc deposition and retention, zinc excretion and immune response in chickens. *British Poultry Science* 40, 108–114.

Morrison, J.N. and Bremner, I. (1987) Effect of maternal zinc supply on blood and tissue metallothionein I concentrations in suckling rats. *Journal of Nutrition* 117, 1588–1594.

Muehlenbein, E.L., Brink, D.R., Deutscher, G.H., Carlson, M.P. and Johnson, A.B. (2001) Effects of inorganic and organic copper supplemented to first-calf cows on cow reproduction and calf health and performance. *Journal of Animal Science* 79, 1650–1659.

Munday, R., Thompson, A.M., Fowke, E.A., Wesselink, C., Smith, B.L., Towers, N.R., O'Donnell, K., McDonald, R.M., Stirnemann, M. and Ford, A.J. (1997) A zinc-containing intraruminal device for faecal eczema control in lambs. *New Zealand Veterinary Journal* 45, 93–98.

NRC (1994) *Nutrient Requirements of Poultry*, 9th edn. National Academy of Sciences, Washington, DC.

NRC (1998) *Nutrient Requirements of Swine*, 10th edn. National Academy of Sciences, Washington, DC.

NRC (2001) *Nutrient Requirements of Dairy Cows*, 7th edn. National Academy of Sciences, Washington, DC.

NRC (2005) *Mineral Tolerances of Animals*, 2nd edn. National Academy of Sciences, Washington, DC.

Neathery, M.W., Miller, W.J., Blackmon, D.M. and Gentry, R.P. (1973a) Performance and milk zinc from low zinc intake in dairy cows. *Journal of Dairy Science* 56, 212–217.

Neathery, M.W., Miller, W.J., Blackmon, D.M., Pate, F.M. and Gentry, R.P. (1973b) Effects of long-term zinc deficiency on feed utilisation, reproductive characteristics and hair growth in the sexually mature male goat. *Journal of Dairy Science* 56, 98–105.

Neathery, M.W., Moos, W.H., Wyatt, R.D., Miller, W.J., Gentry, R.P. and George, L.W. (1980) Effects of dietary aflatoxin on performance and zinc metabolism in dairy calves. *Journal of Dairy Science* 63, 789–799.

Noh, S. and Koo, S.I. (2001) Intraduodenal infusion of lysophosphatidylcholine restores the intestinal absorption of vitamins A and E in rats fed a low-zinc diet. *Experimental Biology and Medicine* 226, 342–348.

Oberleas, D. (1996) Mechanism of zinc homeostasis. *Journal of Inorganic Biochemistry* 62, 231–241.

O'Dell, B.L. and Reeves, P.G. (1989) Zinc status and food intake. In: *Zinc in Human Biology*. ILSI Press, Washington, DC, pp. 173–181.

O'Dell, B.L. and Savage. J.E. (1957) Symptoms of zinc deficiency in the chick. *Federation Proceedings* 16, 394.

O'Dell, B.L., Newberne, P.M. and Savage J.E. (1958) Significance of dietary zinc for the growing chicken. *Journal of Nutrition* 65, 503–508.

Oh, S.H., Nakane, H., Deagan, J.T., Whanger, P.D. and Arscott, G.H. (1979) Accumulation and depletion of zinc in chick tissue metallothioneins. *Journal of Nutrition* 109, 1720–1729.

O'Hara, P.J., Newman, A.P. and Jackson, R. (1960) Parakeratosis and copper poisoning in pigs fed a copper supplement. *Australian Veterinary Journal* 36, 225–229.

Olokosi, O.A., Cowieson, A.J. and Adeola, O. (2007) Age-related influence of a cocktail of xylanase, amylase and protease or phytase individually or in combination in broilers. *Poultry Science* 86, 77–86.

Orr, C.L., Hutcheson, D.P., Grainger, R.B., Cummins, J.M. and Mock, R.E. (1990) Serum copper, zinc, calcium and phosphorus concentrations of calves stressed by bovine respiratory disease and infectious bovine rhinotracheitis. *Journal of Animal Science* 68, 2893–2900.

Ott, E.A., Smith, W.H., Stob, M. and Beeson, W.M. (1964) Zinc deficiency syndrome in young lamb. *Journal of Nutrition* 82, 41–50.

Ott, E.A., Smith, W.H., Stob, M., Parker, H.E., Harrington, R.B. and Beeson, W.M. (1965) Zinc requirement of the growing lamb fed a purified diet. *Journal of Nutrition* 87, 459–463.

Ott, E.A., Smith, W.H., Harrington, R.B., Parker, H.E. and Beeson, W.M. (1966a) Zinc toxicity in ruminants. 4. Physiological changes in tissues of beef cattle. *Journal of Animal Science* 25, 432–438.

Ott, E.A., Smith, W.H., Harrington, R.B. and Beeson W.M. (1966b) Zinc toxicity in ruminants. 1. Effect of high levels of dietary zinc on gains, feed consumption and feed efficiency of lambs. *Journal of Animal Science* 25, 414–418.

Outten, C.E. and O'Halloran, T.V. (2001) Femtomolar sensitivity of metalloregulatory proteins controlling zinc homeostasis. *Science* 292, 2488–2492.

Payne, R.L., Beidner, T.D., Fakler, T.M. and Southern, L.L. (2006) Growth and intestinal morphology of pigs from sows fed two zinc sources during gestation and lactation. *Journal of Animal Science* 84, 2141–2149.

Pitts, W.J., Miller, W.J., Fosgate, O.T., Morton, J.D. and Clifton, C.M. (1966) Effect of zinc deficiency and restricted feeding from two to five months of age on reproduction in Holstein bulls. *Journal of Dairy Science* 49, 995–1000.

Pond, W.G. (1983) The effect of dietary calcium and zinc levels on weight gain and blood and tissue mineral concentrations in growing Columbia- and Suffolk-sired lambs. *Journal of Animal Science* 56, 952–959.

Prasad, A.S. (2007) Zinc: mechanisms of host defence. *Journal of Nutrition* 137, 1345–1349.

Price, J. and Humphries, W.R. (1980) Investigation of the effect of supplementary zinc on growth rate of cattle in farms in N. Scotland. *Journal of Agricultural Science, Cambridge* 95, 135–139.

Roberson, K.D. and Edward, H.M. (1994) Effects of 1,25-dihydroxycholecalciferol and phytase on zinc utilisation in broiler chicks. *Poultry Science* 73, 1312–1326.

Roberson, K.D., Kalbfleish, J.L., Pan, W., Applegate, T.J. and Rosenstein, D.S. (2005) Comparison of wheat bran phytase and a commercially available phytase on turkey tom performance and litter phosphorus content. *International Journal of Poultry Science* 4, 244–249.

Rossi, P., Rutz, F., Anciuti, M.A., Rech, J.L. and Zauk, N.H.F. (2007) Influence of graded levels of organic zinc on growth performance and carcass traits in broilers. *Journal of Applied Poultry Research* 16, 219–225.

Roth, P. and Kirchgessner, M. (1985) Utilisation of zinc from picolinic acid or citric acid complexes in relation to dietary protein source in rats. *Journal of Nutrition* 115, 1641–1649.

Rowe, D.J. and Bobilya, D.J. (2000) Albumin facilitates zinc acquisition by endothelial cells. *Proceedings of the Society for Experimental Biology and Medicine* 224, 178–186.

Sahin, K. and Kucuk, O. (2003) Zinc supplementation alleviates heat stress in laying Japanese quail. *Journal of Nutrition* 133, 2808–2811.

Sahin, K., Smith, M.O., Onderci, M., Sahin, N., Gursu, M.F. and Kucuk, O. (2005) Supplementation of zinc from organic and inorganic sources improves performance and antioxidant status of heat-distressed quail. *Poultry Science* 84, 882–887.

Sandoval, M., Henry, P.R., Ammerman, C.B., Miles, R.D. and Littell, R.C. (1997a) Estimation of the relative bioavailability of zinc from inorganic zinc sources for sheep. *Animal Feed Science and Technology* 66, 223–235.

Sandoval, M., Henry, P.R., Littell, R.C., Cousins, R.J. and Ammerman, C.B. (1997b) Relative bioavailability of supplemental inorganic zinc sources for chicks. *Journal of Animal Science* 75, 3195–3205.

Schryver, H.F., Hintz, H.F. and Lowe, J.E. (1980) Absorption excretion and tissue distribution of stable zinc and ^{65}zinc in ponies. *Journal of Animal Science* 51, 896–902.

Schwarz, W.A. and Kirchgessner, M. (1975) Experimental zinc deficiency in lactating dairy cows. *Veterinary Medical Review* 1/2, 19–41.

Smith, B.L., Embling, P.P., Towers, N.R., Wright, D.E. and Payne, E. (1977) The protective effect of zinc sulfate in experimental sporidesmin poisoning of sheep. *New Zealand Veterinary Journal* 25, 121–127.

Smith, I.D., Grummer, R.H., Hoekstra, W.G. and Phillips P.H. (1960) Effects of feeding an autoclaved diet on the development of parakeratosis in swine. *Journal of Animal Science* 19, 568–579.

Smith, J.C. Jr, McDaniel, E.G. and Chan, W. (1976) Alterations in vitamin A metabolism during zinc deficiency and food and growth restriction. *Journal of Nutrition* 106, 569–574.

Smith, J.W., Tokach, M.D., Goodband, R.D., Nelsson, J.L. and Richert, B.T. (1997) Effects of the interrelationship between zinc oxide and copper sulfate on growth performance of early weaned pigs. *Journal of Animal Science* 75, 1861–1866.

Spais, A.G. and Papasteriadis, A.A. (1974) Zinc deficiency in cattle under Greek conditions. In: Hoekstra, W.G., Suttie, J.W., Ganther, T.T.E. and Mertz, W. (eds) *Trace Element Metabolism in Animals – 2.* University Park Press, Baltimore, Maryland, pp. 628–631.

Spears J.W. (1989) Zinc methionine for ruminants: relative bioavailability of zinc in lambs and effects on growth and performance of growing heifers. *Journal of Animal Science* 67, 835–843.

Spears, J.W. (1996) Organic trace minerals in ruminant nutrition. *Animal Feed Science and Technology* 58, 151–163.

Spears, J.W. and Kegley, E.B. (2002) Effect of source (zinc oxide vs zinc proteinate) and level on performance, carcass characteristics and immune response of growing and finishing steers. *Journal of Animal Science* 80, 2747–2752.

Stake, P.E., Miller, W.J., Gentry, R.P. and Neathery, N.W. (1975) Zinc metabolic adaptations in calves fed a high but non-toxic zinc level for varying time periods. *Journal of Animal Science* 40, 132–137.

Starcher, B. and Kratzer, F.H. (1963) Effect of zinc on bone alkaline phosphatase in turkey poults. *Journal of Nutrition* 79, 18–22.

Stevenson, J.W. and Earle, I.P. (1956) Studies on parakeratosis in swine. *Journal of Animal Science* 15, 1036–1045.

Stevenson, M.H. and Jackson, N. (1984) Comparison of dietary hydrated copper sulfate, dietary zinc oxide and a direct method for inducing moult in laying hens. *British Poultry Science* 25, 505–517.

Sunde, M.L. (1972) Zinc requirement for normal feathering of commercial Leghorn-type pullets. *Poultry Science* 51, 1316–1322.

Sunde, M.L. (1978) Effectiveness of early zinc supplementation to chicks from five commercial egg strains. In: *Proceedings XVI World Poultry Congress, Rio de Janeiro, Brazil*, Vol. IV, p. 574.

Suttle, N.F. (1983) Assessment of the mineral and trace element status of feeds. In: Roberds, G.E. and Packham, R.G. (eds) *Proceeding of the Second Symposium of the International Network of Feed Information Centres*. Commonwealth Agricultural Bureaux, Farnham Royal, UK, pp. 211–237.

Suttle, N.F. and Mills, C.F. (1966) Studies on the toxicity of copper to pigs. 1. The effects of oral supplements of zinc and iron salts on the development of copper toxicosis. *British Journal of Nutrition* 20, 135–148.

Suttle, N.F., Lloyd-Davies, H. and Field, A.C. (1982) A model for zinc metabolism in sheep given a diet of hay. *British Journal of Nutrition* 47, 105–112.

Swinkels, J.W.G.M., Kornegay, E.T., Zhou, W., Lindemann, M.D., Webb, K.E. and Verstegen, M.W.A. (1996) Effectiveness of a zinc amino acid chelate and zinc sulfate in restoring serum and soft tissue zinc concentrations when fed to zinc-depleted pigs. *Journal of Animal Science* 74, 2420–2430.

Todd, W.R., Elvehjem. C.A. and Hart, E.B. (1934) Zinc in the nutrition of the rat. *American Journal of Physiology* 107, 146–156.

Towers, N.R., Smith, B.L., Wright, D.E. and Sinclair, D.P. (1975) Preventing facial eczema by using zinc. *Proceedings of the Ruakura Farmers' Conference*, pp. 57–61.

Tucker, H.F. and Salmon, W.D. (1955) Parakeratosis or zinc deficiency disease in the pig. *Proceedings of the Society for Experimental Biology and Medicine* 88, 613–616.

Underwood. E.J. and Somers, M. (1969) Studies of zinc nutrition in sheep. 1. The relation of zinc to growth, testicular development and spermatogenesis in young rams. *Australian Journal of Agricultural Research* 20, 889–897.

Underwood, E.J. and Suttle, F. (1999) *The Mineral Nutrition of Livestock*, 3rd edn. CAB International, Wallingford, UK.

Vallee, B.L. and Falchuk, K.H. (1993) The biochemical basis of zinc physiology. *Physiological Reviews* 73, 79–118.

Wagner, J.J., Engle, T.E., Wagner, J.J., Lacey, J.L. and Walker, G. (2008) The effects of ZinMet brand liquid zinc methionine on feedlot performance and carcass merit in crossbred yearling steers. *Professional Animal Scientist* 24, 420–429.

Webster, K.A., Prentice, H. and Bishopric, N.H. (2001) Oxidation of zinc finger transcription factors: physiological consequences. *Antioxidants and Redox Signalling* 3, 535–548.

Wedekind, K.J. and Baker, D.H. (1990) Zinc bioavailability of feed grade sources of zinc. *Journal of Animal Science* 68, 684–689.

Wedekind, K.J., Hortin, A.E. and Baker, D.H. (1992) Methodology for assessing zinc bioavailability: efficacy estimates for zinc-methionine, zinc sulfate and zinc oxide. *Journal of Animal Science* 70, 178–187.

Wedekind, K.J., Collings, G., Hancock, J. and Titgemeyer, E. (1994a) The bioavailability of zinc-methionine relative to zinc sulfate is affected by calcium level. *Poultry Science* 73 (Suppl. 1), 114.

Wedekind, K.J., Lewis, A.J., Giesemann, M.A. and Miller, P.S. (1994b) Bioavailability of zinc from inorganic and organic sources for pigs fed corn–soybean meal diets. *Journal of Animal Science* 72, 2681–2689.

Wegner, T.N., Ray, D.E., Lox, C.D. and Stott, G.H. (1973) Effect of stress on serum zinc and plasma corticoids in dairy cattle. *Journal of Dairy Science* 56, 748–752.

White, C.L. (1993) The zinc requirements of grazing ruminants. In: Robson, A.D. (ed.) *Zinc in Soils and Plants: Developments in Plant and Soil Sciences*, Vol. 55. Kluwer Academic Publishers, London, pp. 197–206.

White, C.L., Chandler, B.S. and Peter, D.W. (1991) Zinc supplementation of lactating ewes and weaned lambs grazing improved Mediterranean pastures. *Australian Journal of Experimental Agriculture* 31, 183–189.

White, C.L., Martin, G.B., Hynd, P.T. and Chapman, R.E. (1994) The effect of zinc deficiency on wool growth and skin and wool histology of male Merino lambs. *British Journal of Nutrition* 71, 425–435.

Wichert, B., Kreyenberg, K. and Kienze, E (2002) Serum response after oral supplementation of different zinc compounds in horses. *Journal of Nutrition* 132, 1769S–1770S.

Williams, R.B., McDonald, I. and Bremner, I. (1978) The accretion of copper and of zinc by the foetuses of prolific ewes. *British Journal of Nutrition* 40, 377–386.

Wing, K., Wing, A., Sjostrom, R. and Lonnerdal, B. (1997) Efficacy of a Michaelis-Menton model for the availabilities of zinc, iron and cadmium from an infant formula diet containing phytate. In: Fischer, P.W.F., L'Abbe, M.R., Cockell, K.A. and Gibson, R.S. (eds) *Proceedings of the Ninth International Symposium on Trace Elements in Man and Animals (TEMA 9)*. NRC Research Press, Ottawa, pp. 31–32.

Wright, C.L., Spears, J.W. and Webb, K.E. (2008) Uptake of zinc from zinc sulfate and zinc proteinate by ovine ruminal and omasal epithelia. *Journal of Animal Science* 86, 1357–1363.

Xu, C., Wensing, T. and Beynen, A.C. (1997) The effect of dietary soybean versus skim milk protein on plasma and hepatic concentrations of zinc in veal calves. *Journal of Dairy Science* 80, 2156–2161.

Zago, M. and Oteiza, P.I. (2001) The antioxidant properties of zinc: interactions with iron and antioxidants. *Free Radicals in Biology and Medicine* 31, 266–274.

Zang, P., Duhamel, G.E., Mysore, J.V., Carlson, M.P. and Schneider, N.R. (1995) Prophylactic effect of dietary zinc in a laboratory mouse model of swine dysentery. *American Journal of Veterinary Research* 56, 334–339.

17 Occasionally Beneficial Elements

The 12 minerals discussed so far are all undeniably essential to the health and well-being of farm livestock. For each element, it has not been difficult to demonstrate specific physiological functions, severe disabilities when dietary concentrations are low and adaptive mechanisms that can ameliorate any deficiency. The elements boron, chromium, lithium, molybdenum, nickel, rubidium, silicon and vanadium are grouped as occasionally beneficial elements (OBE) because they can also be regarded as essential, albeit in 'ultratrace' concentrations (<1 mg kg^{-1}) (Nielsen, 1996), although a similar situation applies to potentially toxic elements (PTE) (e.g. arsenic, cadmium, lead; see Chapter 18). The customary evidence for 'essentiality' is threefold:

- the deprived animal fails to thrive and becomes unhealthy;
- the animal shows physiological responses to minute supplements when repleted; and
- the animal has homeostatic mechanisms for modulating retention of the element.

Other ubiquitous and reactive elements with specific functions in lower organisms may eventually have their essentiality to humans and livestock confirmed (Nielsen, 1996, 2002), but the list of known essential minerals has shortened since the last edition of this book. A number of important questions remain to be answered before essentiality is unequivocally demonstrated for all OBE and PTE:

- Was the deficient diet adequate with respect to all other essential elements?
- Were the physiological responses life-enhancing?
- Were the homeostatic mechanisms any more than homeorhetic mechanisms providing defence against excess?
- Where does a nutritional response end and a pharmacological response begin?

Doubts on the first point have led to tin being withdrawn from the list of essential minerals, and call into question the position of others, particularly rubidium (NRC, 2005). The clinical consequences of deprivation of OBE and PTE are rarely specific (e.g. poor growth, low viability of newborn) and the same abnormalities have progressively been attributed to a series of 'newer' ultratrace elements (see Chapter 18). Many of the biochemical consequences of deprivation and responses to supplementation are also shared between 'essential' elements; for example, chromium and zinc both reduce plasma insulin in cold-stressed birds; lack of either lithium or nickel impairs glycolysis. All the above questions are important to farmers who are sometimes tempted to purchase novel supplements because they contain rare elements from the far reaches of the periodic table, but do their stock need more of them on their 'table'? For the most part, OBE are so abundant in the farm environment that natural deficiencies are unlikely to arise.

Boron

Essentiality

Boron is an essential nutrient for plants, but evidence for essentiality to animals has been slow to gain strength. Boron binds strongly to *cis*-hydroxyl groups and may thus form complexes with riboflavin, oestradiol and vitamin D metabolites. For example, by binding with 24,25-dihydroxyvitamin D produced by the microsomal enzyme 24-hydroxylase, boron may up-regulate the enzyme (Miljkovic et al., 2004). This allows pharmacological rather than essential nutritional responses to occur and the boron supplementation of equine feeds is commonly practised in the hope that it will decrease the risk of leg injuries (O'Connor et al., 2001). If boron deprivation occurs, it is most likely to do so in livestock fed grain-based diets.

Poultry

Benefits from boron supplementation in animals were first noted in vitamin D-deficient chicks in 1981 and subsequently confirmed and extended (Hunt et al., 1994). However, the basal diet was exceedingly low in boron (<0.18 mg kg^{-1} dry matter (DM)) and the beneficial concentration so low (1.4 mg kg^{-1} DM) that it would only be found in boron-deficient plants. Supplementation of commercial rations for broilers with 5 or 60 mg kg^{-1} DM marginally increased growth rate, more so in males than females, but significantly improved tibia weight and strength (Rossi et al., 1993). The latter findings have been confirmed (Wilson and Ruszler, 1997) and extended to hens, with the improvement in bone strength occurring sooner in laying than non-laying birds (Wilson and Ruszler, 1998). However, a similar range of boron supplements did not improve egg production in another study (Eren et al., 2004). The addition of 5 or 25 mg B kg^{-1} DM to a corn/soy starter diet of unspecified boron content improved growth rate and feed conversion efficiency (FCE) of broiler chicks, whether the diet was low or adequate in vitamin D_3 (Kurtoglu et al., 2001), but the benefit disappeared on a grower diet that contained 15% wheat. Although plasma calcium and alkaline phosphatase were both lowered by the end of the study, their relationship to performance was unclear. Similar doubts surround the narrowing of the epiphyseal growth plate seen in boron-supplemented groups at 45 days of age (Kurtoglu et al., 2005).

Pigs

Young pigs have been found to eat more food and gain proportionately more weight when 5 or 15 mg boron is added (as sodium borate) to a diet based on ground corn and containing 2.2 mg B kg^{-1} DM. The higher level increased the shear strength of fibula and the phosphorus content of fat-free bone ash, but did not affect calcium or phosphorus retention or plasma calcium, magnesium or phosphorus (Armstrong and Spears, 2001). Continuation of boron supplementation through a first pregnancy had no significant effects on fertility, fecundity or birth weight and no significant effects ($P<0.05$) on bone mineralization (Armstrong et al., 2002). In offspring of these gilts reared on a similar diet, inflammatory responses to an intradermally injected mitogen were muted, but cytokine responses to intramuscular challenge with lipopolysaccharide were increased when compared to offspring from boron-supplemented gilts (Armstrong and Spears, 2003). The essentiality of boron for immunocompetence remains unclear.

Ruminants

An association between poor conception and 'low' serum boron concentrations of 0.1–0.13 mg (9.2–13.6 µmol) l^{-1} has been reported within a beef cow herd (Small et al., 1997), but association is not proof of essentiality.

Sources

Boron concentrations in soils in England and Wales are relatively high and consistent, with 50% of values falling between 20 and 40 mg kg^{-1} DM (Table 17.1). However, only 2.5–5.0% was extractable in hot water, the common test of availability to plants. Concentrations in plants are often lower than those in soils and are particularly low in cereal grains. However, higher levels are found during early cereal

Table 17.1. Median concentrations (range) of occasionally beneficial trace elements found in soils.

Element	Concentration (mg kg^{-1} DM)
B	34 (20–40)
Cr	20 (10–121)
Mo	5 (0.1–20.0)
Li	28[a]
Ni	4 (7.3–70.0)
Rb	4[a]
V	40[a]

[a] Mean values from Eastern Europe (Kabata-Pendias and Pendias, 1992): all other data are for England and Wales (Archer and Hodgson, 1987).

growth, with concentrations falling from 7 to 2 mg B kg^{-1} DM in wheat plants between May and August (Silanpaa, 1982). Boron values tend to be lower in grasses than in legumes (Table 17.2). However, consistently high levels of 38–40 mg kg^{-1} DM have been found in grass hays from Nevada (Green and Weeth, 1977; Weeth et al., 1981). Root crops are rich in boron; turnips, swedes and sugarbeet have been found to contain roughly 10 times more boron (13.5–20.8 mg kg^{-1} DM) than four cereal grains (1.4–1.7 mg kg^{-1} DM) grown at the same site in Finland (Fig. 17.1) (Silanpaa, 1982). The inclusion of soybean meal (SBM) probably accounts for the high levels of boron reported in mixed rations for sheep (32 mg kg^{-1} DM) (Brown et al., 1989) and poultry (9.4 and 15.6 mg B kg^{-1} DM) (Rossi et al., 1993). Bovine milk normally contains 0.09–0.25 mg B l^{-1}, but concentrations greatly increase when boron intakes are increased (Kirchgessner et al., 1967).

Metabolism

There is little evidence of homeostatic control of metabolism at high boron intakes. In cattle, plasma and urine boron are almost linearly related to boron intake over the range 1–45 mg B kg^{-1} live weight (LW) (Weeth et al., 1981) and up to 69% is excreted in bovine urine (Green and Weeth, 1977). The plasma boron level associated with feeding a boron-rich hay (40 mg B kg^{-1} DM) was 2.4 mg l^{-1}. In a balance study with sheep, the faeces was the major route of boron excretion from the basal diet, but urinary excretion became the more important when borate (45 and 175 mg^{-1} head^{-1} day^{-1}) was added (Brown et al., 1989). In broiler chicks, liver and muscle boron increased linearly from <2 to >16 mg B kg^{-1} DM as dietary boron increased to 330 mg kg^{-1} DM (Rossi et al., 1993). In mature horses given a diet consisting predominantly of hay, containing 24 g B kg^{-1} DM, the apparent absorption of boron (AA_B) was 46% but plasma contained only 0.2 mg B l^{-1} after 13 days (O'Connor et al., 2007).

Toxicity

Boron is less well-tolerated by poultry than was once thought (Underwood and Suttle, 1999). A marked weakening of the eggshell occurs when dietary boron is increased from 100 to

Table 17.2. Concentrations (mg kg^{-1} DM) of occasionally beneficial trace elements commonly found in forages and cereals: ranges of means from surveys in different Eastern European countries (from Kabata-Pendias and Pendias, 1992).

Element	Grasses	Legumes	Cereals
B	4.9–7.4	14–78	0.7–7.3
Cr	0.1–0.35	0.2–4.2	0.01–0.55
Li	0.07–1.5	0.01–3.1	0.05
Mo	0.33–1.4	0.5–2.5	0.16–0.92
Ni	0.13–1.1	1.2–2.7	0.22–0.34
Rb	130	44–98	3–4
V	0.1–0.23	0.18–0.24	0.007–0.060

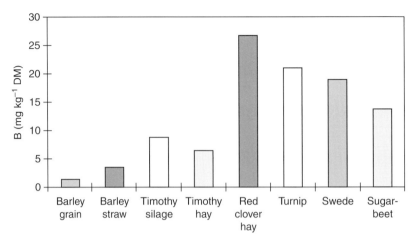

Fig. 17.1. Crops and forage species grown at the same location in Finland show considerable variation in boron concentrations (data from Silanpaa, 1982).

200 mg kg^{-1} DM, accompanied by a reduction in egg yield in laying hens (Fig. 17.2) (Eren et al., 2004). Riboflavin protects animals from boron toxicity (NRC, 2005). Loss of appetite and body weight has occurred in heifers given 150–300 mg B l^{-1} in their drinking water (Green and Weeth, 1977), but the treatment was equivalent to >800 mg B kg^{-1} DM in the diet and drinking water from natural sources is unlikely to contain such high levels. Plasma concentrations >4 mg B l^{-1} indicate abnormally high intakes of boron. Toxicity has occurred following the ingestion of borate fertilizer by cattle and is indicated by widespread increases in tissue levels from <10 to >100 mg B kg^{-1} DM (Puls, 1994). Enrichment of edible products with boron following livestock exposure to excess boron does not present a hazard to the consumer (NRC, 2005).

Chromium

The nutritional biochemistry of trivalent chromium (Cr III) has been reviewed by Vincent (2007).

Essentiality

The essentiality of chromium for mammals was first indicated when supplements of Cr III improved glucose tolerance in rats (Vincent and Stallings, 2007). This property was shared by the organic chromium in brewers' yeast and termed 'glucose tolerance factor' (GTF). Purification of GTF proved difficult, but a low-molecular-weight, chromium-binding protein (chromodulin) that facilitates the uptake of glucose by insulin-sensitive cells has been isolated (Vincent, 2000). The initial observation of Schwarz and Mertz (1959) implies that

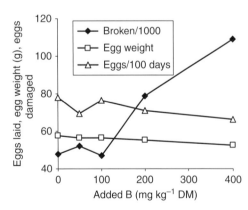

Fig. 17.2. Harmful effects of boron supplementation in laying hens can appear at concentrations >100 mg kg^{-1} DM, taking the form of a decreased egg production rate and higher proportion of damaged eggs. No benefits from smaller supplements have been recorded (data from Eren et al., 2004).

chromodulin can be synthesized in vivo, and chromium is found in GTF-like forms in the livers of mammals given inorganic chromium (Yamamoto et al., 1988). Supplementation of livestock diets with Cr III changes protein, nucleic acid and lipid metabolism (Anderson, 1987), but this could reflect the manifold effects of insulin rather than further biochemical roles for chromium. Ruminants use acetate rather than glucose as a carbon source and this may lower their insulin sensitivity, but a dietary supplement of 1 mg Cr kg^{-1} DM (as amino acid chelate) has been found to cause a 30% increase in the potential for glucose to be used for fat synthesis in adult sheep by increasing the activity of ATP-citrate lyase (Gardner et al., 1998). Responses to Cr III may be important to the twin-bearing ewe, which has a uniquely high requirement for glucose; this has not yet been investigated.

Natural sources

Chromium is far more abundant in soils than in crops. For soils, Puls (1994) gave ranges, based on Canadian experience, of 1–25 mg kg^{-1} DM, but data from England suggest a greater abundance (Table 17.1). Reported values for feedstuffs range from 0.01 to 4.2 mg Cr kg^{-1} DM with cereals relatively poor and legumes relatively rich in chromium (Table 17.2). Significant amounts of soil-borne chromium are ingested by grazing animals. Concentrations of chromium reported in foods decreased enormously towards the end of the last century due to improvements in analytical techniques. Values of 40–60 µg Cr kg^{-1} DM have been reported for single batches of lupins, barley and hay fed to sheep in Australia (Gardner et al., 1998), but mixed diets for livestock usually contain 1–2 mg Cr kg^{-1} DM (e.g. Bryan et al., 2004). Bovine milk contains 55 µg Cr l^{-1} and the level is not increased by dietary supplementation with chromium methionine (CrMet) (Hayirli et al., 2001). Significant amounts of chromium can be added to rations as contaminants in phosphate supplements, which can contain 0.6–1.3 g Cr kg^{-1} DM (Sullivan et al., 1994). The importance of the forms in which chromium is present in the diet has still not been critically examined. Brewers' yeast contains an ethanol-extractable, small-molecular-weight complex (400 Da), while alfalfa contains a much larger complex (2600 Da) that is insoluble in non-polar solvents (Starich and Blincoe, 1983). Similar complexes have been found in wheat (Toepfer et al., 1973).

Metabolism

Little is known about the absorption (A_{Cr}) and metabolism of chromium in any species. The success of chromium oxides, salts and chelates as inert markers in digestibility studies with weaned animals implies that many inorganic chromium sources are poorly absorbed, but the milk-fed animal may absorb chromium more efficiently. Estimates of A_{Cr} are generally low, ranging from 0.5% to 2.8% for Cr III in rats and humans (NRC, 2005), but the mechanisms of absorption remain unclear. Absorbed chromium is assumed to bind to plasma transferrin like injected chromium (Feng, 2007). Insulin activates transferrin receptors in cell membranes and induces a flux of chromium that leads to the formation of holochromodulin from stores of the apoprotein. Holochromodulin in turn facilitates insulin uptake by adipocytes via tyrosine kinase and thus serves as an autoamplifier in insulin signalling (Feng, 2007). Short-term ingestion of two chromium-rich soils, adding 105 and 408 mg Cr kg^{-1} DM to the diet by sheep, raised chromium concentrations in the liver from 0.25 to 0.64 and 0.90 mg kg^{-1} DM, respectively (Suttle et al., 1991), although only a small proportion was apparently absorbed. The presence of higher tissue or urinary chromium in subjects given 'organic' rather than inorganic sources (see NRC, 2005) is not evidence of superior bioavailability in a functional context, particularly when given in 'macro' doses. For example, a synthetic chelate with picolinic acid (CrPic) is highly stable at acid pH and is believed to be absorbed from the small intestine unchanged (Vincent, 2000), but chromium from the core of CrPic must be reduced before it can participate in the insulin signalling process (Vincent, 2000). The chromium methionate (CrMet) complex escapes rumen degradation (Kim et al., 2005) and Cr may be drawn into the host body's methionine pool in the same manner as selenium given as SeMet (Fig. 15.3), if the molecule is absorbed intact.

Responsiveness

The Committee on Animal Nutrition (NRC, 1995) was unconvinced that chromium supplementation is generally likely to benefit livestock. However, the addition of several organic chromium sources to livestock rations is now permitted in the USA. These have been widely studied (Lindemann, 2007), but few trials have clarified the importance of either chromium concentration or form to livestock health and performance. Changes have been reported in an increasing spectrum of metabolic, hormonal and carcass quality indices following supplementation, but results from individual studies can add little to understanding the nutritional significance of chromium because nutritional and environmental factors that induce insulin synthesis will clearly influence responses to Cr III.

Stressed calves

Recently transported feedlot calves have poor appetites and can acquire a variety of respiratory infections that cause high mortality and morbidity (Underwood and Suttle, 1999). Calves usually receive multiple vaccines to protect them from these diseases, and the addition of organic Cr III to the diet has generally improved appetite, early weight gain and humoral immune function (Chang and Mowat, 1992; Moonsie-Shageer and Mowat, 1993; Mowat et al., 1993). In a two-by-two factorial study in which an amino acid chelated chromium source (adding 0.14 mg chromium to a diet containing 0.32 mg Cr kg^{-1} DM) and multiple vaccines were the main treatments, significant improvements were mostly confined to the combined treatment with Cr III and vaccine (Wright et al., 1994). In newly weaned calves, the addition of 0.5 mg organic Cr kg^{-1} DM significantly improved antibody titres after vaccination against infectious rhinotracheitis (Burton et al., 1994), but similar studies failed to confirm this finding (Burton et al., 1993; Kegley et al., 1997b) and the basis for improved survival probably lies elsewhere. Multiple vaccines, particularly those directed against Gram-negative bacteria (e.g. *Pasteurella haemolytica*) are themselves 'stressful' and raise plasma cortisol while lowering blood glucose in recently transported lambs (Suttle and Wadsworth, 1991). Thus, vaccination may add to the stress of transportation, enhancing responses to a cortisol-limiting, insulin-facilitating Cr III supplement. In a dairy herd in Mexico, a remarkable reduction in the incidence of placental retention from an exceptional 56% to 16% ($P<0.01$; $n=25$) has been attributed to supplementation with CrPic (Villalobos et al., 1997). The disorder was attributed to the stress caused by transporting cows to delivery facilities, but its persistence in settled unsupplemented controls suggests other contributory factors, such as vaccination.

It is probably more important to vaccinate animals well in advance of transportation than to supplement them with Cr III.

Normal cattle

In 'unstressed' calves, a supplement of 0.37 mg Cr kg^{-1} DM as CrPic has been reported to lower blood cholesterol and improve glucose tolerance without improving growth, food consumption or FCE (Bunting et al., 1994). Similar responses have been reported in growing heifers given chromium propionate (CrPr, 5–15 mg day^{-1}) (Sumner et al., 2007). In steers, a small CrPic supplement (0.2 mg kg^{-1} DM) had no effects on similar blood metabolites or performance, but reduced serum and liver triglycerides (Besong et al., 2001). Young calves given supplements of 0.4 mg Cr kg^{-1} DM to a milk replacer containing 0.31 mg Cr kg^{-1} DM showed 'intensified' responses to insulin, but biochemical responses to CrCl$_3$ and chromium–nicotinic acid complex (CrNA) were similar and neither source improved growth (see Fig. 17.3) (Kegley et al., 1997a). In a study with CrNA alone, the addition of 0.4 mg Cr kg^{-1} DM only tended to improve growth in steers (10% improvement; $P<0.1$) Kegley et al., 1997b).

Periparturient cows

A supplement of 5 mg Cr mg kg^{-1} DM (as amino acid chelate) has improved the milk

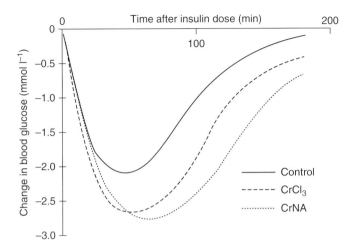

Fig. 17.3. Continuous dietary supplements of chromium chloride (CrCl$_3$) and chromium–nicotinic acid complex (CrNA) both accentuate the hypoglycaemic response of preruminant calves to intravenous insulin (Kegley et al., 1997a).

yield of primiparous cows by 7% and 13% in two experiments (Yang et al., 1996), but was of no benefit to multiparous cows. This suggests that nutritional, physiological and psychological stressors associated with the first lactation increase chromium requirements. The increased milk yield occurred during the first 2 months after parturition and further studies (Subiyatno et al., 1996) have suggested that gluconeogenesis or glycogenolysis are altered by organic chromium. Provision of 40 mg Cr kg^{-1} in a mineral supplement as CrPic from 75 days pre-partum has decreased body-weight loss in 2- to 3-year-old but not older beef cows; pregnancy rates tended to improve (Stahlhut et al., 2005). Daily doses of encapsulated CrMet (0.03–0.12 mg kg^{-1} LW$^{0.75}$) have increased dry matter intake pre- and post-partum in dairy cows, decreasing weight loss and improving glucose tolerance (Hayirli et al., 2001), but daily dosing for 49 days was arguably a stressor. The yields of milk, milk fat and lactose were claimed to show 'quadratic increases' in response to chromium, but negative quadratic responses would be a fairer description, with yields lower than in controls at the maximum level of CrMet (Fig. 17.4) and poorly correlated with plasma insulin response. Similar doses given via the diet to multiparous cows have had only modest effects on markers of carbohydrate metabolism, and effects on hepatic gluconeogenesis in vitro depended on the nature of the dietary carbohydrate source offered to the

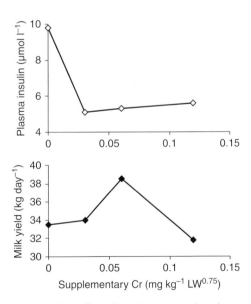

Fig. 17.4. The effect of supplementary chromium, given as Cr methionine, on plasma insulin and milk yield in dairy cows can be extremely sensitive to the precise level added (data from Hayirli et al., 2001). Unless the 'Cr potency' of a commercial ration is known, routine supplementation is unlikely to be of certain benefit.

cows (Smith et al., 2008). Under grazing conditions, CrMet (6.25 mg day^{-1}) has not improved lactation performance, although serum non-esterified fatty acids (NEFA) were decreased peri-partum and supplemented cows tended to conceive earlier (Bryan et al., 2004). Adding 10 mg day^{-1} CrPic to the pre- and immediate post-partum diet of dairy cows significantly increased food intake by 20% long after treatment ceased, but the response in milk yield was much smaller and not significant (McNamara and Valdez, 2005). The effects of chromium on rumen metabolism must be considered since CrPic alters patterns of volatile fatty acids (VFA) production by rumen microbes during in vitro culture (Besong et al., 2001). CrMet, however, 'bypasses' the rumen (Kim et al., 2005) and would presumably have to be dissociated at some later stage to attain biopotency.

In summary, the benefits from Cr III supplementation are often restricted to groups of a particular age or parity, and are of limited duration and economic benefit in peri-parturient cows. Investigations of responses to inorganic chromium and the potency of chromium inherent in commercial rations are required to form a background against which responses to organic chromium supplements can be usefully assessed. When food intake is affected by Cr III supplementation, paired feeding studies are needed to discern the specific effects of chromium on insulin function and lipid metabolism.

Pigs

A suggestion that commercial rations should be supplemented with 'bioavailable' chromium to optimize performance (Page et al., 1993) has not been confirmed in subsequent studies with growing pigs (Ward et al., 1997; Matthews et al., 2001, 2005; Shelton et al., 2003) and was hard to support even from the protagonists' data, which showed inconsistent responses to organic chromium and equal benefits from inorganic chromium (Underwood and Suttle, 1999). In the Page et al. (1993) study, CrPic consistently lowered plasma cholesterol and there was a tendency for pigs to grow more latissimus dorsi muscle and deposit less rib fat when given organic chromium. Subsequent studies have usually featured CrPic, providing 0.2 mg Cr kg^{-1} DM, and occasionally showed related metabolic and re-partitioning effects. However, their direction can depend on the energy density of the diet (Shelton et al., 2003) and they can occur in the absence of an effect on glucose tolerance (Matthews et al., 2001). Furthermore, not all measures of muscling or fatness in pigs are improved by CrPic (Boleman et al., 1995) and in one study only three (loin pH, '21-day purge loss' and carcass length) out of over 20 diverse attributes were changed by CrPic (Matthews et al., 2005). In the sow, CrPic supplementation during pregnancy may change the blood concentrations of isolated metabolic and hormonal markers, including insulin-like growth factor, without affecting the course of pregnancy or viability of offspring (Woodworth et al., 2007). The effects of stressors on the responses of pigs have received little attention. Blood glucose, lactate and cortisol concentrations measured after shipping to and exsanguination at the abattoir have shown no effect of CrPr supplementation (Matthews et al., 2005). Supplementation with CrCl$_3$, CrPic or CrNA, providing 0.2 mg Cr kg^{-1} DM, has not improved immune responses in pigs given a basal diet high in chromium (5.2 mg kg^{-1} DM) (Van Heugton and Spears, 1997). The commercial benefits of Cr III supplementation have yet to be demonstrated in pigs.

Unstressed poultry

The growth of newly hatched turkey poults, given a corn/SBM diet low in crude protein (230 g kg^{-1}), has been improved by 10% with the addition of 20 mg Cr kg^{-1} DM (as CrCl$_3$) and was accompanied by a 60% increase in hepatic lipogenesis (Steele and Roseburgh, 1981). Fermentation by-products with Cr^{3+} (supplying 5 mg Cr kg^{-1} DM) can improve egg quality in hens and protect the egg interior from harmful effects of vanadium (Jensen et al., 1978). Supplements of CrPic (0.8–3.2 mg kg^{-1} DM) have improved the growth and food intake of broilers while lowering blood glucose and NEFA concentrations and changing the distribution of body fat (increased in the liver, decreased in the abdomen) (Lien et al., 1999). Similar responses in terms of broiler performance and blood metabolites have been reported by Sahin et al. (2002), albeit on a diet high in animal fat (approximately 5.0%). In Japanese quail, adding 0.4–1.2 mg Cr as CrPic to a diet containing 1.3 mg inherent Cr kg^{-1} DM has improved food intake, egg production and egg quality, while raising plasma insulin and lowering cortisol concentrations (Sahin et al., 2001).

Stressed poultry

Commercial layers under conditions of cold (Fig. 17.5) (Sahin et al., 2001) and heat stress (Fig. 17.6) (Sahin et al., 2002) exhibit similar biochemical responses to CrPic as unstressed birds. The expected physiological changes in blood glucose, insulin, cortisol, thyroxine and triiodothyronine in response to extreme temperatures are attenuated by CrPic (0.2–0.4 mg kg^{-1} DM). In one study under cold-stress conditions, supplements of CrPic reduced biochemical evidence of peroxidation, while increasing the digestibility of dietary protein and fat (Onderci et al., 2002). However, in a two-by-two factorial study that compared responses to CrPic and zinc (adding 30 mg to the 33 mg Zn kg^{-1} DM in the basal diet), improvements in egg production and blood biochemistry were similar but not additive (Sahin et al., 2001). Responses to chromium (and zinc) in cold-stressed birds were obtained without increases in food

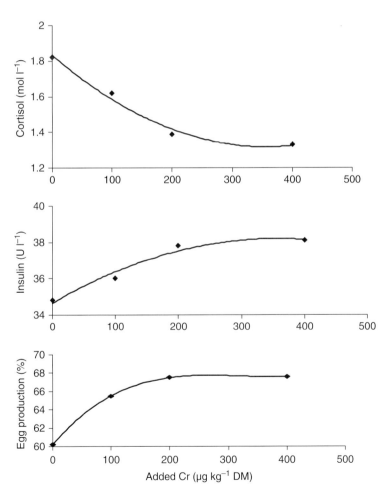

Fig. 17.5. The addition of chromium (given as chelate with picolinate) to the diet of cold-stressed laying hens lowers plasma cortisol and raises both plasma insulin and egg production (data from Sahin et al., 2001). Similar biochemical responses have been found in heat-stressed broiler chicks (Sahin et al., 2002).

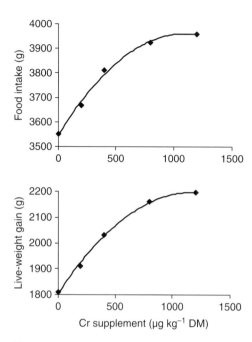

Fig. 17.6. The effect of supplementary chromium on live-weight gain and food consumption in broiler chicks after 42 days' exposure to heat stress (32.8°C) (Sahin et al., 2002).

intake, but the addition of CrPic was only advantageous to cold-stressed birds if the diet contained insufficient zinc. Qualitatively similar responses in blood insulin, glucose, cortisol and cholesterol have been found with vitamin C (250 mg kg^{-1} DM) and CrPic in cold-stressed laying hens (Sahin and Onderci, 2002). There may a common biochemical lesion, a pancreas damaged by peroxidation, whose effects are alleviated by chromium, zinc and/or vitamin C; responses in plasma malondialdehyde and digestibility to Cr III and zinc can be additive (Onderci et al., 2002), whereas those to vitamin C and chromium on plasma insulin may not (Sahin and Onderci, 2002). In summary, supplementary Cr III generally benefits poultry, whether or not they are stressed, but comparisons between inorganic and organic chromium at similar concentrations are required to demonstrate benefits of chelation.

Requirements

NRC (1995) concluded that the chromium requirements of livestock are ill-defined. Little has changed since that review because the concentrations of chromium in basal diets are not always reported and the relative potency of natural, synthetic organic and inorganic sources of chromium remains largely untested. Strenuous exercise, transportation and infection increase losses of chromium in urine and may increase chromium requirements (Anderson, 1987). Gardner et al. (1998) found that exercise in sheep, sufficient to increase energy demands by 9% over a 10-week period, halved chromium concentrations in the liver. However, supplementation of the lucerne–maize diet of heat-stressed sheep with chromium yeast (1 mg Cr kg^{-1} DM) did not alter responses in blood glucose or insulin (Sano et al., 2000). All that can be said is that the chromium concentrations commonly found naturally in pasture or feeds for housed stock (1.0–2.0 mg Cr kg^{-1} DM) have rarely limited performance in livestock other than poultry, and that the level of supplementation required for poultry is <1 mg Cr kg^{-1} DM. For stressed poultry, chromium requirements appear to be higher during heat than cold stress, but the diet, high in fat (6–7%), is also 'heat-stressed', i.e. exposed to high temperature and therefore prone to oxidation.

Status

If inorganic and organic sources of Cr III vary in functionality and metabolism then the assessment of the chromium status of both diet and livestock may present substantial difficulties. Organic sources varied in the extent to which they increased chromium concentrations in different tissues in pigs given 5 mg Cr kg^{-1} DM: CrPic and CrMet caused chromium to accumulate in bone, liver, kidney and ovary, whereas CrPr and CrY only increased levels in bone and kidney (Lindemann et al., 2008). CrPic raised tissue chromium more than CrMet and the authors concluded that CrPic had the by far the highest 'bioavailability'. However:

- differences in tissue Cr do not infer corresponding differences in functional availability (see Chapter 2, p. 30; p. 463);
- the prospect of an inorganic chromium source having a high 'bioavailability' was avoided by not including an inorganic source in the study; and

- no differences were found in metabolic profiles or carcass traits between organic chromium sources.

The adequacy of total chromium concentrations in blood or tissues as a measure of sufficiency has been questioned (Subiyatno et al., 1996), but they have rarely been reported. Just one of the papers cited above (Onderci et al., 2002) measured chromium in serum, with concentrations increasing from 0.12 to 0.15–0.24 µg l^{-1} following supplementation with CrPic. Milk contained 56 µg l^{-1} regardless of whether chromium was added to the diet of cows (Hayirli et al., 2001). Baseline values in porcine bone, kidney, liver, muscle and ovary were 34, 62, 53, 49 and 55 µg kg^{-1} DM, respectively, in Lindemann et al. (2008).

Inorganic chromium as a marker

Compounds such as chromium sesquioxide (Cr$_2$O$_3$) have a long history of use as inert markers. Chromium is almost totally excreted in the faeces and faecal analysis for chromium can provide information on food intake, digestibility of organic matter and the total faecal excretion of other minerals and nutrients when given orally in regular known doses (e.g. in sows, Saha and Galbraith, 1993). Chromium has an affinity for fibrous particles and this has been exploited by impregnating fibrous feeds with chromium (a process known as 'mordanting') when studying the digestion of fibre by ruminants (Uden et al., 1980; Coleman et al., 1984). Information on sites of absorption and rates of passage of digesta can be generated by administering chromium to suitably cannulated or fistulated animals, including grazing animals, given Cr$_2$O$_3$ continuously from an intra-ruminal capsule (Parker et al., 1990). Radioisotopes of chromium in a soluble chelated form (^{75}CrEDTA) have also been used as inert markers (Downes and McDonald, 1964), showing that the chelation of a mineral does not necessarily enhance its absorbability. In the milk-fed animal, significant absorption and urinary excretion of CrEDTA may occur, indicating a temporary enhancement of gut permeability to the molecule (N.F. Suttle, unpublished data).

Toxicity

Bioreduction of Cr^{6+} to the less toxic Cr^{3+} state is effected by many organisms (Starich and Blincoe, 1983), but may generate free radicals and that represents a potential hazard. In the case of CrPic, reduction is essential for biological potency and the risk of causing peroxidative tissue damage has been flagged (Vincent, 2000), but the threat is unlikely to be realized at the low levels of supplementation employed commercially. Chromium toxicity is rare and even soluble sources such as the chloride and chromate appear to be tolerated at concentrations of up to 1000 mg kg^{-1} DM (NRC, 2005) by livestock (Table 17.3). Tissue chromium concentrations are normally low (<0.1 mg kg^{-1} fresh weight (FW)) (Lindemann et al., 2008) and are readily distinguished from the levels associated with exposure to excess chromium (>10 mg Cr kg^{-1} FW) (Puls, 1994). However, disturbances of metabolism, in the form of inhibition of

Table 17.3. Tolerances of livestock towards occasionally beneficial elements, given as soluble salts (NRC, 2005).

Element	Pigs	Poultry	Ruminants
B	Unknown	200[L]	800
Cr (III)[Cr]	Unknown	500 (50[L])	Unknown
Li	25	25	25
Mo	Unknown	100	2–20[Cu]
Ni	250	250	100
V	<200	25 (<5[L])	>200, <400

[Cr] Tolerance of hexavalent (VI) Cr is much higher.
[Cu] Tolerance of Mo can be as little as 2 mg kg^{-1} DM on diets high in S and low in Cu and > 10-fold higher when S is low and Cu high.
[L] Tolerance of layers < growers.

cytochrome P450-linked monooxygenases and hydroxylation of testosterone, have been reported with as little as 50 mg Cr kg^{-1} DM added to the rations of laying hens (Guerra et al., 2002). The inhibition of monooxygenases was more marked with chromium added as an aminoacinate than as $CrCl_3$.

Lithium

The role of lithium in nutrition was reviewed by Mertz (1986).

Essentiality

Lithium was proposed to be an essential dietary constituent for goats in 1981, when research headed by Anke (cited by Mertz, 1986) reported growth retardation, impaired fertility, low birth weights and reduced longevity in goats on a diet containing 3.3 μg Li kg^{-1} DM; supplemented cohorts given 24 μg Li kg^{-1} DM remained healthy. However, similar responses were subsequently obtained with PTE by the same group and they may reflect hormesis rather than essentiality (see Chapter 18). Reduced fertility was also reported in second- and third-generation rats, given a low-lithium diet (5–15 μg kg^{-1} DM), consisting principally of ground yellow corn (67%). Conservation of relatively high tissue concentrations in endocrine organs in the face of depletion (57 μg Li kg^{-1} FW in adrenal, 140 μg Li kg^{-1} FW in pituitary) may indicate an endocrine function for lithium (Patt et al., 1978).

Sources

Lithium ranks 27th in abundance amongst the elements in the Earth's crust. A mean level of 28 mg Li kg^{-1} DM has been reported for soils in Eastern Europe, but values vary widely according to soil type. Concentrations are much lower in crops, particularly cereal grains, while leguminous forages and field beans are richer sources of lithium (Table 17.2). Within food types, lithium can be two- to fourfold richer in those grown on high- than on low-lithium soils, while concentrations in different cereals rank in the order rye > wheat > barley > oats (NRC, 2005).

Metabolism

The metabolism of lithium resembles that of the more abundant alkali metal, sodium. High solubility and low molecular weight allow almost complete absorption by a paracellular route, rapid turnover, extensive urinary excretion and minimal storage (NRC, 2005). Plasma lithium is normally low (<0.02 mg l^{-1}) but not protein bound and is linearly related to intake, a fact that has been exploited through the use of lithium salts as a marker for the intake of supplementary foods at pasture (Vipond et al., 1996); with dietary lithium increased to 162 mg kg^{-1} DM, plasma concentrations rose to 3–6 mg l^{-1}. Any lithium-dependent pathway would have to be protected from competition with sodium and studies of lithium status in ruminants reared under conditions of salinity would be of interest. In rats, natural sources of lithium have been reported to support lower tissue lithium than an inorganic source (lithium carbonate, $LiCO_3$) and were thought to be of lower availability (Patt et al., 1978). However, interactions with other feed constituents may have been responsible.

Toxicity

Farm livestock are rarely exposed to excess lithium, but relatively low concentrations of 50–200 mg Li kg^{-1} DM suppress appetite, with chicks being the most sensitive and cattle the least sensitive species (Table 17.3) (NRC, 2005). Occasional reports of poisoning following exposure to industrial products (e.g. lithium greases) have been reported. The symptoms in mature beef cattle were severe depression, diarrhoea, ataxia and death, and were reproduced by acute oral exposure to 500–700 mg Li kg^{-1} LW (as LiCl) (Johnson et al., 1980). Serum concentrations rose to 40–60 mg Li l^{-1}, while soft-tissue concentrations rose from 5–20 to 50–120 mg kg^{-1} FW with little difference between the sites examined (muscle, brain, heart, liver and kidney). Lower doses (250 mg Li kg^{-1} LW) cause mild diarrhoea and are sufficiently unpleasant to illicit aversive behaviour in cattle (Ralphs, 1997) and sheep (Thorhallsdottir et al., 1990). Aversion to lithium salts has been exploited by training livestock to associate their aversion to lithium with concurrent intake of toxic plants (e.g. tall larkspur), which they subsequently avoid when given a choice under range conditions (Lane et al., 1989). Livestock clearly recover well after brief exposure to sublethal lithium doses, but their aversive behaviour is quickly overridden by social influences such as the presence of untrained cohorts. In sheep, continuous exposure to lithium was necessary to maintain aversion to vermeerbos in South Africa (Snyman et al., 2002). In humans, toxicity has arisen from the use of lithium as an antidepressant by subjects on salt-restricted diets (see Johnson et al., 1980 for citations). It is possible, therefore, that both the toxic and the deficient dietary levels of lithium for livestock are influenced by sodium intake.

Molybdenum

Essentiality

Early nutritional interest in molybdenum was centred upon its profound effects on the copper metabolism of ruminants (see Chapter 11). The first indication of an essential role for molybdenum came from the discovery by de Renzo et al. (1953) that activity of the flavoprotein enzyme, xanthine oxidase (XO), depended upon its molybdenum content. Subsequently, aldehyde and sulfite oxidases were also shown to require molybdopterin cofactors (Johnson, 1997) and reduced activities, sufficient to impair function, have been observed in chicks when deficiency is exacerbated by high intakes of the molybdenum antagonist tungsten (Higgins et al., 1956; Johnson et al., 1974). In tungstate-supplemented chicks, growth is retarded, anaemia develops, tissue molybdenum values fall and the capacity to convert xanthine to uric acid is reduced. The biochemical activity of molybdenum hinges upon its ability to change between the quadri- and hexavalent states giving reduction–oxidation potential, which is linked to electron acceptors (cytochrome c, molecular oxygen, nicotinamide adenine dinucleotide (NAD$^+$)). Inherited defects that reduce the activity of one or all of the above molybdenum-dependent enzymes in humans are life-threatening (Johnson, 1997). Using a diet containing only 24 μg Mo kg^{-1} DM, Anke et al. (1978) reported on what was to become a familiar litany of defects that are alleviated, among other elements, by a small dietary supplement of molybdenum.

Sources

Soils vary widely in molybdenum content (Table 17.1) – with sandy soils at the low extreme and those of marine origin at the high extreme – and in the extent to which they yield the element to growing crops and pastures. Approximately 10% of soil molybdenum is normally extractable, but the proportion rises with soil pH and with it the concentration in crops and pastures (Fig. 2.2, p. 17). Soils can become contaminated with molybdenum by the burning of industrial wastes (up to 930 mg Mo kg^{-1} DM; Raisbeck et al., 2006). Cereals rarely contain >1 mg Mo kg^{-1} DM (Table 17.2), but pasture levels vary widely depending on soil conditions: a median value of 1.1 mg Mo kg^{-1} DM has been reported for 20 improved hill pasture sites in Scotland, but the maximum value was 60 mg kg^{-1} DM (Suttle and Small, 1993), and a cluster of raised values is found in association with outcrops of black shale in Caithness (Suttle, 2008a). The literature regarding pasture species differences is inconsistent, but they are not generally large (Table 17.2). In British Columbia, mean values in forages have been reported to range from 0.9 to 2.6 mg Mo kg^{-1} DM, with corn silage having the least molybdenum, sedge the most and legumes and grasses similar amounts, with intermediate values of 1.8 and 2.0 mg Mo kg^{-1} DM, respectively (Miltimore and Mason, 1971). However, legumes such as red clover may accumulate molybdenum when grown on molybdenum-rich soils (Kincaid et al., 1986). Animal feed sources are mostly low in molybdenum, with the exception of marine products and milk from animals grazing molybdenum-rich pastures.

Metabolism

Absorption

Molybdenum can be well absorbed from hexavalent, water-soluble molybdates (MoO_4) and herbage high in molybdenum (mostly water-soluble) by cattle (Ferguson et al., 1943) and sheep (Grace and Suttle, 1979). However, at high copper to molybdenum ratios in fibrous, high-sulfur diets, ruminants may excrete the majority of ingested molybdenum in faeces (Fig. 17.7) due to the formation of insoluble thiomolybdates (TM) in the rumen (Chapter 11). Metabolism of MoO_4 is quite different in the non-ruminant or milk-fed ruminant, being well absorbed from the stomach (Van Campen and Mitchell, 1965) or abomasum (Miller et al., 1972). Molybdate absorption across the intestinal mucosa is by an active, carrier-mediated process that is shared with and inhibited by sulfate (SO_4) (Mason and Cardin, 1977). Since little sulfur leaves the rumen as SO_4 (see Chapter 9), the antagonism of MoO_4 absorption by SO_4 will only occur naturally in animals without a functional rumen. High sulfur intakes by non-ruminants may reduce their tolerance of low molybdenum intakes.

Cellular activity

Absorbed MoO_4 is normally transported in plasma in the free ionic state but is stored in tissues as molybdopterin, bound to xanthine dehydrogenase (XDH) and aldehyde oxidase in the cytosol and to sulfite oxidase in mitochondrial membranes (Johnson, 1997). The development of anaemia in induced molybdenum deprivation probably reflects the importance of the reduction of XO activity, promoter of ferroxidase activity in both the intestinal mucosa (Topham et al., 1982b) and liver (Topham et al., 1982a). XO activities are inhibited by adding tungstate to rat diets, thereby impairing the release of iron bound to ferritin (see Chapter 13). High activities of XDH are found in the liver, endothelial cells lining blood vessels, macrophages and mast cells (Helsten-Westing et al., 1991), and in the mammary gland in close association with sulfite oxidase (Blakistone et al., 1986). Interconversion between XO and XDH, which can be catalysed by sulfite oxidase, may be important as a cellular source of peroxide and free superoxide radicals. It may thus be significant in membranotropic processes in the udder (Blakistone et al., 1986) and cause muscle damage during intense exercise in humans (Helsten-Westing et al., 1991).

Retention and excretion

Excess molybdenum is predominantly excreted via the kidneys. In the renal tubule, competitive inhibition of MoO_4 reabsorption by SO_4 can increase urinary molybdenum excretion (Bishara and Bray, 1978). The ruminant and non-ruminant will express the same antagonism because sulfur absorbed from the rumen as sulfide or from the intestine as microbial sulfur amino acids is metabolized to SO_4. The marked reduction in molybdenum retention reported by Dick (1956) in sheep given supplementary SO_4 was an artefact of the experiment. Sheep were switched from a low- to a high-sulfur diet, having built up high blood and tissue molybdenum concentrations while receiving excess molybdenum. When SO_4 and MoO_4 are added simultaneously to the diet, molybdenum retention increases when compared with sheep given MoO_4 alone (Grace and Suttle, 1979) and adverse effects mediated by the impairment of copper metabolism are exacerbated.

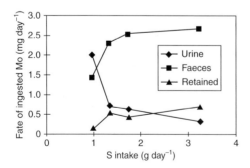

Fig. 17.7. Increases in daily sulfur intake can shift the route of excretion of a dietary molybdenum supplement (5 mg kg^{-1} DM) from urine to faeces while increasing Mo retention in sheep (Grace and Suttle, 1979).

Natural occurrence of molybdenum deprivation

Large areas of molybdenum-deficient soils exist in which yield responses to applications of molybdenum occur in crops and pastures (Anderson, 1956), but the health of livestock is rarely compromised. Chicks have been reported to tolerate diets containing only 0.2 mg Mo kg^{-1} DM (NRC, 2005). A significant growth response to added molybdenum and an improvement in cellulose digestibility have been reported in lambs fed on a semi-purified diet containing 0.36 mg Mo kg^{-1} DM. However, many pastures grazed regularly by sheep and cattle contain less molybdenum yet have no recognized adverse effects on growth or health other than enhanced copper retention in the tissues (see Chapter 11). Induction of XO may be required to trigger the inflammatory response to trauma (Friedl et al., 1989) and variations in dietary molybdenum concentrations within the range encountered by grazing ruminants can at first inhibit (at <5 mg Mo kg^{-1} DM) (Suttle et al., 1992) and then enhance (at >5 mg Mo kg^{-1} DM) (McClure et al., 1999) invasion of the ovine gut mucosa by parasitic nematodes, a perennial problem for most grazing livestock.

Molybdenum toxicity

Tolerance of high dietary molybdenum intakes in farm animals varies mainly with the species, dietary copper to molybdenum ratio, copper status of the animal and dietary concentration of sulfur, with high sulfur intakes being protective in non-ruminants but exacerbative in ruminants (Table 17.3). In ruminants, the same three-way interaction between copper, molybdenum and sulfur that determines susceptibility of the grazing animal to hypocuprosis – involving the formation of triple complexes, Cu TM, in the rumen (see Chapter 11) – determines tolerance to excess molybdenum. This makes it hard and often unnecessary to differentiate a state of molybdenum toxicity from one of induced copper deprivation in copper-responsive cattle grazing molybdenum-rich pastures. The semantics are illustrated in Fig. 17.8.

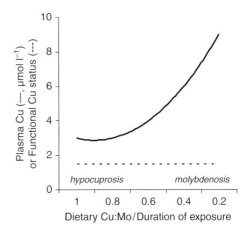

Fig. 17.8. 'Molydenosis' and 'hypocuprosis' should be used to describe clinical disorders that lie at opposite extremes of a common continuous scale. Underlying both conditions is a low functional copper status at some critical locus although the locus may change (e.g. from the hair follicle in hypocuprosis to the intestinal mucosa in molybdenosis); as the Cu:Mo ratio declines, the time taken for symptoms to develop shortens and accumulation of an abnormal trichloroacetic acid insoluble constituent in plasma causes total plasma to paradoxically increase.

Natural occurrence in cattle

Within a few days of being turned on to pastures containing 20–100 mg Mo kg^{-1} DM in Somerset, UK, cattle began to scour profusely and develop harsh, staring coats. Stock were of normal copper status, as judged by blood and liver copper analysis, yet were effectively treated with large oral doses of copper sulfate (CuSO$_4$, 1–2 g day^{-1}) or intravenous injections of 200–300 mg Cu day^{-1} (Ferguson et al., 1938). Much smaller oral supplements given via the drinking water can prevent diarrhoea (Farmer et al., 1982). The laxative effect of such pastures has led to their description as 'teart' and the condition as 'teart scours'. In Somerset, the underlying cause was the molybdenum-rich Lower Lias clay on which the pastures were established. A similar diarrhoea was subsequently described in New Zealand and termed 'peat scours' according to the association with a different soil type, but in that instance the copper status of affected stock was generally low (Cunningham, 1950).

Diarrhoea may be caused by acute, TM-induced copper deprivation in the intestinal mucosa, while low plasma and liver copper only occur after prolonged molybdenum exposure (Suttle, 1991; Fig. 17.8). There is thus a spectrum of copper-responsive conditions that extends from the acute to the subclinical, where affected animals are neither clinically nor biochemically distinguishable from those deprived of copper on molybdenum-free diets: preventative measures are also shared (see p. 283; Raisbeck et al. 2006). If left untreated, exposed cows on teart pastures conceive with difficulty. Male infertility has not been studied, but young bulls exhibit a complete lack of libido with testicular damage and little spermatogenesis when exposed to excess molybdenum under experimental conditions (Thomas and Moss, 1952). Skeletal abnormalities may also develop and appear as 'beading' of the ribs, widening of metatarsal epiphyses and lameness.

'Thiomolybdate toxicity'

In a hypothesis that has yet to pass peer scrutiny, it has been suggested that normocupraemic cattle on pastures or rations with normal dietary copper to molybdenum ratios commonly suffer from TM-induced infertility (McKenzie et al., 1997; Kendall et al., 2001). This hypothesis contradicts the model proposed in Fig. 17.8 and was generated to explain unexpectedly wide variations in the relationship between plasma copper and the oxidase activity of its principal component, caeruloplasmin, in the output from one UK diagnostic laboratory. This variation was alleged to arise from TM-induced inhibition of oxidase activity and the accumulation of trichloroacetic acid (TCA)-insoluble copper. However, neither the natural exposure of cattle to high-molybdenum pastures (Suttle, 2008a) nor the pharmacological exposure of sheep to tetrathiomolybdate (TM4), used in the prevention of chronic copper toxicity (Suttle, 2008b), inhibits oxidase activity. Poor relationships between oxidase activity and copper concentration in plasma are probably method- rather than TM-dependent. Large parenteral doses of TM4 do temporarily lower superoxide dismutase activity in sheep (Suttle et al., 1992) and repeated doses can cause histopathological changes in the pituitary (Haywood et al., 2004). In the treatment of chronic copper toxicity, TM4 must be used with caution and given only to individuals with biochemical evidence of severe liver damage and high liver copper burdens, since the copper acts as a detoxifier of TM4 while being detoxified itself.

Endocrine changes in adult ewes given 24 mg TM4 day^{-1} by continuous intravenous infusion and responses of bovine follicular cells to TM4 in vitro indicate possible mechanisms for molybdenum-induced infertility (Kendall et al., 2003) but the degree to which natural exposure was simulated is unknown. Exposure to molybdenum can be sufficient to raise molybdenum concentrations in serum and liver, while TCA-insoluble copper accumulates, yet does not harm production in sheep or cattle (Raisbeck et al., 2006). There is no specific biochemical test for 'molybdenosis'.

Natural occurrence in other grazing species

The exposure of other grazing animals to molybdenum-rich pastures is far less likely to cause diarrhoea. However, sheep exposed to high levels of dietary molybdenum in artificial diets (25 mg kg^{-1} DM) can develop diarrhoea (Suttle and Field, 1968). In addition, joint abnormalities, lameness, osteoporosis and spontaneous bone fractures have been observed in some lambs grazing pastures heavily fertilized with molybdenum (Pitt, 1976). Goats and yak (Xiao-Yun et al., 2006) can also succumb to molybdenum-induced, copper-responsive disorders, but deer are particularly tolerant (Grace et al., 2005). Tolerance may be related to the rate at which TM are generated in and flow from the rumen (Suttle, 1991). The tolerance of horses, evident from their ability to thrive on 'teart' pastures that severely affect cattle and confirmed by experiments (Pearce et al., 1999), may be explained by the extensive absorption of molybdenum prior to the site of anaerobic fermentation in the caecum. If this is the case, tolerance of molybdenum may be shared by other hind-gut fermenters.

Tolerance in non-grazing species

Pigs remain unaffected by dietary concentrations of 1000 mg Mo kg^{-1} DM, some 20–100 times the level that harms cattle (Underwood, 1977). Growth is inhibited at 200 mg Mo kg^{-1} DM in the chick (NRC, 2005) and at 300 mg Mo kg^{-1} DM in turkey poults (Kratzer, 1952);

higher levels can cause anorexia and weight loss. High tolerance is explained by the absence of a rumen to generate the TM that poison the bovine intestinal mucosa. Differences between non-ruminant species are harder to explain, but they may also involve antagonisms of copper metabolism. Exposure of the rat to molybdenum induces similar TCA-insoluble copper complexes in the plasma to those found in ruminants (N. Sangwan and N.F. Suttle, unpublished data) and these complexes are believed to restrict the tissue uptake of copper in certain circumstances. Molybdenum can be more toxic when given to non-ruminants of low than of high copper status, as is the case with ruminants, but sulfur supplements afford protection whether given as SO_4, high-protein diets, thiosulfate, cysteine or methionine (Underwood and Suttle, 1999). The least-tolerant monogastric species (avian) have low natural circulating levels of the copper-transport protein caeruloplasmin.

Nickel

Essentiality

Non-ruminants

Nickel was first shown to be essential for chicks fed on a highly purified diet under strict environmental control (Nielsen and Sauberlich, 1970). Depigmentation of the shank skin, thickened legs, swollen hocks, growth retardation, ultrastructural liver changes and anaemia have been reported in chicks given a basal diet providing only $2-15\,\mu g$ Ni kg^{-1}; an additional $50\,\mu g$ Ni kg^{-1} alleviated all evidence of deprivation (Nielsen et al., 1975). Nickel supplements also improve growth and fertility in goats (Anke et al., 1991a). The symptoms of deficiency have been confirmed in first-generation rat offspring as growth retardation, impaired reproduction and anaemia, accompanied by lowered liver activities of many enzymes involved in the tricarboxylic acid cycle and biochemical evidence of liver damage (Stangl and Kirchgessner, 1996), but a specific functional role for nickel has not been identified.

The basal diets used in these studies have all been exceedingly low in nickel ($<0.1\,mg\,kg^{-1}$ DM) and a mere $1\,mg$ Ni kg^{-1} DM is usually sufficient to restore normality. However, more nickel is needed to reverse some of the effects of vitamin B_{12} deprivation in pigs; a supplement of $6\,mg$ Ni kg^{-1} DM lowers those indices that are raised (liver iron and serum homocysteine) and raises those that are lowered (serum vitamin B_{12} and folate) by lack of vitamin B_{12} (Stangl et al., 2000). Mechanisms for the vitamin B_{12}-sparing effect of nickel have yet to be elucidated.

Ruminants

Lower organisms undoubtedly require nickel and, since these include the anaerobic rumen microbes, nickel might influence rumen metabolism. Supplementation of diets containing $0.26-0.85\,mg$ Ni kg^{-1} DM with $5\,mg$ Ni (as nickel chloride, $NiCl_3$) has increased the ruminal urease, growth rate and FCE of lambs and steers given diets high in energy and low in protein (Spears, 1984). However, nickel supplementation has also increased urease activity in either rumen liquor or rumen epithelium without affecting growth (Spears et al., 1986), methanogenesis (Oscar et al., 1987) or nitrogen metabolism (Milne et al., 1990). Nickel supplementation can alter VFA production in the rumen, but the changes have been inconsistent (Milne et al., 1990). Other potentially limiting nickel-dependent activities in rumen microbes include SO_4 reduction (Nielsen, 1996), but it may be become necessary to add vitamin B_{12} synthesis to the list. Similar biochemical abnormalities to those found in nickel-deprived rats and pigs have been reported in nickel-deprived goats reared on diets with $<100\,\mu g$ Ni kg^{-1} DM (Anke et al., 1991a), accompanied by non-specific clinical abnormalities also seen with diets low in lithium, molybdenum and some PTE (see Chapter 18).

Sources

Nickel is not abundant in soils and crops (Tables 17.1 and 17.2) and only 5% is present in easily extractable forms (Archer and

Hodgson, 1987). Concentrations of nickel in pasture grasses are lower than in soils, but legumes such as alfalfa contain more (Table 17.2). In a study of wheat seed from 12 locations in North America, Welch and Cary (1975) reported nickel concentrations ranging widely from 0.08 to 0.35, with a low overall mean of 0.18 mg Ni kg^{-1} DM. Oilseed meals are richer sources of nickel (5–8 mg kg^{-1} DM) (NRC, 2005). Bovine milk normally contains about 0.02–0.05 mg Ni l^{-1}, but concentrations are four times higher in bovine colostrum (Kirchgessner et al., 1967), a similar increase to that found for essential minerals (see Chapter 2).

Metabolism

Nickel is normally poorly absorbed (1–5% by calves; O'Dell et al., 1971), but absorption increases during pregnancy, lactation and iron deprivation, suggesting the involvement of active transport mechanisms shared with iron (NRC, 2005). Concentrations of nickel in plasma are low (<0.017 μmol l^{-1}) and it is bound chiefly to albumin, but it is histidine-bound nickel that may facilitate the uptake of nickel by cells. Tissue nickel concentrations vary from organ to organ, being much higher in the kidneys and lung than elsewhere in calves (O'Dell et al., 1971). A similar distribution (2.78 and 0.77 μmol kg^{-1} kidney and liver respectively) is found in rat pups born of dams given adequate nickel; depletion reduced these concentrations by 39% and 44%, respectively (Stangl and Kirchgessner, 1996). Supplementation of a cow's ration with 365 or 1835 mg Ni day^{-1} as nickel carbonate (NiCO$_3$) caused no increase in milk nickel (O'Dell et al., 1970a) but excess absorbed nickel is excreted via the urine rather the faeces (NRC, 2005). Small nickel supplements (5 mg Ni kg^{-1} DM) can cause sixfold increases in bovine kidney nickel to 0.3 mg Ni kg^{-1} DM, suggesting a gross dietary excess (Spears et al., 1986).

Requirements

The low levels of nickel in milk (0.2–0.5 mg kg^{-1} DM) are presumably equal to or exceed the needs of suckled calves for that element. Naturally occurring nickel deprivation in grazing livestock has never been reported and appears unlikely to occur in view of the low requirements of other species and the relatively high nickel concentrations that are commonly present in pasture plants. The relatively high cobalt requirements of ruminants on maize-based diets (p. 236) give pause for thought, in view of the low nickel content of maize (0.2 mg kg^{-1} DM) (NRC, 2005) and new interest in nickel–cobalt interactions.

Toxicity

There is a wide margin between the beneficial and toxic doses of nickel for all species, particularly non-ruminants. The susceptibility of the ruminant may be a reflection of the susceptibility of the rumen microflora to relatively low levels of nickel when administered in soluble forms (e.g. >50 mg Ni kg^{-1} DM as NiCl$_2$), with disturbance of rumen function leading to inappetence but no specific toxicity signs or pathology other than nephritis (O'Dell et al., 1970b). With less soluble forms such as NiCO$_3$, 250 mg kg^{-1} DM is tolerated by ruminants (O'Dell et al., 1971). Poultry and pigs tolerate up to 400 mg Ni kg^{-1} DM as the sulfate or acetate (NRC, 2005). In toxicity, kidney concentrations reach 30–50 mg Ni kg^{-1} DM.

Rubidium

Essentiality

The first evidence that rubidium might be essential to animal life came from application of the 'Anke goat model', which produced the familiar non-specific signs of disorder when the diet contained <0.28 mg Rb kg^{-1} DM (Anke et al., 1993). However, these findings have not been repeatable elsewhere (NRC, 2005).

Sources

Pig feeds contain 2.6–26 mg Rb kg^{-1} DM and components of the human diet also fall within this range (NRC, 2005). Cereals appear to be at the lower end of this range and forage grasses and legumes well above it (Table 17.2), but high levels may reflect contamination with soil that is relatively rich in rubidium.

Metabolism

There is a striking similarity between the metabolism of rubidium and that of potassium and there is evidence of interactions between the two elements. Like potassium, rubidium is extensively absorbed and may share the same active absorption pathway. There is further similarity in that the primary route of excretion is via the kidneys (NRC, 2005). Rats given diets containing 0.5 or 8.1 mg Rb kg^{-1} DM have shown five- to 10-fold increases in blood and tissue rubidium, but reductions in blood and tissue potassium (Yokoi et al., 1994).

Requirements and tolerance

If there is a requirement for rubidium then it is likely to be met by most natural diets. The risk of toxicity is also remote, with the tolerable concentration falling between 250 and 1000 mg Rb kg^{-1} DM (NRC, 2005). Both requirements and tolerance might be affected by interactions with potassium, which is normally present in far higher 'macro' concentrations, particularly in the diets of grazing livestock (see Chapter 7).

Silicon

Essentiality

In a study of chicks reared in a plastic isolator with a filtered air supply, those on a diet containing only 1 mg Si kg^{-1} DM became decidedly unhealthy compared with those supplemented with 250 mg Si kg^{-1} DM (Carlisle, 1986). The growth rate was reduced by 30–50%, wattles did not develop, combs were stunted and the development of skull and long bones was impaired. Silicon is involved in glycosaminoglycan formation in the cartilage and connective tissue. In bone, lack of silicon leads to reduced calcification. Weakening of connective tissues, including cartilage, causes defects at articular surfaces, and wound healing may be impaired in animals deprived of silicon (Seaborn and Nielsen, 2002). Connective tissues such as tendons and organs rich in connective tissue (e.g. aorta and trachea) have high silicon concentrations compared to other tissues and organs (11–17 versus 2–4 mg kg^{-1} FW) and here the role of silicon may be structural rather than functional (see Chapter 1). However, it is highly unlikely that silicon deprivation will occur naturally on crop- or forage-based diets.

Sources

As the second most abundant element in the Earth's crust, silicon is a ubiquitous contaminant of the environment and presents problems of nutritional excess rather than deficiency. Grasses commonly contain silicon in macro element proportions (20–40 g kg^{-1} DM) and the coarser fodder or range species up to twice that range (Fig. 17.9) (Smith and Urquhart, 1975). The leaf is richer in silicon than the stem of the rice plant. Cereal grains are equally rich in silicon, particularly the more fibrous types such as oats (Carlisle, 1986). Leguminous forages contain relatively little silicon (Fig. 17.9), sometimes as little as 1.8 g kg^{-1} DM. Silicon occurs in crops and forages in inorganic form as silica (SiO_2) and silicic acid, and also in organic forms such as pectin and mucopolysaccharides. Most animal products are low in silicon (e.g. eggs contain 3 mg Si kg^{-1} DM) and the most marginal food is milk, with only 1 mg Si l^{-1}. Risks of silicon deficiency in the suckled or milk-fed animal are theoretically possible, but are probably avoided in practice by environmental contamination. Sodium zeolite A, an aluminosilicate, has become fashionable as a food additive in livestock nutrition and as a remedy for arthritic problems in horses (O'Connor et al., 2007).

Metabolism

The analysis of silicon in feeds is complicated by the difficulty of 'digesting' the element

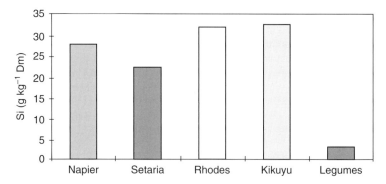

Fig. 17.9. Silicon concentrations are high in tropical grasses, but much lower in legumes (Jumba, 1989).

even with strong acids, so it is hardly surprising that silicon is often poorly absorbed. Sheep given dry gramineous or leguminous diets have been reported to excrete 99% of ingested silicon in their faeces; the absorbed silicon was mostly recovered in urine (Jones and Handreck, 1965). In grazing situations silicon is excreted predominantly via the urine, and silicon in green feeds may be far more absorbable than that in dry feeds (Nottle and Armstrong, 1966). Humans can apparently absorb >50% of the silicon in natural foods (NRC, 2005) and horses 55% (O'Connor et al., 2007); pigs and poultry are probably similar. When aluminium silicate and orthosilicic acid were compared in horses, the latter was better absorbed and retained (apparent absorption (AA_{Si}) of 41%) and additionally improved boron absorption, AA_B increasing from 46 to 66% (O'Connor et al., 2007). Silicon is partially converted to monosilicic acid in the gut and is probably absorbed and transported as such, readily entering tissues and crossing membranes in this free form (Carlisle, 1986). Silicon is uniformly distributed within the cells of soft tissues such as the liver. It is present in plants as small grains of amorphous silica (opal) or phytoliths, some fine enough to be absorbed and excreted in the urine unchanged, while some large enough to become trapped in the lymph nodes (Carlisle, 1986). Therefore, absorption does not guarantee a supply of utilizable silicon. The silicon that is absorbed is rapidly excreted and concentrations in blood do not reflect the absorbed load.

Anti-nutritional effects

Silicon toxicity is not a recognized condition, but high dietary silicon intakes have a variety of adverse effects on health and nutrition. In grazing livestock, the abrasive affects of silicate-rich soil particles may cause excessive tooth wear (Healy and Ludwig, 1965). Complexation of elements such as magnesium, iron and manganese by silicon has been described (Carlisle, 1986) and may partially explain their reduced AA when silicates are added to the diet of ruminants (Smith and Nelson, 1975; Smith and Urquhart, 1975). However, some lowering of AA_{Si} is inevitable on dry diets and on coarse forages containing 6–8% of largely unabsorbable silicon. The high urinary concentrations of silicon (some present as phytoliths) that accompany high silicon intakes by grazing animals may trigger urolithiasis. Risk is, however, governed by the volume of urine excreted and physicochemical properties, such as pH, which influence the initiation and growth of the calculus, rather than silicon intake per se. Thus, urolithiasis in sheep in Western Australia has been associated with inland rather than coastal regions and with the dry season when low water intakes give rise to maximal silica concentrations in the urine (Nottle and Armstrong, 1966). Siliceous calculi consist of polysilicic acid–protein complexes (Keeler, 1963) and these are less likely to form when the urine is high in phosphate and of low pH (Emerick, 1987).

Reciprocal antagonism between silicon and molybdenum has been reported in chicks, whereby supplements of one element reduce blood and tissue concentrations of the other (Carlisle and Curran, 1979). The addition of sodium zeolite to diets can lower the availability of dietary phosphorus, although this may be the result of antagonism from aluminium rather than silicon. Provided that dietary phosphorus is adequate, the addition of sodium zeolite A to the rations of laying hens can improve eggshell strength (Roland, 1990).

Vanadium

Essentiality

Female goats given a diet containing <10 µg V kg^{-1} DM had a higher abortion rate, produced less milk and reared fewer offspring than control goats given 2 mg V kg^{-1} DM (Anke et al., 1991b). In the rat, vanadium deprivation tends to decrease growth while increasing thyroid weight. Diets low in iodine may amplify responses to dietary vanadium (Nielsen, 1996) and those high in goitrogens may have similar effects.

Sources

Vanadium is abundant in soils, particularly in the finer clay fractions (65–200 mg V kg^{-1} DM), but only a small fraction (<10%) is extractable (Berrow et al., 1978) and levels in pasture and crops are usually low (<0.1 mg kg^{-1} DM) (Table 17.2). Natural foodstuffs contain vanadium mostly in the form of vanadyl (VO^{2+}) (NRC, 2005). Since grazing animals cannot avoid consuming soil when they graze and contamination of herbage varies from approximately 2% in summer to 10% or more of dry-matter intake in winter, soil ingestion is a major source of vanadium intake for pastured animals. Free-range hens may also ingest significant amounts of soil vanadium. Levels in the environment may be increasing due to the combustion of vanadium-rich oils (Davison et al., 1997) and use of vanadium-rich dietary phosphate supplements, which contain on average 135 mg V kg^{-1} DM and as much as 185 mg V kg^{-1} DM (Sell et al., 1982; Sullivan et al., 1994). In Minnesota, however, the prolonged use of phosphatic fertilizers has not increased the vanadium (or arsenic, chromium or lead) content of soils or maize crops (Goodroad and Campbell, 1979).

Metabolism

Animals generally absorb very little vanadium and sheep absorb only 1.6% of their dietary vanadium, whether the diet contains 2.6 or 202.6 mg V kg^{-1} DM (Patterson et al., 1986). Intravenously administered dioxovanadium is metabolized differently to the orally dosed element, raising questions as to the physiological significance of in vitro studies with vanadium. The reactivity of vanadium stems from the wide range of oxidation states that the element can enter; movements between the tetravalent and pentavalent state are the major contributor to its reduction–oxidation potential in mammalian tissues (Nielsen, 1996; Davison et al., 1997). Vanadium is readily transported across membranes by non-specific anion channels and binds to proteins such as transferrin and lactoferrin. Vanadate can substitute for phosphate, mimicking the effects of cyclic AMP. As peroxo-vanadate, vanadium influences free-radical generation and mimics insulin (Nielsen, 1996). Tissue concentrations of vanadium are normally low (<10 µg kg^{-1} FW), but increase substantially in the liver and bone when large dietary supplements of metavanadate (VO_3^-) are given (Fig. 17.10). Of more practical significance is the fact that simulated soil-ingestion studies with sheep have shown that the ingestion of as little as 4.9 or 8.3 mg soil V kg^{-1} DM for 41 days can raise kidney vanadium from undetectable levels (<0.2 mg kg^{-1} DM) to means of 0.65 or 1.85 mg kg^{-1} DM (Suttle et al., 1991). Ingested soil vanadium is therefore partially absorbed.

Toxicity

Vanadium has been studied mostly for its toxicological and pharmacological properties. Vanadium toxicity increases as the valency state of the source increases and causes diuresis,

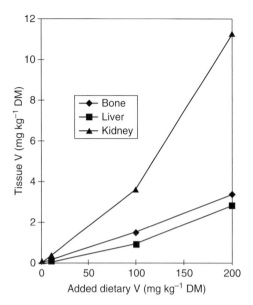

Fig. 17.10. Exposure of lambs to vanadate via the diet cause proportional increases in vanadium in liver and bone. Large increases in kidney V suggest that the urinary route of excretion is important for dealing with dietary excesses (data from Hansard et al., 1978).

natriuresis, diarrhoea and hypertension, with irreversible damage to both the liver and kidneys at high doses (Nechay, 1984). Species vary widely in their susceptibility to vanadium toxicity, with poultry the most and sheep the least susceptible of farmed species (Table 17.3). As little as 5 mg V kg^{-1} DM, added as NH_4VO_3, has been reported to retard the growth of broiler chicks given a corn/SBM diet, but previous work had indicated a tolerance to 13–20 mg V kg^{-1} DM from various sources, including superphosphates (Cervantes and Jensen, 1986). Albumen quality in the egg is particularly sensitive to vanadium supplementation, with as little as 6–7 mg V kg^{-1} DM added as dicalcium phosphate lowering egg albumin content (Sell et al., 1982). Hen body weight and egg production were both reduced during 140 days on a diet containing 50 mg added vanadium as calcium orthovanadate, while 100 mg added V kg^{-1} DM caused up to 56% mortality without specific pathology (Kubena and Phillips, 1983). Recent studies indicate even higher sensitivity to vanadium given as NH_4VO_3 (the common source, used in subsequent citations) in the laying hen. As little as 50 mg V kg^{-1} DM fed for 56 days reduced hatchability and increased the moisture content of hens' faeces (Bressman et al., 2002). Brown eggs were 'bleached' by 15 mg V kg^{-1} DM, but the effect was cancelled out by supplementation with vitamin C (100 mg) (Odabasi et al., 2006).

Growth suppression, mortality, enteritis and cystitis have been recorded in weanling pigs given 200 mg V kg^{-1} DM for 10 weeks (Van Vleet et al., 1981), but lambs tolerated 200 mg added vanadium for at least 84 days – higher levels (400 and 800 mg kg^{-1} DM) caused immediate diarrhoea and anorexia (Hansard et al., 1978). Dietary factors such as high iron, sulfur amino acid and vitamin C concentrations reduce the toxicity of vanadium (Nielsen, 1996) and ruminants may be protected by the reduction of vanadium in the rumen. Tissue concentrations of vanadium are normally relatively low (<0.25 mg kg^{-1} FW) (Puls, 1994), but are increased in lambs given diets with 202 mg V kg^{-1} DM to 3.3, 2.8 and 11.1 mg kg^{-1} FW in bone, liver and kidney, respectively, without signs of toxicity (Fig. 17.10). The increase in muscle vanadium was exceedingly small by comparison (from 0.04 to 0.41 mg kg^{-1} DM).

References

Anderson, R.A. (1956) Molybdenum as a fertiliser. *Advances in Agronomy* 8, 163–202.

Anderson, R.A. (1987) Chromium. In: Mertz, W. (ed) *Trace Elements in Human and Animal Nutrition*. Academic Press, New York, pp. 225–244.

Anke, M., Grun, M., Partschefeld, M. and Groppel, B. (1978) Molybdenum deficiency in ruminants. In: Kirchgessner, M. (ed) *Proceedings of the 3rd International Symposium on Trace Element Metabolism in Man and Animals*. Arbeitskries fur Tierernahrungsforschung, Weihensstephan, Germany, pp. 230–235.

Anke, M., Groppel, B. and Krause, V. (1991a) The essentiality of the toxic elements cadmium, arsenic and nickel. In: Momcilovic, B. (ed.) *Proceedings of the Seventh International Symposium on Trace Elements in Man and Animals (TEMA 7), Dubrovnik*. IMI, Zagreb, pp. 11–6 to 11–8.

Anke, M., Groppel, B. and Krause, V. (1991b) The essentiality of the toxic elements aluminium and vanadium. In: Momcilovic, B. (ed.) *Proceedings of the Seventh International Symposium on Trace Elements in Man and Animals (TEMA 7), Dubrovnik.* IMI, Zagreb, pp. 11–9 to 11–10.

Anke, M., Angelow, L., Schmidt, A. and Gurtler, H. (1993) Rubidium: an essential element for animal and man? In: Anke, M., Meissner, D. and Mills C. F. (eds) *Proceedings of the Eighth International Symposium on Trace Elements in Man and Animals, Dresden.* Verlag Media Touristik, Gersdorf, Germany, pp. 719–723.

Archer, F.C. and Hodgson, J.H. (1987) Total and extractable trace element content of soils in England and Wales. *Journal of Soil Science* 38, 421–431.

Armstrong, T.A. and Spears, J.W. (2001) Effect of dietary boron on growth performance, calcium and phosphorus metabolism and bone mechanical properties in growing barrows. *Journal of Animal Science* 79, 3120–3127.

Armstrong, T.A. and Spears, J.W. (2003) Effects of boron on supplementation of pig diets on the production of tumour necrosis factor-α and interferon-γ. *Journal of Animal Science* 81, 2552–2561.

Berrow, M.L., Wilson, M.J. and Reaves, G.A. (1978) Origin of extractable titanium and vanadium in the horizons of Scottish podzols. *Geoderm* 21, 89–103.

Besong, S., Jackson, J.A., Trammell, D.S. and Akay, V. (2001) Influence of supplemental chromium on concentrations of liver triglycerides, blood metabolites and rumen VFA profile in steers on a moderately high fat diet. *Journal of Dairy Science* 84, 1679–1685.

Bishara, H.N. and Bray, A.C. (1978) Competition between molybdate and sulfate for renal tubular reabsorption in sheep. *Proceedings of the Australian Society of Animal Production* 12, 123.

Blakistone, B.A., Aurand, L.W. and Swaisgood, H.E. (1986) Association between sulfhydryl oxidase and xanthine oxidase in bovine mammary tissue. *Journal of Dairy Science* 69, 2803–2809.

Boleman, S.L., Boleman, S.J., Bidner, T.D., Southern, L.L., Ward, T.L., Pontif, J.E. and Pike, M.M. (1995) Effects of chromium picolinate on growth, body condition and tissue accretion in pigs. *Journal of Animal Science* 73, 2033–2042.

Bressman, R.B., Miles, R.D., Comer, C.W., Wilson, H.R. and Butcher, G.D. (2002) Effect of vanadium supplementation on commercial egg-laying type hens. *Journal of Applied Poultry Research* 11, 46–53.

Brown, T.F., McCormick, M.E., Morris, D.R. and Zerinque, L.K. (1989) Effects of boron on mineral balance in sheep. *Nutrition Research* 9, 503–512.

Bryan, M.A., Socha, M.T. and Tomlinson, D.J. (2004) Supplementing intensively grazed late-gestation and early lactating dairy cattle with chromium. *Journal of Dairy Science* 87, 4269–4277.

Bunting, L.D., Fernandez, J.M., Thompson, D.L. and Southern, L.L. (1994) Influence of chromium picolinate on glucose usage and metabolic criteria in growing Holstein calves. *Journal of Animal Science* 72, 1591–1599.

Burton, J.L., Mallard, B.A. and Mowat, D.N. (1993) Effects of supplemental chromium on immune responses of peri-parturient and early lactation dairy cows. *Journal of Animal Science* 71, 1532–1539.

Burton, J.L., Mallard, B.A. and Mowat, D.N. (1994) Effects of supplemental chromium on antibody responses of newly weaned feedlot calves to immunisation with infectious bovine rhinotracheitis and parainfluenza 3 virus. *Canadian Journal of Veterinary Research* 58, 148–151.

Carlisle, E.M. (1986) Silicon. In: Mertz, W. (ed.) *Trace Elements in Human and Animal Nutrition*, Vol. 2, 5th edn. Academic Press, New York, pp. 373–390.

Carlisle, E.M. and Curran, M.J. (1979) A silicon–molybdenum interaction *in vivo. Federation Proceedings* 38, 553.

Cervantes, H.M. and Jensen, L.S. (1986) Interaction of monensin with dietary vanadium, potassium and protein and its effect on hepatic rubidium and potassium in chicks. *Poultry Science* 65, 1591–1597.

Chang, X. and Mowat, D.N. (1992) Supplemental chromium for stressed growing feeder calves. *Journal of Animal Science* 70, 559–565.

Coleman, S.W., Evans, B.C. and Horn, G.W. (1984) Some factors influencing estimates of digesta turnover rates using markers. *Journal of Animal Science* 58, 979–986.

COSAC (1982) Trace elements in Scottish soils and their uptake by crops, especially herbage. In: *Trace Element Deficiency in Ruminants*. Report of a Scottish Agricultural Colleges (SAC) and Research Institutes (SARI) Study Group. SAC, Edinburgh, pp. 49–50.

Cunningham, I.J. (1950) Copper and molybdenum in relation to diseases of cattle and sheep in New Zealand. In: McElroy W.D. and Glass, B. (eds) *Copper Metabolism: a Symposium on Animal, Plant and Soil Relationships, Baltimore, USA*. Johns Hopkins Press, Baltimore, Maryland, pp. 246–273.

Davison, A., Kowalski, L., Xuefeng, Y. and Siu-Sing, T. (1997) Vanadium as a modulator of cellular regulation the role of redox reactivity. In: Fischer, P.W.F., L'Abbé, M.R. and Cockell, K.A. (eds) *Proceedings of the Ninth International Symposium on Trace Elements in Man and Animals, Banff.* NRC Research Press, Ottawa, pp. 229–233.

de Renzo, E.C., Kaleita, E., Heytler, P., Oleson, J.J., Hutchings, B.L. and Williams, J.H. (1953) The nature of the xanthine oxidase factor. *Journal of the American Chemical Society* 75, 753.

Dick, H.T. (1956) Molybdenum in animal nutrition. *Soil Science* 81, 229–258.

Downes, A.M. and McDonald, I.W. (1964) The chromium–Si complex of ethylene-diamine tetra-acetic acid as a soluble rumen marker. *British Journal of Nutrition* 18, 153–163.

Emerick, R.J. (1987) Phosphate inhibition of protein–polysilicic acid complex formation *in vitro*, a factor in preventing silica urolithiasis. *Journal of Nutrition* 117, 1924–1928.

Eren, M., Uyanik and Kucukersan, S. (2004) The influence of dietary boron on egg quality and serum calcium, inorganic phosphorus, magnesium levels and alkaline phosphatase activity in laying hens. *Research in Veterinary Science* 76, 203–210.

Farmer, P.E., Adams. T.E. and Humphries, W.R. (1982) Copper supplementation of drinking water for cattle grazing molybdenum-rich pastures. *Veterinary Record* 111, 193–195.

Feng, W. (2007) The transport of Cr (III): implications for function. In: Vincent, J.B. (ed.) *The Nutritional Biochemistry of Chromium (III)*. Elsevier, Amsterdam, pp. 121–138.

Ferguson, W.S., Lewis, A.H. and Watson, S.J. (1938) Action of molybdenum in milking cattle. *Nature, UK* 141, 553.

Ferguson, W.S., Lewis, A.H. and Watson, S.J. (1943) The teart pasture of Somerset. The cause and cure of teartness. *Journal of Agricultural Science, Cambridge* 33, 44–51.

Friedl, H.P., Till, G.O., Trentz, O. and Ward, P.A. (1989) Roles of histamine, complement and xanthine oxidase in thermal injury of skin. *American Journal of Pathology* 135, 203–217.

Gardner, G.E., Pethick, D.W. and Smith, C. (1998) Effect of chromium chelavite supplementation on the metabolism of glycogen and lipid in adult Merino sheep. *Australian Journal of Agricultural Research* 49, 137–145.

Goodroad, L.L. and Campbell, A.C. (1979) Effects of phosphorus fertilizer and lime on the As, Cr, Pb and V content of soils and plants. *Journal of Environmental Quality* 8, 493–496.

Grace, N.D. and Suttle, N.F. (1979) Some effects of sulfur intake on molybdenum metabolism in sheep. *British Journal of Nutrition* 41, 125–136.

Grace, N.D., Wilson, P.R. and Quinn, A.K. (2005) Impact of molybdenum on the copper status of red deer (*Cervus elaphus*). *New Zealand Veterinary Journal* 53, 31–38.

Green, G.H. and Weeth, H.J. (1977) Responses of heifers to boron in water. *Journal of Animal Science* 46, 812–818.

Guerra, M.C., Renzulli, C., Antelli, A., Pozzetti, L., Paolini, M. and Speroni, E. (2002) Effects of trivalent chromium on hepatic CYP-linked monooxygenases in laying hens. *Journal of Applied Toxicology* 22, 161–165.

Hansard, S.L.I., Ammerman, C.B., Fick, K.R. and Miller, S.M. (1978) Performance and vanadium content of tissues in sheep as influenced by dietary vanadium. *Journal of Animal Science* 46, 1091–1095.

Hayirli, A., Bremmer, D.R., Bertics, S.J., Socha, M.T. and Grummer, R.R. (2001) Effect of chromium supplementation on production and metabolic parameters in periparturient dairy cows. *Journal of Dairy Science* 84, 1218–1230.

Haywood, S., Dincer, Z., Jasani, B. and Loughram, M.J. (2004) Molybdenum-associated pituitary endocrinopathy in sheep treated with ammonium tetrathiomolybdate. *Journal of Comparative Pathology* 130, 21–31.

Healy, W.B. and Ludwig, T.G. (1965) Wear of sheep's teeth. 1. The role of ingested soil. *New Zealand Journal of Agricultural Research* 8, 737–752.

Helsten-Westing, Y., Sollevi, A. and Sjodin, B. (1991) Plasma accumulation of hypoxanthine, uric acid and creatine kinase following exhaustive rune of differing duration in man. *European Journal of Applied Physiology* 62, 380–384.

Higgins, E.S., Richert, D.A. and Westerfeld, W.W. (1956) Molybdenum deficiency and tungstate inhibition studies. *Journal of Nutrition* 59, 539–559.

Hunt, C.D., Herbel, J.L. and Idso, J.P. (1994) Dietary boron modifies the effects of vitamin D on indices of energy substrate utilisation and mineral metabolism in the chick. *Journal of Bone and Mineral Research* 9, 171–181.

Jensen, L.S., Maurice, D.V. and Murray, M.W. (1978) Evidence for a new biological role of chromium. *Federation Proceedings* 37, 404.

Johnson, J.H., Crookshank, H.R. and Smalley, H.E. (1980) Lithium toxicity in cattle. *Veterinary and Human Toxicology* 22, 248–251.

Johnson, J.I., Rajagopalan. K.V. and Cohen, H.J. (1974) Effect of tungsten on xanthine oxidase and sulfite oxidase in the rat. *Journal of Biological Chemistry* 249, 859–866.

Johnson, J.L. (1997) Molybdenum. In: O'Dell, B.L. and Sunde, R.A. (eds) *Handbook of Nutritionally Essential Mineral Elements*. Marcel Dekker, New York, pp. 413–438.

Jones, L.H.P. and Handreck, K.A. (1965) The relation between the silica content of the diet and the excretion of silica by sheep. *Journal of Agricultural Science, Cambridge* 65, 129–134.

Jumba, I.O. (1989) Tropical soil–plant interactions in relation to mineral imbalances in grazing livestock. PhD Thesis, University of Nairobi, Chapter 5.

Kabata-Pendias, A. and Pendias, H. (1992) *Trace Elements in Soils and Plants*, 2nd edn. CRC Press, Boca Raton, Florida.

Keeler, R.F. (1963) Silicon metabolism and silicon–protein matrix interrelationship in bovine urolithiasis. *Annals of the New York Academy of Sciences* 104, 592–611.

Kegley, E.B., Spears, J.W. and Brown, T.T. (1997a) Effect of shipping and chromium supplementation on performance, immune response and disease resistance of steers. *Journal of Animal Science* 75, 1956–1964.

Kegley, E.B., Spears, J.W. and Eisemann, J.H. (1997b) Performance and glucose metabolism in calves fed a chromium–nicotinic acid complex or chromium chloride. *Journal of Dairy Science* 80, 1744–1750.

Kendall, N.R., Illingworth, D.V. and Telfer, S.B. (2001) Copper responsive infertility in British cattle: the use of a blood caeruloplasmin to copper ratio in determining a requirement for copper supplementation. In: Diskin, M.G. (ed.) *Fertility in The High-Producing Dairy Cow*. Occasional Publication No. 26, British Society of Animal Science, pp. 429–432.

Kendall, N.R., Marsters, P., Scaramuzzi, R.J. and Campbell, B.K. (2003) Expression of lysyl oxidase and effect of copper chloride and ammonium tetrathiomolybdate on bovine ovarian follicle granulosa cells cultured in serum-free media. *Reproduction* 125, 657–665.

Kim, C.H., Park, B.K., Park, J.G., Sung, K.I., Shin, J.S. and Ohh, S.J. (2005) Estimation of rumen by-pass rate of chromium methionine chelates by ruminal bacteria analysis. *Journal of Animal Science and Technology* 47, 759–768.

Kincaid, R.L., Gay, C.C. and Krieger, R.I. (1986) Relationship of serum and plasma copper and caeruloplasmin concentrations of cattle and the effects of whole sample storage. *American Journal of Veterinary Research* 47, 1157–1159.

Kirchgessner, M., Friesecke, H. and Koch, G. (1967) *Nutrition and the Composition of Milk*. Crosby Lockwood, London, p. 129.

Kratzer, F.H. (1952) Effect of dietary molybdenum upon chicks and poults. *Proceedings of the Society for Experimental Biology and Medicine* 80, 483–486.

Kubena, L.F. and Phillips, T.D. (1983) Toxicity of vanadium in female leghorn chickens. *Poultry Science* 62, 47–50.

Kurtoglou, F., Kurtoglou, V. and Coskun, B. (2001) Effects of boron supplementation of adequate and inadequate vitamin D_3-containing diet on performance and serum biochemical parameters of broiler chickens. *Research in Veterinary Science* 71, 183–187.

Kurtoglou, F., Kurtoglou, V., Celik, I., Kececi, T. and Nizamloglu, M. (2005) Effects of boron concentration on some biochemical parameters, peripheral blood lymphocytes and plasma cells and bone characteristics of broiler chicks given diets with adequate or inadequate cholecalciferol (vitamin D3) content. *British Poultry Science* 46, 87–92.

Lane, M.A., Ralphs, M.H., Olsen, J.D., Provenuza, F.D. and Pfister, J.A. (1989) Conditioned taste aversion: potential for reducing cattle loss to larkspur. *Journal of Range Management* 43, 127–131.

Lichten, L.A. and Cousins, R.J. (2009) Mammalian zinc transporters: nutritional and physiological regulation. *Annual Review of Nutrition* 29, 153–178.

Lien, T.-F., Horng, Y.-M. and Yang, K.-H. (1999) Performance, carcase characteristics and lipid metabolism of broilers affected by chromium picolinate. *British Poultry Science* 40, 357–363.

Lindemann, M.D., Cromwell, G.L., Monegue, H.J. and Purser, K.W. (2008) Effect of chromium source on tissue concentration of chromium in pigs. *Journal of Animal Science* 86, 2971–2978.

Mason, J. and Cardin, C.J. (1977) The competition of molybdate and sulfate ions for a transport system in the ovine small intestine. *Research in Veterinary Science* 22, 313–315.

Matthews, J.O., Guzik, A.C., Le Mieux, F.M., Southern, L.L., Fernandez, J.M. and Bidner, T.D. (2005) Effect of chromium propionate on growth, carcass traits and pork quality of growing-finishing pigs. *Journal of Animal Science* 83, 858–862.

Matthews, J.O., Southern, L.L., Fernandez, J.M., Pontif, J.E., Bidner, T.D. and Odgaard, R.L (2001) Effect of chromium picolinate and chromium propionate on glucose and insulin kinetics of growing barrows and on growth and carcass traits of growing-finishing barrows. *Journal of Animal Science* 79, 2172–2178.

McClure, S.J., McClure, T.J. and Emery, D.L. (1999) Effects of molybdenum intake on primary infection and subsequent challenge by the nematode parasite *Trichostrongylus colubriformis* in weaned Merino lambs. *Research in Veterinary Science* 67, 17–22.

McDowell, L.R. (2003) *Minerals in Animal and Human Nutrition.* Elsevier, Amsterdam.

McKenzie, A.M., Illingworth, D.V., Jackson, D.W. and Telfer, S.B. (1997) The use of caeruloplasmin activities and plasma copper concentrations as an indicator of copper status in ruminants. In: Fischer, P.W.F., L'Abb'e, M.R., Cockell, K.A. and Gibson, R.S. (eds) *Proceedings of the Ninth International Symposium on Trace Elements in Man and Animals.* NRC Press, Ottawa, pp. 137–138.

McNamara, J.P. and Valdez, F. (2005) Adipose tissue metabolism and production responses to calcium propionate and chromium propionate. *Journal of Dairy Science* 88, 2498–2507.

Mertz, W. (1986) Lithium. In: Mertz, W. (ed.) *Trace Elements in Human and Animal Nutrition*, Vol. 2, 5th edn. Academic Press, New York, pp. 391–398.

Miljkovic, D., Miljkovic, N. and McCarty, M.F. (2004) Up-regulatory impact of boron on vitamin D function – does it reflect inhibition of 24-hyroxylase? *Medical Hypotheses* 63, 1054–1056.

Miller, J.K., Moss, B.R., Bell, M.C. and Sneed, N. (1972) Comparison of ^{99}Mo metabolism in cattle and swine. *Journal of Animal Science* 34, 846–850.

Milne, J.S., Whitelaw, F.G., Price, J. and Shand, W.J. (1990) The effect of supplementary nickel on urea metabolism in sheep given a low protein diet. *Animal Production* 50, 507–512.

Miltimore, J.E. and Mason, J.L. (1971) Copper to molybdenum ratio and molybdenum and copper concentrations in ruminant feeds. *Canadian Journal of Animal Science* 51, 193–200.

Moonsie-Shageer, S. and Mowat, D.N. (1993) Effects of level of supplemental chromium on performance, serum constituents and immune status of stressed feeder calves. *Journal of Animal Science* 71, 232–238.

Mowat, D.N., Chang, X. and Yang, W.Z. (1993) Chelated chromium for stressed feeder calves. *Canadian Journal of Animal Science* 73, 49–55.

Nechay, B.R. (1984) Mechanisms of action of vanadium. *Annual Reviews of Pharmacology and Toxicology* 24, 501–524.

Nielsen, F.H. (1996) Other trace elements. In: Filer, L.J. and Ziegler, E.E. (eds) *Present Knowledge in Nutrition*, 7th edn. International Life Sciences Institute, Nutrition Foundation, Washington, DC.

Nielsen, F.H. (2002) The nutritional properties and pharmacological potential of boron for higher animals and humans. In: Goldbach, H.E., Rerkasem, B., Wimmer, M.A., Brown, P.H., Thellier, M. and Bell, R.W. (eds) *Boron in Plant and Animal Nutrition.* Kluwer Academic/Plenum, New York, pp. 37–49.

Nielsen, F.H. and Sauberlich, H.E. (1970) Evidence of a possible requirement for nickel by the chick. *Proceedings of the Society for Experimental Biology and Medicine* 134, 845–849.

Nielsen, F.H., Myron, D.R., Givand, S.H. and Ollerich, D.A. (1975) Nickel deficiency and nickel–rhodium interaction in chicks. *Journal of Nutrition* 105, 1607–1619.

Nottle, M.C. and Armstrong, J.M. (1966) Urinary excretion of silica by grazing sheep. *Australian Journal of Agricultural Research* 17, 165–173.

NRC (1995) *The Role of Chromium in Animal Nutrition.* National Academy of Sciences, Washington, DC.

NRC (2005) *Mineral Tolerances of Animals,* 2nd edn. National Academy of Sciences, Washington, DC.

O'Connor, C.I., Nielsen, B.D., Woodward, A.D., Spooner, H.S., Ventura, B.A. and Turner, K.K. (2007) Mineral balance in horses fed two supplemental silicon sources. *Journal of Animal Physiology and Animal Nutrition* 92, 173–181.

O'Dell, G.D., Miller, W.J., King, W.A., Ellers, J.C. and Jurecek, H. (1970a) Effect of nickel supplementation on production and composition of milk. *Journal of Dairy Science* 53, 1545–1548.

O'Dell, G.D., Miller, W.J., King, W.A., Moore, S.L. and Blackmon, D.M. (1970b) Nickel toxicity in the young bovine. *Journal of Nutrition* 100, 1447–1454.

O'Dell, G.D., Miller, W.J., Moore, S.L., King, W.A., Ellers, J.C. and Jurecek, H. (1971) Effect of dietary nickel level on excretion and nickel content of tissue in male calves. *Journal of Animal Science* 32, 769–773.

Odabasi, A.Z., Miles, R.D., Balaban, M.O., Portier, K.M. and Sampath, V. (2006) Vitamin C overcomes the detrimental effect of vanadium on egg pigmentation. *Journal of Applied Poultry Research* 15, 425–432.

Onderci, M., Sahin, N., Sahin, K. and Kilic, N. (2002) Antioxidant properties of chromium and zinc: *in vivo* effects on digestibility, lipid peroxidation, antioxidant vitamins, and some minerals under low ambient temperature. *Biological Trace Element Research* 92, 139–150.

Oscar, T.P., Spears, J.W. and Shih, J.C.H. (1987) Performance, methanogenesis and nitrogen metabolism of finishing steers fed monensin and nickel. *Journal of Animal Science* 64, 887–896.

Page, T.G., Southern, L.L., Ward, T.L. and Thompson, D.L. (1993) Effect of chromium picolinate on growth and serum and carcass traits of growing-finishing pigs. *Journal of Animal Science* 71, 656–662.

Parker, W.J., Morris, S.T., Garrick, D.J., Vincent, G.L. and McCutcheon, S.N. (1990) Intraruminal chromium controlled-release capsules for measuring intake in ruminants: a review. *Proceedings of the New Zealand Society of Animal Production* 50, 437–442.

Patt, E.L., Pickett, E.E. and O'Dell, B.L. (1978) The effect of dietary lithium levels on tissue lithium concentrations, growth rate and reproduction in the rat. *Bioinorganic Chemistry* 9, 299–310.

Patterson, B.W., Hansard, S.L., Ammerman, C.B., Henry, P.R., Zech, L.A. and Fisher, W.R. (1986) Kinetic model for whole-body vanadium metabolism: studies in sheep. *American Journal of Psychology* 251, R325–R332.

Pearce, S.G., Firth, E.C., Grace, N.D. and Fennessy, P.F. (1999) The effect of high pasture molybdenum concentrations on the copper status of grazing horses in New Zealand. *New Zealand Journal of Agricultural Research* 42, 93–99.

Pitt, M.A. (1976) Molybdenum toxicity: interactions between copper, molybdenum and sulfate. *Agents and Actions* 6, 758–769.

Puls, R. (1994) *Mineral Levels in Animal Health*, 2nd edn. Sherpa International, Clearbrook, British Columbia.

Raisbeck, M., Siemion, R.S. and Smith, M.A. (2006) Modest copper supplementation blocks molybdenosis in cattle. *Journal of Veterinary Diagnostic Investigation* 18, 566–572.

Ralphs, M.H. (1997) Persistence of aversions to larkspur in naive and native cattle. *Journal of Range Management* 50, 367–370.

Roland, D.A. (1990) The relationship between dietary phosphorus and sodium aluminosilicate to the performance of commercial leghorns. *Poultry Science* 69, 105–112.

Rossi, A.F., Miles R.D., Damron, B.L. and Flunker, L.K. (1993) Effect of dietary boron supplementation on broilers. *Poultry Science* 72, 2124–2130.

Saha, D.C. and Galbraith, R.L. (1993) A modified chromic oxide indicator ratio technique for accurate determination of nutrient digestibility. *Canadian Journal of Animal Science* 73, 1001–1004.

Sahin, K. and Onderci, M. (2002) Optimal dietary concentrations vitamin C and chromium for alleviating the effect of low ambient temperature on serum insulin, corticosterone and some blood metabolites in laying hens. *Journal of Trace Elements in Experimental Medicine* 15, 153–161.

Sahin K., Sahin, N., Onderci, M., Gursu, F. and Cikim, G. (2002) Optimal dietary concentration of chromium for alleviating the effect of heat stress on growth, carcass qualities and some serum metabolites of broiler chickens. *Biological Trace Element Research* 89, 1–12.

Sahin, N., Sahin, K. and Onderci, M. (2001) Effects of dietary chromium and zinc on egg production, egg quality and some blood metabolites of laying hens reared under low temperature conditions. *Biological Trace Element Research* 85, 47–58.

Sano, H., Konno, S. and Shiga, A. (2000) Effects of supplemental chromium and heat exposure on glucose metabolism and insulin action in sheep. *Journal of Agricultural Science, Cambridge* 134, 319–325.

Seaborn, C.D. and Nielsen, F.H. (2002) Silicon deprivation decreases collagen formation in wounds and bone and ornithine transaminase enzyme activity in liver. *Biological Trace Element Research* 89, 239–250.

Sell, J.L., Arthur, J.A. and Williams, I.L. (1982) Adverse effects of dietary vanadium, contributed by dicalcium phosphate, on albumen quality. *Poultry Science* 61, 2112–2116.

Shelton, J.L., Payne, R.L., Johnston, S.L., Bidner, T.D., Southern, L.L., Odgaard, R.L. and Page, T.G. (2003) Effect of chromium propionate on growth, carcass traits, pork quality and plasma metabolites in growing-finishing pigs. *Journal of Animal Science* 81, 2515–2524.

Silanpaa, M. (1982) *Micronutrient and the Nutrient Status of Soils: A Global Study*. Food and Agriculture Organization, Rome, p. 91.

Small, J.A., Charmley, E., Rodd, A.V. and Freeden, A.H. (1997) Serum mineral concentrations in relation to estrus and conception in beef heifers and cows fed conserved forage. *Canadian Journal of Animal Science* 77, 55–62.

Smith, G.S. and Nelson, A.B. (1975) Effects of sodium silicate added to rumen cultures on forage digestion, with interactions of glucose, urea and minerals. *Journal of Animal Science* 41, 891–899.

Smith, G.S. and Urquhart, N.S. (1975) Effect of sodium silicate added to rumen cultures on digestion of siliceous forages. *Journal of Animal Science* 41, 882–890.

Smith, K.L., Waldron, M.R., Ruzzi, L.C., Drackley, J.K., Socha, M.T. and Overton, T. (2008) Metabolism of dairy cows as affected by prepartum dietary carbohydrate source and supplementation with chromium throughout the periparturient period. *Journal of Dairy Science* 91, 2011–2020.

Snyman, L.D., Schulz, R.A., Kellerman, T.S. and Labuschagne, L. (2002) Continuous exposure to an aversive mixture as a means of maintaining aversion to vermeerbos (*Gergeiria ornativa* O. Hoffm.) in the presence of non-averted sheep. *Onderstepoort Journal of Veterinary Research* 69, 321–325.

Spears, J.W. (1984) Nickel as a 'newer trace element' in the nutrition of domestic animals. *Journal of Animal Science* 59, 823–835.

Spears, J.W., Harvey, R.W. and Samsell, L.J. (1986) Effects of dietary nickel and protein on growth, nitrogen metabolism and tissue concentrations of nickel, iron, zinc, manganese and copper in calves. *Journal of Nutrition* 116, 1873–1882.

Stahlhut, H.S., Whishant, C.S. and Spears, J.W. (2005) Effect of chromium supplementation and copper status on performance and reproduction of beef cows. *Animal Feed Science and Technology* 128, 266–275.

Stangl, G.I. and Kirchgessner, M. (1996) Effect of nickel deficiency on various metabolic parameters of rats. *Animal Physiology and Animal Nutrition* 75, 164–174.

Stangl, G.I., Roth-Maier, D.A. and Kirchgessner, M. (2000) Vitamin B_{12} deficiency and hyperhomocysteinaemia are partly ameliorated by cobalt and nickel supplementation in pigs. *Journal of Nutrition* 130, 3038–3044.

Starich, G.H. and Blincoe, C. (1983) Dietary chromium – forms and availabilities. *The Science of the Total Environment* 28, 443–454.

Steele, N.C. and Roseburgh, W. (1981) Effect of trivalent chromium on hepatic lipogenesis by the turkey poult. *Poultry Science* 60, 617–622.

Subiyatno, A., Mowat, D.N. and Yang, W.Z. (1996) Metabolite and hormonal responses to glucose or propionate infusions in periparturient dairy cows supplemented with chromium. *Journal of Dairy Science* 79, 1436–1445.

Sullivan, T.W., Douglas, J.H. and Gonzalez, N.J. (1994) Levels of various elements of concern in feed phosphates of domestic and foreign origin. *Poultry Science* 73, 520–528.

Sumner, J.M., Valdez, F. and McNamara, J.P. (2007) Effects of chromium propionate on response to an intravenous glucose tolerance test in growing heifers. *Journal of Dairy Science* 90, 3467–3474.

Suttle, N.F. (1991) The interactions between copper, molybdenum and sulfur in ruminant nutrition. *Annual Review of Nutrition* 11, 121–140.

Suttle, N.F. (2008a) Relationships between the concentrations of trichloroacetic acid-soluble copper and caeruloplasmin in the serum of cattle from areas with different soil concentrations of molybdenum. *Veterinary Record* 162, 237–240.

Suttle, N.F. (2008b) Lack of effect of parenteral thiomolybdate on ovine caeruloplasmin activity: diagnostic implications. *Veterinary Record* 162, 593–594.

Suttle, N.F., Brebner, J., Small, J.N. and McLean, K. (1992) Inhibition of ovine erythrocyte superoxide dismutase activity (ESOD; EC1.14.1.1) *in vivo* by parenteral ammonium tetrathiomolybdate. *Proceedings of the Nutrition Society* 51, 145A.

Suttle, N.F. and Field, A.C. (1968) The effect of intake of copper, molybdenum and sulfate on copper metabolism in sheep. I. Clinical condition and distribution of copper in the blood of the pregnant ewe. *Journal of Comparative Pathology* 78, 351–362.

Suttle, N.F. and Small, J.N.W. (1993) Evidence of delayed availability of copper in supplementation trials with lambs on molybdenum-rich pasture. In: Anke, M., Meissner, D. and Mills, C.F. (eds) *Proceedings of the Eighth International Symposium on Trace Elements in Man and Animals, Dresden.* Verlag Media Touristik, Gersdorf, Germany, pp. 651–655.

Suttle, N.F. and Wadsworth, I. (1991) Physiological responses to vaccination in sheep. *Proceedings of the Sheep Veterinary Society* 11, 113–116.

Suttle, N.F., Brebner, J. and Hall, J. (1991) Faecal excretion and retention of heavy metals in sheep ingesting topsoil from fields treated with metal-rich sewage sludge. In: Momcilovic, B. (ed.) *Proceedings of the Seventh International Symposium on Trace Elements in Man and Animals*. IMI, Zagreb, pp. 32-7–32-8.

Suttle, N.F., Knox, D.P., Angus, K.W., Jackson, F. and Coop, R.L. (1992) Effects of dietary molybdenum on nematode and host during *Haemonchus contortus* infection in lambs. *Research in Veterinary Science* 52, 230–235.

Thomas, J.W. and Moss, S. (1952) The effect of orally administered molybdenum on growth, spermatogenesis and testis histology of young dairy bulls. *Journal of Dairy Science* 94, 929–934.

Thorhallsdottir, A.G., Proveuza, F.D. and Ralph, D.F. (1990) Social influences on conditioned food aversions in sheep. *Applied Animal Behaviour Science* 25, 45–50.

Toepfer, E.W., Mertz, W., Royinski, E.E. and Polansky, M.M. (1973) Chromium in foods in relation to biological activity. *Journal of Agricultural and Food Chemistry* 21, 69–73.

Topham, R.W., Walker, M.C. and Calisch, M.P. (1982a) Liver xanthine dehydrogenase and iron mobilisation. *Biochemical and Biophysical Research Communications* 109, 1240–1246.

Topham, R.W., Walker, M.C., Calisch, M.P. and Williams, R.W. (1982b) Evidence for the participation of intestinal xanthine oxidase in the mucosal processing of iron. *Biochemistry* 21, 4529–4535.

Uden, P., Colucci, P.E. and van Soest, P.J. (1980) Investigation of chromium, cerium and cobalt as markers in digesta. Rate of passage studies. *Journal of Science in Food and Agriculture* 31, 625–632.

Underwood, E.J. (1977) *Trace Elements in Human and Animal Nutrition*, 4th edn. Academic Press, London.

Underwood, E.J. and Suttle, N.F. (1999) *The Mineral Nutrition of Livestock*, 3rd edn. CAB International, Wallingford, UK.

Van Campen, D.R. and Mitchell, E.A. (1965) Absorption of Cu^{64}, Zn^{65}, Mo^{99} and Fe^{65} from ligated segments of the rat gastrointestinal tract. *Journal of Nutrition* 86, 120–126.

Van Heugten, E. and Spears, J.W. (1997) Immune response and growth of stressed weanling pigs fed diets supplemented with organic and inorganic forms of chromium. *Journal of Animal Science* 75, 409–416.

Van Vleet, J.F., Boon, G.D. and Ferrans, V.J. (1981) Induction of lesions of selenium–vitamin E deficiency in weanling swine fed silver, cobalt, tellurium, zinc, cadmium and vanadium. *American Journal of Veterinary Research* 42, 789–799.

Villalobos, F., Romero, R.C., Farrago, C.M.R. and Rosado, A.C. (1997) Supplementation with chromium picolinate reduces the incidence of placental retention in dairy cows. Canadian *Journal of Animal Science* 77, 329–330.

Vincent, J.B. (2000) The biochemistry of chromium. *Journal of Nutrition* 130, 715–718.

Vincent, J.B. and Stallings, D. (2007) Introduction: a history of chromium studies, 1955–1995. In: Vincent, J.B. (ed.) *The Nutritional Biochemistry of Chromium (III)*. Elsevier, Amsterdam, pp.1–43.

Vipond, J.E., Horgan, G. and Anderson, D. (1996) Estimation of food intake in sheep by blood assay for lithium content following ingestion of lithium-labelled food. *Animal Science* 60, 513.

Ward, T.L., Southern, T.L. and Bidner, T.D. (1997) Interactive effects of dietary chromium tripicolinate and crude protein level in growing-finishing pigs provided with inadequate pen space. *Journal of Animal Science* 75, 1001–1008.

Weeth, H.J., Speth, C.F. and Hanks, D.R. (1981) Boron content of plasma and urine as indicators of boron intake in cattle. *American Journal of Veterinary Research* 42, 474–477.

Welch, R.M. and Cary, E.E. (1975) Concentration of chromium, nickel and vanadium in plant materials. *Journal of Agricultural and Food Chemistry* 23, 479–482.

Wilson, J.H. and Ruszler, P.L. (1997) Effects of boron on growing pullets. *Biological Trace Element Research* 56, 287–294.

Wilson, J.H. and Ruszler, P.L. (1998) Long term effects of boron on layer bone strength and production parameters. *British Poultry Science* 39, 11–15.

Woodworth, J.C., Tokach, M.D., Nelssen, J.L., Goodband, R.D., Dritz, S.S., Koo, S.S., Minton, J.E. and Owen, K.Q. (2007) Influence of dietary L-carnitine and chromium picolinate on blood hormones and metabolites of gestating sows fed one meal per day. *Journal of Animal Science* 85, 2524–2537.

Wright, A.J., Mowat, D.N. and Mallard, B.A. (1994) Supplemental chromium and bovine respiratory disease vaccines for stressed feeder calves. *Canadian Journal of Animal Science* 74, 287–295.

Yamamoto, A., Wada, O. and Suzuki, H. (1988) Purification and properties of biologically active chromium complex from bovine colostrum. *Journal of Nutrition* 118, 39–45.

Yang, W.Z., Mowat, D.N., Subiyatno, A. and Liptrap, R.M. (1996) Effects of chromium supplementation on early lactation performance of Holstein cows. *Canadian Journal of Animal Science* 76, 221–230.

Yokoi, K.M., Kimura, M. and Itokawa, Y. (1994) Effects of a low-rubidium diet on macro-mineral levels in rat tissues. *Journal of the Japanese Society for Nutrition and Food Science* 47, 295–299.

Xiao-Yun, S., Guo-Zhen, D. and Hong, L. (2006) Studies of a naturally occurring molybdenum-induced copper deficiency in the yak. *The Veterinary Journal* 171, 352–357.

18 Potentially Toxic Elements

Elements that are potentially toxic to animals (PTE) also have important industrial uses and the past industrial removal of PTE from the Earth's crust continues to contaminate some farm environments (e.g. Geeson *et al.*, 1998). Environmental contamination typically causes subclinical exposure, with no overt, adverse effects in livestock but significant accumulations of PTE can be found in animal tissues and produce. The insidious nature of some adverse effects of PTE in animals and man caused concern in the 1980s (Chowdhury and Chandra, 1987; Fox, 1987; Scheuhammer, 1987) and that concern has subsequently increased (World Health Organization, 2000, 2001; NRC, 2005). The perceived risks to human health have set in motion a 'hazard-reduction juggernaut' that has repercussions for animal production, including the imposition of maximum permissible levels (MPL) for PTE in livestock feeds (Table 18.1) and maximum acceptable concentrations (MAC) in edible products. These do little to protect the health of livestock, but occasionally threaten the sustainability and profitability of livestock production. The identification and regulation of contaminated environments is a major challenge.

With the exception of mercury, all of the PTE can, paradoxically, fulfil the criteria of essential elements for livestock (see Chapter 17). The validity of these criteria is re-considered at the end this chapter, along with the apparent essentiality of occasionally beneficial elements (OBE). The risk of a PTE causing toxicity in livestock depends on several factors:

- the uptake of the PTE from soil by plants;
- the inadvertent ingestion of soil by animals;
- the absorbabilities of inherent and contaminant PTE from the diet;
- the presence of dietary factors that enhance or inhibit PTE uptake and retention;
- the existence of predilection sites for accumulation in the body; and
- the duration and pattern of PTE exposure.

The influence of these factors can make nonsense of regulatory limits by disrupting the relationship between apparent PTE exposure and tissue accretion of PTE, which can vary between and within species (Table 18.1).

The soil acts as a major source and reservoir of PTE (Table 18.2) but plant uptake is generally poor and soil ingestion by foraging livestock is often the major determinant of risk to both livestock and consumer. Soil ingestion is highly variable – it is determined by seasonal, husbandry and weather-related factors (see Chapter 2) and usually contributes 2–5% of dry-matter intake (DMI), reaching maximum in outwintered animals. Contamination of pastures with fluorine and lead in an old mining area reached a maximum in winter, peak values in unwashed samples being two or more times those found in washed samples

Table 18.1. Maximum permitted level (MPL), maximum tolerable level (MTL) (NRC, 2005) and recorded ranges of tolerance (see text) for potentially toxic trace elements in complete feeds for ruminants; where different standards are adopted for pigs and poultry, they are given in parentheses.

Element	MPL (mg kg^{-1} DM)	MTL	Range of tolerance (mg kg^{-1} DM)	Factors that influence tolerance
As	2	30	Uncertain	Inorganic > organic
Cd[a]	1	10	3–30 (0.5–1.0)	Pb ↓: Cu, Zn ↑
Fl[b]	150 (100)	40 (150)	40–60[d] (225)	Inorganic < contaminant
Hg (organic)	0.1	2	Uncertain	Inorganic > organic
Pb[c]	5	100 (10)	150–800 (5–10)	Cd ↓

MTL and MPL often bear no relationship to each other and can appear illogical.
[a]The same low MPL is set for cadmium for all classes of livestock to protect the consumer, but non-ruminants are both more vulnerable to Cd exposure and a greater threat to the consumer than ruminants.
[b]More fluorine is allowed in cattle feeds than they can tolerate, yet the more tolerant pig is given better protection.
[c]Cattle and sheep are ten times more tolerant of lead than pigs and the lesser threat to consumers because they accumulate less lead from a given level of exposure, yet a common MPL is decreed.
[d]Sheep are more tolerant than cattle.

Table 18.2. Median (maximal) or ranges of concentrations of potentially toxic elements found in soils (mg kg^{-1} DM at sampling depth 0–7.5 cm).

Element	England and Wales[a]	New Zealand[b]
As	10.4 (140)	4.7 (58)
Cd	0.5 (10.5)	0.4 (1.5)
F	200–400[c]	217–454[d]
Pd	36.8 (2900)	10 (83)
Hg	0.09 (2.12)	–

Data from: [a]Archer and Hodgson (1987); [b]Longhurst et al. (2004); [c]Fuge and Andrews (1988); [d]Loganathan et al. (2001).

(Geeson et al., 1998). Studies that measure PTE concentrations only in herbage cut from contaminated pastures in summer and suggest that levels remain well within those tolerable by livestock (e.g. Gaskin et al., 2003) provide incomplete assessments of the risk both to grazing livestock and humans. Animals that are exposed to PTE go through phases of accretion and dysfunction before clinical toxicity occurs (see Fig. 3.3, p. 44).

Aluminium

Aluminium is the most abundant mineral in most soils so it is hardly surprising that it presents problems of excess and not deficiency to livestock under farming conditions.

Sources

Aluminium constitutes 3–6% of most soils, but concentrations in the soil solution and groundwater remain low (<2.25 mg l^{-1}) because the element is largely present in soils in insoluble siliceous complexes. Concentrations of aluminium in uncontaminated crops and forages are probably in the range 2.5–10.0 mg kg^{-1} DM, judging from levels found in human foods (Greger, 1993). However, trees, ferns and tropical plants are 'aluminium accumulators' and may contain 3–4 g Al kg^{-1} DM (Underwood and Suttle, 1999). The principal source of aluminium exposure for grazing livestock is from soil-contaminated pastures; aluminium may constitute 0.3–1.2% of DM because soil can constitute 10–20% of DM intake (I) (see Chapter 2). The act of grazing itself can increase the degree of contamination of pasture with soil and aluminium (Robinson et al., 1994). High concentrations in the finest soil particles (Brebner et al., 1985) ensure that aluminium, like silicon, is an ubiquitous contaminant of the farm environment and the addition of soil to continuous cultures of rumen microbes raises the soluble aluminium concentration in the effluent (Brebner, 1987). Aluminium may also enter the diet through the use of contaminated mineral supplements (e.g. soft phosphate, 70 g Al kg^{-1} DM), feed-pelleting agents such as bentonite (110 g Al kg^{-1} DM) or 'aids to digestion' such as zeolite (60 g Al kg^{-1} DM). It may enter the environment following the treatment of public water supplies with alum (NRC, 2005).

Metabolism

Less than 1% of ingested aluminium is normally absorbed (A) and retained by weaned animals – even from simple, relatively soluble salts, such as aluminium chloride (AlCl$_3$) – and urinary excretion, which takes precedence over biliary excretion, only becomes important when intakes are excessive (Alfrey, 1986; Greger, 1993; NRC, 2005). Absorption can take place by passive paracellular processes or active processes involving non-specific mucosal metal-binding proteins (NRC, 2005). Organic acids such as citric acid enhance A_{Al} in non-ruminants (NRC, 2005), but the effect of weaker acids produced by the rumen is unknown. As far as ingested soil is concerned, aluminium in coarse particles is probably trapped in the rumino-reticulum, while aluminium in fine particles is rapidly excreted in faeces (Brebner et al., 1985). Thus, major soft tissues normally contain only 2–4 mg Al kg^{-1} DM and the exposure of lambs to 2 g Al kg^{-1} DM for 2 months raised values by no more than 2 mg kg^{-1} DM, with liver the most- and brain the least-affected tissue (Valdivia et al., 1982). Much higher concentrations (10–80 mg Al kg^{-1} DM) have been reported in silicon-rich tissues such as the lung, thymus, aorta and trachea, prompting suggestions of related functions for the two elements (Carlisle and Curran, 1993). However, these findings might reflect common forms (aluminium silicates) and routes (inhalation, ingestion) of entry for the two elements. The anaemia of iron deficiency may increase tissue aluminium by reducing the rate of aluminium turnover rather than increasing A_{Al} (NRC, 2005).

Toxicity

Concerns over the pathogenicity of aluminium in the aged or renal dialysis patient obscure the fact that aluminium, even when ingested in reactive forms such as AlCl$_3$, is of low toxicity to animals because the small proportion of the dietary aluminium intake that is absorbed is well excreted by healthy kidneys. If aluminium does present a hazard to farm livestock, it is likely to arise in free-range animals, both ruminant and non-ruminant.

Non-ruminants

The addition of 1 g Al (as Al$_2$(SO$_4$)$_3$) kg^{-1} DM to the diet of laying hens for 4 months does not reduce egg production and serves as an antidote to fluoride toxicity. However, lower exposure (0.6 g) inhibits calcium absorption in chicks via the regulation of calbindin (see Chapter 4) at the post-transcriptional level and inhibits the calbindin response to calcium depletion (Cox and Dunn, 2001). A note of caution is sounded over the use of citrate to improve phosphorus utilization in pigs and poultry rations (see p. 155) because it may increase A_{Al}.

Ruminants

An antagonism of phosphorus metabolism from the ingestion of aluminium-rich soil (Krueger et al., 1985) has been implicated in phosphorus deficiency in cattle grazing tropical pastures (McDowell, 2003). When >2 g Al kg^{-1} DM (as AlCl$_3$) was added to the diet of lambs, appetite and phosphorus absorption decreased, particularly on a low-phosphorus diet, but no clinical abnormalities developed during 2 months of exposure (Valdivia et al., 1982). The interaction may involve the formation of insoluble and unavailable Al$_2$(PO$_4$)$_3$ complexes in the gut (NRC, 2005). However, no evidence of hypophosphataemia or poor bone mineralization has been found during long-term soil-ingestion studies with diets adequate in phosphorus (N.F. Suttle, unpublished data) or in short-term studies with cattle (Allen et al., 1986). The addition of 1.3 g Al kg^{-1} DM, in the form of soil with high sorptive capacity for phosphorus, increased the apparent absorption (AA) of phosphorus in lambs (Garcia-Bojalil et al., 1988), but adsorbed phosphorus may have been retained on soil particles in the foregut. It is possible that A_{Al} is higher during late gestation or lactation and also for the suckling, but there are no data for livestock. High intakes of aluminium have been implicated in the aetiology of hypomagnesaemic tetany in cattle, and intra-ruminal doses of $Al_2(SO_4)_3$, equivalent to 4 g Al kg^{-1} DM, cause rapid reductions in serum magnesium that precede appetite suppression (Allen et al., 1984). Furthermore, additions of 2.9 mg Al kg^{-1} DM as the citrate reduce plasma magnesium while increasing urinary calcium excretion and causing only a small reduction in appetite (Allen et al., 1986). However, similar concentrations of aluminium given as soil have no effect on plasma magnesium. Grazing livestock can clearly withstand considerable and sustained intakes of soil-borne aluminium (Robinson et al., 1984; Allen et al., 1986; Sherlock, 1989). The absorbability and antagonism of aluminium from tropical shrubs, used as forages, merits investigation. The accumulation of aluminium in edible farm products poses no threat to the human consumer.

Arsenic

Arsenic was once surpassed only by lead as a toxicological hazard to farm livestock (Selby et al., 1977). Most incidents arose from the exploitation of such toxic effects to control pests (rodenticides), common pathogens of livestock (including coccidia and cestodes) and crops (insecticides and fungicides). The incidence of arsenic toxicity in Europe fell markedly following the withdrawal of most arsenic-containing products from the marketplace, although arsenic remains a permitted ingredient of the wood preservative copper chrome arsenic. In view of its long and colourful history as a poison, claims that arsenic was an essential dietary constituent (Nielson 1996) were hard to swallow and have been disputed (NRC, 2005). The recent recognition of arsenic as a carcinogen has greatly complicated the regulatory situation. Traditionally, any trace of a carcinogen is undesirable and so are foodstuffs that might add significant amounts of arsenic to the human diet.

Sources

The natural abundance of arsenic in agricultural soils was low according to one study in England and Wales, with a scatter of high values associated with mining areas (Table 18.2). In a more recent study in south-west England, values as high as 727 mg As kg^{-1} DM were recorded in an area of 722 km^2 where arsenic occurs with other metals, such as tin and copper in geological deposits (Thornton, 1996). Mining and smelting activities, dating back centuries, distributed those metals in the environment. A recent survey in New Zealand found far less soil arsenic (mean 5.4 mg kg^{-1} DM) and no evidence of enrichment associated with farming (Longhurst et al., 2004). Plants usually contain <0.2 mg As kg^{-1} DM in New Zealand, but peat soils, relatively rich in arsenic (approximately 10 mg kg^{-1} DM), support grass with more arsenic than other soils (0.24 mg kg^{-1} DM). Cereal grains contain little arsenic (<0.2 kg^{-1} DM) but marine sources are much richer, with fish meals containing 2–9 mg As kg^{-1} DM (NRC, 2005) and seaweeds or their derivatives 35 mg As kg^{-1} DM. Inorganic arsenic has been added to pig and poultry rations as a growth stimulant (e.g. Proudfoot et al., 1991). The MAC of arsenic in drinking water in the European Community is 50 µg As l^{-1}, but 'cancer scares' have led authorities in the USA to reduce that level to 10 µg As l^{-1}. In arseniferous areas, levels in groundwater can be measured in milligrams (NRC, 2005)! For their part, the European Communities (2002) have reduced permissible levels in air-dry animal feeds to 2 mg As kg^{-1}, making exceptions for some forage by-products (<4 mg As kg^{-1}). However, low arsenic intakes may predispose to colonic cancer by causing hypomethylation (NRC, 2005). Balancing the needs of animal production and human health is an issue returned to in Chapter 20.

Metabolism

Like most elements that exist primarily in the anionic state, arsenic is well absorbed (>90%) from inorganic forms by non-ruminants

(NRC, 2005) but lambs absorbed only 0.45 of a labelled oral dose of $^{75}AsCl_3$ (Beresford et al., 2001). The relative rates of tissue arsenic accretion were ranked in the following order: kidneys > liver > muscle (Beresford et al., 2001). The bioavailability of arsenic ingested as contaminated soil is relatively low (Casteel et al., 2001). Organic forms of arsenic such as those in fish meals are also highly absorbed, but arsenic is well excreted in the urine and tissue accretion of arsenic is therefore slow, occurring mainly in the liver, kidneys, and skin and its appendages (Selby et al., 1977). An early literature review (Doyle and Spaulding, 1978) reported arsenic concentrations of 0.02–0.19 µg g^{-1} wet matter (WM) for the liver, 0.01–0.15 µg g^{-1} WM for the kidneys and 0.01–0.11 µg g^{-1} WM for muscle (multiply by 13.35 for µmol kg^{-1}) of cattle, pigs and sheep. Higher values for poultry (0.70, 0.28 and 0.09 µg g^{-1} WM, respectively) were attributed to the widespread use of arsenicals in poultry feeds. The implementation of brief withdrawal periods prior to slaughter prevents such increases in tissue arsenic. The high arsenic concentrations produced in the liver and thigh muscle of broiler chickens (1.23 and 0.41 µg g^{-1} WM) by feeding arsanilic acid at 'growth stimulant' levels (90 mg kg^{-1} DM) for 42 days are reduced to background concentrations (0.14 and 0.23, respectively) 7 days after withdrawing the additive (Proudfoot et al., 1991). Significant differences in arsenic concentrations between muscles were reported, with levels in breast 60% or more lower than in thigh muscle. More importantly, no growth stimulation was evident by the end of the study. Therefore, the benefits of adding arsenicals to poultry rations is questionable and their use as growth promoters is banned in most countries.

Toxicity to livestock

Cases of arsenic toxicity are now extremely rare. Symptoms vary with the rate and duration of exposure and the source of arsenic involved (Bahri and Romdane, 1991). Water-soluble inorganic salts such as arsenites and arsenates are the most toxic, particularly the trivalent forms, while the insoluble elemental form is non-toxic (Arnold, 1988; NRC, 2005). There are three mechanisms by which arsenic disturbs normal metabolism: impaired methylation through inhibition of sulfhydryl enzyme systems; oxidative stress though inhibition of keto acid oxidation; and antagonism of other elements (NRC, 2005).

Acute toxicity

Accidental exposure to large oral doses of inorganic arsenic causes acute toxicity in which the primary symptom is gastroenteritis. A rapid drop in blood pressure causes collapse, accompanied by muscular twitching, convulsions and gastrointestinal haemorrhage in exposed sheep (McCaughey, 2007). Confirmation of arsenic poisoning is provided by the presence of high arsenic levels in the tissues and body fluids and also in the gut contents if exposure has been recent. In poisoned horses, arsenic concentrations were higher in liver than kidney (11–14 and 4.2–6.5 mg kg^{-1} fresh weight (FW), respectively; Pace et al., 1997), some 30-fold greater than normal (Puls, 1994). Where the route of exposure has primarily involved the lungs or skin, the focus of pathology shifts to the liver and kidneys, which will both show severe necrosis.

Chronic toxicity

Chronic toxicity is most commonly caused by over-exposure to organic arsenic sources of medium (e.g. phenylarsonic compounds) to low (e.g. arsanilic acid) toxicity in arsenical acaricides, rodenticides and feed additives. Early signs are those of neurological disorder, such as incoordination, swaying and ataxia ('drunken hog syndrome'), but affected animals remain alert and continue to eat and drink. Later, extreme agitation may be shown by pigs, which scream with their noses pressed to the ground when they are disturbed. Demyelination of some peripheral nerves may be evident histologically (Kennedy et al., 1986). The biochemical confirmation of arsenic poisoning depends on the species and the characteristics of arsenic exposure, notably pattern and source (Puls, 1994).

Causes

Arsenical sheep dips have posed a problem because the product becomes trapped in the fleece and absorbed through the skin (McCaughey, 2007). There are a few areas, such as south-west England (Thornton, 1996), where the disturbance of arsenic-rich deposits during mining has led to pollution of the soil, atmosphere or groundwater, increasing the risk of toxicity in grazing livestock. In the case of groundwater, the disturbance of natural drainage systems by pipe-laying operations can further change the distribution of arsenic in the affected area. For other livestock, problems can still arise in countries (such as the USA) that permit the use of organo-arsenicals at high doses for the control of infections and at low doses as growth stimulants. For example, arsanilic acid is used at 45–100 mg As kg^{-1} DM as a growth stimulant for pigs and at 200–250 mg As kg^{-1} DM for the control of swine dysentery, while 400–2000 mg As kg^{-1} DM can cause chronic toxicity (NRC, 2005). When pigs have been given the organo-arsenical roxarsone, 53.4 mg As kg^{-1} DM was toxic (Rice et al., 1985). Excessive intakes of arsenic can result from errors in feed formulation, extended duration of treatment or the pharmacological use of arsenic in the treatment of dehydrated and debilitated animals.

Cadmium

Cadmium is a highly reactive and toxic element that is sparsely distributed in most agricultural ecosystems (Pinot et al., 2002). Once absorbed by animals or humans, however, cadmium is poorly excreted and increasing efforts are being made to limit the entry of cadmium into the human food chain. Of particular concern is the contamination of pastures with cadmium derived from fertilizers and its subsequent fate in grazing livestock (Loganathan et al., 2008).

Sources

Soils, crops and contaminants

The soils of England and Wales have a low median cadmium value (Table 18.2), similar to that found in New Zealand (Bramley, 1990; Longhurst et al., 2004), but in both countries some soils have become enriched threefold by applications of cadmium-rich inorganic (e.g. Loganathan et al., 2008) or organic fertilizers (e.g. Brebner et al. 1993) and dispersal of wastes from the mining and smelting of metals such as zinc and lead. The uptake of cadmium by plants is generally poor, particularly from clay soils, and forages and crops grown on normal soils usually contain <1 mg Cd kg^{-1} DM. Cadmium is strongly retained in the topsoil and, if contamination continues, soil cadmium concentrations slowly increase. Plant cadmium concentrations also rise, but the increase depends on soil pH, plant species and the part of the plant sampled (Van Bruwaene et al., 1984). Uptake of cadmium by sheep grazing pastures on acidic, sandy soils is higher than that on alkaline, clay soils (Morcombe et al., 1994b).

Superphosphates

The most important exogenous sources of cadmium are superphosphate fertilizers. These vary in concentration from <5 to 134 mg Cd kg^{-1}, depending on country or region of origin (Bramley, 1990; Sullivan et al., 1994). The regular use of cadmium-rich Pacific sources of superphosphate in Australia and New Zealand has led to increases in soil and pasture cadmium sufficient to raise kidney cadmium in grazing lambs above the MAC (>1 mg kg^{-1} FW) (Bramley, 1990; Morcombe et al., 1994a). In New Zealand, soil cadmium is correlated with soil phosphorus, indicating the importance of phosphorus fertilizers as a major source of cadmium enrichment (Longhurst et al., 2004). Pasture cadmium can be contained within acceptable limits by avoiding cadmium-rich sources of superphosphate and probably by liming, which reduces cadmium uptake by cereals (Oliver et al., 1996). The ingestion of fertilizer and topsoil during grazing adds to the body burden of cadmium in the grazing animal (Stark et al., 1998).

Municipal sewage sludges

Sewage sludges contain variable and occasionally excessive cadmium concentrations (up to

20 mg kg^{-1} DM) and bans on dispersal to sea in some countries have increased dispersal on agricultural land (Hill *et al.*, 1998a). In Europe, guidelines aim to restrict the accumulation of cadmium in sludge-amended soils to <3 mg kg^{-1} DM (sampling depth, 20 cm) (European Communities, 2002). Under the flooded rice culture practised in the Po valley in Italy, soil cadmium concentrations have risen by up to 6% per annum (Van Bruwaene *et al.*, 1984).

Metabolism

There is a vast literature on the biochemical responses of animals to cadmium, but most of it has little or no nutritional significance because of the high dietary concentrations used or the total bypassing of the protective gut mucosa (by parenteral administration) during experiments.

Absorption

Absorption of cadmium (A_{Cd}) by weaned animals is low in *all* species. In ruminants, AA_{Cd} is <1% for inorganic cadmium salts (Neathery *et al.*, 1974; van Bruwaene *et al.*, 1982, 1984; Houpert *et al.*, 1995). With sludge-amended soil (Suttle *et al.*, 1991) or sewage sludge (Hill *et al.*, 1998a) providing the cadmium AA_{Cd} is not significantly different from zero, although accretion of cadmium in tissues indicates that some absorption must take place. Inorganic cadmium sources added to milk are relatively well absorbed (Bremner, 1978), but AA_{Cd} is still only 4% in neonatal pigs (Sasser and Jarboe, 1980) and cadmium is not transferred naturally via the dam's milk (Suttle *et al.*, 1997). An extraordinarily high general A_{Cd} of 15% for cattle cited by the NRC (2005) is highly misleading and unsupported by the literature. The cadmium in superphosphates is of lower availability to pigs than that in cadmium chloride (CdCl$_2$) (King *et al.*, 1992). The addition of phytase to the diet (Rimbach *et al.*, 1996) and deficiencies of minerals, including calcium, iron and zinc (Kollmer and Berg, 1989), increase A_{Cd}, while excesses (including manganese) reduce (NRC, 2005).

Retention and maternal transfer

The small proportion of dietary cadmium that is absorbed after oral pulse doses of ^{109}Cd is avidly retained: of the 0.75% absorbed by cows after 14 days, 17% is still present in the tissues after 131 days; the 0.15–0.5% absorbed by sheep has a half-life for elimination from the body of 100–150 days (Houpert *et al.*, 1995). In a short-term (4-day) study with Japanese quail, the proportion of an oral cadmium dose retained increased with the size of the dose (Scheuhammer, 1987), but this may reflect disruption of control mechanisms at high doses (NRC, 2005). The principal sites of cadmium retention are the gastrointestinal tract, liver and kidneys (Neathery *et al.*, 1974; Houpert *et al.*, 1995); in the kidneys, the half-life is measured in years. In a 191-day study with lambs, liver cadmium was not linearly related to dietary cadmium, which ranged up to 60 mg kg^{-1} DM (Doyle *et al.*, 1974). Control of cadmium retention is achieved largely by metallothionein (MT), a cysteine-rich protein that is the major cadmium-binding protein in the body and transports cadmium from the liver to the kidneys. Exposure to cadmium induces the synthesis of MT at many sites, including the gut mucosa and liver, in what is believed to be a protective role, but cadmium–MT is nephrotoxic (Bremner, 1978). Cadmium is not readily transferred across either ovine (Mills and Dalgarno, 1972; Suttle *et al.*, 1997) or bovine (Neathery *et al.*, 1974; Smith *et al.*, 1991a) placenta and mammary glands or across the avian oviduct (Sell, 1975; Stoewsand *et al.*, 2005).

Interactions

Cadmium shares important biochemical properties with other metals, such as binding to MT (shared with copper and zinc; see Chapter 11), divalent metal transporter 1 (with iron, manganese and lead; see Chapter 1) and calbindin (with calcium and aluminium; see Chapter 4). These elements are thus often mutually antagonistic, with exposure to cadmium lowering copper status and exposure to zinc lowering cadmium status (see Chapter 3) (Bremner, 1978). The addition of phytate (5 g kg^{-1} DM) to a rat diet supplemented with 5 mg Cd kg^{-1} DM has been reported to increase liver cadmium threefold in what was believed to be a secondary response to a lowering of A_{Ca} and induction of calbindin in the gut mucosa (Rimbach et al., 1996). Concurrent exposure of pigs to cadmium (1.0 or 2.5 mg kg^{-1} DM) and lead (5, 10 or 25 mg kg^{-1} DM) results in higher tissue cadmium than exposure to cadmium alone (Phillips et al., 2003). Cadmium also interacts with other minerals through other mechanisms. The administration of selenium lowers the toxicity of cadmium by shifting the distribution of tissue cadmium from MT towards high-molecular-weight proteins (Bremner, 1978). Raising dietary sulfur from a moderate (1.9 g kg^{-1} DM) to a very high level (5.9 g kg^{-1} DM) has been found to reduce cadmium accumulation in the liver and kidneys of sheep by 60% (Smith and White, 1997). Molybdenum was less effective and the two antagonists did not act additively, providing a contrast with the synergistic antagonism of copper (see Chapter 11).

Chronic toxicity in livestock

Primary toxicity

The listed mechanisms of primary cadmium toxicity are many and varied and include disruption of the cell reduction–oxidation cycle and calcium signalling (Pinot et al., 2002; NRC, 2005), but they have mostly been studied under conditions of acute exposure, sometimes in cell cultures. Puls (1994) gives normal, high and 'toxic' levels of cadmium for different species; those of agricultural interest are summarized in Table 18.3. Blood cadmium is only marginally increased above normal (>0.01 μg l^{-1}) and is analytically problematic and diagnostically unhelpful in assessing cadmium toxicity, but excretion of MT in urine is indicative of cadmium exposure. In avian species, dietary concentrations of 75 mg Cd kg^{-1} DM commonly cause kidney damage and delayed maturation of the testes in males (Scheuhammer, 1987). In the laying hen, exposure to diets containing 48 (Leach et al., 1979) and 60 mg Cd kg^{-1} DM (Sell, 1975) has reduced egg production. With diets or animals of normal or high copper and zinc status, relatively high cadmium concentrations are required to cause toxicity, the major abnormality being nephrotoxicity. A diagnosis of primary cadmium toxicosis is confirmed by elevated kidney cadmium and histological evidence of kidney damage (Scheuhammer, 1987). Diagnostic problems arise because the symptoms of secondary cadmium toxicity involving interactions with copper and zinc are indistinguishable from those caused by straightforward deficiencies of copper and

Table 18.3. The likelihood of cadmium toxicity in farm livestock can be assessed from cadmium concentrations in the diet, liver or kidneys. Normal (N), high (H) and toxic (T) ranges are given (chiefly from Puls, 1994).

Species	Cd level	Diet (mg kg^{-1} DM)[a]	Liver (mg kg^{-1} FW)[a]	Kidney (mg kg^{-1} FW)[a]
Cattle and sheep	N	0.1–0.2	0.02–0.05	0.03–0.10
	H[b]	0.5–5.0	0.1–1.5	1.0–5.0
	T	>50	50–160	100–250
Pigs	N	0.1–0.8	0.1–0.5	0.1–1.0
	H[b]	1.0–5.0	1.0–5.0	2.0–5.0
	T	>80	>13	<270
Poultry	N	0.1–0.8	0.05–0.5	0.1–1.5
	H[b]	1.0–5.0	1.0–50	2.0–10.0
	T	>40	>15	>70

[a]Multiply by 8.9 to obtain units in mmol kg^{-1} DM.
[b]High dietary cadmium levels may lead to unacceptably high concentrations in offal. Maximum acceptable concentrations vary from country to country, but are often 1 mg Cd kg^{-1} FW and likely to fall. Cadmium may also induce copper or zinc deficiencies if the respective dietary concentrations are marginal.

zinc. The level of exposure required for their manifestation is lower but highly variable and susceptibility varies from species to species (see Chapter 2).

Secondary toxicity: ruminants

In early studies employing high dietary cadmium (>40 mg kg^{-1} DM), the symptoms of cadmium toxicity were often similar to those of zinc deprivation and were prevented by zinc supplementation (Powell et al., 1964). They included loss of appetite, poor growth, retarded testicular development and parakeratosis in sheep, yet liver and kidney zinc increased (Doyle and Pfander, 1975). Later studies with lambs using diets marginal in zinc showed that growth was retarded by a less severe cadmium challenge (3.4 mg Cd kg^{-1} DM) and prevented by a small zinc supplement (150 mg Zn kg^{-1} DM) (Campbell and Mills, 1979; Bremner and Campbell, 1980): secondary zinc deprivation had been induced. Similar cadmium exposure of pregnant ewes and/or lambs (3.0–3.4 mg Cd kg^{-1} DM) using a diet adequate in zinc induced anaemia, and impaired bone mineralization, loss of wool crimp, abortion and stillbirths – all signs associated with copper deprivation (see Chapter 11), although supplementary zinc (750 kg^{-1} DM) was still protective (Mills and Dalgarno, 1972; Dalgarno, 1980): secondary copper deprivation had been induced. Housed lambs tolerate 15 mg Cd kg^{-1} DM (30–60 mg retards growth) when given a diet that maintains high liver copper concentrations (Doyle et al. 1974). In studies in cattle, a supplement of 5 mg Cd kg^{-1} DM fed throughout pregnancy has been reported to reduce liver copper in the newborn calf by 29%, while 1 mg Cd kg^{-1} DM was sufficient to reduce liver copper in the dam by 40% (Smith et al., 1991a).

The high susceptibility of grazing ruminants to copper deprivation (see Chapter 11) makes them more vulnerable than other species to the antagonistic effects of cadmium.

Secondary toxicity: non-ruminants

Interactions between cadmium, zinc and copper can occur in non-ruminants. When rats on a diet marginal in copper (2.6 mg kg^{-1} DM) were given one of four cadmium levels at each of three zinc levels, each antagonist was most inhibitory towards copper when the other was *not* added (Fig. 18.1) (Campbell and Mills, 1974). Cadmium was by far the more potent antagonist, with a supplement of 6 mg Cd kg^{-1} DM having the same inhibitory effect as 1000 mg Zn kg^{-1} DM. Supplee (1961)

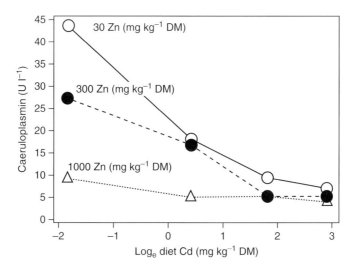

Fig. 18.1. When rats are concurrently exposed to two copper antagonists (cadmium and zinc) there is a negative interaction between the antagonists, each being less antagonistic towards copper in the presence than in the absence of the other (data from Campbell and Mills, 1974).

demonstrated antagonism between cadmium and zinc in turkey poults, with 2 mg Cd kg^{-1} DM inducing hock and feather abnormalities on a low zinc but not a high zinc diet (10 and 60 mg kg^{-1} DM, respectively). The higher zinc level did not fully protect against 20 mg added Cd kg^{-1} DM. Tolerance will be determined by ratios (Cd:Zn and/or Cd:Cu) rather than cadmium concentration per se.

Tolerance by livestock

Non-ruminant species are generally less tolerant of cadmium exposure than ruminants, but are fortunately less likely to be exposed in agricultural practice. Limits of tolerance are hard to define because they are determined by dietary factors and the criteria of harmfulness.

Pigs

Studies with pigs illustrate the difficulties of setting limits on tolerance for cadmium (these difficulties apply to most PTE). With experimental exposure to Cd(NO$_3$)$_2$ alone and growth as probably the most sensitive index of harm, the limit of tolerance lies between 0.5 and 1.0 mg kg^{-1} DM – but with concurrent exposure to lead acetate, the tolerable limit lies between 0 and

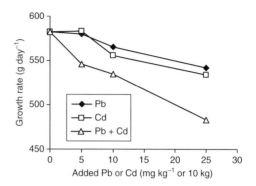

Fig. 18.2. The tolerance of added cadmium and lead by growing pigs is greatly reduced when exposure to them is simultaneous (from Phillips et al., 2003).

0.5 mg kg^{-1} DM (Fig. 18.2) (Phillips et al., 2003). In a further experiment, the exposure of sows to 1 mg cadmium with 10 mg Pb kg^{-1} DM for 6 months was clearly harmful to their offspring (Fig. 18.3) (Phillips et al., 2003). NRC (2005) gives a much higher tolerable limit for all species of 10 mg Cd kg^{-1} DM. One possible reason for the sensitivity of the growing pigs in Fig. 18.2 to cadmium is the use of wheat and barley to provide 50% of the cereals in the diet and wheat bran as carrier for cadmium: the phytase content of the diet would be higher and the phytate content lower than that in the

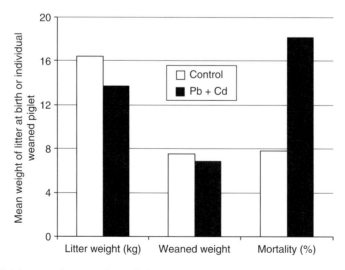

Fig. 18.3. The intolerance of sows to 6 months' simultaneous exposure to cadmium and lead (1 and 10 mg kg^{-1} DM, respectively) as reflected by reduced litter weight at birth and the reduced weight and increased mortality of survivors at weaning (from Phillips et al., 2003).

widely used corn/soybean meal diet. If free-range sows forage on sludge-amended soils then they will probably be concurrently exposed to cadmium and lead. However, absorbabilities would be lower (Casteel *et al.*, 2001) and the experiments of Phillips *et al.* (2003) therefore represent a 'worst-case scenario'.

Poultry

There is no evidence from outward appearances that broilers cannot tolerate up to 10 mg Cd kg^{-1} DM (NRC, 2005) and the tolerance of laying hens is probably raised by the high calcium levels they receive, but studies on phytase-rich diets would be of interest. Estimates of tolerance depend on the criteria of normality used. Japanese quail fed a diet of 60% wheat grown on sewage sludge-amended soils have shown induction of hepatic microsomal enzymes when compared with those given untreated wheat (Stoewsand *et al.*, 2005). Although kidney cadmium was raised, the ultrastructure of the kidney was normal. The treated wheat contained only 1.94 mg Cd kg^{-1} DM and it is possible that another component of sewage sludge 'upset' the liver, but the probability of cadmium-induced hepatotoxicity is extant.

Grazing livestock

Because cadmium is poorly taken up by plants, the major notional risk of toxicity occurring in grazing or foraging livestock is via the ingestion of soils enriched with cadmium from inorganic (superphosphate) or organic (sewage sludge or 'biosolids') fertilizers. However, in the many recent studies of experimental PTE exposure cited below, no ill health or loss of production has been reported in sheep or cattle.

Livestock as sources of cadmium exposure in humans

The introduction of regulations to limit the entry of cadmium into the human food chain means that the exposure of livestock to cadmium-contaminated soils assumes economic importance through the unacceptability of carcasses for human consumption (Loganathan *et al.*, 2008).

Intensity of animal exposure

In simulated soil-ingestion studies with lambs given grass pellets containing 10% soil for 84 days, kidney cadmium increased linearly and liver cadmium increased curvilinearly with soil cadmium concentration (Fig. 18.4), with a significantly higher accretion in the liver from a soil lower in iron and higher in calcium (Cassington, shown) than at another (Royston, not shown) site. However, this was probably a 'worst-case' simulation. In grazing studies on sites where sewage sludge has been applied for several years, liver and kidney cadmium have shown little (Wilkinson *et al.*, 2001) or no increase in lambs (Rhind *et al.*, 2005), but soil cadmium was not increased by the application of sludge in the latter study and soil ingestion reached a maximum of only 5% over winter.

Duration of exposure in sheep

With cumulative poisons such as cadmium, the risk of chronic toxicity (and also the hazard to the consumer) increases with duration of exposure as well as dietary cadmium concentration. Kidney cadmium increases linearly with exposure time in lambs grazing pastures on soils contaminated with fertilizer cadmium (Petersen *et al.*, 1991) and may exceed MAC (Langlands *et al.*, 1988; Petersen *et al.*,

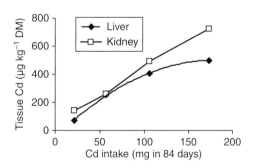

Fig. 18.4. The effects of cumulative cadmium intake from a sludge-amended soil on the concentrations of cadmium in the liver and kidney of lambs (data for Cassington soil from Hill *et al.*, 1998b).

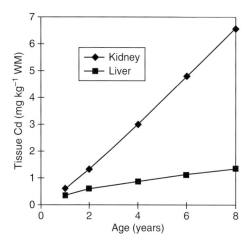

Fig. 18.5. The continuous exposure of sheep to pastures enriched with cadmium from superphosphates in Western Australia causes linear increases in cadmium concentrations in the kidneys and liver over time. Values in the kidneys can exceed maximum acceptable concentrations for human consumption (Peterson et al., 1991).

1991) but cadmium accumulates to a lesser extent in liver (Fig. 18.5) (Petersen et al., 1991) and hardly at all in muscle. Long-term exposure studies suggest that adaptive mechanisms limit cadmium retention in liver. When ewes were exposed for 3 years to diets containing soils enriched with cadmium from sewage sludges (dietary cadmium up to 0.4 mg kg^{-1} DM), liver cadmium concentrations reached a plateau after about 6 months and the soil with the most Cd had a relatively small cumulative effect (Fig. 18.6) (Stark et al., 1998). Curvilinear increases in liver cadmium over time (28 months) have also been reported in lambs grazing pastures of slightly raised cadmium content (0.5 mg Cd kg^{-1} DM) (Lee et al., 1996). Nevertheless, chronic toxicity is more likely to occur in aged breeding stock or milking herds than in young stock slaughtered for meat consumption. Higher liver and kidney cadmium have been reported in ewes than in their lambs and this was partly attributed to duration of exposure (Rhind et al., 2005).

Duration of exposure in cattle

In one study, liver cadmium in heifers showed no further increase when exposure to 5 mg Cd kg^{-1} DM (added as $CdCl_2$) continued from 394 days (parturition) and throughout lactation for a further 160 days (Smith et al., 1991b), but this may reflect the negative

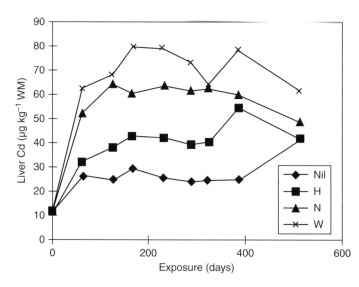

Fig. 18.6. Early high rates of cadmium accumulation in the liver of lambs were not sustained during long-term exposure via three ingested soils (H, N and W), enriched with cadmium from sewage sludge to varying extents (1.7, 3.9 and 7.3 mg Cd kg^{-1} DM, respectively). Homeostatic adjustment to loading with Cd is implied and confirmed by the relatively large, final effect of the soil with the least Cd (W) on liver Cd accretion (Stark et al., 1998).

effect of lactation on cadmium accretion rather than adaption to exposure. Five- to eightfold increases in kidney and liver cadmium have been reported in cattle exposed to pastures treated with sewage sludge by irrigation and soil incorporation for 8 years (Fitzgerald et al., 1985). Cadmium showed larger proportionate increases than any other element, but no clinical or histopathological abnormalities were recorded and the maximal kidney cadmium level recorded (32 mg kg^{-1} FW) remained well below the level usually associated with kidney damage in humans (200 mg kg^{-1} FW). Exposure to cadmium from contaminated phosphates was apparently responsible for 24%, 33% and 55% higher cadmium concentrations in the kidney, liver and mammary tissue of dairy cows reared under a conventional rather than an 'organic' management system on the same farm (Olsson et al., 2001). However, removal of older cows that entered the 'organic' herd as milkers removed the difference: the primary effect was on animal longevity and there was a correlation of kidney cadmium with both age and milk yield.

Exposure of consumers

Cadmium in the kidney is largely present as MT. It is sufficiently heat stable to survive cooking, of similar availability to CdCl$_2$ and largely accumulates in the kidneys of the consumer. The removal of kidneys from carcasses at the slaughter facility is ordered by some authorities (NRC, 2005) but surveys show considerable variation in animals of similar age from the same farm (e.g. in porcine kidney Cd; Linden et al., 2003). Current MAC are likely to remain and even decrease and all sources of cadmium and factors that increase accumulation must therefore be carefully monitored in the farm ecosystem and practical steps taken to minimize them. However, the chances of a person consuming enough liver or kidney of consistently high cadmium content to harm the human kidney seem remote. The fate of liver and kidney cadmium after exposure has ceased has not been studied in livestock, but remedial measures of the kind advocated for lead (see later) are worth exploring.

Fluorine

Chronic endemic fluorosis in humans and farm animals was identified early in the last century in several countries (e.g. Jacob and Reynolds, 1928). Severe fluorosis in cattle during and after volcanic eruptions in Iceland has an even longer history, and the early literature on severe fluorine toxicity was reviewed by Roholm (1937). Industrial contamination of herbage and the use of fluorine-containing phosphates as mineral supplements increased the incidence of chronic fluorine toxicity in livestock and stimulated interest in the distribution of fluorine in soils, plants and animal tissues, the susceptibility of different animal species and the prevention and control of chronic fluorosis. As the principal sources of fluorine in the farm environment became recognized and controlled, the emphasis of research shifted briefly to verification of the essentiality of fluorine in the diet, not only for dental health but also for growth.

Sources

Pastures and crops

Most plant species have a limited capacity to absorb fluorine from the soil, in which most is present as calcium fluoride (CaF$_2$). Pastures and forages are therefore characteristically low in fluorine unless they have been contaminated by the deposition of fumes and dusts of industrial origin or by irrigation with fluorine-rich (often geothermal) waters (Shupe, 1980). Since the soil is usually far richer in fluorine (Table 18.2; >200 kg^{-1} DM) than the plant, ingestion of fluorine-rich soil on overgrazed pastures can contribute significantly to fluoride intakes. An analysis of uncontaminated pastures in England revealed a range of 2–16 mg F kg^{-1} DM about a mean of 5.3 mg F kg^{-1} DM (Allcroft et al., 1965). In 107 'clean' samples of lucerne hay from areas throughout the USA, levels ranged widely from 0.8 to 36.5 mg F kg^{-1} DM about a median of 2.0 mg F kg^{-1} DM (Suttie, 1969). Cereals and other grains and their by-products usually contain only 1–3 mg F kg^{-1} DM (McClure, 1949) and uncontaminated feeds are safe for farm livestock.

Rock phosphate supplements

The principal sources of fluorine for livestock are commercial feeds that contain fluorine-rich phosphate supplements. Thus, 10% of 168 samples of dairy feed from seven different US states contained >30 mg F kg^{-1} DM, and some samples had over 200 mg F kg^{-1} DM (Suttie, 1969). Different fluoro-apatite rock phosphate (RP) sources vary widely in their fluorine content, depending on their origin. European and Asian sources of RP contain 3–4% fluorine, whereas Pacific and Indian Ocean island deposits usually contain only half as much. The island deposits can apparently be used safely in the rations of cows, pigs and poultry (Snook, 1962). High-fluorine RPs can be injurious to livestock when used over long periods (Phillips *et al.*, 1934) and are now routinely thermally 'defluorinated' without reducing phosphorus availability (see Chapter 6). A variety of chemically processed phosphorus sources were used by the feed industry and they varied widely in phosphorus to fluorine ratio (Table 18.4). In the USA, a minimum acceptable phosphorus to fluorine ratio of 100:1 has been set to minimize the risk of fluorosis. Sources with high phosphorus to fluorine ratios are generally the more expensive (Thompson, 1980).

Animal by-products

The bones of mature animals, even in the absence of abnormal fluorine exposure, are far higher in fluorine than the soft tissues. Bone meals can therefore constitute a significant source of fluorine for farm animals, typically containing 0.7 g F kg^{-1} DM, towards the lower end of the range given in Table 18.4. Variable and occasionally low phosphorus to fluorine ratios coupled with high costs make bone meal an unattractive source of phosphorus (Thompson, 1980). Meat meal or tankage is a significant source of fluorine only when it contains a high proportion of bone; some commercial meat meals contain as much as 200 mg F kg^{-1} DM. Other components of

Table 18.4. A comparison of phosphorus sources in terms of their fluoride (F) concentration, P:F ratio and relative cost as phosphorus sources (after Thompson, 1980).

Source	F (g kg^{-1} DM)	P:F ratio	F provided in adding 2.5 g P kg^{-1} DM (mg kg^{-1} DM)	Cost[a] relative to DCP
'Dicalcium phosphate' (DCP)	1.4–1.6	132	19	100
Precipitated DCP	0.5	360	7	94
Defluorinated rock phosphate (DFP)	1.6	112	22	94
Phosphoric acid (wet process)	2.0	118	21	72
Phosphoric acid (furnace process)	0.1	>2000	<1.25	156
Sodium phosphate	0.1	>2000	<1.25	200
Ammonium polyphosphate	1.2	120	21	107
Monoammonium phosphate				
Feed grade	1.8	133	19	107
'Low F'	0.9	266	9	128
Fertilizer grade	22.0	10	250	77
Diammonium phosphate				
Feed grade	1.6	125	20	113
Fertilizer grade	20.0	10	250	88
Bone meal[b]	**0.2**–3.5	35–**6000**	71–**0.4**	250
Curacao rock phosphate	5.4	26	96	86
Colloidal (soft) phosphate	12.0–14.0	6–7	361	39
Triple superphosphate	20.0	10.5	240	67
Fluoride rock phosphate (RP)	37.0	3.5	711	13

[a]Costs per unit of P at 1979 prices.
[b]Values towards bold limit are the more typical.
[c]See Appendix 5 for formula.

animal origin are invariably low in fluorine because the soft tissues and fluids of the body rarely contain >2–4 mg F kg^{-1} DM. Milk and milk products contain even less fluorine: cow's milk contains 0.1–0.3 mg F l^{-1} or 1–2 mg F kg^{-1} DM, and levels are little affected by fluorine intake.

Drinking water

Chronic fluorosis is enzootic in sheep, cattle, goats and horses in parts of Turkey, India, Australia, Argentina, Africa and the USA as a consequence of the consumption of waters abnormally high in fluorine, usually from deep wells or bores. Surface waters from such areas commonly contain <1 mg F l^{-1}, whereas bore waters often contain 5–15 mg F l^{-1} and as much as 40 mg l^{-1} when evaporation has occurred in troughs or bore drains before consumption by stock (Harvey, 1952). Elsewhere, drinking water is not normally a significant source, containing just 0.1–0.6 mg F l^{-1}.

Industrial and agricultural contamination

In parts of North Africa a chronic fluorosis, known as 'darmous', occurs from the contamination of the herbage and water supplies with high-fluorine, phosphatic dusts blown from RP deposits and quarries. The further industrial processing of such phosphates, aluminium reduction, brick and tile production and steel manufacturing also release significant amounts of fluorine into the atmosphere and forages downwind from production sites become contaminated by fumes and dusts (Oelschlager et al., 1970). The introduction of stringent controls on industrial emissions has greatly reduced fluorosis problems of this nature (e.g. Lloyd, 1983), but they can still occur (Patra et al., 2000). The prolonged use of phosphatic fertilizers containing 15–40 mg F kg^{-1} DM in Australia and New Zealand has raised soil fluorine to 217–454 mg kg^{-1} DM and the ingestion of such soils as a herbage contaminant by grazing sheep and cattle represents a potential risk for fluorine toxicity (Loganathan et al., 2008). Past mining activity can raise soil concentrations to >1000 mg kg^{-1} DM (Geeson et al., 1998).

Metabolism

The hazard presented by a particular fluorine source is determined not only by the quantity ingested but by the way it is metabolized.

Absorption

The availability of fluorine in natural feeds is high, as it is for other halides. Hay fluorine is as available as that in soluble sources such as sodium fluoride (NaF) (see Table 18.5) (Shupe et al., 1962), and 0.76 of the of fluorine from lucerne pellets has been reported to be apparently absorbed (AA_F) by sheep, slightly more than was absorbed from NaF (Table 18.6) (Grace et al., 2003a,b). Fluorine in less soluble compounds, such as bone meal, CaF_2, cryolite (Na_3AlF_6), RP and defluorinated RP (DFP), is less well absorbed. Fluoride retention in the bones of growing sheep is much lower from DFP than from dicalcium phosphate (Hemingway, 1977). Suttie (1980) found that fluorine in RP, DFP and dicalcium phosphate was 70%, 50% and 20% as available as that in NaF, as judged by skeletal fluorine accretion in lambs given diets containing 60 mg F kg l^{-1} DM as each source for 10 weeks. The AA_F in two fluorine-enriched soils from New Zealand ranged from 0.32 to 0.44 in sheep and cows (Grace et al., 2003a,b) (Table 18.5). The availability of fluorine from fertilizer-contaminated soils is thus similar to that of the source of contamination, fertilizer phosphates, but less than that of NaF. Flume and brick dusts also contain fluoride of relatively low avalability (Oelschlager et al., 1970).

Post-absorption

A significant proportion of A_F is excreted via the urine (e.g. 20–30%) (Grace et al., 2003a). The kidneys of exposed animals contain higher fluorine concentrations than any other soft tissue, although still <3 mg kg^{-1} DM in studies with cattle (Shupe, 1980) and sheep (Grace et al., 2003a). Retention is largely confined to the skeleton, but there is a negative exponential relationship between fluorine intake and bone fluorine after a fixed exposure time, at

Table 18.5. Mean alkaline phosphatase activity and fluoride content of the bones of groups of four heifers given different amounts and forms of fluoride (F) for 588 days (Shupe et al., 1962).

	F content (mg kg^{-1} DM)			Phosphatase activity[a]	
Treatment	Diet	Metatarsal	Ribs[b]	Metatarsal	Ribs[b]
Low-F hay	10	344	423	73	85
Low-F hay + CaF$_2$	69	931	1287	104	116
Low-F hay + NaF	68	1880	2753	146	248
High-F hay	66	2130	2861	163	191

[a]mg P hydrolysed in 15 min g^{-1} bone.
[b]The 11th, 12th and 13th ribs.

Table 18.6. Contrasting apparent absorption (%) of fluoride (F) from lucerne pellets, with or without added sodium fluoride (NaF), and soils contaminated by fertilizer F when fed to young sheep or cows (from Grace et al., 2003a,b). The tolerance for soil F ingestion by grazing livestock is underestimated by standards of tolerance based on studies with NaF (see Table 18.1)

	Source of F			
Species	Lucerne	+ NaF	+ Low-F soil	+ High-F soil
Sheep	76 (9)	69 (69)	32 (69)[a]	41 (86)
Cows	61 (110)	–	44 (450)	43 (1560)

[a]Daily fluorine intakes (mg) given in parentheses.

least in the ribs of sheep (Masters et al., 1992). The relationship between duration of exposure at a fixed intake and bone fluorine in cattle also follows the law of diminishing returns (Fig. 18.7) (Shupe, 1980), as does the relationship with serum fluorine in sheep (Grace et al., 2003a) and cows (Grace et al., 2003b). Exposure of the pregnant and lactating animal raises blood fluorine in the neonate and milk fluorine in the dam, but the increases are far less than those seen in the dam's bloodstream. The addition of 30 mg F l^{-1} to the drinking water of ewes has been reported to raise their plasma fluorine from <0.03 to 0.15–0.64 mg (7.9–33.7 μmol) F l^{-1}; levels in the newborn lamb rose from 9.0 to 300 μg (0.5–15.8 μmol) F l^{-1} and those in the ewe's milk from 0.13 to 0.40 mg (6.8–21.1 μmol) F l^{-1} (Wheeler et al., 1985).

Toxicity

Fluorine is largely a cumulative poison and clinical signs of toxicity may not appear for many weeks or months in animals ingesting moderate amounts. The mechanisms of chronic toxicity are not clearly understood, but may involve enzyme inhibition and the ability of fluoride ions (F$^-$) to substitute for hydroxide (OH$^-$) ions (NRC, 2005). Animals are protected by two physiological mechanisms: fluorine excretion in the urine and deposition in the bones or eggs. During this latent period before toxicity becomes apparent in cattle, neither milk production (Suttie and Kolstad, 1977) nor the digestibility and utilization of nutrients is significantly depressed (Shupe et al., 1962). There may, however, be transient subclinical changes in sensitive organs such as the testes and the development of spermatozoa may be impaired in poultry (Mehdi et al., 1983) and rabbits (Kumar and Susheela, 1995). Fluorine deposition in the skeleton proceeds rapidly at first and then more slowly until a saturation stage is eventually reached. Beyond this saturation point, which is marked by values of 15–20 g F kg^{-1} dry bone in cattle (30–40 times greater than normal) (Table 18.8) and 2.3 g F kg^{-1} in hens (five times greater than normal), 'flooding' of the susceptible soft tissues with

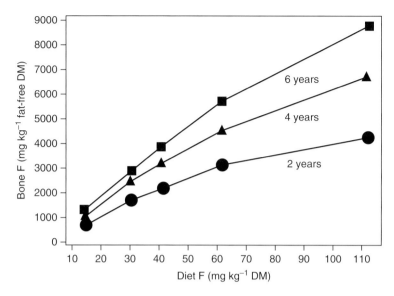

Fig. 18.7. Fluorine concentrations in the skeleton of dairy cows increase with dietary F concentration and duration of exposure, but both responses are curvilinear (data from Shupe, 1980).

fluorine occurs, metabolic breakdown takes place and death ensues. Chronic skeletal abnormalities are usually detected before this terminal stage is reached.

Tolerance to dietary fluorine

Direct signs of fluorosis are rare in newborn or suckling animals because placental and mammary transfer of fluorine is limited (Forsyth *et al.*, 1972; Wheeler *et al.*, 1985). General undernutrition tends to accentuate the toxic effects of fluorine (Suttie and Faltin, 1973). Whatever the measure of harm, be it clinical or subclinical, tolerance of dietary fluorine depends largely on the following:

- the species and age of the animal;
- the chemical form of fluorine ingested;
- the severity of contamination;
- the duration and continuity of exposure; and
- whether the skeleton is in a state of net calcium accretion or resorption.

Mink

Mink are the most resilient species of farmed animal studied. They can tolerate approximately 600 mg F kg^{-1} DM (as NaF) of a fish/liver/chicken diet for over 1 year, but 1000 mg F kg^{-1} DM causes unthriftiness, hyperexcitability and mortality in adult females and their offspring as well as dental and skeletal lesions (Aulerlich *et al.*, 1987).

Poultry

Poultry are also a tolerant species and maximum safe dietary levels of 300–400 and 500–700 mg F kg^{-1} DM as RP have been reported for growing chicks and laying hens, respectively (Gerry *et al.*, 1949). Similar tolerance levels have been reported for growing female turkeys, but in the young male poult 200 mg F kg^{-1} DM as NaF decreased weight gain (Anderson *et al.*, 1955). Increased proportions of dietary fat can enhance the growth-retarding effect of excess fluorine in chicks (Bixler and Muhler, 1960), possibly by delaying gastric emptying (McGown and Suttie, 1974). The distribution of fluorine in hens exposed to NaF at two rates is shown in Table 18.7. The lower level (100 mg F kg^{-1} DM) caused large increases in tibial fluorine but little accumulation elsewhere and no fall in egg production. The higher level (1300 mg kg^{-1} DM) caused fluorine to accumulate in the

Table 18.7. The effect of the level of fluoride (F) exposure, as sodium fluoride (NaF), over 112 days on the concentrations of fluoride in body fluids, tissues and products of hens (Hahn and Guenther, 1986).

Dietary F (mg kg^{-1} DM)	Plasma[a] (mg l^{-1})	(mg kg^{-1} fat-free DM)[b]				Egg	
		Tibia	Liver	Kidney	Muscle	Yolk	Shell
16	0.7	538	5.2	3.4	4.0	4.3	3.9
100	2.8	2247	5.5	3.8	4.0	3.3	12.7
1300	10.1	2600	19.2	31.8	6.7	18.4	307.2
1300 + 1040 aluminium	6.0	2228	10.5	11.9	5.5	7.5	55.5

[a]Ionic F.
[b]Multiply by 52.63 for µmol kg^{-1} fat-free DM.

eggshell, kidney, liver, egg yolk, plasma and muscle (in decreasing order) (Hahn and Guenther, 1986). An early loss of feathers occurred as though moulting had been induced, slight diarrhoea was induced and appetite and egg production decreased by 50% during the 112-day trial, but mortality did not increase. The tolerance of the laying hen may be attributable to the deposition of fluorine in each egg that is laid, but hatchability may be impaired.

Pigs

Pigs are tolerant of fluorine exposure, with no abnormalities reported when young female pigs have been fed up to 450 mg F kg^{-1} DM as NaF through two pregnancies (Forsyth et al., 1972). Accumulation of fluoride was noted in their newborn and 4-week-old offspring, particularly if calcium in the maternal diet had been reduced (from 12 to 5 g kg^{-1} DM), but there was no 'generation build-up'.

Sheep

Downward revision of the safe fluorine allowances for breeding ewes was suggested by Wheeler et al. (1985), who found evidence for marginal toxicity on treatments equivalent to 25 mg added F kg^{-1} diet. The widely used figures set by successive NRC committees (e.g. 2005) are 60 mg F kg^{-1} DM for ewes and 150 mg kg^{-1} DM for fattening lambs. Subsequent work has supported a higher tolerance for lambs (e.g. Milhaud et al., 1984) but in Grace et al.'s (2003a) study, 125 mg F kg^{-1} DM was tolerated for 55 days when given as enriched soil.

Cattle

The limit of tolerance for dairy cows is approximately 40 mg F kg^{-1} DM when ingested as NaF. Signs of fluorosis have appeared within 3–5 years at 50 mg F kg^{-1} DM from this source (Suttie et al., 1957). However, in a study beginning with young calves and lasting for 7 years, tolerance for NaF was no more than 30 mg F kg^{-1} DM (Shupe et al., 1963b). Tolerance of dairy cows to the fluorine in RP and CaF$_2$ (and presumably to soil fluorine) is approximately twice that to the fluoride in NaF, with each source providing 65 mg F kg^{-1} DM, but there is little difference in toxicity between NaF and fluorine-contaminated hay (Shupe et al., 1962). Tolerance to fluorine in DFP may be five times that to NaF, the source used experimentally to define tolerances. Thus, while many phosphorus fertilizers appear to provide discomfortingly high additions of fluorine to cattle diets, they are probably safe and can only be improved upon at considerable cost (Table 18.4). In a short-term (90-day) exposure study in male buffalo, tolerance was placed between 30 and 60 mg F kg^{-1} DM (as NaF) on the basis of changes in plasma calcium (↓) and phosphorus (↑) induced at the higher level (Madan et al., 2009).

Clinical signs of chronic fluorosis in ruminants

The need to combine clinico-pathological information and biochemical data in a flexible framework when diagnosing mineral imbalances is nowhere more apparent than in the diagnosis of fluorosis in dairy cattle. Table 18.8 summarizes the exhaustive studies of Shupe (1980) and illustrates the distinctive contribution a particular criterion of fluorine exposure can make:

- incisor score indicates the severity of past daily exposure;
- fluorine in milk or blood indicates the current severity of daily exposure;
- fluorine in bone and urine, molar and hyperostosis score indicate the duration as well as severity of exposure.

Each criterion can vary widely between individuals, with clinico-pathological features (i.e. the most important in welfare and economic terms) the most variable. The criteria in Table 18.8 are broadly applicable to sheep.

Dental lesions

In young animals exposed to excess fluorine *before* the eruption of their permanent teeth (this varies from 12 months of age for the first incisors to 30 months of age for the last molars to appear in sheep), the teeth become modified in size, shape, colour, orientation and structure. This is because fluorine impairs the function of ameloblasts – the enamel-forming cells – and the abnormal enamel matrix fails to calcify (Shearer *et al.*, 1978a,b; Suttie, 1980). The incisors become pitted and blackened with erosion of the hypoplastic enamel, the molars become abraded and there may be exposure of the pulp cavities due to fracture or wear. The uneven molar 'table' impairs mastication of food and is particularly debilitating (Suttie, 1980).

Skeletal abnormalities

The development of skeletal abnormalities during fluorosis was reviewed by Shupe (1980) and Suttie (1980) and there is little new information. The exposure of immature stock to excess fluorine can decrease endochondral bone growth, but the lesion in adult stock is predominantly one of excessive periosteal bone formation (hyperostosis) and the affected bones are often visibly enlarged and

Table 18.8. A guide to the diagnosis of fluorosis in dairy cattle based on chronic experimental exposure studies and field cases (after Shupe, 1980).

	Age (years)	Fluorosis state				
		Normal	Marginal	Mild	Moderate	Severe
F in diet (mg kg^{-1} DM)[a]	–	>15	15–30	30–40	40–60	60–109
Incisor score	–	0–1	0–2	2–3	3–4	4–5
F in milk (mg l^{-1})[a]	–	<0.12	<0.12	0.12–0.15	0.15–0.25	>0.25
F in blood (mg l^{-1})[a]	–	<0.30	<0.30	0.3–0.4	0.4–0.5	0.5–0.6
F in rib bone[b]	2	0.4–0.7	0.71–1.60	1.6–2.1	2.1–3.0	3.0–4.2
(g kg^{-1} fat-free DM)	4	0.7–1.1	1.14–2.40	2.4–3.1	3.1–4.5	4.5–6.6
	6	0.65–1.22	1.22–2.80	2.8–3.8	3.8–5.6	5.6–8.7
F in urine (mg l^{-1})[a]	2	2.3–3.8	3.8–8.0	8.0–10.5	10.5–14.7	14.7–19.9
	4	3.5–5.3	5.3–10.3	10.3–13.3	13.3–18.5	18.5–25.6
	6	3.5–6.0	6.0–11.3	11.3–14.8	14.8–21.0	21.0–30.1
Molar score	2	0–1	0–1	0–1	0–1	0–3
	4	0–1	0–1	0–1	1–2	1–4
	6	0–1	0–1	0–1	1–3	1–5
Periosteal hyperostosis	2	0	0–1	0–1	0–2	0–3
score	4	0	0–1	0–1	0–3	0–4
	6	0	0–1	0–2	0–4	0–5

[a]Multiply by 52.6 to obtain values in µmol l^{-1} or kg^{-1} DM.
[b]To obtain approximate values in rib ash, multiply by 1.67 (assuming 60% ash in sample); since rib and tail vertebrae values are similar on an ash basis, the same factor is required for generating values from fat-free DM to vertebral ash; metacarpal and metatarsal bones contain 25% less F than rib on a fat-free DM basis.

have porous, chalky white surfaces instead of an ivory sheen. In cross-section, the shaft is often irregularly overgrown and lacking its normal compact appearance. These abnormalities can occur at any age and are not necessarily accompanied by dental lesions. Exostoses of the jaw and long bones become visible in severely exposed animals and the joints may become thickened and ankylosed due to calcification of tendons at their points of attachment to the bone. Stiffness and lameness then become apparent, movement is difficult and painful and foraging is therefore restricted. Growth may become subnormal and weight losses may occur, together with a reduction in milk production and fertility, secondary to the reduced feed consumption (Phillips and Suttie, 1960). These manifestations of fluorosis are less prominent in young and pen-fed animals. The poor lamb and calf crops characteristic of fluorosis areas arise primarily from mortality of the newborn due to the impoverished condition of the mothers, rather than from a failure of the reproductive process itself (Harvey, 1952). Thyroid enlargement and anaemia have been associated with fluorine exposure in the field (Hillman et al., 1979), but they have not been seen in experiments with cattle (Suttie, 1980), pregnant ewes (Wheeler et al., 1985) or mink (Aulerlich et al., 1987).

Biochemical evidence of fluorine exposure

Bone and tooth composition

In unexposed adult animals, the whole bones commonly contain 0.2–0.5 g F kg^{-1} and values rarely exceed 1.2 g kg^{-1} fat-free (ff) DM. Normal teeth contain much less fluorine than bone (<0.2 g kg^{-1} ff DM), with higher levels in the dentin than in the less metabolically active enamel and marked variations in concentration within the fluorotic tooth (Shearer et al., 1978b). Fluorine is incorporated more rapidly into active areas of bone growth than into static regions (epiphyses > shaft), into cancellous bones (ribs, vertebrae and sternum) faster than compact cortical bones (metacarpals) (see Table 18.5) and into the surface areas (periosteal and endosteal) of the shaft in preference to the inner regions. Differences in fluorine concentrations within and among bones (and teeth) and between results expressed on different bases can influence assessments of exposure. For example, it has been concluded from the lack of effect of short-term exposure to soil fluorine on metacarpal fluorine (fat and organic matter not removed) that the fertilizer-enriched soils of New Zealand present no fluorosis hazard to the dairy cow (Grace et al., 2003b). However, removal of fat and organic matter raises bone fluorine, and values in the metacarpus (ff DM) are relatively low – approximately 50% of values in the tail bone, although highly correlated with them (Suttie, 1967). The aged (7-year-old) cows studied by Grace et al. (2003b) came from an area of fluorine-enriched soils and had background metacarpal fluorine values ranging from 625 to 1625 mg kg^{-1} DM; fluorine in rib ff DM could have been two- to threefold higher, changing the assessment of risk to mild/marginal (Table 18.8). In dairy cattle, experimental fluorine toxicosis has been associated with values in excess of 5.5 g F kg^{-1} ff DM in compact bone and 7.0 g kg^{-1} DM in cancellous bone, with concentrations between 4.5 and 5.5 g F kg^{-1} DM indicating a marginal fluorine status (Suttie et al., 1958). Bovine tail vertebrae and rib contain similar fluorine values on an ash basis and values of 4–6 g F kg^{-1} ash have been associated with lameness in the UK (Burns and Allcroft, 1966). Tail bone samples can be obtained by biopsy (Burns and Allcroft, 1962). The toxic thresholds for cancellous bones of sheep have been placed at 7 g F kg^{-1} ff DM or 11 g F kg^{-1} ash (Puls, 1994). The calcium to phosphorus ratio of fluorotic bone is normal, the magnesium content increased and there may be a precipitation as CaF_2.

Urine

In sheep and cattle not exposed to excess fluorine the urine rarely contains >10 mg F l^{-1}. However, long-term fluorosis experiments with dairy cows have reported the following urinary concentrations: normal animals, <5 mg F l^{-1}; borderline toxicity, 20–30 mg F l^{-1}; and overt toxicity, >35 mg F l^{-1} (Suttie et al.,

1961; Shupe et al., 1963a). Urinary fluorine can come from the skeleton after animals have been removed from a high-fluorine supply, particularly during periods of bone resorption (e.g. during early lactation). High urinary fluorine therefore reflects either current or previous fluorine exposure and can be easily measured using an ion-specific electrode. Urine samples are best taken in the morning on more than one occasion and results should be expressed on a specific gravity or creatinine basis (Shupe, 1980).

Blood

Fluorine is higher in plasma than in whole blood and reflects the current rate of fluorine ingestion rather than cumulative fluorine intake in sheep (Milhaud et al., 1984; Grace et al., 2003a) and cattle (Grace et al., 2003b). During their formative period, the teeth are sensitive to small changes in plasma fluorine: when concentrations approach 0.5 mg l^{-1} or more, severe dental lesions appear in young cattle; at values between 0.2 and 0.5 mg l^{-1}, less-severe damage occurs; while below 0.2 mg l^{-1}, few adverse effects are apparent (Suttie et al., 1972). In a long-term study of dairy cows that were continuously or periodically exposed to high fluorine intakes, plasma values increased from 0.1 to 1.0 mg l^{-1} (Suttie et al., 1972). Good correlations between fluorine intake and serum fluorine in short-term studies led Grace et al. (2003a,b) to conclude that the latter is a useful index of fluorine status, but this statement requires qualification. The relationship between serum fluorine and bone fluorine is curvilinear, with serum fluorine reaching a plateau while fluorine continues to accumulate in bone (Grace et al., 2003a). Serum fluorine therefore reflects the fluorine status of the current diet rather than the cumulative effect of fluorine exposure. Furthermore, plasma fluorine can show marked diurnal variations and may only reflect recent fluorine exposure. Sheep appear to tolerate higher plasma fluorine than cattle. In lambs given up to 2.5 mg F kg^{-1} live weight (LW) in a milk diet, plasma values reached 0.6 mg (31.5 μmol) F l^{-1} with no obvious signs of ill health (Milhaud et al., 1984), while in pregnant ewes a similar level was associated with only marginal evidence of toxicity (Wheeler et al., 1985). However, the sheep studies were both of relatively short duration (4–8 months). The exposure of 14-month-old lambs for 55 days to 60 mg F kg^{-1} DM caused no dental or skeletal abnormalities, although serum fluorine was maintained at about 0.3 mg l^{-1} for the last 30 days (Grace et al., 2003a).

Other indices

Serum alkaline phosphatase activity is increased in fluorotic chicks (Motzok and Branion, 1958) and cows (Shupe et al., 1962), but increases can occur in exposed buffalo without evidence of clinical disease (Madan et al., 2009). Alkaline phosphatase activity was increased in the bones of heifers exposed to various sources (Table 18.5) of fluoride.

Continuity of intake

The importance of continuity of fluorine intake and age of the animal was illustrated when sheep consuming artesian bore water containing 5 mg F l^{-1} developed severe dental abnormalities and other signs of fluorosis in the hot climatic conditions of Queensland (Harvey, 1952). However, in the cooler conditions of South Australia, where little water is drunk during the wet winter months, no ill effects were seen in mature sheep given 20 mg F l^{-1} as NaF in the drinking water (Peirce, 1954). During periods of low fluorine intake, the exchangeable skeletal fluorine stores are depleted and fluorine is excreted in the urine. When compared on the basis of total yearly intake, skeletal fluorine storage can be similar for continuous and intermittent exposure (Suttie, 1980). However, the latter can be the more harmful because of rapid increases in tissue fluorine during periods of high intake (Suttie et al., 1972). With short-term exposure to high intakes (approximately 90 mg F kg^{-1} DM), systemic reactions such as weight loss and unthriftiness due to decreased appetite can arise in dairy cattle (Suttie et al., 1972). The seasonal peak in fluoride ingestion in winter by the in-lamb ewe grazing pastures contaminated by past mining can be followed by outbreaks of lameness and stiffness in lambs (Geeson et al., 1998).

Treatment and prevention of fluorosis

Both the dental and skeletal lesions of chronic fluorosis are essentially irreversible, the latter being beyond the scope of remodelling. If livestock cannot be removed from the source of exposure, steps can be taken to delay the progression of the disease. Aluminium salts protect against high fluorine intakes in sheep (Becker et al., 1950) and cattle (1% as aluminium sulfate (AlSO$_4$)) (Allcroft et al., 1965), apparently through reducing fluorine absorption from the intestinal tract. Similar additions of AlSO$_4$ have greatly reduced fluorosis in hens given 1300 mg F kg^{-1} DM and reduced fluorine concentrations at most sites, particularly the eggshell, liver and kidney (Table 18.7). Borate and silicate supplements have limited the affects of excess fluorine on the skeleton of pigs (Seffner and Tuebener, 1983).

Essentiality

Nearly 70 years ago, the discovery that the incidence of dental caries in human populations was significantly higher where water supplies were virtually free from fluoride than in areas where the water contained 1.0–1.5 mg F l^{-1} focused attention on the beneficial as well as the toxic effects of fluorine. Dental caries does not present a health problem in farm animals and there is as yet no evidence that fluorine performs any specific essential function. However, the provision of fluorine has been found to stimulate the growth of rats fed on a purified diet in plastic isolators with minimum atmospheric contamination (Schwarz and Milne, 1972; Milne and Schwarz, 1974). In the latter study, potassium fluoride was added at 2.5 mg F kg^{-1} DM to a diet containing <0.04 mg F kg^{-1} DM. Others have failed to confirm the need for dietary fluorine for optimum growth (Maurer and Day, 1957; Doberenz et al., 1964). Controversy continued with the report that female mice on a diet containing 0.1–0.3 mg F kg^{-1} DM developed progressive infertility in two successive generations, accompanied by severe anaemia in mother and offspring, unless they were given 50–200 mg F l^{-1} as NaF in the drinking water (Messer et al., 1974). Other workers have reported no such effects (Weber and Reid, 1974). Tao and Suttie (1976) suggested that the apparent essentiality could have been due to a pharmacological effect of fluoride in improving iron utilization in a marginally iron-sufficient basal diet, since fluoride might enhance the intestinal absorption of iron in rats (Ruliffson et al., 1963). All natural diets are likely to provide sufficient fluorine to meet any dietary requirement.

Lead

Lead poisoning is one of the most frequently reported causes of poisoning in farm livestock, with cattle the most commonly affected species (Blakley, 1984) and some 50 incidents reported annually in the UK (VIDA, 2007). Young calves are particularly vulnerable. In western Canada, 50.9% of lead poisoning cases in bovines were in animals <6 months of age and cases were more common in dairy than in beef breeds (Blakley, 1984). These effects of species, breed and age are attributable to the access cattle are inadvertently given to point sources of lead (e.g. old batteries, tins of paint) in the vicinity of farm buildings, the trapping of heavy objects in the adult reticulum and the lack of a functional and protective rumen microflora in the youngest calves. Cases of lead poisoning are mostly acute in form and the 20,000 cases that once occurred in cattle every year, worldwide, were seen as an inevitable consequence of the vast amounts of lead that were redistributed from the Earth's crust each year by industrial processes (1.5×10^6 tons) (R.P. Botts, cited by Bratton, 1984). The effects of chronic lead exposure on tissue lead in farm livestock are receiving increasing attention due to concerns over the subclinical effects of raised blood lead in young children. Lead is commonly regarded as a cumulative poison with risk primarily determined by the source of lead and duration of exposure (NRC, 2005), but this is far from true.

Sources

Soil lead varies widely and concentrations are much higher in the UK than in New Zealand (Table 18.2). Mining activity (Moffat, 1993; Geeson et al., 1998) and the dispersal of sewage sludge (Suttle et al., 1991; Hill et al., 1998a,b; Gaskin et al., 2003; Tiffany et al., 2006) can raise soil lead to $\geq 500\,mg\,kg^{-1}$ DM. Although soils can contain appreciable concentrations of lead in EDTA-extractable forms (Archer and Hodgson, 1987), lead is poorly taken up by plants and concentrations in pastures and crops are generally $<0.3\,mg\,kg^{-1}$ DM (Longhurst et al., 2004). The principal threat to livestock and the consumer comes from the soil, as livestock consume soil while grazing or foraging on contaminated land (Abrahams and Steigmajer, 2003). Until the addition of lead to petrol was banned in 1985, automotive exhaust fumes caused local accumulations beside major roads and contributed to widespread atmospheric deposition, which still affects vast areas. These include the upland peat soils of Scotland, in which levels of $>50\,mg\,Pb\,kg^{-1}$ DM have frequently been recorded (Bacon et al., 1992). A continuing source of anthropogenic contamination is the dispersal of municipal sewage sludge on agricultural land (Hill et al., 1998a,b). When lead-contaminated pasture is ensiled, downward migration of lead in the silo can lead to fourfold increases in lead concentration in the lowermost layers, sufficient at $119\,mg\,Pb\,kg^{-1}$ DM to cause clinical lead toxicity in a dairy herd (Coppock et al., 1988). Calcium supplements are sometimes rich in lead (Ross, 2000).

Availability

Effect of dietary composition

Sheep and rabbits apparently absorb (AA_{Pb}) only 1% of the lead they ingest (Blaxter, 1950) and rats only 0.14% (Quarterman and Morrison, 1975). A negative AA_{Pb} has been reported for lambs given grass nuts (Hill et al., 1998a). The high AA_{Pb} of 15% that has been recorded for a large lead supplement ($1000\,mg\,kg^{-1}$ DM from lead acetate (PbAc)) (Pearl et al., 1983) in sheep may have been a characteristic of the diet fed: it consisted of 56% corn, contained no long roughage, was low in calcium and probably low in sulfur. Raising dietary calcium from 2.5 to $5.0\,g\,kg^{-1}$ DM reduced AA_{Pb} to 12.8% on the lead-supplemented diet and from 11.0% to 5.7% on the basal diet with no added lead. A qualitatively similar effect of dietary calcium had been reported previously (Morrison et al., 1977). The most likely explanation for the calcium effect is that some lead is absorbed inadvertently through its affinity for calbindin (Fullmer et al., 1985), which is downregulated by calcium supplementation (see Chapter 4). Lead absorption can also be decreased by raising dietary calcium or phosphorus in avian species (Scheuhammer, 1987). Lead also binds to divalent metal transporter 1, which is induced by iron deprivation (see Chapter 13) and cadmium exposure, explaining why AA_{Pb} increases in iron deficiency (Morrison and Quarterman, 1987) and tissue accretion of lead is raised in pigs by giving cadmium simultaneously (Phillips et al., 2003).

There is an inverse relationship between dietary sulfur and A_{Pb} that is only manifested in ruminants. In lambs, raising dietary sulfur from low ($0.7\,g\,kg^{-1}$ DM) to moderate ($2.3\,g\,kg^{-1}$ DM) or high ($3.8\,g\,kg^{-1}$ DM) levels increased the mean survival time from 6 to 15 or 30 weeks, respectively, when $200\,mg\,Pb\,kg^{-1}$ DM was added as PbAc to their diet (Quarterman et al., 1977). Precipitation of unabsorbable lead sulfide (PbS) in the sulfide-rich rumen at the higher sulfur intakes probably reduced the toxicity of lead.

Lead is thus generally very poorly absorbed by non-ruminant and ruminant species alike. Large proportional reductions in AA_{Pb} are caused by increases in dietary calcium, cadmium and lack of iron. In ruminants, diets low is sulfur raise AA_{Pb}.

Duration of exposure and dietary concentration

The decrease in AA_{Pb} noted with increasing duration of exposure or dietary lead concentration (Fick et al., 1976) may be caused by the trapping of lead by divalent metal transporter 1 in the intestinal mucosa and its subsequent sloughing into the gastrointestinal tract in the same way that excess iron is lost (see Chapter 13).

Source

The sources of adventitious lead encountered by livestock can vary greatly in availability. A soil enriched by lead-mining activity had a relative availability (RA_{Pb}) of 58–78% when compared with PbAc in young pigs, depending on the base for calculation (Casteel et al., 1997). A subsequent standardized comparison of different contaminated soils found that their RA_{pb} varied from 3 to 86% in the pregnant pig (Casteel et al., 2001). The RA_{Pd} in sewage and harbour sludges is about half that of PbAc when added to the diet of lambs (Van der Veen and Vreman, 1986; cited by NRC, 2005). Although the lead in sewage sludge (Stark et al., 1998), like that in roadside soils, (N.F. Suttle, unpublished data) is extensively extracted by EDTA, it had an AA_{Pb} of only 10–30% when added to grass nuts and fed to lambs (Hill et al., 1998a). However, the absolute AA_{Pb} attributed to a given source will depend on the composition of the diet that has been contaminated (p. 511) and the lead in the grass nuts had an AA_{Pb} of zero!

Metabolism

Transport and tissue distribution

Some 90% of absorbed lead is taken up by red blood cells, while the remainder is largely bound to albumin (NRC, 2005). Eventually, lead is released from erythrocytes but remains firmly bound to cytosolic-binding proteins and MT. Patterns of lead accretion vary widely from tissue to tissue. Blood and liver lead show early rapid increases in lambs at a constant daily rate of exposure before reaching a plateau (Pearl et al., 1983; Brebner et al., 1993). During medium-term exposure (7 weeks), most lamb tissues retain diminishing proportions of the ingested lead as dietary concentrations increase, with the liver showing a more marked curvilinearity than the brain or bone, for example (Fig. 18.8) (Fick et al., 1976). Animals adapt to lead exposure, and blood and liver lead are poor indices of long-term risk of toxicity and body lead burden.

Kidney lead shows distinctive changes during extreme lead exposure, increasing markedly with dietary levels of 1000 mg Pb kg^{-1} DM (Fig. 18.8). Strong linear increases in kidney lead with duration of exposure (Stark et al., 1998) confirm the limits of adaptations elsewhere. Lead accumulates in bone more than in soft tissues at lower levels of exposure (Fig. 18.8), particularly at growth points in the immature animal and at times of rapid bone turnover in the mature animal. Four- to fivefold greater increases in bone lead have been reported in laying than in non-laying birds exposed to lead (Finley and Dieter, 1978). Calcium and phosphorus supplements delay the release of lead from the skeleton when exposure to lead has ceased (Quarterman et al., 1978).

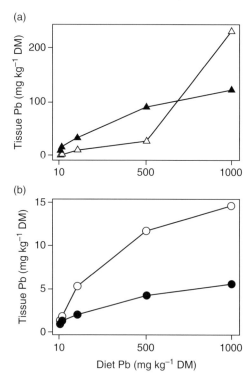

Fig. 18.8. Tissue lead concentrations increase with dietary lead concentration and duration of exposure, but both responses are curvilinear (see text for the latter effect). Limits to liver accumulation are particularly marked and partition may shift towards the kidney at very high exposure levels (data from Fick et al., 1976). △, kidney; ▲, bone; ○, liver; ●, brain.

Maternal transfer

When cows (Pinault and Milhaud, 1985) or ewes (Carson et al., 1974; Suttle et al., 1997) are exposed to lead during gestation, raised liver or blood lead in the newborn provides evidence of the placental transfer of lead. Tissue lead accretion is higher in gestating than barren pigs but higher in the mother than in her unborn litter (Casteel et al., 2001). Raised milk lead indicates that maternal transfer will continue during lactation and, while increases are small, lead is well absorbed from milk (Morrison and Quarterman, 1987) and a risk to the suckling or other consumer may arise.

Toxicity

Acute lead poisoning

Acute lead toxicity is characterized by gastrointestinal haemorrhage and anaemia, together with liver necrosis and kidney damage; high lead concentrations may be found in digesta and faeces. Diagnosis of lead toxicity is confirmed by the presence of high kidney lead (Table 18.9) and histological evidence of kidney damage at post-mortem. Surviving cohorts are likely to have raised blood lead (Table 18.9) and early dysfunction is indicated by a fall in erythrocyte aminolaevulinic acid dehydratase (ALAD) (Rice et al., 1985; Bratton et al., 1986) and a rise in zinc protoporphyrin (ZPP) concentrations. However, ALAD activity is slightly reduced by heavy exposure to other metals such as cadmium and selenium.

Chronic lead poisoning

In chronic lead exposure, lesions of the alimentary tract and anaemia are not seen. However, osteoporosis may accompany hydronephrosis, particularly in young lambs (Butler et al., 1957; Quarterman et al., 1977). The first visible sign of malaise is loss of appetite. 'Lead lines' (probably porphyrin pigments arising from erythrocyte breakdown) may eventually be seen in the gingiva adjoining the teeth. Suckling lambs in a former lead-mining area have been found to be particularly vulnerable to lead-induced osteodystrophy between 6 and 10 weeks of age (Butler et al., 1957) and subsequent work in the same area has shown that blood lead peaks at that time (Fig. 18.9) (Moffat, 1993). A similar peak in blood lead has been reported in the offspring of housed ewes on constant lead intakes for over 1 year and was associated with a peak in milk lead in mid-lactation (Suttle et al., 1997). The mobilization of lead from a demineralizing maternal skeleton (see Chapter 4) may be responsible. The risks of lead toxicity to both

Table 18.9. Guildelines for assessing lead exposure and likelihood of lead toxicity in livestock from Pb concentrations in diet, blood and tissues; normal (N), high (H) and toxic (T) levels are given.

Group	Status	Diet (mg kg^{-1} DM)[a]	Bone (mg kg^{-1} DM)[a]	Liver (mg kg^{-1} FM)[a]	Kidney (mg kg^{-1} FM)[a]	Blood (mmol l^{-1})[b]
Cattle	N	1–6	1–7	0.1–0.5	0.1–0.5	0.005–0.25
and	H[c]	20–1000	30–75	0.8–2.0	0.7–4.0	0.5–1.5
sheep	T[d]	>2000	>75	>8.0	>20.0	>2.0
Pigs	N	2–8	1?	0.2–0.5	0.2–1.2	0.2–1.5
	H[c]	20–750	100–500	5.0–35.0	5.0–25.0	1.5–4.8
	T[d]	>750	>500	>35.0	>25.0	>4.8
Poultry	N	1–10	<50	0.1–5.0	0.1–5.0	0.2–1.0
	H[c]	20–200	150–400	5.0–18–0	5.0–20.0	1.0–2.0
	T[d]	>200	>400	>18.0	>20.0	>2.0

[a] Multiply by 4.826 to obtain values in μmol kg^{-1} DM or FM.
[b] Multiply by 207.2 to obtain values in μg l^{-1}.
[c] Values within the 'high' category indicate unusually high levels, but not high enough to harm animal health.
[d] Values in 'toxic' category will cause harm but not necessarily death.

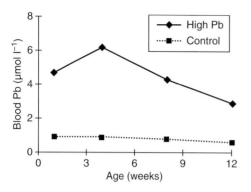

Fig. 18.9. Outbreaks of osteodystrophy in suckling lambs in former lead-mining areas are preceded by high blood lead concentrations. The efficient absorption of Pb from milk and ingested Pb-rich soil may be responsible (Moffat, 1993).

the mother and lamb may be heightened at that time. Conversely, factors that increase bone mineralization, such as dietary supplements of calcium and phosphorus, may reduce lead toxicity partly by increasing the safe capture and 'dilution' of lead in an increased volume of newly formed bone.

Tolerance

In both acute and chronic toxicity, maximum tolerable oral doses of lead vary widely in all species and are determined by:

- developmental status;
- source;
- species; and
- dietary composition.

Tolerance in non- or pre-ruminants

Tolerance is minimal in quail chicks. In a study in which chicks were given 1 mg Pb kg^{-1} DM from hatching, egg production was reduced at sexual maturity (Edens and Garlich, 1983). When exposed to lead after sexual maturity, ten times more lead was required to retard growth; and when laying hens were exposed towards the end of lay, 200 times more lead was required to lower egg production. The difference may be largely due to loss of lead in the eggshell and yolk, which results in lower blood lead in hens than in cockerels given the same exposure (Mazliah et al., 1989). Laying ducks have tolerated 12–24 mg Pb day^{-1} (given by capsule as lead nitrate (PbNO$_3$)) (Jeng et al., 1997), a level probably equivalent to approximately 120–240 mg kg^{-1} DM. As little as 10 mg Pb kg^{-1} DM has reduced the growth rate in weanling pigs (Fig. 18.2) and harmed the offspring of exposed sows (Fig. 18.3); with concurrent exposure to cadmium, tolerance was roughly halved (Fig. 18.2).

A dose of 5 mg Pb kg^{-1} LW day^{-1} as PbAc killed milk-fed calves within 7 days and as little as 1 mg caused mortality after longer exposures (Zmudzki et al., 1984), confirming the protective influence of a functional rumen. Calves given lead carbonate (PbCO$_3$) while on a milk replacer diet tolerated 2.6 mg kg^{-1} LW day^{-1} for 7 weeks (Lynch et al., 1976a), but succumbed to 3.9 mg kg^{-1} LW day^{-1} after 12 weeks (Lynch et al., 1976b). High tolerance may have been due to the less-soluble source used, but this may have been more representative of the sources encountered on farms.

The MTL is probably <80 mg kg^{-1} DM for PbAc in milk replacers based on casein, but probably higher on those based on vegetable protein.

With pigs and poultry, trends to reduce supplementation with inorganic calcium, phosphorus, zinc and manganese are likely to increase lead accretion in edible produce if dietary lead remains unchanged.

Tolerance in ruminants

Weaned ruminants are extremely tolerant of lead exposure, and a dose that kills milk-fed calves (1 mg Pb kg^{-1} LW day^{-1}) is apparently harmless to weaned calves of the same age (Zmudzki et al., 1984). Calves given 1000 mg Pb kg^{-1} DM as lead sulfate (PbSO$_4$) have shown only a slight reduction in appetite and growth after 4 weeks (Neathery et al., 1987); weaned lambs suffered no adverse effects of similar exposure after 6 or 12 weeks and no reduction in diet digestibility (Fick et al., 1976; Pearl et al., 1983). However, Quarterman et al. (1977) reported rapid weight loss in some individuals after 6 weeks with 200 mg Pb kg^{-1} DM and a moderate sulfur intake. In dairy cows exposed to lead, blood lead peaked at 350 μg l^{-1} and did so sooner when exposure began during lactation than during gestation (9 versus 13 weeks). Tolerance may be reduced during early lactation when the net resorption of bone mineral (see Chapter 4) releases lead previously stored and limits the fresh deposition of lead. The risk posed by anthropomorphic sources of lead will be greatly influenced by the composition of the diet that they contaminate. Reductions in the atmospheric deposition of sulfur are likely to increase lead accretion from soil-contaminated pasture at a given level of lead contamination.

Subclinical effects

Publicity given to the alleged effects of small increases in blood lead on the learning ability and behaviour of children in the 1960s prompted a study of the effects of chronic exposure on ewes in terms of lamb behaviour and ewe reproduction. Lead exposure commenced 1 month prior to mating and continued for 26 weeks, during pregnancy and into lactation (120 or 230 mg elemental lead orally day^{-1}). No behavioural defects were found in their lambs, despite increases in mean blood lead from 64 to 302 µg l^{-1} (Van Gelder et al., 1973), but similar levels of exposure throughout gestation raised blood lead in lambs to 250 µg l^{-1} and slowed their learning behaviour (Carson et al., 1974). In a study with calves (Lynch et al. 1976b), the normal age-related increase in ALAD was totally inhibited by as little as 0.65 mg Pb kg^{-1} LW day^{-1} and was accompanied by a rise in blood lead to 911 µg l^{-1}, although calf growth was unimpaired. Early reductions in ALAD were reported in cows given mixed sources of lead equivalent to 100 mg Pb kg^{-1} DM with no harmful effects on live weight or milk yield (Pinault and Milhaud, 1985). Increases in ZPP concentrations were also found, indicating that lead exposure had been sufficient to inhibit ferrochelatase activity and affect haem synthesis.

Livestock as sources of lead exposure for humans

The problems are essentially the same as those outlined for cadmium earlier (p. 499).

Risk assessment

Lead does not accumulate in muscle during commonly encountered levels of lead exposure (Stark et al., 1998) and there is no risk to humans in consuming carcass meat from livestock exposed to lead. There is, however, a small possible risk from the consumption of offal and a greater risk that carcasses may be rejected because lead levels in offal exceed national MAC (2 and 1 mg kg^{-1} FW in liver and kidney, respectively, in Europe). Chronic exposure to lead may thus raise kidney lead to levels deemed unacceptably high for human consumption. Simulation of high levels of ingestion (100 g kg^{-1} DM) of soils containing 592 and 841 mg Pb kg^{-1} DM by lambs has caused mean kidney lead values to rise to 9.0 and 9.7 mg Pb kg^{-1} DM (Brebner et al., 1993), similar to those attained after feeding 100 mg Pb kg^{-1} DM (as PbAc) (Fick et al., 1976), albeit for a longer period (160 versus 84 days). The accumulation of lead was much lower in a parallel study, but was linearly related to lead intake in both the kidney and liver (Fig. 18.10). No increase in kidney lead occurred in cattle grazing on sludge-amended soils, containing 79–135 mg Pb kg^{-1} DM, for 8 years (Fitzgerald et al., 1985), or in ewes and lambs exposed to much lower levels of soil lead (Rhind et al., 2005). The much higher AA_{Pb} in sewage sludge (p. 512) than in sludge-amended soils implies that risks to the consumer will be reduced by allowing adequate

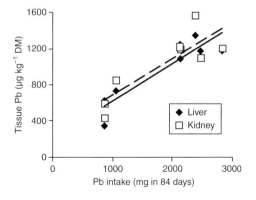

Fig. 18.10. The effects of cumulative lead intake from a sludge-amended soil on the concentrations of lead in the liver and kidney of lambs (data for two soils and two inclusion rates from Hill et al., 1998b).

Residual effects of exposure

Natural exposure to excess lead will often be episodic rather than continuous and it is therefore important to know what happens to tissue lead once the source of exposure is removed. The half-life of tissue lead varies widely between tissues. In sheep, the half-life is shortest in the kidney and spleen (approximately 45 days), intermediate in blood, liver, muscle and heart (approximately 90 days) and longest in the brain (approximately 150 days). Bone lead is exceptional in showing further substantial *increases* 45 days after the cessation of exposure, but a fast decline thereafter (Pearl et al., 1983). Where lead exposure is known to have occurred, the feeding of an acidogenic, grain-based diet for 6 weeks should reduce liver and kidney lead to acceptable levels. Simply transferring animals to uncontaminated pasture prior for a few months will also reduce body lead burdens.

Mercury

The 1970s saw an explosion of interest in the toxicity of mercury following events in Minamata Bay, Japan. Industrial discharges of inorganic mercury into the bay were converted to organic (methylated) mercury by lower marine organisms and entered the human food chain in toxic amounts for a coastal community that was heavily dependent on fish in the diet. Stringent restrictions were placed on acceptable mercury levels in food (0.5 mg kg^{-1} DM by the Food and Drug Administration) and renewed interest was taken in the residual effects of the long-standing exploitation of fungicidal and other toxic properties of the element in agriculture. The methylation of mercury is not, however, a modern phenomenon, but goes back to the dawn of creation. It is a fundamental property of anaerobic microorganisms, expressed in any environment that is naturally or unnaturally endowed with mercury. Puls (1994) has published an extensive bibliography on the subject and here only the essential features pertaining to the toxicity of mercury to livestock and their possible importance as a source of mercury in the human diet will be considered.

Sources

The sources of mercury that pose threats to livestock or enter the food chain via livestock do not originate in the farm environment. Soils contain little mercury (Table 18.2) and pastures and crops even less (<0.1 mg kg^{-1} DM). The only mercury-rich component in normal diets is likely to be fish meal. However, the use of mercury-based fungicides for dressing cereal seed raises mercury concentrations to potentially hazardous levels (approximately 20 mg Hg kg^{-1} DM) and such materials will cause toxicity if fed to cattle (Boyd, 1985) or other species. The accidental consumption of mercurous chloride, intended for the treatment of club root in brassica crops, has caused the deaths of dairy heifers (Simpson et al., 1997). The topical administration of mercury-containing drugs for the treatment of skin infections has also resulted in deaths from mercury poisoning.

Metabolism

Absorption

The metabolism of mercury is dominated by the contrasting behaviours of inorganic and organic forms and differences between ruminants and non-ruminants. Inorganic forms are poorly absorbed, according to one source (AA_{Hg} 1–3%) (Kostial et al., 1978) but not another (5–15% of intake) (Clarkson, 1987). Very young, milk-fed mammals absorb mercury efficiently even from inorganic sources (30–40% of intake) (Kostial et al., 1978). Organic forms are fat-soluble and much more absorbable, although rumen microbes can demethylate methyl mercury (Kozak and Fosberg, 1979). Neathery et al. (1974) recorded an AA_{Hg} of 59% for radiolabelled methyl mercury chloride in lactating dairy cows, and values as high as 90% have been reported for phenyl derivatives (Clarkson, 1987).

Distribution

Mercury has an affinity for sulfhydryl groups, which may explain its tissue distribution, toxicity and ability to interact with other micronutrients (Scheuhammer, 1985). Absorbed mercury is transported primarily in association with glutathione-rich erythrocytes, which contain 20 times more mercury than plasma in exposed animals. In plasma, mercury binds to sulfhydryl groups in albumin. Mercury is avidly retained in tissues by similar associations, although there is interconversion between organic and inorganic mercury forms (NRC, 2005). Little mercury appears in urine (1.1%) or milk (0.17%) and most (72%) is found in muscle (Neathery et al., 1974). Inorganic mercury has an affinity for cysteine-rich molecules such as MT, from which it can displace bound zinc and cadmium (NRC, 2005). MT probably affords protection from inorganic mercury toxicity, but methyl mercury does not react with MT in this way.

Toxicity

Tolerance

Most incidents of mercury poisoning involve exposure to organic sources. When weanling pigs have been given 0.5, 5.0 or 50 mg Hg kg^{-1} DM in either inorganic or organic form, the lowest level in the organic form (CH_3HgCl) caused fatty liver degeneration and the highest was lethal after 3–4 weeks; in comparison, the highest inorganic level caused only subclinical (fatty liver) disorder after rearing to 82 kg LW (Chang et al., 1977). The tolerance of mercury in raw fish was higher than anticipated at 3–8 mg Hg kg^{-1} DM. No signs of toxicity were visible clinically or at post-mortem and there were no teratogenic effects. Fish is rich in selenium and it has been suggested that this attenuates the threat from mercury in fish products (Ganther et al., 1972). The tolerance of poultry to organic forms of mercury is said to be fivefold higher than for inorganic forms (1 versus 0.2 mg kg^{-1} DM) (NRC, 2005). Poultry are particularly susceptible to inorganic mercury toxicity: as little as 2 mg Hg kg^{-1} DM has impaired testicular development in young quail and 8 mg Hg kg^{-1} DM has depressed the fertility of eggs produced by young hens (Scheuhammer, 1987). The addition of selenium to the diet affords substantial protection from inorganic and organic mercury toxicity in poultry and there is a mutual antagonism between these elements (El-Begearmi et al., 1977). The mercury–selenium interaction may be mediated by the induction of the cysteine-rich selenoprotein P (see Chapter 15) (also Chang and Suber, 1982). However, vitamin E supplements also afford protection against mercury toxicity, suggesting that exposure to mercury creates an oxidative stress at the membrane level.

Symptoms

The symptoms of acute toxicity are similar in all farmed species exposed to organic mercury: anorexia, inability to drink, blindness, ataxia, hypersensitivity to people and sudden movements, recumbency and death. In heifers acutely poisoned by mercuric chloride ($HgCl_2$, 300–350 mg kg^{-1} DM) the clinical symptoms were bloody diarrhoea, excessive thirst, salivation, extreme depression and an unsteady gait (Simpson et al., 1997). Post-mortem examination usually reveals generalized congestion and haemorrhagic gastroenteritis. In chickens there is loss of appetite and signs of wing and muscle weakness; demyelination has also been reported (Scheuhammer, 1987).

Diagnosis

Chang et al. (1977) found high levels of mercury in the hair (7.1 mg kg^{-1} DM), kidney (1.0 mg kg^{-1} DM) and liver (0.6 mg kg^{-1} DM) in affected pigs, but no elevation of levels in the central nervous system. Higher mercury concentrations have been found in clinically normal pigs exposed to inorganic mercury than in those poisoned by organic mercury. Tissue mercury concentrations must therefore be interpreted cautiously and in conjunction with pathological evidence. Exceedingly high concentrations of mercury were found in bovine kidneys by Simpson et al. (1997). Diagnostic guidelines have been presented by Puls (1994).

'Newer Essential Elements': Hormesis or Artefact?

Evidence for 'essentiality' in goats

The suggestion that arsenic, of all elements, might be an essential dietary constituent for the normal development and health of livestock caused quite a stir when it was made at a conference in Dubrovnik in 1991. Female goat kids had been given a diet containing <35 µg As kg^{-1} DM and compared with a control group given 350 µg As kg^{-1} DM. Those on the lower arsenic diet grew less well, were less likely to conceive, had more abortions and produced less milk to nourish their young, which showed developmental abnormalities (Anke et al., 1991a). The experiments were repeated 13 times, leaving little doubt that arsenic is an essential element for goats in the chosen highly purified, roughage-free diet. However, similar experiments had been conducted with cadmium and nickel and small supplements of both elements averted the same apparent symptoms of deprivation. In a companion paper (Anke et al., 1991b), aluminium and vanadium were claimed to be essential on the same grounds. The final elements to be added to the essential list for goats were fluorine and rubidium in 1997 (Underwood and Suttle, 1999).

Evidence for essentiality of potentially toxic elements in other species

Young rats maintained under rigorously clean conditions and fed highly purified diets containing <200 µg Pb kg^{-1} DM have been reported to grow more slowly than a lead-supplemented group (Schwarz, 1974) and these findings have been repeated by others (Nielsen, 1996). Disturbances of iron metabolism and anaemia were prominent features in the lead-depleted rat. Artificially reared piglets given a diet containing 31 µg Pb kg^{-1} DM gained 15% less weight than those supplemented with 800 µg Pb kg^{-1} DM. Impaired lipid metabolism was suspected (Kirchgessner et al., 1991).

The theory of hormesis

The concept that PTE with no recognized function in mammals or birds can be essential to their development remains hard to accept, but Paracelsus, writing in the 16th century, observed that most toxic substances are beneficial when administered in small doses. This idea was developed into a concept of 'chemical hormesis' two centuries later by the German scientist Schulz (Stebbing, 1982). An attempt has been made to standardize dose–response curves in terms of the strength of evidence for hormesis (Calabrese and Baldwin, 1998). Inevitably, the evidence for hormesis in mammalian responses is weak and incomplete compared with that obtainable with, for example, yeast cultures because of the small number of doses used (usually two at the low extreme and three at most at the high extreme in separate studies). As for a general mechanism, it has been suggested that over-compensation to small harmful stimuli may be beneficial (van der Woude et al., 2005).

Two possible mechanisms for hormesis

In keeping with their predominantly pathogenic qualities, both lead and cadmium stimulate cells of the immune system at low concentrations in vitro (0.2–1.0 mg l^{-1} and 50–500 nmol l^{-1}, respectively) (Borella and Giardino, 1991). If such mitogenic properties are induced by small dietary supplements of PTE, the adjuvant effect might improve the resistance of livestock to miscellaneous infections, improving longevity and fertility in long-term studies. Mammals have symbiotic relationships with their gut microflora and these are particularly important in the ruminant. Low-dose 'benefits' to the addition of PTE or OBE (see Chapter 17) in particular microbial species might have secondary beneficial effects, without participating in any essential metabolic pathway in the host. For example, it is the bactericidal effects of fluorine that are responsible for its beneficial effects on dental caries in childhood when added at low concentrations to drinking water.

Artefacts of experiments

In embarking on the goat studies cited above, the researchers faced a dilemma. They were stepping into the unknown and did not know how many undiscovered 'essential' elements lay ahead and so they took the precaution of adding nearly every element from the periodic table to their basal diet, from beryllium to zirconium. Given the prolonged exposure often necessary to produce clinical symptoms of 'deficiency', it is possible that harmful responses to these little-studied elements were countered by the additions of PTE or OBE.

Apparent benefits from supplements of potentially toxic elements

The evidence for essentiality cited so far has involved supplementation in ultratrace (µg) amounts. The addition of 30 mg Al kg^{-1} DM to the low-aluminium diet of day-old chicks has been found to increase their growth rate by 22% over the first 18 days of life, but clinical abnormalities in unsupplemented chicks were not reported (Carlisle and Curran, 1993). A tendency for the growth of pigs to be increased by the addition of 50 mg Hg kg^{-1} DM as HgCl$_2$ to the diet has been attributed to bactericidal activity akin to that underlying the growth response to antibiotics (Chang et al., 1977). The growth of broiler chicks has also been improved by cadmium supplementation (Bokori and Fekete, 1996), but the response may again have been pharmacological. Lead supplements have increased egg production in hens in a long-term study, but only by accelerating the reproductive cycle and increasing the overall time spent in lay (Mazliah et al., 1989). When given in trace amounts (mg), PTE may have pharmacological effects that are not proof of essentiality.

Conclusion

In mineral nutrition, the important distinction to be made for each OBE and PTE is whether it yields specific benefit or is an experimental artefact. The nettle must be grasped because the drive to reduce human exposure to PTE seems to know no end, and could conceivably impoverish the diet of consumers if some PTE are truly beneficial. This point is addressed in the final chapter.

References

Abrahams, P.W. and Steismajer, J. (2003) Soil ingestion by sheep grazing the metal enriched floodplain soils on mid-Wales. *Environmental Geochemistry and Health* 25, 17–24.

Alfrey, A.C. (1986) Aluminium. In: Mertz, W. (ed.) *Trace Elements in Human and Animal Nutrition.* Academic Press, New York, pp. 399–413.

Allcroft, R., Burns, K.N. and Hebert, C.N. (1965) *Fluorosis in Cattle. 2. Development and Alleviation: Experimental Studies.* Animal Disease Surveys Report No. 2, Part 2. HMSO, London, 58 pp.

Allen, V.G., Robinson, D.L. and Hembry, F. (1984) Effects of ingested aluminium sulfate on serum magnesium and the possible relationship to hypomagnesaemic tetany. *Nutrition Reports International* 29, 107–115.

Allen, V.G., Horn, F.P. and Fontenot, J.P. (1986) Influence of ingestion of aluminium, citric acid and soil on mineral metabolism of lactating beef cows. *Journal of Animal Science* 62, 1396–1403.

Anderson, J.O., Hurst, J.S., Strong, D.C., Nielsen, H., Greenwood, D.A., Robinson, W., Shupe, J.L., Binns, W., Bagley R. and Draper, C. (1955) Effect of feeding various levels of sodium fluoride to growing turkeys. *Poultry Science* 34, 1147–1153.

Anke, M., Groppel, B. and Krause, V. (1991a) The essentiality of the toxic elements cadmium, arsenic and nickel. In: Momcilovic, B. (ed.) *Proceedings of the Seventh International Symposium on Trace Elements in Man and Animals (TEMA 7), Dubrovnik.* IMI, Zagreb, pp. 11-6–11-8.

Anke, M., Groppel, B. and Krause, V. (1991b) The essentiality of the toxic elements aluminium and vanadium. In: Momcilovic, B. (ed.) *Proceedings of the Seventh International Symposium on Trace Elements in Man and Animals (TEMA 7), Dubrovnik.* IMI, Zagreb, pp. 11-9–11-10.

Archer, F.C. and Hodgson, J.H. (1987) Total and extractable trace element content of soils in England and Wales. *Journal of Soil Science* 38, 421–431.
Arnold, W. (1988) Arsenic. In: Seiler, H.G. (ed) *Handbook on Toxicity of Inorganic Compounds*. Marcel Dekker, New York, pp. 79–93.
Aulerlich, R.J., Napolitano, A.C., Bursian, S.J., Olson, B.A. and Hochstein, J.R. (1987) Chronic toxicity of dietary fluorine to mink. *Journal of Animal Science* 65, 1759–1767.
Bacon, J.R., Berrow, M.L. and Shand, C.A. (1992) Isotopic composition as an indicator of origin of lead accumulation in surface soils. *International Journal of Environmental Analytical Chemistry* 46, 71–76.
Bahri, L.E. and Romdane, S.B. (1991) Arsenic poisoning in livestock. *Veterinary and Human Toxicology* 33, 259–264.
Becker, D.E., Griffith. J.M., Hobbs, C.S. and MacIntire, W.H. (1950) The alleviation of fluorine toxicosis by means of certain aluminum compounds. *Journal of Animal Science* 9, 647.
Beresford, N.A., Crout, N.M. and Mayes, R.W. (2001) The transfer of arsenic to sheep tissues. *Journal of Agricultural Science* 136, 331–334.
Bixler, D. and Muhler, J.C. (1960) Retention of fluoride in soft tissues of chickens receiving different fat diets. *Journal of Nutrition* 70, 26–30.
Blakley, B.R. (1984) A retrospective study of lead poisoning in cattle. *Veterinary and Human Toxicology* 26, 505–507.
Blaxter, K.L. (1950) Lead as a nutritional hazard to farm livestock. II. Absorption and excretion of lead by sheep and rabbits. *Journal of Comparative Pathology* 60, 140–159.
Bokori, J. and Fekete, S. (1996) Complex study of the physiological role of cadmium. IV Effects of prolonged exposure of broiler chickens to cadmium. *Acta Physiologica Hungarica* 44, 57–74.
Borella, P. and Giardino, A. (1991) Lead and cadmium at very low doses affect *in vitro* immune response of human lymphocytes. *Environmental Research* 55, 165–177.
Boyd, J.H. (1985) Organomercuric poisoning in fat cattle. *Veterinary Record* 116, 443–444.
Bramley, R.G.V. (1990) Cadmium in New Zealand agriculture. *New Zealand Journal of Agricultural Research* 33, 505–519.
Bratton, G.R. (1984) Laboratory diagnosis of lead poisoning in cattle: a reassessment and review. *Veterinary and Human Toxicology* 26, 387–392.
Bratton, G.R., Childress, M., Zmudski, J., Womac, C., Rowe, L.D. and Tiffany-Castiglioni, E. (1986) Delta aminolaevulinic acid dehydratase (EC 4.3.1.24) activity in erythrocytes from cattle administered low concentrations of lead acetate. *American Journal of Veterinary Research* 47, 2068–2074.
Brebner, J. (1987) The role of soil ingestion in the trace element nutrition of grazing livestock. PhD thesis, Imperial College, University of London.
Brebner, J., Thornton, I., McDonald, P. and Suttle, N.F. (1985) The release of trace elements from soils under conditions of simulated rumenal and abomasal digestion. In: Mills, C.F., Bremner, I. and Chesters, J.K. (eds) *Proceedings of the Fifth International Symposium on Trace Elements in Animals and Man*. Commonwealth Agricultural Bureaux, Farnham Royal, UK, pp. 850–852.
Brebner, J., Hall, J., Smith, S., Stark, B., Suttle, N.F. and Sweet, N. (1993) Soil ingestion is an important pathway for the entry of potentially toxic elements (PTE) from sewage sludge-treated pasture into ruminants and the food chain. In: Allan, R.J. and Nriagu, J.O. (eds) *Proceedings of the Ninth International Conference on Heavy Metals in the Environment*, Vol. 1. CEP, Edinburgh, pp. 446–449.
Bremner, I. (1978) Cadmium toxicity. *World Review of Nutrition and Dietetics* 32, 165–197.
Bremner, I. and Campbell, J.K. (1980) The influence of dietary copper intake on the toxicity of cadmium. *Annals of the New York Academy of Sciences* 355, 319–332.
Burns, K.N. and Allcroft, R. (1962) The use of tail bone biopsy for studying skeletal deposition of fluorine in cattle. *Research in Veterinary Science* 3, 215–218.
Burns, K.N. and Allcroft, R. (1966) *Fluorosis in Cattle: Occurrence, Diagnosis and Alleviation*. Report of the Fourth International Meeting, World Association for Buiatrics, pp. 94–115.
Butler, E.J., Nisbet, D.I. and Robertson, J.M. (1957) Osteoporosis in lambs in a lead mining area. 1. A study of the naturally occurring disease. *Journal of Comparative Pathology* 67, 378–398.
Calabrese, E.J. and Baldwin, L.A. (1998) Hormesis as a biological hypothesis. *Environmental Health Perspectives* 106 (Suppl. 1), 357–362.
Campbell, J.K. and Mills, C.F. (1974) Effects of dietary cadmium and zinc on rats maintained on diets low in copper. *Proceedings of the Nutrition Society* 33, 15A–17A.

Campbell, J.K. and Mills, C.F. (1979) The toxicity of zinc to pregnant sheep. *Environmental Research* 20, 1–13.

Carlisle, E.M. and Curran, M.J. (1993) Aluminium: an essential element for the chick. In: Anke, M., Meissner, D. and Mills, C.F. (eds) *Proceedings of the Eighth International Symposium on Trace Elements in Man and Animals*. Verlag Media Touristik, Gersdorf, Germany, pp. 695–698.

Carson, T.L., Van Gelder, G.A., Karas, G.C. and Buck, W.B. (1974) Slowed learning in lambs prenatally exposed to lead. *Archives of Environmental Health* 29, 154–156.

Casteel, S., Cowart, R.P., Weis, C.P., Henningsen, G., Hoffman, E., Brattin, W.J., Guzman, R.E., Starost, M.F., Payne, J.T., Stockham, S.L., Becker, S.V., Drexler, J.W. and Turk, J. (1997) Bioavailability of lead to juvenile swine dosed with soil from the Smuggler Mountain NPL Site of Aspen, Colorado. *Fundamental and Applied Toxicology* 36, 177–187.

Casteel, S., Evans, T., Turk, J., Basta, N., Weis, C., Henningsen, G. and Hoffman, E. (2001) Refining the risk assessment of metal-contaminated soils. *International Journal of Hygiene and Environmental Health* 203, 473–474.

Chang, C.W.J. and Suber, R. (1982) Protective effect of selenium on methyl mercury toxicity: a possible mechanism. *Bulletin of Environmental Contamination and Toxicology* 29, 285–289.

Chang, C.W.J., Nakamura, R.M. and Brooks, C.C. (1977) Effect of varied dietary levels and forms of mercury on swine. *Journal of Animal Science* 45, 279–285.

Chowdury, B.-A. and Chandra, R.K. (1987) Biological and health implications of toxic heavy metal and essential trace element interactions. In: *Progress in Food and Nutrition Science*, Volume II. Permagon Press, Oxford, UK, pp. 55–113.

Clarkson, T.W. (1987) Mercury. In: Mertz, W. (ed.) *Trace Elements in Human and Animal Nutrition*, Vol. 7. Academic Press, New York, p. 417.

Coppock, R.W., Wagner, W.C. and Reynolds, R.D. (1988) Migration of lead in a glass-lined, bottom-loading silo. *Veterinary and Human Toxicology* 30, 458–459.

Cox, K.A. and Dunn, M.A. (2001) Aluminium toxicity alters the regulation of calbindin-D28k protein and mRNA expression in chick intestine. *Journal of Nutrition* 131, 2007–2013.

Dalgarno, A.C. (1980) The effect of low level exposure to dietary cadmium on cadmium, zinc, copper and iron content of selected tissues of growing lambs. *Journal of Science in Food and Agriculture* 31, 1043–1049.

Doberenz, A.R., Kurnick, A.A., Kurtz, E.B., Kemmerer, A.R. and Reid, B.L. (1964) Effect of a minimal fluoride diet on rats. *Proceedings of the Society for Experimental Biology and Medicine* 117, 689–693.

Doyle, J.J. and Pfander, W.H. (1975) Interactions of cadmium with copper, iron, zinc and manganese in ovine tissues. *Journal of Nutrition* 105, 599–606.

Doyle, J.J. and Spaulding, J.E. (1978) Toxic and essential elements in meat – a review. *Journal of Animal Science* 47, 398–419.

Doyle, J.J., Pfander, W.H., Grebing, S.E. and Pierce, J.O. (1974) Effect of dietary cadmium on growth, cadmium absorption and cadmium tissue levels in growing lambs. *Journal of Nutrition* 104, 160–166.

Edens, F.W. and Garlich, J.D. (1983) Lead-induced egg production in Leghorn and Japanese quail hens. *Poultry Science* 62, 1757–1763.

El-Begearmi, M.M., Sunde, M.L. and Ganther, H.E. (1977) A mutual protective effect of mercury and selenium in Japanese quail. *Poultry Science* 56, 313–322.

European Communities (2002) Directive 2002/32/EC of The European Parliament and the Council. *Official Journal of the European Commission* L140/10–L140/21.

Fick, K.R., Ammerman, C.B., Miller, S.M., Simpson, C.F. and Loggins, P.E. (1976) Effect of dietary lead on performance, tissue mineral composition and lead absorption in sheep. *Journal of Animal Science* 42, 515–523.

Finley, M.T. and Dieter, M.P. (1978) Influence of laying on lead accumulation in Mallard ducks. *Journal of Toxicology and Environmental Health* 4, 123–129.

Fitzgerald, P.R., Peterson, J. and Lue-Hing, C. (1985) Heavy metals in tissues of cattle exposed to sludge-treated pasture for eight years. *American Journal of Veterinary Research* 46, 703–707.

Forsyth, D.M., Pond, W.G. and Krook, L. (1972) Dietary calcium and fluoride interactions in swine: *in utero* and neonatal effects. *Journal of Nutrition* 102, 1639–1646.

Fox, M.R.S. (1987) Assessment of cadmium, lead and vanadium status of large animals as related to the human food chain. *Journal of Animal Science* 65, 1744–1752.

Fuge, R.G. and Andrews, M.J. (1988) Fluorine in the UK environment. *Environmental Geochemistry and Health* 10, 96–104.

Fullmer, C.S., Edelstein, S. and Wasserman, R.H. (1985) Lead-binding properties of intestinal calcium-binding protein. *Journal of Biological Chemistry* 260, 6816–6819.

Ganther, H.E., Goudie, C., Sunde, M.I., Kopecky, M.J., Wagner, P., Ott, S.-H. and Hoekstra, W.G. (1972) Selenium: relation to decreased toxicity of methylmercury added to diets containing tuna. *Science* 175, 1122–1124.

Garcia-Bojalil, C.M., Ammerman, C.B., Henry, P.R., Littell, R.C. and Blue, W.G. (1988) Effects of dietary phosphorus, soil ingestion and dietary intake level on performance, phosphorus utilisation and serum and alimentary tract mineral concentrations in lambs. *Journal of Animal Science* 66, 1508–1519.

Gaskin, J.W., Brobst, R.B., Miller, W.B. and Tollner, E.W. (2003) Long-term biosolids application effects on metal concentrations in soil and Bermudagrass forage. *Journal of Environmental Quality* 32, 146–152.

Geeson, N.A., Abrahams, P.W., Murphy, M.P. and Thornton, I. (1998) Fluorine and metal enrichment of soils and pasture herbage in the old mining areas of Derbyshire, UK. *Agriculture, Ecosystems and the Environment* 217–231.

Gerry, R.W., Carrick, C.W., Roberts, R.E. and Hauge, S.M. (1949) Raw rock phosphate in laying rations. *Poultry Science* 28, 19–23.

Grace, N.D., Loganathan, P., Hedley, M.J. and Wallace, G.C. (2003a) Ingestion of soil fluorine: its impact on the fluorine metabolism and status of grazing young sheep. *New Zealand Journal of Agricultural Research* 46, 279–286.

Grace, N.D., Loganathan, P., Hedley, M.J. and Wallace, G.C. (2003b) Ingestion of soil fluorine: its impact on the fluorine metabolism of dairy cows. *New Zealand Journal of Agricultural Research* 48, 23–27.

Greger, J.L. (1993) Aluminium metabolism. *Annual Review of Nutrition* 13, 43–63.

Hahn, P.H.B. and Guenther, W. (1986) Effect of dietary fluoride and aluminium on laying hen performance and fluoride concentration in blood, soft tissue, bone and egg. *Poultry Science* 65, 1343–1349.

Harvey, J.M. (1952) Chronic endemic fluorosis of Merino sheep in Queensland. *Queensland Journal of Agricultural Science* 9, 47–141.

Hemingway, R.G. (1977) Fluorine retention in growing sheep: a comparison between two phosphorus supplements as sources of fluorine. *Proceedings of the Nutrition Society* 36, 82A.

Hill, J., Stark, B.A., Wilkinson, J.M., Curran, M.K., Lean, I.J., Hall, J.E. and Livesey, C.T. (1998a) Accumulation of potentially toxic elements by sheep given diets containing soil and sewage sludge. 1. Effect of type of soil and level of sewage sludge in the diet. *Animal Science* 67, 73–86.

Hill, J., Stark, B.A., Wilkinson, J.M., Curran, M.K., Lean, I.J., Hall, J.E. and Livesey, C.T. (1998b) Accumulation of potentially toxic elements by sheep given diets containing soil and sewage sludge. 2. Effect of the ingestion of soils historically treated with sewage sludge. *Animal Science* 67, 87–96.

Hillman, D., Bolenbaugh, D.L. and Convey, E.M. (1979) Hypothyroidism and anaemia related to fluoride in dairy cattle. *Journal of Dairy Science* 62, 416–423.

Houpert, P., Mehennaoni, S., Joseph-Enriquez, B., Federspiel, B. and Milhaud, G. (1995) Pharmokinetics of cadmium following intravenous and oral administration to non-lactating ewes. *Veterinary Research* 26, 145–154.

Jacob, K.D. and Reynolds, D.S. (1928) The fluorine content of phosphate rock. *Journal of Official Agricultural Chemists* 11, 237–258.

Jeng, S.L., Lee, S.J., Liu, Y.F., Yang, S.C. and Liou, P.P. (1997) Effects of lead ingestion on concentrations of lead in tissues and eggs of laying Tsaiya ducks in Taiwan. *Poultry Science* 76, 13–16.

Kennedy, S., Rice, D.A. and Cush, P.F. (1986) Neuropathology of experimental 3-nitro-4-hyroxyphenyl-arsonic acid toxicosis in pigs. *Veterinary Pathology* 23, 454–461.

King, R.H., Brown, W.G., Amenta, V.C.M., Shelley, B.C., Handson, P.D., Greenhill, N.B. and Wilcock, G.P. (1992) The effect of dietary cadmium intake on the growth performance and retention in growing pigs. *Animal Feed Science and Technology,* 37, 1–7.

Kirchgessner, M., Plass, D.L. and Reichlmayr-Lais, A.M. (1991) Lead deficiency in swine. In: Momcilovic, B. (ed.) *Proceedings of the Seventh International Symposium on Trace Elements in Man and Animals (TEMA 7), Dubrovnik.* IMI, Zagreb, pp. 11-20–11-21.

Kollmer, W.E. and Berg, D. (1989) The influence of a zinc-, calcium- or iron-deficient diet on the resorption and kinetics of cadmium in the rat. In: Southgate, D.A.T., Johnson, I.T. and Fenwick, G.R. (eds) *Nutrient Availability: Chemical and Biological Aspects.* Royal Society of Chemistry Special Publication No. 72, Cambridge, UK, pp. 287–289.

Kostial, K., Jugo, S., Rabar, I. and Maljkovic, T. (1978) Influence of age on metal metabolism and toxicity. *Environmental Health Perspectives* 25, 81–86.

Kozak, S. and Forsberg, C.W. (1979) Transformation of mercuric chloride and methyl mercury by the rumen microflora. *Applied Environmental Microbiology* 38, 626–636.

Krueger, G.L., Morris, T.K., Suskind, R.R. and Widner, E.M. (1985) The health effects of aluminium compounds in mammals. *CRC Critical Reviews in Toxicology* 13, 1–24.

Kumar, A. and Susheela, A.K. (1995) Effects of chronic fluoride toxicity on the morphology of ductus epididymis and the maturation of spermatozoa in the rabbit. *International Journal of Experimental Pathology* 76, 1–11.

Langlands, J.P., Donald, G.E. and Bowles, J.E. (1988) Cadmium concentrations in liver, kidney and muscle in Australian sheep and cattle. *Australian Journal of Experimental Agriculture* 28, 291–297.

Leach, R.M. Jr, Wang, K.W. and Baker, D.E. (1979) Cadmium and the food chain: the effect of dietary cadmium on tissue composition in chicks and laying hens. *Journal of Nutrition* 109, 437–443.

Lee, J., Rounce, J.R., MacKay, A.D. and Grace, N.D. (1996) Accumulation of cadmium with time in Romney sheep grazing ryegrass–white clover pasture: effect of cadmium from pasture and soil intake. *Australian Journal of Agricultural Research* 47, 877–894.

Linden, A., Olsson, I-M., Bensryd, I., Lundh, T., Skerfving, S. and Oskarsson, A. (2003) Monitoring cadmium in the food chain from soil via crops and feed to pig blood and kidney. *Ecotoxicology and Environmental Safety* 55, 213–222.

Lloyd, M.K. (1983) Environmental toxicity. In: Suttle, N.F., Gunn, R.G., Allen, W.M., Linklater, K.A. and Wiener, G. (eds) *Trace Elements in Animal Production and Veterinary Practice*. British Society of Animal Production Occasional Publication No. 7, Edinburgh, pp. 119–124.

Loganathan, P., Hedley, M.J., Wallace, G.C. and Roberts, A.H.C. (2001) Fluoride accumulation in pasture forages and soils following the long-term application of phosphorus fertilisers. *Environmental Pollution* 115, 275–282.

Loganathan, P., Hedley, M.J. and Grace, N.D. (2008) Pasture soils contaminated with fertilizer-derived cadmium and fluorine: livestock effects. *Reviews in Environmental Contamination and Toxicology* 192, 29–66.

Longhurst, R.D., Roberts, A.H.C. and Waller, J.E. (2004) Concentrations of arsenic, cadmium, copper, lead and zinc in New Zealand pastoral topsoils and herbage. *New Zealand Journal of Agricultural Research* 47, 23–32.

Lynch, G.P., Jackson, E.D. and Smith, D.F. (1976a) Responses of young calves to oral doses of lead. *Journal of Dairy Science* 59, 1490–1494.

Lynch, G.P., Smith, D.F., Fisher, M., Pike, T.L. and Weinland, B.T. (1976b) Physiological responses of calves to cadmium and lead. *Journal of Animal Science* 42, 410–421.

Madan, J., Puri, J.P. and Singh, J.K. (2009) Growth, feed conversion efficiency and blood profile of buffalo calves consuming high levels of fluoride. *Tropical Animal Health and Production* 41, 295–298.

Masters, D.G., White, C.L., Peter, D.W., Purser, D.B., Roe, S.P. and Barnes, M.J. (1992). A multi-element supplement for grazing sheep. II. Accumulation of trace elements in sheep fed different levels of supplement. *Australian Journal of Agricultural Research* 43, 809–817.

Maurer, R.L. and Day, H.G. (1957) The non-essentiality of fluorine in nutrition. *Journal of Nutrition* 62, 561–573.

Mazliah, J., Barron, S., Bental, E. and Reznik, I. (1989) The effect of chronic lead intoxication in mature chickens. *Avian Diseases* 33, 566–570.

McCaughey, W. (2007) Inorganic and organic poisons. In: Aitken, I.D. (ed.) *Sheep Diseases*. Blackwell Scientific, London.

McClure, F.J. (1949) Fluorine in foods: survey of recent data. *Public Health Reports* 64, 1061–1074.

McDowell, L.R. (2003) Aluminium, arsenic, cadmium, lead and mercury. in: *Minerals in Animal and Human Nutrition*. Academic Press, New York, pp. 355–356.

McGown, E.L. and Suttie, J.W. (1974) Influence of fat and fluoride on gastric emptying time. *Journal of Nutrition* 104, 909–915.

Mehdi, A.W.R., Al-Soudi, K.A., Al-Jiboori, N.A.J. and Al-Hiti, M.K. (1983) Effect of high fluoride intake on chicken performance, ovulation, spermatogenesis and bone fluoride content. *Fluoride* 16, 37–43.

Messer, H.H., Armstrong, W.D. and Singer, L. (1974) Essentiality and function of fluoride. In: Hoekstra, W.G., Suttie, J.W., Ganther, H.E. and Mertz, W. (eds) *Trace Element Metabolism in Animals – 2*. University Park Press, Baltimore, Maryland, pp. 425–437.

Milhaud, G., Cazieux, A. and Enriquez, B. (1984) Experimental studies on fluorosis in the suckling lamb. *Fluoride* 17, 107–114.

Mills, C.F. and Dalgarno, A.C. (1972) Copper and zinc status of ewes and lambs receiving increased dietary concentrations of cadmium. *Nature, London* 239, 171–173.

Milne, D.B. and Schwarz, K. (1974) Effect of different fluorine compounds on growth and bone fluoride levels in rats. In: Hoekstra, W.G., Suttie, J.W., Ganther, H.E. and Mertz, W. (eds) *Trace Element Metabolism in Animals – 2*. University Park Press, Baltimore, Maryland, pp. 710–714.

Moffat, W.H. (1993) Long term residual effects of lead mining on man and grazing livestock within a rural community in Southern Scotland. PhD thesis, University of Edinburgh, pp. 152–214.

Morcombe, P.W., Petterson, D.S., Masters, H.G., Ross, P.J. and Edwards, J.R. (1994a) Cadmium concentrations in kidneys of sheep and cattle in Western Australia. 1. Regional distribution. *Australian Journal of Agricultural Research* 45, 851–862.

Morcombe, P.W., Petterson, D.S., Ross, P.J. and Edwards, J.R. (1994b) Soil and agronomic factors associated with cadmium accumulations in kidneys of grazing sheep. *Australian Veterinary Journal* 71, 404–406.

Morrison, J.N. and Quarterman, J. (1987) The relationship between iron status and lead absorption in rats. *Biological Trace Element Research* 14, 115–126.

Morrison, J.N., Quarterman, J. and Humphries, W.R. (1977) The effect of dietary calcium and phosphate on lead poisoning in lambs. *Journal of Comparative Pathology* 87, 417–429.

Motzok, I. and Branion, H.D. (1958) Influence of fluorine on phosphatase activities of plasma and tissues of chicks. *Poultry Science* 37, 1469–1471.

Neathery, M.W., Miller, W.J., Gentry, R.P., Stake, P.E. and Blackmon, D.M. (1974) Cadmium[109] and methyl mercury metabolism, tissue distribution and secretion into milk in dairy cows. *Journal of Dairy Science* 57, 1177–1184.

Neathery, M.W., Miller, W.J., Gentry, R.P., Crowe, C.T., Alfaro, E., Fielding, A.S., Pugh, D.G. and Blackmore, D.M. (1987) Influence of high dietary deed on selenium metabolism in dairy calves. *Journal of Dairy Science* 70, 645–652.

Nielsen, F.H. (1996) Other trace elements. In: Filer, L.J. and Ziegler, E.E. (eds) *Present Knowledge in Nutrition*, 7th edn. International Life Sciences Institute, Nutrition Foundation, Washington, DC.

NRC (2005) *Mineral Tolerances of Animals,* 2nd edn. National Academy of Sciences, Washington, DC.

Oelschlager, W.K., Loeffler, K. and Opletova, L. (1970) Retention of fluorine in bones and bone sections of oxen in respect of equal increases in fluorine in the form of soil, flue dust from an aluminium reduction plant and sodium fluoride. *Landwirsch Forsch* 23, 214–224.

Oliver, D.P., Tiller, K.G., Conyers, M.K., Slattery, W.J., Alston, A.M. and Merry, R.H. (1996) Effectiveness of liming to minimise uptake of cadmium by wheat and barley grain grown in the field. *Australian Journal of Agricultural Research* 47, 1181–1193.

Olsson, I.-G., Jonsson, S. and Oskarsson, A. (2001) Cadmium and zinc in kidney, liver, muscle and mammary tissue from dairy cows in conventional and organic farming. *Journal of Environmental Monitoring* 3, 531–538.

Pace, L.W., Turnquist, S.E., Casteel, S.W., Johnson, P.J. and Frankeny, R.L. (1997) Acute arsenic poisoning in five horses. *Veterinary Pathology* 34, 160–164.

Patra, R.C., Dwivedi, S.K., Bhardwaj, B. and Swarup, D. (2000) Industrial fluorosis in cattle and buffalo around Udaipur, India. *Science of the Total Environment* 253, 145–150.

Pearl, D.S., Ammerman, C.B., Henry, P.R. and Littell, R.C. (1983) Influence of dietary lead and calcium on tissue lead accumulation and depletion, lead metabolism and tissue mineral composition in sheep. *Journal of Animal Science* 56, 1416–1426.

Peirce, A.W. (1954) Studies on fluorosis of sheep. 2. The toxicity of water-borne fluoride for mature grazing sheep. *Australian Journal of Agricultural Research* 5, 545–554.

Petersen, D.S., Masters, H.G., Spiejers, E.J., Williams, D.E. and Edwards, J.R. (1991) Accumulation of cadmium in sheep. In: Momcilovic, B. (ed.) *Proceedings of the Seventh International Symposium on Trace Elements in Man and Animals (TEMA 7), Dubrovnik.* IMI, Zagreb, pp. 26-13–26-14.

Phillips, C., Gyori, Z. and Kovacs, B. (2003) The effect of adding cadmium and lead alone or in combination to the diet of pigs on their growth, carcase composition and reproduction. *Journal of the Science of Food and Agriculture* 83, 1357–1365.

Phillips, P.H. and Suttie, J.W. (1960) The significance of time in intoxication of domestic animals by fluoride. *Archives of Industrial Health* 21, 343–345.

Phillips, P.H., Hart, E.B. and Bohstedt, G. (1934) *Chronic Toxicosis in Dairy Cows due to the Ingestion of Fluorine*. Research Bulletin Wisconsin Agricultural Experiment Station No. 123.

Pinault, L. and Milhaud, G. (1985) Evaluation of subclinical lead poisoning in dairy cattle. *Veterinary Pharmacology and Toxicology* 61, 715–124.

Pinot, F., Kreps, S.E., Bachelet, M., Hainault, P., Bakonyi, M. and Polla, B.S. (2002) Cadmium in the environment: sources, mechanisms of biotoxicity. *Reviews in Environmental Health* 15, 299–323.

Powell, G.W., Miller, W.I., Morton, J.D. and Clifton, C.M. (1964) Influence of dietary cadmium level and supplemental zinc on cadmium toxicity in the bovine. *Journal of Nutrition* 84, 205–211.

Proudfoot, F.G., Jackson, E.D., Hulan, H.W. and Salisbury, C.D.C. (1991) Arsanilic acid as a growth promoter for chicken broilers when administered via either the feed or drinking water. *Canadian Journal of Animal Science* 71, 221–226.

Puls, R. (1994) *Mineral Levels in Animal Health*, 2nd edn. Sherpa International, Clearbrook, British Columbia, Canada.

Quarterman, J. and Morrison, J. N. (1975) The effects of dietary calcium and phosphorus on the retention and excretion of lead in rats. *British Journal of Nutrition* 34, 351–362.

Quarterman, J., Morrison, J.N., Humphries, W.R. and Mills, C.F. (1977) The effect of dietary sulfur and of castration on lead poisoning in lambs. *Journal of Comparative Pathology* 87, 405–416.

Quarterman, J., Morrison, J.N. and Humphries, W.R. (1978) The influence of high dietary calcium and phosphate on lead uptake and release. *Environmental Research* 17, 60–67.

Rhind, S.J., Kyle, C.E. and Owen, J. (2005) Accumulation of potentially toxic metals in the liver tissue of sheep grazed on sewage sludge treated pastures. *Animal Science* 81, 107–113.

Rice, D.A., Kennedy, S., McMurray, C.H. and Blanchflower, W.J. (1985) Experimental 3-nitro-4-hydroxyphenylarsonic acid toxicosis in pigs. *Research in Veterinary Science* 39, 47–51.

Rimbach, G., Pallauf, J., Brandt, K. and Most, E. (1996) Effect of phytic acid and microbial phytase on Cd accumulation, Zn status and the apparent absorption of Ca, P, Mg, Fe, Zn, Cu and Mn in growing rats. *Annals of Nutrition and Metabolism* 39, 361–370.

Robinson, D.L., Hemkes, O.J. and Kemp, A. (1984) Relationships among forage aluminium levels, soil contamination on forages and availability of elements to dairy cows. *Netherlands Journal of Agricultural Science* 32, 73–80.

Roholm, K. (1937) *Fluorine Intoxication*. H.K. Lewis, London.

Ross, E.A. (2000) Lead contamination of calcium supplements. *Journal of the American Veterinary Medical Association* 284,1425–1429.

Ruliffson, W., Burns, L.V. and Hughes, J.S. (1963) The effect of fluoride ion on ^{59}Fe iron levels in the blood of rats. *Transactions of the Kansas Academy of Science* 66, 52.

Sasser, L.B. and Jarboe, G.E. (1980) Intestinal absorption and retention of cadmium in neonatal pigs compared to rats and guinea pigs. *Journal of Nutrition* 110, 1641–1647.

Scheuhammer, A.M. (1987) The chronic toxicity of aluminium, cadmium, mercury and lead in birds: a review. *Environmental Pollution* 46, 263–295.

Schwarz, K. (1974) New essential trace elements (Sn, V, F, Si): progress report and outlook. In: Hoekstra, W.G., Suttie, J.W., Ganther, H.E. and Mertz, W. (eds) *Proceedings of the Second International Symposium on Trace Element Metabolism in Animals*. University Press, Baltimore, Maryland, pp. 355–380.

Schwarz, K. and Milne, D.B. (1972) Fluorine requirements for growth in the rat. *Bioinorganic Chemistry* 1, 331–336.

Seffner, W. and Tuebener, W. (1983) Antidotes in experimental fluorosis in pigs: morphological studies. *Fluoride* 16, 33–37.

Selby, L.A., Case, A.A., Osweiler, G.D. and Mayes, H.M. (1977) Epidemiology and toxicology of arsenic poisoning in domestic animals. *Environmental Health Perspectives* 19, 183–189.

Sell, J.L. (1975) Cadmium in the laying hen: apparent absorption, tissue distribution and virtual absence of transfer into eggs. *Poultry Science* 54, 1674–1678.

Shearer, T.R., Kolstad, D.L. and Suttie, J.W. (1978a) Bovine dental fluorosis: histologic and physical characteristics. *American Journal of Physiology* 212, 1165–1168.

Shearer, T.R., Kolstad, D.L. and Suttie, J.W. (1978b) Electron probe microanalysis of fluorotic bovine teeth. *American Journal of Veterinary Research* 39, 1393–1398.

Sherlock, J.C. (1989) Aluminium in foods and the diet. In: Massey, R.C. and Taylor, D. (eds) *Aluminium in Food and the Environment*. Royal Society of Chemistry Special Publication, London, pp. 68–76.

Shupe, J.L. (1980) Clinicopathologic features of fluoride toxicosis in cattle. *Journal of Animal Science* 51, 746–757.

Shupe, J.L., Miner, M.L., Harris, L.E. and Greenwood, D.A. (1962) Relative effects of feeding hay atmospherically contaminated by fluoride residue, normal hay plus calcium fluoride, and normal hay plus sodium fluoride to dairy heifers. *American Journal of Veterinary Research* 23, 777–787.

Shupe, J.L., Harris, L.E., Greenwood, D.A., Butcher, J.E. and Nielsen, H.M. (1963a) The effect of fluorine on dairy cattle. 5. Fluorine in the urine as an estimator of fluorine intake. *American Journal of Veterinary Research* 24, 300–306.

Shupe, J.L., Miner, M.L., Greenwood, D.A., Harris, L.E. and Stoddard, G.E. (1963b) The effect of fluorine on dairy cattle. 2. Clinical and pathologic effects. *American Journal of Veterinary Research* 24, 964–979.

Simpson, V.R., Stuart, N.C., Munro, R., Hunt, A. and Livesey, C.T. (1997) Poisoning of dairy heifers by mercurous chloride. *Veterinary Record* 140, 549–552.

Smith, G.M. and White, C.L. (1997) A molybdenum–sulfur–cadmium interaction in sheep. *Australian Journal of Agricultural Research* 48, 147–154.

Smith, R.M., Griel, L.C., Muller, L.D., Leach, R.M. and Baker, D.H. (1991a) Effects of dietary cadmium chloride throughout gestation on blood and tissue metabolites of primigravid and neonatal dairy cattle. *Journal of Animal Science* 69, 4078–4087.

Smith, R.M., Griel, L.C., Muller, L.D., Leach, R.M. and Baker, D.H. (1991b) Effects of long term dietary cadmium chloride on tissue, milk and urine mineral concentrations of lactating dairy cows. *Journal of Animal Science* 69, 4088–4096.

Snook, L.C. (1962) Rock phosphate in stock feeds. The fluorine hazard. *Australian Veterinary Journal* 38, 42–47.

Stark, B.A., Livesey, C.T., Smith, S.R., Suttle, N.F., Wilkinson, J.M. and Cripps, P.J. (1998) *Implications of Research on the Uptake of PTEs from Sewage Sludge by Grazing Animals*. Report to the Department of the Environment, Transport and the Regions (DETR) and the Ministry of Agriculture, Fisheries and Food (MAFF). WRc, Marlow, UK.

Stebbing, A.R.D. (1982) Hormesis – the stimulation of growth by low levels of inhibitors. *Science of the Total Environment* 22, 213–234.

Stoewsand, G.S., Telford, J.N., Anderson, J.L., Bache, C.A., Gutenmann, W.H. and Lisk, D.J. (2005) Toxicological studies with Japanese quail fed winter wheat grown on municipal sludge-amended soil. *Archives of Experimental Contamination and Toxicology* 13, 297–301.

Sullivan, T.W., Douglas, J.H. and Gonzalez, N.J. (1994) Levels of various elements of concern in feed phosphates of domestic and foreign origin. *Poultry Science* 73, 520–528.

Supplee, W.C. (1961) Production of zinc deficiency in turkey poults by dietary cadmium. *Poultry Science* 40, 827–831.

Suttie, J.W. (1967) Vertebral biopsies in the diagnosis of bovine fluoride toxicosis. *American Journal of Veterinary Research* 28, 709–712.

Suttie, J.W. (1969) Fluoride content of commercial dairy concentrates and alfalfa forage. *Journal of Agricultural and Food Chemistry* 17, 1350–1352.

Suttie, J.W. (1980) Nutritional aspects of fluoride toxicosis. *Journal of Animal Science* 51, 759–766.

Suttie, J.W. and Faltin, E.C. (1973) Effects of sodium fluoride on dairy cattle: influence of nutritional state. *American Journal of Veterinary Research* 34, 479–483.

Suttie, J.W. and Kolstad, D.L. (1977) Effects of dietary fluoride ingestion on ration intake and milk production. *Journal of Dairy Science* 60, 1568–1573.

Suttie, J.W., Miller, R.F. and Phillips, P.H. (1957) Studies of the effects of dietary NaF on dairy cows. 1. The physiological effects and the developmental symptoms of fluorosis. *Journal of Nutrition* 63, 211–224.

Suttie, J.W., Phillips, P.H. and Miller, R.F. (1958) Studies of the effects of dietary sodium fluoride on dairy cows. 3. Skeletal and soft tissue fluorine deposition and fluorine toxicosis. *Journal of Nutrition* 65, 293–304.

Suttie, J.W., Gesteland, R. and Phillips, P.H. (1961) Effects of dietary sodium fluoride on dairy cows. 6. In young heifers. *Journal of Dairy Science* 44, 2250–2258.

Suttie, J.W., Carlson, J.R. and Faltin, E.C. (1972) Effects of alternating periods of high- and low-fluoride ingestion on dairy cattle. *Journal of Dairy Science* 55, 790–804.

Suttle, N.F., Brebner, J. and Hall, J. (1991) Faecal excretion and retention of heavy metals in sheep ingesting topsoil from fields treated with metal rich sludge. In: Momcilovic, B. (ed.) *Proceedings of the Seventh International Symposium on Trace Elements in Man and Animals (TEMA 7), Dubrovnik*. IMI, Zagreb, pp. 32-7–32-8.

Suttle, N.F., Brebner, J., Stark, B., Sweet, N. and Hall, J.W. (1997) Placental and mammary transfer of lead and cadmium by ewes exposed to lead and cadmium enriched, sewage-sludge treated soils. In: Fischer, P.W.F., L'Abbé, M.R., Cockell, K.A. and Gilson, R.S. (eds) *Proceedings of the Ninth International Symposium on Trace Elements in Man and Animals (TEMA 9), Banff.* NRC Research Press, Ottawa, pp. 168–170.

Tao, S. and Suttie, J.W. (1976) Evidence for a lack of an effect of dietary fluoride level on reproduction in mice. *Journal of Nutrition* 106, 1115–1122.

Thompson, D.J. (1980) Industrial considerations related to fluoride toxicity. *Journal of Animal Science* 51, 767–772.

Thornton, I. (1996) Sources and pathways of arsenic in the geochemical environment. In: Appleton, J.D., Fuge, R. and McCall, G.J.H. (eds) *Environmental Geochemistry and Health.* Geological Society Special Publication No. 113, London, pp. 153–162.

Tiffany, M.E., McDowell, L.R., O'Connor, G.A., Martin, F.G., Wilkinson, N.S., Percival, S.S. and Rabiansky, P.A. (2002) Effects of residual and reapplied biosolids on performance and mineral status of grazing steers. *Journal of Animal Science* 80, 260–269.

Underwood, E.J. and Suttle, F. (1999) *The Mineral Nutrition of Livestock*, 3rd edn. CAB International, Wallingford, UK.

Valdivia, R., Ammerman, C.B., Henry, P.R., Feaster, J.P. and Wilcox, C.J. (1982) Effect of dietary aluminium and phosphorus on performance, phosphorus utilisation and tissue mineral composition in sheep. *Journal of Animal Science* 55, 402–410.

Van Bruwaene, R., Gerber, G.B., Kirchmann, R. and Colard, J. (1982) Transfer and distribution of radioactive cadmium in dairy cows. *International Journal of Environmental Studies* 9, 47–51.

Van Bruwaene, R., Kirchmann, R. and Impens, R. (1984) Cadmium contamination in agriculture and zootechnology. *Experientia* 40, 43–50.

Van der Woude, H., Alink, G.M. and Rietjens, I.M. (2005) The definition of hormesis and its implications for *in vitro* and *in vivo* risk assessment. *Critical Reviews in Toxicology* 35, 603–607.

Van Gelder, G.A., Carson, T., Smith, R.M. and Buck, W.B. (1973) Behavioural toxicologic assessment of the neurologic effect of lead in sheep. *Clinical Toxicology* 6, 405–418.

VIDA (2007) Table 3 in *Veterinary Investigation Surveillance Report, 2007.* Veterinary Laboratories Agency, Weybridge, UK.

Weber, C.W. and Reid, B.L. (1974) Effect of low-fluoride diets fed to mice for six generations. In: Hoekstra, W.G., Suttie, J.W., Ganther, H.E. and Mertz, W. (eds) *Trace Element Metabolism in Animals – 2.* University Park Press, Baltimore, Maryland, pp. 707–709.

Wheeler, S.M., Brock, T.B. and Teasdale, D. (1985) Effects of adding 30 mg fluoride/l drinking water given to pregnant ewes and their lambs upon physiology and wool growth. *Journal of Agricultural Science, Cambridge* 105, 715–726.

Wilkinson, J.M., Hill, J. and Livesey, C.T. (2001) Accumulation of potentially toxic elements in the body tissues of sheep given diets containing soil and sewage sludge grazed on grassland given repeated applications of sewage sludge. *Animal Science* 72, 179–190.

World Health Organization (2000) Lead. In: *Safety Evaluation of Certain Food Additives and Contaminants. Fifty-third meeting of the Joint FAO/WHO Expert Committee on Food Additives (JECFA), Food Additives Series 44.* WHO, Geneva, pp. 273–312.

World Health Organization (2001) Cadmium. In: *Safety Evaluation of Certain Food Additives and Contaminants. Fifty-third meeting of the Joint FAO/WHO Expert Committee on Food Additives (JECFA), Food Additives Series 46.* WHO, Geneva, pp. 247–305.

Zmudzki, J., Bratton, G.R., Womac, C. and Rowe, L.D. (1984) The influence of milk diet, grain diet and method of dosing on lead toxicity in young calves. *Toxicology and Applied Pharmacology* 76, 490–497.

19 Design of Supplementation Trials For Assessing Mineral Deprivation

The purpose of this chapter is to bridge the gaps in knowledge that complicate assessments of mineral deprivation and the extent to which they limit livestock production.

Uncertainties

For most minerals – major and trace – there are uncertainties regarding:

- the effective mineral supply (i.e. amount ingested, absorbed and utilized);
- the minimal demand, when mineral losses can be reduced and stores redistributed;
- the impact of abrupt physiological and environmental changes on demand;
- rate-limiting metabolic pathways, when supply fails to meet demand; and
- correlations between conventional clinical biochemistry criteria and the activity of rate-limiting pathways.

The soundest diagnosis of mineral deprivation is provided by a clinical or production response to supplements of the mineral or minerals that are thought to be lacking (Suttle, 1987; McDowell, 1992; Judson, 1996). Guidelines for the design and interpretation of supplementation trials have been laid down to make fuller use of resources and data (Clark et al., 1985; Grace, 1994; Underwood and Suttle, 1999), but can still be improved upon. Countless trials have been destined to confirm the null hypothesis before any data have been recorded because of inadequate numbers and the failure to consult a statistician before starting. The essential requirements for a good supplementation trial are given here.

Controls

The benefits from supplementation cannot be gauged unless sufficient numbers of animals are left unsupplemented as controls (e.g. Suttle and Phillippo, 2005; p. 449). 'Before and after treatment' comparisons are notoriously unreliable because any spontaneous recoveries (e.g. improved coat or fleece condition in spring; improved lamb growth with acquisition of immunity to gut parasites), new constraints (e.g. drought) or concurrent changes in husbandry (e.g. use of natural instead of artificial insemination) can give false-positive or -negative results.

Control bias

Mineral supplementation per se constitutes a husbandry change and leaving animals untreated may indirectly benefit the treated cohorts if the source of major nutrients is shared and mineral deprivation causes loss of appetite. For example, if cobalt deprivation limits lamb growth by restricting appetite and individually treated (parenteral or oral) cohorts are being fed below appetite, treated animals will have access to more feed than if all lambs were treated. Thus, if one member of a suckling twin pair or half of the lambs in a flock grazed at high stocking rate is left untreated, treated animals may outperform untreated controls partly by default and the response to treatment becomes exaggerated. Sometimes poorly performing controls are withdrawn from trials on welfare grounds (e.g. Grace et al., 2003; Gruner et al., 2004). If they are simply treated and left in the flock, bias will be reduced. Such 'control bias' can be reduced by leaving only a small fraction (e.g. 10%) of the flock untreated or by treating both members of a pair of suckling twins while leaving a minority of pairs untreated (e.g. Suttle and Small, 1993). On experimental stations, treated and control groups can be housed or grazed separately in replicated groups (Masters and Peter, 1990; Whelan et al., 1994), but this tactic is generally impractical under farm-trial conditions.

Adequate Group Size

For statistical significance

The group sizes needed to demonstrate statistically significant differences in performance can be calculated from previous records of variability in the chosen criterion. Whether the variables are continuous (e.g. live weight (LW)) or discrete (e.g. mortality), the appropriate analysis (e.g. analysis of variance or chi-square) can be run on simulated datasets. The effect of group size on the probability of discounting as non-significant a 5% improvement in LW with

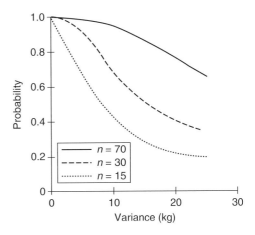

Fig. 19.1. Minimum acceptable group sizes for dose–response trials can be calculated from data on the variance of the response trait and the anticipated benefit from treatment. Here, the assumed benefit is one of 2 kg in live-weight gain in lambs; unless variance is low, large groups are required to correctly reject the null hypothesis (from White, 1996).

different degrees of background variation is illustrated in Fig. 19.1. Wu and Satter (2000) estimated that a group size of 250 was needed in their experiments with dairy cows to detect a 10% improvement in milk yield when no account was taken of pre-trial yield! Although such a goal is small, it may well be economically significant and representative of the maximum benefit likely to be attained when mineral deprivation is marginal (Kumagai and White, 1995). Table 19.1 illustrates the importance of group size on the attainment of statistical significance for a discrete variable such as calf mortality or ewe infertility. Even when the incidence of the event in question is high (40%)

Table 19.1. A hypothetical illustration of the group size needed to obtain a significant improvement in the incidence of a disorder (e.g. mortality or infertility) with a particular treatment (for $P < 0.05$, chi-square must be >3.84).

	Hypothetical group		Group size					
	Control	Treated	10	20	40	80	160	320
	Incidence (%)		Chi-square					
Experiment A	40	20	0.95	1.91	3.81	7.62	15.24	30.48
Experiment B	10	5	–	0.36	0.72	1.44	2.88	5.76

and is halved by the hypothetical treatment, groups of over 40 animals are needed for statistical significance by the chi-square test (>3.8 for $P < 0.05$ with 1 degree of freedom). When performance is good (10% incidence) but still improvable (the incidence can be halved to 5%), over 200 animals are needed in each group to prove the point. Figure 19.2 demonstrates the importance of large group size in lamb mortality trials. A proportional hazards model has now been applied to bovine infertility data to increase the use made of limited numbers of cows (Black and French, 2004) but this may exaggerate treatment effects by returning 'early conceivers' to the pool of data and treating them as newcomers of equal vulnerability to the single 'hazard', infertility.

For economic significance

The contrast between statistical and economic significance is highlighted by discrete, all-or-nothing variables. For example, a non-significant ($P > 0.05$) increase of 5% in the calf or lamb crop every other year would be welcomed by most farmers and would repay the outlay on treatments faster than a consistent significant 5% improvement in growth. In fact, under farm-trial conditions, the use of whole flocks and pooling data from several flocks has detected a small but significant response in ewe fertility to parenteral selenium, despite a mean blood selenium status well above normal on all farms and the absence of significant responses within farms (Munoz et al., 2009).

Unavoidable and avoidable constraints

Group sizes are subject to some unavoidable constraints such as flock or herd size, available grazing area or housing, handling facilities and the need to return stock to their farm environment without undue disruption. However, it may be neither necessary nor advisable to return stock on the same day that they are gathered. Gathering animals the day prior to recording or treatment and allowing them to fast overnight might reduce errors in weighing and blood analysis due to the recent ingestion of food while increasing the number of animals that can be sampled in a working day. Other constraints may be entirely avoidable. For example, group size is sometimes restricted by the excessive number of observations made on or samples taken from each individual and by the time and cost of subsequently processing and analysing all of the samples. Priority should be given to providing sufficient animals to confirm or deny responsiveness in the performance criterion. The number of blood samples needed to attain statistical significance for a given biochemical test is often smaller than that required for a performance criterion. Expensive tests (e.g. for methylmalonyl acid) can be restricted in number if there is no prior production response. There are increasing pressures to reduce the number of animals used in experiments on welfare grounds. 'Restricted' or 'licensed' procedures (e.g. blood sampling) need be applied to only a small proportion of those subjected to unrestricted farm procedures (e.g. weighing). Repeated subsampling of a small sentinel subgroup allows seasonal trends to be followed, but relationships to health or performance in the

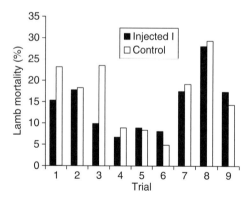

Fig. 19.2. The importance of sample size in dose–response trials. Improvements in lamb survival were found following the parenteral supplementation of ewes with iodine pre-mating on three out of nine farms in New Zealand (from Clark et al., 1998). A large group size of 500 allowed a small improvement to be detected in trial 4 but poor relationships with markers of mean iodine (I) status were reported. This is hardly surprising with only ten ewes or lambs sampled on each farm and only one in eight ewes benefiting from supplementation. I supply was a minor contributor to variation in lamb mortality.

main group may be disappointing (Fig. 19.2). Values recorded for the same animal on several occasions are not independent and must be analysed statistically as 'repeated measures' (St-Pierre and Jones, 1999).

Accurate Measurement

Statistical significance is determined by the ratio of variation due to fixed effects (e.g. of treatment, litter and pen) to unaccountable, random variation. Random variation can be reduced by rigorous attention to the detail and consistency of all procedures and observations. Under statutory pressure, the principles of good experimental and laboratory practice are becoming increasingly adopted. While some requirements may seem pedantic, it takes little time to regularly check the accuracy of both field (e.g. weighings) and laboratory (e.g. analyses) results against known standards and to adjust either the equipment or results accordingly. Electronic recording and data transference is more reliable than the human ear, hand and memory. All statistical inputs should be double-checked and attention given to the outliers that are routinely 'flagged' by modern statistical packages. The discarding of biologically impossible weight gains, for example, is duty not deceit; so, too, is the reanalysis of duplicate samples with statistically verified discordance.

Repetition

Repetition of observations, samplings or trials can provide a safeguard against diurnal, day-to-day, seasonal or year-to-year variations.

Diurnal variation

Awareness of diurnal fluctuations is a prerequisite for deciding and adhering to an optimum time of observation (e.g. before or after feeding, morning or afternoon milking). If the sequence of a prolonged set of samplings is repeated (e.g. in individually penned animals) then the sequence might usefully be recorded and entered as a variable in the statistical model.

Day-to-day variation

Repeated observations (e.g. body weights on successive days) are increasingly used to improve the reliability of the beginning and end points of an observation period. In heavily fleeced sheep, for example, LW can be increased by several kilograms by heavy rainfall prior to a particular weighing.

Seasonal variation

Seasonality affects the growth of crops and pastures, the physiological processes of animals and the life cycles of pathogens. It is therefore a feature of many farm trials. Seasonal influences can be sought by dividing an experiment into periods, although period effects are not necessarily synonymous with true season effects; for example, they may indicate the waning efficacy of a treatment or differences in composition between batches of diet. Knowledge of the seasonality of responses to mineral supplementation should influence the timing of any treatment and the method of supplementation. When different treatments are applied sequentially to the same group, responses can become confounded with seasonal influences.

Year-to-year variation

For the farmer, the repeatability of responses to supplementation from year to year is of paramount importance. A classic example of year-to-year variation in response to a mineral supplement was provided by Lee (1951) in early studies of coast disease in South Australia. Positive responses to cobalt supplementation were obtained in only eight of 13 years, and the severity of deprivation varied in 'responsive' years from marginal growth depression to 100% mortality at the same site. In a sequential study of growth responses in lambs to copper, cobalt and selenium in successive years (Suttle and Small, 1993), the significance of copper responsiveness may have been underestimated by year-to-year fluctuations, with mean plasma copper being highest in the year that copper was under test. Year-to-year variation is exploited commercially when tales of 'the good years' are more readily told than those of 'the bad years' by sellers of mineral supplements.

Making Use of Individual Variation

The typical presentation in marginal mineral disorders is one of irregular performance rather than consistent failure, but that variation may be only partly attributable to variations in mineral status (Table 19.2). Instead of being regarded as a nuisance, individual variation can be used as part of experimental design through covariance analysis to reduce group size, redefine diagnostic limits and form the basis of a cost–benefit analysis. More can be done than simply pairing or stratifying the population initially according to a production parameter such as LW or milk yield. Before a growth trial even begins, a correlation between past growth and initial mineral status is worth looking at, although trends would only emerge if growth was being restricted by the last two factors listed in Table 19.2 or if maternal performance had been restricted by deprivation of the same mineral that their offspring lacked. Similar approaches can be taken with all parameters of production. Figure 19.3 illustrates the variation found in a small experimental herd of dairy cows and the futility of conducting an experiment without allocating such animals according to pre-trial performance or past lactation records.

Covariance with early mineral-dependent performance

If pre-treatment performance has been restricted by mineral deprivation then random variations in subsequent performance should be reduced by using the initial ranking or prior growth rate as a covariate and the relationship should be *negative* (i.e. the poorest initial performers show the greatest response to supplementation). Thus, in lactating cows on an alfalfa diet that was low in sodium, individuals with the poorest yields prior to treatment showed the largest responses to sodium, while previously good yielders failed to respond (Joyce and Brunswick, 1975). It would be surprising if the use of conception rate records from previous matings as covariates does not greatly increase the chances of detecting and quantifying the effects of suspected mineral deprivation on fertility, but this tactic has rarely been used. Significantly different relationships between trial performance and mineral status have been found *within* treated or untreated groups that did *not* differ significantly in their mean trial performances. As shown in Table 19.3, the influences of copper supplementation on the early growth of lambs only became significant when initial LW was used as a covariate. This indicates that the population distribution contained a 'tail' of responsive animals that alone benefited from supplementation. However, differences in regression were not found later in the season (6–12 weeks) when growth was

Table 19.2. Seven possible reasons for poor growth in a group of young growing animals on first inspection, listed in order of probability.

Order	Reason for poor growth
1	Too little digestible, utilizable non-mineral nutrients provided by the diet
2	Maternal undernutrition (in the case of newly weaned or suckled offspring)
3	Parasitic, microbial or viral infection
4	Environmental stressors (e.g. wind, temperature, water supply)
5	Low genetic potential
6	Too little of one mineral in the diet
7	Too little of two or more minerals in the diet

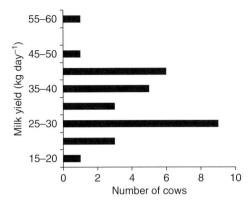

Fig. 19.3. An example of pre-trial variation in milk yield in a herd of multiparous Holstein cows. Unless cows are allocated to treatments on the basis of past lactation performance, the chances of detecting significant responses to a mineral supplement can be negligible (Voelker et al., 2002).

slower and the supplement delivered less copper (Suttle, 2002). Such populations are marginally responsive and a cost–benefit analysis can be applied to ascertain whether all, none or part of the population merits treatment.

Covariance with early mineral-independent performance

If individual performance prior to treatment is restricted only by factors other than supply of the supplemented mineral (reasons 1–5 in Table 19.2), the relationship with subsequent performance is likely to be *positive* when the influence continues (e.g. poor milk supply, poor genotype, chronic infection) but *negative* when it does not continue (e.g. weaning, acute infection) and compensatory growth is expressed. Whatever mechanism underlies a significant correlation, covariance analysis will improve the chances of obtaining a significant response to treatment. In all cases where the final measure of performance is *proportional* to performance prior to treatment, the old practice of calculating improvement by the difference between the initial and final value (e.g. LW gain) will be *less* effective than covariance analysis.

Non-linear covariance

Few biological relationships are linear over the complete range of two covariates. In the series of trials to which Fig. 19.4 refers (Suttle and Small, 1993), no significant growth responses to parenteral selenium were found, but contrasting relationships with initial LW indicated that parenteral treatment with selenium had made a difference on some farms. Whereas the final LW of untreated lambs was highly and linearly correlated to initial LW (Fig. 19.4a) the relationship in selenium-treated lambs was much more complex, with selenium appearing to boost the growth of lightweight lambs but not of the heaviest lambs (Fig. 19.4b). Curvilinear relationships within groups between LW and serum vitamin B_{12} have been used to confirm the marginal range for serum vitamin B_{12} in lambs (Grace et al., 2003). Covariance analysis should

Table 19.3. Treatment with copper oxide (CuO) needles failed to improve the average weight gain of a group of 25 lambs compared to their untreated twins on molybdeniferous pasture (50–60 mg Mo kg^{-1} dry matter). However, within-group regressions of live weight (kg) after 6 weeks on initial live weight were different (unpublished data from the trials described by Suttle and Small, 1993).

CuO	Live-weight gain (g day^{-1})	Regression (kg units)	
		Intercept (± SD)	Coefficient (± SD)
0	173[a]	8.1[b] (± 2.19)	0.95[d] (± 0.162)
+	191[a]	1.1[c] (± 2.27)	1.51[e] (± 0.168)
	21.6 (SED)		

SD, standard deviation; SED, standard error of the difference.
Means with different superscripts within columns differ significantly.

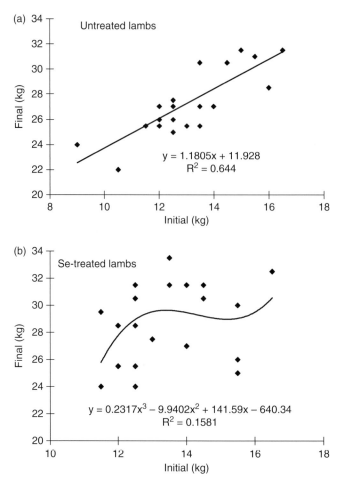

Fig. 19.4. The relationship between initial and 12-week live weight of individual suckling lambs in the last year of 3 years in which the possibility of first copper, then cobalt and finally selenium deprivation was examined on numerous Scottish hill farms (Suttle and Small, 1993). While significant improvements in live-weight gain were not found in groups given parenteral Se on any of 11 farms, intervention with Se clearly affected the growth of treated lambs on this farm (a), disrupting the linear relationship between final and initial live weight found in untreated lambs (b).

always initially allow for the possibility that a relationship may be non-linear.

Making Use of Group Responses

The relationship between mean improvement in group performance and a measure of mean mineral status in dose–response trials has been used to predict responses to treatment on 'unknown' farms from the same index of status. For example, the relationship between plasma vitamin B_{12} concentrations and average LW benefit after treating lambs with cobalt or vitamin B_{12} can be used to improve the interpretation of plasma vitamin B_{12} results (see Chapter 10) as well as to predict benefit in terms of performance (Fig. 19.5), but the outcome will differ from that based on individual values (Grace et al., 2003). The exponential nature of the relationship is partly explained by the fact that as vitamin B_{12} status declines, more and more individuals show larger and larger benefits; however, it may also partly reflect control bias. The same approach has been used with selenium (Fig. 19.6) and emphasizes the continuous

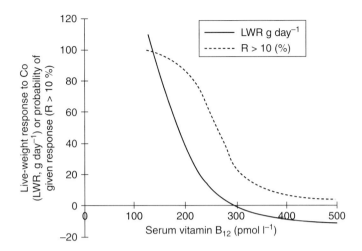

Fig. 19.5. Collation of data from cobalt supplementation trials with lambs in New Zealand has allowed the interpretation of serum and liver vitamin B_{12} concentrations to be given an economic base (Clark et al., 1985, 1989), but that base varies within and between countries.

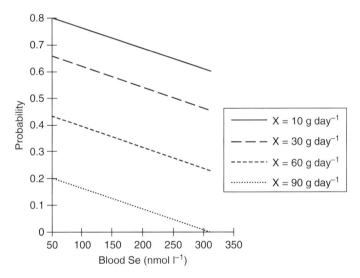

Fig. 19.6. The collation of data from large numbers of dose–response trials can allow prediction of the likely benefits from supplementation in other areas from a common measure of trace element status. Here, the probability of different growth responses in cattle is predicted from blood selenium (Langlands et al., 1989).

nature of the distribution of responses and the need for marginal bands in classifying responsiveness (see Chapter 3). In the trials illustrated, as blood selenium declined linearly below 316 nmol (25 μg) l^{-1}, so the probability of obtaining a growth response in calves increased, but large responses (90 g day^{-1}) remained improbable. The same research group has used a 'bent-stick' model to predict responses in wool growth from blood selenium (Langlands et al., 1991). These simple strategies allow the animal to be the arbiter of 'deprivation' and begin to address all of the uncertainties set out at the start of this chapter.

Choosing the Most Responsive Individuals

It may not always be necessary or preferable to conduct trials with groups that are representative of the whole herd or flock. In New Zealand, for example, selenium supplements rarely produce significant improvements in milk yield despite a generally low blood and herbage selenium status (Wichtel, 1998). However, Knowles et al. (1999) have recorded a 30% increase in milk-fat yield in a herd of unexceptional mean blood selenium status on pastures of unexceptional selenium content. By selecting the 10% of the herd with the lowest blood selenium concentrations and allocating them according to previous milk yield, the researchers greatly increased the chances of achieving a positive response to selenium. Such results cannot be extrapolated to the whole herd or to other herds, but second-order questions such as 'what fraction of the herd was selenium responsive?' and 'which cows should be treated?' can be addressed once selenium deprivation in the 'tail' of a population has been demonstrated. The 1-day-old chick is a popular model for demonstrating responses to minerals because large numbers of chicks can be mustered and multifactorial designs accommodated. However, the immaturity of the digestive system may mean that any responses observed are not representative of those obtainable throughout the majority of a bird's life. Thus, the capacity of phytase supplements to free phytate phosphorus from cereals and vegetable protein sources may have been greatly exaggerated by studies with chicks in the first 3 weeks of life (Fig. 4.5, p. 62). This model is therefore popular with manufacturers of phytase supplements, but misleading to environmentalists.

Choosing the Best Treatment(s)

The choice of treatments for diagnostic trials is governed by different criteria from those that a farmer might routinely use and there are various options to be considered. Where possible, animals should be supplemented individually rather than as a group on the grounds of maintaining the certainty and uniformity of the treatment. The product used experimentally may be more expensive than that used in practice in order to guarantee efficacy.

Sustained efficacy

The chosen supplement must be capable of alleviating the suspected mineral deprivation throughout the period of risk, yet present no risk of excess when first administered. For example, in the series of trials referred to in Fig. 19.5, injections of vitamin B_{12} were preferred to cobalt bullets for raising vitamin B_{12} status, but the chosen dose (1 mg) was inadequate when repeated in lambs that had grown from 20 to 30 kg between doses during the first 6 weeks of the experiment. The vitamin B_{12} status of the untreated lambs declined steadily and responses were more likely to have been obtained after the second dose, had it been adequate. The failure to scale selenium injections to LW may have contributed to the wide variation in lamb growth evident in Fig. 19.4. The scaling of repeated doses to body weight in growing animals is, therefore, recommended. Mineral-responsive disorders can only be diagnosed by deploying treatments of a type and frequency that achieve sustained alleviation of deficiencies in the animal.

The case for multiple treatments

The first decision that must be made is whether to provide more than one mineral in the supplement. The decision may depend on the weight of evidence from previous analyses of soil, diet, blood or tissues, but, given the uncertainty of the respective relationships with animal performance, there is much to commend the 'blunderbuss' approach of initially providing a mixture of all essential minerals (necessarily by diet for the major elements) and measuring their combined impact on performance (White et al., 1992). The major advantage of the multi-element approach is that it can indicate whether *any* mineral is likely to be limiting performance while also indicating the order of the benefit that is likely to arise once any constraints are lifted (Kumagai and White, 1995).

The case against multiple treatments

Several difficulties arise with multiple treatments:

- No evidence will be provided as to which particular mineral component (or components) is responsible for any improvements in health and performance.
- Part of the flock or herd must remain unsupplemented as controls (i.e. put at risk of multiple deprivations) for diagnostic purposes, an unpopular decision at farm level.
- Uptake of a free-access mineral mixture (the common method of multiple supplementations) may vary substantially within treated groups (see Chapter 3).
- Minerals not needed in the mixture may be provided to such an excess that performance is impaired, masking responses to any mineral that is lacking.

The case for and against single treatments

Single treatments can provide an unequivocal answer to a simple question: is the supplemented mineral currently limiting performance? In the case of trace elements, the mineral supplement can be given individually by mouth or by injection, allowing the effects of a known dose to be assessed. That question will remain unanswered if the chosen method of treatment is inadequate for individuals, but safeguards can be incorporated into the design: scaling dose carefully to LW; re-dosing if criteria of mineral status fall into or below the 'marginal' category; and discarding data for exceptional individuals that show no early evidence of improved mineral status on the assumption of maladministration. The disadvantages of single treatments are: the cost of failure to select the single most-limiting element at an early stage; and underestimating responsiveness when other elements limit performance. Examples of the single-treatment approach are given in the relevant sections of the chapters dealing with specific minerals.

The case for clustered treatments

Compromises can be reached by increasing the number of treatment groups to three: a fully supplemented group (A); a group deprived of only one mineral (B); and an unsupplemented group (C). Then, A – C gives the total response to minerals and B – C gives the minimal response to the selectively withdrawn mineral. If the most limiting macro-mineral is likely to be sodium, this would be clearly demonstrated by the contrast between Na + Ca + P + Mg versus Ca + P + Mg versus 0, with the minerals given as continuous dietary supplements. Similarly, the most limiting trace element might be ascertained by selectively withdrawing copper, cobalt or selenium from the set Cu + Co + Se given by mouth or injection (Suttle, 1987). Some workers have provided salt for control groups and looked for increments in performance following the addition of further minerals by mouth (e.g. phosphorus, cobalt) or injection (e.g. copper, iodine, selenium) in all possible combinations (McDowell *et al.*, 1982). However, the responses to individual elements may be so small that only combined supplements give significant responses (McDowell *et al.*, 1982; Masters and Peter, 1990).

Re-examining unresponsive populations

The failure of either multiple, clustered or single mineral treatments to improve performance in a field trial does not mean that the potential for response should be dismissed for all time. The next step is to investigate other causes for the poor performance that prompted the supplementation trials in the first instance: lack of digestible organic matter and chronic parasitic infections are the most likely alternative causes of ill-thrift in grazing livestock (Appendix 1). If such constraints are identified and removed, mineral supplies may become truly limiting at the new, increased level of performance. However, prior evidence of marginal mineral status should be obtained before re-embarking on costly and laborious supplementation trials.

References

Black, D.H. and French, N.P. (2004) Effects of three types of trace element supplement on the fertility of diary herds *Veterinary Record* 154, 652–658.

Clark, R.G., Wright, D.F. and Millar, K.R. (1985) A proposed new approach and protocol to defining mineral deficiencies using reference curves: cobalt deficiency in young sheep is used as a model. *New Zealand Veterinary Journal* 32, 1–5.

Clark, R.G., Wright, D.F., Millar, K.R. and Rowland, J.D. (1989) Reference curves to diagnose cobalt deficiency in sheep using liver and serum vitamin B_{12} levels. *New Zealand Veterinary Journal* 37, 7–11.

Clark, R.G., Sargison, N.D., West, D.M. and Littlejohn, R.P (1998) Recent information on iodine deficiency in New Zealand sheep flocks. *New Zealand Veterinary Journal* 46, 216–222.

Grace, N.D. (1994) *Managing Trace Element Deficiencies*. New Zealand Pastoral Agricultural Research Institute, Palmerston North, New Zealand, pp. 63–68.

Grace, N.D., Knowles, S.O., Sinclair, G.R. and Lee, J. (2003) Growth responses to increasing doses of microencapsulated vitamin B_{12} and related changes in tissue B_{12} concentrations in cobalt-deficient lambs. *New Zealand Veterinary Journal* 51, 89–92.

Gruner, T.M., Sedcole, J.R., Furlong, J.M., Grace, N.D., Williams, S.D., Sinclair, G.R. and Sykes, A.R. (2004) Changes in serum concentrations of MMA and vitamin B_{12} in cobalt-supplemented ewes and their lambs on two cobalt-deficient properties. *New Zealand Veterinary Journal* 52, 17–128.

Joyce, J.P. and Brunswick, I.C.F. (1975) Sodium supplementation of sheep and cattle. *New Zealand Journal of Experimental Agriculture* 3, 299–304.

Judson, G.J. (1996) Trace element supplements for sheep at pasture. In: Masters, D.G. and White, C.L. (eds) *Detection and Treatment of Mineral Problems in Grazing Sheep*. ACIAR Monograph No. 37, Canberra, pp. 57–80.

Knowles, S.O., Grace, N.D., Wurms, K. and Lee, J. (1999) Significance of amount and form of dietary selenium on blood milk and casein selenium concentrations in grazing cows. *Journal of Dairy Science* 82, 429–437.

Kumagai, H. and White, C.L. (1995) The effect of supplementary minerals, retinol and α-tocopherol on the vitamin status and productivity of pregnant Merino ewes. *Australian Journal of Agricultural Research* 46, 1159–1174.

Langlands, J.P., Donald, G.E., Bowles, J.E. and Smith, A.J. (1986) Selenium excretion in sheep. *Australian Journal of Agricultural Research* 37, 201–209.

Langlands, J.P., Donald, G.E., Bowles, J.E. and Smith, A.J. (1989) Selenium concentration in the blood of ruminants grazing in northern New South Wales. 3. Relationship between blood concentration and the response in liveweight of grazing cattle given a selenium supplement. *Australian Journal of Agricultural Research* 40, 1075–1083.

Langlands, J.P., Donald, G.E., Bowles, J.E. and Smith, A.J. (1991) Subclinical selenium insufficiency. 1. Selenium status and the response in liveweight and wool production of grazing ewes supplemented with selenium. *Australian Journal of Experimental Agriculture* 31, 25–31.

Lee, H.J. (1951) Cobalt and copper deficiencies affecting sheep in South Australia. Part 1. Symptoms and distribution. *Journal of Agriculture, South Australia* 54, 475–490.

Masters, D.G. and Peter, D.W. (1990) Marginal deficiencies of cobalt and selenium in weaner sheep: response to supplementation. *Australian Journal of Experimental Agriculture* 30, 337–341.

McDowell, L.R. (1992) *Minerals in Animal and Human Nutrition*. Academic Press, New York, p.18.

McDowell, L.R., Bauer, B., Galdo, E., Koger, M., Loosli, J.K. and Conrad, J.H. (1982) Mineral supplementation of beef cattle in the Bolivian tropics. *Journal of Animal Science* 55, 964–970.

Munoz, C., Carson, A.F., McCoy, M.A., Dawson, L.E.R., Irwin, D., Gordon, A.G. and Kilpatrick, D.J. (2009) Effects of supplementation with barium selenate on the fertility, prolificacy and lambing performance of hill sheep. *Veterinary Record* 164, 265–271.

St-Pierre, N.R. and Jones, L.R. (1999) Interpretation and design of non-regulatory on-farm trials. *Journal of Animal Science* 77 (suppl. 2J), 177–182.

Suttle, N.F. (1987) The absorption, retention and function of minor nutrients. In: Hacker, J.B. and Ternouth, J.H. (eds) *Nutrition in Herbivores*. Academic Press, Sydney, Australia, pp. 333–362.

Suttle, N.F. (2002) Meeting the copper requirements of ruminants. In: Garnsworthy, P.C. and Cole, D.J.A. (eds) *Recent Advances in Nutrition – 1994*. Nottingham University Press, Nottingham, UK, pp. 173–188.

Suttle, N.F. and Phillippo, M. (2005) Effect of trace element supplementation on the fertility of dairy herds. *Veterinary Record* 156, 155–156.

Suttle, N.F. and Small, J.N.W. (1993) Evidence of delayed availability of copper in supplementation trials with lambs on molybdenum-rich pasture. In: Anke, M., Meissner, D. and Mills, C.F. (eds) *Proceedings of the Eighth International Symposium on Trace Elements in Man and Animals, Dresden.* Verlag Media Touristik, Gersdorf, Germany, pp. 651–655.

Underwood, E.J. and Suttle, F. (1999) *The Mineral Nutrition of Livestock*, 3rd edn. CAB International, Wallingford, UK.

Voelker, J.A., Burato, G.M. and Allen, M.S. (2002) Effects of pretrial milk yield on responses of feed intake, digestion and production to dietary forage concentration. *Journal of Dairy Science* 85, 2650–2661.

Whelan, B.R., Barrow, N.J. and Peter, D.W. (1994) Selenium fertilizers for pastures grazed by sheep. II. Wool and liveweight responses to selenium. *Australian Journal of Agricultural Research* 45, 875–886.

White, C.L. (1996) Understanding the mineral requirements of sheep. In: Masters, D.G. and White, C.L. (eds) *Detection and Treatment of Mineral Problems in Grazing Sheep*. ACIAR Monograph No. 37, Canberra, pp. 15–30.

White, C.L., Masters, D.G., Peter, D.W., Purser, D.B., Roc, S.P. and Barnes, M. (1992) A multi-element supplement for grazing sheep. I. Intake, mineral status and production response. *Australian Journal of Agricultural Research* 43, 795–808.

Wichtel, J.J. (1998) A review of selenium deficiency in grazing ruminants 2: towards a more rational approach to diagnosis and prevention. *New Zealand Veterinary Journal* 46, 54–58.

Wu, Z. and Satter, L.D. (2000) Milk production and reproductive performance of dairy cows fed two concentrations of phosphorus for two years. *Journal of Dairy Science* 83, 1052–1063.

20 Minerals and Humans

So far, this book has considered mineral nutrition almost exclusively from the viewpoint of livestock as consumers of minerals, the primary concern being to avoid constraints of either inadequate or excessive mineral supplies on production. This final chapter deals with the interface between animals and humans. There are several facets at which either the mineral nutrition of livestock affects humans or human activity affects the mineral nutrition of livestock. Four facets are of particular concern:

- the impact of the mineral nutrition of livestock on the environment;
- the impact of climate change on the mineral nutrition of livestock;
- manipulating the composition of edible products to enhance their mineral value and sensory appeal to the consumer; and
- reducing perceived 'mineral hazards' to human health.

The Mineral Footprint

A significant proportion of the minerals that are consumed by livestock is voided in urine and faeces, and this proportion increases as the margin of mineral supply over mineral requirement increases. Three driving forces lead to the over-provision of minerals on farms:

- Farmers generally follow a precautionary principle, based on the belief that their stock cannot get too much of an essential mineral.
- Suppliers of minerals and compound feeds exploit that belief while holding a similar one, anxious to avoid all risk of their product failing to optimize production.
- Lack of specialized knowledge of mineral nutrition in the feed industry leads to lack of confidence in estimates of requirement.

Excessive mineral supplies pollute the environment and the chosen or default policy of mineral nutrition on a given farm leaves behind a 'mineral footprint' (e.g. Hristov et al., 2006, 2007). This legacy provokes regulatory and legislative reactions that will increasingly constrain animal production (see Chapter 18) (Meyer and Mullinax, 1999).

Maximum permissible limits

The European Community (EC) and national authorities (e.g. the Food and Drug Administration) have set maximum permitted levels (MPL) for mineral concentrations in foodstuffs in order to protect livestock, the consumer and/or the environment. These follow similar precautionary principles to those adopted by producers. For essential minerals, the authorities set MPL at levels several times greater than minimum requirements; feed compounders add minerals to attain concentrations closer to the MPL than the requirement, operating on their 'precautionary principle' and regarding the MPL as an optimal rather than the intended maximum *acceptable* level or level of last resort. Table 20.1 compares current practice for supplying minerals to ruminants in the UK with mineral requirements and the latest MPL. Mineral provision from the concentrate alone would exceed requirement for nearly all elements and classes of livestock, yet the mixed ration would normally meet most minimum requirements. Similar over-generous trace mineral specifications can be found in compound feeds for pigs and poultry. Manufacturers are clearly drawn by the MPL rather than animal need and this leads to unnecessarily large 'mineral footprints'. Furthermore, many trace elements are potent oxidants, capable of inducing deleterious changes in the ration, particularly during prolonged storage at high temperatures. The very existence of MPL, which are there to protect the environment, has the opposite effect in practice – raising mineral concentrations in compound feeds far above the needs of livestock.

The need for sub-sets in maximum permitted levels

As currently employed, MPL ignore large genetic and dietary effects in both mineral needs and tolerances amongst animals. Thus, MPL can sometimes protect the environment at the expense of livestock or protect the majority of livestock at the expense of the minority. For example, breeds of sheep vary widely in their ability to absorb copper and some, such as the Texel, can accumulate potentially hepatotoxic copper concentrations on diets far below the MPL for copper (Suttle *et al.*, 2002). Because it is hard to attain a safe copper level for vulnerable breeds, feed compounders had adopted the practice of routinely adding copper antagonists such as iron, molybdenum, sulfur and zinc to their rations to reduce the risk of chronic copper poisoning. However, the MPL for zinc has been lowered by the EC (2003) directive from 350 to 150 mg kg^{-1} dry matter (DM) on environmental protection grounds, and this may make it impossible to protect sheep such as Texels from chronic copper poisoning via the 'antagonist cocktail' route. The constraint could be offset by

Table 20.1. Current mineral supplementation practice in Europe (from responses to a livestock feed industry questionnaire; A. Packington, personal communication) indicates application of precautionary principles, with trace mineral provision via complementary feeds generally exceeding maximum requirements (*values in italics*): the significant but variable contribution of minerals from forages (assumed to constitute 50% of total ration) is ignored.

Contribution to whole ration from complementary feed	Dairy (milking)	Calf rearer	Beef	Ewe	Maximum requirement[a]	MPL
Calcium (g kg^{-1} DM)[c]	4.0 *(5.3)*	5.5 *(8)*	7 *(11)*	6.5 *(4.3)*	3–11	
Phosphorus	2.5 *(2.8)*	2.5 *(4.4)*	2.5 *(2.7)*	3 *(3.2)*	1–4.4	
Copper (mg kg^{-1} DM)[d]	20	10	10	0	5–*28*	40(17)[b]
Cobalt	0.25	0.25	0.25	0.5	0.05–*0.15*	2.2
Manganese	10	30	30	40	15–*25*	170
Iodine	2.5	2.5	2.5	2	0.1–*1*	6
Selenium	0.4	0.25	0.25	0.35	0.05–*0.15*	0.6
Zinc	70	40	40	50	15–*40*	170

MPL are stated on an 'as fed' or 88% DM basis and have been divided by 0.88 to allow a comparison with provision and requirement on a 100% DM basis.
[a]The average requirement is skewed towards the lower limit of the range, the maximum rarely applies throughout a cycle of production and temporary dietary shortfalls can usually be tolerated.
[b]Addition of copper to sheep feeds is forbidden in European Community countries and a lower MPL is set to lower risks of copper poisoning.
[c]Concentration contributed by the complementary feed, including minerals added to it.
[d]Concentration contributed exclusively by the supplementary minerals in complementary feeds.

increasing the customary molybdenum supplement from 2 to 3 mg Mo kg^{-1} DM) in concentrates for Texels, and limiting the inclusion rate to 60% of the total ration. However, sensible nutritionally based MPL for sheep would be cast in terms of a Cu:Mo ratio of >4.0, rather than separate concentrations of Cu and Mo.

Copper and zinc footprints on pig farms

The use of copper as a growth stimulant in pig nutrition (see Chapter 11) created the problem of how to safely dispose of copper-rich excreta. In one study, the annual soil application for 8 years of slurry from pigs given 251 mg dietary Cu kg^{-1} DM increased soil extractable (DTPA) copper linearly with the application rate (Payne et al., 1988). Although grain and leaf-tissue copper in corn grown on the treated land was not increased, copper may enter the food chain via soil ingestion in mixed farming systems. A survey of copper and zinc in the livers of cattle in north-west Spain found a positive correlation between liver copper and the density of pig-rearing units (Lopez-Alonso et al., 2000). Concerns about this 'copper footprint' have prompted authorities in Europe to halt the common practice of adding 250 mg Cu kg^{-1} DM to all rations by imposing MPL of 170 and 25 mg Cu kg^{-1} DM in the growing and fattening stages, respectively (EC, 2003). A similar problem arises if dietary zinc is used 'supranutritionally' to reduce the incidence of diarrhoea in early weaned pigs (see Chapter 16). Where such pharmacological responses to minerals are obtainable, the underlying cause of the problem should be identified instead of relying on extravagant mineral supplies as a 'quick fix'. In the case of diarrhoea in piglets, the cause of intestinal hypersecretion should be identified and the stimulus countered rather than use excess zinc to inhibit mucosal secretion. In the above survey in Spain (Lopez-Alonso et al., 2000) no evidence of zinc accumulation in cattle was found, but liver zinc is a poor indicator of exposure (see Chapter 16).

Copper footprints on dairy farms

In a recent case of over-provision of copper in a UK dairy herd, copper supplements were being used on a grand scale (Table 20.2). Pregnant cows were given a total mixed ration containing added copper closer, at 35 mg kg^{-1} DM, to the MPL (50 mg kg^{-1} DM) than to the copper requirement (5–10 mg kg^{-1} DM) for much of the year. Calves were reared on a milk replacer containing 10 mg added Cu kg^{-1} DM, a common industry practice that ignores – as does the EC MPL of 30 mg Cu kg^{-1} DM – the very high efficiency with which copper is absorbed from milk (see Chapter 11). Replacement heifers were weaned on to a ration supplemented with coccidiostat as well as copper, ignoring the fact that the medication would defaunate the rumen and greatly increase copper absorbability. Dry cows were as generously supplemented with copper as their lactating cohorts. Cumulative over-provision of copper led to calves being born with livers containing 13–17 mmol Cu kg^{-1} DM. Some of these calves died with signs of hepatopathy after a few weeks, the liver apparently poisoned by a copper supplementation policy that met EC guidelines! The ignorance of feed formulators of important details of copper nutrition,

Table 20.2. Simultaneous routes of copper supplementation employed in a UK dairy herd and the hepatotoxic concentrations that were a cause for concern in newborn calves (actual copper requirements are given in parentheses). (A. Hunter and N.F. Suttle, unpublished data.)

	Cu (mg kg^{-1} DM)
Dry cow	30–50 (5–10)
Milking cow	40–60 (5–10)
Calf milk replacer	10 (1–2)
Calf rearer	20 (5–8)
Liver copper in newborn calves	13–17 mmol kg^{-1} DM

the eagerness of farmers to follow the precautionary principle at every opportunity and the weaknesses of the MPL system also leave behind a colossal 'copper footprint' as tonnes of copper are annually dispersed around farms in policies unlikely to benefit herd health. Bovine copper toxicity is of growing concern in the UK (Livesey, 2002) and now almost as prevalent as lead poisoning (see Chapter 11) but it is hardly accidental.

Phosphorus footprints

Among the major minerals, the problem posed by over-feeding phosphorus has long been recognized (Joengbloed and Lenis, 1998). Not only do farmers and feed compounders regularly include more phosphorus than national authorities recommend (e.g. dairy cows, Hristov et al., 2006; pigs, NRC, 1998; turkeys, Godwin et al., 2005), but also the recommendations themselves are generally excessive (Chapter 6). The problem is accentuated by the over-feeding of calcium (e.g. in dairy cows, Chapuis-Lardy et al., 2004) and the fact that phosphorus is excreted in highly labile forms even on dairy farms (Toor et al. 2005; McDowell et al., 2008). The survey data for current phosphorus supplementation practice in the UK (Table 20.1) confirms that phosphorus is generally being over-fed to most all classes of ruminant that receive complementary feeds.

Reducing the Phosphorus Footprint

Genetic manipulation of plants and microbes

Biotechnology has been applied to the problem of reducing the 'phosphorus footprint' by improving the utilization of plant phytate (Py) phosphorus (PyP) by 'genetic engineering'. A first generation of microbial and fungal phytases showed limited capacity to degrade PyP when added as dietary supplements to pig and poultry rations and organisms were modified to improve the pH optima and stability of phytases in both the pelleted ration and gut (see Chapter 6). In parallel studies, lines of genetically modified corn, low in Py but not phosphorus, were developed and found to provide more available phosphorus than normal corn, although the margin of improvement was far greater in one pig study (Spencer et al, 2000) than in another (Sands et al., 2001) in which the feed was fed as a mash. Low-Py barley (Veum et al., 2002) and soybean (Dilger and Adeola, 2006) also provide more available phosphorus than normal varieties. However, low-Py mutations of barley give poor yields and there are problems to be overcome before they become commercially viable (Bregitzer and Raboy, 2006).

Genetic manipulation of animals

Pigs have been genetically modified to express microbial phytase in their saliva and these breeds are able to utilize PyP better than conventional pigs (Golovan et al., 2001). More recently, poultry have been modified to overexpress a recombinant avian phosphatase with high activity towards PyP, a development that has been claimed to offer something akin to a 'golden bullet' solution to the problem of environmental phosphorus pollution (Cho et al., 2006); however, extensive commercial application remains a distant prospect. Any genetic modification of plant, ration or animal that meets the need for absorbable phosphorus (A_p) (on a concentration basis) during periods of peak demand could provide excess phosphorus at times of low demand (e.g. finishing pigs and laying hens) and simply re-route surplus phosphorus from the faeces to urine while increasing its lability (see Chapter 6). If the two technologies were concurrently adopted then it could become increasingly difficult to keep genetically modified animals away from genetically modified feeds, and the combination could lead to the hyper-excretion of phosphorus.

Dietary manipulation for pigs and poultry

There are many complementary ways of *immediately* reducing phosphorus footprints that do not require costly supplements or new technologies:

- Partial replacement of corn with a 'low-phosphorus' feed (e.g. cassava, sugarbeet pulp) in rations for animals with low phosphorus demands (e.g. finishing broilers, poults and pigs, hens).
- Feeding grain by-products of the distillery and biofuel industries, which are of much higher 'phosphorus value' than the grains themselves (Pederson et al., 2007).
- Reducing calcium supplements in rations to the minimum required (see Chapter 4).
- Blending corn and/or soybean meal with cereals and cereal by-products that are rich in phytase activity (e.g. laying hens, Scott et al., 2000; turkey poults, Roberson et al., 2005).
- Adding water to rations prior to feeding to stimulate endogenous phytase activity (see Chapter 6) (Godwin et al., 2005).
- Avoiding pelleting, which lowers the 'phosphorus value' of phytase-rich feeds (see Chapter 6).
- Substituting animal protein for soybean protein, since its phosphorus is three times more available (see Chapter 6).

Some of these measures have been tested in a pig experiment in which inorganic phosphorus was partially replaced with wheat middlings during growth and all or most mineral and vitamin supplements were withdrawn for the last 27 days of fattening (Shaw et al., 2006). Serum osteocalcin and pyridinoline, indicators of osteoblast activity and bone resorption (see Chapter 4), were increased and bone density and strength of the metacarpal were both reduced, suggesting that removing bone-forming minerals such as inorganic phosphorus is a risky procedure, but the incidence of bone fractures at slaughter did not increase. Furthermore, the addition of up to 30% wheat middlings should have allowed reductions in both total phosphorus and calcium below the generous levels of the NRC (1998) that were fed prior to withdrawal (see Chapter 4) and the abrupt reduction from 6.1 to 3.8 g Ca kg^{-1} DM and from 5.2 to 3.4 g P kg^{-1} DM for the last 27 days was neither wise and nor necessary.

Dietary manipulation in ruminants

It has been suggested that dairy farmers in Ontario, Canada, could save themselves $20 per cow and spare the environment 1.3 kt P per annum by reducing the phosphorus concentration in feed from 4.1 to 3.5 g kg^{-1} DM (Kebreab et al., 2008). A similar suggestion had been made to Idaho farmers (Hristov et al., 2006) and both are valid as far as they go. However, no harm is done by reducing phosphorus in the pre-calving diet to 2.1 g kg^{-1} DM (Petersen et al., 2005), and the calculations presented here (see Chapter 6) suggest that 2.5 g kg^{-1} DM is sufficient during lactation – in other words, the savings to both farmer and environment can be more than doubled! Not only is the addition of inorganic phosphorus unnecessary on most dairy farms, but its use may decrease the degradation of PyP (McDowell et al., 2008).

Reducing Other Mineral Footprints

The 'low-tech' approach can reduce other footprints.

Feeding to minimum mineral requirements

All mineral footprints can be reduced by feeding minerals at levels that meet the best estimates of maximum individual requirements (Table 20.1). Throughout this book, efforts have been made to derive low average requirements and identify factors that increase them so that proportionate allowances can be made for 'adverse mineral circumstances'. Implementation of the proposed requirements will be assisted by changes in current commercial and legislative practice. The removal of all supplementary trace minerals from a fattening ration for pigs has only minor effects on performance and carcass quality, and not all are negative (Shelton et al., 2004).

Maximum desirable limits

The tendency for MPL to inadvertently drive essential mineral concentrations far beyond animal needs, as mentioned earlier, could be countered by adding a further tier to regulations: that of the maximum desirable limit (MDL), based on satisfactory mineral allowances (see Chapter 1). The juxtaposition of MDL and MPL, and particularly the numerical gap between the two levels, would remind farmers and feed compounders that the MPL is warning against excess and is *not* a nutritional target.

Chelated minerals – a false economy

Suppliers of chelated minerals have leapt on the environmental protection bandwagon by arguing that the use of such minerals can reduce mineral footprints; that feeding hyperavailable mineral sources at reduced dietary concentrations can maintain performance in both nutritional and supranutritional contexts, while reducing the dispersal of mineral in animal wastes. If chelation was a guarantor of hyperavailability then this would indeed be a valid approach, but there is little or no evidence that it is (see Chapter 2). Undaunted and assuming what remains to be proven, sponsored experimenters have compared faecal copper excretion by pigs given copper sulfate ($CuSO_4$) at growth-stimulatory levels to that by pigs given chelated copper at lower levels (Veum *et al.*, 2004). The predictable result is lower faecal copper excretion with chelated copper, but the linear relationship between dietary and faecal copper concentrations across sources (Fig. 20.1) indicates no difference in availability and that similar environmental protection could be achieved by reducing the permitted level of $CuSO_4$ supplementation. Similarly, it has been suggested that the pharmacological benefits of adding 2–3 g Zn kg^{-1} DM as ZnO on the performance of early weaned, scouring piglets could be attained with less environmental impact by feeding less zinc in a chelated form (Mullan and D'Souza, 2005). However, faecal zinc excretion is determined by dietary zinc concentration rather than source (Case and Carlson, 2002) and neither inorganic nor a variety of organic sources improved the performance of weanling pigs when fed at the lower level of 0.5 g Zn kg^{-1} DM (Hollis *et al.*, 2005).

Footprints from mineral sources must be compared over similar ranges of mineral concentrations until consistent hyperavailability has been demonstrated for a given chelated or organic mineral source. The tendency of commercially sponsored trials to be designed to give a commercially desirable answer rather than to address a crucial question (e.g. of availability) is a disturbing trend, particularly given the lack of independent research on applied mineral nutrition (see Preface).

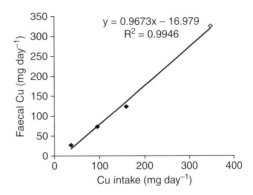

Fig. 20.1. An attempt to justify the use of chelated copper supplements (solid symbol) as growth stimulants for pigs rather than copper sulfate ($CuSO_4$, open symbol) (data from Veum *et al.*, 2004). Assuming that the chelate contains hyperavailable Cu, much lower dietary Cu levels are used and faecal Cu excretion is reduced, but extrapolation through a common regression line suggests that similar proportions of Cu are absorbed from the two sources and that the cheapest way to protect the environment is to use less $CuSO_4$.

Effects of Climate Change on Mineral Nutrition of Livestock

Regardless of whether the industrial and domestic activity of humans is responsible, climate change and the steps taken to reverse it will affect the mineral nutrition of livestock.

Changes in the growing season

Higher temperatures and heavier spring and autumn rainfalls are extending the grazing season in temperate zones. Those mineral disorders that are associated with the sudden or prolonged exposure to lush pastures (e.g. hypomagnesaemic tetany (see Chapter 5) and copper deprivation (see Chapter 11)) and associated increases in pasture infection with nematode larvae may be exacerbated, as may other disorders (e.g. sodium deprivation (see Chapter 8)). By contrast, shifting rainfall distribution from already arid zones will extend areas of salinity and aridity, with attendant problems for the mineral nutrition of livestock (see Chapter 7). In addition, enforced feeding of imported grains, feeds and by-products will change certain mineral supplies in ways to which native breeds may find it hard to adapt.

Increased production of biofuels

As energy-dense crops are increasingly grown for fuel production, soaring grain prices will lead to the decreased use of cereals in livestock feeding, particularly for ruminants. The livestock feed gap will be partly filled by an increased use of by-products from the biofuel industry. Corn distillers' solubles from modern ethanol plants contain phosphorus of higher availability than that of corn itself (Whitney et al., 2006; Pederson et al., 2007). Some products are largely unknown quantities in terms of mineral nutrition; others, such as bagasse and rapeseed meal, are more familiar and current levels of use have not presented recognizable or insurmountable problems in mineral nutrition. However, increased use is likely to be made of rapeseed meal and with it may come problems associated with the increased ingestion of goitrogens and their transfer to milk. These problems will be of particular concern in low-selenium (see Chapter 15) and low-iodine areas (see Chapter 12).

Reducing methane emissions by ruminants

Ruminants are significant contributors of the 'greenhouse gas' methane to the environment and attempts are being made to manipulate the rumen microflora by pharmacological and genetic means. One avenue is to reduce methane output from protozoa by defaunating the rumen. Rumen protozoa have a major influence on sulfur metabolism and their deletion will enhance the capture of sulfuride as sulfur amino acids in microbial protein. There will be secondary effects on trace mineral availability, both good and bad: the availability of copper will increase, reducing the need to supplement but increasing the risk of copper poisoning, particularly in sheep (see Chapter 11). The toxicity of cadmium and lead may be increased in defaunated animals, since their precipitation as insoluble sulfurides is part and parcel of ruminants' tolerance of those elements. The effects of methane inhibitors on the synthesis of B vitamins, including vitamin B_{12}, do not appear to have been studied.

Enhancing the 'Mineral Value' of Animal Foods in Human Diets

A new area of research has opened up since the last edition of this book was published, served by dedicated institutes and journals. This research is focused on the life-sustaining properties of micronutrients in foods consumed by humans and the associated possibilities for enhancing both 'functional' and market value by changing methods of food production (Welch and Graham, 2004). Particular interest surrounds the ability of selenium at supranutritional levels to reduce susceptibility to carcinogens (Rayman, 2005; Irons et al., 2006). There is substantial commercial interest from an agricultural industry that is anxious to reverse a tide of adverse publicity on the life-threatening qualities of foods, such as the link between saturated fat intake and coronary heart disease. Lucrative markets for trace element supplements for humans testify to concerns that modern diets are inadequate, although some claimed benefits of these supplements, such as improved fertility, vitality and resistance to infection, may be illusory. The temptation to over-provide minerals is as strong in human as in animal nutrition and notes of caution are sounded here.

Enhancing thyroid function

Goitre is such a conspicuous and specific sign of iodine deprivation that its regional appearance can prompt interventions to increase dietary iodine intakes, sometimes with spectacular success. A high incidence of goitre in Denmark led to a nationwide scheme in 1988 to increase iodine intakes by arranging for table salt and bread to be iodized. Surprisingly, the incidence of hypothyroidism rose, particularly in areas that had previously been relatively free from goitre (Laurberg et al., 2006). Daily iodine intakes varied widely from area to area due to variations in iodine supply from the drinking water and, when the national supply of iodine was increased, some individuals from high-iodine areas were exposed to an excess of iodine and thyroid activity was suppressed. The iodine supply can be improved by other means, such as feeding supranutritional levels of iodine to the dairy cow, because there is a linear relationship between dietary and milk iodine (see Chapter 12). However, there are those who think that levels in milk are already too high (Phillips et al., 1998). Mass medication to improve micronutrient-dependent functions is fraught with problems because of variations in individual micronutrient intakes as well as vocal minorities who do not want either their food or water supplies tampered with.

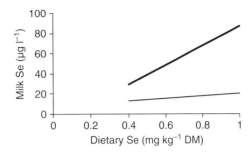

Fig. 20.2. When selenium yeast is added at high levels to the diet of dairy cows (bold line), the increase in milk Se is far greater than when the same amount of Se is added as selenite or a 'metalosate' (amino acid complex) (Givens et al., 2004).

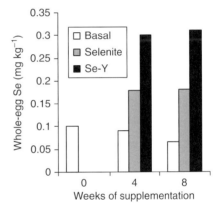

Fig. 20.3. The supplementation of the ration of laying hens with selenium yeast (Se-Y) at the maximum level permitted in the USA (+0.3 mg kg^{-1} DM) raises the Se concentration in the whole egg threefold and is more effective than the same amount of Se added as selenite. Note that egg Se declines significantly with time when hens are fed a basal diet that contains 0.1 mg Se kg^{-1} DM (data from Utterback et al., 2005).

Enhancing antioxidant status

The hypothesis has been advanced that oxidant stress is a contributory factor in many diseases (Surai, 2006). Since one component of antioxidant defence, glutathione peroxidase (GPX) activity, is enhanced simply by raising selenium intake (Lei et al., 1998; see Chapter 15) and human selenium intakes have declined in the UK (Givens et al., 2004) and are low in Northern Europe, China, New Zealand and elsewhere (Rayman, 2000; Combs, 2001), there are notional health benefits from raising selenium intakes. The supplementation of livestock rations with selenium was authorized in Sweden in 1980 and in the USA in 1985 and selenium fertilization of cereal crops was promoted in Finland in 1995. The supplementation of rations with sources rich in selenomethionine (SeMet), such as selenium yeast, has also been advocated because it raises milk selenium in dairy cows (Fig. 20.2: Juniper et al., 2006), egg selenium in hens (Fig. 20.3) and muscle selenium in lambs (Juniper et al., 2008) more effectively than the same amount of inorganic selenium. Most of the additional selenium in milk (Phipps et al., 2008) and muscle (Juniper et al., 2008) is present as SeMet and the partition of selenium in the egg shifts from the yolk to the albumen when Se yeast is used instead of selenite (Leeson et al., 2008). The 'success' of SeMet is attributable to its intact absorption and resistance to degradation but availability for

GPX synthesis in the supplemented animal is relatively poor (see Chapter 15), and turnover rate in muscle is slow (Juniper et al., 2008). Whether SeMet enrichment of the human diet raises GPX activity to the same extent as enrichment with selenite or is equally available at crucial intracellular sites of oxidant stress, such as the pathogen-engorged phagocyte, has not been studied, probably for fear that it may be inferior. If selenium bioavailability in and from the tissues of SeMet-supplemented animals is relatively poor then total tissue or blood selenium will provide inaccurate measures of the status or value to animals or consumers; the increased use of organic selenium will change the rules for diagnosis (see Chapter 15).

Selenium and cancer prevention

The relationship between selenium nutrition and cancer incidence has been reviewed by Rayman (2005). An inverse relationship between the selenium status of forage crops and cancer mortality has been reported across states of the USA and selenium supplementation of the human diet with SeMet has reduced skin cancer incidence in controlled trials. The protection against colonic cancer afforded to rats by a nutritional supplement (0.1 mg Se kg^{-1} DM) may be related to enhanced DNA methylation, evident in the liver and colon (Davis et al., 2000). However, at a supranutritional level (2.0 mg Se kg^{-1} DM) other protective effects may occur. Pre-neoplastic crypt abnormalities are induced more readily in the colons of mice with genetically impaired selenoprotein expression than in wild-type mice, and high-level selenium partially counters the effect while increasing the accumulation of selenium in non-protein compounds (Irons et al., 2006). It has been suggested that methylated selenium compounds with cancer-protecting properties accumulate when excess selenium is fed.

Selenium source and cancer prevention

The induction of GPX activity in selenium-depleted, Caco-2 cells has been reported to be faster with selenium added as selenite or methyl-selenocysteine than as SeMet (Zeng et al., 2008). As in the nutritional context, it was suggested that SeMet is diverted into non-selenoprotein synthesis. SeMet is no more effective than selenite for promoting DNA methylation (Davis et al., 2000). Any benefit from selenium given as SeMet (or selenium yeast) rather than as inorganic selenium seems unlikely, and efficacy may be poor when rates of protein accretion or secretion are high (e.g. infants, adolescents and nursing mothers; Fig. 15.3, p. 386). Natural selenium accumulators with atypical profiles of selenium speciation (e.g. certain brassicas and vetches) may be of higher nutraceutical value than SeMet. Furthermore, the association of value with a particular 'exotic' species would allow consumers to choose for themselves a dietary component that might increase carcinogen tolerance.

Enhancing iron status

Iron-deficiency anaemia is the most prevalent trace-element-responsive disorder of humans, affecting as many as 100 million people worldwide, most of them adult women or infants (Beard, 2008). The basic problem is one of low availability of iron from cereal sources, particularly maize (see Chapter 13), and this offers a major opportunity for improving the 'functional iron status' of human diets. Vegetable protein sources can be modified to lower their content of iron antagonists, notably Py, or increase their content of available iron (White and Broadley, 2009). However, enhancing iron nutrition increases susceptibility to malaria (Prentice, 2008) and there is an obvious need to reduce the possibility of infection before releasing 'high-iron genotypes' into the localities of at-risk populations. One simple dietary modification is to consume a vitamin-C-rich drink or supplement prior to eating, but the possible value of consuming vitamin-C-rich milk from animals given green feeds does not appear to have been explored.

Enhancing the Sensory Value of Foods

During the interval between the slaughter of an animal and the cooking of it either whole or as a joint, autolytic changes take place that affect the taste, smell and visual appearance of the product. The process is essentially one of oxidative deterioration and consequences such as change of colour (e.g. accumulation of metmyoglobin) (Ryu et al., 2005), exudation ('watery pork') and taint (Bou et al., 2005) can have a negative effect on retailer and consumer valuation of meat products. Manufacturers of mineral supplements with potential antioxidant properties, including zinc chelates and SeMet, have been quick to search for evidence that feeding them to livestock enhances the sensory value of meat, measuring everything from pH and shear strength to 'tastiness' in blindfolded consumers in the hope of positive outcomes similar to those found with vitamin E supplements. However, these trials have been conducted with selenium given in combination with vitamin E (Ryu et al., 2005). Improvements in storage characteristics have generally been small (Bou et al., 2005). For example, a supranutritional inorganic selenium supplement produces only minor improvements in the oxidative stability of chicken meat (Ryu et al., 2005). An organic selenium supplement that raised the selenium level in a pig ration from 0.18 (high) to 0.48 mg kg^{-1} DM (very high) reduced 'drip loss' significantly and more effectively than inorganic selenium (Mateo et al., 2007), but whether the reduction of 0.8% in drip loss, only one-third of the total loss, justifies 'medication' of the pork is questionable.

Putting a price on functional and sensory value

The enhancement of functional or sensory value will usually incur added production costs and will only be practised commercially if those costs are more than recouped in the marketplace. Furthermore, enhanced functional value will have to be measurable and measured as part of quality assurance schemes in any sustainable commercial venture. Alternatively, governments might sponsor the production of 'functionally valuable' products if their mass consumption is likely to produce cost-saving improvements in the nation's health. In either case, far more evidence of real health benefits is required before the enhancement of functional value becomes either a commercially viable proposition or the subject of a national intervention policy. Major purchasers of meat could perhaps recognize superior keeping quality in the produce of particular suppliers and offer a premium price.

New regulations on additives to animal feeds

Given the need to avoid micronutrient excess as well as deprivation, there will be a need to regulate disparate attempts to improve the functional and sensory value of our foods. There is already evidence of substantial variation in the trace mineral intakes provided by edible cuts of meat within Switzerland (Gerber et al., 2009). The same paper indicated that the early approval of selenium supplementation in the USA has probably contributed to threefold higher selenium concentrations in US than in Swiss beef. Some consumers are already eating – wittingly or unwittingly – bread, eggs and bacon enriched with selenium, and the milk on their breakfast table may also be enriched with selenium from selenium yeast. But how do consumers know and when should those in the habit of self-medication stop taking their selenium-rich pills? There is an urgent need for international regulatory standards to constrain powerful commercial interests that make substantial profits from selling mineral supplements for livestock that neither animals nor consumers need.

Conflict between Anti-nutritional and Nutritional Properties

Selenium and polyunsaturated fatty acids

Concern about the adverse effects of saturated fats in dairy and meat products on human health has stimulated interest in desaturating animal foods by feeding livestock more unsaturated fat. It is not difficult to producer softer, less saturated fat even in ruminants, in which unsaturated dietary fat can be protected from saturation in the rumen. However, high polyunsaturated fatty acid (PUFA) intakes increase the demand for vitamin E and increase the risk of selenium-responsive disorders in livestock (see Chapter 15). It is probably only a matter of time before high PUFA intakes are recognized as potentially harmful to human health, increasing humans' requirements for selenium!

Phytate – friend or foe?

The enhanced utilization of plant phosphorus from cereals and protein sources selected for *low Py* concentrations has been hailed as a breakthrough in terms of environmental protection from phosphorus pollution, but are there possible unforeseen side effects that could harm human health? Questions must be asked when the benefits of *high Py* intakes to humans are being investigated. Dietary *Py* supplements have anti-neoplastic activity against a range of tumours (Singh et al., 2004; Zhang et al., 2005) and protect tissues from calcification (Grases et al., 2004). *Py* is found also in mammalian tissues and cell membranes, where incorporation into phospholipids may influence cell signalling and angiogenesis (Vucenik et al., 2004). Populations that rely heavily on grains in their diets have low prevalences of bowel and certain other cancers. The extensive replacement of 'wild-type' cereals with low-*Py* varieties would lower *Py* intakes from 'Western' diets with possible adverse effects on human health, including increased cancer incidence. This is a striking example of the potential impact of a change in the mineral nutrition of livestock upon humans and the need for human and animal nutritionists to re-establish a dialogue.

Conflict Between Health Hazard Reduction and Animal Nutrition

The current approach is one of hazard reduction rather than risk assessment; in other words, to lower the concentration of a potentially harmful mineral at all costs, rather than to conduct a risk (to humans)–benefit (to livestock) analysis.

Hazard labelling for mutagenic or carcinogenic but essential minerals

Toxicologists and politicians can and increasingly do take their roles as protectors of the human population to extremes. For example, the identification of cobalt as a carcinogen (National Toxicology Program, 1998) has led to the introduction of a new EC Regulation (No. 1272/2008) that requires a skull and crossbones symbol to be placed on bags of feed containing >0.01% added cobalt salts with a warning 'may cause cancer by inhalation' – no concession is made for pelleted diets! The real risk, if any, is to those involved in the manufacturing process and not to the livestock farmer, yet consideration is being given to the energy (and nutritionally) inefficient option of using synthetic vitamin B_{12} as a feed supplement. The evidence of carcinogenicity comes from experiments with mice exposed semi-continuously (5 days a week) for months on end to an aerosol containing $3 mg Co m^{-3}$, a volume that contains enough cobalt to keep an adult sheep nourished for a month! It is probably only a matter of time before other trace elements join the list of carcinogens, creating further difficulties for those who formulate and manufacture animal foodstuffs with no credible impact on cancer risk.

Maximum permissible levels for potentially toxic elements

The case for reducing human exposure to potentially toxic elements via the food chain once seemed unarguable, but the idea that all elements, even lead and cadmium, are beneficial in small doses (see Chapter 18) is causing debate amongst toxicologists that has yet to impinge upon those who set MPL. Could the drive to reduce human exposure to lead and cadmium eventually be harmful to health? On these and other matters there is a need to restore a balance, which until now has been determined entirely by simplistic 'hazard-reduction' standards. As for the potentially toxic elements, MPL should be based on the cumulative exposure (amount of the element over time) likely to cause the element in the kidneys at slaughter to reach maximum acceptable concentrations in the growing and breeding animal, rather than on a single concentration.

the problem of environmental pollution from pig and poultry farms and enhance the availability of iron, copper, manganese and zinc.

Permissible sources

In the wake of the BSE scare in the UK, the use of animal protein sources in animal feeds was banned in Europe to protect consumers from a perceived health hazard. The ban has so far been only partially relaxed. The possibility of using meat and bone meal to reduce the need for inorganic phosphorus supplementation is thus prohibited, although it could lessen

Conclusion

There is clearly a need for dialogue between those whose priority lies in the domain of public health and those whose priority is animal nutrition. Unless each side listens and learns from the other, the problems of adequate but safe mineral nutrition for animals and humans are likely to increase in number and severity.

References

Beard, J.L. (2008) Why iron deficiency is important in infant development. *Journal of Nutrition* 138, 2523–2528.
Bou, R., Guardiola, P., Barroeta, A.C. and Codony, R. (2005) Effect of dietary fat sources and zinc and selenium supplements on the consumption and consumer acceptability of chicken meat. *Poultry Science* 84, 1129–1140.
Bregitzer, P. and Raboy, V. (2006) Effects of four independent low-phytate mutations on barley agronomic performance. *Crop Science* 46, 1318–1322.
Case, C.L. and Carlson, M.S. (2002) Effect of feeding organic and inorganic sources of additional zinc on growth performance and zinc balance in pigs. *Journal of Animal Science* 80, 1917–1924.
Chapuis-Lardy, L., Fiorini, J., Toth, J. and Dou, Z. (2004) Phosphorus concentration and solubility in dairy faeces: variability and affecting factors. *Journal of Dairy Science* 87, 4334–4341.
Cho, J., Choi, K., Darden, T., Reynolds, P.R., Petite, J.N. and Shears, S.B. (2006) Avian multiple inositol polyphosphate phosphatase is an active phytase that can be engineered to help ameliorate the planet's 'phosphate crisis'. *Journal of Biotechnology* 1, 248–259.
Davis, C.D., Uthus, E.O. and Finley, J.W. (2000) Dietary selenium and arsenic affect DNA methylation in vitro in Caco-2 cells and *in vivo* in rat liver and colon. *Journal of Nutrition* 130, 2903–2909.
Dilger, R.N. and Adeola, O. (2006) Estimation of true phosphorus digestibility and endogenous phosphorus loss in growing pigs fed conventional and low-phytate soybean meals. *Journal of Animal Science* 84, 627–634.
EC (2002) European Communities Directive 2002/32/EC of The European Parliament and the Council. *Official Journal of the European Commission* L140/10–L140/21.
EC (2003) European Communities Directive No. 1334/2003 of The European Parliament and the Council. *Official Journal of the European Commission* L187/11–L187/15, Annex.
Gerber, N., Brogioli, R., Hattendof, B., Scheeder, M.R.L., Wenk, C. and Gunther, D. (2009) Variability in selected trace elements of different meat cuts determined by ICP-MS and DRC-ICPMS. *Animal* 3, 166–172.
Godwin, J.L., Grimes, J.L., Christensen, V.L. and Wineland, M.J. (2005) Effect of dietary phosphorus and phytase levels on the reproductive performance of large white turkey breeder hens. *Poultry Science* 84, 485–493.
Golovan, S.P., Meidinger, R.G., Ajakaiye, A., Cottrill, M., Wiederkehr, M.Z., Barney, D.J., Plante, C., Pollard, J.W. and Fan, M.Z. (2001) Pigs expressing salivary phytase produce low-phosphorus manure. *Nature Biotechnology* 19, 741–745.
Grases, F., Perello, J., Prieto, R.M., Simonet, B.M. and Torres, J.J. (2004) Dietary myo-inositol hexaphosphate prevents dystrophic calcifications in soft tissues: a pilot study in Wistar rats. *Life Sciences* 75, 11–19.

Hill, G.M., Mahan, D.C., Carter, S.D., Cromwell, G.L., Ewan, R.C., Harrold, R.L., Lewis, A.J., Miller, P.S., Shurson, G.C. and Veum, T.L. (2001) Effect of pharmacological concentrations of zinc oxide with or without the inclusion of an antibacterial agent on nursery pig performance. *Journal of Animal Science* 79, 934–941.

Hollis, G.R., Carter, S.D., Cline. T.R., Crenshaw, T.D., Cromwell, G.L., Hill, G.M., Kim, S.W., Lewis, A.J., Mahan, D.C., Miller, P.S., Stein, H.H. and Veum, T.L. (2005) Effects of replacing pharmacological levels of dietary zinc oxide with lower dietary levels of various organic zinc sources for weanling pigs. *Journal of Animal Science* 83, 2123-2129.

Hristov, A.N., Hazen,W. and Ellsworth J.W. (2006) Efficiency of use of imported nitrogen, phosphorus, and potassium and potential for reducing phosphorus imports on Idaho dairy farms. *Journal of Dairy Science* 89, 3702–3712.

Hristov, A.N., Hazen,W. and Ellsworth J.W. (2007) Efficiency of use of imported magnesium, sulfur, copper and zinc on Idaho dairy farms. *Journal of Dairy Science* 90, 3702–3712.

Irons, R., Carlson, B.A., Hatfield, D.L. and Davies, C.D. (2006) Both selenoproteins and low molecular weight selenocompounds reduce colon cancer risk in mice with genetically impaired selenoprotein expression. *Journal of Nutrition* 136, 1311–1317.

Joengbloed, A.W. and Lenis, N.P. (1998) Environmental concerns about animal manure. *Journal of Animal Science* 76, 2641–2648.

Juniper, D.T., Phipps, R.H., Jones, A.K. and Bertin, G. (2006) Selenium supplementation of lactating dairy cows: effect on selenium concentration in blood, milk, urine and feces. *Journal of Dairy Science* 89, 3544–3551.

Juniper, D.T., Phipps, R.H., Ramos-Morales, E. and Bertin, G. (2008) Selenium persistency and speciation in the tissues of lambs following the withdrawal of a high-dose selenium-enriched yeast. *Animal* 2, 375–380.

Kebreab, E., Odongo, N.E., McBride, B.W., Hanigan, M.D. and France, J. (2008) Phosphorus utilization and environmental pollution and economic implications of reducing phosphorus pollution from Ontario dairy farms. *Journal of Dairy Science* 91, 241–246.

Laurberg, P., Jorgensen, T., Perrild, H., Ovesen, L., Knudsen, N., Pedersen, I.B., Rasmussen, L.B., Carle, A. and Vejberg, P. (2006) The Danish investigation on iodine intake and thyroid disease, DanThyr: status and perspectives. *European Journal of Endocrinology* 155, 219–228.

Leeson, S., Namkung, H., Caston, L., Durosoy, S. and Schlegel, P. (2008) Comparison of selenium levels and sources and dietary fat quality in diets of broiler breeders and laying hens. *Poultry Science* 87, 2605–2612.

Lei, X.G., Dann, H.M., Ross, D.A., Cheng, W.-S., Combs, G.F. and Roneker, K.R. (1998) Dietary selenium supplementation is required to support full expression of three selenium-dependent glutathione peroxidases in various tissues of weanling pigs. *Journal of Nutrition* 128, 130–135.

Livesey, C.T., Bidewell, C.A., Crawshaw, T.R. and David, G.P. (2002) Investigation of copper poisoning in adult cows by the Veterinary Laboratories Agency. *Cattle Practice* 10, 289–294.

Lopez-Alonso, M.L., Benedito, J.L., Miranda, M., Castillo, C., Hernandez, J. and Shore, R.F. (2000) The effect of pig farming on copper and zinc accumulation in cattle in Galicia (North-West Spain). *The Veterinary Journal* 160, 259–266.

Mateo, R.D., Spalholz, J.E., Elder, R., Yoon, I. and Kim, S.W. (2007) Efficacy of dietary selenium sources on growth and carcass characteristics of growing-finishing pigs fed diets containing high endogenous selenium. *Journal of Animal Science* 85, 1177–1183.

McDowell, L.R. (1992) *Minerals in Animal and Human Nutrition*. Elsevier, Amsterdam, p.18.

McDowell, R.W., Dou, Z., Toth, J.D., Cade-Menun, B.J., Kleinman, P.J.A., Soder, K. and Saporito, L. (2008) A comparison of the phosphorus speciation and potential bioavailability in feed and feces of different dairy herds using ^{31}P nuclear magnetic resonance spectroscopy. *Journal of Environmental Quality* 37, 741–752.

Meyer, D. and Mullinax, D.D. (1999) Livestock nutrient management concerns: regulatory and legislative overview. *Journal of Animal Science* 77 (Suppl. 2J), 51–62.

Mullan, B. and D'Souza, D. (2005) The role of organic minerals in modern pig production. In: Taylor-Pickard, J.A. and Tucker, L.A. (eds) *Redefining Mineral Nutrition*. Nottingham University Press, Nottingham, UK, pp. 89–117.

National Toxicology Program (1998) NTP toxicology and carcinogenesis studies of cobalt sulfate heptahydrate (CAS No. 10026-24-1) in F344/N and B6C3F1mice (inhalation studies). *National Toxicology Program Technical Report Series* 471:1, 1–268.

NRC (1998) *Nutrient Requirements of Swine*, 10th edn. National Academy of Sciences, Washington, DC.

Pang, C., Huang, H., Zhao, Y., Qin, S., Chen, F. and Hu, Q. (2007) Effect of selenium source and level in hen's diet on tissue selenium deposition and egg selenium concentrations. *Journal of Agricultural and Food Chemistry* 55, 1027–1032.

Payne, G.G., Martens, D.C., Kornegay, E.T. and Lindemann, M.D. (1988) Availability and form of copper in three soils following eight annual applications of copper-enriched swine manure. *Journal of Environmental Quality* 17, 740–746.

Pedersen, C., Boersma, M.G. and Stein, H.H. (2007) Digestibility of energy and phosphorus in ten samples of distillers dried grains with solubles fed to growing pigs. *Journal of Animal Science* 85, 1168–1176.

Peterson, A.B., Orth, M.W., Goff, J.P. and Beede, D.K. (2005) Periparturient responses of multiparous Holstein cows fed different dietary phosphorus concentrations prepartum. *Journal of Dairy Science* 88, 3582–3594.

Phillips, D.I.W., Nelson, M., Barker, D.J.P., Morris, J.A. and Wood, T.J. (1988) Iodine in milk and the incidence of thyrotoxicosis in England. *Clinical Endocrinology* 28, 61–66.

Phipps, R.H., Grandison, A.S., Jones, A.K., Juniper, D.T., Ramos-Morales, E. and Bertin, G. (2008). Selenium supplementation of lactating dairy cows: effects on milk production and total selenium content and speciation in blood, milk and cheese. *Animal* 2, 1610–1618.

Prentice, A.M. (2008) Iron metabolism, malaria and other infections: what is all the fuss about? *Journal of Nutrition* 138, 2537–2541.

Rayman, M.P. (2000) The importance of selenium in human health. *Lancet,* 356, 233–241.

Rayman, M.P. (2005) Selenium in cancer prevention: a review of the evidence. *Proceedings of the Nutrition Society* 64, 527–542.

Roberson, K.D., Kalbfleish, J.L., Pan, W., Applegate, T.J. and Rosenstein, D.S. (2005) Comparison of wheat bran phytase and a commercially available phytase on turkey tom performance and litter phosphorus content. *International Journal of Poultry Science* 4, 244–249.

Ryu, Y.-C., Rhee, M.-S., Lee, K.-M. and Kim, B.-C. (2005) Effects of different levels of dietary supplemental selenium on performance, lipid oxidation and color stability of broiler chicks. *Poultry Science* 84, 809–815.

Sands, J.S., Ragland, D., Baxter, C., Joern, B.C., Sauber, T.E. and Adeola, O. (2001) Phosphorus bioavailability, growth performance and nutrient balance in pigs fed high available phosphorus corn and phytase. *Journal of Animal Science* 79, 234–2142.

Scott, T.A., Kampen, R. and Silversides, F.G. (2000) The effect of phosphorus, phytase enzyme and calcium on the performance of layers fed wheat-based diets. *Canadian Journal of Animal Science* 80, 183–190.

Shaw, D.T., Rozeboom, D.W., Hill, G.M., Orth, M.W., Rosenstein, D.S. and Link, J.E. (2006) Impact of supplement withdrawal and wheat middling inclusion on bone metabolism, bone strength and the incidence of bone fractures occurring at slaughter in pigs. *Journal of Animal Science* 84, 1138–1146.

Shelton, J.L., Southern, L.L., LeMieux, F.M., Bidner, T.D. and Page, T.G. (2004) Effects of microbial phytase, low calcium and phosphorus, and removing the dietary trace mineral premix on carcass traits, pork quality, plasma metabolites, and tissue mineral content in growing-finishing pigs. *Journal of Animal Science* 82, 2630–2639.

Singh, R.P., Sharma, G., Mallikarnjuna, G.U., Dhanalakshmi, S., Agarwal, C. and Agarwal, R. (2004) *In vivo* suppression of hormone-refractory prostate cancer growth by inositol hexaphosphate. *Clinical Cancer Research* 10, 244–250.

Spencer, J.D., Allee, G.L. and Sauber, T.E. (2000) Phosphorus bioavailability and digestibility of normal and genetically modified low-phytate corn for pigs. *Journal of Animal Science* 78, 675–681.

Surai, P.F. (2006) *Selenium in Animal and Human Health*. Nottingham, Nottingham University Press, UK.

Suttle, N.F., Lewis, R.M. and Small, J.N. (2002) Effects of breed and family on rate of copper accretion in the liver of purebred Charollais, Suffolk and Texel lambs. *Animal Science* 75, 295–302.

Toor, G.S., Cade-Munn, B.J. and Sims, J.T. (2005) Establishing a linkage between phosphorus forms in dairy diets, faeces and manures. *Journal of Environmental Quality* 34, 1380–1391.

Utterback, P.L., Parsons, C.M., Yoon, I. and Butler, J. (2005) Effect of supplementing selenium yeast in diets of laying hens on egg selenium content. *Poultry Science* 84, 1900–1901.

Veum, T.L., Ledoux, D.R., Bollinger, D.W., Raboy, V. and Cook, A. (2002) Low-phytic acid barley improves calcium and phosphorus utilisation and growth performance in growing pigs. *Journal of Animal Science* 80, 2663–2670.

Veum, T.L., Carlson, M.S., Bollinger, D.W. and Ellersieck, M.R. (2004) Copper proteinate in weanling pig diets for enhancing growth performance and reducing faecal copper excretion compared with copper sulfate. *Journal of Animal Science* 82, 1062–1070.

Vucenik, I., Passaniti, A., Vitolo, M.I., Tantivejkul, K., Eggleton, P. and Shamsuddin, A.M. (2004) Anti-angiogenic activity of inositol hexaphosphate (IP6). *Carcinogenesis* 25, 2115–2123.

Welch, R.M. and Graham, R.D. (2004) Breeding for micronutrients in staple food crops. *Journal of Experimental Botany* 55, 353–364.

White, P.J. and Broadley, M.R. (2009) Biofortification of crops with seven mineral elements often lacking in human diets–iron, zinc, copper, calcium, magnesium, selenium and iodine. *New Phytologist* 182, 49–84.

Whitney, M.H., Shurson, G.C., Johnston, L.J., Wulf, D.M. and Shanks, B.C. (2006) Growth performance and carcass characteristics of grower-finishing pigs fed high-quality corn distillers dried grain with solubles originating from a modern Midwestern ethanol plant. *Journal of Animal Science* 84, 3356–3363.

Zeng, H., Botnen, J.H. and Johnson, L.K. (2008) A selenium-deficient Caco-2 cell model for assessing differential incorporation of chemical or food selenium into glutathione peroxidase. *Biological Trace Element Research* 123, 98–108.

Zhang, Z., Song, Y. and Wang, X-L. (2005) Inositol hexaphosphate-induced enhancement of natural killer cell activity correlates with suppression of colon carcinogenesis in rats. *World Journal of Gastroenterology* 11, 5044–5046.

Appendix 1

Appendix 1. A scheme for the differential diagnosis of mineral disorders with non-specific signs. Proceed vertically with a given line of investigation, starting with 1 and switching to 2 and 3 if the results are normal or negative.

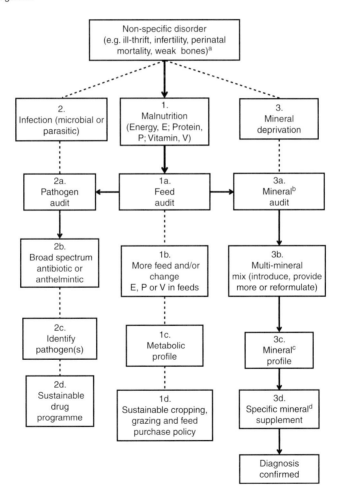

[a]First rule out exotic agents, e.g. plant or industrial poisons.
[b]See Appendix 2 for guidance on macromineral audit.
[c]See Appendices 3 and 4 for guidance on profile interpretation.
[d]Some animals must be left untreated.

Appendix 2

Appendix 2. Mean macromineral composition of major feedstuffs sampled in the UK by the Ministry of Agriculture (standard deviations in parentheses) (MAFF, 1990). By following the diagnostic flow diagram (Appendix 1), the data can be used to conduct a 'mineral audit'. Farm values for a *major* ration constituent that lie >1 standard deviation below the mean for that feed (or above for an antagonist such as potassium) strengthen the possibility of mineral imbalance. The exception is for sodium, where coefficients of variation are generally high.

	Ca	Mg	P	K	Na	S	Ratio of S to N[a]
Roughages							
Grass – fresh (242)	5.4 (1.7)	1.6 (0.56)	3.0 (0.68)	25.8 (6.6)	2.5 (2.1)	2.2 (0.6)	0.088
Grass – hay (128)	5.2 (2.5)	1.4 (0.52)	2.3 (0.75)	20.7 (5.3)	2.1 (1.7)	—	—
Grass – silage[b] (180)	6.4 (2.0)	1.7 (0.54)	3.2 (1.7)	25.8 (6.8)	—	—	—
Grass–dried (112)	7.4 (4.9)	1.8 (0.47)	3.3 (1.1)	26.0 (8.0)	2.8 (1.6)	3.5 (0.3)	0.12
Lucerne – hay (5)	15.6 (1.8)	1.7 (0.27)	3.1 (0.66)	27.3 (5.0)	0.6 (0.07)	—	—
Lucerne – silage (8)	17.6 (3.4)	1.8 (0.23)	3.0 (0.35)	24.6 (4.0)	1.3 (0.95)	—	—
Lucerne – dried (50)	15.0 (2.6)	2.3 (0.65)	3.0 (0.8)	25.4 (8.3)	1.3 (0.8)	—	—
Clover silage (10)	16.7 (5.5)	2.3 (0.75)	3.1 (0.66)	27.4 (7.6)	0.3 (0.1)	—	—
Maize silage (26)	4.3 (2.0)	2.2 (0.69)	2.6 (1.2)	12.3 (4.1)	0.3 (0.2)	—	—
Fodder beet (10)	2.8 (2.4)	1.6 (0.30)	1.8 (0.3)	17.5 (4.8)	0.3 (0.4)	—	—
Kale – all varieties (10)	13.1 (1.4)	1.6 (0.2)	4.1 (0.4)	29.3 (3.2)	1.2 (0.6)	—	—
Straw – barley[c] (17)	3.8 (1.8)	0.7 (0.31)	1.1 (1.10)	16.0 (6.5)	1.8 (2.1)	2.0 (0.4)	0.33
Straw – oat (6)	3.9 (1.2)	0.9 (0.31)	0.9 (0.24)	17.9 (2.4)	1.8 (2.1)	2.0 (0.4)	0.19
Straw – wheat (70)	3.9 (1.1)	0.9 (0.9)	0.8 (0.35)	10.2 (3.7)	4.8 (0.6)	1.3 (0.4)	0.24
Concentrates							
Barley[d] (56)	0.9 (0.6)	1.2 (0.2)	4.0 (0.46)	5.0 (0.7)	0.3 (0.4)	1.5 (0.3)	0.073
Maize (16)	0.1 (0.1)	1.3 (0.1)	3.0 (0.3)	3.5 (0.2)	—	1.6 (0)	0.10
Oats[d] (27)	0.9 (0.2)	1.0 (0.1)[e]	3.4 (0.5)	5.0 (0.9)	0.2 (0.1)	1.9 (0.1)	0.11
Sorghum (5)	0.3 (0.3)	1.1 (0.1)	2.8 (0.5)	3.7 (0.14)	0.5 (0.0)	—	—
Wheat (37)	0.6 (0.2)	1.1 (0.1)	3.3 (0.4)	4.6 (0.4)	0.1 (0.06)	1.6 (0.08)	0.078
Cottonseed meal (9)	2.1 (0.21)	5.8 (0.4)	8.9 (1.1)	15.8 (0.8)	0.2 (0.06)	5.0 (0.32)	0.083
Fish meal, white (7)	56 (6.0)	2.3 (0.3)	38 (14.5)	10.2 (1.3)	11.2 (1.5)	—	—
Linseed meal (5)	3.4 (0.1)	5.4 (0.1)	8.7 (0.1)	11.2 (0.1)	0.7 (0.0)	5.0	0.083
Maize-germ meal (9)	0.2 (0.04)	2.1 (0.7)	2.8 (1.2)	—	—	4.9	0.046
Maize-gluten feed (22)	2.3 (3.6)	4.1 (0.7)	9.3 (1.2)	12.5 (2.7)	2.6 (1.4)	4.1	0.065
Palm-kernel meal (7)	2.4 (0.3)	3.0 (0.4)	6.2 (0.6)	6.9 (1.2)	0.2 (0.1)	3.2	0.118
Rapeseed meal (17)	8.4 (2.7)	4.4 (0.5)	11.3 (1.5)	14.3 (2.2)	0.4 (0.3)	16.9 (1.7)	0.26
Soybean meal[f] (12)	3.9 (1.6)	3.0 (0.2)	7.4 (0.4)	25.0 (1.0)	0.16 (0.05)	4.6	0.058
Sunflower meal (6)	4.8 (1.4)	5.8 (0.5)	10.8 (1.9)	17.1 (1.8)	1.0 (1.2)	—	—

Continued

Appendix 2. Continued.

	Ca	Mg	P	K	Na	S	Ratio of S to N[a]
By-products							
Wheat bran (10)	1.1 (0.2)	6.2 (2.7)	12.6 (3.0)	–	–	–	–
Rice-bran meal (8)	0.9 (1.0)	3.8 (4.1)	17.4 (1.7)	5.6 (5.9)	0.24 (0.13)	–	–
Brewers' grains (20)	3.5 (1.4)	1.7 (0.4)	5.1 (1.0)	0.6 (0.95)	0.3 (0.30)	3.0 (0.17)	0.084
Distillers' grains[g] (7)	1.7 (0.32)	1.7 (0.5)	9.6 (0.8)	10.2 (0.8)	1.2 (1.2)	–	–
Citrus pulp (13)	14.6 (2.4)	1.1 (0.5)	1.1 (0.1)	–	–	3.7	0.084
Sugarbeet pulp[h] (14)	7.6 (2.0)	1.1 (0.2)	0.8 (0.2)	18.2 (1.9)	–	–	–
Cassava meal (5)	2.2 (0.6)	1.4 (0.6)	0.9 (0.1)	8.1 (1.5)	0.6 (0.25)	18.1 (0.6)	0.127
Feather meal (5)	5.6 (0.1)	0.4 (0.04)	3.1 (0.1)	1.5 (0.0)	1.4 (0.0)	18.1	0.127

[a]The ratio for S to N is more important and the means far less variable than data for S (ratio standard deviation not given in database).
[b]Clamp.
[c]Untreated.
[d]Winter and spring.
[e]Winter.
[f]Extracted.
[g]Barley.
[h]Molasses.

Appendix 3

Appendix 3. Marginal bands for the interpretation of macro-mineral concentrations in blood or bone biopsy samples from chronically deprived weaned ruminants. Mean values should normally exceed these bands; the nearer they are to the lower limit or the higher the proportion of individual values falling within the band, the greater the likelihood of mineral deprivation.

Animal	Plasma (mmol l^{-1})					
	Ca	Mg	P	K	Na	Sulfate
Cattle	1.8–2.0	0.50–0.75	1.0–1.5	<2.5	124–135	<0.3
Sheep	1.8–2.0	0.60–0.75	1.0–1.5	2.4–4.0	140–145	<0.3
	Bone ash (g kg^{-1} DM)					
Calf/cow	500–600	<12[a]	500–600 (rib cortex)	–	–	–
Lamb	0.2–0.3	<7[a]	200–300 (vertebra)	–	–	–

[a] mg g^{-1} ash.

Appendix 4

Appendix 4. A summary[a] of marginal bands for diagnostic criteria indicating a possible risk of trace element disorder[b] in ruminants. Mean plasma and tissue values for a randomly sampled population falling within these bands indicate the probability of disorder, while 10% of individual values falling within the bands indicates possible benefits from supplementation, particularly if other possible causes of ill-thrift have been eliminated (see Appendix 1).

Sample	Cu	Co	I[c]	Se	Zn
Diet (mg kg⁻¹ DM)	S:Cu, 500[d]–1000 Fe:Cu, 50[d]–100 Mo:Cu, 0.3[d]–1	0.05–0.10[e]	Summer, 0.10–0.15 Winter, 0.20–0.30	0.025–0.05[e]	10–20
Plasma (l⁻¹)	Neonate, 2–4 μmol Others, 3–9[f] μmol	B_{12}, 230–350 pmol[g] B_{12}, 336–500 pmol[g] MMA, 5–15 μmol	Thyroxine, 40–60 nmol Thyroxine, 20–30 nmol Total iodine, 30–40 μg	250–500 nmol 150–250[d] nmol	– 6–9 μmol
Tissue (kg⁻¹ FW)	Neonate liver, 260–1050 Others, 30–100 μmol	Liver B_{12}, 280–340 nmol	Thyroid weight, 0.4–0.9 g kg⁻¹ LW Thyroid I, 1.2–2.0 g kg⁻¹ DM Milk I, 70–100 μg l⁻¹	Liver, 800–900 nmol Liver, 250–450[e]	Bone, 0.77–1.08 mmol kg⁻¹ fat-free dry rib
Other	Fleece/hair, 31–62 μmol kg⁻¹ DM	Soil, extractable Co <0.4 mg kg⁻¹ DM Soil, extractable Mn >1000 mg kg⁻¹ DM Soil, pH <5 or >6 Milk B_{12}, 250–500 pmol l⁻¹		Erythrocyte GPX 30–50 U g⁻¹ Hb or 10–16 U ml⁻¹ cells (assay temp. 37°C)	Fleece/hair, 1.15–1.54 mmol kg⁻¹ DM

DM, dry matter; FW, fresh weight; GPX, glutathione peroxidase (an important indicator of functional selenium status).
[a]It is impracticable to cover in a single table the many factors that can affect the reliability of a single measure of trace element status, but bands depend on species and assay conditions; see Chapter 15.
[b]For the diagnosis of iron and manganese deprivation, rarely encountered in weaned animals, see Chapters 13 and 14, respectively. In sheep, for example, the interpretation of plasma selenium is different for barren and pregnant ewes (see Chapter 15) and trace element deprivation in early pregnancy may have residual effects on perinatal mortality and lamb growth that are not detectable from later sampling. Readers are advised to consult the relevant section in each chapter.
[c]Exposure to goitrogens is a major cause of iodine deprivation and may influence the reliability of biochemical markers of iodine and selenium status in ways that have yet to be investigated (see Chapters 12 and 15). The link between iodine and selenium metabolism may mean that more dietary selenium is needed in winter than in summer, as is the case for iodine.
[d,e]Species differences in susceptibility, cattle and sheep, respectively, being the ruminants most likely to suffer at relatively mild exposure to copper and cobalt or selenium deprivation. The high-yielding dairy cow may need higher plasma selenium than other animals to avoid selenium deprivation.
[f]Plasma must be used, not serum.
[g]Bovine assays often underestimate.

Appendix 5

Appendix 5. Common ingredients of mineral supplements in Europe and their mineral content prior to mixing. *Source*: International Association of the European (EU) Manufacturers of Major, Trace and Specific Feed Mineral Materials (EMFEMA), Brussels.

Mineral	Source	Formula	Content (%) Theoretical	Content (%) Actual	Other elements
Ca	Calcium carbonate	$CaCO_3$	40	37–39.5	
	Oyster shells	$CaCO_3$	40	c.37	
Mg	Magnesium oxide	MgO		50.5–52.0	
	Magnesium sulfate	$MgSO_4 \cdot 7H_2O$		10.0	13 S
	Magnesium chloride	$MgCl_2 \cdot 6H_2O$		12.0	35 Cl
	Magnesium hydroxide	$Mg(OH)_2$		36–38	
	Magnesium phosphate	$MgHPO_4 \cdot nH_2O$		24–28	13–15 P
Na	Marine salt	$NaCl$		39.2	
	Vacuum-dried salt	$NaCl$		38.9–39.2	
	Rock salt	$NaCl$		36.4	
	Bicarbonate	$NaHCO_3$		27.0	
	Sodium sulfate (anhydrous)	Na_2SO_4		32.0	22 S
P	'Di' calcium phosphate (anhydrous)	$CaHPO_4$		18–20.5	25–27 Ca
	'Di' calcium phosphate (dihydrate)	$CaHPO_4 \cdot 2H_2O$		17–18.6	23–25 Ca
	'Mono-di' calcium phosphate	$CaHPO_4 \cdot Ca(H_2PO_4)_2 \cdot H_2O$		20–22	19.5–23.0 Ca
	'Mono' calcium phosphate	$Ca(H_2PO_4)_2 \cdot H_2O$		22–22.7	15–17.5 Ca
	Phosphate-containing $CaCO_3$	$Ca(H_2PO_4)_2 \cdot xH_2O + CaCO_3$		18	21 Ca
	Calcium–magnesium phosphate	$(Ca,Mg)PO_4 \cdot nH_2O$		18	15 Ca; 9 Mg
	Na–Ca–Mg phosphate	$(Na,Ca,Mg)PO_4 \cdot nH_2O$		17.5	7.5–8.5 Ca; 4.5–5.5 Mg; 12.0–14.5 Na
	Monoammonium phosphate	$NH_4H_2PO_4$		26	11 N
	Monosodium phosphate	$NH_2PO_4 \cdot nH_2O$		24	20 Na
S	Calcium sulfate	$CaSO_4 \cdot 2H_2O$		18.5	23.3 Ca
Co	Cobaltous sulfate heptahydrate	$CoSO_4 \cdot 7H_2O$		21	
Cu	Copper sulfate	$CuSO_4 \cdot 5H_2O$		25	
Zn	Zinc sulfate	$ZnSO_4 \cdot H_2O$		35	
	Zinc oxide	ZnO		75	
Fe	Ferrous sulfate monohydrate	$FeSO_4 \cdot H_2O$		29.5–30.0	
	Ferrous sulfate heptahydrate	$FeSO_4 \cdot 7H_2O$		19.5	
Mn	Manganous oxide	MnO		52–63	
	Manganous sulfate monohydrate	$MnSO_4 \cdot H_2O$		31–32	
Mo	Sodium molybdate dihydrate	$Na_2MoO_4 \cdot 2H_2O$		39.5–40	
I	Calcium iodate monohydrate	$Ca(IO_2)_2 \cdot H_2O$		61.6	
	Sodium iodate	NaI		84.7	
	Potassium iodide (may be stabilized with stearates)	KI		76.5	
Se	Sodium selenite	Na_2SeO_3		45	
	Sodium selenate	Na_2SeO_4		41.8	

Appendix 6

Appendix 6. Chemical symbols, atomic numbers and atomic weights of elements of nutritional significance and their grouping in the periodic table.

Name	Symbol	Atomic number	Atomic weight	Group	Name	Symbol	Atomic number	Atomic weight	Group
Aluminium	Al	13	26.98	IIIb	Mercury	Hg	80	200.59	IIb
Arsenic	As	33	74.92	Vb	Molybdenum	Mo	42	95.94	VIa
Barium	Ba	56	137.34	IIa	Nickel	Ni	28	58.71	VIII
Boron	B	5	10.81	IIIb	Nitrogen	N	7	14.007	Vb
Cadmium	Cd	48	112.40	IId	Oxygen	O	8	15.9994	VIb
Calcium	Ca	20	40.08	IIa	Phosphorus	P	15	30.974	Vb
Carbon	C	6	12.011	IVb	Potassium	K	19	39.102	Ia
Cerium	Ce	58	140.12	IIIa	Rubidium	Rb	37	85.47	Ia
Caesium	Cs	55	132.91	Ia	Ruthenium	Ru	44	101.1	VIII
Chlorine	Cl	17	35.453	VIIb	Selenium	Se	34	78.96	VIb
Chromium	Cr	24	52.00	VIa	Silicon	Si	14	28.09	IVb
Cobalt	Co	27	58.93	VIII	Silver	Ag	47	107.870	Ib
Copper	Cu	29	63.54	Ib	Sodium	Na	11	22.9898	Ia
Fluorine	F	9	19.00	VIIb	Strontium	Sr	38	87.62	IIb
Hydrogen	H	1	1.0080	VIIb	Sulfur	S	16	32.064	VIb
Iodine	I	53	126.90	VIIb	Technetium	Tc	43	(99)	VIIa
Iron	Fe	26	55.85	VIII	Tin	Sn	50	118.69	IVb
Lanthanum	La	57	138.91	IIIa	Titanium	Ti	22	47.90	IVa
Lead	Pb	82	207.19	IVb	Tungsten	W	74	183.85	VIa
Lithium	Li	3	6.939	Ia	Vanadium	V	23	50.94	Va
Magnesium	Mg	12	24.312	IIa	Zinc	Zn	30	65.37	IIb
Manganese	Mn	25	54.94	VIIa					

Index

anorexia
    fluorine-induced 40
    lambs, beef cattle and dairy cows 145
    toxicity 199
    zinc deprivation 428
ataxia
    manganese deprivation 365–366
    myelination 258
    types 258

bone disorders 259
boron
    essentiality
        pigs and ruminants 460
        poultry 460
    metabolism 461
    source
        concentration, soil and plant 460–461
        crops and forage species 462
        soybean meal (SBM) 461
    toxicity
        eggshell, weakening 461
        laying hens 462
        riboflavin 462
bush-sickness 223

cadmium
    copper and zinc, antagonisms 33
    exposure, livestock
        animals and 499
        cattle 500–501
        consumers 501
        sheep 499–500
    metabolism
        absorption 495
        biochemical properties,
            metals 495–496
        retention and maternal transfer 495
        selenium and molybdenum 496
    non-ruminant, secondary toxicity
        rats and copper antagonists 497
        zinc level 498
    primary toxicity
        diagnostic problems 496–497
        species 496
    ruminants, secondary toxicity 497
    source
        municipal sewage
            sludges 494–495
        soil, crops and contaminants 494
        superphosphates 494
    tolerance
        grazing livestock 499
        pigs 498–499
        poultry 499
calcium
    acute deprivation features
        hypocalcaemia, ewes 75
        milk fever see milk fever
        transport tetany 75
    calbindin 491
    chronic deprivation see chronic calcium
        deprivation
    concentrates 59
    deprivation see calcium deprivation
    disorder diagnosis
        acute 80
        chronic 80–81

calcium (continued)
  forages
    leaf 59
    values 59
  functions
    bone growth and mineralization 55
    eggshell formation 55
    non-skeletal 54–55
  high dietary concentrations 370
  horses, forage absorbability
    alfalfa 61
    depleting effects, oxalate-rich roughages 60–61
  manganese, nutritive value 359–360
  metabolism *see* metabolism, calcium
  milk and milk replacers
    phytate 59
    protein source effect 58–59
  milk fever, acute disorders *see* milk fever
  and phytate 433–434
  pigs, absorbability
    phytate, AA_{Ca} 62
    phytate phosphorus 62
  poultry, absorbability
    broilers fed diets 61, 62
    saturated fat supplements 62
    vitamin D_3 61–62
  requirements
    affecting factors 62–63
    dietary concentration 63
    egg production, poultry 66–68
    horses 64–65
    meat production, poultry 66
    pigs 65–66
    sheep and cattle 63–64
  ruminants, forage absorbability
    alfalfa 59–60
    corn silage 60
    ryegrass/clover 60
  toxicity
    calcium soaps of fatty acids (CSFA) 84
    hypercalcaemia and growth retardation 84
  treatment and prevention, chronic disorder *see* calcium chronic disorders
  zinc toxicity 450
calcium acute disorders
  milk fever 80
    grazing herds 75–76
    housed herds 76–77
  plasma levels 80
  status 80
  treatment and prevention, milk fever
    grazing herds 82
    housed herds 81–82
    hypocalcaemia, ewe and goats 82–83
    parenteral 81
calcium chronic disorders
  hypocalcaemia 80
  inappetence 225
  laying hen 80–81
  treatment and prevention, milk fever
    grazing herds 82
    housed herds 81–82
    hypocalcaemia, ewe and goats 82–83
    parenteral 81
calcium deprivation
  acute, blood changes
    at calving 68
    metabolic 68
    milk fever 68, 71
  bone changes, adult
    ash, spongy and compact 73
    caudal vertebrae and metacarpal 73
  bone changes, young
    baby pigs and lambs 72
    chicks 72–73
    feeding diets, lamb 72
    lambs 72
    mineralization 71–72
  chronic, blood changes
    phosphorus 68
    plasma calcium 68
  laying hens, blood and bone changes
    shell calcification 74
    skeleton 74
  parasitism and bone mineralization
    demineralization 74
    lambs 73–74
cardiovascular disorders 260
cattle
  blood and plasma selenium 401–402
  cadmium 500–501
  copper poisoning 291
  copper requirement
    gross 270
    net 268
  deprivation, phosphorus 148–149
  fluorine 506
  iron requirement 340
  molybdenum 473–474
  phosphorus net requirement
    factorial models, AFRC 140
    lactation 138
    roughage 138
  potassium requirements
    calves 171–172
    dietary concentration 172
    lactation 171
  rate-limiting function, cobalt anomalies, lipid metabolism 235–236

plasma homocysteine 235
propionate load hypothesis 236, 237
requirements
  calcium 63–64
  chloride 190–191
  magnesium 101, 102
  manganese 362
selenium
  deprivation 383–384
  requirements 393, 394
serum, vitamin B_{12}
  human blood samples 242
  lambs and calves 242, 243
  radioisotope dilution (RID) method 242
sodium requirement
  milk yield 190
  saliva 190
tetany 106
zinc requirements 437–439
chloride deprivation
  biochemical consequences
    abomasal parasitism 197
    lactating cows 197
  eye defects 184
  lactating cows 184
chromium
  essentiality
    mammals 462
    ruminants 463
  inorganic, marker
    mordanting 469
    radioisotopes 469
  metabolism
    holochromodulin 463
    methionate complex 463–464
  natural sources 463
  normal cattle 464
  periparturient cows
    Cr III supplementation 466
    methionine 465
    milk yield 464–465
  pigs 466
  poultry
    stressed 467–468
    unstressed 466–467
  requirements 468
  status
    CrPic 468–469
    inorganic and organic 468
  stressed calves 464
  toxicity 469–470
chronic calcium deprivation
  bone abnormalities
    rickets, osteomalacia and osteoporosis 77
    tibial dyschondroplasia 77–78
  egg yield and quality 78

grain-feeding
  calcium deprivation 79
  rickets 79
grazing livestock
  cows, osteodystrophic diseases 79
  vitamin D status, sheep 79–80
growing birds
  bone disorders 78–79
  dietary calcium and dihydroxycholecalciferol effects 79
  leg weakness see tibial dyschondroplasia
  rickets, broiler chicks 78
growth and survival 77
laying hens
  bone fractures 79
  shell quality 79
milk yield depression 78
prevention
  hatched chicks 83
  housed ruminants 83–84
  non-ruminants 83
teeth 78
chronic copper poisoning (CCP)
  absorbability estimation 262–263
  cattle and goat
    chelate 291–292
    diagnosis and prevention 292
    parenteral injections 291
    subclinical hepatotoxicity 292
  haemolytic crisis, sheep 47
  pigs and poultry
    dietary concentrations 293
    haemolytic crisis 292–293
  risk reduction 541
  sheep
    description 288–289
    diagnosis 289–290
    pathogenesis 289
    prevention 291
    subclinical hepatotoxicity 289
    treatment 290–291
chronic selenosis
  biochemical manifestations
    accretion phase 411
    dysfunction phase 411
    marginal bands 412
  grazing livestock
    blind staggers 410
    primary lesions 410
  pigs and poultry 410
  prevention and control
    dietary treatments 413–414
    soil treatments 413
coast disease 223

cobalt
    action mode 223
    anaemia 225
    bush-sickness 223
    carcinogen 550
    chronic disorders 225
    coast and wasting disease 223
    deprivation *see* cobalt deprivation
    Grand Traverse disease 223
    lecksucht 223
    Nakuruitis 223
    requirements
        grass and cereals 237
        milk 238
        species 236
    salt sickness 223
    sources
        concentrates 227
        forages 226–227
        milk and milk products 227
        soil 227
    toxicity 247
    vitamin B_{12} *see* vitamin B_{12}
cobalt deprivation
    biochemical changes 232
    depletion and deficiency 231
    disease susceptibility
        Mycobacterium avium 226
        Ostertagia circumcincta 226
        Trichostrongylus vitrinus 226
    fatty liver diseases
        features 226
        WLD *see* white-liver disease
    hair colour loss 225
    infertility and perinatal mortality 226
    lipid metabolism
        biochemical abnormalities 234, 235
        MMA CoA 234
        WLD 234
    methylation dysfunction
        formiminoglutamic acid (FIGLU) 233–234
        methionine synthesis 233
        plasma homocysteine 233, 234
        S-adenosyl methionine (SAM) 233
    prevention and control
        dietary supplements 245
        non-ruminants 247
        oral and parenteral
            supplementation 245–246
        parenteral/oral dosing 245
        pasture fertilizers 245
        sources 246
        toxins, protection 246–247
    propionate dysfunctions, ruminants
        age-based diet 232, 233
        blood falls and intermediary MMA 232
        plasma MMA 232
    rate-limiting function
        cattle 235–236
        sheep 235
    responsive disorders
        biochemical criteria 240
        cattle, serum vitamin
            B_{12} 242
        dietary and liver cobalt 241
        dysfunction indicators 245
        functional B_{12}, serum 242–243
        indicators, impaired propionate
            metabolism 243–244
        liver and serum vitamin B_{12} 241–242
        methylation, indicators 244
        milk, vitamin B_{12} 243
        vitamin B_{12} assessment 240
    rumen microbes, propionate dysfunctions
        Selenomonas ruminantium 231
        succinate levels 231, 233
    ruminants
        energy substrate effect 240
        geographical distribution 239
        species, breed, physiological state and
            husbandry 239–240
    subclinical consequences
        appetite and feed conversion
            efficiency 238–239
        wool growth rate 239
cobalt pine 223
copper
    biochemical consequences, ruminants
        deficiency 274
        depletion 273–274
        disorder 275
        dysfunction 274–275
    chelated supplements 545
    chronic poisoning
        cattle and goats 291–292
        pigs and poultry 292–293
        sheep 288–291
    deprivation in non-ruminants
        horses 286
        pigs 286–287
        poultry 287
    enzyme functions
        cellular respiration 256–257
        genes controlling 256
        iron transport 257
        mammalian tissues 255–256
        oxidants protection 257
    feed, ruminants
        absorption 262
        availability 262–263
        forages, interactions 263–265
    footprint, farms
        dairy 542–543
        pig 542

growth promoter
    pigs 287
    poultry 288
    sources 288
metabolism
    biliary secretion and urinary
        excretion 272–273
    description 271–272
    transport and cellular uptake 272
molybdenum exposure, ruminants
    blood, abnormal distribution 275–276
    reproduction and growth 276–277
    thiomolybdates 277
MT synthesis 441
non and pre-ruminants, sources
    milk and milk replacers 267
    solid feeds 267–268
non-ruminant requirements
    horse 271
    pigs 271
    poultry 271
poisoning 541
responsive disorder, ruminants
    diagnosis 279–283
    genetic factors 279
    plant factors 278
    prevention and control 283–285
    seasonal factors 278–279
    soil factors 277–278
ruminants, deprivation, clinical signs
    anaemia 258
    ataxia 258
    bone disorders 259
    cardiovascular disorders 260
    depigmentation 258–259
    diarrhoea 259
    growth retardation 260–261
    infectious diseases 260, 261
    infertility 260–261
    microbial infections 260
    tissue disorders 260
    wool and hair abnormalities 258
ruminants, gross requirements
    cattle 270
    deer and goat 270
    diet influence 268–269
    genotype influence 268–269
ruminants, net requirements 268
sources and antagonists
    feeds composition 267
    grasses 266
    pasture 266–267
    soil condition and fertilizer 265–266
copper oxide wire particles (CuoWP)
    dosing 285
    newborn calves 291
covariance analysis, mineral deprivation

mineral-dependent performance
    cost–benefit analysis 533
    pre-treatment performance 532–533
mineral-independent performance 533
non-linear
    parenteral selenium 533
    relationships 533–534

dietary sources
    iron
        animal 335
        forages 335
        grains and seeds 335
    manganese 356–357
    phosphorus
        cereals 124
        feeds 123
        grazing 125–126
        herbage 125
        milk and animal by-products 124
        phytase and phosphatase, plant
            source 125
        plant protein and by-products 124–125
        roughages and forage crops 126
    potassium
        absorbability 170
        feeds 169–170
        forages 169
    selenium 384–385
    sodium and chloride
        concentrates 185–186
        drinking water 186–187
        forages 184–185
dietary supplementation
    copper
        effects 289
        inorganic sources 283–284
        organic sources 284
    phosphorus 153

elements, grouping 564

feed conversion efficiency (FCE)
    boron content 460
    cobalt deprivation 238–239
    copper status 284
    growth retardation 277
    mica 114
    phosphorus
        and calcium, effects 132
        deprived pigs 147
    selenium 394
feeds, mineral forms
    absorbability, *in vivo*
        apparent 27

feeds, mineral forms (*continued*)
    isotope-free methods 29
    isotopes 28–29
  accessibility, *in vitro*
    amino acid–trace element stability 26–27
    bioaccessibility 26
    simulated digestion 27
  food constituents interaction
    dietary balance 31, 32
    mechanisms 32–33
    mineral value and interactions 31–32
    outcome 33
  functionality measurement
    retention 30–31
    zinc and phosphorus 31
  iron and manganese 25–26
  mineral cycles
    crops/livestock 33–34
    phosphorus pool sizes and flows 34
    swayback, newborn lambs 34
  phytate 26
  retainability
    partial 30
    whole-body 29–30
  value measurement
    accessibility and absorbability 26
    bioavailability 26
fertilizers
  magnesium 110–111
  nitrogenous and potassic
    legume growth and herbage yields 18
    reseeding 18
  pasture 245, 284
  phosphorus
    concentrations 18
    herbage 17–18
  potassium 108, 112–113
  sulfur deprivation 216
  superphosphate 494
  trace element and sulfur
    gypsum 18
    molybdenum 18
    zinc and selenium 18
  use 152–153
  zinc 429, 430, 445
fluorine
  absorption
    hay fluorine 503
    mean alkaline phosphatase activity 504
  biochemical evidence
    blood 509
    bone and tooth composition 508
    continuity 509–510
    serum alkaline phosphatase activity 509
    urine 508–509
  chronic fluorosis, ruminants
    dental lesions 507
    diagnosis 507
    mineral imbalance 506–507
    skeletal abnormalities 507–508
    treatment and prevention 510
  essentiality
    dental caries 510
    pharmacological effect 510
  post-absorption
    plasma fluorine 504
    retention 503
  sources
    animal by-products 502–503
    drinking water 503
    industrial and agricultural contamination 503
    pastures and crops 501
    and phosphorus 502
    rock phosphate supplements 502
  tolerance, dietary
    cattle 506
    mink 505
    pigs 506
    poultry 505–506
    sheep 506
  toxicity
    mechanism 504
    skeleton 504–505
free-access mineral mixtures
  common salt (NaCl) 49
  consumption, individual and seasonal variation 49
  housed and grazing livestock 83

goats
  chloride requirement 191
  copper
    poisoning 291–292
    requirements, ruminant 270
  hypocalcaemia 82–83
  PTE essentiality 518
  requirements
    magnesium 101
    manganese 362
    selenium 393–394
    zinc 440
goitrogens
  deiodinase inhibitors 313
  genetic selection 313
  goitrogenicity
    CG and PTU-like 313
    precursor concentration 313
  iodine requirements
    inorganic 315
    organic 315
  metabolism

glucosinolates 313–314
   SCN and TSH secretion 314
thiouracil-type, countering 326
thyroperoxidase inhibitors
   glucosinolates 313
   myrosinase 312
   thiocyanates 312–313
Grand Traverse disease 223
gross requirements, magnesium
   goats 101
   horses 101, 103
   pigs
      baby 103
      older and fattening 103
   poultry
      growth rate, broiler chicks 103
      laying hens 103
   sheep and cattle
      dietary concentrations 99, 101
      forage/roughage concentrate
         mixture 101, 102
      hypomagnesaemia 101
gross requirements (GR), minerals
   dietary trials 9
   dose–response relationship 8
   factorial models
      advantage 9
      lactating ewe, calcium 8

horses
   calcium requirement
      carcass and milk 64
      faecal endogenous losses 64–65
      thoroughbred foal/lactating mare 65
   copper
      deprivation 286
      requirements 271
   excretion, phosphorus
      lactating mares 128
      urine 127–128
   forage calcium absorbability 60–61
   requirements
      magnesium 101, 103
      phosphorus 141
      selenium 395
   zinc
      requirements 440
      value, feeds 434
humans and minerals
   anti-nutritional and nutritional properties
      phytate 550
      selenium and polyunsaturated fatty
         acids 549–550
   climate changes
      biofuels 546
      growing season 545–546

      methane emissions, ruminants 546
   footprint
      chelated copper supplements 545
      dairy farms, copper 542–543
      farms 540
      phosphorus 543
      pig farms, copper and zinc 542
      reduction 544–545
   health hazard reduction and animal nutrition
      MPL, toxic elements
         550–551
      mutagenic/carcinogenic 550
      permissible sources 551
   MPL
      precautionary principles 540–541
      sub-sets 541–542
      supplementation practice 541
   phosphorus footprint reduction, manipulation
      pigs and poultry, dietary 543–544
      plants and microbes, genetic 543
      ruminants, dietary 544
   sensory value enhancement, food
      additives and regulations 549
      inorganic and organic selenium 549
      price 549
      slaughter 548–549
   value, animal foods and diet
      antioxidant status 547–548
      iron status 548
      micronutrients 546
      selenium and cancer prevention 548
      thyroid function 546–547
hypomagnesaemic tetany
   aluminium intake 492
   causative/associated factors 109
   control, dusting pastures 111
   dairy cows 106–107
   diagnosis
      calcium concentrations 109
      lactation and phosphorus status
         effects 110
      plasma and serum magnesium 109
   dietary change, pasturage
      composition and rumen
         characteristics 108
      plasma magnesium recovery 108
   direct intervention, prevention
      magnesium fertilizers 110–111
      mineral sources 112
   milk diet 93
   model 69
   outwintered beef cattle 107
   parturition, dietary change
      housed dairy cow 108–109
      potassium 109
   potassium fertilizers
      pastures 108

hypomagnesaemic tetany (*continued*)
    plasma magnesium and crude
        protein 108
    prevention methods, indirect
        dietary change, turnout 113
        genetic selection 114
        potassium fertilizers 112–113
        predictors 112
        transport stress 113–114
    risk 76, 177
    sheep 107
    signs 93
    transport 109
    treatment
        magnesium lactate injection 109–110
        serum magnesium 110
    weather
        grass temperature 108
        spring conditions 108

IDD *see* iodine deficiency disorders
indirect prevention methods,
        hypomagnesaemic tetany
    dietary change turnout 113
    genetic selection 114
    potassium fertilizers, use
        feeds 113
        herbage 112–113
        slurry dispose 113
    predictors
        plasma and urinary magnesium 112, 113
        potassium and calcium ratio 112
    transport stress
        pigs 114
        sheep and cattle 113–114
iodine
    cereals and oilseed meals 311
    deficiency disorders (IDD)
        see iodine deficiency disorders
    deprivation *see* iodine deprivation
    farm animals 307
    fish meals 311
    forages
        botanical species 310
        plant 310
        soil 310
    function
        gene transcription 307
        leptin production 308
        metabolic rate 307–308
        thyroxine 307
    goitre 306, 307
    goitrogens *see* goitrogens
    impaired deiodination
        ID2 activity 317–318
        pigs, RSM 318
        propylthiouracil, oral doses 318
        T4:T3 ratios, serum 317
    metabolism
        concentration, cow milk 309
        recycling 308
        thyroidal iodine recycling 308–309
    milk and milk products
        colostrum 312
        minerals 312
    requirement
        goitrogens *see* goitrogens
        non-ruminants 314
        ruminants 314–315
    soil
        pasture 310–311
        pelleted lucerne diet 310
    status, SeRD 406
    T4 *see* thyroxine
    thyroid activity and basal metabolic rate
        306
    thyroid dysfunction *see* thyroid dysfunction
    toxicity
        pharmacological side effects 326
        tolerance 326
    water
        drinking 311
        sea 311–312
iodine deficiency disorders (IDD)
    biochemical confirmation 324
    bloodstream
        protein-bound iodine (PBI) 322
        serum 322–323
    ewe milk, thyroid enlargement 324
    genetic factors 321
    geochemical factors
        goitre incidence 320
        plant-available soil iodine 320
    prevention and control methods
        continuous 324–325
        discontinuous 325
        slow-release 325
        thiouracil-type goitrogens 325
    secondary
        mustard cake 321
        selenium supply 321
        thyroid enlargement, pigs 321, 322
    soil–plant interactions 320
        grazing animal 320
        outwintered ewes 320–321
    T3 assays
        calves 323
        selenium-induced 323
    T4 assays
        newborn lambs and calves 323
        plasma 323
        serum 323
    thyroid

goitre 322
weight 322
urine and milk 323–324
iodine deprivation
    deficiency
        bloodstream, serum T4 316–317
        hyperplasia, cuboidal epithelium 316
    depletion
        oxidative stress 316
        thyroid 316
    dysfunction 317
iron
    absorption
        affecting factors 337
        mucosal uptake 336
        retention 335–336
        retention and serum ferritin
            concentrations 337
        serosal transfer 336–337
    –copper interaction 263–264
    cosmetic 50
    deficiency anaemia 548
    deprivation see iron deprivation
    dietary sources
        animal 335
        forages 335
        grains and seeds 335
    functions
        flavoprotein enzymes 334–335
        haemoglobin (Hb) 334
        myoglobin 334
    herbage 267
    manganese and 362
    and molybdenum antagonisms 32
    nutritive value, feeds
        affecting factors 337
        ascorbate, relative biological value
            (RBV) 337
        haem iron 338
        milk iron 337–338
        non-haem iron 338
    requirements
        haemoglobin (Hb) and red cell
            mass 338–339
        pigs 339–340
        poultry 340
        sheep and cattle 340
    soil 148, 152
    storage
        ferritin 341
        haemosiderin 341
    transport
        caeruloplasmin 341
        copper function 257
        haem and non-haem sources 340–341
        hepcidin 341
        transferrin 341

iron deprivation
    anaemia
        hypochromic and microcytic 343
        piglets 343
    cardiac hypertrophy 343
    cognitive dysfunction 344
    deficiency
        anaemic piglets 342
        blood haemoglobin and mean cell volume
            (MCV) 341–342
        haemoglobin and myoglobin 342
        hypoferraemia 342
        serum 341
    depletion 341
    diagnosis
        differential 346
        serum and liver iron 346
        whole blood criteria 345–346
    dysfunction
        biochemical changes 343
        metallothionein concentrations 343
    hypothyroidism 343–344
    infections
        host defence mechanisms 345
        lactoferrin UIBC 345
        microbial pathogens 345
        susceptibility 344
    lamb and calf anaemia
        versus piglets 344–345
        twin calves 345
    piglet anaemia 344
    prevention
        inorganic sources 347–348
        oral route 347
        organic sources 348
    risk 347
    toxicity
        diagnosis 349–350
        dietary, tolerance 349
        interactions 349
        mechanisms 348
        sources 348–349
        vitamin E and polyunsaturated fatty
            acids 348
    treatment
        iron-dextran/dextrin-ferric oxide
            injection 346–347
        piglet anaemia 347
    veal production 345
iron-responsive disorders
    in humans 348
    serum and liver
        ferritin 346
        TIBC and UIBC 346
    whole blood criteria
        haemoglobin (Hb) values 345–346
        sex 346

lead
    acute poisoning 513
    availability
        dietary composition 511
        exposure/dietary
            concentration 511–512
        source 512
    chronic poisoning
        calcium and phosphorus, diet 514
        osteodystrophy, suckling lambs
            513–514
    livestock sources, human exposure
        residual effects 516
        risk assessment 515–516
    maternal transfer 513
    poisoning 510
    sources 511
    tolerance
        non/pre-ruminants 514
        ruminants 514–515
        subclinical effects 515
    transport and tissue distribution
        accretion 512
        kidney 512
lecksucht 223, 225
lithium
    essentiality and sources 470
    metabolism 470
    toxicity
        aversive behaviour 471
        symptoms, beef cattle 470
livestock, mineral status
    appendages 46
    balance 46
    biochemical indicators, mineral excess
        chronic copper poisoning 47
        deprivation 46–47
    clinical and pathological changes
        diseases 41
        fluorine-induced anorexia 40
        mild mineral imbalance 40
    deficits, over-correction
        copper and iron 50
        phosphorus and magnesium 50
    deprivation, biochemical indicators
        macro-model 41
        nano-model 42
        theme variation 41–42
    deprivation differentiation see mineral
            deprivation
    erythrocyte
        diet, fluctuations 45
        enzyme activities 45
    food selection
        diet 39–40
        grazing 39
    functional forms and indices 46

    hair, hoof, fleece/feathers 46
    imbalance correlation
        holistic approach 47
        supplementation 47–49
    imbalance, self correction
        appetite 50
        magnesium 51
        sodium-deficient 50–51
    secretions
        preparatory mouth washes 45
        saliva 45
    sources
        inorganic versus organic 49–50
        supplement components 49
    tissues
        kidneys 46
        liver and bone 45–46
    urine and faeces 45
    whole blood
        erythrocyte 44–45
        plasma and serum 44
luteinizing hormone (LH) 276

magnesium
    absorption, rumen see rumen, magnesium
            absorption
    availability
        grass and alfalfa hays 100
        monogastric species 100
    biochemical changes, deprivation
        deficiency 105
        depletion 104
        dysfunction 105
        non-ruminant 105–106
        plasma 103–104
    concentrations, UK livestock feeds 94
    deprivation, clinical signs
        cattle and sheep, tetany 106
        subclinical disorder 106
    forages
        concentration 93
        legume 93–94
        and potassium concentration, season
            effect 93
    functions
        microsomes 92
        non-enzymic 92
        versus potassium 92
        rumen microorganisms 92–93
    and heat stress 114
    hypomagnesaemia and tetany see
            hypomagnesaemic tetany
    lack, diet 92
    and meat quality
        glycolytic changes, muscle 114
        pork 114

milk and milk replacers
    absorption 93
    concentration 93
potassium contrasting effects 95
and potassium effect, pastures 93
ruminant feeds, availability
    animal species and potassium
        interaction 98
    dietary influences 100
    diet type and potassium 98–100
    genetic variation 100
    potassium effects 98
secretion and absorption, rumen
    fluxes, hind gut mucosa 96
    potassium 96–97
serum 75
soluble and pH 94–95
toxicity
    homeostatic mechanisms 114–115
    ruminants 115
manganese
    bone development 355
    and cobalt concentrations, plants 17
    –copper interaction 264–265
    deprivation *see* manganese deprivation
    dietary sources
        forages 357
        protein supplements 357
        seeds and grains 356–357
    glycosyltransferase 356
    metabolism
        absorption, transport and cellular
            uptake 362
        excretion 363
        tissue distribution and storage 363–364
    nutritive value, feeds
        absorption control 357–358
        calcium and phosphorus 359–360
        forages 360
        grains 359
        milk and milk products 358–359
    organic supplements
        effect, locus of 368–369
        response indices 369–370
        source purity 369
        test conditions 369
    pyruvate carboxylase 355–356
    and reproduction 356
    requirements
        cattle 362
        pigs 361–362
        poultry 360–361
        sheep and goats 362
    superoxide dismutase 356
    toxicity
        pigs and poultry 371
        ruminants 370–371

manganese deprivation
    ataxia 365–366
    biochemical changes
        bone alkaline phosphatase 364
        pathophysiological, sequence 365
    blood and plasma 366
    control 368
    functional manganese 367–368
    integument 367
    marginal bands 368
    occurrence, ruminants 366
    reproductive disorders 366
    skeletal abnormalities
        goats 365
        perosis 365
    tissues 366–367
maximum permitted levels (MPL)
    precautionary principles 540–541
    PTE 550–551
    subsets
        copper levels 541
        sheep 542
    supplementation practice 541
mercury
    industrial discharges 516
    metabolism
        absorption 516
        distribution 517
    sources 516
    toxicity
        diagnosis 517
        symptoms 517
        tolerance 517
metabolism, calcium
    absorption control
        calbindin role 55
        hormones 55
        mechanism 56
    excretion
        dry matter intake (DMI) 58
        faecal versus dietary calcium, pig 58
        G-linked calcium ion receptors 57–58
    hormonal control, skeleton
        nuclear receptors 57
        serum osteocalcin concentrations 57
    skeleton, calcium movement
        bone mineralization 56
        immature bodies 56–57
metabolism, copper
    biliary secretion 272–273
    deprivation 271
    transport and cellular uptake 272
    urinary excretion 273
microbial protein synthesis (MPS)
    feeds 208
    and sulfur capture 209
    sulfur requirements 211–213

milk
- bovine, manganese 357
- cobalt 227
- colostrum 385
- composition 23–24
- diet effect
  - cobalt/vitamin B_{12} 24
  - copper and iodine deficiencies 24
  - selenium 24–25
- fever see milk fever
- iodine
  - deprivation 320
  - thyroid enlargement 324
  - and urine 323–324
- iron 337–338
- lactation effect, parity and species
  - ash content 24
  - bovine colostrum 24
  - offspring growth rate 24
- magnesium concentration 101
- manganese 357–359
- and milk products
  - cobalt 227
  - iodine 312
- and milk replacers
  - calcium 58–59
  - copper 267
  - magnesium 93
  - phosphorus 128
- phosphorus
  - and animal by-products 124
  - deprivation 147
- production 236
- products 430
- selenium deprivation, yield 384
- vitamin B_{12} 243
- yield depression 78
- zinc availability 435

milk fever
- description and symptoms 74
- grazing herds
  - body fatness 76
  - disease costs, UK 75–76
- housed herds
  - dietary calcium 76–77
  - magnesium and phosphorus 77
  - potassium concentrations 76
- hypocalcaemia 74–75
- parenteral treatment
  - hypocalcaemia 81
  - mastitis and ketosis 81
- prevention
  - grazing herds 82
  - housed herds 81–82
  - parenteral methods 81
- serum magnesium 75
- see also potassium

mineral deprivation
- biochemical indicators nano-model
  - macro-mineral metabolism 42
  - zinc deprivation 42
- biochemical markers 42–43
- dysfunction criteria
  - methylmalonic acid (MMA) 43
  - mRNA 43
- functional forms and indices 46
- macro-model, biochemical indicators
  - diet 41
  - phases 41
- marginal value delineation
  - biochemical events, chronic exposure 44
  - pathophysiological changes 43
  - three-tier system 43
- theme variations, biochemical indicators
  - calcium 41
  - iron 41–42
  - selenium 42

mineral deprivation, supplementation trials
- accurate measurement 531
- control
  - bias 528–529
  - treatment comparisons 528
- group responses
  - blood selenium 535
  - data collation 535
  - vitamin B_{12} 534
- group size
  - dose–response trials 529
  - economic and statistical significance 529–530
  - unavoidable and avoidable constraints 530–531
- individual variation
  - mineral-dependent performance, covariance 532–533
  - mineral-independent performance 533
  - non-linear covariance 533–534
  - poor growth reasons 532
- repetition, variation
  - day-to-day 531
  - diurnal 531
  - seasonal 531
  - year-to-year 531–532
- responsive individuals 536
- treatment choice
  - clustered 537
  - multiple 536–537
  - single 537
  - sustained efficacy 536
  - unresponsive populations 537
- uncertainties 528

mineral requirements
- affecting factors, calcium 62–63
- breeding stock

lifetime production 10
reproduction and lactation 9–10
complexity 2
criteria
copper and zinc 9
pigmentation and keratinization, wool 9
disagreement
dietary mineral imbalances 10
estimate, methods and
assumptions 10–11
discoveries 1
disorders 555
essentiality 2–3
expression
dietary concentrations 11
phosphorus, hens 11
functions
calcium and phosphorus 3
cobalt 4
copper 3
metalloenzymes 3
metalloproteins 3
regulatory 2–3
regulatory proteins/peptides 3–4
structural, physiological and catalytic 2
gross requirements (GR)
dietary trials 9
factorial models 8–9
horses 64–65
mean macromineral composition 557–558
metabolism
calcium accretion and resorption 4, 5
digestive process 4
turnover rates, tissues 4
net
maintenance 4–6
production 6–8
work and reproduction 6
two-way flow, pools 5
mineral supplementation
continuous direct methods
free-access mineral mixtures 49
hand/mechanically fed stock 48–49
discontinuous direct methods
oral dosing 48
slow-release oral 48
indirect methods
cobalt/selenium, fertilizers 47–48
phosphorus 48
ingredients 563
urea 47
molybdenum
biochemical consequences 275–277
and copper
absorbability 263
antagonists, pasture 266
and sulfur, forages 263

deficient soil 18
deprivation, natural occurrence 473
essentiality and sources 471
and iodine, milk 24, 29
and iron, antagonisms 32
metabolism
absorption 472
cellular activity 472
retention and excretion 472
pasture, treatment 48
toxicity
cattle 473–474
grazing species 474
molybdenosis and hypocuprosis 473
non-grazing species, tolerance 474–475
thiomolybdate toxicity 474
urinary excretion 210
MPL *see* maximum permitted levels

Nakuruitis 223
natural sources, minerals
analytical and experimental precision
diets, composition 20
milling 20
animal by-products
macro-mineral concentrations 25
meat and fish 25
skimmed and buttermilk 25
atmospheric inputs
soil pH 20
sulfur 20
chromium 463
drinking water
iodine concentrations 21, 23
sodium and chlorine 23
sulfur concentrations 21
feeds *see* feeds, mineral forms
ingestion, soil
daily intake 20
toxic elements 20
maturity and season
factors, internal and external 19
fluctuations 19
phosphorus and potassium
concentrations 19–20
selenium concentration 19
wet- and dry-forages 19
milk
composition 23–24
diet effect 24–25
lactation stage, parity and species 24
plant genotype *see* plant genotype, mineral
sources
representative values, feeds
by-products, crop 21
crops and forages 21

natural sources, minerals (*continued*)
    industrial process effect 23
    UK and USA 22
    soil and fertilizers effects
        cobalt and manganese concentration 17
        grazing animals, deficiencies 17
        inadequate supplies, plant 16–17
        nitrogenous and potassic 18
        phosphorus 17–18
        plants versus animals 17
        soil pH 17
        trace element and sulfur 18
    zinc 429–430
net requirements, minerals
    maintenance
        faecal endogenous loss 6
        non-productive livestock 4, 6
    production
        calcium requirements 7
        milk 7–8
        weight gain/milk yield/fleece growth 6–7
        white and red meat 7
    reproduction
        fetal accumulation pattern 7
        twin-bearing ewe 6
    work
        energy maintenance 6
        oxygen consumption 6
nickel
    essentiality
        non-ruminants 475
        ruminants 475
    metabolism 476
    requirements and toxicity 476
    sources 475–476
non-ruminants
    aluminium 491
    cadmium toxicity 497–498
    chronic calcium deprivation, prevention 83
    copper
        deprivation 286–287
        requirements 270–271
    dietary amino acids 206
    iodine requirements 314
    magnesium deprivation 105–106
    nickel essentiality 475
    phosphorus
        absorption 127
        deprivation, prevention and control 154–156
    selenium
        absorption 386
        blood and plasma 402
        organic 408–409
    vitamin B_{12} 247
    zinc
        absorbability 434–435
        deprivation 445–446
        value 433
non-ruminants, phosphorus deprivation
    inorganic supplementation 154
    phytase supplementation
        fungal 154
        microbial 155
        poultry rations 155
    vitamin D supplements
        broiler chicks 155
        chick growth rate and tibia ash 155
        citrate 155–156
        phytase 155

occasionally beneficial elements (OBE)
    boron
        essentiality 460
        metabolism 461
        sources 460–461
        toxicity 461–462
    chromium
        essentiality 462–463
        inorganic, as marker 469
        metabolism 463–464
        natural sources 463
        requirements 468
        responsiveness 464–468
        status 468–469
        toxicity 469–470
    lithium
        essentiality and sources 470
        metabolism 470
        toxicity 470–471
    median concentration, soil 461
    molybdenum
        deprivation, natural occurrence 473
        essentiality and sources 471
        metabolism 472
        toxicity 473–475
    nickel
        essentiality 475
        metabolism and requirements 476
        sources 475–476
        toxicity 476
    rubidium
        essentiality 476
        metabolism 477
        requirements and tolerance 477
        sources 476–477
    silicon
        anti-nutritional effects 478–479
        essentiality and sources 477
        metabolism 477–478
    vanadium
        essentiality and sources 479
        metabolism 479
        toxicity 479–480

organic supplements
   manganese
      chelate, lower solubility 369
      effect, locus of 368–369
      methionine 369
      response indices 369–370
      source purity 369
   selenium
      forms 408
      non-ruminants 408–409
      ruminants 409
OWLD see white-liver disease in ovines

phosphorus
   absorption
      mucosa and kidneys 126–127
      ruminants and non-ruminants 127
      small intestine 126
   availability
      calcium effect, pigs see pigs
      calcium effect, poultry 131
      experiment 133
      milk and milk replacers 128
      pig feeds 129–130
      poultry feeds 130–131
      ruminant feeds 128–129
      value assesment 133
   deprivation
      biochemical consequences 141–145
      clinical manifestations 145–147
      diagnosis, grazing animals 150–152
      non-ruminants, prevention and control 154–156
      occurrence 148–150
      prevention and control, ruminants 152–153
      subclinical manifestations 147–148
   dietary sources see dietary sources, phosphorus
   excretion
      horse see horses
      pig see pigs
      poultry see poultry
   footprint reduction, manipulation
      dietary 543–544
      genetic 543
   functions
      bones and teeth 123
      glycolytic metabolism, erythrocytes 123
      organic bone matrix 123
      osmotic and acid-base balance 123
   and lactation effects 110
   nutritive value, manganese 359–360
   recycling and excretion, ruminants 127
   requirements
      boars, gilts and sows 136
      estimation 137–138
      pig see pigs
      poultry see poultry
   ruminants, requirements
      absorption coefficients 139
      cattle, net see cattle
      estimation 139
      horses see horses
      net, sheep see sheep
      phased feeding 139–140
   ruminants versus nonruminants 122
   sources comparison 502
   toxicity
      PO_4 ion 156
      urinary calculi 156
phosphorus deprivation
   bones abnormalities 147
   deficiency 142–143, 146
   depletion
      demineralization 142–143, 146
      rumen and saliva 142
   diagnosis, grazing animals
      bone assessment methods, in vivo 150–151
      bone criteria 150
      faecal and urine 152
      plasma inorganic 150
      post-mortem assessment, mineralization 151–152
   dysfunction
      bone, breaking and shear strength 144
      growth rates, pig 144
      pathophysiological changes 141
   egg yield and quality, reduction
      calcium 148
      grain-based diets 148
   in grazing cattle 148–149
      concentrations, herbage 148–149
      pasture productivity and milk yield 148
      roughages, urea 149
      sheep 149
   impaired feed utilization 147
   milk yield depression
      pregnancy and early lactation 147
      transitional diet 147–148
      weaning weight 148
   mineralization, bone 142
   nematodiasis, bone demineralization 148
   pica
      bone-chewing, cows 146
      consumption 145–146
   plasma, fluctuations
      bone ash and serum 142
      dietary concentration 142
      inflows and outflows 141–142
      ruminants 142

phosphorus deprivation (*continued*)
   poor appetite
      anorexia, lambs 145
      yearling cattle 145, 146
   in poultry and pigs
      broiler chicks, mortality 149
      hypophosphataemic rickets 149–150
   reproductive disturbances
      infertility 146
      puberty/pregnancy age 146–147
      subnormal fertility 146
phosphorus footprint reduction
   dietary manipulation
      pigs and poultry 543–544
      ruminants 544
   genetic manipulation
      animals 543
      plants and microbes 543
pigs
   acid–base balance 176–177
   boron 460
   calcium
      absorption coefficient 65
      effects 131–133
      factorial estimates 65
      feeds, absorbability 61
      phytases 66
      phytate phosphorus 65–66
      provision 65
   chromium 466
   copper
      deprivation 287–288
      poisoning, chronic 292–293
   dietary oxidants 405
   feeds, phosphorus availability
      absorption coefficients 129
      blending maize 130
      cereals 129
      dietary calcium concentrations 130
   gross requirements, phosphorus
      AA_p 136
      factorial model 135–136
      test, growing pig 135
   iron requirement
      bovine and synthetic milk 339
      lactation 339–340
      non-haem 338
   magnesium 103
   magnesium requirement *see* gross requirements, magnesium
   manganese
      requirement 361–362
      toxicity 370–371
   non-haem iron 338
   phosphorus
      absorption 130
      and calcium deprivation 131–133

      deprivation 149–150
      excretion 127
      requirements 133–136
   potassium 170
   selenium 394, 410
   sodium and chloride requirement
      factorial model 192
      lacting sow 191–192
   zinc requirements
      factorial estimates 436
      feeding trials 436
plant genotype, mineral sources
   browse species
      Balanites aegyptiaca 16
      leaves 16
   environment adaptation
      Astragalus 14
      saline soil, salt bushes and blue-bushes 14
   grains and seeds
      lupin seeds 16
      manganese concentrations 16
   grasses and forages
      natrophilic and natrophobic species 15
      selenium concentrations, Florida 15–16
   legumes
      calcium concentrations 14
      Lolium perenne and *Trifolium* 14–15
polioencephalomalacia (PEM) 218
potassium
   absorption
      non-ruminants 172–173
      ruminants 173
   and acid–base balance
      DCAD and milk fever 176
      dietary cation–anion differences (DCAD) 175–176
      forage-based diets 175
      heat stress 176
      pigs and poultry 176–177
   antagonism 95
   deprivation
      biochemical consequences 174
      clinical consequences 175
      occurrence 175
   dietary sources
      absorbability 170
      carbohydrate 169
      cereals and non-protein 169–170
      concentration, pastures and foodstuffs 170
      feed analysis 168
      forages 169
   excretion 173–174
   functions
      concentrations 168
      electrical potential 168

and sodium distributions 169
soft tissues 168
magnesium
   absorption 95
   animal species 98
   concentrations 93
   and diet type 98–100
   fertilizers 108, 112–113
membrane transport 173
primary toxicity 177
requirements
   cattle 171–172
   pigs and poultry 170–171
   sheep 172
secretion
   milk 174
   ruminants 174
toxicity, secondary
   calcium 177–178
   and magnesium interaction 177
   prevention 178
   sodium metabolism 177
potassium deprivation
   biochemical consequences
      depletion 174
      serum, urine and diet 174
   clinical consequences 175
   occurrence
      corn-based diets 175
      dry swards, ageing and leaching 175
potentially toxic elements (PTE)
   aluminium
      contaminated mineral supplement source 490–491
      metabolism 491
      non-ruminants 491
      principal source 490
      ruminants 491–492
   arsenic
      accidental exposure 493
      acute toxicity 493
      causes 494
      chronic toxicity 493
      metabolism 492–493
      sources 492
      toxicity symptoms 493
   artefacts, experiments 519
   cadmium
      absorption 495
      human exposure 499–501
      metabolism 495–496
      non-ruminants, secondary toxicity 497–498
      primary toxicity 496–497
      retention and maternal transfer 495
      sources 494–495
      tolerance 498–499

concentrations 490
essentiality
   goats 518
   young rats 518
fluorine
   chronic fluorosis, ruminants 506–508, 510
   essentiality 510
   exposure, biochemical evidence 508–510
   metabolism 503–504
   sources 501–503
   tolerance 505–506
   toxicity 504–505
hormesis
   mechanisms 518
   theory 518
lead
   availability 511–512
   livestock, exposure 515–516
   metabolism 512–513
   sources 511
   tolerance 514–515
   toxicity 513–514
livestock factors 489, 490
mercury
   levels, food 516
   metabolism 516–517
   sources 516
   toxicity 517
soil 489
supplement benefits 519
poultry
   absorbability, calcium 61–62
   boron 460
   cadmium 499
   calcium effect, phosphorus retention
      broiler chicks 131
      phytase and vitamin D 131
   calcium, meat production
      dietary, broiler 66
      hatched broiler chicks 66
      inorganic phosphorus 66
   chromium 466–468
   chronic selenosis 410
   copper
      chronic poisoning 292–293
      deprivation 287
      growth promoter 288
      requirements 271
   dietary oxidants 405
   egg production, calcium
      dietary effects 67
      estimation 66–67
      leghorn hens 67
   feeds, phosphorus availability
      broiler chicks 131
      cereal 131

poultry (continued)
    dietary calcium effects 130, 131
    laying hens and newly hatched
        chicks 130
  fluorine 505–506
  gross requirements, phosphorus
    broiler chicks 136–137
    egg production 137
    leghorn chicks 136
    turkey poults 137
  iron 340
  magnesium 103
  magnesium requirements *see* gross
        requirements, magnesium
  manganese toxicity 371
  non-haem iron, chicks 338
  organic zinc sources 448
  phosphorus
    excretion 127
    requirement 133–134
  phosphorus footprints 543–544
  potassium requirements
    chickens 171
    dietary conditions 170–172
    turkey 171
  requirements
    iron 340
    manganese 360–361
    sodium and chloride 192, 193
  selenium 394–395
  zinc requirements
    chick and poult growth 436–437
    egg production 437
PTE *see* potentially toxic elements

RBV *see* relative biological value
relative biological value (RBV)
  assessment 431, 432
  influence factors 432–433
  iron ascorbate 337
  zinc and phytate effect 431–432
  ZnO versus $ZnSO_4$ 447–448
rubidium
  essentiality 476
  metabolism 477
  requirements and tolerance 477
  sources 476–477
rumen, magnesium absorption
  anions, effects 96
  dietary protein effect, fatty acids 96
  dietary sodium effects, salivary flow
    magnesium uptake 96
    urinary excretion 96
  metabolic events 94
  potassium, antagonism
    concentration 95

    contrasting effects 95
    dietary supplement 95
  sheep and cattle 94
  soluble and pH
    sheep 94–95
    ultrafilterable magnesium 95
ruminan feeds, magnesium availability
  animal species and potassium
    goats, AA_{Mg} 98
    sheep versus cattle 98
  diet type and potassium
    cattle 98–99
    cereal and alfalfa silages 99–100
    concentration, cattle ration 98–99
    sheep fed forages 98
  genetic variation
    milk and serum concentrations 100
    sire, urinary excretion 100
  influences, values 100
  potassium effects
    dietary concentration 98, 99
    dried grasses 98
    magnitude 98
ruminants
  absorption, cadmium 495
  boron concentrations 460
  cadmium toxicity 497
  calcium
    forage absorbability 59–60
    housed, prevention 83–84
  chronic fluorosis 506–508
  cobalt deprivation 239
  copper
    deprivation 257–262, 273–275
    molybdenum exposure 275–277
    prevention and control,
        deprivation 283–285
    requirements 268–270
    responsive disorders 277–283
    value 262–265
  dietary oxidants 404–405
  dysfunctions, propionate metabolism 232
  iodine 314–315
  lead, tolerance 514–515
  macro-mineral concentration interpretation
    559
  magnesium
    availability 98–100
    deprivation 115
  manganese toxicity 370–371
  methane emissions, reduction 546
  nickel 475
  non-haem iron 338
  phosphorus
    absorption 127
    feeding 139
    feeds 128

recycling and excretion 127
requirements 139
phosphorus deprivation
  dietary supplementation 153
  fertilizers 152–153
  mineral sources 154
  supplementation, direct and indirect 153
phosphorus metabolism 491–492
selenium
  absorption 386–387
  organic 409
trace element disorder 561
zinc
  deprivation 446–447
  organic sources 448–449
  see also sulfur
ruminants, responsive disorders, copper
  diagnosis
    blood and erythrocyte 281–282
    clinical signs 279
    conditions 279
    dietary 279–280
    hair and fleece 282
    liver 280–281
    neutrophil 282
    plasma and caeruloplasmin 281
  factors
    genetic 279
    plant 278
    seasonal 278–279
    soil 277–278
  prevention and control
    antagonisms 285
    dietary supplementation 283–284
    genetic selection 285
    oral and parenteral dosage 284
    oral supplementation 284–285
    pasture fertilizers 283

salivary phosphorus
  absorption 127
  secretion 127
salt sickness 223
selenium
  absorption
    methionine and cysteine 386
    non-ruminants 386
    ruminants 386–387
    selenite and selenate 385
  calcium and manganese, fetal accumulation 7
  chronic selenosis *see* chronic selenosis
  deprivation *see* selenium deprivation
  dietary sources
    animal protein 385
    cereals and legume seeds 385
    colostrum and milk 385

    forages 384, 385
  excretion 389
  functional tests, availability
    chicks 389–390
    dose–response profile 391
    selenoamino acids 391
    selenomethionine (SeMet) 390
  functions
    glutathione peroxidases 378–380
    iodothyronine 5′-deiodinase 380
    selenophosphate synthetase
      (SPS)-2 381
    thioredoxin reductases 380–381
  grains and pastures 18
  indices
    disorder 400
    dysfunction 400
    GPX, blood and serum 399
  maternal transfer 389
  non-functional indices
    cattle and deer 401–402
    milk 402
    pre/non-ruminants 402
    sheep 400–401
    tissue 402, 403
  non-specific tests
    enhanced transfer 391
    milk, increase 392
  organic supplements
    forms 408
    non-ruminants 408–409
    ruminants 409
  pathways, metabolism
    methionine 387
    transsulfuration 387–388
  requirements
    cattle and sheep 393
    goats 393–394
    hierarchy 392–393
    horses 395
    pigs 394
    poultry 394–395
  retention
    dietary concentration 388
    liver 388–389
    tissue 389
  SeRD *see* selenium-responsive disorders
  tolerance
    lactation 413
    livestock 411–412
    source 412
    species differences 412–413
  toxicity
    accumulator plants 410
    acute 414
    vegetation 409
  tracers 392

selenium deprivation
   biochemical consequences
      antioxidant defence, dysfunction 397
      deficiency 395–397
      depletion 395
      dysfunctional deiodination 397
      livestock 396
   clinical disorders 381
   immunological consequences
      cellular immunity 398
      specificity 397
   non-specific SeRD
      blood 383
      lowered resistance 383
      neurodegenerative 382
      peri-natal and post-natal mortality 382
      reproductive 382
   prevention and control
      dietary supplementation 407
      parenteral supplementation 407–408
      slow-release oral methods 408
      soil and foliar treatment 407
   subclinical
      cumulative effects 384
      growth retardation 383–384
      milk yield 384
      wool production 383
selenium-responsive disorders (SeRD)
   diagnosis 398
   dietary oxidants
      pigs and poultry 405
      ruminants 404–405
   endogenous oxidants
      heat stress 406
      myopathic lesions 405
      prolonged heat stress 406
   high dietary sulfur 406–407
   iodine status and cold stress 406
   non-specific
      blood 383
      lowered resistance 383
      neurodegenerative 382
      peri-natal and post-natal mortality 382
      reproductive 382
   soil and feeds
      forages 403
      geochemistry and mapping 402
   toxins 406
   vitamin E
      colostrum 403–404
      forage conservation 403
      inclusion rates 404
      relative contributions 404
SeRD *see* selenium-responsive disorders
sheep
   blood and plasma selenium 400–401
   phosphorus net requirement
      factorial models, AFRC 140
      genetic variation 138
      roughage factor 138
   rate-limiting function, cobalt
      biochemical dysfunction 235, 236
      growth cessation and plasma homocysteine 235
   requirements
      calcium 63–64
      iron 340
      manganese 362
      potassium 172
   selenium 393
   serum, vitamin B_{12}
      lambs 241
      plasma 241–242
   sodium and chloride requirements
      grazing lambs and ewes 189
      obligatory faecal losses 190
      saliva 189–190
   zinc requirements
      factorial models 438–439
      feeding trials 438
      field experience 437–438
silicon
   anti-nutritional effects
      reciprocal antagonism 478–479
      urolithiasis 478
   essentiality and sources 477
   metabolism
      absorption 478
      digesting 477–478
slow-release oral methods 48
sodium and chloride
   absorption
      dietary and endogenous 187
      large intestine 187
   and acid–base balance 197–198
   deprivation
      biochemical and physiological signs 192–194, 197
      diagnosis 194–195
      occurrence 195–196, 198
      prevention 197
   dietary sources
      concentrates 185–186
      drinking water 186–187
      forages 184–185
   excretion
      dietary excesses 189
      kidneys 188
      potassium supplementation 188–189
   functions
      amino acid and glucose 183
      osmotic skeleton 182–183
   heat stress 189
   hormonal influences 188

membrane transport
    cellular level 187
    mechanisms 187–188
palatability 187
requirements
    cattle 190–191
    goats 191
    pigs 191–192
    poultry 192
    sheep 189–190
salt toxicities
    plants 199
    saline drinking water 199
    sodium hydroxide-treated feeds 200
    supplements 198–199
secretion
    ruminants 188
    skin 188
water deprivation 189
sodium deprivation
    deficiency
        buffalo heifers 194
        concentration, urine 193
        conservation mechanism 193
        mammary gland 193–194
    depletion 192–193
    diagnosis
        saliva 194–195
        salt craving 194
        urine output 195
    dysfunction
        haemoconcentration 194
        laying hens 194
    occurrence
        climatic factors 196
        conditions 195
        dietary 195–196
        disease factors 196
        heat stress 198
        pasture potassium 196
        spontaneous turkey cardiomyopathy 198
        transport 196
    pica 183
    prevention
        energy metabolism 197
        milking cows and ewes 197
        pig and poultry feeding 197
    water consumption 183–184
sulfur
    deprivation *see* sulfur deprivation
    excretion
        molybdenum 210
        urine 210
    functions
        glutathione 207
        hormones 206
        metallothionein 206–207
        sulfhydryl (SH) 206
    high dietary 406–407
    homeostasis, metabolism
        saliva 210
        tissue protein, catabolism 210
    nutritive value, feeds
        MPS 208
        plant tannins 208
    post-rumenal metabolism
        methionine 210
        microbial capture 210
    rumen microbiota, metabolism
        copper 210
        inorganic sources 209
        MPS and 209
        primary flows 209
        rumen-degradable protein (RDP) 209
    ruminants, requirement
        application and tests 212–213
        definition 211
        estimating system 211–212
        hair and wool production 211
        live weight and wool growth rate effects 211
    ruminants, sources
        drinking water 207–208
        foodstuffs 207
        forages 207
    toxicity
        compounds 218
        direct effects 217–218
        effects, indirect 218
sulfur deprivation
    animal factors
        gut nematodes 214
        nitrogen 214–215
        stocking systems 214
    biochemical abnormalities
        hypoalbuminaemia 213
        plasma 213–214
    diagnosis
        blood and liver GSH concentrations 215
        seasonal fluctuations 215
        sulfide sulfur level 215
    plant factors
        pastoral systems 214
        spear grass 214
        wool production and live-weight 214
    prevention and control
        fertilizers 216
        inorganic sources, nutritive value 216–217
        sources 217
        supplements, direct 216
    symptoms 214

tetrathiomolybdate (TTM)
        copper–molybdenum–sulfur antagonism 290
        lambs 276
        liver copper accretion 275–276
        pharmacological doses 276
thyroid dysfunction
        embryo and fetal development, impaired
                brain maturation 319
                thyroid dysfunction 319
        goitre 318
        impaired fertility
                lactation 318
                male 318–319
        integument disorders
                methylthiouracil, cashmere
                        goats 319–320
                pigs 319
        milk yield, low 320
        post-natal mortality and growth retardation
                cold stress 319
                hypothermia 319
thyroxine (T4)
        activation
                ID1 309–310
                selenium deprivation 309–310
        assay 323
        synthesis and storage
                environmental temperature 309
                thyroperoxidase 309
tibial dyschondroplasia (TD)
        broilers 78
        dietary calcium and dihydroxycholecalciferol
                effects 79
        grain-based diets 79
tissue disorders 260
total iron-binding capacity (TIBC)
        plasma 341
        and UIBC, serum iron 346

unsaturation iron-binding capacity (UIBC)
        iron-free apotransferrin 341
        lactoferrin, milk 345
        and TIBC, serum iron 346

vanadium
        essentiality and sources 479
        metabolism 479
        toxicity
                dietary factors 480
                valency state 479
vitamin B_{12}
        absorption
                bile salts 230
                features 230
        Cbl-transporting proteins 229

        deprivation, pigs 475
        excretion 231
        liver 241
        milk 243
        mucosal transfer, transport and cellular
                uptake 230
        production 229
        rumen production 228
        secretion 231
        serum
                cattle 242
                sheep 241–242
        storage
                concentration, cobalt-deprived
                        sheep 230
                liver 230
                plasma 230
        synthesis and cobalt availability
                energy and protein substrates
                        effect 228
                intake effect 227–228
                source, effect 228
                species 228
                synthesis, in vitro 228–229

wasting disease 223
white-liver disease (WLD)
        features 226
        polioencephalomalacia 225–226
        signs 225
        symptoms 239
white-liver disease in ovines (OWLD)
        cobalt deprivation, lipid and vitamin E 235
        pathogenesis 234

zinc
        absorbability enhancement,
                non-ruminants 434–435
        availability, milk products 435
        biochemical manifestations, deprivation
                deficiency 442–443
                depletion 442
                disease resistance 443–444
                disorder 443
                dysfunction 443
                heavy metals, accumulation 444
        deprivation, clinical consequences
                anorexia 428
                reproductive disorders 428–429
                skeletal disorders 428
                skin and appendages, abnormalities 428
        deprivation occurrence
                non-ruminants 445–446
                ruminants 446–447
        dietary, requirement derivation 435–436

disorder, prevention and control
    grazing livestock 447
    housed livestock 447
    organic and inorganic
        sources 447–449
feed phytate
    mathematical modelling 433–434
    RBV assessment, problems 432, 433
functions
    antioxidant defence 427–428
    appetite control 427
    fat absorption 427
    gene expression 427
    structural and functional integrity 426
    'zinc-finger' 426–427
homeostasis
    cellular uptake 440–442
    plasma zinc, control 440
    storage 442
metabolism
    absorption 431
    excretion and retention 430
parakeratosis 426
pharmacological responses
    early-weaned pigs, growth 449
    facial eczema 450
    organic versus inorganic zinc 450
RBV, feeds
    assessment 431, 432
    effect, dietary zinc
        concentration 432–433
    phytate effect 431–432
    response index effect 432, 433

requirements
    goats and horses 440
    pigs 436
    poultry 436–437
    sheep and cattle 437–439
responsive disorders, diagnosis
    indices 445, 446
    plasma and serum 444
    serum metalloproteins and
        enzymes 444–445
sources
    concentrates 430
    forages 429–430
    milk and milk products 430
toxicity 450
value, horse feeds 434
zinc deprivation
    biochemical manifestations
        deficiency 442–443
        depletion 442
        disease resistance 443–444
        disorder 443
        dysfunction 443
        heavy metals, accumulation 444
    clinical consequences
        anorexia 428
        reproductive disorders 428–429
        skeletal disorders 428
        skin and appendages, abnormalities 428
    occurrence
        non-ruminants 445–446
        ruminants 446–447
    vulnerable cells 42